Historical & Cultural Astronomy

The Historical & Cultural Astronomy series includes high-level monographs and edited volumes covering a broad range of subjects in the history of astronomy, including interdisciplinary contributions from historians, sociologists, horologists, archaeologists, and other humanities fields. The authors are distinguished specialists in their fields of expertise. Each title is carefully supervised and aims to provide an in-depth understanding by offering detailed research. Rather than focusing on the scientific findings alone, these volumes explain the context of astronomical and space science progress from the pre-modern world to the future. The interdisciplinary Historical & Cultural Astronomy series offers a home for books addressing astronomical progress from a humanities perspective, encompassing the influence of religion, politics, social movements, and more on the growth of astronomical knowledge over the centuries.

The Historical & Cultural Astronomy Series Editors are: Wayne Orchiston, Marc Rothenberg, and Cliff Cunningham.

Juan Antonio Belmonte • José Lull

Astronomy of Ancient Egypt

A Cultural Perspective

With Contributions by Terry Mahoney

Springer

Juan Antonio Belmonte
Instituto de Astrofísica de Canarias and
Universidad de La Laguna
La Laguna, Spain

José Lull
Departament de Ciències de l'Antiguitat i
de l'Edat Mitjana, Instituto de Estudios del
Próximo Oriente Antiguo (IEPOA)
Universitat Autònoma de Barcelona
Bellaterra, Barcelona, Spain

ISSN 2509-310X ISSN 2509-3118 (electronic)
Historical & Cultural Astronomy
ISBN 978-3-031-11828-9 ISBN 978-3-031-11829-6 (eBook)
https://doi.org/10.1007/978-3-031-11829-6

Cover Image - Front cover: Close-up of the recently restored astronomical ceiling of the Pronaos of the Temple of Dendara. Photo by José Lull.

This Springer imprint is published by the registered company Springer Nature Switzerland AG
The registered company address is: Gewerbestrasse 11, 6330 Cham, Switzerland

Michael Hoskin (1930–2021)
IN MEMORIAM

A nuestras chicas
Marga & Sol, Mari Carmen & Valeria
The true stars of our life

Foreword

Herodotus said that 'Egypt is a gift of the Nile'. And while this is certainly true, because without the river and its annual flood there would be no Egypt, it is also true that the ancient Egyptians incorporated every aspect of their environment into their peculiar way of life: not only the river, its banks, animals and plants, but also the mountains, the desert and the sky, most notably—but not only—the sun. All these factors, and of course many others, were interpreted in a most peculiar way by the people who inhabited this land for nearly three millennia and who left behind some of the most fascinating examples of human interaction with the skyscape that we are still unveiling today.

It is a commonplace in our days that our Western society has turned its back to the sky in everyday life and we long for a past in which light pollution, and a more rural way of life kept us more closely connected to the rhythms of the sky. However, the evocative role of the heavens and the discoveries of astronomers make it front-page news. One may also find examples of the close engagement of our everyday life with the sky, such as the importance we apparently give to the chance alignment of some streets in cities such as Manhattan where, twice each year the 'cosmic' phenomenon of the setting sun causes a people collapse and traffic jams even today.

Astronomers today are often asked what astronomy is useful for. The puzzled astronomer then struggles to provide an answer in terms of the utility of astronomy today in developing tools for observations, for the processing of images, and other technological techniques that may have helped develop some of the fancy tools we benefit from in our everyday life.

This question was easier to answer in the past, when the sky, as the final object of study of astronomy, then had many different 'uses'. It is clear that the recurrent rhythms of the sky, such as the rising or setting of the sun, the moon and its phases or cycles (or those of other heavenly bodies), helped in acquiring a sense on the flow of time, first by following those rhythms and later on, when the need arose, by establishing more nuanced and refined timekeeping systems, such as calendars.

The precise regularity of the rising and setting of heavenly bodies always happens in similar ways in the landscape. The rising of a given body always occurs in a preferred direction, and not in others, while the setting takes place in an area where no risings occur. This set of circumstances helps define directions in the landscape. In other words, it helps to organize the space by orientating it. In a sense, if Bender (1998) says that time builds the landscape, it can be shown that the definition of preferred directions—the orientation—helps to establish such a concept of landscape.

But there is a final way in which the sky can also aid our understanding of our surroundings by helping us in organizing or receiving our explanations for everything that is around us. The sky and the events, the items and figures we can see in the constellations, for example, even today act as a metaphysical source and repository. All the images of giants, animals, queens, and heroes in the classical constellations come from a particular attempt to find a sense in those figures or to project onto the skies our interpretation of how everything worked.

Things were quite different in the past. The ancient Egyptians attempted to understand their environment by comparison and analogies, creating through these means symbols taken from nature and creating narratives in which the sky played a prominent role. The life of the ancient Egyptians, as we may understand from their own records, revolved around the prominent role played by the sun.

Despite the large number of works on the importance of the sky in the Egyptian culture, one is astonished to find one of the giants in the study of ancient astronomy, Otto Neugebauer in his *The Exact Sciences in Antiquity* (1969: 71) declaring, 'Astronomy played a uniformly insignificant role in all periods of Egyptian history'. The present book clearly shows otherwise.

In my view, Neugebauer's statement stems from the restrictive use of the word 'astronomy' to mean celestial mechanics. On that limited basis, it is true that we do not find that kind of knowledge in Egypt. However, if we broaden the definition of astronomy to incorporate anything that has to do with skywatching, the picture changes completely.

This last way of 'employing' the sky is in fact of most relevance for the Egyptians, as for many other cultures worldwide, as it provides a handle on their world-view. It was from the heavens, in particular from the sun, that the rulers—the Pharaohs—obtained the sanction of their government, as we shall see in this book. To keep justice on earth, the *Ma'at* (the cosmic order) it was necessary to perform the correct rituals at the right site and at the appropriate timing, and most probably facing in the right directions.

When one approaches the Egyptian culture, one is usually first overwhelmed by a mass of information on the intricate politics of the Pharaohs, their dynasties, and the successions. One may become lost in admiration of the sheer grandeur of their architectural achievements. The magnificent decoration of their tombs will also surely evoke images of a dynamic and prosperous society.

But it is when coming closer to this society and its culture that one can see how the Egyptians lived, how they understood their life, and how they interacted with their environment. In this aspect, we find the sky permeated by several aspects concerning this society. The task of attempting such an understanding of any society commonly falls to what has been termed 'Astronomy in Culture' (e.g. Iwaniszewski, 2009; Ruggles, 2011).

We may investigate the meaning of the sky for other societies from two points of view. One method is to use an *etic* approach from the outside of that society and from our point of view in order to try to interpret and comprehend their world-view. The opposite—*emic*—approach tries to understand that sky from within that society. In this sense, a complete understanding of the meaning of the sky for an ancient culture is impossible, but the present volume illustrates how our understanding of the different knowledges, understandings, and complex reflections upon the sky of the Egyptians have evolved over the last century.

From an *etic* point of view, an historian of science or a present-day astronomer, might like to know what the Egyptians saw when they looked at the sky. Research over the past two centuries has revealed the importance to the Egyptians of not only the sun but also the moon, certain stars, and the planets, as the authors have noted in these pages, along with how they mapped the sky and how they might have used the regularities in the movements of the sun and moon to define calendars and time-keeping devices, and the importance of Sirius, the brightest star in the sky in heralding the most important moment in their annual cycle, the flooding of the Nile.

However, the authors also present, from an *emic* point of view, our current understanding on how such knowledge was generated and transmitted in Ancient Egypt. They present us a body of priests that might have been charged, among other possible duties, with being aware of signals in the sky that announced when a given ritual needed to be performed, especially by the 'Lord of the Offerings', the King himself, as son of Re. Who did these observations, and how they performed them, are also the subject of this research, including their attributes as a possible social elite, their garments and tools, their ambitions, and their passions. It is of interest to know how they used such astronomical terminology as 'enclosed by the Duat' possibly to mean that a celestial body was in conjunction with the sun, or 'troop' to mean 'star cluster'.

We also discover how Re travelled across the sky day and night sailing in his boats, stopping at mooring posts (the hours) along the way, and how this was viewed by the priests by day and night; how such a concept helped them to make and keep appointments or regulate working hours in the tombs of the Valley of the Kings; what the stars and planets represented for them, and how they conceived the sun, the moon, and the stars as manifestations of several gods. We may wonder at the astonishing lack of records of solar eclipses for a society where the sun was constantly monitored. Was it lack of interest? Are we missing or misinterpreting relevant information? Or did, perhaps, the Egyptians feel a kind of taboo for these events that prevented their sages from keeping records of them?

We also find out how the living needed maps of the sky during the night to guide them in their journey to the Netherworld, with narratives to explain the images in the sky. We observe the soul of the deceased Pharaoh journeying towards the Imperishable Stars to become one of them, and how the living attempted to secure a similar fate with ceremonies and instruments (ritual adzes) that resembled those star patterns in the sky; how several of those images appeared throughout their history and evolved, and how concepts from other countries finally permeated this society in the same way that we find Mesopotamian concepts in the late Ptolemaic and Roman periods, for example.

The present volume also describes how the ancient Egyptians encapsulated their understanding of the world around them in the many buildings they erected, mostly in relation to the Nile, but also, as the authors highlight and has been put forward especially over the last two decades, notably in relation to the skies, in particular the sun and perhaps some stars. As images of the Cosmos, the temples not only represented a miniature of the World but also needed to face the correct direction so that rituals could be performed at the right time (as mentioned above) while facing the deities. In fact, the Egyptians themselves tell us that they knew how to do this. 'He made these great and perfect monuments for all eternity. He knew how to orientate them towards the Horizon of the Sky'.

One final way in which astronomy is 'useful', given the regularities of the movements of the skies, is that we can model how the sky was seen at any moment in the past with a fair degree of accuracy. This is something that no other environmental science can yet achieve with this level of certainty (although there are of course always uncertain areas!). This may help the Egyptologist to date the several documents written by the Egyptians at the time of certain astronomical events. As the authors show, this is not straightforward. Astronomical dating is an innovative and clever way of using the orientation of buildings as cultural information on the Egyptian calendar and astronomy to provide a new handle on Egypt's past.

In this sense, this volume presents a still image of a rapidly evolving subject that touches on a large number of different branches of Egyptology. It is true that this is not the first book on this subject, as the lengthy bibliography attests, and this will surely not be the final word in many of the subjects treated in the book; it will, however, provide the reader with a balanced understanding of the latest research in the field. The volume also deals with the various controversies in all these subjects in a very open and frank way.

In summary, the present volume is a compendium of multidisciplinary research on the heavens in Ancient Egypt, but from an interdisciplinary and transdisciplinary viewpoint. In fact, as the authors disclose in their Introduction, it is most fortunate that this compendium is presented by two Egyptologists with different backgrounds: an astronomer with studies in Egyptology, and an Egyptologist with a deep knowledge of astronomy. If the key to doing interdisciplinary research is trying to combine several fields in a balanced way, one could not wish for a better foundation than such a collaboration as this.

To conclude, the reader will discover a complex society with an elaborate image of the world that we can figure out thanks to the work of the many researchers that the authors have referenced in these chapters. In my opinion, it is after reading books like this that one really catches a glimpse of the people of the past. Enjoy it!

President of the European Society
for Astronomy in Culture (SEAC)
Instituto de Ciencias del Patrimonio, INCIPIT-CSIC
Santiago de Compostela, Spain

A. César González-García

References

Bender, B. (1998). *Stonehenge: Making space*. Berg.

Iwaniszewski, S. (2009). Por una astronomía cultural renovada. *Complutum, 20*(2), 23–37.

Ruggles, C. L. N. (2011). Pushing back the frontiers or still running around the same circles? 'Interpreting archaeoastronomy' thirty years on. In C. L. N. Ruggles (Ed.), *Archaeoastronomy and ethnoastronomy: Building bridges across cultures* (IAU Symposium 178) (pp. 1–18). Cambridge University Press.

Preface, by Way of Introduction

This book was inspired by necessity. Although they had known of each other for some time, the authors first met nearly two decades ago when they realized that their approach to Egyptian skies was extremely similar. This brought them to their first collaboration (Lull & Belmonte, 2006). Much more importantly, however, they eventually came to realize that they were a sort of mirror image of each other.

Juan Antonio Belmonte was a professional astronomer at an astrophysical research centre working in stellar physics who has since moved on to exoplanets, the Instituto de Astrofísica de Canarias. Since his teens, however, he had had a passion for ancient Egypt and her culture and in the 1990s he initiated a joint research project on archaeoastronomy in which Egyptian astronomy and chronology played a role. This marked a point of inflexion in his research and academic carrier.

José Lull was a professional Egyptologist, trained at Tübingen University, who had read his PhD thesis on the troublesome period of the end of the New Kingdom and the beginning of the Third Intermediate Period. He had been a member of several archaeological missions in Egypt and was now teaching Egyptology in the Universitat Autònoma de Barcelona. However, since his teens, he had had a passion for astronomy, being a dedicated and skillful amateur astronomer, having built his own observatory at home. Sooner than later, it was obvious that he would turn his attention to the skies of ancient Egypt.

The result of these complementary natures had been two books: *La astronomía en el antiguo Egipto* (Lull, 2004) and *Pirámides, templos y estrellas: astronomía y arqueología en el Egipto antiguo* (Belmonte, 2012). The former included an extensive analysis of the sources and a solid set of new hypotheses. The latter was the first essay dedicated to aspects of ancient Egyptian cultural astronomy. These volumes are well known in Spanish-speaking Egyptological circles but passed largely unnoticed in the rest of the world. Spanish is not a language Egyptologists are used to. Actually, something very similar happened to the extraordinary *Astronomia Egizia* of Pierangelo Mengoli (2006), which is hardly referred in academic circles, possibly because it is written in Italian, a language Champollion dominated but most modern transalpine Egyptologists do not. This needs to change!

The authors decided that a new book should be written. A book in which our combined knowledge of ancient Egyptian astronomy could be captured in a simple but scholarly manner, specially dedicated to explaining the interaction over three millennia between astronomy and the civilization of ancient Egypt. Hence the subtitle of the book: 'a cultural perspective'. The end result is the volume readers have now in their hands. It is not meant solely for Egyptologists and historians of science but also for anyone interested in the culture of ancient Egypt and how this people interacted with their environment, including the sky, to produce such a sophisticated culture.

Our work has been built on the work of others. We too have stood on the shoulders of giants and, relied on the labours of generations of Egyptologists who, since the pioneering works of Champollion (1842), Brugsch (1883) and Borchardt (1917), have inspired a plethora of dedicated researchers. We could do nothing without that bible of astronomy in Egyptology, *Egyptian Astronomical Texts*, the work of a decade (1960–1969) of toil of two giants, Otto Neugebauer and Richard Parker. This work and Parker's *Calendars* (1950) have reigned in solitude for several decades until a new generation of Egyptologists took over the discipline. A reference work is Marshall Clagett's *Calendars, Clocks and Astronomy* (Clagett, 1995), in which several 'paradigms' were questioned for the first time in a dialectic way.

A new generation of researchers appeared in the final decades of the last century. Rolf Krauss, Anthony Spalinger, Leo Depuydt, and Christian Leitz (among others), whose works will be profusely referred to throughout the volume, have dedicated substantial effort to improving our knowledge of the ancient Egyptian heavens and time-reckoning systems. More recently, Rita Gautschy, Alexandra von Lieven, and Joachim Quack must be included among the giants of whose shoulders we have availed ourselves.

Archaeoastronomical studies were a blank space in Egyptology, and this would still be the case if Gerard Hawkins in the 1970s and Edwin Krupp in the 1980s had not continued the pioneering work of Sir Norman Lockyer in his *Dawn of Astronomy* at the end of the 19th century. The recent inspiring works of Luc Gabolde should not be forgotten. These names represent but a small sample of the scholars who have been our inspiration and leitmotiv in writing this book. The bibliography recalls the vast majority of their names, activities, and results.

As we intend to demonstrate in this volume, astronomy played a most important role in ancient Egyptian culture as the basic tool for a perfect orientation in both time and space. Without observation of the sky, the Pharaohs' subjects would have been unable to create the 'most intelligent calendar ever developed' (Neugebauer *dixit*), to measure time by day and night, and to orientate their buildings following premises established by cult necessities. Modern western culture owes these aspects of our civilization to the ancient Egyptians' world-view.

However, astronomy has played no role at all in international recognition of the ample heritage of ancient Egypt, in spite of the wonders inscribed in UNESCO's World Heritage list (Belmonte, 2010). Egyptologists have continuously used astronomy in one aspect or another of their research, but without realizing it and without appreciating its importance until very recent times. We hope this book will go some way towards remedying this oversight.

One of the wonders just referred to where astronomy's role has been successfully popularized relates to the Egyptian pyramids.

It is well-known that the pyramids of Egypt, particularly those of the 4th Dynasty kings Khufu, Khaefre, and Menkaure—Herodotus' Cheops, Kefren, and Micerinos—erected on the Giza Plateau probably around 4500 years ago, are orientated with 'surprising' precision towards the four cardinal points (Petrie, 1883: 39; recently confirmed using modern techniques by Nell & Ruggles, 2014). This has been repeated ad nauseam in books, magazines, press, and TV documentaries. We deliberately emphasized 'surprising' because a corollary often expressed is how such a 'primitive' people could have been able to obtain such a precise orientation. When it is not claimed the pyramids are simply the works of beings from outer space, the conclusions drawn often involve the existence in the distant past of great long-disappeared civilizations, such as Atlantis, or other not so exotic but no less controversial explanations.

The Orion Correlation Theory (OCT) and its corollaries are an interesting example of this dynamic. The story started with a curious hypothesis (Bauval, 1989) and developed into a full theory in 'The Orion Mystery' (Bauval & Gilbert, 1994), which culminated in 'The Egypt code' (Bauval, 2006), a decade later. The debate recently came full circle with a hypothesis that attributes the origins of Pharaonic civilization to a mysterious star-ceremonialist African people from the distant past who possessed great astronomical skills, of whom the great Imhotep would be their most recognized heir and swan song representative (Bauval & Brophy, 2013).

This volume does not deal with arcane mysteries, secret codes, or mysterious peoples, but with the ancient Egyptians, persons who read and wrote, calculated, built splendorous monuments, interacted with their neighbours, and indeed observed the skies, because they themselves have told us they did it. They left a series of unsolved questions for us to answer, including how and why they orientated the pyramids so precisely.

The answer does indeed lie in natural and social sciences. It is much simpler than the reader might imagine, and is 'written' in the stars. And when we say written, this is literally true: this magnificent orientation was able to be obtained by the almost obsessive interest of the Egyptians in the stars, an interest that is already reflected in the Old Kingdom, not only in the *Pyramid Texts* (Faulkner, 1969; Allen, 2015), but also in the very names of these buildings:

The pyramid [whose name] is 'Djedefre is a Sehedu star'.

This is the name of the funerary monument of King Djedefre, eldest son and successor of Khufu, which suggests that when this pharaoh died he would have become a *Sehedu* star. This term sometimes appears in the *Pyramid Texts* but its meaning is not yet clear, having often been translated as ‘firmament’ and not as the name of a particular star, or even being identified as one of the Egyptian denominations for the planets, although this is far from proven (see Chap. 4). It has occasionally been suggested, through an alternative translation, that it might not be Djedefre himself but his pyramid that would be a stone representation of a *Sehedu* Star, but there are indications that the first reading is by far the most likely (Fischer, 1996).

The stars, perhaps even the sun, seem to be the basic explanation of the ‘surprising’ orientation referred to above, and we can show that this knowledge did not arise from zero. It was no less than the worthy heir, and supreme achievement, of a tradition that had begun centuries before, at the dawn of Pharaonic civilization, when the first kings of Egypt were buried at Abydos (see Chap. 6).

However, the first kings of the Second Dynasty, in particular Hotepsekhemuy, Ninetjer and Raneb, the latter being the first to include the name of the sun god Re in his Horus name, were not buried at Abydos. Instead, they choose a new royal cemetery in Saqqara to the south-west of the huge mud brick mastabas of the nobles and members of the royal family of the 1st Dynasty. These early mastabas had their axes orientated in a lax north–south direction, possibly dictated by the local topography of the cliff overlooking the Nile Valley as seen from the site. However, a substantial change was introduced at the beginning of the new dynasty when kings began the construction of large underground chamber complexes that would possibly have been covered by superstructures of an equivalent category. Unfortunately, these buildings had already been dismantled in ancient times to make room for new constructions (in particular the pyramid complex of the 5th Dynasty King Wenis) and only the underground chambers and entrance corridors have survived to the present.

Figure 1 shows the entrance to the substructures of the tomb of Hotepsekhemuy, founder of the Second Dynasty. This is a monument visitors pay little attention to, despite its importance. The substructure of Hotepsekhemuy has an azimuth of 1½°, while the accesses to the royal tombs of Abydos followed a not too regular intercardinal pattern with tombs indistinctly accessed from the north-east or north-west. Hotepsekhemuy’s tomb is possibly the first structure built in Egypt with the intention of orientating it more or less precisely to geographical north, and perhaps to the domain of the imperishable stars, the *Ikhemu Sekiu*, as defined in the *Pyramid Texts* (see Fig. 1).

There was a royal domain related to the funerary complex with the very suggestive name, ‘Horus, the rising star’ (Figure 1) that could be interpreted in two ways. On the one hand, it could be construed as Horus (that is, the king), once deceased, ascends to the heavens towards the sphere of the *Ikhemu Sekiu*. Alternatively, it could mean that when the king died, he identified himself with the Morning Star, i.e. the planet Venus, the celestial manifestation *par excellence* of the god Horus at that time (Krauss, 2016a), which ascends in the eastern sky. Both interpretations are possible in the context of the *Pyramid Texts* because both were post-mortem destinations of the Pharaoh, either as one more star of the Imperishables or as the Morning Star, son of Sah and Sopdet (see Chap. 4).

Fig. 1 Access dromos to the substructure of the Hotepsekhemuy complex in Saqqara (left). This was orientated northwards with an azimuth of 1½° and possibly to the realm of the *Ikhemu Sekiu*, the Imperishable Stars of the later *Pyramid Texts* (right). In particular the image is a snapshot of the texts within the burial chamber of Teti, where the *Ikhemu Sekiu* are mentioned in the last column while the constellation Sah is mentioned in the first. The small quadrant to the left is the name of a royal domain related to the complex: xA sbA Hrw 'Horus, the rising star'. (Diagram and images by the authors)

The northern orientation of the underground complex leans towards the first hypothesis, but the second cannot be ruled out. Actually, it is more than likely that both were possible in the context of the astral eschatology of the time, whose maximum exponents are certainly the *Pyramid Texts*. These were written later (they are first documented in Wenis's burial chamber to be exact), but almost certainly reflect older traditions. This meridian orientation towards the north was followed, with greater or lesser success, by other royal monuments of the 3rd Dynasty in Saqqara, in particular the complex of Netjerkhet (Belmonte, 2012: 176–181), but reached its paroxysm during the 4th Dynasty.

Since the turn of this century, published literature on the ancient Egyptian pyramids—the 4th Dynasty ones in particular—and astronomy has reached unaffordable proportions. There are articles and even books written by specialists, based on qualified measurements and serious historical sources. These try to contextualize the almost perfect orientation of many of these monuments and will be discussed in due course, although briefly.

It is an established fact that the plan of the Egyptian pyramid complexes is such that most of these monuments are orientated to within a few degrees of the four cardinal points, reaching notable precision for a few cases. This is rigorously true for those of the Old Kingdom with one notable exception, and less so for the Middle Kingdom (see Figure 2). A few methods that might have aided the Egyptians in achieving such a goal have been suggested, and have been reviewed and analysed by Glen Dash (2015). There is one procedure using the shadow cast by the sun—with variants—and three stellar methods. These involve the pole star, a circumpolar star and simultaneous transit—with several alternatives, respectively. Dash (2015) has shown that most of these methods could achieve the necessary precision, but the question is whether the ancient Egyptians used them, or even had the appropriate tools or knowledge to accomplish these tasks.

The north (cardinal) orientation was brought to near perfection in the pyramids of Dahshur and Giza (see Table 1) and would thus continue, with greater or lesser accuracy, in the reigns of all the builders of the great pyramids of the 4th, 5th, and 6th Dynasties. There was only one exception (see Fig. 2): the pyramid complex of Teti, the first king of the 6th Dynasty, with an azimuth of almost 81°, instead of the normative *c.* 90°. We discuss this anomaly and its chronological implications in Chap. 7. In general, however, the builders looked for the north more or less precisely.

Table 1 shows that there seems to be an improvement in the precision over time from Men Sneferu to Akhet Khufu, followed by a decline afterwards. Interestingly, Dorner's (1981) measurements for his PhD thesis (unfortunately never duly published or translated) of the descending passageway for Sneferu's pyramids at Dahshur suggest that these were orientated north very precisely, even more than at Khufu's pyramid at Giza. This fact is rarely taken into account and is often ignored by scholars dealing with the topic, including the first author in the past (see below), but it is indeed most relevant for any informed and comprehensive discussion of the topic.

Which came first, the chicken or the egg? With regard to the pyramid faces, scholars may reach the certain conclusion that pyramid orientations improved with time until reaching near perfection in the largest and most complex pyramid ever built, that of Akhet Khufu. If instead, however, the corridors are taken into account, the conclusions reached may be totally different: the great builder Sneferu had reached the best precision in the first true pyramids he erected (Kha Sneferu and

Table 1 The orientation of the pyramids of the 4th Dynasty

Pyramid	Corridor	N side	E side	S side	W side	Average
Men Snefru	−21.6	−35.4	−20.6	−23.6	−18.1	−24.4
Kha Sneferu Resy	−1.0	−7.5	−17.3	−4.2	−11.8	−10.2
Kha Sneferu	+3.0		−8.7			−8.7
Akhet Khufu	−5.8	−2.9	−3.4	−3.7	−4.6	−3.6
Wer Khaefre	−5.6	−5.2	−6.0	−5.7	−6.0	−5.7
Netjer Menkaure	13.3	16.8	12.4	13.0		14.1

The first column gives the name of the complex. The first numerical column stands for the orientation (in arc minutes) of the descending passageway, the corridor entering the pyramid on its northern side. The following four columns represent the deviation of the respective faces to the cardinal points, and finally the average of the four measurements, where available. See the text for further discussion. Data adapted from Dash (2015, and references therein)

Fig. 2 (continued) The lower panel excluded Netjerkhet and Teti as outliers. Note how the orientation improves during the 4th Dynasty, is a bit chaotic for the 5th, and improves back and stabilized from the reign of Wenis onwards, only to worsen again during the 12th Dynasty. T stands for the solar temples of Userkaf and Niuserre. The main outcome, however, is that whichever method was used after Menkaure, it almost always produced a clockwise (positive) deviation from due north. (Diagram by the authors with data taken from Belmonte and Shaltout (2009: Appendix II))

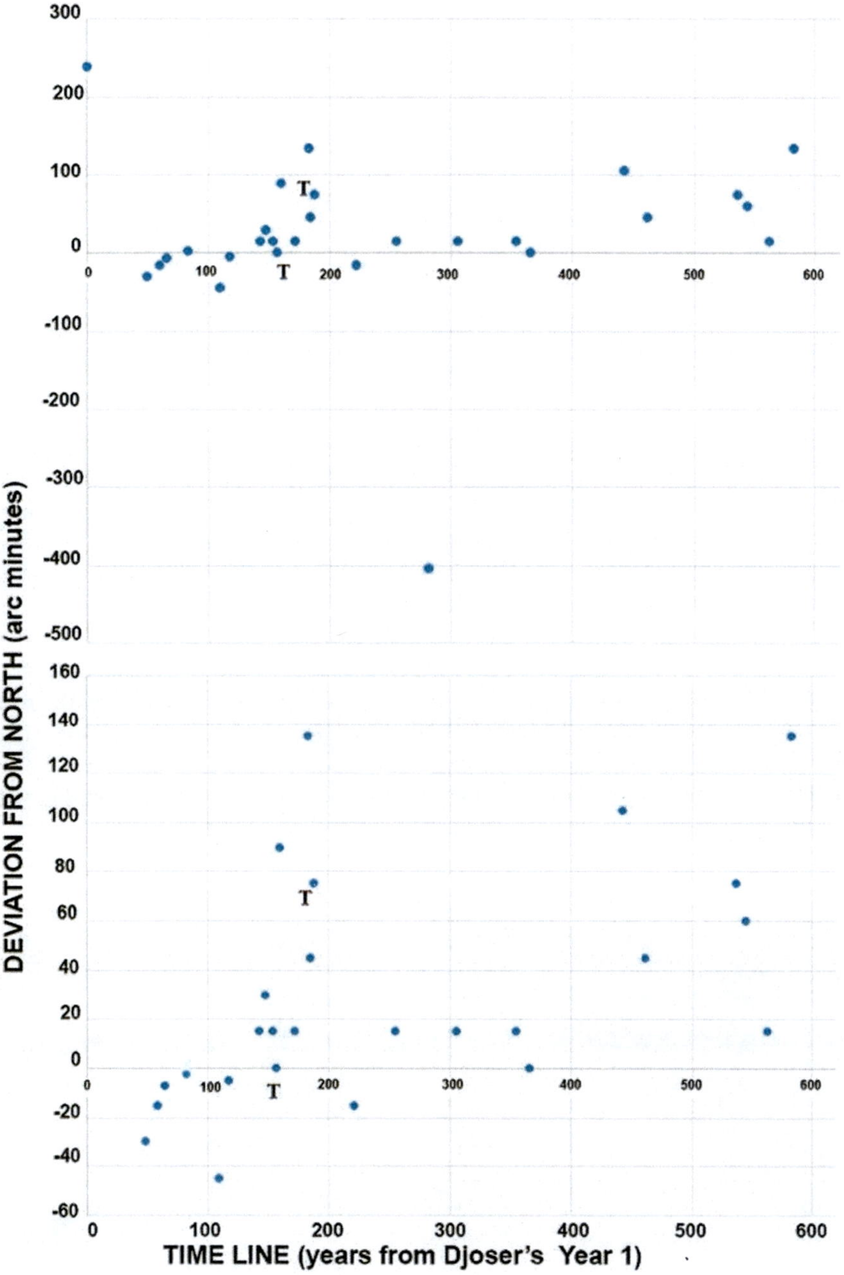

Fig. 2 Orientation of the pyramid complexes of the Old and Middle Kingdoms, as measured from the orientation of their funerary temples, versus time. The data have an average precision of ¼° (i.e. 15′). To avoid chronological issues, the zero point is Netjerkhet year 1, although the length of the reigns has been taken from Hornung et al. (2006, see Table 7.1) and could be slightly biased.

Kha Sneferu Resy; see Chap. 6). Meidum, which was started as a step pyramid, may offer a different framework. The dilemma is not an easy one to solve.

Figure 2 also illustrates what happened later, after Khufu's reign. The Djedefre and Khaefre pyramids break the rules. During the 5th Dynasty, pyramid and solar temple complexes do not seem to follow any regular pattern or custom. The dispersion is huge, but with an interesting outcome: the orientations of the complexes always deviate clockwise, with the exception of Djedkare. The same can be argued for the end of the Old Kingdom and the pyramids of the Middle Kingdom. This fact remains unexplained, although a not too precise solar orientation in winter or spring (see below) may account for it. Pyramids of the late 6th Dynasty were built in standard, almost industrial, proportion using 100 cubits as a base unit. Orientations look similar if an outlier (Teti) is disregarded. Finally, during the Middle Kingdom, the clockwise deviation is maintained but no logic seems to be hidden behind a random distribution.

In any case, the question remains of the method by which such precision, or imprecision, could be achieved (Lull, 2004: 287–302).

In western tradition, we are used to solar orientations whenever we want to establish the cardinal directions (in the absence of a compass). The sun, was very soon considered by Egyptologists as a good option as the target for orientating the pyramids. The astronomy required is simple.

Chatley (1940) supported the idea of a determination of the north–south axis through the shortest shadow produced at noon by a gnomon. Neugebauer (1980) proposed that the previously designed and sculpted pyramidon could play the role of the gnomon (see Figure 3). However, the shadow cast would be quite imprecise in both instances. Isler (1989: 197–206) proposed a variant of the 'Indian Circle' to establish the first perpendicular and hence the meridian line to improve the shadow sharpness by means of a device similar to a *bay* (see Fig. 3 and Chap. 2).

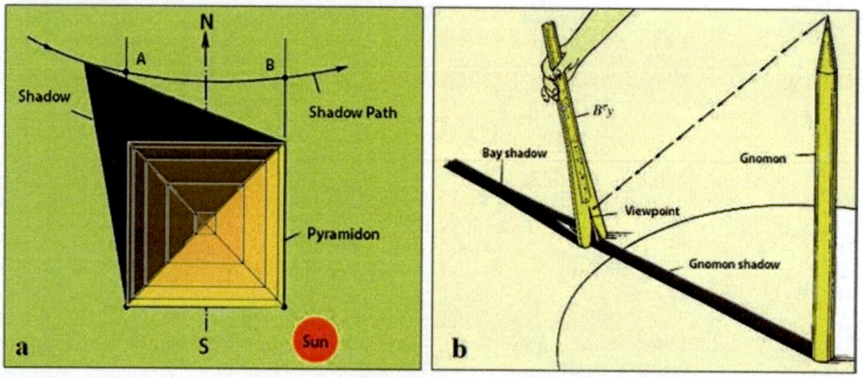

Fig. 3 The sun as an observing target. (**a**) A sketch of Neugebauer's method to align pyramids. The line connecting A and B would be the first perpendicular and hence the E–W line. (**b**) Since the definition of a sun shadow border is always a problem, Isler proposed using a *bay* as an auxiliary tool to improve shadow sharpness and hence accuracy. (Adapted from Lull (2004: 290–292), inspired by Neugebauer (1980) and Isler (2001))

A similar method has been defended much more recently by Dash (2015) who even suggested that the ancient Egyptians may have used the date of the equinoxes (when any shadow of the sun would form a straight line in the E–W direction) to determine the orientation (Dash, 2017). However, as Belmonte (2021) has argued, there are serious doubts whether the Egyptians ever knew about or used the equinox for any purpose before the Hellenistic period. Hence, the use of a solar method is far from proven, although it would be the only one to provide a systematic clockwise deviation (see above) when the solar declination increases during the day in winter and spring. But deviations are great, and considerable freedom must have been granted to very imprecise observers. A final shortcoming of solar observations is that they would be useless for fixing the precise orientation of the descending passageways of the pyramids.

Because of the problems arising from using the sun as a tool, most researchers since Charles Piazzi Smyth in the 19th century have therefore preferred a method using stellar alignments. This Astronomer Royal for Scotland was responsible for the presence of batteries of huge telescopes on the tops of high mountains throughout the world (Smyth, 1858), but his theories regarding Khufu's pyramid were also the original sin that generated a long chain of fringe ideas about this building (Smyth, 1864).

Petrie (1883: 211–12) was the first to suggest using the pole star at the time of the pyramids (Thuban, α Draconis) to get the meridian line. The idea was to use the maximum eastern and western digression (or elongation) of the star and get the north direction by bisecting these positions. Dash (2015) tested this hypothesis with modern facilities, and a 10 m tall plumb line, using Polaris on two different nights separated by 20 days. He got quite a precise result. The question is whether the ancient Egyptians were at all aware of Thuban's maximum elongations. Thuban is only a 3.6 magnitude star, dimmer than Polaris (2.5 mag), and precise observations might have been far more difficult. In any case, this is a method that warrants consideration, especially when the orientation of the descending passageways are to be taken into account.

Lightboy (2020) has recently re-enacted this idea by suggesting that the so-called trial passages located at Giza close to Khufu's pyramid could have been used as a device to observe Thuban's maximum elongations in a dark, collimated, and secure environment with no wind to disturb the long plumb line (see Figure 4). This hypothesis, which has used all the facilities of modern 3D modelling techniques (often called experimental archaeology), is indeed attractive and very challenging. However, our personal impression is that scholars should not put all their confidence in models produced at desks with sophisticated technologies and should be aware that the ancient Egyptians did not have these tools. As the first author's former PhD student, Andrea Rodríguez-Antón likes to say, 'nothing compares to fieldwork!', and she knows what she is talking about after dealing with plans, remote sensing, 3D models, and on-site measurements of more than one hundred Roman sites (e.g. Rodríguez-Antón et al., 2018).

The question would now be how to extend a baseline produced at the trial passages to the rest of the pyramid complex. Recent work by astronomer Henk Spruit

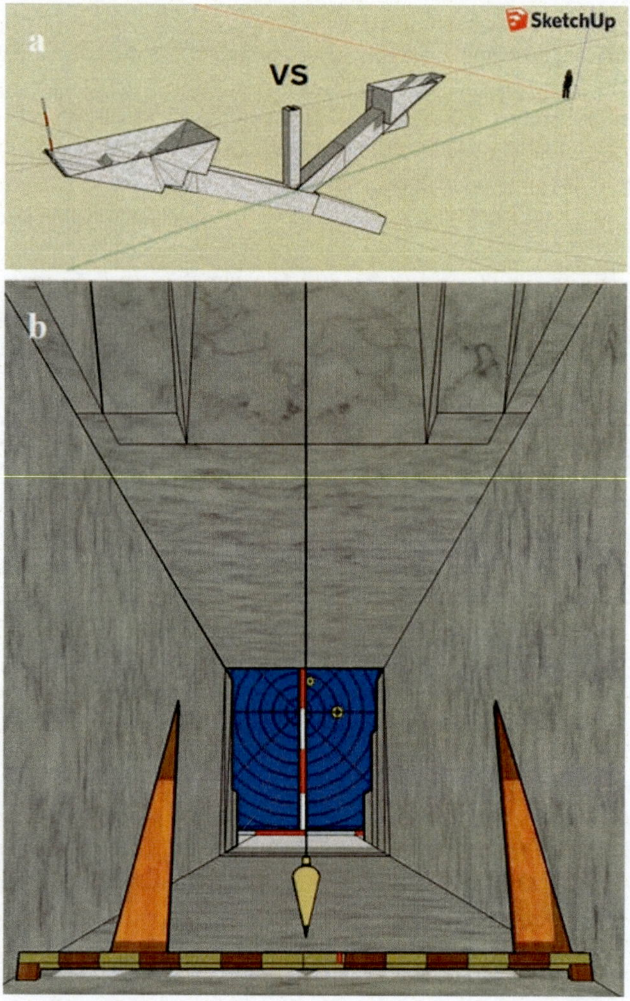

Fig. 4 (**a**) General arrangement of the so-called trial passages of the Khufu complex at Giza. There are three entrances into the lower passages, with two sloped passages and one vertical shaft (VS). According to Lightboy's (2020) suggestive hypothesis, these structures could have been used as a device to find the meridian line through observation of the pole star, Thuban, within a secluded dark environment. (**b**) The vertical shaft (VS) would have saved the long plumb line required from critical weather conditions. (Images kindly provided by David Lightboy, adapted from Lightboy (2020: figs. 2 & 6))

(in preparation), who has kindly given us access to his most recent hypotheses, can help with this. He suggests that a couple of poles such as those used in the stretching of the cord ceremony (Belmonte et al., 2009) may suffice. He has physically tested a procedure, perfectly available to ancient Egyptian topographers, by means of which he could extend a 5 m long baseline to a distance of 230 m with a precision better than one arc minute (1′). For longer distances and parallel lines, the errors would increase but would still remain within reasonable limits. This challenging experiment could help us to understand why the two huge pyramids at the Giza Plateau had such similar numbers and precision in their alignments.

Many authors, however, have not been happy with the pole star method in the past. Hence, Edwards (1993), one of the greatest specialists in the study of the pyramids, suggested in the 1950s the use of a circumpolar or nearly circumpolar star rising and setting on an artificial horizon as a reliable and precise method of determining the meridian line by bisecting the angle formed by both positions (see Figure 5). This proposal was quite intelligent but suffered from a formal defect

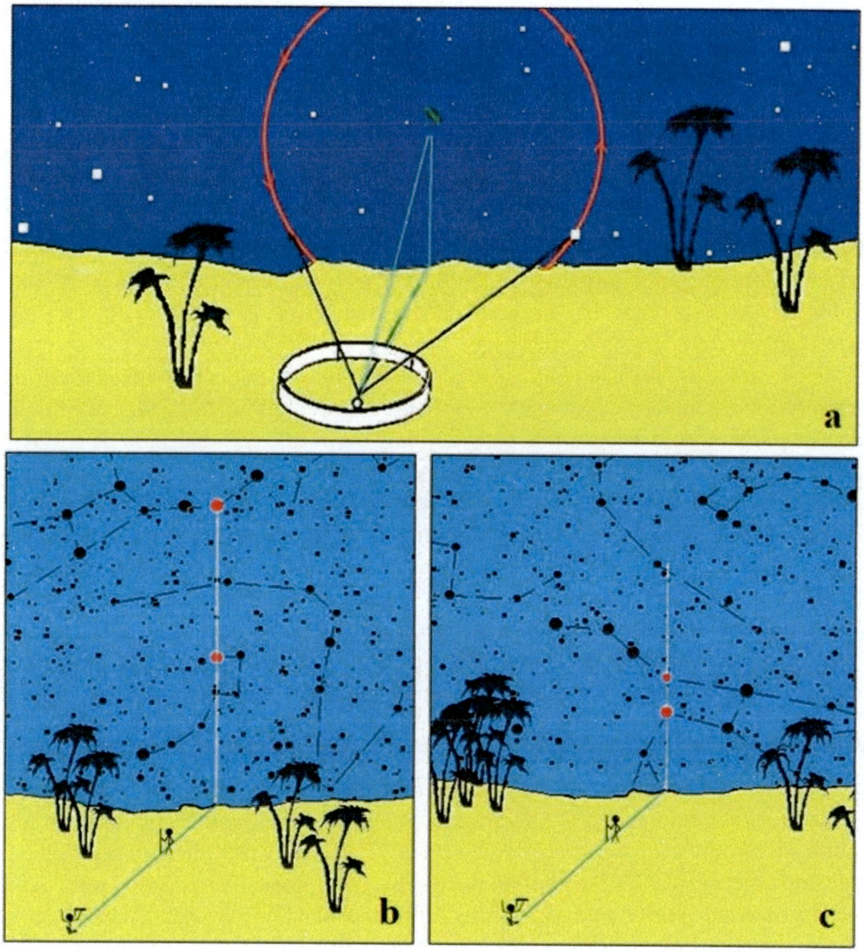

Fig. 5 Three different methods of orientation of the pyramids using stars circling the pole. (**a**) The circumpolar star model as advocated by Edwards (1993). A circumpolar or nearly circumpolar star is observed rising and setting on an artificial horizon, while the bisector of the angle of the two positions gives us the meridian line. (**b**) Spence's (2001) model for the simultaneous transit of two stars on opposite sides of the celestial pole. (**c**) An alternative model with two stars on the same side of the pole, as proposed by Belmonte (2001). See the text for further details. (Diagram by the authors adapted from Lull (2004))

owing to the lack of any textual or archaeological support—artificial flat horizon devices have never been found in Egypt. Because of the prestige of that author, however, this model has been continuously referred to in the Egyptological literature.

Actually, this was the most accepted method for decades, even after Haack (1984) discovered that errors in the orientation of the pyramids of the 4th Dynasty seemed to follow a certain temporal evolution—from lower to higher precision, then decreasing again—rather than showing a random pattern as would be expected from Edwards's model. Therefore, Haack suggested the use of the rising and setting position of a certain star in opposite directions of the horizon (in particular β Sco). These locations would move over time on the horizon owing to the precession of the equinoxes. However, it is easy to show that it is very difficult to determine the position of a star rising and setting above the horizon, even in the case of the brightest luminaries, with an error of less than 1° (see Chap. 6), so Haack's hypothesis was never taken into account.

However, ideas are never completely lost. At the end of the last century, just over two decades ago, the British Egyptologist and historian of architecture Kate Spence rediscovered Haack's trend and presented the theory of the 'simultaneous transit' of two stars as a paradigm to solve the enigma of pyramid orientation and, in passing, promote a new chronology that was supposed to be revolutionary. This work was published in the last issue of *Nature* of the 20th century, and hence it was given wide coverage.

The hypothesis consisted of the use of the (celestial) meridian transit of a pair of stars, one of them at its upper and the other in its lower culmination, whose vertical line of union would necessarily pass through the north celestial pole at a given moment (see Fig. 5). Because of precession, in a period of time before and after that instant, this method would have given increasingly imprecise alignments over time (see also Kolbe, 2008). After performing numerous tests, Spence proposed the pair formed by Mizar (ζ UMa) and Kochab (β UMi). The method included the unique corollary that Khufu's great pyramid was aligned around 2478 BC, at least 80 years later than the lowest dates admitted at the time by Egyptologists (see Table 7.1).

In its day, the authors found this work tremendously inspiring. Belmonte (2001) accordingly proposed an alternative version of this theory.

At this point, it would perhaps be interesting to ask what the Egyptians of the 4th Dynasty tell us in this regard. There are reports from the 1st Dynasty onwards of the existence of the ceremony of the stretching of the cord in which the king and the goddess Seshat determined the main axis (actually the four corners) of a construction by tensioning a rope between two poles. To do this, the Pharaoh observed a celestial object (a star or asterism) in the position diametrically opposite him. There are no contemporary references dating from the 4th Dynasty concerning which astronomical objects were used in this ceremony. However, the references we have, from the Ptolemaic period, repeatedly mention Meskhetyu, one of the celestial adzes, and a post-mortem destiny of the king's soul (Belmonte et al., 2009, and references therein) as the appropriate celestial target. The Egyptians identified Meskhetyu, in the form of a bull's foreleg, with the asterism of the Plough in the constellation of Ursa Major (see Chap. 4).

Taking into account all these premises, Spence's hypothesis could be seen from a different point of view. This idea was inspired by how Polaris is today located in the sky by extending northwards the line joining Merak and Dubhe (β UMa and α UMa, respectively).

Coincidentally, in the first half of the third millennium BC, the line joining another pair of stars in the asterism of the Plough, Phecda (γ UMa) and Megrez (δ UMa), pointed more or less towards the pole star of that epoch. That star was Thuban (α Draconis). Belmonte (2001) used this coincidence for a new working hypothesis: the orientation, following the meridian line, could be achieved by observing the meridian transit of this pair of stars (see Fig. 5). In the process, he discovered that the idea had already been suggested by the astronomer Bedrich Polák decades ago, although he had not verified it mathematically (Polák, 1952: 219; see also Lull, 2004: 295–296). His discovery had passed largely unnoticed west of the Iron Curtain.

This hypothesis collected the best of Haack's and Spence's hypotheses but in a simpler way. It made use of historical information and particularly simplified the observation of the phenomenon, either through the poles mentioned in the stretching of the cord ceremony, or through the use of more sophisticated instruments, such as a *merkhet* or the symbol of the deity Seshat (see Chap. 2), although these would only be useful at lower culmination (see Figure 6).

Moreover, this working hypothesis posed no apparent chronological problems. This last point is especially delicate since the maximum precision would have been achieved around 2562 BC and, consequently, Khufu's pyramid could have been aligned at dates close to 2556 BC, with Meskhetyu in lower culmination (see Chap. 7). Spence's controversial dates may be quarantined. However, Khaefre's pyramid would be observed with the asterism in upper culmination.

In the decade after Belmonte postulated his hypothesis, he realized the enormous difficulty of observing the upper transit of a star at an angular height of more than 45° (or alternatively a zenith distance below that figure) in the sky (he tested this personally at Teide Observatory) when trying to fix an alignment, not so for timekeeping. In this context, it is difficult to imagine an instrument that allows an accuracy of a few minutes of arc under such uncomfortable circumstances, certainly not in the Egyptian sphere. Therefore, Belmonte (2012: 188) favoured the use of only the lower meridian transit of Phecda and Megrez as the astronomical phenomenon observed for the orientation of the great pyramids of Giza, and perhaps other buildings of the 4th Dynasty (see Fig. 6).

This was the case, even though it implies having to abandon the correlation between error in the alignment and the epoch of commencement of construction that had permitted venturing a chronology for the kings of the 4th Dynasty. However, through this new prism, a second, perhaps much more interesting, possibility arose: the near contemporaneity of the orientation process of the two great pyramids, those of Khufu and Khaefre (see Chap. 6). If Khufu's pyramid was the first monument to be aligned, a date of *c.* 2556 BC would be obtained. If it was Khaefre's pyramid or either of the two pyramid descending passages (see Table 1) that were the first structures to be aligned, a date of *c.* 2544 BC would have resulted instead. An error of

Fig. 6 Diagram showing different sketches associated with the stretching of the cord ceremony and how this could be related to the orientation of the great pyramids at the Giza Plateau. The sign of Seshat, as represented in Old Kingdom reliefs (upper row), suggests something more sophisticated, maybe a transit instrument (see also Fig. 2.16). The ceremony was indeed performed in Dahshur, where relief fragments of it have been found in Sneferu's valley temple. The simultaneous lower culmination of Phecda and Megrez could be used as the target for orientating the pyramids in the interval 2550 ± 10 BC (*c.* 2556 BC if Khufu's is considered, *c.* 2466 BC if Khaefre's sector was the one selected). The orientation of most of the elements in the pyramid complex (the other pyramid, minor pyramids, mastaba fields) could have been obtained by simple trigonometric and topographic procedures. (Diagram by the authors, courtesy of IAC-SMM)

plus or minus 4 years in either direction is indeed appropriate considering human capabilities (see Dash, 2015). Hence, if Belmonte's model was correct, a date of 2550 ± 10 BC could be used as a reference time framework for the apogee of the 4th Dynasty.

Variants of the simultaneous transit model have been produced in recent decades. For example, the Ukrainian scholar Alexander Puchkov defends a hypothesis with a peculiar stellar configuration involving the upper culmination of Meskhetyu.[1] This needs dates for the 4th Dynasty and the entire Old Kingdom that are 200 years older than most recent hypotheses making use of astronomical and C14 dates would support (see Chap. 7).

[1] See https://www.academia.edu/41240818/_Stretching_of_the_cord_ceremony_for_astronomical_orientation_of_the_Old_Kingdom_pyramids_2019, for a preprint of Puchkov's as yet unpublished work.

The authors' impression is that, with new and improved programs simulating the sky, such as Stellarium, new 3D modelling facilities, and the amelioration of remote sensing techniques available to both scholars and the interested lay researcher, we shall be presented with more and more sophisticated models for explaining the orientation of the Egyptian pyramids. This does not mean that a final universally accepted solution will be reached. Quite the contrary, researchers will tend to favour their own paradigms. The authors have selectively done just this for Giza and they may indeed be wrong.

However, our contention is that this will remain an open question unless substantial new evidence rises from beneath the sands of the Egyptian desert. Surprises are always possible as amply demonstrated by the recent discovery of papyrus fragments of Khufu's pyramid accountancy in a barren spot of the Red Sea coast (Taller & Lehner, 2021).

Other aspects of astronomical symbolism in the pyramid that the authors have dealt with elsewhere will not be treated in this volume, as the authors have independently covered a few of these topics in earlier works (e.g. Lull, 2004: 302–322). Belmonte (2012: 200–205) discussed the controversial issue of the air-shafts in the Great Pyramid as hypothetical stellar channels. Krauss (2019) has recently presented a very skeptical approach to the topic.

Before concluding this theme, it is perhaps worth mentioning the ongoing discussion concerning the slopes of the pyramids and how this was related to the *seked*, the Egyptian unit used to define this magnitude (Rossi, 2004). Belmonte (2012: 188–194, and Table 5.1) discussed this topic and how these slopes could have represented a solar symbolism (see Chap. 6). Interestingly, in the Middle Kingdom, when standard values of the *seked* were in use, a stellar symbolism could also be inferred directly from the text (see Figure 7). This opens up perspectives for the future.

This book is the fruit of both authors' joint passion for astronomy and Egyptology. It aims to address a series of open questions of the type that fascinate seekers of unsolvable enigmas concerning the relationship of the ancient Egyptians with the Universe. We shall see that most of these apparent enigmas can be easily solved and indeed contextualized within Pharaonic culture. This is therefore a volume on the Egyptian world-view.

Chapter 1, 'Cosmogonies: the formation of the ancient Egyptian Universe' deals with those aspects of Egyptian religion that are directly and unquestionably connected to the sky. The solar cults and the different—often contradictory—cosmogonies that rely on the origin of the Universe will be discussed. This is needed in order to understand what then follows.

Chapter 2, 'Egyptian astronomers throughout history', deals with the people doing astronomy in ancient Egypt, and describes the different roles Egyptian astronomers might assume, from hour watchers to dedicated star observers and excellent architects.

Chapter 3, 'Measuring time: ancient Egyptian star, water, and solar clocks', is the most technical chapter of all. We take a tour of the different ways that the Egyptians used to measure time. This is the chapter where we will be mostly perched on the 'shoulders of giants', since the Egyptological evidence is abundant.

Fig. 7 The Egyptian *seked* depicted as the slope of one of the faces of the pyramidon of Amenemhat III's pyramid at Dahshur. In this case, a 5/7 *seked* is defined, i.e. 5 cubits in the horizontal per 7 cubits in the vertical. While the eastern face offers a canonical solar symbolism (notice the winged sun), possibly related to sunrise, the hieroglyphic text written in the northern face is unexpectedly related to the southern constellations. King Nima'atre accordingly ascends to the sky to place himself as a star higher than the constellation of Sah. This star could be Betelgeuse (see Chap. 4). The northern face of the pyramid would supposedly have acted as a device (perhaps as a symbolic ramp or stair) to enable such an event to take place. Image by the authors, courtesy of the Egyptian Museum in Cairo.

Chapter 4, 'Mapping the stars: the skies of ancient Egypt', is where the constellations, stars and other celestial bodies of ancient Egypt will be presented, all within the framework of a highly developed astral eschatology. We visit beautiful astronomical ceilings, such as Senenmut's celestial diagram that will serve as a master model for our hypotheses, and we try to reveal the keys hidden at the famous Zodiacs of Dendara, along with many other issues.

Chapter 5, 'The Calendar of ancient Egypt: a gift of the Nile', will focus on the genesis and evolution of one of the greatest inventions of the ancient Egyptians, their 365-day civil calendar. The importance and the keys of this splendid device will be analysed in perspective. We will also study the festivals and their impact on the study of chronology, one of the great controversies of modern Egyptology.

Those expecting a convoluted debate on inextricable arcane matters will probably be disappointed.

Chapter 6, 'Egyptian land- and skyscapes', will focus on expounding the best and most spectacular results obtained by the Egyptian–Spanish Mission on 'Archaeoastronomy of Ancient Egypt' in the 2000s, upgraded and updated by the authors' own research in the field over the last decade. A series of case studies on the most important pyramid complexes and great temples of Egypt will be examined. An excursus to ancient Kush will also be discussed. The interaction between Egyptian sacred landscapes, using the Nile as a paradigm, and solar, lunar and stellar skyscapes will be emphasized.

Chapter 7, 'Astronomy and Chronology', discusses the various chronologies that have traditionally been proposed for the history of ancient Egypt. These are contrasted with the most recent and contradictory analyses on the topic (a fraught enterprise indeed). The peculiarities of each of these proposals, along with their advantages and disadvantages are debated. Astronomical dating is confronted with its nemesis, C14 dating.

The volume ends with a brief glossary that should be useful for the inexperienced reader and there is an extensive bibliography. The latter has not been organized by chapters in order to avoid repetitions, optimize space, and follow Springer's 'Historical and Cultural astronomy' series tradition. The interested reader can use these references in search of more detailed or accurate information.

This is not a standard volume where we are going to focus on every single detail of the hieroglyphic and historical sources available, or offer detailed descriptions of the monuments on which we have worked, unless it is necessary for following the argument. This work does, however, reflect the various facets in which astronomy has proved useful as a tool at the service of culture. The Egypt of the Pharaohs used it for varied purposes and in diverse areas, such as time reckoning, architecture, religion and eschatology, royal power, and even the economy. We hope you will enjoy it.

La Laguna, Spain
Bellaterra, Barcelona, Spain

Juan Antonio Belmonte
José Lull

Acknowledgements

This book has been written with great enthusiasm. Not in vain, it reflects the efforts of more than a quarter century in Herodotus' Gift of the Nile by the part of the authors: Juan Antonio Belmonte Avilés and José Lull García. We have studied the sources and analysed the texts. We have also carried out fieldwork in almost every corner of the country, from the Delta to the Sudanese border, and from the distant Gilf al-Kebir and the oases of the Western Desert, Siwa included, to the mountains of the Eastern Desert and the Sinai Peninsula and beyond. This has led us to visit more than a hundred archaeological sites.

Juan's travelling companions in part of this adventure during the 2000s were his Egyptian colleagues: the late astronomer Mosalam Shaltout, and the Egyptologist Magdi Fekri. Without them, the decade-old Egyptian–Spanish Mission for the archaeoastronomy of ancient Egypt would have been an impossible task. Thank you for making Juan feel at home in Egypt on so many occasions. The Mission's support by the Supreme Antiquities Service, its inspectors, its local representatives, and its former Secretary General, Zahi Hawass, was also indispensable.

For the analysis of many texts and iconographic sources, Juan has been fortunate to have the invaluable collaboration of his hieroglyph teacher, the Egyptologist Miguel Ángel Molinero. It is a great advantage to deal with the often devilish Egyptian astronomical texts in their original language. Juan's fellow students Noemí and Dani, helped make that learning much more enjoyable. To work with Giulio Magli has been inspiring. He is always full of smart, innovative ideas. Carmen Pérez Die, Juan's *mudira* at Herakleopolis Magna, has been an inspiration over the last decade. Thank you for sharing your precious wisdom with him. Juan first met Maitane Urrrutia Aparicio after a lecture he gave on Nefertiti nearly a decade ago. Now she is his brilliant PhD student, always open to discussion, and has been responsible for some of the graphics presented here. Her co-supervisor, César González García is an irreplaceable presence in Juan's list of scholars of greatest confidence. He has done us the huge favour of writing the foreword in record time, considering his skills, duties and work overload as one of the most reputed cultural astronomers worldwide *¡Gracias César!*

José's passion since childhood for astronomy and astronomical observation has led him, as an Egyptologist, to Egyptian astronomy. In this astronomical journey, the *Agrupación Astronómica de La Safor* (Gandía, Valencia) was of the greatest importance for innumerable years. The many hours of debate between good friends and colleagues, and the many nights of observation will always stay in his memory that he longs to share again with the same intensity.

Regarding Egyptology, José will always remember his years at Tübingen, where the *Schloss Hohentübingen* and the *Ägyptologisches Institut der Universität Tübingen* became his second home. Frau Prof. Dr. Ingrid Gamer-Wallert and her family welcomed him as a son, and he will always keep them in his heart. These were unforgettable and very productive years, where learning with her, and with lecturers such as Herr Prof. Dr. Wolfgang Schenkel, Prof. Dr. Shafik Allam, and Dr. Farouk Gomaa was fundamental to his career.

Back in Spain, José participated heartily in the great Egyptological project created, through titanic effort, by Josep Cervelló at the *Universitat Autònoma de Barcelona*. Both are now friends, collaborators and companions in victory and defeat. José has continued to work until today within the framework of the UAB and it is there that he has continued, among other things, with his studies on ancient Egyptian astronomy and, above all, with the teaching of this discipline.

The authors are much indebted to the German Egyptologist and chronologist Rolf Krauss. He has often been an excellent critic of our work and has always allowed us to preview his works even before they were published. His knowledge, blended with his irony, have frequently been a breath of fresh air. The pages of this book will certainly reflect some of his ideas, past and present, always controversial, but indeed fascinating.

Rita Gautschy has also been a window open to the world. Her original training as a stellar physicist has parallels with Juan's so they have always understood each other very well. Her math skills are legendary and are proving useful to many Egyptologists, including the authors. Egyptologist Luc Gabolde has always been keen to provide the authors with his most recent hypotheses. We have endeavoured to ensure that his ideas are faithfully reflected in this volume.

The authors' mother tongues are Spanish (Juan) and Spanish and Valencian (José), and their English is not always as perfect as they would wish. Hence, Terry Mahoney's input has been fundamental. His long decades of skill as an editor of astronomy texts have proven to be absolutely vital in improving the quality of our manuscript. There are not enough words to say thank you. *¡Mil gracias Terry!*

When Juan and José first discussed the idea of writing a book in English together, they wondered who would publish it. After Juan's participation in the *Handbook of Archaeoastronomy and Ethnoastronomy* in 2015 and the *Handbook of Exoplanets* in 2018, Springer seemed a reasonable option. We do appreciate the interest shown in our manuscript first by Springer's Editor Maury Solomon and, after Maury's retirement, by her efficient successor Hannah Kaufman. Thank you, Hannah, for your understanding of the difficult family and academic conditions that a pandemic can impose upon the lives of two stressed researchers. Thank you for your patience and confidence. Ramon Khanna, our new editor, and Dinesh Vinayagam, production editor of this volume, helped in ensuring that the final product was the best it could be.

The research presented in this volume was done at a cost. The authors acknowledge support from the State Research Agency (AEI) of the Spanish Ministry of Science and Innovation (MICINN), and the European Regional Development Fund (ERDF) under grants with reference AYA2015-66787-P 'Orientatio ad Sidera IV' and PID2020-115940GB-C21 'Orientation ad Sidera V', and from the internal IAC project P310793 'Arqueoastronomía'.

Notwithstanding, Mar has shared Juan's passions for adventure and distant cultures since their lives crossed almost by chance so long ago. Although she has not always accompanied him, she was his travel companion in their first unforgettable encounter with the Nile three decades ago. Juan's parents always allowed their children to be extravagant, because that is what many people think Juan's job is, and his siblings have never stopped supporting him in his often crazy enterprises. The new sun of Juan's life, his daughter Sol, must forgive his father for being so often absent, either abroad or at the office, and for not devoting to her all the time her luminous character deserves. Like the sun, she was born in peace at the winter solstice of 2009. The future is hers.

Mari Carmen has always been José's great support and knows very well how astronomy and Egyptology are for him like the oxygen he breathes. She has known this well since that Leonid shower of November 2002, which they experienced under a celestial vault decorated with countless stars in an almost unbearable cold. Now, in addition, she has enjoyed reliving with him his childhood dream, to build an astronomical observatory. All the love, joy, and generosity that Mari Carmen transmits to him have helped José to move so far forward with enthusiasm.

José's parents have always been by his side, being his great reference. They have always encouraged him to fight to follow the path he chose. To them, he owes everything and all that he is. Valeria, the daughter of José and Mari Carmen, is an immense treasure that has brought great happiness and joy to her family. José will always be in debt to her and her mother, for all the time he has had to give to his work instead of sharing it with them, even though that time with them would undoubtedly have brought him much more happiness.

Juan and José first contacted and knew of each other's parallel advances in the discipline because Professor Michael Hoskin asked the former to review, for the *Journal of the History of Astronomy*, the PUV 2004 book of the latter. Hence, our beloved Michael is somehow responsible for this volume. It was with great sadness that we learnt of his death while the final pieces of this book were being assembled. An extraordinary colleague, the best of mentors, and a dear friend, he will always remain in our memory.

La Laguna and Gandía
December 2021

Juan Antonio Belmonte and José Lull

Contents

Chapter 1
Cosmogonies: The Formation of The Ancient Egyptian Universe

Ancient cultures developed beliefs around issues as complex as the origin of the world, placing a select group of gods and animated forces as vital elements of that initial ignition that ultimately led to the creation and development of the known world.

In ancient Egypt there was not one but several cosmogonic, parallel and diachronic theories to explain the origin of the world. When comparing them, we observe differences and even variants among them over time, but despite the differences from one to another, common elements stand out, especially in terms of the characteristics of the pre-creation world from which space and time arose.

1.1 Cosmogonic Traditions

1.1.1 Heliopolitan Cosmogony

Of all the Egyptian cosmogonic theories, the Heliopolitan one stands out owing to its antiquity and the abundant references to it. In this creation theory Atum acts as demiurge, although it is possible that in archaic times the falcon god Horus had played this important role (Clagett, 1992: 279). The main source of information on the Heliopolitan cosmogony is found in the *Pyramid Texts*[1], the oldest known version of which is the one that appears on the inside walls of the pyramid of Wenis (*c.* 2320 BC), the last king of the 5th Dynasty, but it is also documented in a large number of later sources, including the well-known *Coffin Texts* and the *Book of Coming Forth by Day* (formerly known as *Book of the Dead*). In these sources the initial protagonist god, the one acting as demiurge, is a solar god, Atum, although he is also sometimes referred to as Re or Khepri.

[1] For *Pyramid Texts*, *Coffin Texts* and the *Book of Coming Forth by Day* we use the following abbreviations: PT, CT and BD, respectively.

J. A. Belmonte, J. Lull, *Astronomy of Ancient Egypt*, Historical & Cultural Astronomy, https://doi.org/10.1007/978-3-031-11829-6_1

Atum alone is the creator god, *ḫpr ḏs.f* ('by himself') and is the first of the Heliopolitan ennead. His loneliness is also a reflection of an inertness time (Pires, 2019: 146). In the *Coffin Texts* (CT 306), with parallels in chapter 79 of the *Book of Coming Forth by Day* (Allen, 1960: 154), its creative facet is recognized on numerous occasions: 'Hail to you, Atum, you who made the sky and created what exists, who came forth from the earth, who created seed, Lord of all things, who fashioned the gods' (Faulkner, 1994: I, 224).

But before Atum came into existence, the Egyptians describe a pre-creation space that, in some way, tries to capture the idea of emptiness, although not in a strict sense, since that space contains the essence (although inactive) of the demiurge. According to Assmann (2019), 'the pre-world is, according to Egyptian thought, neither chaos nor a gaping void, but rather an embryonic pleroma'. This pre-creation realm of the world is expressed by a primordial ocean, the Nun, characterized by its extreme darkness, its stillness, and its silence. It should be borne in mind that although the pre-creation world is often defined as chaotic, it should in no way be viewed as negative or destructive. In fact, PT 607 §1701 (M 352) and 627B (N 32) indicate how the king is born from the Nun and how he is protected by Nun from the 'gods of disorder'. As Popielska-Grzybowska (2017: 19) points out:

> Passage through Nu or existence in Nu augured evolution of a being, a change so vividly crucial in the progressive metamorphoses of the pharaoh. Thus, in the *Pyramid Texts* the emphasis is put on living and not destruction. It seems obvious that immersion into primordial non-existence, which simultaneously was the beginning of every element of life, was indispensable to undergo a final metamorphosis into the creator Atum and hence assume a final form.

That is why the chaotic image of the Nun, as something negative, does not correspond to the Egyptian reality, since the references to the passage through the Nun in the funerary context are always propitious.

The serpent Apophis, a well-known symbol of chaos that every day tries to disrupt order by attacking the solar boat, is not documented in the Old Kingdom, it appears only from the 9th Dynasty onwards (as *ʿ3pp* in a tomb of el-Moalla; Morenz, 2004), not being mentioned in the pre-creacional scope. This detail also shows that the Egyptian vision of the Nun is quite positive, since from it Atum could create himself and thus begin the ordering of the created world.

The *nun* offers a variety of transcriptional variants (*nww*, *njw*, *nw*, *nnw,* or *nwnw*) (Gardiner, 1957: 530 W24). In the *Pyramid Texts* of King Wenis the most usual variant is , while in immediate examples it is usual , and with Pepi II the determinative of 'god' () is added. As Allen (1988: 4) points out, on occasion, some written variants seem to seek a similarity with the verb *njnj* ('be inert'), since being inert is ultimately a quality of the Nun. The word, according to Allen (1988: 4), could come from an old form for 'water'. In Coptic it is ⲛⲟⲩⲛ, translated as 'abyss' or 'depth' (Crum, 2000: 226), a word whose phonetics has led to the more widespread use of the form Nun.

The determinative of heaven for Nun () is documented in PT 585 §1583a (Nt 7). Subsequently, both and are common forms. These written variants are very significant, since they use the sky/sky and water signs as determinative, which also offers clues about some of the qualities of the Nun, such as its association with the watery and celestial element.

In CT 76 it is said, *ḫpr.n Jtm jm m ḥḥw m nww m kkw m tnmw* ('Atum came into being in chaos, in the Abyss (Nun), in darkness and in gloom'; Faulkner, 1994: I, 78). Here *ḥḥw* and *nww* are associated with the aquatic, while *kkw* and *tnmw* relate to darkness, so that the characteristics of the pre-creation world are clearly defined by watery and total darkness.

As Meeks and Favard-Meeks (1994: 29–30) point out, although the texts say that nothing existed, the abstraction of nothingness did not fit into Egyptian thought, so pre-creation space was also described by everything that did not exist. Seen this way, it was the very negation of existence, a state of not being. Bickel (1994: 31), however, opines: *l'absence de composants n'est que le corollaire de l'etat primordial, non sa definition* ('the absence of components is merely the corollary of the primordial state, not its definition').

In PT 486 §1040 a–d (P 338) we read:

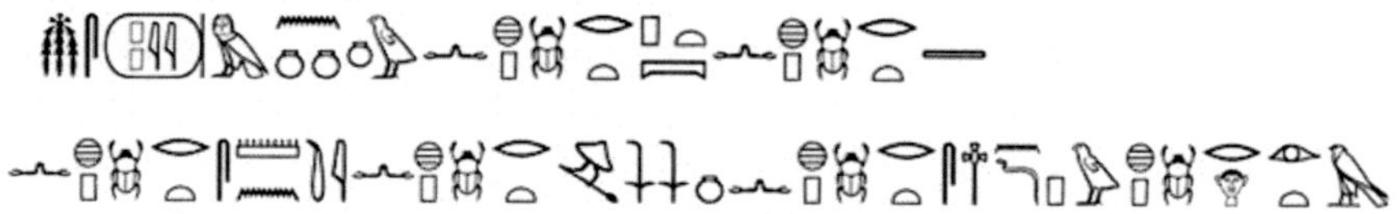

§ 1040 a–b *msjw Ppy m nww nj ḫpr.t pt nj ḫpr.t tꜣ*
§ 1040 c–d *nj ḫpr.t smntj nj ḫpr.t ḫnnw nj ḫpr.t snḏ pw ḫpr ḥr [jrt] ḥrw*

> Pepi was born in Nu  (Allen, 2013: IV, 165) when the sky had not yet come into being, when the earth had not yet come into being, when the establishment (of the world) had not yet come into being, when disturbance had not yet come into being, when the fear that came into being because of Horus's [eye] had not yet come into being. (Allen, 2005: 133)

In the previous fragment, belonging to the version of King Pepi, the late Pharaoh acts as Atum. Before anything had been created the essence of Atum was therefore in the Nun. In his pre-creation state Atum himself is defined by his serpent form, for this form may reflect the pre-existential state. According to Hornung (1983: 178), this serpent must represent non-existence, to which the creator god himself will arrive at the end of time, thus being an element of pre-existence.

In the case of Atum, his name (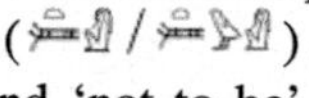) comes from the verb  *tm*, which means both 'to complete' and 'not to be', the name of the god probably meaning 'the completed one' (Allen, 1988: 9), although there are several proposals that have been made about its meaning (Myśliwiec, 1979: 78–83). On the other hand, Anthes (1983: 129) postulated that his name should mean 'the one who has been completed by absorbing the other first living beings', so we could interpret this to mean that Atum is all the vital essence that remained inactive in the Nun.

A whole series of small descriptions provide clues about Atum's state before his own existence. In CT 80 it is said, *je flotte étant entièrement engourdi, mes membres (?) étant inertes* ('I float entirely sluggishly, my members (?) being inert'; Bickel, 1994: 48), while in CT 132 it is clarified, *c'est moi, celui qui a rompu son repos au temps de Noun* ('it is I that disturbed my repose in the time of Noun'; Bickel, 1994: 35), after being in that state of inertia alone, in the Nun. Therefore, the parts of Atum were inert in the Nun, in the time of Nun (*m rk nww*); that is, in the time of pre-existence. This 'time of Nun' is also mentioned in CT 321 (IV, 146c) when Hapy indicates that its antiquity dates back to the time of Nun (Bickel, 1994: 128). In CT 76; on the other hand, the place where Atum was developed is described as a site characterized by / *ḥhw* ('flood'), / *nw* ('waters'), *tnmw* ('chaos'), and *kkw* ('darkness').

In the *Coffin Texts* (CT 714) a temporal concept marks the beginning of time, linked to the process of creation:

> I am Nu , the Sole One who has no equal, and I came into being yonder on the great occasion of my flood, when I came into being (…) I am he who originated in the Abyss, and see, the Chaos-god came out of me; see, I am hale. I brought my body into being through my power; I am one who made myself, and I formed myself at my will according to my desire. (Faulkner, 1994: II, 270)

Although the text seems to convey Nun's words, it is very possible that it is actually Atum himself who is speaking, so that following Bickel (1994: 58), the beginning of the text could be amended as, *Je suis [celui qui est sorti du?] Noun* ('I am [that which was born of?] Noun'). The concept *sp wr n mḥt.j* ('the great occasion of my flood') refers just to the moment when the god came into existence, so it must be considered the beginning of time.

Atum, after emerging from the Nun as *benben* ('primordial hill'), is the creator of the first pair of gods, Shu and Tefnut. In PT 600 §1652–53 (N 359) this process is explained as follows: 'Atum Beetle! You became high, as the hill; you rose as the *benben* in the Benben Enclosure in Heliopolis. You sneezed Shu and spat Tefnut' (Allen, 2005: 269), but in PT 527 §1248–49 (P 475): 'Atum is the one who came into being as one who came (with penis) extended in Heliopolis. He put his penis in his fist so that he might make orgasm with it, and the two twins were born, Shu and Tefnut' (Allen, 2005: 164).

Also in the *Book of Coming Forth by Day* (BD 15 A3), following the Heliopolitan doctrine, Atum is recognized as the demiurge. It is also pointed out that with his creation time began:

> Thou art lord of sky and earth, who made the stars above and mankind below, sole God, who came into being at the beginning of time, who made the lands and created common people, who made the deep and created the inundation, who made the water and gave life to what is in it, who fashioned the mountains and brought into being man and beast. (Allen, 1974: 18)

The means used for creation has many variations. In CT 75 (Faulkner, 1994: I, 337–340) it is said that Shu was created 'from the flesh of the self-created god', 'by his wish', 'exhaled from his nostril'. In CT 1130 (Faulkner, 1994: VII, 465), on the other hand, it is said that it was from his sweat that the gods arose: 'Words spoken by him whose names are secret, the Lord of All (…) I created the gods from my sweat, and mankind from the tears of my eye' (Faulkner, 1994: III, 167). The creation of humankind from tears  *jw rmṯ m rmwt jrt.j* ('people are the tears of my eye') probably seeks assimilation with the word used precisely to designate people (Mathieu, 1986).

In a hieratic papyrus of the 19th Dynasty (pTurin 1993) known as *The Legend of Isis and the Name of Re*, of which part of the same text is also preserved in the Chester Beatty XI r° papyrus (Gardiner, 1935), it is Re who appears as demiurge and creator of all that is known:

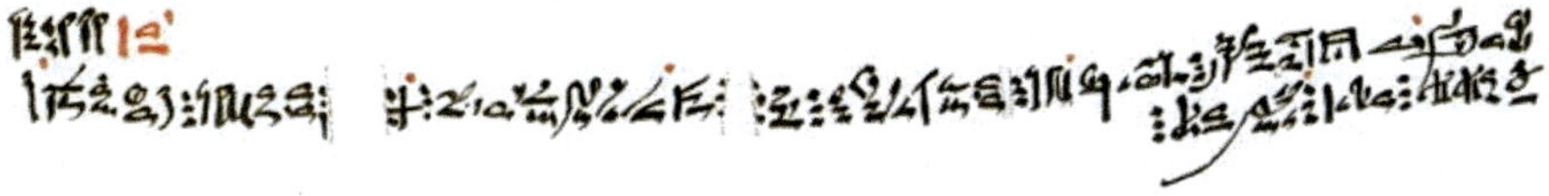

> Spell of the divine god, who came into being by himself, who made heaven, earth, water, the breath of life, fire, gods, men, flocks, herds, reptiles, birds, and fish, the kingship of gods and men altogether, with limits beyond numerous years, […] and with numerous names. (Ritner, 2003: 33; Rossi & Pleyte, 1869–1876: pl. CXXXI, l. 12–14).

Atum, Re, and Khepri are solar gods, forms of the same god and for that reason there are texts in which one or the other appears as demiurge. Thus, in the hieratic papyrus BM 10188, dated in the year 12 of Alexander IV (311 BC), is preserved *The Book of Knowing the Creations of Re and of felling Aapep* (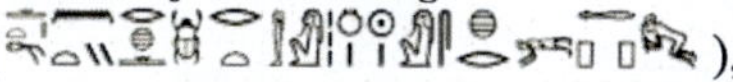), where Khepri acts as demiurge.

Thus spake the Lord of All after he had come into being: It was I who came into being as Khopri. When I came into being, 'Being' came into being, and all beings came into being after I came into being; manifold were the beings which came forth from my mouth ere the sky had come into being, ere the earth had come into being, ere the ground and reptiles had been created in this place. I created (some) of them in Nun I as Inert Ones when I could as yet find no place where I could stand. I considered (?) in mine heart, I surveyed with my sight, and I alone made every shape ere I had spat out Shu, ere I had expectorated Tefenet, ere there had come into being any other I who could act with me. I planned with mine own heart and there came into being a multitude of forms of living creatures, namely the forms of children and the forms of their children. I indeed made excitation with my fist, I copulated I with mine hand, I spat with mine own mouth; I spat out Shu, I expectorated Tefenet, and my father Nun brought them up, mine Eye following after them since the aeons when they were far from me. After I had come into being as sole god, there were three gods in addition to myself. I came into being in this land and Shu and Tefenet rejoiced in the Nun, in which they were. They brought back to me mine Eye with them after I had united my members; I wept over them, and that is how men came into being I from the tears which came forth from mine Eye, for it was wroth with me when it returned and found that I had made another in its place, having replaced it with the Glorious (Eye). So I promoted it in my face, and when it exercised governance over I this entire land, its wrath died away (??), for I had replaced what had been (?) taken from it. I came forth from the roots, I created all reptiles and all that exists among them. Shu and Tefenet begat I Geb and Nut, and Geba and Nut begat Osiris, Horus Mekhantenirti, Seth, Isis, and Nephthys from the womb, one after the other, and they begat their multitudes in this land. (Faulkner, 1937a: 172, 1933: 60–61, l. 26:21–27:5)

This text summarizes the entire creative process from the origins of the solar god in the Nun, in his Khepri form, to the formation of his offspring, thus completing the Heliopolitan ennead and all other known beings. Reptiles appear not only at the time the primordial hill came into existence but even earlier, in the Nun, as inert (that is, as yet undefined) essential elements that would come to life during the process of creation. The image of the serpents is thus very evocative, representing the pre-creation element from which creation can later emerge. On the other hand, the epithet that the god receives as 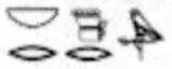 *Nb r ḏr* ('Lord of All') but literally translated as 'Lord to the Limit', already appears in CT 534 (VI, 131) as . It can have both a spatial and a temporal connotation. The limit, in both senses, takes the domain of the creator god to the boundary of the Nun. It is an epithet used by other gods of a demiurgic character, including the goddess Hathor in Dendara 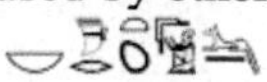 *nbt* (*r*) *ḏr jmnt sštꜣ,* 'Lady (to) the Limit, hidden of image' (Richter, 2012: 211).

Later, the same papyrus explains again the process through which existence came to be. In the same way as in the previous fragment, the beginning of the explanation is a complex play of words where sound, in its magical-religious aspect, should play an important role:

> Thus spake the Lord of All: When I came into being, 'Being' came into being. I came into being in the form of Khopri who came into being on the First Occasion; I came into being in the form of Khopri when I came into being, and that is how 'Being' came into being'. (Faulkner, 1937b: 41, 1933: 69, l. 28: 20–21; see also, Sauneron & Yoyotte, 1959: 49–50)

Next, the solar divinity insists, in his role as the primordial god, that nothing was done before he existed, and that he himself was the one who created the primordial gods:

> I was more primaeval (?) than the primaeval ones whom I had made; I was the most primaeval (?) of the primaeval ones, and my name was more primaeval (?) than theirs (?), (for) I made primaeval time and the primaeval ones. (Faulkner, 1933: 69, l. 28: 21, 1937b: 41)

New words of interest appear in the text we have just seen. On the one hand, 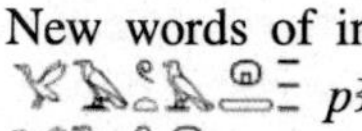 *pꜣwtjw* (the 'primaeval ones'), the oldest gods, and, on the other,  *pꜣwt* ('primaeval time'), which will also be used to refer to the moment of regeneration in the funerary sphere.

As indicated in CT 80, Atum, being in the Nun, could not find a place to stand: *sk wj wꜥ.kwj ḥnꜥ Nww m nnwt n gm.n.j bw ꜥḥꜥ.j jm n gm.n.j bw ḥms.j jm* ('When I was alone with the Nun, as inert, unable to find a place to stand'). The first terrestrial element that appears from the Nun, in the initial creative process, is the space in which Atum is placed. In CT 79 this space is said to be offered by Mehet-Weret, who came into existence in order to give Atum a resting place (Bickel, 1994: 65).

The Berlin 3047 papyrus (Reymond, 1969: 58) defines this site as the *sꜣ ṯꜣw* ('ground'), which must have been in the middle of the Nun. In PT 600 §1652–53 (N 359) it is said of him: 'Atum Beetle! You became high, as the hill; you rose as the *benben* in the Benben Enclosure in Heliopolis', which is why Atum is identified with the *ḳꜣꜣ* ('(primordial) hill'), with the *bn* or *bnbn*. This assimilation is also described in the *Book of Coming Forth by Day* (BD 79), when Atum is said to be the one 'who made the sky, who created what exists; who emerged as land'.

The Heliopolitan ennead (see Figure 1.1) is a reflection of the parts of the created universe. If Atum (or Re or Khepri) is the demiurge whose essence was already in the Nun, the first couple created, Shu and Tefnut, will constitute divinities linked to the atmosphere or the air. The name *šw* (Shu) possibly means 'void' or 'emptiness', but the etymology of *tfnt* (Tefnut) is unclear (Allen, 1988: 9). The next couple, Geb and Nut, will be associated with the earth and the celestial realm, respectively.

The separation of heaven, earth, and the Nun is already indicated, albeit very briefly, in the *Pyramid Texts*. Differently from other components of the created world, however, the reason for this separation is explained, although not in the oldest documents. Thus, for example, in PT 519 §1208c (P 467) it is said, 'the sky was parted from the land and the gods went off to the sky' (Allen, 2005: 160) and in PT 627 §1778 (M 32) the text speaks of 'him who sweeps the sky away from the earth and Nu' (Allen, 2005: 244).

However, the *Book of the Heavenly Cow* (Hornung, 1982), the oldest (and incomplete) version of which dates from the time of Tutankhamun, explains precisely the moment when Nut became the vault of heaven. According to this book it was Re who, fed up with mankind and wanting to decide what to do, summoned Shu, Tefnut, Geb, and Nut, along with the fathers and mothers who were with him in the Nun, as well as the god (Nun) with his servants (*shentyu*). When Re

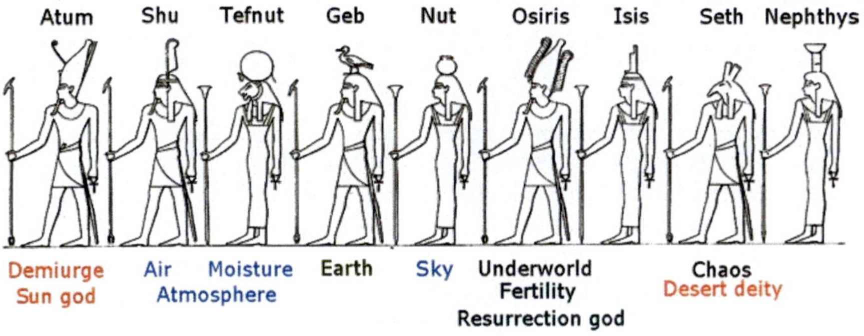

Fig. 1.1 The Heliopolitan ennead. (Adapted from Wilkinson, 2000a, b: 82)

refers to Nun, he says of him that he is the oldest of the gods from which he came into existence, but when Nun describes Re he considers him his son and his superior:

> His Majesty became aware of the plot being contrived against him by mankind, and so his Majesty, l.p.h., said to those who were his retinue, 'Please summon to me my Eye, Shu, Tefnut, Geb, and Nut as well as the fathers and mothers who were in my company when I was in Nun, and also the god Nun, bringing his courtiers/along with him (…). It is within Nun, at the place where I originated, that I shall return.
>
> So these gods were brought [right away], and these gods were (positioned) on either side of him, touching the ground with their foreheads in His Majesty's presence, so that he might state his problem in the presence of the father of the eldest ones, the maker of mankind and king of humanity. They then said unto His Majesty, 'Speak to us so that we may learn about it'. Then said Re to Nun, 'O eldest god in whom I originated and you ancestral gods, see, mankind, who originated from my Eye, has contrived a plot against me. Tell me what you would do about this, since I am seeking (a solution). I cannot slay them until I have heard what you might have to say about this. (Simpson, 2005: 290)

However, Nun still plays a very important role, as before a weary and weak Re he takes the initiative in order to raise Re to the celestial realm, lifting Nut and placing him on his back:

> The Majesty of Nun replied, 'O my son Re, the god mightier than the one who produced you and greater than those who created you, stay put on your throne! The fear of you is great; your Eye shall proceed against those who conspire against you' (…)
>
> The Majesty of this god (Re) then said to the Majesty of Nun, 'My body is weak for the first time. I won't wait until another gets to me'. Then said the Majesty of Nun, '(My) son Shu, [your] eye shall serve [your] father (Re) as his protection. (My) daughter Nut, you shall place him [upon your back]'. Nut replied, 'But how, my father Nun?' Said Nut, 'Don't [⌜be silly⌝], Nun'. [And so] Nut became [a cow]. Then the Majesty of Re [placed] himself upon her back (…).
>
> This god then said to Nut, 'It is in order that I may be uplifted that I have placed myself on your back'. 'What's this?' asked Nut. And so she came to be there in both the heavens. The Majesty of this god [said], 'Stay far away from them! Lift me up! Look at me!' And so ⌜she⌝ became the sky. Then the Majesty of [this] god was visible within her. She ⌜said⌝, 'If only you would provide me with a multitude!' [And so the ⌜Milky Way⌝] came into being.
>
> His Majesty, l.p.h., said, 'Peaceful is the field here'. And so the Field of Offerings came into being. 'I shall make vegetation grow/in it.' And so the Field of Rushes came into being. 'I shall provide them with everything.' And so ⌜planets⌝ and stars (came into being). Then Nut started trembling because of the height. So the Majesty of Re said, 'If only I had millions supporting her!' And so the Infinite Ones came into being.
>
> The Majesty of Re said, 'O my son Shu, place yourself under (my grand)daughter Nut and watch over for me the two groups of Infinite Ones there so that they may live in the twilight. Place her over your head and nurture her'. (Simpson, 2005: 290–293; see also Clagett, 1992: 539–540; Hornung, 1982: 41–42)

What is explained in the previous text is perfectly reflected in the iconography that appears in the tomb of Sethy I (see Figure 1.2). The goddess Nut, in the form of a cow, is supported by Shu while Hehu pairs (the 'Infinite Ones') prop up the four legs of the cow, so that she does not lose balance (see Figure 1.3). The four legs, like four pillars, can be linked to the four cardinal points, since in New Kingdom inscriptions the goddesses of the four cardinal points play an important role in the arrangement of the sky. On the other hand, these Hehu are already mentioned in CT 76 (II, 1) and in CT 78 (II, 19–20), which are said to be those 'who guard the path of Nut which

Fig. 1.2 The heavenly cow in the tomb of Sethy I in KV 17. (Drawn by María Hidalgo Portillo)

Fig. 1.3 Detail of the solar ships, and of Shu and the *hehu* under the heavenly cow in the tomb of Sethy I. (Image by courtesy of María Lull)

is under Atum' (Faulkner, 1994: 81). Stars are represented on the skirt of the cow, on the level where the solar boat sails. These are the *jḫjḫw pw sbꜣw immerleuchtenden Sterne* ('ever-luminous stars') referred to in the text (translated by Wente as "planets' and stars'). In the *Pyramid Texts* these stars appear as *ꜣḫꜣḫw*.

The etymology of the name of Nut is complex but very interesting. The name of the goddess appears as in the first mentions in the *Pyramid Texts*, with the feminine anthropomorphic determinative from the example of Ibi. Already in the *Coffin Texts* we find the inclusion of the symbol of water (Billing, 2002: 11), which gives an idea of one of the aspects that define the sky of Nut. For Barta (1973: 100–101), Nut could have emerged as a female counterpart to Nun. However, Allen (1989: 16) denies this possibility observing that in the *Pyramid Texts* no variant of the name of Nut is found with three *nw* signs (W24), as Nun is written, so it can hardly be her feminine form.

The separation of heaven and earth by the god Shu is also shown in the iconography of *The Fundamentals of the Course of the Stars* (formerly known as the *Book of Nut*), as seen in the cenotaph of Sethy I at Abydos (Frankfort, 1933: pl. 81; *EAT* I: 40 fig. 21) and at the tomb of Ramesses IV in the Valley of the Kings (Hornung, 1990: pl. 68). In fact, it is an oft-repeated iconographic motif, sometimes with curious variations (see Figure 1.4).

Other texts of a funerary nature also make mention of this separation. Thus, for example, in the *Book of the Imiduat* (*sš nj ꜥt jmnt* 'Book of the Hidden Chamber') it is said that Shu 'separated heaven from earth in the primordial darkness' (Hornung, 1963: 188; te Velde, 1977: 161–170).

Fig. 1.4 Nut, Shu, and Geb. https://commons.wikimedia.org/wiki/File:Geb_and_Nut03.png. (Author: E. A. Wallis Budge (1857–1937). Public domain, via Wikimedia Commons)

The creation of heaven appears to be associated with the solar god, as Re or Atum, in the *Coffin Texts*. In CT 306 (IV, 60) Atum is he 'who made the sky (*pt*) and created what exists', and in CT 648 (VI, 270) 'created the mountains and knit the firmament together' (Faulkner, 1994: I, 224; II, 223).

1.1.2 Memphite Cosmogony

Another important cosmogonic tradition is the one in which the god  Ptah of Memphis is the protagonist of creation (see Figure 1.5). This cosmogonic tradition is little known until the New Kingdom, although already at this time Ptah appears in

Fig. 1.5 Standing figure of Ptah, Late Period, MET Museum Accession Number 26.7.881. (Public domain image, courtesy of the Metropolitan Museum of New York)

many texts as the creator god. Thus, for example, in the papyrus Harris I (BM EA9999/2), from the beginning of the reign of Ramesses IV, he is described as the oldest of the gods:

> *Gegrüßet seiest du, der du groß und uralt bist, Tatenen, Vater der Götter! Du ältester Gott des Anbeginns, der die Menschen baute und die Götter schuf, Der das Werden begann als erster Urgott, alles, was kam, entstand nach ihm. Der den Himmel schuf nach dem Plan seines Herzens und ihn hochhob mit Hilfe des Luftraums; der die Erde gründete mit seiner eigenen Schöpfung und sie umgibt als Urwasser und Großes Grün; der die Unterwelt schuf und die Leichname ruhen läßt und den Sonnengott an ihnen vorbeiziehen läßt, um sie zu bewahren als Herrscher – er lebe, sei heil und gesund – der Zeit und Herr der Dauer.* (Assmann, 1999: 447)

> Hail to thee, thou who art great and timeless, O Tatenen, father of the gods! Thou most ancient god of the beginning, who man and the gods created; thou who began becoming as the first primordial god, everything that came, came after him.; thou who created the firmament according to his heart's desire and raised it with the help of the air; thou who fashioned the earth through his own will and endowed it with its first water and fertile verdure; thou who created the underworld so that the departed may rest there in peace and the sun god may pass them by, to preserve them as lord – life, prosperity, and health – of time and lord of duration.

Also clear, in this sense, is the text of the Carlsberg Glyptothek stela Nr. 897 from Copenhagen, where thought and word characterize the way in which Ptah begins creation:

> *Ptah, der Große südlich seiner Mauer, Tatenen in den Mauern, der erlauchte Gott des Anbeginns. Der die Menschen formte und die Götter gebar, Urgott, der das Leben der Menschheit schafft. Was in seinem Herzen gesagt wird, sieht man entstehen, der das, was noch nicht ist, vorhersagt und das, was ist, bedenkt; nichts entsteht ohne sein Wissen.* (Assmann, 1999: 503)

> Ptah-the Great-South-of-his-Wall, Tatenen in the Wall, illustrious god of the beginning. He who formed man and brought forth the gods, the most ancient god, who created the life of mankind. That which was spoken in his heart was came into being; that which is not was foretold, and that which is was premeditated; nothing could be without his knowledge.

In the *Onomasticon of Amenope*, written by the scribe of the sacred books in the House of Life, Imenemipet, whose most complete copy dates from the 22nd Dynasty (Papyrus Moscow 169), the scribe author indicates his wish that this work serves 'for instruction of the ignorant and for learning all things that exist: what Ptah created, what Thoth copied down' (Gardiner, 1947: 2*); that is, he recognizes Ptah as the creator.

The Memphite cosmogony shows some differences with respect to the Heliopolitan tradition. In the first place, it must be considered that while in the previous case the protagonist is a solar god, Atum (Re or Khepri), in the Memphite doctrine it is a chthonic god, Ptah. On the other hand, the method used in the creative process also introduces certain changes with respect to the Heliopolitan model.

One of the main sources of knowledge of the Memphite cosmogony is provided by the Berlin 3048 papyrus. This papyrus contains on its recto a hymn to the god Ptah (Wolf, 1929) and a prayer to Ramesses IX, but in its verso is a series of contracts that belong to the time of Takelot II or III (Theban 23rd Dynasty), which seems to be the time when the papyrus was written.

In the Berlin 3048 papyrus, the qualities of Ptah as demiurge are described, with emphasis on the fact that it was through thought and word that things were created according to the god's desire. He is also described, however, as a craftsman who uses his hands to shape things. In his original form he is remembered as Ptah-Tatenen: 'Ptah, father of the gods, Tatenen, the oldest of the primordial gods', 'Ptah-Tatenen, great god (…) father of the parents of the gods', who constitutes the primordial hill from which the universe began to take shape. Tatenen, a form that takes centre stage in these texts, comes to personify the primordial hill that emerged from the primordial ocean (Schlögl, 1980: 71–72). Ptah–Tatenen, therefore, is an original form of Ptah, self-created and creator of the gods. Ptah is 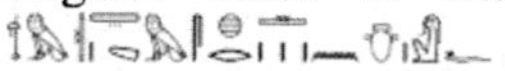 *ḥm tꜣ m sẖrw n jb.f* ('he who made the earth according to the plans of his heart'):

> *Gegrüßt seist Du, Ptah, Vater der Götter, Tanen, Ältester der Urgötter (…) der die Götter am Leben erhält (…) der sich selbst gezeugt hat, als noch kein Gewordenes geworden war; der die Erde gebildet hat nach seinem Herzen; dessen Gestalt (von selbst) entstand; Du bist es, der alles, was ist, gebildet hat; der gezeugt und gestaltet hat, was existiert (…).*
>
> *Sei gegrüßt angesichts Deiner Urgötter, die Du gemacht hast, nachdem Du entstanden warest als Gott. Leib, der seinen Leib selbst gebaut hat, bevor der Himmel entstand, bevor die Erde entstand, bevor das Wasser hervorkam. (…) Du hast keinen Vater, der Dich gezeugt hat, als Du entstandest, Du hast keine Mutter, die Dich geboren hat. Du Dein eigener Chnum! Du Gerüsteter, der gerüstet hervorkam! Du bist aufgestanden auf dem Lande während seiner Müdigkeit, von der es sich nachher zusamengerafft hat, indem Du warest in Deiner Gestalt des Tanen, in Deinem Wesen des Zusamenfügers der beiden Länder. Was Dein Mund gezeugt und Deine Hände geschaffen haben, Du hast es aus dem Urwasser herausgenommen. Das Werk Deiner Hände ist Deiner Schönheit angeglichen.* (Wolf, 1929: 18–19, 26)

> Hail to thee, Ptah, father of the gods, Tanen, oldest of the ancient gods (…) from whom the gods received life (…), who begat himself when nothing had yet come into being; who built the earth after his own heart; whose form was (of itself) created; thou art all that exists, all that has been built; who begat and gave shape to what is (…).
>
> We hail thee in the presence of the primaeval gods that you created after you came into being. Body that built itself before the heavens were formed, before the earth was formed, before the waters came forth. (…) Thou hast no father that made thee, thou camest into being of thyself, thou hast no mother that bore thee. Thou art thine own Khnum! Thou camest out fully armed! Thou stoodst erect on the land during its torpor, from which it has since arisen, by your taking the form of Tanen, in your guise as the joining of both lands. What thy mouth hath issued and what thine hands hath made thou hast taken from the primaeval water. The work of thine hands is a reflection of thy beauty.

Another remarkable text of the Memphite cosmogony, generally known as 'Memphite Theology', is the so-called Stone of Shabaka (BM 408) (Erman, 1911; El-Hawary, 2010). In its introduction it is said, 'this writing was copied out anew by his majesty in the House of his father Ptah-South-of-his-Wall, for his majesty found it to be a work of the ancestors which was worm- eaten, so that it could not be understood from beginning to end' (Lichtheim, 1975: 52). With this, the author of the text probably intended to date the composition of the work in a much older period, thus giving it greater relevance. In fact, the dating of the original composition has been a subject of discussion among specialists, with dates being proposed as old as the archaic era (Sethe, 1928: 2–5), the 19th Dynasty (Schlögl, 1980: 110–117), or Shabaka's own time, as proposed by Junge (1973).

In this text the primordial forms of the god Ptah are enunciated, the first being identified as Ptah-on-the-great-throne. However, the next two forms are related to Nun, to Ptah-Nun and Ptah-Nunet; i.e. to the masculine and feminine manifestations of Nun, respectively. In any case, it is specified that these last two forms are the ones that engendered Atum, so this discourse is intended to relegate the Heliopolitan to a secondary level within creation:

> The gods who came into being in Ptah:
>
> Ptah-on-the-great-throne [...]; Ptah-Nun, the father who [made] Atum; Ptah-Naunet, the mother who bore Atum; Ptah-the-Great is heart and tongue of the Nine [gods]; [Ptah] [...] who bore the gods; [Ptah] [...] who bore the gods; [Ptah] [...]; [Ptah] [...] Nefertem at the nose of Re every day.
>
> There took shape in the heart, there took shape on the tongue the form Atum. For the very great one is Ptah, who gave [life] to all the gods and their kas through this heart and through this tongue. (Lichtheim, 1975: 54)

The text of the Stone of Shabaka insists on the primordiality of Ptah, as god of the beginning, above all others and creator of all others:

> Thus it is said of Ptah: 'He who made all and created the gods'. And he is Ta-tenen, who gave birth to the gods, and from whom every thing came forth, foods, provisions, divine offerings, all good things. Thus it is recognized and understood that he is the mightiest of the gods. Thus Ptah was satisfied after he had made all things and all divine words. (Lichtheim, 1975: 55)

Although from a much later chronology (first half of the 2nd century AD), reference can also be made to the papyrus PSI Inv. D7 (a). The text (Smith, 2002: 18) has numerous gaps, but the context in which Ptah began creation with the use of the word is still clear. The papyrus says that Ptah stood on the sand (*šꜥ3*), which emerged from the Nun. Also, in the *Book of Fayum* (Beinlich, 1991: 257, 294) the sand is the soil that appeared in the middle of the sea. A thicket of reeds was formed there before the ground appeared.

The idea of the thicket of reeds as the first 'ground' on which to settle or perch is also repeated in the cosmogony that appears inscribed in the temple of Horus at Edfu (Reymond, 1962: 83–85). It is referred to there as the first support (*wṯst*) found by the god, located at the edge of the water, in the place where creation will stabilize (Finnestad, 1985: 43).

Finally, in another quality attributed to Ptah, it should also be remembered that in the cosmogony that appears in the temple of the god in Karnak Ptah is identified as *ḳm3 swḥt pr n Nwn* ('the creator of the egg that appeared in the primeval ocean'; Parker & Lesko, 1988: 170).

Another text that has sometimes been associated with the Memphite cosmogony is the demotic papyrus Berlin 13603 (Erichsen & Schott, 1954), although, as Quack (2018a, b, c) indicates, there does not seem to be any relationship between the two; rather, the papyrus seems to follow the Hermopolitan cosmogonic model.

1.1.3 Theban Cosmogony

In the Theban cosmogony the primordial god is *Jmnw* (Amon or Amun; see Figure 1.6), whose name comes from the verb *Jmn* ('be hidden'). According to Allen (1988: 48), it should be best understood as 'Self-Concealing'. In Thebes, this god began to have greater prominence from the Middle Kingdom, since until that moment Montu was the most prominent god of the Theban nome. Proof of this growing prominence is reflected in the name of some of the pharaohs of the 12th Dynasty, called Amenemhat 'Amun is at the front'. Despite the 'late' role of Amun, some mention of this god already appears in the *Pyramid Texts*. The oldest (PT 301 §446c; Allen, 2013: III, 83) comes from the pyramid of Wenis, where (Amun) and his feminine form (Amunet) are cited, as primordial gods, along with *njw* or *nwj* ('the watery one', Niu) and *nwnt* (Nunet).

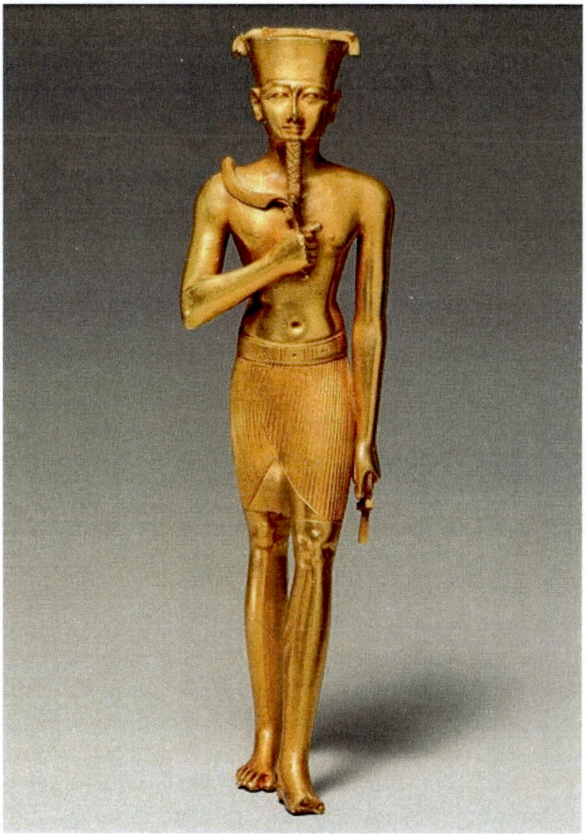

Fig. 1.6 Statuette of Amun, Third Intermediate Period, MET Accession Number: 26.7.1412. (Public domain image, courtesy of the Metropolitan Museum of New York)

In a text from the 18th Dynasty, from the tomb of Kheruef (TT 192), Amun is described as *nṯr nṯry ḫpr ḏs.f ḥry-tp jr ntt nb pr m Nwn ḫꜥ m jtn pꜣwty* ('divine god, who came into being of himself, the master who has made all that is, who came forth from Nun appearing as the solar disk, the primeval one'; Epigraphic Survey, 1980: 67 pl 65).

One of the best examples of this cosmogony is shown in the Leiden I 350 papyrus, from the time of Ramesses II:

> You concealed yourself as Amun, at the head of the gods. You made your development into Ta-tenen, in order to cause the original ones to be born from your first original state (…) You began development with nothing, without the world being empty of you on the first occasion. All gods are developed after you (…)
>
> You emerged first, you began from the start. Amun, whose identity is hidden from gods, oldest elder, more distinguished than these (…) who appeared in the sun, from the waters (…) light was development on the first occasion (…) he began speaking in the midst of silence (…) he began crying out while the world was in illness (…)
>
> Who began development on the first occasion. Amun, who developed in the beginning, whose emanation is unknown, no god developing prior to him, no other god with him to tell his appearance, there being no mother of his from whom his name was made, and no father of his who begot him so as to say 'It is I'. Who smelted his egg by himself. Power secret of birth, creator of his (own) perfection. Divine god, who developed by himself and every god developed since he began himself. (Allen, 1988: 50–52)

Amun is also described as the demiurge, the one who was already before the beginning and was developed from the beginning, alone, without any god being able to see him, since none was with him. As we saw in PT 301, Amun is already mentioned as one of the primordial gods, but it is in Thebes where he acquires maximum prominence, becoming equivalent to the Atum of Heliopolitan cosmogony. It is in the New Kingdom that this leap is consolidated and developed.

1.1.4 Hermopolitan Cosmogony

The most obvious difference between the Hermopolitan cosmogony and other Egyptian cosmogonic traditions resides in the fact that in this it is not a single god but eight, the  (Khemenyu), those who are present at the beginning of creation. This cosmogony has traditionally been associated in Egyptology with the ancient city of Hermopolis  *ḫmnw* (Khemenu), the city of 'Eight', but the Khemenyu, as such, are not documented until the 18th Dynasty in an inscription of Hatshepsut in Speos Artemidos as  (*Urk.* IV: 389, 3), whereas the city's Khemenu name has been known since the Old Kingdom (Zivie-Coche, 2016: 59). In fact, it is possible that the eight gods and the name of the city are not related at all in their origin. They are well documented in Thebes, both in the New Kingdom and especially in Ptolemaic times. There, in Djeme (Medinet Habu) it was believed that they were buried next to his father Kematef ('he who has completed his time'), so the hill of Djeme was also known as the 'Netherworld of Kematef'.

The full series of names of the eight gods would not be seen until the Saite period, as they appear for the first time in the tomb of Benaty, in the oasis of Bahariya. We can also see them listed in late cosmogonic texts, such as the Carlsberg 302 papyrus (Smith, 2002: 32). Once defined (Sethe, 1929: taf. I), the group was formed by (Nun) and (Nunet), as an expression of the primordial ocean, (Hehu) and (Hehet), symbol of the vast space, (Kek) and (Keket), the most complete darkness, and (Amun) and (Amunet), as expression of the hidden and unknown. This series of gods does nothing but represent many of the basic characteristics of the pre-creation world, being established by pairs of divinities consisting of a male and a female entity, thus combining both creative elements and also differentiating them in this way from the solitary androgynous demiurge. In the iconography, documented only from the 26th Dynasty onwards, as in a naos of Amasis (Louvre D 29) (Piankoff, 1933: 167) (see Figure 1.7) and later documents (see Figure 1.8), these deities usually appear in the guise of male frog heads and female serpent heads.

An attempt has generally been made to trace the existence of the 'Hermopolitan' cosmogony to the *Pyramid Texts*, since in PT 301 (W 206) the pairs formed by Nun and Nunet, and Amun and Amunet are mentioned (Allen, 2005: 55). But even when so referred to, they remain inserted into the framework of Heliopolitan cosmogony. In later contexts, it is to Amun(-Re), especially in hymns of the Ramesside period, or to Thoth to whom the eight are subordinated. Thus, for example, in the magical papyrus Harris (pBM 10042; Leitz, 1999: 38–39, pl. 17) is described the drawing of a tetracephalous god Amun surrounded on the right and left by the eight gods, perhaps in the shape of a baboon (Zivie-Coche, 2008: 73) in an attitude of adoration. This image of adoration of the solar god by the eight is also documented in late iconography, such as in reliefs from the temple of Hibis in the El-Kharga oasis (with

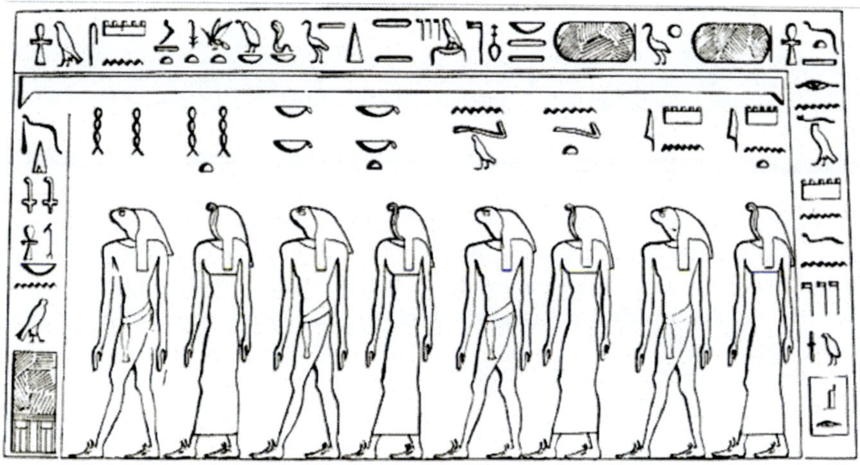

Fig. 1.7 The gods of the ogdoad on an Amasis period naos, Louvre D 29. (After Piankoff, 1933: 167 fig. 7)

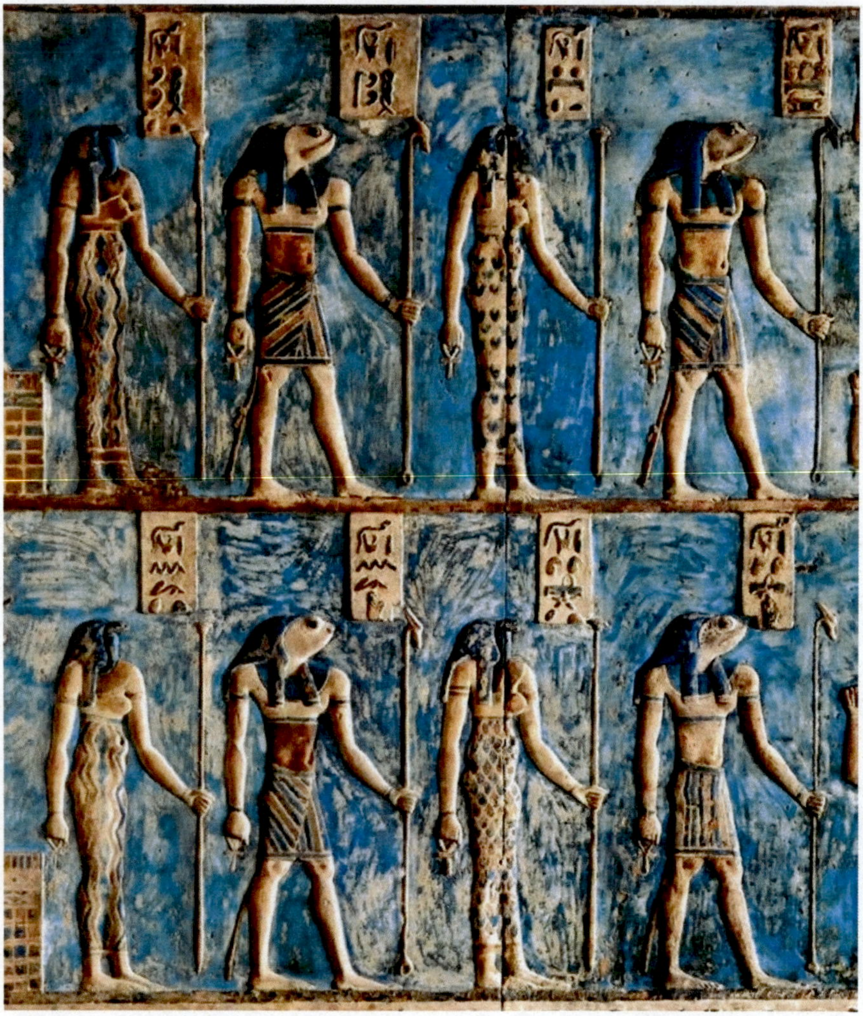

Fig. 1.8 The gods of the ogdoad in the temple of Dendara. (Image by courtesy of Txema Sanz)

the heads of serpents and frogs; Davies, 1953: pl. 4). In any case, as Zivie-Coche (2008: 75) points out, the transition from the form of baboons to that of frogs and serpents probably also marks a milestone in the evolution of the eight, from being mere worshippers of the rising sun into direct participants in their creation as primordial gods.

In Karnak, at the gate of Ptolemy III Evergetes (Bab el-Amara), on the axis facing the temple of Khonsu, some inscriptions offer clues about the ogdoad:

> *L'Ogdoade façonnée dans le Noun, les pères et les mères qui font la lumière, nés à Thèbes, conçus au Mur Blanc, qui mettent au monde le disque solaire dans la grande terre irriguée primordiale, qui naviguent avec lui jusqu'au château de leur naissance pour qu'il prenne*

possession de la royauté de Celui qui crée le souffle (…). Ils (les Huit) sont à ses côtés dans sa caverne du Noun, vivant éternellement et indestructiblement (…)

Les premiers Primordiaux, les enfants de Irta, mâles et femelles, qui créent la lumière, façonnés dans l'Opé par leur père Tatenen dans son officine de l'oeuvre de la Première Fois. Quand ils flottent dans leur district, l'île de l'embrasement, ils font advenir Rê dans le lotus, l'Ogdoade qui adresse des adorations à son héritier car il éclaire la terre pour elle. Ils rentrent dans le Noun auprès de leur père qui les a créés et ils pénètrent dans la Douat de Kematef. Rê les rejoint pour proclamer sa royauté tandis qu'ils lui annoncent ce qui adviendra. (Zivie-Coche, 2016: 74–75)

The Ogdoad formed in the Nun, their mothers and fathers who provide the illumination, born in Thebes, conceived at the White Wall, who give the solar disc to the world placed above the great primordial irrigated land, who sail with him to the palace of their birth that he may take take possession of the kingship of He who gives breath (…). They (the Eight) flank him, eternally and indestructibly alive, in the cavern of the Nun (…).

The first Primordials, children of Irta, males and females, who, giving light, formed in the Opet by their father Tatenen in his guise as the egg of the First Time.When, floating in their place, the island of flames, they make Re appear in the lotus, the Ogdoad, who give adoration to his heir (for he prepares the earth for them). They enter the Nun by the side of their father who created them and they enter into the Duat of Kematef. Re joins them to proclaim his kingship while they announce to him what is about to happen.

In a way that is not explained to us in any text, the ogdoad, according to late versions, brought forth the appearance of the sun on the primordial hill in  *jw nsjsj/nsrsr* ('island of flames'; Sethe, 1929: 49; Kees, 1942) in the 'first time'. It should be noted, however, that the phrase 'island of flames' is usually associated with the deceased. In fact, in CT 75 (I, 367) it is the place where people to be reproduced through Shu are found, and in CT 80 (II, 40) it is the place where the grain god goes when he transforms into Osiris.

On the other hand, the term *sp tpj* ('first time' or 'first occasion') is usually associated with the beginning of time at the moment of creation, but it is a word that does not appear in ancient sources such as the *Pyramid Texts* or the *Coffin Texts*. Expressions, however, that may have a similar meaning are used, such as *sp wr* ('the great time' or 'the great occasion') in CT 714.

In any case, expressions such as *mj sp tpj* ('like the first time') and *ḏr sp tpj* ('since the first time'), or in reference to the god, *twt nṯr nṯrj ḫpr m sp tpj* ('you are the divine god who came into being at the first occasion'), or in relation to the place where the creative event took place. Therefore, *st n sp tpj* ('the place of the first occasion') express beyond doubt the value that is given to the 'first time' as a milestone of temporal reference to be conceived as the beginning. It should in any case be noted that in Egyptian mythology there is no myth that properly refers to the creation of time (Servajean, 2007: 47).

The latest research seems to show that the creation of the cosmogonic myth of the ogdoad as creator gods is relatively recent, there being no possibility of dating the myth as far back as the middle of the first millennium BC (Zivie-Coche, 2016).

One view on the origin of the universe that has been linked to Hermopolitan cosmogony (Lefebvre, 1923) is creation from a 'cosmic' egg (*swḥt*) located on top of the primordial hill (Caminos, 1975: 1185). From the breaking shell emerged a bird, the Great Cackler (*ngng wr*), possibly associated with Atum himself, who was the one that brought light to that primordial world. This birth from an egg, like so many other cosmogonic aspects, was integrated into the regenerative beliefs of the funerary world so that most of the references (Maravelia, 2019: 31–33) that we have about it in the *Pyramid Texts*, *Coffin Texts*, and *Book of Coming Forth by Day* are linked to it.

In CT 686 (VI, 315 f–h), even in a funerary context, the egg is associated with a primordial phase before ‘he who is in Heliopolis’ (Atum) was seen. In addition, in CT 820 (VII, 211m) the egg is related to the protection of Nun. However, beyond the embryonic and life-generating meaning of the egg and its relationship with the solar god, the oldest sources are not clear. Regarding the ogdoad, it would be from the New Kingdom and Late Period when the idea of the primordial egg would develop on the primordial hill of Hermopolis.

A text from the temple of Khonsu in Karnak, known as the ‘Khonsu Cosmogony’, seeks to associate the protagonists of the Theban, Memphite, and Hermopolitan cosmogonies. The main Theban god is said to be the ‘father of the fathers of the Ogdoad’; in his form of Ptah ‘he fertilized the egg and the eight came into existence from it in the district around the Ogdoad (…) (Thus) came into existence the Ogdoad… consisting… its four evils and a wife for each one’. The text indicates that ‘Tanen had created the Ogdoad in Thebes. They were water-borne (to) the Island of Flames, and (thus) came into existence his form, the first primeval one of the “Great Flood”. They adore the eldest among the gods in the portal of Upper Egypt, the great god who came into existence on the first occasion, appraising before him, rejoicing in Nun, (while) darkness circulates before them’ (Parker & Lesko, 1988: 170).

1.1.5 Esna Cosmogony

In the temple of Esna there are records of two main cosmogonies. One is linked to the goddess Neith, so although she appears in Esna she must relate more appropriately to Sais. According to this cosmogony, Neith, as an androgynous being, was recognized as ‘the father of the fathers, the mother of the mothers, the one god who became two gods’ (von Lieven, 2014: 20). She was the one who created Re when he swam in the form of Nile perch in the Nun. There are also references of the egg laid by Neith, from which the sun came out (Sauneron, 1962; Guilhou, 2007). In another text is said that she ‘gave birth to all of the stars (*ḫ3bs*), they arise and then they set’ (Klotz, 2012: 32).

Both the temple of Esna and the inscriptions on its walls are from the Ptolemaic and Roman periods, so the version we have here of the cosmogony of Esna is late. In fact, one of the most important texts in which the demiurge's actions are explained is from Trajan’s time. In this cosmogony the demiurge is Khnum (see Figure 1.9), the ram who shaped the gods and men, he is *nb-nḥp grg.n.f t3 m r3-ʿwj.fj* (‘the lord of the potter’s wheel, the one who with the work of his own hands has made the earth’). To better appreciate the character of this tradition, which although preserved

Fig. 1.9 Khnum and Menhit in a wall relief in the temple of Khnum in Esna. (Image by courtesy of Artiom Gizun)

in late texts must go back further into Antiquity, there is no better way than to read some passages of texts 378 and 394 of Esna (Sauneron, 1968), where it is explained how Khnum was born and created what exists:

> You are the Lord of Esna, the god of the potter's wheel, the one who turned the gods, the one who shaped man and animals (...); there is no god that can equal him, the most beloved ram, the one who made what exists and what does not exist, the one who gave birth to the gods and gave birth to the goddesses.
>
> You are the august god who came into existence in the beginning, the one accompanied by a goddess, the *uraeus*; you are the mysterious one whose form no one knows, the secret god whose appearance no one knows, the one who came out of the Nun, who appears on the

horizon like a flame, the one under whose feet the flood emanates from two caverns (...). You illuminate the Two Lands with your two eyes. His right eye is the Sun and his left eye is the Moon; and he has created beings (in heaven) and on earth, in the Duat and in the Nun.

Greetings to you, Khnum-Re, Lord of Esna, Ptah-Tatenen, who has designed the primordial gods, great god, who came into existence at the beginning, splendid ram, at the first time. He raised the earth and sustained the sky, and there he diffused the light in the form of resplendence (...).

The most important among the important ones, the one who is greater than all the gods, mysterious in appearance, the one who is important above the gods, modeler of the modelers, the oldest of the primordial gods, father of parents, mother of mothers, the one who created superior beings and inferior beings, the august ram who made the rams, Khnum who made the Khnums.

Of Khnum it is said that he came into existence in the beginning, in the first time, and as father and mother of all the gods he is his own predecessor and ancestor, the most ancient and primordial of all. It is interesting to note that, as a primaeval god, it is said of him that, no one knows his shape or appearance. This is exactly the same kind of description that we find of the god Atum, when he still retained the serpent form in the pre-creation world. This, then, is a characteristic of the demiurge, since in his primordial form no other god accompanied him in the Nun and no one, therefore, could observe him or know his form. Very differently from the creator gods, Khnum uses an original way of performing the act of creation, by means of the potter's wheel.

The various preserved texts, whether in the most important religious compendia or the magical papyri, show us a fair number of strangely interrelated traditions that, on many occasions, cannot easily be associated without entering into open contradiction. Perhaps, though, this is also one of the main attractions of Egyptian cosmogony since, depending on the place of origin, the prominence of creation is resolved by one god or another, and in one way or another. In any case, all Egyptian cosmogonic myths have a common starting point, the creation from one or more entities that are activated in the Nun, the inert and dark primordial ocean that existed before creation.

1.2 Structure of the Egyptian Universe

What was the universe like according to Egyptian thought? This is a difficult question, the answer to which can only be based on the descriptions, often abstract and even contradictory, that we find mainly in religious and funerary texts.

A key element of the Egyptian universe that preceded its existence and the course of time began is the Nun, that primordial ocean characterized by its dark, inert, and silent waters. The Nun, therefore, is not extinguished with creation; in fact, it sets the scene for the work of demiurge, being both pre-creation and present at the same time.

For Bonnet (1952: 536), the Egyptians thought that the primordial water was located between the sky and the undersky, holding the earth as a floating disc.

According to Mercer (1952: 60–61 pl. VI), 'this Nun apparently became an all-encircling ocean, like the Greek Okeanos, with the coming into being of the universe; and the universe was conceived as contained within the bounds of that abyss (…). Within the circular Nun was the universe, which came into being perhaps gradually, for the mound was the first to appear and then Atum, which seems to have been a manifestation of Nun as the first god'. Hornung (1956: 29), for his part, stated:

> *Der Nun ist am Himmel, auf der Erde und in den Tiefen der Unterwelt. Er umringt, dem gleich, die Erde, aber er ist ebenso tief unten, als Grundwasser und Quelle der Nilüberschwemmung, wie er überhaupt das fruchtbare Wasser und mit ihm den positiven Aspekt des Chaos bezeichnet. Der Regen vom Himmel kommt, doch erfüllt er allgegenwärtig auch die Unterwelt. Und er ragt endlich so lebendig in die Schöpfung hinein, daß man auch mit Schiffen auf ihm fahren kann.*

> The Nun is in the heavens, on earth, and in the depths of the underworld. It likewise surrounds the earth, but is also deep beneath it as groundwater and the source of the Nile, as it guides the fruit-bearing water and with it the positive aspect of chaos. The rain comes from heaven, but it is omnipresent in the underworld as well. It towers at last so alive in the creation that it can be sailed on.

For Allen (2005: 9), 'The Egyptians (…) saw the world itself as a finite space bounded by land and sky; these two elements were personified respectively as the god Geb and the goddess Nut, whose genders reflect the Egyptian words for "land" (masculine *tꜣ*) and "sky" (feminine *pt*). This world was thought to exist within an infinite ocean, called Nu ("Waters"), which was kept from engulfing the earth by the atmosphere, incorporated in the god Shu, whose name means both "void" and "dry". The sky was seen as the surface of the cosmic ocean where it met the atmosphere, and the Sun's daily journey through the sky therefore required a boat, known as the Dayboat'.

As Bickel (1994: 31) points out, *avec le ciel et la terre, le Noun constitutes une troisième grande région de l'univers* ('together with the earth and sky, the Nun constitutes a third great region of the universe'). The separation of these three main parts of the Egyptian universe is already indicated in PT 627B §1778b (N32) ( *ḏsr pt r tꜣ nww*) and in CT 682. In this case, the separative preposition *r* seems to give a small clue concerning the relationship existing between each of these parts (Bickel, 1994: 186), since in the text it is situated between 'heaven', on the one hand, and 'earth' and 'the Nun' on the other. It would seem, then, that when the separation occurs, it is the sky that moves away from the group formed by 'earth' and 'the Nun', with which a more direct relationship could be understood between the latter, either because the Nun remains under the terrestrial region, or because the Nun surrounds it. For this reason, Bickel concludes, 'it is necessary to indicate a cosmology that places the Nun *pas tout autour du monde creé, mais bien du côté ou en-dessous de la terre.*[2] However, if we look at other texts, this structure is not so evident; rather, it seems that the Nun surrounds the other regions.

[2] 'Not completely surrounding the created world but rather adjoining or beneath the earth'.

An understanding of the structure of the Egyptian universe can only be established from interpretation of cosmogonic texts, from which emerge a series of basic elements (such as the Nun, the sky, the Duat, and the earth), and from the funerary texts, where there is a profusion of descriptions of these parts; however this hinders rather than facilitating understanding, rendering the relationship between each of the parts more complex (and often contradictory).

If we move away from the religious sphere (be it cosmogonic or funerary), references to this kind of question are almost non-existent. An example to take into account is the *Onomasticon of Amenemope*, a kind of 'encyclopaedia' consisting of a list of Egyptian words, grouped by lexical fields, written at the beginning of the 21st Dynasty. If we observe the terms that it introduces for the lexical field referring to the sky (Gardiner, 1947: 4*–13*) we find—in addition to the word *pt* ('sky'), *jtn* (solar 'disc'), *j'ḥ* ('moon'), and *sb3* ('star')—the names of five constellations of the boreal sky: *s3ḥ* ('Sah'), *msḫtjw* ('Meskhetyu'), *j'nw* ('Anu'), *nḫt* ('Nekhet'), and *rrt* ('Reret'). No more words are introduced. For atmospheric phenomena on the other hand it includes *ḳrj* ('storm cloud'), *ẖ3ẖ3* ('tempest'), and others related to light, darkness and water, among these ('Nun'). However, the author of this list, Amenemope, does not offer additional information on other words, thus eschewing those that appear more frequently in religious texts.

1.2.1 Nun and Nunet

In the *Fundamentals of the Course of the Stars* (*snṯ šmt nt sb3w*, formerly called *Book of Nut*; von Lieven, 2007), preserved in the cenotaph of Sethy I at Abydos, to a lesser degree in the tomb of Ramesses IV in the Valley of the Kings, and in its most complete version (but without iconography) in the Carlsberg I papyrus (Lange & Neugebauer, 1940), we find clues about as to some parts of the Egyptian universe were structured.

In the Carlsberg I papyrus, written in hieratic and demotic around AD 144, almost 1500 years after the first known version of the time of Sethy I, some of the descriptions about remote areas of the universe are preserved with a little more development, as in the so-called L text (*EAT* I, pl. 46):

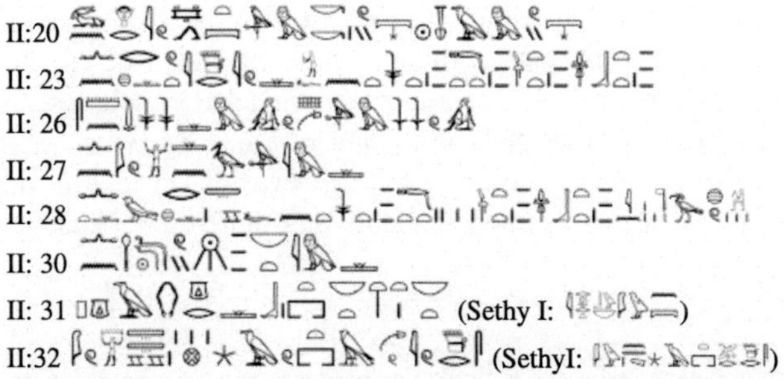

Wnn ḥrw pt tn[3] *m kkw smꜣw*
Nn rḫ.tw ḏrw.f n rsj mḥtj jmntj jꜣbtj
smn nn m Nnw m nnyw
Nn šwj bꜣ jm
Jwtj rḫw tꜣ.f n rsj mḥtj jmntj jꜣbtj jn nṯrw ꜣḫw
Nn ḥḏḏwj (ḥḏwt) nbt jm
pgꜣ ẖr (jr grt bw nb šw n pt šw n tꜣ
dꜣt pw r ḏr.s – Sethy I)

The top of this sky is in the primeval darkness
Its borders to the south, north, west and east are not known.
These are established in the Nun as weary ones
The *ba* does not rise there
Without his land being known, towards the south, north, west and east, by the gods and *akhu*.
There is no kind of light there
As for every place empty of heaven and empty of earth,
this is the Duat to its limit. [4]

The word *kkw smꜣw*, translated by von Lieven as *Urfinsternis* ('primordial darkness') is for Wilson (1997: 1091–1092) a darkness outside the world, prior to creation itself. This word appears in the *Book of the Imiduat* and in the *Book of Gates*, to designate the hours. In the 12th hour of the *Imiduat* of the Tjaihorpata sarcophagus, Shu is also said to be he 'who divides Nun from the land in the uniform darkness (*kkw smꜣw*)' (Manassa, 2007: 373). On the other hand, as *ḫntj kkw smꜣw*, it is used as an epithet for Osiris (Hornung, 1956: 30; also, Bonano, 2017).

If we analyse this description, we may draw some conclusions. On the one hand, in the limits of the sky, above, there is only darkness of unlimited extent. In the explanation that appears in the Carlsberg I papyrus it is said that the outer part of the sky, being a place not visited by Re, is in complete darkness. On the other hand, in an inscription from the time of Hatshepsut (*Urk* IV, 248: 16–17) the same words are used to refer to the limit of the world:

rdj.sn tꜣš.ṯ r wsḫ n pt
r ḏrw kkw-smꜣw

They placed your border across the width of the sky
to the edge of complete darkness.

---

[3] Von Lieven (2007: 59 n. 263)

[4] At this point we follow Rotsch's translation: *Was aber jeden leeren Ort des Himmels und leeren der Erde angeht, so ist das die Dat bis zu ihrer Grenze*. See Rotsch, 2019: 11; Clagett (1995: 374) translated: 'Moreover every place there [in the Abyss] is devoid of sky (i.e. the region where the sun and stars travel), [while] the entire Duat is devoid of lands (i.e. the countries of the living?)'.

On the other hand, it is clear that the Nun is an even darker place than the Duat, since, as the demotic explanation of pCarsberg I indicates, 'It happens that the darkness there is much more than that which is in the Duat'. Therefore, although the sky is illuminated by the sun and the stars, there is an upper and bordering part characterized by darkness. The unknown extremes of this dark area are connected with the Nun, this being a dark place that is totally beyond not only human knowledge but also that of the gods and the *akhu*.

According to this description, it seems to be understood that there is a spatial relationship between the Nun and the sky, even though the first is in a more remote and unknown area, and between them there is an empty region (the Duat) also characterized by primal darkness. Obviously, this description of the Duat differs from that space full of obstacles and beings described profusely in the *Books of the Hereafter*, so it is understandable that, on many occasions, it is a really complex matter to assimilate the basic cosmogonic conceptions, even with their variants, with their homologues in the funerary world.

At this point, we must remember that it is no coincidence that one of the most common written variants of Nun, , carries the determinative of 'sky', since there is a spatial relationship between the two. Let us remember, on the other hand, that the name of Nut could mean, 'of the waters', although the etymology is uncertain.

The image that most comes to mind has sometimes been compared with the description of the creation of heaven that appears in the Bible:

> And God said, Let there be a firmament in the midst of the waters, and let it divide the waters from the waters.
> And God made the firmament, and divided the waters which were under the firmament from the waters which were above the firmament: and it was so. (Genesis 1: 6–7)

Some texts mention four main parts of the universe. This can be seen, for example, in an inscription of the festival calendar that appears on the outside of the south wall of the funerary temple of Ramesses III in Medinet Habu, where they are mentioned in this order in a dedication to the god Amun: heaven, earth, Nun, and Duat (*pt, t3, nww, dw3t*) (Epigraphic Survey, 1934: 138 col. 42). In a hymn to Ptah written in the Great Harris Papyrus, from the time of Ramesses IV, there is also a brief enumeration of the parts of the world (Grandet, 2005: pl. 44, 4–5): *Jr pt m ḳm3.n jb. f sʿẖ sw m sṯsw-šw snnt t3 m jr.nf ḏs.f pẖr sw m nwn w3ḏ-wr jr dw3t* ('(Tatenen) (...) who made heaven after the conception of his heart (mind), who raised it above the atmosphere (Shu), who founded the earth by being himself, who surrounded it with the Nun and the sea[5], who made the Duat'). As we see, regarding the Nun, its position is indicated: 'who surrounded it (the earth) with the Nun and the sea'. Therefore, although the Nun is beyond the sky, it also surrounds the terrestrial world.

According to Allen (1988: 4), the idea transmitted by the Egyptians is that of a 'limitless ocean of dark and motionless water, within which the world of life floats

[5] However, Grandet (2005: 284) prefers the translation *entourée du flot de l'Océan* ('surrounded by the flood of the ocean') instead of 'surrounded with it the Nun and the ocean'. Breasted's (1906: 163) rendition was 'surrounded it with Nun, and the sea'.

as a sphere of air and light'. However, it seems rather that the Egyptians understood those waters to surround the entire world known to gods, *akhu*, and humans. As the Nun exists both under the earth and above the sky, that is why the sources of the Nile were believed to originate from a resurgence of the Nun. The sky, in contrast, acted as a vault ( ) capable of protecting them from the Nun.

In the Great Hymn to Aten it is explained how the flood waters originate from the Duat, while those from rain come from the celestial realm (Davies, 1908: pl. XXVII col. 10): 'The flood (*ḥʿpj*) from heaven is for the foreign people (…) while the flood (*ḥʿpj*) comes from the Duat for Ta-Meri' [6]. If the waters are ultimately resurgences of the Nun, they cross the Duat but also the sky, so they are proof that the Nun surrounds both spaces.

In the third year of Osorkon III's reign, at the beginning of the 8th century BC, there was a great flood (Daressy, 1896: 181–186) that caused extensive damage to Thebes. The text referring to this event is found at the bottom of a wall in the SE corner of the courtyard of the Temple of Amenhotep III at Luxor. In this text the water of the Nun is directly associated with the flood, as though it were capable of gushing from the depths:

> The flood water of the Abyss (*nwn*) welled up in the first (?) […] this entire land. As at the beginning of time (*sp tpj*), it attained the bordering mountains, […] this land being in its power like the sea. There was no canal made by man to repel its intensity, everyone being as fleas in his city, its raging high […] as heaven, every temple of Thebes like swamps. (Ritner, 2009: 417)

Also in pBerlin 3048 (Text E), from the Third Intermediate Period, we read: (He is) *Der den Nil* (*ḥʿpj*) *aus seinem Quelloch hervorbringt* (…) *in seinem Namen 'Nun'* (*nww*) (…) *Der die Überschwemmungswasser* (*nww*) *zum Himmel steigen* ('He is the one who brings forth the Nile (*ḥʿpj*) from its springs (…) in its name of 'Nun' (*nww*), the one that the floodwaters (*nww*) raises to heaven'; Knigge, 2006: 170).

The iconography, on the other hand, shows a relationship between the Nun and Nut, or between the Nun and the place where the solar boat navigates on its nocturnal journey once the sun has set below the western horizon.

We have a good example in a Late Period coffin kept in the Musée du Louvre (E 21611), belonging to an individual named Pasherienaset (Étienne, 2009: 92). Around the iconographic set there is a list of swampy areas or *pḥww*. As a central element Nut, who raises her arms in the direction of a large *djed* pillar, is surrounded by Re and Atum, under which are the personifications of Ankh and Djed on the one hand and under two other divinities, Neheh and Djet, usually understood as expression of cyclical and linear time on the other. Just below Nut's feet, twenty-four zigzagging lines topped by the *ankh* sign connect with a drooping-breasted figure, a representation of Nun similar to that offered by Hapy as a reflection of the fertile

[6] Clagett's (1992: 570) translation is: 'You made Hapy (the Nile god) in dat (the Netherworld), you bring him [as flood waters] when you will, to nourish the people; you made a heavenly Hapy (i.e. rain) descend for them; he makes waves on the mountains like the sea'.

and life-giving waters of the Nile. In this way too there is a relationship between Nun and Hapy.

In the Kawa V stela, from the time of the Nubian king Taharqa, it is explained how in his year 6 there was a great and exceptional flood. The flooded landscape was compared to the Nun:

> As the time of the inundation (*ḥwyt*) of the Nile (*ḥ*ʿ*pj*) came, it overflowed greatly every day and did so for many days. The inundation was of a cubit per day; it penetrated the hills of the South Land and was over the mound of the North Land. The land was like the Nun (*nwn*), an inert (water), the land was not discernible from the river. (Gozzoli, 2009: 237; Ritner, 2009: 543).

To the Egyptians, the image of the lands flooded by the Nile reminds them of the supposed image of the Nun because of the extension of the waters and their stillness: ('The land was like the Nun, an inert (water)'). In texts of the 19th and 20th Dynasties (Barguet, 1950) it is also said that Hapy (the divinity who represents the Nile or the flooding of the Nile) is the beloved of Nun:

> Long live the good god, beloved of Nun, Hapy, father of the gods and of the ennead that is in the water of the flood.

This 'ennead' is made up of gods (Barguet, 1950: 58 n.1), mentioned in the temple of Edfu (Rochemonteix, 1892: 491), among which Nuu (*Nnj*), Niu, Hehu, Shu, Thoth, and Ptah stand out for their connection with the primordial world.

It must be remembered, on the other hand, that in Egyptian religious thought, this connection between the waters of the Nun and the flood was evident in specific parts of the country where water supposedly gushed from caves. In the *Pyramid Texts*, as in PT 250 §268 (W 161), the *tpḥt nw* ('Cavern of Nun') is already mentioned.

In terrestrial topography, the best-known is the one located at the first cataract, beautifully reproduced at Hadrian's Gate (see Figure 1.10) in Philae, but the texts also mention others, especially that of Kher-Aha (Aja Sánchez, 2015: 156–166) near Heliopolis. This is where it was said of the Nubian king Piye that *j*ʿ*j ḥr.f m jtrw nt Nw* ('his face was washed in the river of Nun') (Ritner, 2009: 487; Grimal, 1981a, b: 36* l. 102).

On the other hand, it should be noted that, even in the *Pyramid Texts*, there already appears the idea of an access or door to the Nun, as in PT 272 §392a (W 179), ʿ*rrwt nt nw* 'portal of Nu' or in PT 585 (Nt 7) 'upper gates of Nu' (Allen, 2005: 50, 312), for example. Let us bear in mind, once again, that it is not only the pre-creation space; its function from the religious or funerary point of view is also evident, since the late king passes through the Nun in order to be reborn. In this sense, for example, a papyrus of the Middle Kingdom (Borghouts, 1989: 135; Pepin, 1989) indicates that the Nun possesses the doors that give access to the celestial vault.

Fig. 1.10 The Cavern of Hapy at Philae. (Image by the authors)

In the Great Hymn to Hapy, of which a good number of texts remain, especially from the Ramesside period, it is said that he  'is in the Duat, heaven and earth, on their supports' (*m dw3t pt t3 ḥr sẖnwt.f*) (Van der Plas, 1986: 40). It is interesting to observe the consequence of the translation, because if *sẖnwt.f* 'its supports' refer to those of Hapy (= Nun); consequently, he appears as a key piece of the universe, holding heaven and earth or, at least, as Van der Plas (1986: 96) points out, expressing a notion of dependency. In any case, once again, it is clear that the water of the Nun surrounds all parts of the world.

In Graeco–Roman texts this idea is maintained, as we see in Dendara (Chassinat, 1935: 235.4–5): 'Hapy came to create everything that exists, the sky, the earth, the Duat, the water and the mountains'. In the temple of Horus in Edfu (Finnestad, 1985: 90), for example, a salutation is made to Nun, Flood and Hapy, precisely because of the common elements they present: 'Hail to you, O Nun!, Hail to you, O Flood!, Hail to you, O Hapy!, father of gods!'.

Certainly, in many Egyptian texts, especially since the New Kingdom, an assimilation was made between the Nun and the floodwaters or Hapy (Rotsch, 2005; Aja Sánchez, 2015: 128). It should be borne in mind, however, that this idea already seems to be present in the *Pyramid Texts*. For example, in PT 338 §551 (T 148) was written: 'Go to Nu ( ), travel off to the flood (*ꜣgbj*)' (Allen, 2005: 74); and in PT 493 §1062 (P 376): '[I will look at Nu] and appear at New-Year's days' (Allen, 2005: 143), a fragment that hints at the relationship between Nun and the ideal beginning of the year in which the flood began in I Akhet.

This idea was also transmitted, with its reinterpretations, during the Graeco–Roman era. Thus, both Cicero, in *De Natura Deorum* (II, 22), and Apuleyus, in *Metamorphoses* (XII, 11) explain that Nun means 'born of the Nile'. Even in the 5th century AD the relationship between Nun and the flood was commented on by Horapollo in his *Hieroglyphica* (I. 21): 'To express the "rising of the Nile", which the Egyptians call Nun and translated means "Young", sometimes they draw a lion, other times three large vessels and others the sky and the earth flowing water' (Horapolo, 2011: 70). The drawing of the lion to which Horapolo refers undoubtedly has to do with the constellation of Leo, since that is where the sun was located in part of the summer, coinciding with the flood or, as Pliny the Elder explains in his *Naturalis historia* (V-, X), when the flood was most noticeable.

Regarding the position of the Nun in the structure of the universe, another image of great interest is that offered in the final scene of the *Book of Gates*, a religious composition that has its oldest example in the tomb of Horemheb in the Valley of the Kings (KV 57).

There are many examples of this scene, including that of the cenotaph of Sethy I in Abydos (Murray, 1904: pl. XIII) in relief, or that of the Papyrus of Lady Anhai (BM 10472) of the 20th Dynasty (Budge, 1899: 8). Also, in the calcite sarcophagus of Sethy I we find a complete representation of the end of the *Book of Gates* (Sharpe, 1864: pl. 15; see Figure 1.11). In this image the solar boat carries the god Khepri, who drags and supports the solar disc. The inscription says: 'This god takes a seat in the morning bark (with) the gods, in whose midst he is'. Above, Nut, the goddess of the sky, reaches out with her arms for said disc, to receive it, as the accompanying text indicates: 'This is Nut, who receives Re'. The goddess, depicted in an inverted position, extends from the head of Osiris. According to Hornung (1999: 65–66), this position means, 'the inversion of the Sun's course, which will now once again run in the opposite direction from its course through the netherworld' represented by Osiris, at the top, curved like a circle possibly symbolizing the continuous cycle of regeneration in the Hereafter (the *ẖrt nṯr* of the Egyptians). The inscription next to Osiris reads: *Wsjr pw šn.f dw3t* ('This is Osiris, who surrounds the Duat').

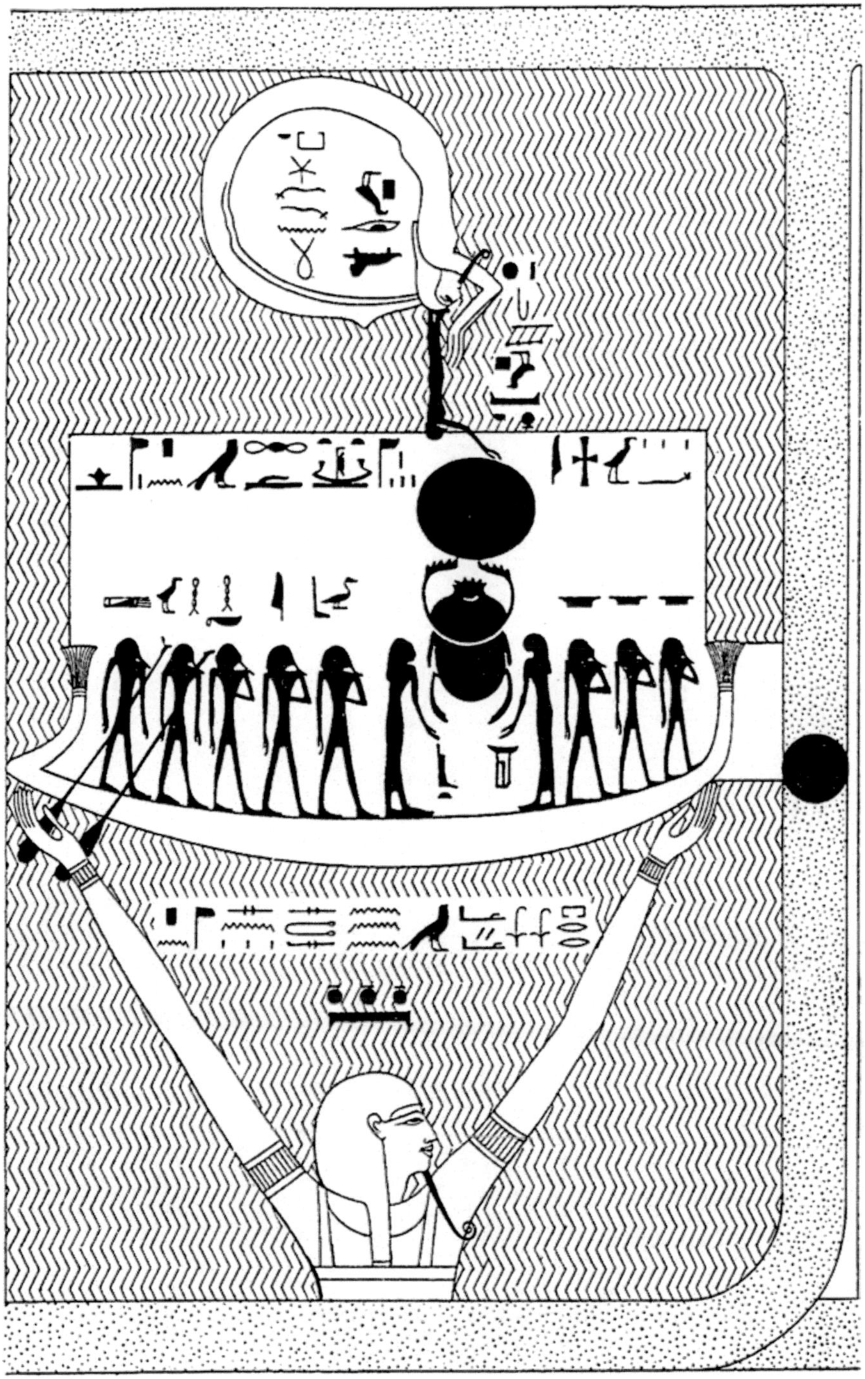

Fig. 1.11 Final scene of the *Book of the Gates* in the calcite sarcophagus of Sethy I. (After Sharpe, 1864: pl. 15)

In the lower part of the scene, on a larger scale, we see Nun as the personification of the primordial waters, raising his arms to support the solar boat (accompanied by the text 'These arms come out of the water, they lift this god. Nun'). In this iconography, therefore, some of the main elements of the Egyptian cosmos are combined, such as the Nun, the sky (*pt*), and the Duat.

In the Khonsumes papyrus Nun is also seen holding the solar boat (Piankoff & Rambova, 1957: vI 49 fig. 33, vII n° 30; see Figure 1.12). The nine columns of text remaining to the left of this scene provide additional information (Piankoff & Rambova, 1957: vI 211). It explains how Re 'is towed by the souls among whom he is', in order to 'pass by the Cavern of Nun', to 'pass the darkness of the mysterious abodes'. However, unlike the other representations at the end of the *Book of Gates*, the entire scene is framed by a circle of water. For Troy (1986: 17), this water circle would represent Nun's feminine aspect, proof of his androgynous nature, through an iconography whose meaning she compares with Gardiner's Aa1 sign (◍), identified as the placenta. For Molinero Polo (2003: 530) 'by representing that Nun with

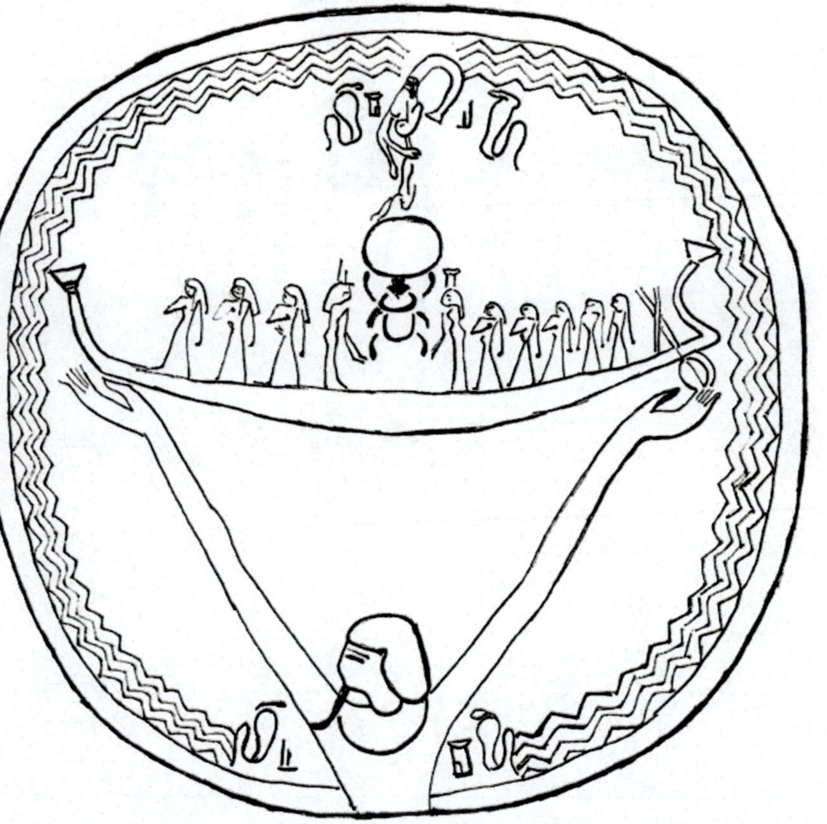

Fig. 1.12 Final scene of the *Book of the Gates* Khonsumes papyrus. (Drawn by the authors after Piankoff & Rambova, 1957: 49 fig. 33)

a circular perimeter, the drawer captured, as Troy has recognized, the generative capacity of the god. But this is the figure of an idea, not a location plan or topographic image. This is much better defined by the texts, when they say that its limits are not known, and therefore cannot be drawn'.

However, in our opinion, the image of Nun could be read in two different ways; on the one hand, from the inherent symbolism of the watery circle, as explained by Troy, and on the other as an expression of the position of the Nun in the structure of the universe. Let us note that only the upper half of the body of Nun is seen emerging from the waters. Both here and in the remaining scenes, perhaps the infinity of his extension is expressed in this way, since it is not possible to see his entire figure. The order in which the three elements have been represented is also significant: Nun, solar boat, Nut, Osiris (touching the waters and with the inscription which tells that he surrounds the Duat). The Nun surrounds the entire Duat and the entire sky, necessarily as a sphere, since, if it were a circle, it could not encompass the celestial and subterranean levels beyond the earth at the same time.

Another reference to the location of the waters of the Nun with respect to the earth is found in a hymn dedicated to Khnum in the temple of Esna, where we read: *sꜣp.nf tꜣ -pn ẖr- Nnw šn wr pẖr wr [ḥr-] pẖr .f* (*Unser Land hat er (Khnum) unter den* Nun *gesetzt, der Schen-Ozean und der Peher-Ozean umgeben es*[7]; Knigge, 2006: 298).

In the *Pyramid Texts* another word appears to designate a zone of the universe: *nnt / nnwt / njwt / nwnt* (also with its personification as divinity, Nunet), defined by Allen (2005: 443) as 'the counterpart of Nut in the Duat'. As we can see, the determinative of sky was inverted, which is why it has been translated as *Gegenhimmel* ('countersky'; Hannig, 1997: 416) or 'nether sky' (Allen, 1989: 12), an opposition with respect to the *pt* sky that is also perceived in PT 215 §149b (written in Wenis and in Pepi I).

However, the undersky must be directly related to the Nun, as both are mentioned together on several occasions in the *Pyramid Texts*. Thus, in PT 222 §207b (W 155) the text says:

> (Unis), you shall release your cleansing to Atum in [Heliopolis; you shall descend] with him, that you may decide the needs of the undersky () and stand up in the places of Nu (*njw / nwj*)'. (Allen, 2005: 40)

That this undersky is spatially related to the Nun seems to be shown by the fact that they are mentioned together on some occasions. Thus, in PT 301 §446a (W 206), 'Nu and Undersky (*njw ḥnꜥ nnwt*), you pair of the gods, who joined the gods with their shadow' (Allen, 2005: 55).

On the other hand, the text PT 571 §1466 (P 511) is also significant:

> The mother of Pepi became pregnant with him, who is in the undersky (), and this Pepi was given birth by his father Atum, when the sky had not yet come into being, when the earth had not yet come into being, when people had not yet come into being, when the gods had not yet come into being, when death had not yet come into being. (Allen, 2005: 179)

---

[7] 'He (Khnum) has placed our land below the Nun, the Shen-Ocean and the Peher-Ocean being circumvented.'

Although this text refers to the rebirth of the deceased king, it may give clues as to the undersky's position. The (re)birth of the king before heaven and earth existed (as a cyclical repetition of what were the original cosmogonic events) takes us to a remote place from which Atum, the demiurge, emerged. This association with the primordial state also appears in P 557: 'He of the undersky commended you, [he of] Nu commended you [to…], when the sky had [not] yet come into being, when the earth had not yet come into being, when canals had not yet been dug, when towns had not yet been founded' (Allen, 2005: 196). On the other hand, in PT 606 §1691b (M336), it is indicated that if Shu and Tefnut are placed to the east and west, respectively, and Nun and Undersky to the south and north, respectively (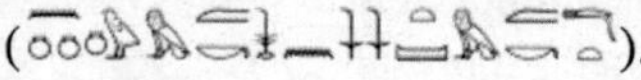).

Allen (1989: 13) understood the undersky as 'an ocean of water lying beneath the earth'. In fact, the concept of the undersky as a lower realm, as opposed to the upper one of the *pt* sky, seems to be clear in the *Pyramid Texts* as well. Thus, in PT 215 §149 (W 148): 'when you demand to go up to the sky, you go up (…) when you demand to go down to the undersky, you go down'. On the other hand, Molinero Polo (2003: 539–540) concludes that, with respect to Nun, 'since he was conceived as primordial waters, infinite and without limits, which left no free space that could have been occupied by his *páredros* (Naunet), she was conceived incorporated in him, forming part of that primeval ocean. Perhaps because of that characteristic of identical element, but in a certain way, because of its gender, different, they explained her as opposed to Nun, in an inverted position'.

The exact location of the undersky is not absolutely clear, since while some references, such as those we have cited above, seem to associate it with the Nun (concerning which it must also be taken into account that in PT 570 §1446a (M) it is written with the determinative of inverted sky: ), others indicate its relationship with the imperishable stars. Thus, although in various recitations it is indicated that gods inhabit the undersky, in PT570 §1456 (P 510, also in N 599) they are called  *nṯrw njwtjw jẖmw-skjw* 'imperishable gods of the undersky'. It seems, then, that the undersky is a feminine part of the Nun, so as a primal world it is useful for regeneration, also becoming a place of passage or transition that allows its relationship with the imperishable stars.

1.2.2 *The Duat*

As mentioned above, an important part of the structure of the Egyptian universe is the Duat, generally translated as *Unterwelt* ('underworld'; *Wb* V, 415). Funerary texts, especially from the New Kingdom, profusely describe the characteristics of this region.

Duat is generally written *dꜣt* in the *Pyramid Texts*, although, according to Allen (1989: 22), *dwꜣt* would be its complete form following the variant  that appears several times in the pyramid of Wenis (PT 247 §257c, PT 252 §272a, PT 263 §341c). The word uses as determinative a sign that represents a star within a circle. According to Hornung (1975: 994) it could originally have represented the *Reich der Sterne* ('star empire') or *Sternenkreis* ('starry realm').

The Duat, like the sky, is also linked to the watery element, since it contains lakes and canals, and the solar retinue also navigates on it. Its access seems to be in the west, descending (PT 688 §2084–2085). The door of the Duat is also comparable to that of *ꜣkr* (Aker), since while in Pepi's version of PT 483 §1014a *mdw tꜣ wn rt dꜣt* is read in those of Merenre and Pepi II, *dꜣt* is replaced by *ꜣkr*. Aker's space is described in the *Book of Earth*, in the New Kingdom, as an underworld.

However, as Molinero Polo (2003: 580–587) warns, the image of the Duat that appears so clearly in the New Kingdom is not so clear in the references to it in the *Pyramid Texts*, where even on one occasion, PT 437 §802c (P), uses the determinative of heaven. When talking about the Duat, there are numerous references to the sky, the *akhet*, stars, Sah and Sopdet, so it is possible that the location and meaning of the Duat has evolved from the Old Kingdom to more recent times.

Although in most of the *Books of the Hereafter* of the New Kingdom the image of the Duat is that of an underground world, in *The Fundamentals of the Course of the Stars* we have already seen how the limit of the sky is associated with the Duat. On the other hand, other texts in the same compedium completely associate the Duat with heaven, since it is through the mouth of Nut that the Duat is accessed:

> When the Incarnation of this god enters her mouth, inside the Duat, it stays open after he sails inside her, so that these sailing stars may enter after him and come forth after him. (Allen, 1988: 2; *EAT* I: pl. 49)

1.2.3 The Sky

The most common word to express the concept of sky in Egyptian is *pt*, although in Egyptian terminology there are a large number of words to express parts of the sky or types of sky (Molinero Polo, 2003: 547–559). The determinative of heaven () undoubtedly expresses in part how the Egyptians conceived this region of the world. The sign, usually painted blue (because of its relation to water) is flat at the top, although it can also be represented curved in iconography (as in the Stele of Lady Taperet and many others) adapting the shape of the celestial vault as Nut (Calmettes, 2003: 10, 13). At the ends it is finished off by two downward points, which represent the supports (two from the Egyptian perspective, four in fact) on which it is supported.

In the *Pyramid Texts* (PT 510 §1143b (P 449) the pillars of heaven are already mentioned: *jtj Mry-Rꜥw pt jwnw.s ꜣḫꜣḫw.s* ('Meryre will acquire the sky, its pillars and its stars'), but in representations it is common to see the *pt* sign supported by *wꜣs* ('sceptres'). It is possible that the cardinal points correspond to the places where the Egyptians placed these pillars, although this is not clearly expressed in the PT. Late representations (such as the zodiac of Dendara) do show, for example, the goddesses of the cardinal points holding up the sky, so they could be assimilated into those pillars.

The sky, in the Egyptian conception, is a place dominated by water. Thus in PT 685 §2063a (N 519) we read about the 'waters of life in the sky'. The gods navigate in it. It is also the place where the stars are located (Allen, 1989: 3–4), for which there are numerous references. Thus, in PT 434 §785d the text speaks about the gods who, like stars, navigate through Nut:

> As you have acquired [every god] with you in his Sea-Mullet boat, that you might make (each of) them a star, as She of a Thousand Bas, and they might not go away from you as stars, you should not let Pepi be away from you, in your identity of the above ( ). (Allen, 2005: 104)

In this passage the goddess Nut is also referred to as *ḥrt* (P), with determinative of sky in various examples (M, N), (Nt). The meaning of this appellation can be translated as 'above', as Allen (2005: 104) does, 'sky' (Faulkner, 1969: 143), *Himmel/Himmelskuppel* ('heaven/heaven's dome'; Hannig, 1997: 555) or *ciel éloigné* ('elongated sky'; Calmettes, 2008: 3). In PT 370 §645d she appears as (P, M), and also with the determinative of divinity.

In the *Pyramid Texts* there are references to the *ḳbḥw* ('fresh water'), written with the determinative of sky in PT 452 §841b (P, Nt) (/) and 876a (P), 1327b (P), 1765c (Ib). These waters are in the sky, even in PT 214 §138b (W) we find the *ḳbḥw sb3w* ('fresh water of the stars'). According to Allen (1989: 8) it 'may not be so much a name for the celestial domain as a general description of its nature'.

Although there are many divinities linked to the *pt* ('sky'), the most significant is Nut, the goddess of the sky *par excellence*. In this relationship, it could not be by chance that the word 'heaven' is feminine in Egyptian (her name appears written with a feminine determinative for the first time in the pyramid of Ibi; Jéquier, 1935: pl. 16 f. V). Thus, according to Allen (1989: 16), 'the distinction between Nut and the sky (*pt*) in the *Pyramid Texts* therefore seems to be mainly one of personification'. Barta (1973: 100–101) understood Nut as a feminine form of Nun, also taking into account her relationship with the aquatic.

In fact, although Allen (1988: 5, 1989: 16) does not share this hypothesis, he does consider that there must be a relationship between the name Nut and the root *nwj* ('waters'), from which the *nisbe* formed from the masculine name would come *nwjt* '([she] of the Abyss' or 'of the waters'). However, he also indicates that the variant (with the determinative N18), with which her name is written in the pyramid of Neith (PT 593 §1629a), could refer to the original character of his name, *nwt* ('oval'), surrounding the body of Osiris (Allen, 1993: 25), which within the pyramid would correspond to the sarcophagus containing the deceased (very clear in PT 364 §616).

In more recent texts, such as the papyrus of Henuttawy (British Museum EA 10018, 21st Dyn.; Niwinski, 1989: 200 fig. 74), Nut appears with her body decorated with stars and in an arched position in the usual way, but ithyphallic (see Figure 1.13). The inscription that accompanies her image recognizes her, according to Billing (2002: 15) as 'She who created the sky, who created the earth, who created the Netherworld' (*jr pr jr t3 jr d3t*; also Calmettes, 2008: 4–5), although the full text seems to start with the reference to 'Osiris, Foremost of Westerners'.

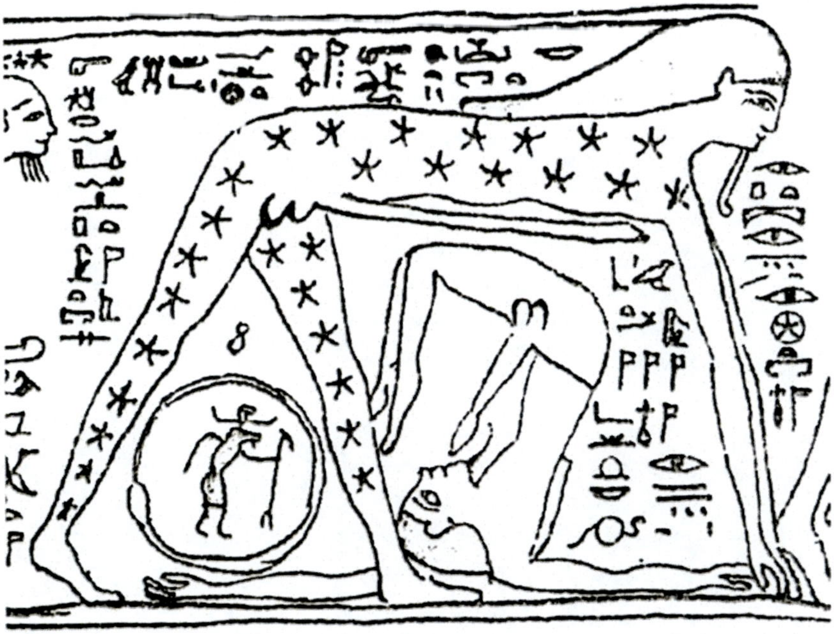

Fig. 1.13 An ithyphallic Nut, celestial goddess, separated from an ithyphallic Geb, terrestrial god. (After R. V. Lanzone, 1883, *Dizionario di mitologia egizia*. Turin: Litografia fratelli Doyen, tav. 159)

The sky is divided into various sectors and regions. In addition to the areas linked to the four cardinal points, mentioned on numerous occasions in the texts, the upper part or zenith is designated as *wpt pt* (*Wb* I 297, 21), as is clear from the description given in PT 456 § 854 (N 356). A region of the sky of great importance to the Egyptians is the *ꜣḫt* ('horizon'). The *ꜣḫt nt pt* ('horizon of the sky'; PT 586 §1582b) is found in its eastern part (PT 263 §341) and is a place of rebirth.

The sky has several entrances (Figure 1.14). Four gates are mentioned in PT 529 §1252, one at each cardinal point. Other texts, however, provide different information. Thus, for example, the *Book of the Day* speaks of four doors on the eastern side (Roulin, 1996: 44). Assmann (1970: 33–34) considers that the gates of the Nun and of heaven through which the access between certain spaces and others takes place must in fact be reduced to a single door.

In both PT 469 §907b (P 319b) and PT 584 (P 560) it is said that the 'doors of *bjꜣ* that are in the *sḥdw* ( ) must be opened'. The *sḥdw,* or *sḥdw*-sky, is a reference to the 'starry sky', as proposed by authors such as Barta (1981a, b: 83) and Allen. This, therefore, is the night sky. And the term *bjꜣ* refers to (meteoric) iron, which is why the Egyptians saw iron gates in the access to heaven.

In the *Pyramid Texts*, as Wainwright (1932a, b) noted almost a century ago, there are numerous mentions of iron in a celestial context. Recently, from a semiotic study of the sign N41 (Almansa-Villatoro, 2019: 77) with which *bjꜣ* is written, it has

Fig. 1.14 The portals of the horizon, under the sky, from the tomb of Sennedjem. Deir el-Medina. (Image by the authors)

been proposed that the sky, as understood by the Egyptians, could be seen as an iron container filled with water. In addition, the *Pyramid Texts* talks about *bjꜣ*-iron cords, bones of the king, chair, limbs of the king, doors of the sky, prow of Sokar-boat (Almansa-Villatoro, 2019: 75). Even in PT 684, the Imperishable Stars are assimilated into the iron bones of the king.

A proof adduced for this is the reference that appears in PT 257 (W 168): 'Unis will acquire the sky and splits its *bjꜣ* (iron)' (*jṯ Wnjs pt pšn.f bjꜣ.s*), *bjꜣ* translated as 'iron' by Faulkner (1969: 67) and 'basin' by Allen (2005: 45). Also, PT 509 (P 462) 'I ascend to the sky, I cross over the iron (firmament), I traverse the *Ḥzꜣw*-water, I demolish the ramparts of Shu' (Faulkner, 1969: 184), and 'he will go forth to the sky. He will cross the Basin for life and authority, travel across the fierce lake, one who topples Shu's walls' (Allen, 2005: 158). Likewise, there are references to iron walls that must be traversed or broken (Almansa-Villatoro, 2019: 76 n. 34).

According to Lesko (1991: 117), the Egyptian firmament (*bjꜣ*) was made of iron, so that when a piece of it became detached it fell in the form of an iron meteorite, leaving, in that iron structure, a hole which,when illuminated by the sun from behind during its night journey, made the stars appear.

The expression *bjꜣ n pt* was used by the Egyptians, at least since the Ramesside period, to designate meteoric iron. The literal translation 'iron from the sky' is very curious, as it could imply that necessarily at some point the Egyptians must have been able to observe a meteor that, beyond leaving its ephemeral and showy trail in the atmosphere, had hit the ground to become a meteorite so that its remains came to be associated with the entire phenomenon, from its celestial origin. Such a situation is not at all common. Shooting stars are very frequently observed, but observing fireballs far less so, although an assiduous observer may observe several during a lifetime.

However, observing a meteorite that survives atmospheric friction and then impacts the surface of the earth (its impact site being located) is a tremendously rare ocurrance. Despite this, at some point the Egyptians must have observed one, thus discovering the celestial origin of meteoritic iron. A possible candidate might be the meteorite that landed 5000 years ago near Gebel Uweinat to form the Gebel Kamil crater (Urbini et al., 2012), next to Egypt's present border with Sudan.

1.2.4 The Earth

Of course, a fundamental part of the Egyptian universe is the earth, *tꜣ*, whose divine personification is Geb. In the *Pyramid Texts* it is generally written as 𓇿 (as in most examples of the Wenis pyramid texts) or 𓇾; that is, with the representation of the extent of the earth with three dots (sometimes two or four, which represent the abundant sand) on the underside.

It is evident that in the Egyptian conception the earth is the centre of the universe. In the upper part of the lid of a sarcophagus from the 30th Dynasty, belonging to the priest of Mut Wereshnefer, kept in the Metropolitan Museum in New York (MMA 14.7.1B) (Ransom, 1914), we have an exceptional example of a representation of the world with the earth as the centre (see Figure 1.15).

In the image stands out the arched figure of Nut, the celestial goddess, covering several series of circles. The innermost circle presents a succession of emblems that symbolize the *nomoi* or provinces of Egypt, 22 corresponding to Upper Egypt and 17 to Lower Egypt, according to the traditional list (Gardiner, 1944: 38–39, pl. V.1). A small fragment of another document is preserved at the Yale Peabody Museum of Natural History (Inv. 61), which originally must have been similar to Wereshnefer's (Clère, 1958: fig. 2; Scott, 1986: 154–155). Even though it is a small fragment of the entire composition, this inner circle shows information that is not detailed in Wereshnefer's example, since each nome is associated with the sign [hieroglyph] or [hieroglyph]; that is, with the chapels *pr wr* or *pr nw*, distinctive of Upper and Lower Egypt, respectively. In the Yale example, the inner circle on which the *nomoi* emblems rest was represented by several lines that could perhaps show the Nile (Clere, 1958: 45).

The motifs included in this circle appear to have no direct relationship to the sector of the *nomoi* or to the land of Egypt. In the upper register there is a winged solar disc and a motif that recalls the 'awakening of Osiris'. In the next register is a row of 19 stars, and in the third, 15 figures facing each other (seven on one side and eight

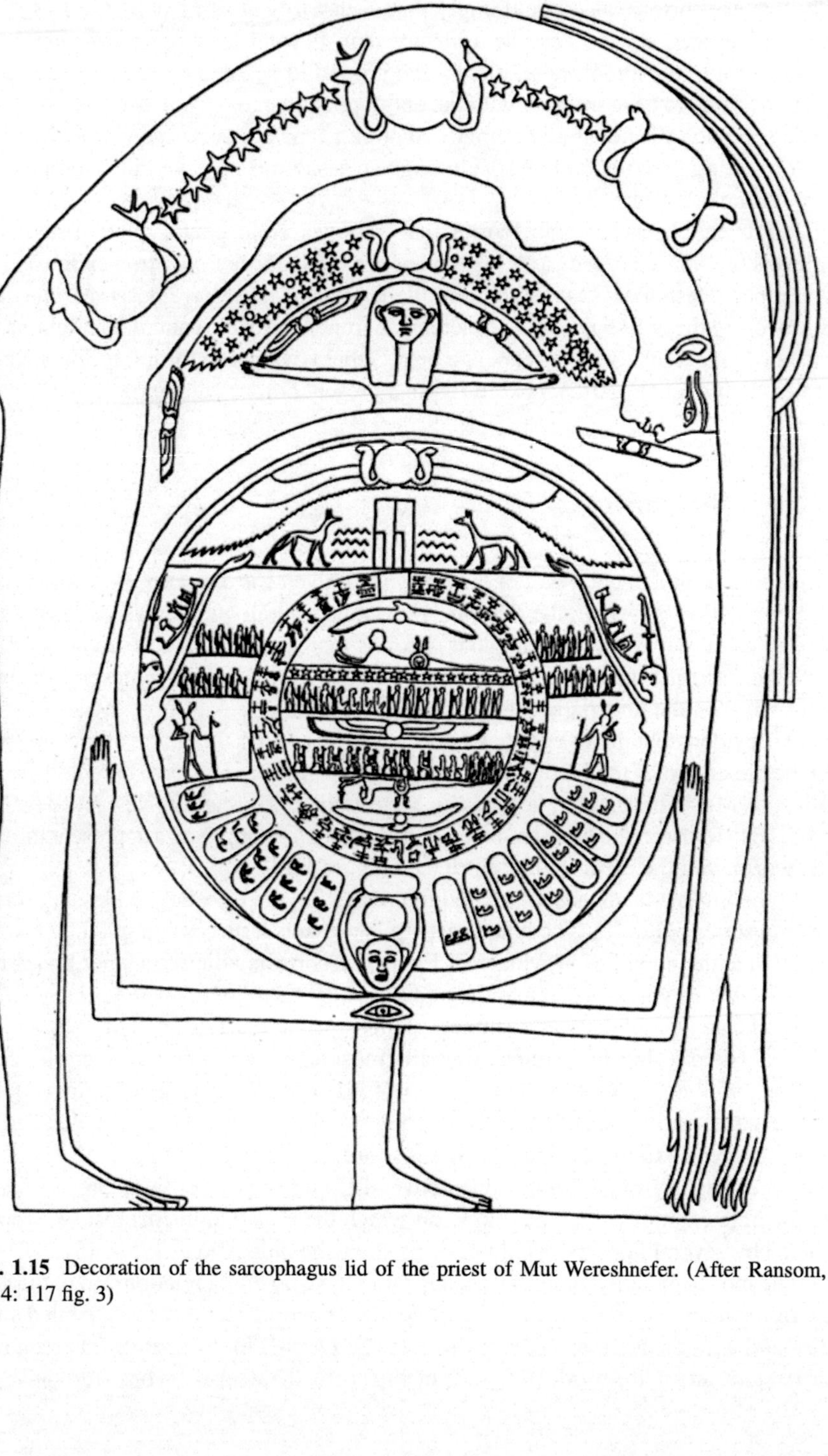

Fig. 1.15 Decoration of the sarcophagus lid of the priest of Mut Wereshnefer. (After Ransom, 1914: 117 fig. 3)

on the other, the latter mumiform). In the fourth register the winged sun appears again, this time with its wings fully spread. In the fifth, there are two other opposing groups of divinities. Finally, the sixth register is (inverted) similar to the first. In short, if on the one hand we see up to three solar motifs, on the other, we have, on two occasions, what seems to be an Osirian motif, as well as those series of divinities.

For this reason, Clère (1958: 46) assumed that it could refer to the Duat or the undersky, in the same way as Calmettes (2017: 40), who believes, *Il s'agit probablement d'une représentation de la Douat ou du «monde de l'au-delà», en tout cas d'un espace lié à la transformation et à la régénération* ('It probably concerns a representation of the Duat or the "afterworld"; in any case a space linked to transformation and regeneration'). There might still be room here for the Nun, if that can explain the traces of water in the Yale example that have survived in this part.

Concentric to the first circle is a second one, inside which we also observe a set of very illustrative iconographic motifs. Occupying the upper part of the interior of this second circle there again appears a winged solar disc with two *uraei*, under which we see the signs . Interpretation of this group is not simple, the animals could be interpreted as the western and eastern desert areas (Calmettes, 2017: 30), as a representation of the eastern and western 'souls', as in the *Book of the Day* (Antelme & Rossini, 2007: 202).

On the right and left, the goddesses of the west and east stand out, respectively, identified thanks to the signs (,) that they carry on their heads. They raise their arms, on which are placed representations of the solar boat, possibly that of the night and that of the day. In front of the chest of each cardinal goddess there appears, inside a box, a feathered divinity. Thanks to the Yale fragment, it is easier to understand the meaning of these figures. It is, on the one hand, the god Ha (which is the one we see at Yale), a god who as 'Lord of the west', is linked to the Libyan desert, the western one (therefore situated on the side of the goddess who represents the west). The records with small human figures associated with this god are also much more developed in the Yale example. Apparently, it is associated with foreign peoples, who would have been organized according to the cardinal orientation.

On the other hand, on the opposite side and in contrast to Ha, we must suppose that Sopdu is located in association with the oriental desert and therefore related to the position of the goddess that represents the oriental region (Clère, 1958: 39). Below, completing the inner part of this circle, a series of ovals with the inscription *ḫ3swt* ('foreign countries') is systematically repeated. This second circle therefore shows us all the land that borders Egypt, identified by its *nomoi* in the first or inner circle.

Above the second circle, attached to it, appears a frontal representation of Nut covered by a solar disc with arched wings full of stars. Below these wings and on either side of Nut's head, other smaller winged discs are shown. These surely reflect the day and night sun, left and right, respectively. This image of Nut (Calmettes, 2017: 41) is quite significant, since its arrangement is reminiscent of a representation belonging to the first hour of the *Book of the Day*, as seen in the tomb of Ramesses VI (Piankoff, 1954: fig. 73). Here, Nut, flanked by Isis and Nephthys, is shown in frontal view holding a disc inside which a child appears with his finger in his mouth; that is, Nut is showing herself pregnant with the young and regenerated sun that will be born again. Nut's role as a life-producing goddess is described by Billing (2002).

At the other end of the circle appears, also in frontal vision, a male figure with a disc on his head, with raised arms supporting an oval shape. According to Calmettes (2017: 46) this could be the solar god as demiurge. Linking him to the sun actually seems very likely. Thus, for example, in the 11th hour of the *Book of Gates*, in the tomb of Ramesses VI (Piankoff, 1954: pl. 58), in reference to the sun there appears a boat with a head looking straight ahead, accompanied by the inscription *ḥr pn jḫ*(*w*) ('This face is made to flourish').

Holding the entire series of concentric circles, arms appear as a *kꜥ* sign, supported on the ground by a pair of legs, perhaps an abbreviated image of Shu (Calmettes, 2017: 43). For Von Lieven (2007: 153–155 §70), however, the legs should instead relate to the goddess Nut. Finally, surrounding the iconography described above, a large bent figure of the goddess Nut stands out. The inside of her torso is decorated with three solar discs with *uraei*, linked by stars. It also highlights the detail of the representation of a winged solar disc at the level of her mouth and another at the level of her genitals, thus expressing the circuit of the sun from her setting to her rebirth.

The elements that we have mentioned do not exactly reflect a physical description of the parts of the universe as understood by the ancient Egyptians, since cosmogonic and funerary elements also appear in it. However, the arrangement of the parts does, in essence, reflect their physical distribution. Egypt is the centre of the universe; therefore, the *nomoi* that constitute it form the central circle and this is surrounded by foreign countries and the limits of the earth. However, the central space of the circle, as we have commented, obeys another part of the world, perhaps the Duat, and everything is finally surrounded by the celestial world, represented by Nut.

1.3 The Structure of the Universe Reflected in Egyptian Architecture

Cosmogonic visions are also reflected in Egyptian architecture. The step pyramids of the 3rd Dynasty, with its first example in the pyramid of Netjerkhet and perhaps with the precedent of the stepped building represented by mastaba 3038 of Saqqara, could be understood not only as a recreation of the primordial hill, whose largest exponents will come in the 4th Dynasty, but also as ladders to access heaven, to which there are numerous references in the *Pyramid Texts*. For example, in PT 572 §1474 (M391) we read: 'He will fetch for you the gods who belong to the sky, having assembled for you the gods who belong to the earth, and they shall put their arms under you, having made a ladder for you so that you might mount on it to the sky. The starry sky's door will be opened to you' (Allen, 2005: 234).

Fig. 1.16 Corbelled vault in the red pyramid of Sneferu at Dahshur. (Image by the authors)

The Sneferu pyramid in Meidum, completed in a final phase of construction as a smooth-faced pyramid, shows an interior with a turriform core and sloping layers in the style of the preceding pyramids. It is the first pyramid in whose interior, in the sepulchral chamber, we observe a corbel vault by approximation of courses. It cannot be ruled out that this type of vault, beyond its architectural function, could also suggest the idea of a stairway to heaven (see Figure 1.16).

On the other hand, although it was not until Wenis that five-pointed stars were shaped for the first time on the gabled roofs of the sepulchral chambers of the pyramids to reflect the celestial sphere of Nut, we may suspect that before these decorative motifs some sepulchral chambers have already sought this association with the celestial world. Thus, it is quite possible that both the vault of the burial chamber of Menkaure (see Figure 1.17) and that belonging to Shepseskaf, from the end of the 4th Dynasty, try to assimilate to the vaulted shape of the sky. It is true that this innovation did not have immediate continuity, but this proposal does not cease to be valid for that reason. We are entitled to think, for example, that it is not by chance that the first astronomical ceiling in the Valley of the Kings (KV 17, tomb of Sethy I) appears in the burial chamber and on a vaulted ceiling in order to recreate the shape of the sky.

The astronomical alignments made during the ritual of the stretching of the cord are a reflection of that existing connection with the celestial world, with the world of the gods, but even in the structure of the temples one may discern an architectural planning that seeks to assimilate cosmogonic concepts recognized by all Egyptians.

The temples, in ancient Egypt, are conceived not as places of worship to which believers must go but as the house of the god. That is why, in Egyptian, the priests are referred ti as the *ḥmw nṯr* (literally, the 'servants of the god'). Taking this into

Fig. 1.17 Burial chamber, with vaulted ceiling, of the Pyramid of Menkaure in Giza. https://upload.wikimedia.org/wikipedia/commons/3/38/17_menkaure_burial_chamber.jpg. (Author: Jon Bodsworth. Public domain, via Wikimedia Commons)

account, it is more logical to assume that this house of the god, symbolically, must express fundamental cosmic principles, principles in which the god must also be involved. This is seen from the beginning, since the primordial hill was symbolically elevated in the oldest Egyptian temples, as in the archaic temple of Nekhen, where a huge mound of earth and sand within the *temenos* had to simulate the primordial hill (Wilkinson, 2000a, b: 76).

An Egyptian temple is the central and most important unit of a larger area defined by a perimeter wall which in classical architecture is known as the *peribolos*. This wall, therefore, is the barrier that separates the temple and its *temenos* from the outside world; in other words, it distinguishes order from chaos. From this wall, the main elements of the temple will be a reflection of the cosmos, of its creation.

If we examine the perimeters of Egyptian temples, especially those that are well preserved from the Graeco–Roman period, we observe a characteristic that is repeated: the walls were not built with horizontal courses of adobe but in sections of concave and convex courses that seem to represent waves (see Figure 1.18). According to Chévrier (1964) this method of building had a functional purpose, since it allowed the dilations of the sectioned wall to be lower than if it had been built in horizontal courses and in a single section. But it also had a symbolic function, since the waves recreate the waters of the Nun, that watery pre-creation space characterized by darkness, silence, and inertia. The wall of the *peribolos*, in the cosmic symbolism of the temple, will therefore mark the limit that the darkness (or chaos) of Nun can reach. Everything that remains within the *peribolos*, therefore, will belong to the space of creation.

We should also bear in mind that in Egyptian temples it is common to gain access through a quay connected by channels to the Nile. The road up to the entrance of the temple surrounded by the *peribolos* could then be protected by sphinxes, protective animals. Not in vain does it come from the outside world, and although the Nile is the great benefactor of Egypt, with Hapy as the deity of the desired flood, in origin these waters could come from the Nun.

Once we pass through the entrance of the *peribolos*, we also find more monumental doors or pylons in the direction of the main temple. The architecture of the pylons seems to start in Thebes in the late 11th Dynasty, in a temple built by Sankhkare Montuhotep on the so-called Thoth hill or Crown of Thebes. From there, especially in the time of the New Kingdom, this type of construction would become characteristic. The pylons form an access consisting of two thick towers with sloping walls joined by a bridge, creating a shape reminiscent of the 𓈌 *3ḫt* ('horizon') sign. Neither is the adoption of this form accidental, since it could represent the horizon through which the sun rises each day, which in turn is a reason for birth, regeneration, and creation. On the lintel of the door, especially in Graeco–Roman times, we can see the solar disc. On the other hand, the pair of obelisks that precede the pylon, as observed in the example of the Luxor temple, may refer to the two horizons, since they were sometimes dedicated to the morning or evening sun god (Wilkinson, 2000a, b: 77–78).

The frontal decoration of the pylons is equally interesting from the point of view of the symbolic, since it is common to see the iconography of the god who delivers the weapon with which the pharaoh must annihilate the foreign prisoners who, caught by their hair, appear kneeling in front of him waiting for the fatal blow. In the created world there are also chaotic elements, so this image expresses the order imposed by the pharaoh, champion of maat on earth, against the chaos represented

Fig. 1.18 *Peribolos* wall of the Ptolemaic temple of Deir el-Medina. (Image by courtesy of Mónica Moritz)

by his enemies. It is an apotropaic motif (Hall, 1986: 16). In Karnak, the second pylon (raised at the beginning of the 19th Dynasty) is attached to a large stone wall whose exterior façade presents a succession of military themes on its two flanks, but not on the inside. The image, then, is quite clear. It may also symbolize the outer limit of Egypt (Molinero Polo, 2000: 86–87), exposed to other peoples.

Behind the first pylon, we gain access to an open-air court. In the Temple of Horus in Edfu, a formidable inscription more than 300 metres in length from the time of Ptolemy X describes the name of this courtyard: the 'Place in which Seftekh (Apophis) is Defeated, the enemy of Harakhte' (Kurth, 2004a, b: 66) or *St-sḫr-Sftḫ ḫftj n 3ḫtj* (Finnestad, 1985: 65). Here, in the great open-air courtyard, the sun is relentless. It represents the victory of creation that repeats itself every day.

After the courtyard, we access the first hypostyle room, full of columns, which can be papyrus-, loti-, or palmiform. At their base, these are decorated with plant motifs, open petals. On the walls of this room, in the plinth, it is also common to find plant motifs. Thus are recreated the marshes that surround the primordial hill, whose distinctive element in the temple will be the naos of the god. The ceiling of this room, on the other hand, is usually decorated with astronomical motifs.

From the first hypostyle room to the naos where the god resides is recreated the ascent to the primordial hill. On the one hand, the ceiling will go down in level, on the other, gentle ramps will make us climb, almost without perceiving it. At the same time, the atmosphere will get darker and darker. In this way, when we stand in the *sancta sanctorum* we are higher and, in addition, the ceiling (decorated with stars) is closer to our heads. This, of course, is intended to show that with the ascent to the primordial hill the sky is closer (Figure 1.19).

The *sancta sanctorum*, especially in late examples such as Dendara and Edfu, constitutes a separate building with an ambulatory. The most important element that we find inside is the naos, since it is there where, in its statue form, the titular god of the temple resides (see Figure 1.20). Its shape is also significant. If we look at the example of the temple of Horus in Edfu, its upper part is pyramidal in shape, thus also resembling the primordial hill. The god is therefore in his place of the first time, at the top of the primordial hill, closer to heaven. In fact, both the doors of the *sancta sanctorum* and those of the naos itself are called the *gates of heaven*.

1.4 The End of the Egyptian Universe

As we have seen, creation from the pre-creational world, the Nun, is a recurring motif in Egyptian cosmogonic texts, even with its different versions and variants. The origin and repetition of creation is the event which, also because of its funerary significance (due to the implicit regeneration that it entails), seems to interest the Egyptians the most. However, in many texts we can also find eschatological references to the end of the universe. These mentions are of great interest, since they give us an almost unexpected vision, taking into account the volume of texts that tell us about the permanence of the gods.

In PT 254 §277–279 (W 165) we find a first reference to the destructive process, still in the form of a curse, a format that appears in many other and varied texts:

> O lord of the Akhet, make a place for Unis. If you do not make a place for Unis, Unis will make a curse on Father Geb, (saying): 'The earth has no spokesman; Geb has no guard', and anyone Unis finds in his way he will devour. The pelican (*ḥnwt*) will prophesy, the sunshine-

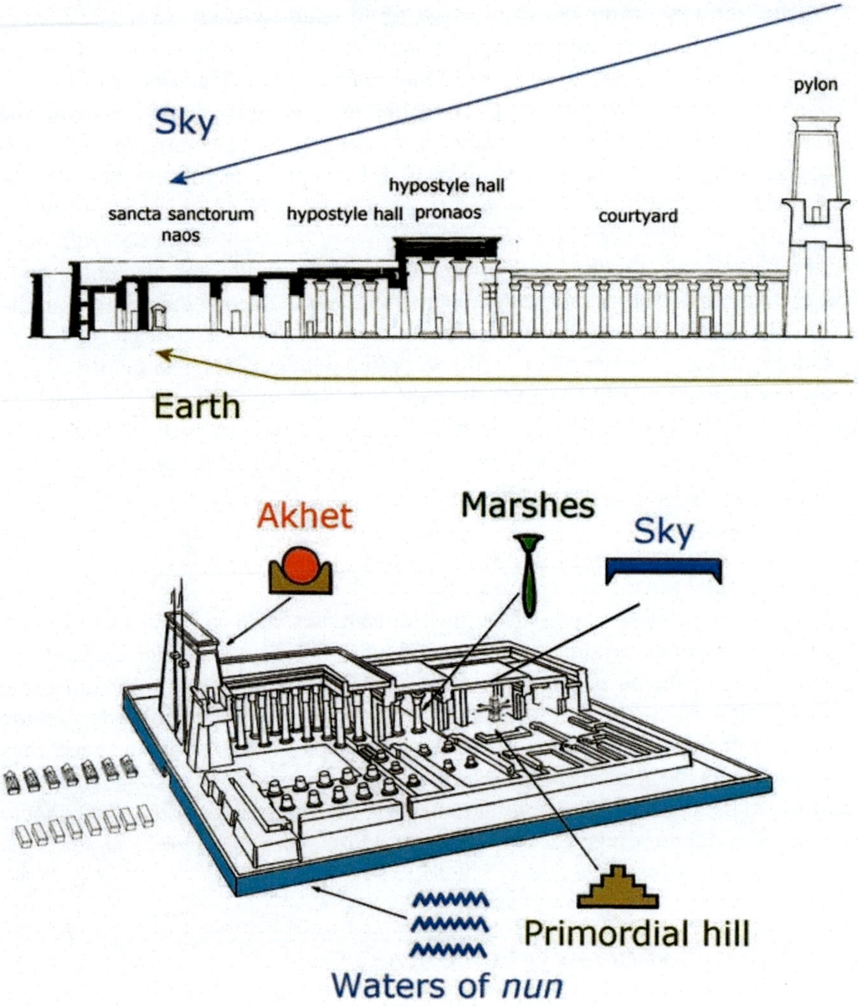

Fig. 1.19 The temple as a reflection of the cosmos. (Adapted from Kurth, 2004a, b; Molinero Polo, 2000: 85)

> bird will come forth, the Great One will stand up, and the Enneads will speak, (saying): 'The earth is completely dammed up, for the mountain ranges on either side of the river have been joined and the two banks have been united, the roads have been made inaccessible to those who would pass, and the terraces have been destroyed for those who would go up. (Allen, 2005: 43)

The mentioned pelican, in the guise of a solar bird, symbolizes the place through which the sun enters and leaves the visible world. The pelican lives on the edge of the universe, so that in this way he is aware of what may happen at the end of time. Thus, when it is said that 'the mountain ranges on either side of the river have been joined and the two banks have been united' it is intended to express that in the end,

Fig. 1.20 Naos of the Temple of Horus at Edfu in the Holy of Holies. (Image by courtesy of María Lull)

space will fold over on itself, until it disappears. It is the same idea that appears in pHarris VII, No. 2–4: 'I will cause the earth to go down into the primeval water, South will become North and Earth will be overturned' (El-Sebaie, 2000: 19).

A similar reference is found in the *Coffin Texts* (CT 619 VI, 231), this time the *benu* bird prophesying the destruction:

> I will strike the Westerners as Geb, father of the gods, the phoenix will prophesy, the shining one will go forth, the Great One will arise, the Ennead will speak, the earth will not open, Geb will not speak, Re will not be on high, the skins of the scorched ones (?) will be joined, and they will unite the river banks (…). (Faulkner, 1994: II, 202)

The *benu* bird, sometimes translated as the phoenix, is at the beginning of time but also has the ability to indicate what the final process will be like at death or at the end of the world. In fact, in chapter 17 of the *Book of Coming Forth by Day* the *benu* is shown as the keeper of the inventory of what exists: 'I am that benu-heron which is in Iunu the keeper of the inventory of what exists' (Quirke, 2013: 56).

We can also trace the end as being associated with a destructive process contemplated by Atum, the demiurge, and Osiris in the *Coffin Texts* (CT 1130 VII, 467–468) when we read: 'I have passed myriads of years between myself and yonder Inert

One (*wrḏ-jb* 'tired of heart'), the son of Geb; I will sit with him in the one place, and mounds will be towns and towns will be mounds; mansion will desolate mansion' (Faulkner, 1994: III, 168). It is a return to the beginning of creation, where the demiurge returns to its pre-existence form. In other texts that return to initial chaos is also recreated (Schott, 1959: 319–330). In the words of Kákosy (1963: 19), *der Weltuntergang sei eine sozusagen "negative" Wiederholung der Schöpfung* ('[that] end of the world be, as it were, the "negative" repetition of creation').

In other types of texts, such as the *Admonitions*, there are also descriptions that refer to the end of the world. In the Ipuwer Papyrus (pLeiden I 344 recto), for example, which, although dating from the 19th Dynasty, traces its origin to the Middle Kingdom, in a context of disorder that refers to a generalized crisis of society, it is said: 'Would that there might be an end of men, no conception, no birth! O that the earth would cease from noise, and tumult be no more!' (Gardiner, 1909: 44). The silence of the earth is a synonym for annihilation, back to the Nun.

There can also occur in texts of a ritual nature a description of the instability and chaos in the world that might come if the rituals that favour the maintenance of order, of maat, are not carried out correctly. In this sense, a good example is the Salt 825 papyrus (pBM 10090), dated to the 30th or 31st Dynasty (Fermat, 2010: 21).

In the *Book of Gates* it is noted how the gods can also fall into Duat. The creator god has the virtue of being able to take with him not only humans but also gods to the world of the dead. Thus, in the first record of the first hour it is said:

> *Die aus Re entstanden sind, aus seinem Glanzauge, die aus seinem Auge hervorgegangen sind.* Er hat ihnen den Vergorgenen Platz zugewiesen, zu dem entdrückt sind die Menschen und die Götter, alle Vieh und alles Gewürm, das diesern grosse Gott geschaffen hat. (Hornung, 1997: 197)
>
> 'That originate from Re, from his gleaming eye, that are brought forth from his eye. He has assigned to them the hidden place to which men and gods are consigned, all cattle and worms that this great god has made.'

But this death is necessary, since the passage of the sun through the Duat implies rebirth. For this reason, in pBerlin 3049 it is said that Re is 'the one who rejuvenates daily, without reaching his end' (Hornung, 1983: 162).

Death, according to chapter 154 of the *Book of Coming Forth by Day*, will reach everyone: 'as does every god, every goddess, as all birds and all fish, as all snakes and all worms, as every flock, entirely, entirely' (Quirke, 2013: 383). In the *Book of Caverns*, another of the great religious compositions of the New Kingdom that appears in its first complete version in the tomb of Ramesses IV, it is also observed how death is not the end but the necessary transition for cyclical rebirth: *j'entre dans la terre de laquelle je suis sorti et que j'ai revisé ma première naissance*[8] (Piankoff, 1942: 6).

But of all the documents that we can analyse, undoubtedly the clearest, regarding the end of the world, is chapter 175 of the *Book of Coming Forth by Day*:

[8] 'I return to the earth whence I came and where I witnessed my first birth'.

> What is a span of life?, Says (Osiris).
> You are to have millions of millions, a lifespan of millions.
> when I have had him send out the elders,
> and I shall indeed destroy all I made,
> and this land shall turn into Nun,
> as a floodwater (*ḥwḥw*), as its original condition.
> I alone am to remain, with Osiris,
> when I have transformed myself into other snakes,
> which men do not know, which gods do not see. (Quirke, 2013: 438)

This descriptive and particularly interesting text shows a conversation between the god of the dead, Osiris, and the demiurge himself, Atum. In response to Osiris's question (*jšst pw ꜥḥꜥ m ꜥḥꜥ*), Atum tells him that his remaining time, although very long, is limited. Atum will destroy everything created to return to the Nun, as the text says, to its original state. But it should be noted that although BD 175 warns that both Atum and Osiris will remain in the Nun, it is also said that Atum will transform into serpents. This indicates that even Atum himself will not remain in his created godly form but will pass into a pre-existential form defined by those serpents. Atum, both before creation and in the return to the Nun after the end of the world, appears in the form of a serpent, as a reflection of that state of pre-existence.

However, in both PT and CT it seems that Atum is also shown in that form: *psḥ.n Tm mḥ.nf rf ꜥnn.f ꜥnn* ('Atum has bitten and has filled his mouth, and he coils up'; Faulkner, 1994: II, 271; CT 714 VI, 346).'The one Atum has bitten has filled his mouth, coiling all up' (Allen, 2005: 53; PT 284 §425 (W 190)).

Therefore, this chapter of the *Book of Coming Forth by Day* establishes that in the distant future the known universe will be destroyed to return to the Nun, the dark, inert, and silent pre-creation place, in the original form 'which men do not know, which gods do not see' (*nn rḫ sn rmṯ nn mꜣ sn nṯrw*). Let us remember that the latter is a quality of the demiurge. In its pre-creation state in the Nun its essence, its force, remains inactive, latent, until it comes into existence before any other god has done. As noted in the pLeiden I 350, 'no other god with him to tell his appearance of him' (Allen, 1988: 52).

In late texts it is also observed that the return to the Nun, as a transition, supposes a renewal. Thus, in reference to the eight primordial gods, a text in the propylon of the temple of Khonsu (Bab el-Amara gate of Ptolemy III) tells us, 'they are at his side in the cavern of Nun, living for eternity, not knowing destruction', 'they enter the Nun with their father who created them' (Brotto, 2016: 241, 245).

Also of great interest is a hymn dedicated to Osiris found in the temple of Opet in Karnak, from the Ptolemaic period:

> *On dit de lui dans les écritures: celui qui dure après le temps neheh, parce qu'il est l'Un qui reste ensemble avec la Majesté de Rê, tandis que la terre sera dans le flot et sera submergée par l'océan primordial comme avant son origine et tandis qu'il n'y aura pas de dieu et pas de déesse qui, eux aussi, se retransformeront en serpent.* (Assmann, 2003: 121).

> 'It is written of him: he that endures after the *neheh* time, for he is the One who remains together with the majesty of Re, while the earth shall be flooded and shall be submerged in the primordial ocean as before its birth, and there shall be no god and no goddess, for they too shall be transformed into serpents.'

The Egyptians did not have any words to express anything as abstract or of such complete meaning as our word 'time' (Assmann, 1983: 193), despite having a large number of words linked to time. However, it is the combination of two words, *ḏt djet* and *nḥḥ neheh* (Assmann, 1975), that comes closest to this concept. Thus, a passage from the *Coffin Texts* says: *Quant à tout ce qui existe, c'est neheh et djet*[9] (Bickel, 2003: 45). The Egyptians made constant use of these two words to express the idea of time/ eternity (Assmann, 2003: 111): *djet* and *neheh*, greatly simplifying their meaning, could respectively correspond on the one hand to a linear and immutable eternity and on the other to a cyclical and regenerative form.

The text of the temple of Opet again refers to the permanence of Re (instead of Atum) and Osiris, in the form of a serpent, in the Nun, in the pre-creation stage, no other god being able to do so. As Assmann (2003: 121) points out, this hymn contributes the idea of eternity after time, since after time *neheh*, with all its cycles, will come the lasting continuity of *djet*. On the other hand, it is noteworthy that while *neheh* time is related to the creative process, *djet* time is absent in this regard. However, neither *djet* nor *neheh* are used as elements present in the Nun (Hornung, 2001: 12).

When the Egyptians refer to a very distant time, they use the term *jswt* ('the old time') or *p3wt* 'ancient times', the moment of creation being the well-known *tp spj* 'first occasion'. Before creation, before the first time, there is no time. A text by Ramesses II, the Rhetorical Stela from Abu Simbel, conveys this idea by stating that before the first time there was no past *mj tpj sp.f nn wn ḫr-ḥ3t* (*KRI* II: 312 l.12; Loprieno, 2003: 131).

The return to the initial pre-creation state will lead to a new beginning, as in the first time. Therefore, beyond the daily cycle of regeneration of the solar god, abundantly expressed in a great variety of texts and iconography, there is also a cycle of (re)creation on a much longer timescale, since it encompasses the universe itself. The Egyptian universe is cyclical, as also endorsed by an inscription from the Ptolemaic temple of Dendara, in which the creator is said to be 'the one who comes into existence after the end of cyclical time and does not disappear' (Daumas, 1987: 152).

It is interesting to recall a work by the Alexandrian poet Claudius Claudian (Derchain, 1956: 4), from the early 5th century AD. In *The Praises of Stilicho*, Claudian speaks of a serpent that, in itself, embodies this cyclical and inexorable aspect:

> *Est ignota procul, nostraeque impervia genti,*
> *Vix adeunda Deis, annorum squalida mater,*
> *Immensi spelunca aevi, quae tempora vasto*
> *Suppeditat revocatque sinu: Complectitur antrum,*
> *Omnia qui placido consumit numine, serpens*
> *Perpetuumque viret squamis, caudamque reductam*
> *Ore vorat, tacito relegens exordia lapsu.* (Claudiani, 1824: 699–700)

[9] 'That which exists shall be neheh and djet.'

> 'Far away, unknown, beyond the range of our race,
> scarce to be approached by the gods, is the dark mother of the years,
> the cave of immeasurable time, which in its vast interior
> spawns and countermands the ages: A snake surrounds the grotto,
> peaceful in spirit, swallowing everything
> and eternally rejuvenating itself with its scales,
> but swallows its tail with its head thrown back, and silently returns to the beginning.'
> (Assmann, 2019: 27)

What this text of Claudian conveys is perfectly comparable to what we have seen in BD 175. The place where the serpent lives is unknown, like the Nun, and the serpent that bites its tail symbolizes cyclical eternity in its process of continuous renewal. On the other hand, this iconographic motif, although it is known by the Greek word οὐροβρόρος, (*ouroboros*, 'tail-devouring'), has its origin in ancient Egypt.

The oldest *ouroboros* depicted in Egyptian iconography appears on the outside of the right side of Tutankhamun's second golden chapel (Piankoff, 1951: pl. IV; 1949). Actually, two were represented, one surrounding the head of a mummiform figure and a similar one surrounding the feet of the figure. On both sides of the deceased's head there is the same inscription *jmn wnwt* 'He who hides the hours'. A double circle surrounds the head, and inside it appears the serpent that bites its tail, accompanied by an inscription in 'enigmatic writing' that identifies it: *mḥn* Mehen, more usually written .

Mehen (Wilkinson, 2003: 222–223) is the serpent that usually appears in the Mesketet or nocturnal boat of the sun, surrounding, as a protective element, the cabin in which the solar god is found. It appears in some of the *Books of the Hereafter* of the New Kingdom, such as the *Book of the Night*, the *Book of Gates*, and the *Book of the Imiduat*. We can also see it in some sarcophagi of kings from the Ramesside period, such as Merenptah (Assmann, 1972: fig. 1; Sourouzian, 1989: pl. 36 b) and Ramesses III.

There is an even more interesting image of the *ouroboros*, owing to the clear spatio–temporal connotations with which it is associated (Assmann, 2011: 54–56, fig. 6), in the papyrus of Lady Heruben (Piankoff & Rambova, 1957: vI, 22 fig. 3, 73, vII n° 1) of the 21st Dynasty (pCairo 133; see Figure 1.21), similar to a scene from pBerlin 3148 (Schott, 1938: pl. 6). A solar disc appears to be surrounded by the *ouroboros* and the embrace, from its upper part, by a pair of arms that must belong to Nut. In the lower part, the disc is supported by a *bucranium*, which according to Reemes (2015: 204–205) must be recognized as a representation of Hathor as a lady of the west.

However, in similar scenes (Uppsala VM 228 coffin; Billing, 2002: 218 and fig. A22) the *bucranium* is decorated on the inside with stars, so it seems to be better related to the celestial realm, with the heavenly cow or the goddess Bat. The solar disc, painted red on the papyrus, contains in its interior the anthropomorphic representation of the boy with a finger in his mouth that symbolizes the young rising sun. In fact, he is shown sitting on a cushion that resembles the *akhet*, the horizon. Also supporting the disc are lions pointing in opposite directions. It is about Ruty, the 'two lions', generally associated with the eastern and western horizons, which in chapter 17 of the *Book of Coming Forth by Day* represent the *sf* ('yesterday')

Fig. 1.21 *Ouroboros* on the Heruben Papyrus. (Drawn by the authors after Piankoff & Rambova, 1957: 22 fig. 3)

and the ★𓅱𓄿𓇳 *dw3w* ('tomorrow'), as we can see, for example, on the 19th Dynasty Ani papyrus (pBM EA 10470), where both lions support the *akhet* and are covered by *pt*, the sign of heaven (Budge, 1913: pl. 7). In BD 17 it is stated, 'Yesterday is Osiris, Tomorrow is Re (…) as for perpetuity (*neheh*), that is day; as for eternity (*djet*), that is night' (Allen, 1974: 27–28; see Figure 1.22).

Overall, therefore, the iconography provided by this scene is very complete. The *ouroboros* protects the solar disc but is at the same time related to the *djet* and *neheh* time, with the continuous cycle of renewal of the sun, and also with the beginning of the world (the first time) as expressed by Niwiński (1981: 46 fig. 1) in a diagram showing the significance of the *ouroboros*. Similarly, there is a spatial dimension, since the cycles are framed between heaven and earth, with the necessary passage of the sun through the Duat, so that the main elements of the universe remain implicit.

Fig. 1.22 Yesterday and tomorrow on Ani's papyrus, BM EA 10470. (After Budge, 1913: pl. 7)

Chapter 2
Egyptian Astronomers Throughout History

Recognizing astronomers in Egyptian documentation is not always an easy task. We must bear in mind that in most cases they are individuals who are not identified in either the iconography or the statuary with a specific clothing or associated with instruments that we may consider proper to the astronomer.

In these conditions, on many occasions we therefore have the option of identifying this kind of specialist only thanks to the meaning of the functions with which they are presented in the documents, provided of course that these responsibilities offer clues as to their relationship with the celestial vault. For this reason, the study of titles or functions is very important in helping to identify them throughout the history of Egypt.

2.1 Astronomers and Horologists During The Pharaonic Period

2.1.1 Astronomers from the Old Kingdom to the Middle Kingdom

According to Strudwick (1985: 159–160), the oldest Egyptian astronomer that we can identify most clearly in written sources was (Tjenti), an individual from the mid-6th Dynasty (*c.* 2300 BC). On the lintel of the access door to his tomb, excavated by Mariette (1889: 148–149) in North Saqqara, were written his main titles (see Figure 2.1): '*Iripat*, *sem*, *hatia*, grand vizier, superior lector priest, sole companion, scribe of the divine book', and after these titles Tjenti also appears designated as *ḥrj-sšt3 n pt* ('superior of the secret of the sky'), *m33 št3 pt* ('who sees the secret of the sky'). Tjenti's relationship with observing the sky seems clear, although we cannot conclude what specific tasks he had as an observer of the secrets of the sky.

J. A. Belmonte, J. Lull, *Astronomy of Ancient Egypt*, Historical & Cultural Astronomy, https://doi.org/10.1007/978-3-031-11829-6_2

Fig. 2.1 Tjenti inscription, 6th Dynasty. (After Mariette, 1889: 149)

Another of the titles with which Tjenti is presented is *wr mꜣw*, generally translated as 'Great of the Seers'.[1] The title has been identified as dating from at least the 1st Dynasty (Cervelló-Autuori, 2011: 1133) and in the 3rd Dynasty one of its bearers was the famous Imhotep. It was also held by the main priest of the solar god in Heliopolis, although on this issue there also been some disagreement. Helck (1954: 95–97), for example, after verifying the relationship between the priesthood of Heliopolis and the leaders of mining expeditions, considered that the 'seers' or 'observers' where not related with the celestial world but with stones and metals (in the sense of 'seekers'). But it has also been interpreted by some authors in its astronomical sense (Roeder, 1917: 4; Borchardt, 1935: 99; Quirke, 2001: 106), as an observer of the hourly stars, the sun, or in relation to the possible discovery in Heliopolis of the relationship between the heliacal rising of Sirius and the calendar. Thus, Sethe (1918: 65) translated it as *Ober(stern)seher* ('star superintendant'), although Neugebauer (1983: 177) spoke of *petitio principii* to deny this relationship. Actually, with the information we have, we cannot affirm that the *wr mꜣw* has among his functions observation of the sky, but, as we shall see, this title is associated on some occasions[2] (as in the case of Tjenti) with others directly related to the astronomers, so we cannot totally deny this possibility.

There has also been speculation about the possibility that the title , *ḥrj sštꜣ n kkw* ('superior of the secret of darkness'; Fischer, 1978: 56; i.e. 'he who is privy to the secret of the darkness'), held by an individual of the 6th Dynasty, refers to an astronomer (Fissolo, 2001: 16), but this seems very doubtful since, in this case, we cannot necessarily link *kkw*, the darkness, with the night sky.

A function that possibly identifies an astronomer, at least in those cases in which it does not refer to security personnel, is that of *wršy* 'watcher', 'sentry' (Faulkner, 1988: 65), *Wache*, *Wächter*, *Tagesdiensthabener* ('guard', 'watchman', 'day-shift guard'; Hannig, 2003: 365), a word documented since the Old Kingdom and that would come from the verb *wrš den Tag verbringen* ('to spend the day'; Hannig, 1997: 206; Gunn, 1921: 108–109). In fact, in the Abusir papyri (Posener-Kriéger, 1976: 29–31), from the funerary temple of King Neferirkare

[1] But other translations of this title have also been proposed. See Jones (2000b: 386–387); Moursi (1972: 147–154); Nuzzolo and Krejčí (2017: 367–369).

[2] Such is the case of the priest Nyankhnesut (6th Dynasty) who, among other titles, uses "Superior of the secret of the sky" and "Great of the Seers". See Priskin, 2018: 31.

Kakai (5th Dynasty, *c.* 2450 BC), we find inscriptions indicating the presence of  ('watchers on the terrace (of the temple)'). If we look at it, the determinative used to designate the watchmen, , represents the sign of the sky from which a star hangs.

This determinative seems quite significant, because although the role of the watchman or observer is not described (so we do not know what the objective of their surveillance is), it is likely that it would have been linked to observation of the night sky. This observation is also carried out from the best place to do it, which is none other than the terrace, the highest part of the building. Undoubtedly, there would be watchers who would work during the daytime, observing the movement of the sun, and others who would work at night, observing the movement of the stars, planets, and the moon.

ntjw m rsw tp ḥwt
tp-rsj tp-mḥtj

'those who stand as watchers on the terrace (of the temple)
south extreme, north extreme'

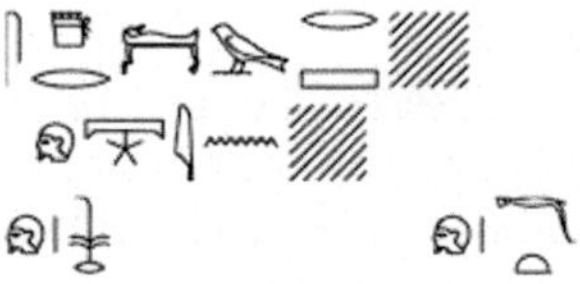

sḏr wrš [...]
tp grḥ jn [...]
tp-rsj tp-mḥtj

'spend the night and spend the day [...]
beginning of the night by [...]
south extreme, north extreme'

On the other hand, in the Abusir papyri it is indicated how these stargazers were distributed on the terrace of the temple: 'those who are as watchmen on the terrace of the temple, south end: 4 (watchmen), north end: 3 (watchmen)'. This arrangement could indicate that the watchers were thus distributed to determine the passage of the stars through the central meridian. In another fragment of the papyrus we read: 'spend the night and spend the day [...] at the beginning of the night at [...] extreme south, extreme north'. Both the words 'extreme north' and 'extreme south' are written in red ink, heading two columns of text. In this way, in each column and in relation to each end or each part of the terrace, the number of 'guards' that should be available were indicated.

Thus, while in table II of the papyrus it is said that four guards were placed on the south side and three in the north, in table V there are two in the southern part and three in the northern part of the temple terrace. An arrangement in facing pairs could recall the representations of the Ramesside stellar clocks in which the body of a person served as a reference for the observer who looked for stellar culminations. A reference such as a pole fixed vertically to the ground would have been equally useful and probably more accurate.

During the Middle Kingdom we also find the words *jrj-ʿt n wršw* ('keeper of the watchers/observers (of the hours of the day)'; Ward, 1982: 58 n° 466 'Hall-keeper of the Watcher of Day-light hours'); *ȝtw n wršw* ('Attendant of the watchers/observers (of the hours of the day)'; Ward, 1982: 6 n° 5 'Attendant of the Watcher (of daylight hours)'; and Wild, 1969: 123 *intendant de la surveillance-du-jour* ('daytime surveillance attendant')). We see this last expression, for example, in the statue of Renseneb ( ). For Wild, this function would not be limited to religious concerns but in the civil sphere would be a state-run service.

There are researchers, citing the well-known example of Sinuhe (Koch, 1990: 18),[3] who believe that these expressions might have a military meaning (Quirke, 2004: 110). However, we must bear in mind that this last example mentions the *wršyw tp-ḥwt* ('watchmen on the terrace') or *wršy tp jnb* ('watchman on the wall'), who might have noticed Sinuhe's presence during his flight from Egypt. Certainly, in this case, the context suggests security guards and hence a military sense, but some combinations of functions might lead us to think that, at least in some cases, they could be expressions used to designate Egyptian astronomers. This can be verified, for example, in the inscriptions on the Berlin 7286 stela from Abydos of the 13th Dynasty (Simpson, 1974: pl. 28, ANOC 18.3; Roeder, 1913: 192). In this stela, belonging to the family and colleagues of Wernebkemui, an individual named Hetepi is designated as *jrj-ʿt n wršw wnwtj* ('keeper of the watchers/observers, the (observer) of the hour'). The combination of the first title with that of the *wnwtj* ('(observer) of the hour', which is the most often used word in much of Egyptian history to designate the astronomer) could justify the claim that the first, at least in certain cases, had astronomical connotation. On the other hand, although the term *wnwtj* "the (observer) of the hour" is not documented in the Old Kingdom, the term *wnwt* for "priestly service" (*Wb* I, 317: 1) is known for that period, a term that used as a determinative, and not coincidentally, a star. This only strengthens the idea that the Egyptian astronomer or horologist, as a measurer of the hours, could have had an important role in the temple from very early on, regulating the times when certain rituals should be performed.

The word *wnwtj*, mentioned in the example above, appears in the Middle Kingdom. Two examples from this period are seen in the Louvre C33 stela (Gayet, 1886: pl. LVIII), where a reference is made to the *jrj-ʿt wnwtj* ('keeper of the (observer) of the hour'; Ward, 1982: 58 n° 463 'Hall-keeper of an Astronomer'), and in the Louvre C5 stela (Gayet, 1886: pl. IX), in which the *wnwtj* ('(observer) of the hour') is mentioned (Ward, 1982: 86 n° 713, 'Astronomer', 'Hour-watcher'). As we can see, the writing presents some variants at this time, to which we must add (Stela CG 20140)[4] and (Martin, 1971: pl. 33 No. 17; Mace, 1922: pl. III No. 11), also during the 12th Dynasty.

[3] Papyrus Berlin 3022, l. 18-19 + Berlin 10499, R44–45.

[4] In the funerary stele of Ikhernofret (CG 20140), of the 12th Dynasty, this variant of writing appears in the title . See, Lange and Schäfer (1902: 166).

In Egyptian, the word 'hour' is written *wnwt* (*Wb* I, 316) or , so that the word , the '(observer) of the hour', is a *nisbe* form of 'hour.' In this sense, the *wnwtj* would correspond to the ὡρολόγος ('horologist') that will appear in the documentation of the Graeco–Roman period. Already in the *Pyramid Texts* mention is made of the hours of the night, for example in PT 320 §515a (W 225) (Allen, 2013: PT 320; Sethe, 1908: Spruch 320, 515a):

> *ḏd mdw ḏsr.n Wnjs grḥ sb.n Wnjs wnwt*[5]
> 'Recitation: Unis has cleared the night, Unis has sent off the hours'[6]

We may assume that observation of the hours of the day and night was the main task of this sky watcher, taking into account that the combination of the functions of 'keeper of the watchmen/observers of the day', and that of 'the (observer) of the hour', which we mentioned in the Berlin 7286 stela, is significant (Quirke, 2004: 36).

Assuming that at that time, by the combination of titles with which they are presented, these horologists are priests, they are also recognized by their assignment during the Middle Kingdom to the *wnwtjw ḥwt-nṯr,* '(observers) of the hour of the temple' (Ward, 1982: 86 n° 714 'Hour-watchers of a Temple'), as shown, for example, in stela CG 20538 (; Lange & Schäfer, 1908: 149) or in Berlin 1204 (; Roeder, 1913: 1204 p. 173). These astronomers should have as their main task measurement of the hours of the day and night in order to mark with precision the times at which certain rituals should be performed, so they could also have had functions directly linked to these rituals.

2.1.2 New Kingdom Astronomers

In those enclosures where there were several main temples, it is possible that there were teams of astronomers, working independently for these temples. In the annals of Thutmose III are mentioned *wnwtjw* (Gabolde & Gabolde, 2015: 62, south

[5] The reading of *wnwt* as a plural of 'star' is not certain (Hannig, 1997: 197). but it should be noted that in the 19th Dynasty and in Late period is found the verb *wnwn* (also) 'to observe the stars' (Wb I, 318: 11), whence *wnwnw*, in Ptolemaic times. would be the 'observer of the stars' (*Wb* I, 318: 12); the most common form of 'star' is *sb3* (abbreviated); another word used for star, also documented from the Middle Kingdom, is *gnḫt* (Wb V, 177); likewise, with several variants, *ḫ3b3s* 'stars' (*Wb* III, 230; Hannig, 1997: 582) documented from the *Pyramid Texts* to the Late period. On the other hand, it should be noted that the hieroglyphic sign representing the star (sign N14) is documented from the reign of Adjib, 1st Dynasty.

[6] Faulkner (1969: 101), translated: 'The King has cleared the night, the King has dispatched the hours', and Allen (2005: 61): 'Now that Unis has swept away the night and Unis has sent off the hour-stars'. However, recently, Shmakov (2012: 173) proposed another interpretation for the first sentence: 'Unis has lit up the night'.

wall, col. 12 N), but in a text (*Urk* IV, 1263: 7–8) of the Akhmenu reference is made to the astronomers who worked for Amun-Re in the Akhmenu on the one hand and for the temple of Amun-Re on the other, thus evidencing the existence of groups linked to the main temples in the Karnak precinct:

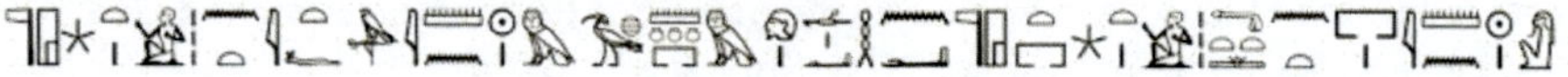

But not all astronomers were priests or solely linked to temples. Thus, in the 18th Dynasty stela Florence 6371 (cited as Florence 1776 in the designation of Schiaparelli in Gardiner, 1947: 62*; Bosticco, 1965: pl. 35; see Figure 2.2) is mentioned an *sš n ꜥt wnwt n Pr-ꜥ3 wnwty ḥr tp-ḥwt ꜥḥ,* 'scribe of the department of the hour of the pharaoh, the (observer) of the hour on the terrace of the palace', who was not a priest. On this stela Černý (1963: 173) read as *sb3y,* making the word out to be a *nisbe* of *sb3* ('star')[7] instead of 'hour'.

Along with in the New Kingdom we see how new variants of the word *wnwtj* appear. Only in the tomb of Nakht (TT52) (Davies, 1917; see Figure 2.3), which was *wnwtj n Jmnw* ('astronomer of Amun') do we find several ways of writing this word: (*Ibid.*, 125), (*Ibid.*, pl. VIII, XXI), (*Ibid.*, pl. XVIII), (*Ibid.*, pl. XXII),[8] (*Ibid.*, pl. XIV fig. 6 statue), (*Ibid.*, pl. VIII, XVIII), (*Ibid.*, pl. XI-XII), (*Ibid.*, pl. XIV), and (*Ibid.*, fig. 8 funerary cone). In the inscriptions on his tomb and associated objects (statue, funerary cones: Zenihiro, 2009: Davies and Macadam #157, p. 96) Nakht makes use, first of all, of the title of *wnwtj* (*n Jmnw*), 'astronomer (of Amun)', sometimes followed by the title 'scribe'. That is, he gives absolute preponderance to his role as astronomer in the temple of Amun, thus becoming quite an exceptional example. Despite this, we do not see Nakht wearing a special dress in any of the scenes in which he is represented in his tomb, or see him associated with astronomical observing instruments such as the *bay* or the *merkhet*, of which we shall mention later. Quite the opposite occurs with another example from the 18th Dynasty, that of Aanen, Second Priest of Amun.

In ancient Egypt, priests could wear a variety of garments, not only according to their rank but also according to their occupations. In the case of the priest-astronomers, one of the most exceptional and beautiful preserved examples is that of Aanen, son of Tuya (and, therefore, brother of the Great Royal Wife of Amenhotep III, Tiye), as identified in a diorite statue from the time of Amenhotep III. This is preserved in the Egyptian Museum in Turin (Turin 5484) (Kozloff & Berman, 1993: 210–211; see Figure 2.4). The stars that decorate the panther skin with which the

[7] The word is rendered in the same way in statue CG 482 of the Middle Kingdom. See, Borchardt, 1925: 63; also in GC 20725, see Lange and Schäfer 1908: 355; another variant, , is found in CG 20524. *Ibidem*, 125.

[8] In this case it could be a dithograph, although in Neo-Egyptian we can find the form *wnwntj.*

Fig. 2.2 The stela of the astronomer Tetiankh (18th Dynasty). Museo archeologico nazionale di Firenze, E6371. (Drawing made by María Hidalgo Portillo after Bosticco, 1965: pl. 35)

Fig. 2.3 Painting of the tomb TT 52, belonging to the *wnwtj*-astronomer Nakht (18th Dynasty). (After Davies, 1917: pl. XXI)

Fig. 2.4 Aanen, Second Priest of Amun (18th Dynasty). Museo Egizio di Torino E5484. Image by Naty Sánchez. (Courtesy of the Egyptian Museum in Turin)

torso is covered relate this individual unequivocally to observation of the celestial vault.

Furthermore, in an inscription on the back of the statue is included among Aanen's other titles *ẖrj-ḥbt rḫw sšm n pt* ('lector-priest who knows the behaviour of heaven'). In this case, neither in his tomb (TT120) nor in other known documents belonging to him do we see Aanen associated with the title of *wnwtj*. However, both his peculiar clothing and the use of the epithet 'the one who knows the behaviour of heaven' make it clear that he is a priest-astronomer. His role as 'lector-priest' also shows his closeness to the most secret knowledge of the rituals of the temple. On the other hand, it is interesting to note how Aanen also makes use of the title *wr mꜣw m ḥwt-sr* ('Great of the seers in the House of the Prince'), perhaps linked, among his other functions, to observation of the sky.

At the time of Ramesses II, the well-known High Priest of Amun (HPA) Bakenkhonsu I has, among his titles, one that links him to observation and knowledge of the celestial vault. He was *ḥrj-sšt3 m pt t3 dw3t* ('superior of the secret in the sky, the earth and the Duat'; *KRI* III, 294: 5), a title that we find again shortly after (written) with another HPA called Bakenkhonsu II (Boraik, 2007: 122 fig. 1) during the reign of Setnakht-Ramesses III, but also in examples from the 21st Dynasty. Bakenkhonsu I was furthermore *wr m3w n Rꜥw m W3st* ('Great of the seers of Re in Thebes'; Kees, 1949: 43–433).

In other documents from this period we also find new variants of *wnwtj*, such as (Janssen, 1961: 15; *KRI* II, 812: 15; Lesko, 2002: 101), *wnwt(y)*, 'priesthood, lay priest-hood, watchman' (pLeiden I 350 v°, from the time of Ramesses II), and *ḥrj wnwtj(w)*, 'chief of the astronomers', as has been more recently found in a stela from the early Ramesside period from Saqqara (Daoud, 1993: 262 fig. 1).

2.1.3 Astronomers of the Third Intermediate Period and Late Period

At the beginning of the 21st Dynasty new variants of *wnwtj* appear, such as (pGolenischeff) (Gardiner, 1947: I, no. 134, p. 61 *; III, pl. VIIIA l.11) and (pHood/BM 10202) (Gardiner, 1947: I, no. 134, p. 61 *; III, pl. XV l. 12) or, more exactly, and *jmj wnwt* 'who is on the hour.' This form, although typical of the time, is already found in an example from the Middle Kingdom: *jmj wnwt* (Ward, 1982: 9 n° 25). Similarly, in Neo-Egyptian we can find the word *wnwntj* (Hannig, 1997: 198).

Also from the 21st Dynasty comes a precious example, that of the priest of Amun Ankhefenkhonsu (Trauckener, 1993: 83–93; Lull, 2008a: 73), who inscribed his titles on the doorjambs of his house located next to the sacred lake of Karnak. Some of the titles mentioned link him to his role as an observer of the celestial vault: *wḥꜥ snṯ n ꜥ3wj pt* ('who explains the *snṯ* of the two gates of heaven'), *rḫ sšt3 Nwt* ('who knows the secret of Nut'). Regarding the first title, it should be noted that the expression *ꜥ3wj pt* (the 'two gates of heaven') served to refer to the doors of the chapel (Černý, 1948: 120) behind which stood the statue of the god in the temple.

In that sense, it could be argued that the title of Ankhefenkhonsu refers to this part of the interior of the temple. However, it may also refer to his astronomical knowledge of horizontal observing of risings and settings. It is enough to remember that, in the so-called Chamber of Re of the funerary temple of Ramesses III in Medinet Habu, there is a representation of the first hour of the *Book of the Night* in which the nocturnal boat of the solar god is located between two doors below the symbol of the sky (*pt*). He is thus crossing the two gates of heaven when entering the western horizon (Epigraphic Survey, 1963: pl. 422).

This iconography is associated with an inscription that reads: 'standing on the western mountain, in the *mesketet* (night boat) by this god, entering the western horizon, opening the secret doors dragging the western *bau*' (Brovarski, 1977: 108).

For this reason, taking into account that Ankhefenkhonsu is also the one 'who knows the secret of Nut', and with this he probably refers to his general knowledge of the celestial vault, we may attribute an astronomical meaning to the first title. Other titles of Ankhefenkhonsu could also have an astronomical meaning, such as *m3w m Jpt swt*, '(chief?) of the seers in Karnak'.

Tjanefer (Sousa, 2018: 453), a Fourth Priest of Amun in Karnak, of the 21st Dynasty (as Ankhefenkhonsu contemporary of HPA Painedjem II) possesses, among his numerous priestly titles, that of 'superior of the secret in the sky, the earth and the Duat' (as Bakenkhonsu I) and 'who opens the two gates of heaven in Karnak', in the same way as Nesiamon (Sousa, 2018: 387), also buried in Bab el-Gasus, or Ankhefenmut, from the end of the Third Intermediate Period, which was 'who opens the two gates of heaven in Karnak' (Zayed, 1968: pl. 7).

In the 25th Dynasty, the Dream Stela of the Kushite king Tanutamun tells us that when this king arrived in Karnak he was received by the 'priest of the great *snṯ*' and the *wnwtjw ḥwt nṯr nt Jmnw-Rˁw* ('astronomers of the temple of Amun-Re'; *Urk* III, 65:13; Grimal, 1981b: 10 l. 12, pl. II). Whatever the 'great *snṯ*' is, we see that this priest is followed by astronomers, perhaps because there is a relationship between the two. In this sense, it should be noted that Ankhefenkhonsu (21st Dynasty) was he 'who explains the *snṯ* of the two gates of heaven', but also the ('keeper of the great *snṯ*'). For Trauckener (1993: 90), the great *snṯ* would be:

> *Un document associant inventaires de matériel et prescriptions de culte et de fonctionnement de l'institution voire une liste du personnel ou du moins un organigramme. Le préposé au grand-sn.t. serait alors un prêtre chargé de la tenue et de la mise à jour de cet important document.*
>
> 'A document associating inventories of cult matters and prescriptions of the functioning of the institution with even a list of staff or at least an organigram. The officer in charge of the great *sn.t.* would therefore be a priest charged with the keeping and updating of this important document'.

The '*snṯ* of the two gates of heaven' might also refer to astronomical documents.

A stela, belonging to an individual of the 25th or 26th Dynasty named Ankhhor (Dhennin, 2012: pl. VI) from the city of Mefkat indicates that he was *jmj-r3 wnwntjw rḫ pr(t) spdt*, 'inspector of the astronomers, the one who knows the (heliacal) rising of Sirius', or *ḥrj wnwntj(w)*, 'chief of the astronomers'. The inscription that contains this stela is very interesting, as it shows the existence of a family that for several generations had astronomers among its members. On the other hand, the reference to the titles and must also be related to that of the astronomer (as Dhennin suspected), because a son of Ankhhor called Hersenef was a , and this last title shows a clear relationship between and knowing the heliacal rising of Sirius.

Astronomers are sometimes associated with specific gods or temples. A Serapeum stela, from the 25th Dynasty, refers to a *jmj wnwt n pr Ptḥ* ('astronomer of the domain of Ptah'; Vercoutter et al., 1968: 117 (149)). Wahibre's stela (Martin, 1979: 50 n° 162, pl. 45) from the 26th Dynasty mentions an *wnwtj (n) 3st* ('astronomer of Isis'), and the stela Florence 2502 (Bosticco, 1972: 21–22, n° 11) shows us a *ḥrj jmj(w) wnwt n Wsjr* ('chief of the astronomers of Osiris') in Abydos. We also find the *wnwtj n ḥrw bḥdtj* ('astronomer of Horus Behedeti'; Spiegelberg, 1904: 157–158). We see, therefore, that the temples usually had to have attendants in charge of the measurement of the hours in order to be able to keep proper schedules for the delivery of offerings and other important rituals in the day-to-day running of the temple. The people in charge of the teams of astronomers would be the *ḥrj jmj(w) wnwt.*

2.2 Basic Instruments of Observation

As horologists, Egyptian astronomers made use of a variety of devices to measure the passage of time, including inflow and outflow clepsydras, sundials and shadow clocks, and tables of hourly stars. However, their repertoire of observing instruments, properly speaking, was very limited. Two of them, the *bay* and the *merkhet,* are well known, not only because of their mention in numerous inscriptions but also because some of them have come down to us. Actually, they are sighting instruments used specifically to verify alignments. Their combined use by astronomers, preferably from high vantage points (see Figure 2.5), is suggested in a well-known illustration by Sloley (1931: pl. XVII: 1), although observations would be feasible not only to the south but also to the north and were possibly better helped with poles or other kinds of fixed references. We should recall that the titles with which some astronomers are designated specify their location on the terraces of temples or palaces.

Fig. 2.5 View looking south from the terrace of the Temple of Hathor at Dendara. (Image courtesy of Marc Mateos Comaposada)

The *bay* (*bay*), according to a specimen that has survived to this day, is a palm rib with a groove in the sheath part. It was supposedly held with an outstretched arm in order to make alignments. It has sometimes been assumed that by means of the *bay*, in order to align with the central meridian, the astronomer could align himself with a second astronomer who had to be at a certain distance to his south or north. With no fixed and stable tripod, pier, or mount for the *bay,* or with no fixed reference other than observation through the *bay*, this method could not have yielded very remarkable results. Therefore, the presence of the second reference astronomer could be perfectly avoided, with the added benefit of greater precision, by using a perfectly aligned fixed pole.

In some astronomical ceilings, right next to the representation of the boreal constellations, a divinity raises and extends its arm horizontally, holding an object that could be well a *bay* or a similar sighting instrument (Lull, 2006–2009). He is represented in the 26th Dynasty tomb of Pedamenope (*EAT* III: pl. 19; see Figure 2.6) and in an early Ptolemaic astronomical ceiling of Tuna el-Gebel (*EAT* III, fig. 14; see Figure 2.7). Both, the shape of this object and the position of the god's arm,

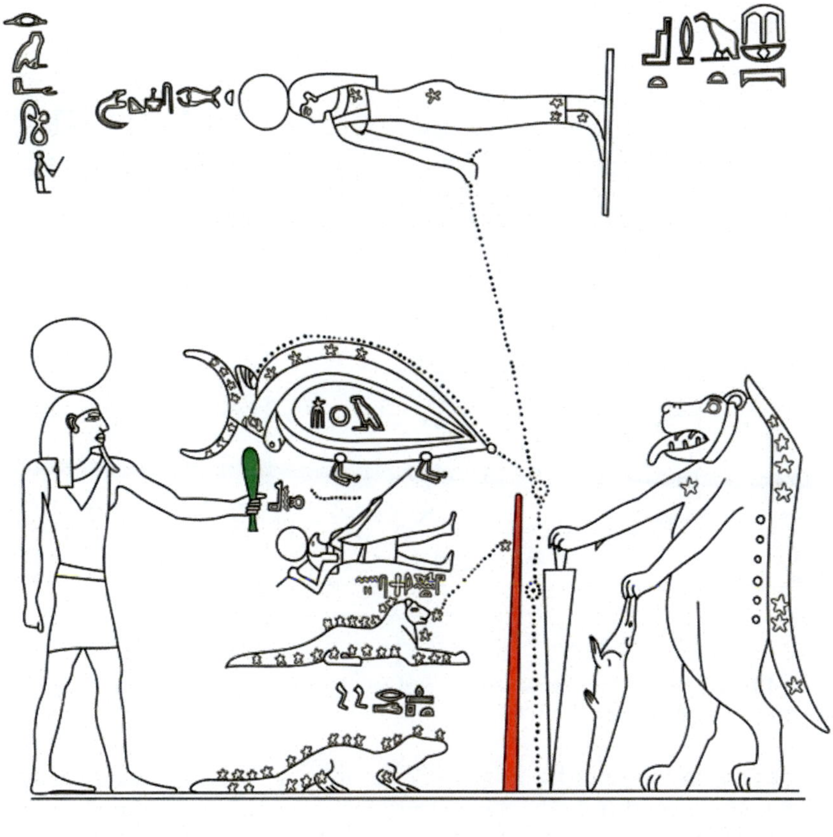

Fig. 2.6 Detail of the astronomical ceiling from the tomb of Pedamenope (26th Dynasty). A *bay* is marked in green, and in red a complementary astronomical device. (Drawing made by María Hidalgo Portillo after *EAT* III: pl. 19)

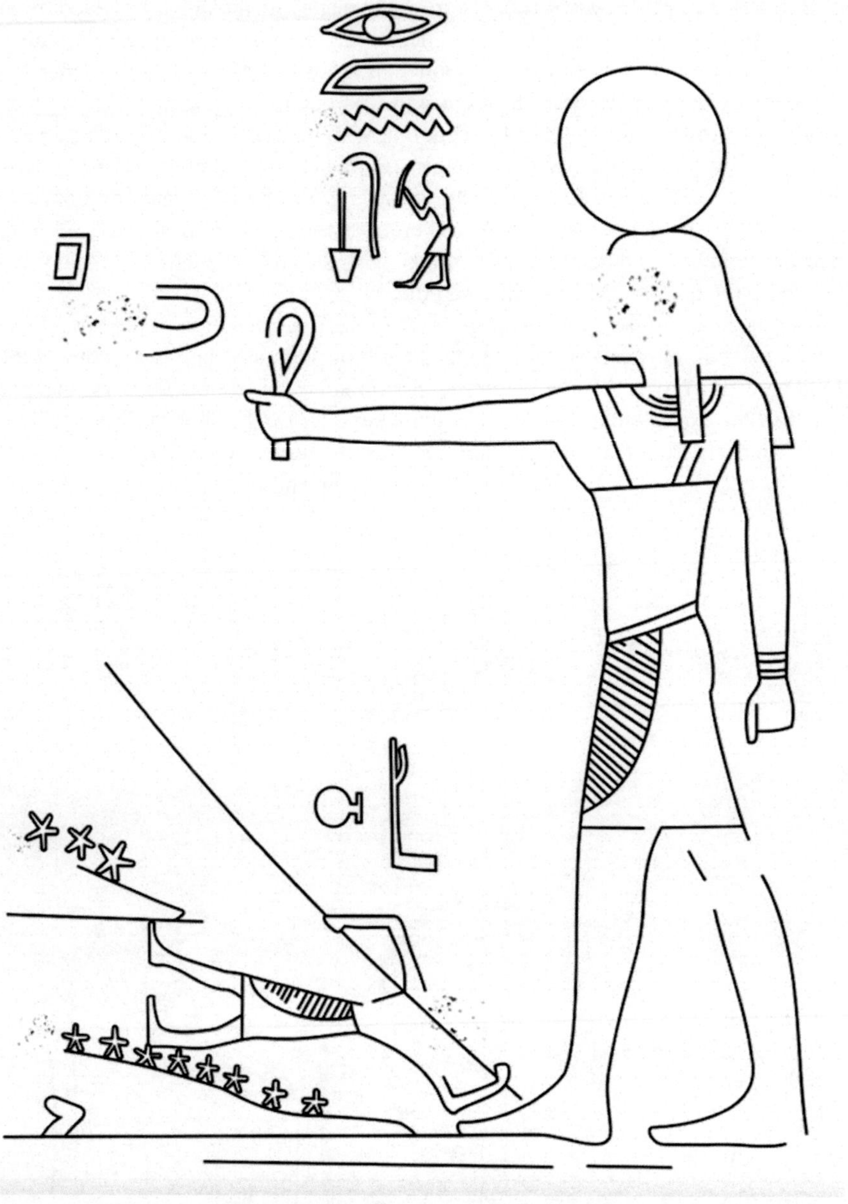

Fig. 2.7 Detail of the early Ptolemaic astronomical ceiling of Tuna el-Gebel. (Drawing made by María Hidalgo Portillo from *EAT* III: fig. 14)

which is extended horizontally, make it difficult to imagine that it could have been used as a weapon or that it is held in a threatening way. However, it is true that in other astronomical ceilings this divinity, called *jr-m-ꜥwꜣ* (Iremawa), whose name means 'the one that acts violently', lives up to its name by clearly holding a knife, as in the tomb of Ramesses VI (*EAT* III: pl. XIII) and other examples. There is a possibility that, what was originally a knife, was reinterpreted later by the Egyptians themselves as a *bay*, especially taking into account the astronomical context in which that divinity appears.

A real example of both a *bay* and a *merkhet*, are preserved in the Berlin Ägyptisches Museum. The instruments originally belonged to an astronomer named Horu, dated in the 26th Dynasty. The *bay* (ÄMB 14084; see Figure 2.8) contains a hieroglyphic inscription (Borchardt, 1899: 10) which reads as follows:

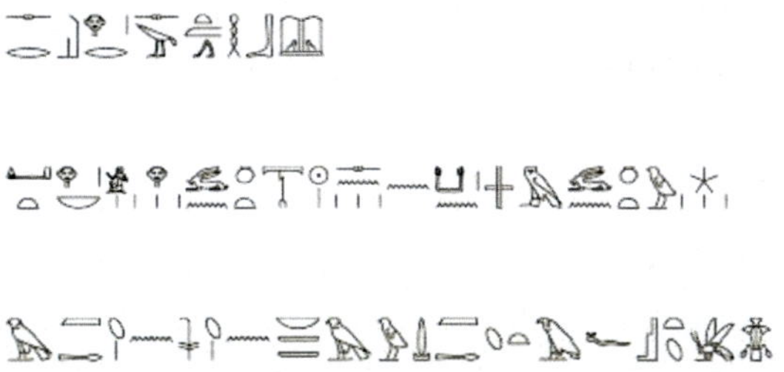

'Attend to the monitoring of the celebration and that all the men are placed in their hours, for the *ka* of the astronomer Horu, justified, son of the son of the King of the Two Lands, Horudja, justified, son of his mother, Asetakhbit'.

In this inscription, although no kind of precise information on the mode of employment or use of the *bay* is offered, we may deduce the ultimate purpose was to set the hours so that the duties of the priests and temple staff were assigned at the appropriate time.

In some of the so called Piankhy blocks discovered in the temple of Mut at Karnak at the end of the 19th century and, in fact, issued under Psametik I at the beginning of the 26th Dynasty (Broekman, 2012), a large shipment of *bꜥj n jmj-wnwt* ('*bay* of the astronomer') is mentioned (Benson & Gourlay, 1899: 373). This shows that, although very few instruments of this type have survived, they were in very common use and the number of astronomers had to be considerable, among other things because their function as horologists was absolutely vital and necessary for the proper functioning of the rituals of the temple.

Fig. 2.8 *Bay* belonging to the astronomer Horu (26th Dynasty). (Courtesy of the Staatliche Museen zu Berlin, Ägyptisches Museum und Papyrussammlung, Inv. No. ÄM 14084. Image by Margarete Büsing)

Fig. 2.9 *Merkhet* belonging to the astronomer Horu (26th Dynasty). (Courtesy of the Staatliche Museen zu Berlin, Ägyptisches Museum und Papyrussammlung, Inv. No. ÄM 14085, Photo: Margarete Büsing)

The 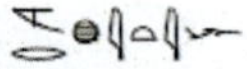 *mrḫyt merkhet*, on the other hand, is an instrument formed by an elongated bar with a head from which a plumb line hangs. Horu's *merkhet* (ÄMB 14085) (see Figure 2.9), which is only 11.5 cm long, has an inscription along its length that is equally interesting and representative of the knowledge of its original owner:

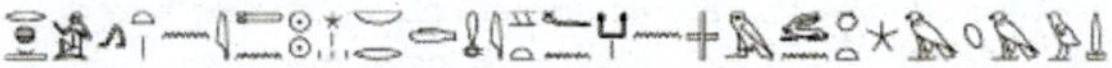

'I know the movement of the two discs (the sun and the moon) and of all the stars towards their corresponding place, for the *ka* of the astronomer Horu, son of Horudja'.

Borchardt, instead of the dual form of *jtn* 'disc', interpreted 'sun' and with a question mark 'moon' (the second disc). However, according to George (1974: 103), it properly refers to the two (daytime and nighttime) suns, while for Magdolen (2001), the word *jtn.wj* (the dual of *jtn*), would actually refer to the two positions of the sun when rising and setting, assuming its use as a clock, since the Egyptians also developed shadow clocks, with the same shape of the *merkhet*, but with a scale in the horizontal plane.

A *merkhet* preserved in the Science Museum in London (Inventory No.1929-585), belongs to Bes, a priest of Horus and son of Khonsuirdis (see Figure 2.10). Another and older of the few *merkhet* that has survived (Oriental Institute Chicago 12144) bears the name of Tutankhamun (Larson, 1992). In iconography, however, it is difficult to find these instruments. An exception is found on a stela (Florence, 2502) from the Late period preserved in the Museo Archeologico Nazionale di Firenze (Bosticco, 1972: 21–22 n° 11), in which a man can be seen carrying a *bay* (see Figure 2.11), and in the iconography of the Ptolemaic cube statue TR 25/10/17/5 (see Figure 2.12), where an astronomer called Horemheb is showed holding a *merkhet* shadow clock with the left hand and a *bay* and a long three-pronged topped stick that must be another sighting instrument.[9] Another statue from the Ptolemaic period (Scott, 1992:

[9] Images of this statue can be seen at: https://www.ifao.egnet.net/bases/cachette/?biblio=Birk%3A2014&os=9#galerie (seen: 25/5/2021).

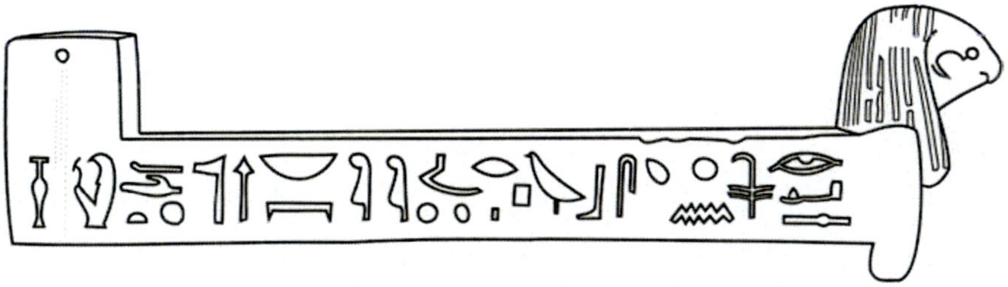

Fig. 2.10 *Merkhet* belonging to the priest of Horus, Bes (26th Dynasty). (Drawing made by María Hidalgo Portillo from a piece preserved in the Science Museum in London, Inv. No.1929-585)

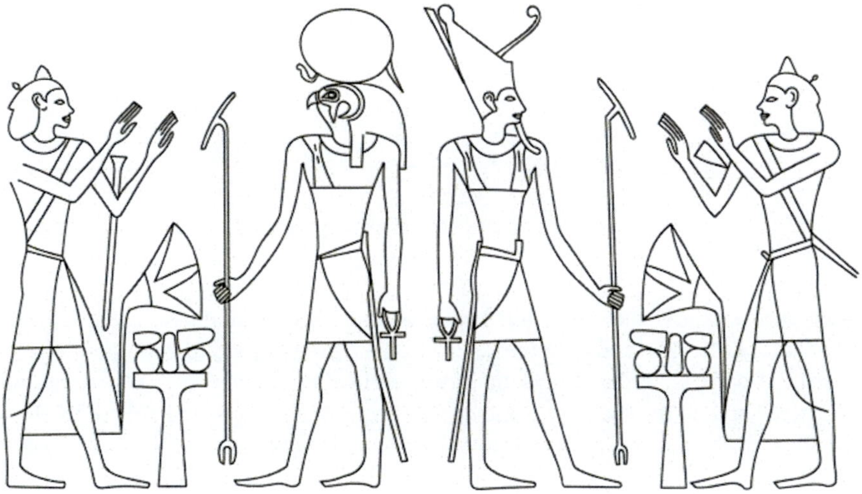

Fig. 2.11 Man holding a *bay*. Museo Archeologico Nazionale di Firenze, Stele Florence 2502, Late period. (Drawing made by María Hidalgo Portillo after Bosticco, 1972: n° 11)

53–55 n° 28; Fissolo, 2001: 20; Birk, 2014: Abb. 2), belonging to an astronomer named Imhotep,[10] shows him holding a complete set of astronomical instruments, including a *merkhet* shadow clock under his left arm, a long three-pronged topped stick, a *bay*, and perhaps a *merkhet* with his left hand (see Figure 2.13).

The use of the *bay* and the *merkhet* could not provide highly accurate results in the performance of its main task, observation of the meridian. This would be so unless, as previously mentioned, N-S alignment posts had been placed on the temple terraces with specific observing points marked on the ground, or even stable pedestals in which to insert the *bay* to free it from the unconscious movement of the

[10] A beautiful 3D Model of the statue of the Astronomer Imhotep (RAFFMA EG.01.008.2020) can be seen at: https://sketchfab.com/3d-models/statue-of-ptolemaic-astronomer-raffma-artifact-068de0851def43febfeb1ad8cbdf4d59 (seen: 1/4/2021).

Fig. 2.12 The astronomer Horemheb holding astronomical instruments. (Drawing made by María Hidalgo Portillo from an IFAO photograph of Statue Cairo TR 25-10-17-5)

astronomer's arm. If such measures had been taken, they would have achieved better results (Lull, 2004a: 64; Couprie, 2011: 48). It is possible that the long three-pronged topped stick instrument that appears in the relief on the cube statue of Horemheb and on the statue of Imhotep had been used as a complement to the *bay* and *merkhet*. In fact, it is quite possible that simple actions similar to the aforementioned were taken by Egyptian astronomers to improve the precision of their observations using such simple instruments.

Although these are the best-documented instruments, the Egyptians could have used other astronomical instruments. A pointed object such as a gnomon or reference post, represented in some Egyptian astronomical ceilings, might have been used as an observing instrument (Lull, 2003: 7–13, 2004a: 235–241, 2006–2009: 47–56; see Figures 2.6 and 2.14). According to the representations we have, there might have been two main variants of this instrument. It could have been a simple wooden pole of a certain height, with the uppermost extreme being pointed, truncated, or rounded. A second variant might have consisted, however, of two poles separated at the base but converging at the upper extreme, thus creating an instrument that, perhaps, could have had a hanging plumb line. The above-mentioned instruments might have been used to calculate the moment of culmination of circumpolar stars or to sight the central meridian. As an ancillary instrument, the observer might have made use of a *bay* or some other useful sighting instrument.

Likewise, the symbol that crowns the head of the goddess Seshat (Belmonte et al., 2009a: 207–210; see Figure 2.15) has also been interpreted as a kind of groma or topographic instrument with a viewer represented according to the perspective of

Fig. 2.13 Statue of the astronomer Imhotep holding his astronomical instruments. RAFFMA EG.01.008.2020. (Image by Bryan Kraemer, courtesy of the Robert and Frances Fullerton Museum of Art, San Bernardino, California)

Egyptian art (see Figure 2.16). The seven radii of the Seshat rosette have been extrapolated to a horizontal wheel with eight radial spokes, the eighth spoke being presumably hidden in the flat representations by the vertical pole. This part of the instrument would become an essential element of the proposed mechanism like the one offering four possible directions (one for each diameter) at any given point for the orientation of the main axis of a certain building. The semicircular arc on this radial element, surmounted by a pair of parallel strokes, then becomes an independent but vital element of the instrument. It would define a sighting device, or eyepiece, in the style of the classical hieroglyph depictions of the *merkhet*.

Fig. 2.14 An astronomical instrument, shaped like an isosceles triangle, has been marked in red. A vertical line coloured green could symbolize the central meridian. These were represented among the boreal constellations in the Ramesseum (19th Dynasty). (Drawing by María Hidalgo Portillo after Wilkinson (1991: 153 fig. 3))

Other instruments that might have been used by the Egyptian astronomer, such as diagonal stellar clocks, inflow and outflow clepsydras, sundials and shadow clocks, tables of hourly stars, etc., are all instruments for measuring or tables for calculating the time, but not observing instruments as such. All these, then, were the known working tools of the Egyptian astronomer.

2.3 Precedents of Astrology Before The Graeco–Roman Period

What undoubtedly characterizes Egyptian astronomers of the Ptolemaic and Roman periods is their rapid adaptation to astrology, this becoming, as is clearly shown in the documentation, an important task. How was astrology associated with traditional Egyptian astronomical practice?

The first thing that must be highlighted is that the divinatory character derived from the interpretation of the events taking place in the celestial vault, although little known to us, was already part of the cultural heritage of Pharaonic Egypt. This

Fig. 2.15 The Goddess Seshat crowned by a peculiar sign that could represent a topographical instrument. Temple of Kom Ombo. (Image by the authors)

Fig. 2.16 This sequence of images illustrates the hypothetical use of the sign of Seshat as a topographical instrument. The upper register shows (left) a relief of a *gromaticus* in a funerary stela discovered in the excavations of Pompey and (right) an idealized reconstruction of it. The middle panel shows the seba parasol as represented in the mastaba of Tiy in Saqqara, demonstrating how the Egyptian 'aspective' operated. The three-dimensional reconstruction of the device

is not usually taken into account in its full dimension, so that the appearance of astrology in Egypt has usually been interpreted as a completely new cultural import. However, in ancient Egyptian documentation we can find evidence which, although scarce, clearly demonstrates the existence of this type of belief.

On the astronomical ceiling of the funerary temple of Ramesses II, above the list of decans, the southern stars augur interesting outcomes to the king (Kákosy, 1982: 187; *EAT* III, pl. 5):

> 'He (Re-Horakhthy) will give the rise of Isis-Sothis on the heaven at the dawn of the New Year's day. She (Sothis) foretells to you (innumerable) years and sed-festivals and Niles. There will be no obstacle to them. The star of every decad (decans) will come forth for you to lengthen all your years… You will sail like Orion on the heaven, your lifetime is its lifetime'.

Although, as we have argued, the Egyptian astronomer acts mainly as an horologist who knows the sky, it should be noted that the divinatory character associated with stargazing can be clearly perceived. This is even the case in some Ramesside texts, such as the so-called Stele of Israel dating from the time of King Merenptah (*c.* 1209 BC) (Brunner, 1988: 225–226; Quack, 2002a: 168; Jambon, 2013: 158–161),

 [11]

> *Nts jn.s pḥwy n ḫftyw.s ḫrw m n3 s3w n n3y.sn sb3w rḫyw r.sn nbw m nw r ṯ3w*
> 'He is the one who will bring the end of his enemies, say those who scrutinize their stars and those who know all their formulas by watching the winds'.

The preceding text shows that the Egyptians of the Ramesside period already believed in the capacity that certain people could have in predicting future events from observation of the stars or certain atmospheric phenomena. Quack (2018a: 74) suspects that it possibly refers to an eclipse, since in Mesopotamia (Koch-Westenholz, 1995: 98) the direction of the wind during an eclipse determined the prediction.

Without the slightest doubt, this text, although exceptional in the Egyptian documentation of the third and second millennium BC, has enormous importance, since it demonstrates a reality very little known at the time: divination from the observation of astronomical phenomena. The fact that this inscription is part of a monumental stela is also very important, as it somehow indicates that it was a standard practice.

Fig. 2.16 (continued) (lower panel, right) follows the same rules which we are proposing for the 'instrument' (sign) of Seshat. The lower left panel shows a relief from the solar temple of Niuserre at Abu Ghurob, where the sign appears as a portable object, and (right) the sign transformed into a real object in which the seven radial spokes appearing in the iconography of the goddess would be turned into an eight-spoked movable 'wheel', when changing the two-dimensional version of the representation into a three-dimensional image. (Adapted from Belmonte et al. (2009a: 208 fig. 7.14–15, 209 fig. 7.16); diagrams courtesy of SMM/IAC)

[11] *KRI* IV, 16: 2–4.

On the other hand, from the Ramesside period there are certain clues that observation of the heliacal rise of Sirius might have served divinatory purposes (von Lieven, 1999: 103–104; Quack, 2017: 191).

In Graeco–Roman times, in fact, a notable divinatory element of astrology in Egypt would have Sirius as its main protagonist, a star of great importance from the religious and economic point of view throughout the entire history of Egypt. This is shown, for example, in the demotic papyrus Cairo 31222 (Hughes, 1951: 256–264), which deals with the *nꜣ sẖnjw* (*n*) *spdt* ('the influences of Sirius'), as indicated at the beginning of the papyrus, or in the demotic papyrus Berlin P 23888+Wien D 9906b, c recto (Stadler, 2018).

The ancient Egyptians also related a series of divinities with the birth of the individual and his destiny (Greenbaum, 2016: 97–98). These include Shai (Quaegebeur, 1975), the main god linked to predestination, but also to the Seven Hathors, which authors such as Thausing (1939: 67) have even associated with the seven main stars of the Pleiades star cluster.

Herodotus (II: 82), who visited Egypt in the 5th century BC, during the first Persian rule, leaves us a description of the prediction that the Egyptians made about the life of a person on the day of his birth:

> 'I pass to other inventions of the Egyptians. They assign each month and each day to some god; they can tell what fortune and what end and what disposition a man shall have according to the day of birth. This has given material to Greeks who deal in poetry. They have made themselves more omens than all the other nations together; when an ominous thing happens they take note of the outcome and write it down; and if something of a like kind happens again they think it will have a like result'. (Herodotus, 1920: 367, 369)

According to Dieleman (2003a: 139), referring to Egyptian horologues, 'the second task of hour-priests was to determine for each day of the year whether it would bring good or ill fortune to businesses or persons'. This form of divination is that pointed out by Herodotus and, in fact, has a long and very old tradition in Egyptian culture. However, it does not correspond to astrology but to hemerology (Leitz, 1994b; von Lieven, 2016a: 31–41), so it is clear that it does not suppose a precedent directly linked to astral birth divination. Even in spite of this, perhaps the genethliacal astrological practice could probably have been adopted more easily in Egypt thanks to deep-rooted traditions like the one previously mentioned.

Inside the bubastide portal of Karnak, in a well-known text from the late 9th century BC called the *Chronicle of Prince Osorkon*, we also see how a celestial event, in this case a lunar eclipse, was a clear harbinger of ill fortune (Lull, 2006a). The text, dated in the year 15 IV Shemu 25 of Takelot II, reflects, in fact, what would usually have happened when a lunar eclipse occurred:

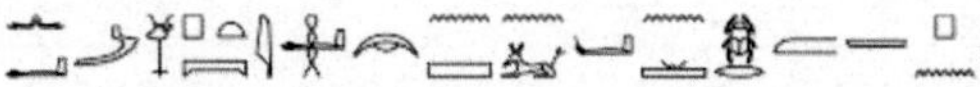 [12]

'(Although) the sky did not swallow the moon a great fight took place in this land'.

[12] Epigraphic Survey 1954: pl. 21 col. 7; Caminos 1958: 88-90.

The same expression used in this text, *n ˁm pt jˁḥ* ('that the sky does not swallow the moon'), is found in the *Ritual to repel evil*, whose oldest source is the papyrus BM 10252 (*Urk* VI, 123, 11 and 12, Louvre 3129), of the 30th Dynasty. The association between a lunar eclipse and a disastrous event is clearly revealed in this ritual (Jambon, 2012: 147–148).

On the other hand, already in some oracular decrees (Edwards, 1960: 24; Quack, 2016: 238) of the Third Intermediate Period, it is clear that it was considered that certain stars could exert an influence on people. On the ceiling of the tomb of Osorkon II in Tanis (22nd Dynasty), where the decans are represented with lion and serpent heads, one of them is designated as *ˁnḫw š (ȝw)* ('living ones of fate'; Kákosy, 1982: 165). This name could suggest a relationship between the decanal stars and fate (von Lieven, 2000: 181). A relationship between the decans and the decades of the year calendar is also observed, for example, in papyrus Cairo JE 89131–6, of the 26th Dynasty (von Bomhard, 2008: 185).

A papyrus dating from the 1st to 2nd century AD that we could qualify as exceptional is pBM 10651 (Quack, 2017: 193, 204–205). It is written in hieratic (unusual at that time) and refers to the Egyptian decans and to their ordering by decades, rather than according to the zodiacal constellations (which appear in Babylonian astrological texts at the end of the 5th century BC; Steele, 2018: 99). It uses as main divinatory references eclipses of the moon and the sun and the risings of Mercury. This has led us to suppose that its original version could date back to the Saite period. Another papyrus that could reflect an older Egyptian tradition is Carlsberg 695 (Quack, 2017: 192), in which the size and brightness of the star play a role in astral divination.

In another document prior to the Ptolemaic period, a naos from the time of king Nectanebo I (378–361 BC) from Saft el-Henna, known as *Naos of the Decades* (see Fig. 2.17), there appear elements that have been described as astrological (Brennan, 2017: 7). Von Bomhard (2008: 105) has pointed out, 'The texts are of a character that is often much more akin to mythology than to astrology, but their perusal provides a better understanding of how one gradually glides towards the other'.

The five images that characterize the record of each decan in this monument seem to be associated with the kind of information provided by the text of the *Fundamentals of the Course of the Stars* (Lull, 2004a: 126). Von Bomhard (2008: 189, 2011) has wondered 'if that ancient book of observations did not already contain the seeds of the later foundations of astrology, reproduced on Shu's monument (Naos of the Decades)', since 'the monument could reflect or precede the astrological systems that foretell the future from the changing position of the planets, the sun and the decans according to the hours'. Astrological papyri such as the hieratic BM 10651 or the Greek Oxyrhynchus 465 (Quack, 2010: 178–179), both from Roman times, have been compared to some extent with the naos (von Bomhard, 2011: 130).

In any case, Leitz (1995: 49–50) has proposed that the naos texts could come from the same source as the *Salmeschiniaka*. This is known only from a few fragments of papyrus written in Greek, which constitute the earliest hermetic astrological treatise mostly recognized as Egyptian (Bohleke, 1996: 17; Quack, 2018b:

Fig. 2.17 Naos of the decades. Reign of Nectanebo I, 30th Dynasty. The left and right flanks were discovered on the site of Canopus by the archaeological mission of IEASM headed by Franck Goddio. (Photograph courtesy of Christoph Gerigk©Franck Goddio/Hilti Foundation)

99–100), and whose model was placed by Gundel (1936a: 49) in the third century BC. In a reference, contributed by Hephaestion of Thebes, the astrological influence of the decans according to their position is precisely expressed (Gundel, 1936b: 410).

It is from the Ptolemaic period when the documentation clearly shows the introduction of new functions and words to designate the Egyptian astronomer, since he begins to be linked more assiduously to new tasks in the astral divinatory field; that is, to astrology.

2.4 From Astronomers to Astrologers in the Transition to the Graeco–Roman Period

Astronomers, as an integral and indispensable part of the staff of Egyptian temples, continued their work in Graeco–Roman times. Numerous mentions of this period are also preserved, both in ritual texts embodied in temples that have reached us in excellent condition, such as Edfu or Esna, as well as in some statues or, mainly, in papyri and ostraca written in demotic and Greek.

As in the previous Pharaonic period, in the Graeco–Roman period we find many astronomers associated with a god or a temple. Among them, we can discern the *ḥrj jmj(w) wnwt n Jmnw-Rꜥw* 'chief of the astronomers of Amun-Re' (Mogensen, 1918: 85 pl. XXVII, 85–86 pls. XXVII–XXVIII); the *jmj wnwt n ẖnm* 'astronomer of Khnum' (Johnson, 2001: 118); the *jmj wnwt n ḥrj-šf* 'astronomer of Herishef' (early Ptolemaic, pLille 51; Hughes & Jasnow, 1997: 14 Z), and the *ḥrj jmj(w) wnwt n Jmnw-Re nsw nṯrw* 'chief of the astronomers of Amun-Re, king of the gods' (Spiegelberg, 1912: 44–45 'Oberhoroskopen').

In demotic we also have (second half of the Ptolemaic period), the 'astronomer of the domain of Sobek' (Daressy, 1919: 277–278); the *jmj wnwt n sbk* 'astronomer of Sobek' (in pOI 25255 and in pOI 25256, from the beginning of Ptolemy III);[13] the *jmj wnwt n ḥwt-nṯr n tꜣ-nb-tꜣ-ntj* 'astronomer of the temple of Tebtunis' in Ptolemaic times (pCairo 30607 and 34662; Arlt, 2008: 20), and others. The temples, therefore, continue to have attendants charged with the measurement of the hours.

Other words and variants that appear at this time are *rḫ pt* ('who knows the sky'; Depauw, 2001: 1–2) in the 3rd or 2nd century BC; in Edfu (Chassinat, 1931: 346); , and *jmw pt* ('astronomer'; Quack, 2004: 16; Sauneron, 1959: 36–41 read *bꜣḳ pt*) in Esna (Sauneron, 1968: 9 num. 197, 162 num. 265, 195 num. 284). The astronomer is also identified in Ptolemaic times as *wnwnw* ('observer of the stars'; *Wb* I, 318), or as *jmj wnwt* ('who is in the hour'; Wilson, 1997: 69), abbreviated (Mogensen, 1918: 87) and in demotic (Vittmann, 1998: 323–324) the form *pꜣ jmj wnwt* continues to appear (Erichsen, 1954: 90).

A statue from the early Ptolemaic period (Cairo CG 680; El-Sayed, 1980: 235, *Supérieur -de la détermination- de l'heure d'Amon au quatrième phylé*[14]) belonging to an individual named Nesmin, identifies him, among his other titles ('servant of Horus and scribe of Amun'), as *ḥrj (jmjw) wnwt (n) Jmnw* 'Chief of the astronomers of Amun' or *ḥrj (jmjw) wnwt n Jmnw sꜣ-4* 'Chief of the astronomers of Amun' of the fourth phyle. Also, in a statue of a man called Horemheb (Cairo TR 25/10/17/5) we find the form *ḥrj (jmjw) wnwt (n) Jmnw* or *ḥrj (jmjw) wnwt n Jmnw* 'chief of the astronomers of Amun'; the last one also in another statue of the same individual (JE 43652)[15] and rare variants as and (Birk, 2014: 81). Also in the statue of Imhotep (Scott, 1992: 54 n° 28), for example, we find the form *ḥrj (jmjw) wnwt n Jmnw* 'chief of the astronomers of Amun'.

In the same way we know, thanks to the inscriptions found by Schiaparelli on two jars in the temple of Deir el-Medina (Botti, 1967: n° 2a–b and 8a–b), two

[13] Hughes and Jasnow 1997: Chicago Hawara Papyrus 1 (365–464 BC) pl. 3, Chicago Hawara Papyrus 7a (245 BC) pl. 40, Chicago Hawara Papyrus 8 (243 BC), pl. 48.

[14] 'Senior keeper of the hour of Amun during the fourth *phyle*.'

[15] Images of the statue can be seen at: https://www.ifao.egnet.net/bases/cachette/?ids=436&os=1#galerie (seen: 20/3/2020).

horologists from the time of Ptolemy V and Ptolemy VIII from Djeme (the town developed in Medinet Habu), both called Psenminis.

In both the Dakka and Philae temples there are late demotic and hieroglyphic graffiti that mention individuals linked to astronomical knowledge, albeit in relation to the kingdom of Kush. In Dakka, for example, a hieroglyphic text (Griffith, 1937: 27–28 Dak 30) belonging to an individual named Hornakhtyotef II (Török, 2009: 461), who was high priest of Thoth at this place, defines him as:

ḥm-nṯr n spdt jmj-rꜣ mšꜥ n jꜥḥ wꜥb n pꜣ sbꜣ 5 ꜥnḫw nty rḫ pꜣ wš n jr ꜥb n rꜥ jꜥḥ
'Priest of Sothis, general of the moon, *wꜥb*-priest of the five living stars (the five visible planets), who knows the moment of the eclipse of the sun and moon'.

This Hornakhtyotef II was also 'ritualistic chief of the king of Kush', so here we can also see how, in many cases, there is a relationship between someone who is a maximum connoisseur of the temple texts and at the same time an astronomer.

Also in Philae, at Hadrian's Gate, a graffito speaks of the 'royal scribes of Kush who know the rises of the five planets and find the moment of the eclipse of the sun and moon'. This was a time towards the second half of the 3rd century AD and still within the Meroitic world, in which the Blemmies dominated the region (Griffith, 1937: 112 Ph 410, also 121 Ph 421).

Of course, during the Ptolemaic period the *wnwtj* continues his work as an horologist. Thus, for example, in the Temple of Horus in Edfu we find 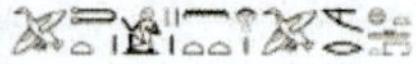 *pꜣ ṯt ntt ḥr pꜣ mrḫt* ('the team of those in charge of the *merkhet*'; Chassinat, 1931: 222), made up of those in charge of the indication of the hours and who use the *merkhet* for this purpose. In texts from Esna (Sauneron, 1959: 38) the importance of his work is clearly shown:

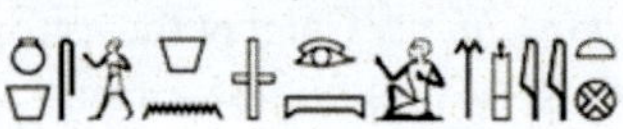

'The (time) signal is given by the astronomer at Esna'

'When the fourth hour of the day comes, the astronomer of this temple will come and give the signal to the atelier's servers'.

In a piece known as Dattari statue (Brooklyn Museum 52.89; Guermeur, 2005: 305; see Figure 2.18), belonging to a priest of Amun from the 30th Dynasty, is expressed the importance of the *merkhet* and the clepsydra in measuring the hours. And precisely with a 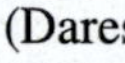 *merkhet* or shadow clock appears in his statue (Cairo TR 25/11/18/3) a priest and 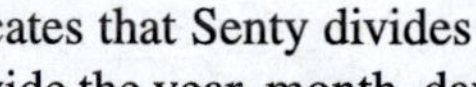 ('astronomer of the domain of Sobek') called Senty (Daressy, 1919: 277–278). The inscription on this statue (Winkler, 2016: 273 n. 93) indicates that Senty divides the hours according to the will of the god, knows how to divide the year, month, day, hour, minute (Altmann-Wendling & Stockhusen, 2020: 111 n.46), and the orbits of the stars, observing their movements, predicting what is to come (that is, the future). In the absence of a new publication

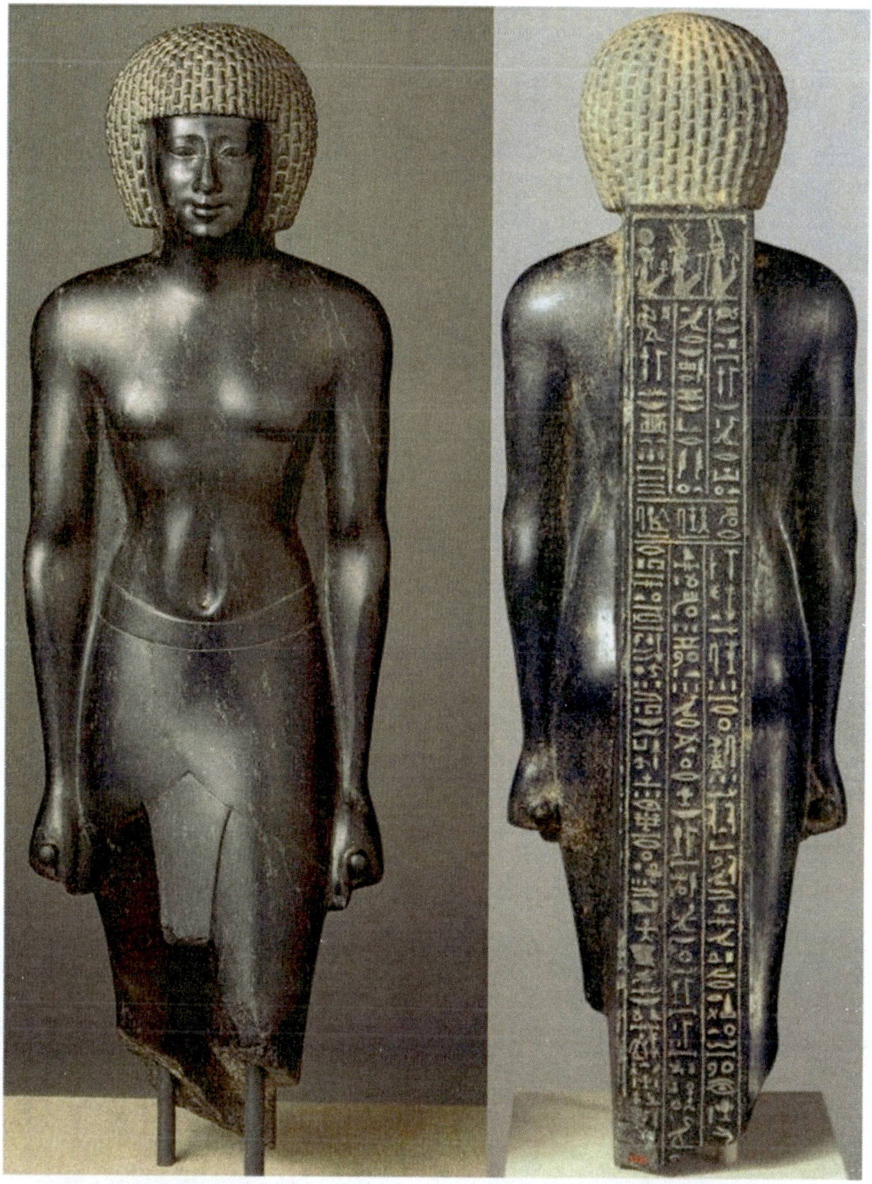

Fig. 2.18 Dattari statue, belonging to a priest of Amun from the 30th Dynasty, front and back. Brooklyn Museum, Charles Edwin Wilbour Fund, 52.89. Creative Commons-BY. (Photograph courtesy of the Brooklyn Museum)

on this inscription, it seems to be deduced that at this time the functions of the Egyptian astronomer also began to be associated with those of the astrologer. It is said of Senty that his work included 'announcing to man his future, telling him about his youth and his death' (Clagett, 1995: 490).

Fig. 2.19 Horkhebi statue (Cairo Museum, JE 38545). 30th Dynasty or early Ptolemaic period. (Adapted from Razanajao, 2006: pl. IX)

A more complete example is provided by another statue dated between the 30th Dynasty and the Ptolemaic period (Kamal, 1906: 239–240), belonging to an astronomer named Horkhebi, or Hor(en)tabat according to a new reading (Altmann-Wendling & Stockhusen, 2020: 107 n.26). Of this statue (Razanajao, 2006: pl. IX; see Figure 2.19), which originally could have been about 80 cm high, two fragments are known. One was found by Petrie, although its whereabouts are unfortunately currently unknown, and the other is the one that was found in 1905, currently in the Cairo Museum (Cairo JE 38545).

This is a document of undoubted importance to verifying the transition, in terms of astronomical knowledge and its applications, from the ancient *wnwtj*-astronomer to the Ptolemaic *wnwtj*, even though this title is not specified for him (Daressy, 1916: 2–3; Derchain, 1989: 74–89; Clagett, 1995: 495–496; Dieleman, 2003a, erratum; Lull, 2004a: 54–59; Razanajao, 2006: 84–86, pl. VIII; Altmann-Wendling & Stockhusen, 2020: 103–115).

Horkhebi is a *wb3-jb m mdw-nṯr* ('expert in sacred texts'), so he is a priest with access to the papyri kept in the temple, where there was a wide range of texts, from rituals to those of astronomical content and from other disciplines. As an astronomer, Horkhebi is *m33 m33wt nb m pt m t3* ('observer of everything that can be observed in the sky and on earth') and, as such, is a *wb3-jb n m33 ḫ3b3s* ('expert in stargazing').

In the main text, arranged in three columns at the back of the statue, we can read:

> '(He is) the one who announces the rising(s) and setting(s) at their time, with the gods who predict the future, for which he has purified himself on their rising days (heliacal risings), when the (decan) *3kh* is next to *Benu* above them, so that he can satisfy them with his predictions; the one who observes the culmination of every star in the sky [...], the one who announces the heliacal rising of Sirius at the beginning of the year and observes her on her first day of the festival, having calculated her course [...], the one who observes what she does every day, so that everything she has planned is in his charge; the one who knows what goes to the north and south of the solar disc, making known all its omens and predictions, so that he can say what they cause when they arrive at their time; the one who divides the hours of the day and the night without making mistakes during the night [...], one who is wise in every phenomenon observed in the sky, who has waited for them, instructed in their winds and their omens'.

Horkhebi has a great knowledge of the celestial vault. He knows the times of the risings and settings of the stars (mainly, we understand, those of the decanal or hourly stars), the sun, and the moon. He knows the culmination of every star, that is, the moment when it reaches its greatest height above the southern horizon, coinciding with its transit through the central meridian. Without a doubt, of all the heliacal risings with which he dealt, the most transcendent was that of the star Sirius, with which he was very well acquainted. Sirius was taken into account not only in temple and administrative realms but also in agricultural society, as it roughly marked the beginning of the flood, a vital period for the economy of the whole of Egyptian society. He also knows the extreme positions of the sun on the horizon ('what goes north and south of the solar disc'), that is, the summer and winter solstices. Horkhebi

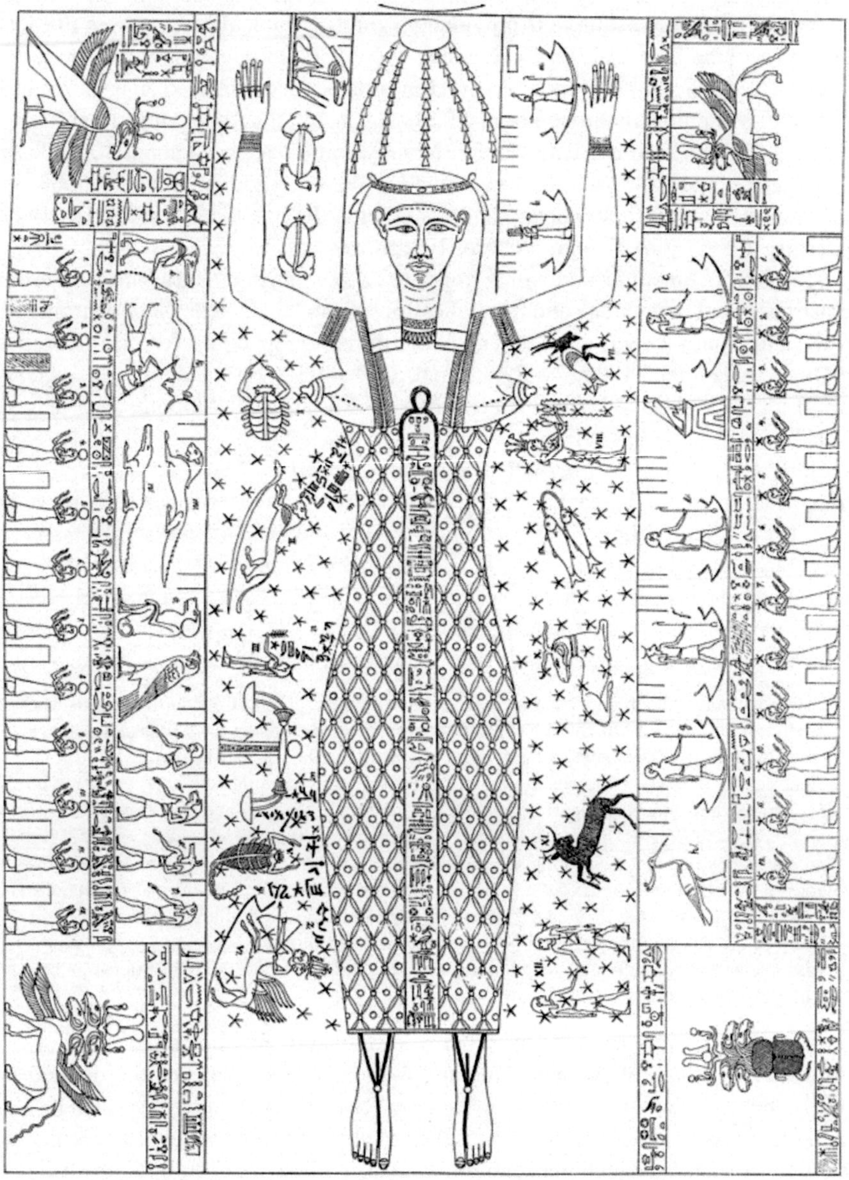

Fig. 2.20 Coffin of Heter. Roman period, 2nd century AD. (After Brugsch, 1862: pl. XVII)

is also 'the one who divides the hours of the day and the night', as this is his main mission as an horologist.

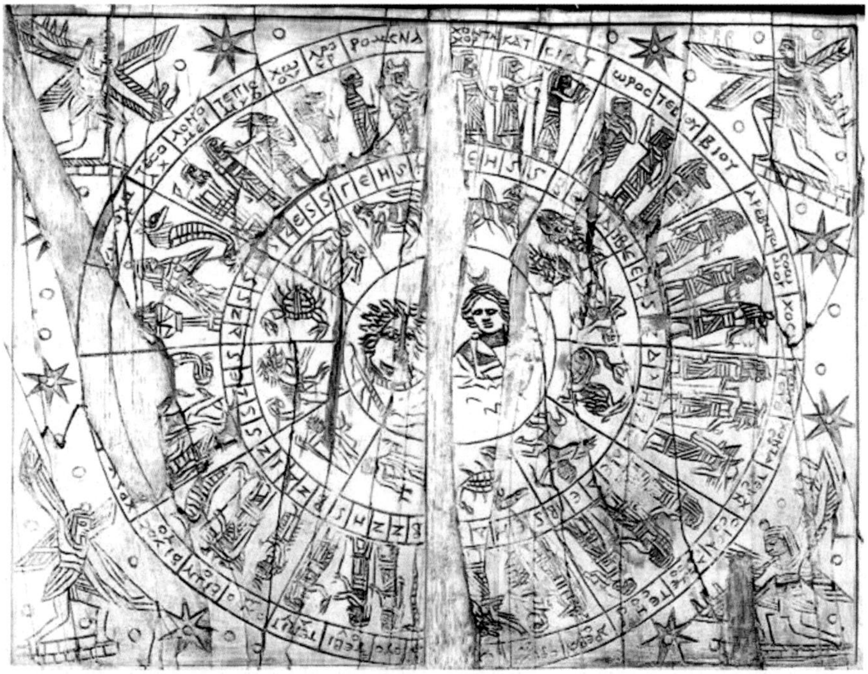

Fig. 2.21 Astrological board from Grand. Musée d'Archéologie National, MAN 83675. (Adapted from Nenna, 2003: 376 fig. 7)

However, the Horkhebi inscription also offers signs of an astrological nature, although the correct translation of the text and its interpretation is not absolutely clear. For example, the following part:

šsr ꜥnḫ(w) ḥtp(w) r nw.sn ḥnꜥ nṯrw sr(w) ḫntj
'(He is) the one who announces the rising(s) and setting(s) at their time, with the gods who predict the future'.

Here Neugebauer and Parker (*EAT* III, 214) understood that rising and setting should refer to the signs of the zodiac, while the 'gods' would be the planets, thus establishing the astrological relationship. However, given that the preceding text indicates that he is an 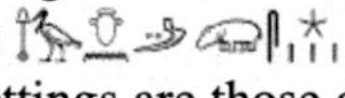 ('expert in stargazing'), it is logical to think that the risings and settings are those of these stars, hourly decans, or stars in general. On the other hand, the 'gods who predict the future' could be the decans themselves, seen as divine entities linked to the prediction of the future, as is clear in Egyptian astral predictions prior to the development of horoscopic astrology in Egypt.

Fig. 2.22 One of the four Stobart Tablets, M11467c, reverse and obverse. (Courtesy of National Museums Liverpool, World Museum)

Immediately below, we have another part of the text that has usually served to indicate this astrological relationship, although since Daressy (1916: 2–3) first published it, it has been translated in various ways (*EAT* III: 214; Derchain, 1989: 76; Clagett, 1995: 495; Dieleman, 2003b: e .; Lull, 2004a: 54–55; Quack, 2018b: 106; Altmann-Wendling & Stockhusen, 2020: 104–105):

wʿb.n.f n.sn m hrww.sn n pr 3ḫ r-gs bnw ḥr.sn

'for which he has purified himself on their rising days (heliacal risings), when the (decan) Akh is next to Benu above them'.

Generally, this has been understood in the sense that  *3ḫ* refers to the decan  *3ḫwj* (Akhwy) and that  *bnw* (Benu) refers to the planet Venus. This being the case, and taking into account that *3ḫwj* is one of the three decans that divide Pisces (as we see in Esna A), it is logical to think that the relationship between Pisces and Venus would follow the Mesopotamian tradition, in which Venus is in Pisces in exaltation—since in Pisces there is the *hypsoma* of Venus (*EAT* III: 204 fig. 32; Rochberg, 1998: 47–50). However, Quack (2018b: 107) believes that the translation is incorrect, and also that the *hypsoma* does not exactly fit that decan; if that is so, thit particular influence of Babylonian astrology would disappear.

Finally, it is also worth analysing a last sentence of the main text which has been translated in various ways, depending on the different readings that have been given to some of the signs: 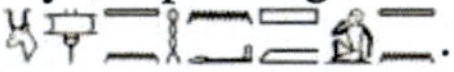. The Neugebauer and Parker translation (*EAT* III: 215), 'skilled with respect to their conjunction(s) and their regular movement(s)', follows the reading *šs3 (m) ẖnm(w).sn ḥnᶜ gsgs(w).sn*. Brunner (1988: 228), on the other hand, proposed: *Erfahren in ihren Winden (ṯ3w) und deren Wehen (?),*[16] thus inviting the possibility that it made reference to an omen from the observation of the winds, as clearly appears on the Stele of Israel. Jansen-Winkeln (1998: 10), however, reads *šs3 (m) nfj.sn ḥnᶜ ᶜšm.sn* as *Kundig (in) ihrem Anfachen (Aufleuchten) und ihrem Verlöschen.*[17]

Finally, Ritner (1993: 36–37 n. 167) has taken into account other parallels for the last word of the sentence, and has translated: 'diagnosing their wind(s) and their omens'. More recently, Quack (2018b: 107) and Altmann-Wendling and Stockhusen (2020: 105), have proposed the reading *šs3 (m) ṯ3w(.w).sn ḥnᶜ šm. (w).sn, der erfahren ist (mit) ihren Winden und ihren Omina* ('experienced is (with) their winds and their omens'). In our opinion, this is the best option.

However, the text is not entirely explicit about who the omens may refer to. Are they predictions for Egypt or the king of Egypt, or are they also made for individuals? The only part of Horkhebi's text that seems to offer any clue to this indicates that he is *wp(w) mdw ḫr nb t3wj* ('who opens his speech to the lord of the Two Lands'). Moyer (2011: 240) interprets this to mean that Horkhebi's astrological predictions are directed at the king (and by extension to Egypt), thus denying the possibility that he could make individualized predictions such as those documented in the late Ptolemaic period in genethliacal astrology. In this sense, they would be judicial or mundane astrological predictions like those provided by papyri such as BM 10660 and 10661 (Andrews, 1992) from the Ptolemaic period. They affect territories and kings (in fact BM 10661 even mentions the death of the king of Persia if conditions are met). However, it is not clear that Horkhebi's predictions are necessarily unique to judicial astrology. In the case of Senty, the text warned that this astronomer was able to predict the future of a person (through the stars).

It seems (or made more evident), then, that to the Egyptian priest-astronomer tasks related to astral divination are now being added. Diodorus of Sicily, who visited Egypt towards the end of the first rule of Ptolemy XII, admired the wisdom of Egyptian astronomers, and the zeal and thoroughness with which they recorded the positions of the stars. Diodorus (I: 81) describes how in his time they were already influenced by astral divination:

> 'For the positions and arrangements of the stars as well as their motions have always been the subject of careful observation among the Egyptians, if anywhere in the world; they have preserved to this day the records concerning each of these stars over an incredible number

[16] 'Experienced in their winds (ṯ3w) and labours (?).'

[17] 'Expert in their lighting up and their extinction.'

> of years, this subject of study having been zealously preserved among them from ancient times, and they have also observed with the utmost avidity the motions and orbits and stoppings of the planets, as well as the influences of each one on the generation of all living things—the good or the evil effects, namely, of which they are the cause. And while they are often successful in predicting to men the events which are going to befall them in the course of their lives, not infrequently they foretell destructions of the crops or, on the other hand, abundant yields, and pestilences that are to attack men or beasts, and as a result of their long observations they have prior knowledge of earthquakes and floods, of the risings of the comets, and of all things which the ordinary man looks upon as beyond all finding out.'

For Sauneron (1960: 66) at this time there was a separation between the horologist (ὡρολόγος) and the astrologer-*horoskopos* (ὡροσκόπος), although the existing documentation does not really seem to show such a separation, but quite the contrary.

According to Winkler (2016: 272), although it has been suggested that the words astrologer (ἀστρολόγος), astronomer (ἀστρονόμος), mathematician (μαθηματικός) and horologist (ὡρολόγος) designate astronomers in the Greek documentation of Egypt, it is not ruled out that they were busy in relation to genethliacal astrology. Even the *hierogrammateus* (ἱερογραμματεύς) or 'scribe of the sacred books' (Otto, 1905: 89–91) is related with astronomy.

As Jones (1994) points out: the 'astronomical competence of the Egyptian priests is found in two Hellenistic Greek papyri, pHibeh I.27 from the third century and pPar 1 from the second century BC, which have almost identical passages stating that the ἀστρολόγοι and ἱερογραμματεῖς, that is the 'star-watchers' and the 'priest-scribes', use a combination of lunar months, the 365-day Egyptian year, and stellar risings to establish the dates of religious festivals'.

Perhaps the *pastophoroi* (παστοφόροι) (see functions in Thomas, 2014: 122–126) could be associated with the study of the stars (Winkler, 2016: 274), though it even seems that they were linked to the interpretation of dreams (Renberg, 2016: 719–721). Vandorpe and Clarysse (2019: 417) classify the *pastophoros* as 'door-keeper, supervising the temple area open to the public' belonging to the lower priesthood. However, it is not clear that the *pastophoroi* belong to the group of priests since Greek sources such as pGnomon §82 expressly indicate that 'for *pastophoroi* it is no permitted to identify themselves as priests' (Renberg, 2016: 720 n. 7).

Therefore, authors such as Hoffmann and Quack (2014) believe that the *pastophoroi* from Greek sources are the *Vorhangträger* ('curtain holders') or *Vorhangshochheber* ('curtain raisers'), *jemand, der als Pförtner des Haus bzw. den Tempel bewacht und dem Ein- oder Ausgehenden den Türvorhang zur Seite oder hochhebt* ('someone who guards the house or the temple and lifts the door curtain to the side or up raise the door curtain to those entering and leaving'), corresponding to the *jrj-ꜥꜣ der zur Tür Gehörige* ('the one belonging to the door') of the demotic sources, so that although they are linked to the chores in the temple they are not priests *im Sinne einer präzisen Terminologie* ('in the sense of precise terminology'). Renberg (2016: 720) defines the *pastophoroi* as a 'nebulous group of low-level cult officials (...) who appear to have focused on serving the needs of the public'. For her part, Escolano-Poveda (2020: 77) concludes: 'Although part of the temple personel, the *pastophoroi* were not priestly figures'.

Similarly, in an ostracon (OMM 1156) from the 1st century AD, found in Narmuthis, the word *khronokrator* (χρονοκράτωρ) has been understood to designate another specialist in astrology (Hoffmann, 2000: 46), although it is highly doubtful that it refers to a function (Dieleman, 2003a: 277). Finally, the term *genethlialogos* would refer to the astrologer.

The fusion of the astronomer and the astrologer, in the religious sphere, is clearly perceived in the interesting description of a procession in honour of the god Osiris in Alexandria, carried out by Clement of Alexandria (1st–2nd century AD) in his *Stromata* (Dieleman, 2003a: 278; Roberts et al., 2007: 488; *Stromata* VI: 4):

> 'For first advances the Singer (...) And after the Singer advances the horoskopos (ὡροσκόπος), who is holding his insignia, the horologion (ὡρολόγιον) and the astronomical palm leaf (φοίνιξ ἀστρολογίας) in his hand. He must have the astrological books of Hermes, which are four in number, always in his mouth. Of these, one is about the order of the fixed stars that are visible, and another about the conjunctions and luminous appearances of the sun and moon; and the rest respecting their risings. Next in order advances the sacred Scribe (ἱερογραμματεύς), with wings on his head, and in his hand a book and rule, in which were writing ink and the reed, with which they write. And he must be acquainted with what are called hieroglyphics, and know about cosmography and geography, the position of the sun and moon, and about the five planets'.

The aforementioned horoskopos (ὡροσκόπος) must be taken as the equivalent of the horologue (ὡρολόγος) or the *wnwtj*, '(observer) of the hour', of ancient Egyptian texts. Clement of Alexandria indicates that he carries a horologion (ὡρολόγιον) and an astrological (astronomical) palm (φοίνιξ ἀστρολογίας), instruments that must correspond to the *merkhet* clock and *bay*, better known in the astronomy of the Pharaonic period. Furthermore, he refers to the existence of four 'astrology' books, of which only three are mentioned with the *horoskopos*: one on the fixed stars that are visible; one on the conjunctions and luminous appearances of the sun and the moon; and another on their risings. The fourth book could be the one that, according to the text (and perhaps because of confusion), is carried by the sacred scribe (ἱερογραμματεύς), a book on the 'position of the sun, the moon and the five planets'.

However, in another translation Dieleman (2003b: 138) gives the four books to the *horoskopos*: 'the first is about the arrangement of the fixed stars, the second about the movements of the sun and the moon and the five planets, the third about the encounters and illuminations of the sun and the moon, and the last about the rising stars'. In summary, both the instruments and the books, with astronomical ephemerides, are a reflection of a complete compendium of astronomical knowledge acquired during centuries of observations of the celestial vault by ancient Egyptian observers, now also enhanced by transfers of Mesopotamian origin. A clear distinction between astrology and astronomy did not arise until the 18th century.

Porphyry, a 3rd century AD philosopher, citing the words of Chaeremon the Stoic, writes in *De abstinentia ab esum animalum* (IV, 8), that 'The true method of philosophizing, likewise, was preserved by the prophets, by the hierostolistae, and the sacred scribes, and also by the horologi, or calculators of nativities. But the rest of the priests, and of the pastophori, curators of temples, and ministers of the Gods, were similarly studious of purity, yet not so accurately, and with such great continence, as the priests of whom we have been speaking' (Porphyry, 1823: 143). That is, it provides a list of people, with different positions, linked to the temple that in some way reminds the members of the procession mentioned by Clement of Alexandria.

The papyrus scrolls containing all this information were to be kept in the archives of the temples (Ryholt, 2019). In the specific case of the 'house of the book' (*pr mḏꜣt*) of the Ptolemaic temple of Horus in Edfu, the titles of at least 31 works have survived inscribed on its walls, including several on astronomy (Clagett, 1992: 35–36). One is entitled 'Knowledge of the periodic returns of the sun and the moon' (*rḫ nmtt n ḥꜣytj*), while another bears the title 'The government of the periodic returns of the stars' (*ḥḳꜣ nmtt sbꜣw*; Alliot, 1949: 148).

These two papyrus scrolls had to include long lists of observations that would allow the calculation of the risings, settings, or transits of all these celestial bodies, since the calculation of these positions was of great importance in determining the hours, decades, months, etc. In other words, it was essential to monitor the time, be able to measure it, and be able to order the celebrations or religious services at the ritually established moments. These books could be the same ones mentioned by Clement of Alexandria. There is no doubt that all the astronomical information contained in these 'books' could also be used for astrological work.

Likewise, the Carlsberg I papyrus, in its commentaries in demotic of the *Fundamentals of the Course of the Stars*, cites as reference works a long series of books related to astronomy of which we only know the titles. These are 'The description of the movements of the stars'; the *mꜣꜣ jtn,* '(Book) of Seeing the Disk'; 'The (Book) of the sky', and so on (Clagett, 1995: 366–367; von Lieven, 2007: 284ss).

In Egypt, demotic astrological treatises date mainly from Roman times, but fragments from the Ptolemaic period have been found at Tebtunis (Quack, 2016: 238). Although most are unpublished, there is evidence of a large number of demotic astrological treatises of various kinds.

While papyri, such as pCairo 31222 (Hughes, 1951), in which the rising of Sirius is associated with the position of a planet (or the sun or the moon) in a zodiacal constellation, correspond to the so-called universal (καθολικόν) astrology, others, such as pBerlin 8345 from the 1st century AD (Hughes, 1986: 53–69), correspond to genethliacal (γενθλιαλογικόν) astrology, with such predictions as:

> *Wer geboren wurde, während Venus im Deszendenten war, wird viel Leid aufgrund einer Frau haben, und ihm wird übler Leumund zuteil werden, während er ein Knabe ist (?). Wenn er alt geworden ist, wird sein Herz glücklich sein mit seiner Lage. Er wird Glück haben in dem, wozu [er auszieht] in allen Dingen.* (Quack, 2008: 368–370)

> 'Who was born while Venus was in the descendant would know great sorrow over a woman and an evil spell be cast over him while he is a boy (?). When he comes of age, his heart will rejoice at his position. He will be fortunate in what [he moves] in all things'.

In this case, they are forecasts in which the position of the planet at the time of the individual's birth prevails, taking into account the position in one of the 12 houses of the dodecatropos (δωδεκάτροπος), a system that is absent in Mesopotamian astrology but makes its appearance in Egypt, a great centre of Hellenistic astrology. Another kind of astrological text, typical of Egypt, uses the 36 Egyptian decans as a birth reference (Quack, 2016: 237), a group into which pFlorence 8 would fall (*EAT* III, 252).

Sixty demotic horoscopes are catalogued, mainly on ostraca from Athribis, Luxor, Medinet Habu, Thebes, Karnak, and Medinet Madi. The oldest horoscope known in Egypt dates from 44 and 38 BC (Quack, 2018b: 100 n.183; Escolano-Poveda, 2022), during the reign of Cleopatra VII, and is written in demotic (oAshmolean DO 633). Beyond its age, it is interesting to note that the information it provides is very complete for horoscopic astrological practice:

> Year 8 of the Queen, fourth month of Peret (Pharmouthi),
> day 22, seventh (hour) of the day
> Sun (in) Taurus 4°, (Terms of) Venus
> Moon (in) Gemini 20° 30′, (Terms of) Mars; Libra 6°, (Terms of) Saturn
> Year 13, first, he adds
> Saturn (in) Sagittarius, his own Terms (scil. Saturn), he kills
> Jupiter] (in) Aquarius 4°, (Terms of) Mercury, (he) adds
> Mars] (in) Pisces 19°, (Terms of) Mercury, (he) adds
> Venus in Taurus] 7°, his own Terms (scil. Venus), (he) adds
> Mercury in .] 10°, (Terms of) Venus, (he) adds
> Ascendant (in) Leo 19°; Descendant in] Aquarius 19°;
> (Lake of the) Sky (in) Taurus [17°; (Lake of the) Duat
> (in) Scorpio 17°;
> Provision of Life in Virgo; Brother in] Pisces; Good Fate [in Cancer;
> Bad Fate (in) Capricorn;] Life (in) Aries; [Death (in)
> Libra; Good Daimon (in) Pisces; Bad Daimon (in) Virgo. (Escolano-Poveda, 2022: 59)

It offers the position of the sun, moon, and planets (Mercury is missing), and the four cardinal elements; that is, the horoscope or ascendant (point of the horizon through which the ecliptic rises, *rꜥ-ḫꜥ* 'Place of Ascension'), the descendant (point of the horizon where it is hidden, *rꜥ-ḥtp* 'Place of Rest'), the *mesouranema* (*medium coeli*, 'middle of the sky', i.e. the culmination of the ecliptic, *šy n pt* 'Lake of the Sky') and the *hypogeion* (*imum coeli*, 'bottom of the sky', lower culmination of the ecliptic, *šy n dwꜣt* 'Lake of the Duat').

In addition, demotic/hieratic signs are used to express names of planets, zodiacs, and a number of astronomical words (Bohleke, 1996: 20; Escolano-Poveda, 2022). The oldest written in Greek dates from 9 BC (Dieleman, 2003a: 280).

At this time both astrologers and astronomers can be priests of the temple, although it should be clear that their functions are shared. The Roman Marcus Manilius, author at the beginning of the 1st century AD of the Astronomicon, a Latin didactic poem in five books, indicates that priests were among those who contributed to the development of astrology (I, 47). It is also worth remembering that, at this time in which Hellenistic astrology is going to develop, astronomy and astrology will go hand in hand. Let us think that Claudius Ptolemy himself (2nd century AD) is not only the author of the famous Almagest, purely astronomical, but also of the Tetrabiblos or Apotelesmatiká, an astrological text that uses the astronomical knowledge expressed in the Almagest to make predictions.

On the other hand, although many of the astrological documents are written in Greek, rather than demotic, it seems that both Egyptian and Greek were used by the priests (Jones, 1994: 25–52; Winkler, 2016: 277–278; Quack, 2018b: 76–78). It is worth reproducing the words of Quack (2016: 238–239) when he says: 'It would be a great mistake to differentiate between the languages (and scripts) in a way which attributes everything Greek to a 'Greek' culture distinct from the Egyptian one, and perhaps even constructs an opposition between a more 'scientific' Greek and a more 'superstitious' Egyptian tradition. The fallacy of such an approach can be highlighted by looking in more detail at the manuscripts themselves'.

Of course, as astrologers they would make use of specific methods to carry out their predictions, although the basis of their calculations was served by astronomical ephemerides. In this sense, it must be taken into account that papyri that contain information of an astronomical nature (i.e. without divinatory predictions), such as the hieratic papyrus Berlin 8279, from the middle of the 1st century AD, which includes planetary positions through the zodiac for the Augustan period, or the famous 2nd century AD demotic Carlsberg 9 papyrus with its lunar cycle, may also have served a purely astrological purpose (Bohleke, 1996: 24–26).

Equally noteworthy is the demotic pBerlin 13146+13147 from Abusir el-Meleq (Neugebauer et al., 1981: 312–327; Steele, 2000: 3.2). This papyrus, based on the Egyptian civil calendar, makes a list of the 2 days in the year in which lunar eclipses could occur (coinciding, of course, with the ascending or descending node), between the years 84 and 73 BC, indicating in which sign of the zodiac it occurred (since the nodes regress with respect to the ecliptic, fulfilling a cycle of 18.6 years) and sometimes including positions of planets in the zodiac.

However, it should be noted that the dates follow a Greek scheme (Jones, 1999a: 14), as they are calculated according to the fourth Callippic cycle (cycles of 76 years, equivalent to 940 synodic months, counted from the summer solstice of 330 BC). On the back, it also has calculations of equinoxes and solstices. Despite the references by classical authors to the number of eclipses recorded by the ancient Egyptians, this is the oldest eclipse record preserved, although in the words of its editors, 'in a style reminiscent of contemporary Babylonian texts'. On the other hand, this papyrus is also one of the oldest proofs of the introduction of the zodiac into Egypt. Neither, in this case, are there any indications of astrological use, since the relationship is purely astronomical, but the introduction of the zodiac may also indicate that these calculations could have served for further astrological purposes.

In relation to purely astrological texts, there is a great variety, a reflection of the various currents and methods that were used. However, we also find simple notes taken by the astrologers in order to have the personal data of the clients or individuals for whom they were going to make forecasts. The demotic pVienna D 6005 (2nd century AD) (Hoffmann, 1995: 22–26), for example, shows a list of three people for whom the year, month, season, day, and time of birth are indicated, evidently for astrological purposes, so that the specialist could do his calculations and provide a forecast of the future of each of these people.

From the beginning of the 4th century AD there is a papyrus (pOxy Astron. 4262) which shows the fortune of an individual, taking into account the time of his birth on 18 April, AD 300, and the position of the stars at that time:

> Good fortune. Nativity. Year 16 of Diocletian, Pharmuthi 23, hour 4 of the day. Ascendant, Mars in Gemini; Saturn in Leo; moon in Virgo; Jupiter in Capricorn; Sun, Mercury, Venus in Aries. Good luck! (Jones, 1999a: 403, 2009: 347)

This example, therefore, shows a basic prediction, although it ignores the method used to reach that conclusion.

Astrology was particularly widespread in Egypt in the first two centuries of our era. It is not uncommon to find coffins in which the goddess Nut is represented on the inside of the lid following the ancient Egyptian tradition but now surrounded by zodiacal constellations with an explanation, in demotic, of the position of the planets at the time of the birth of the deceased. A good example is the coffin of Heter (see Figure 2.20), now lost (Brugsch, 1862; *EAT* III, 93–95 pl. 50), an individual born in AD 93. The demotic inscriptions that appear between the constellations of Leo and Sagittarius indicate the position of the planets at the time of his birth, that is, his horoscope: Jupiter and Saturn in Leo, at the end of Virgo Mars, the ascendant and Mercury in Scorpius, and Venus in Sagittarius.

In some cases (oMedinet Madi 1156), there remains even the price that the astrologer charged the client:

> In case you are one who foretells fate regarding the dominant heavenly body of a man, you will make ten obols minus the costs that were made at your expense and those that are at your expense today also and those that you will acquire with them as well. (Dieleman, 2003a: 277)

There are even instances of the care that the astrologer had to take so as not to be discovered at the time of making the astrological chart in the temple (oMedinet Madi 82) (Gallo, 1997: 86–87 n° 82: 1–7).

To know the position of the stars at the time of a person's birth or at any other time, the astrologer necessarily had to make calculations of an astronomical ephemeris, or make use of almanacs or astronomical tables (Jones, 1999b: 302–305), such as the Greek Oxy. LXI. 4175 papyrus, one of the earliest, with data from 24 BC (Jones, 2000a) from which that information or part of it could be obtained.

In order to predict the future for their clients, we know that Egyptian astrologers of the Roman period used diagrams and boards (πίνακες, possibly the πίναξ mentioned by authors such as Thrasyllus and Critodemus) in which, in addition, the Egyptian decans were related to the signs of the zodiac, and on which they placed stones to represent the planets, sun, and moon (on the word πίναξ, see Greenbaum & Ross, 2015). In oMedinet Madi 423 from Narmuthis (Medinet Madi), 2nd century AD, the tools used by the astrologer are a great πίναξ or board, a book of the year 3 of Petosiris, and a horoscope book (Menchetti, 2009: 231).

The last indigenous king of Egypt, Nectanebo II, was traditionally recognized as a great magician and astrologer (Matthey, 2012). In the *Roman d'Alexandre*, whose oldest preserved version in Greek dates from the third century AD (Anonymous, 1991: 38, I: 4), he is shown using a board for queen Olympia, the mother of

Alexander the Great. The board was divided into three concentric zones: an outer one with a representation of the 36 decans, an intermediate one with the 12 signs of the zodiac, and an inner one with the sun and the moon. On the board were placed eight pieces carved in different types of stones and gems that represented the five visible planets, the sun, the moon and the horoscope. Although this description is literary, some of these boards have survived.

The oldest of those found to date (in a Croatian cave), made of ivory, is from the 1st century AD (Forenbaher & Jones, 2011). The best-preserved, however, is the Grand board (Abry, 1993; see Figure 2.21), discovered in 1967–1968 in the temple of Apollo Grannus at Grand dans les Vosges (ancient Andesina), now in the Musée d'Archéologie National Saint-Germain-en Laye (MAN 83675). It was made of elephant ivory in two pieces, as a diptych, and was deliberately destroyed around AD 170. Similarly, the papyrus pWashington University inv. 181 (2nd–3rd century AD) contains a magnificent reference to this kind of board (Evans, 2004: 4).

2.4.1 *Two Egyptian Astrologers in the Tradition*

Many ancient authors, referring to the origin and development of astrology, made special mention of two Egyptian sages named Petosiris and Nechepso/s (Gundel & Gundel, 1966: 27–36). According to Pliny the Elder (*Nḥ* II: 21; 1855: 54–55), both came to calculate the dimensions of the cosmos up to the orbit of Saturn:

> The Egyptian calculation, which was made out bý Petosiris and Nechepsos, supposes that each of the lunar orbit (which, as I have said, is the least) consists of little more than 33 stadia; in the very large orbit of Saturn the number is double; in that of the sun, which, as we have said, is in the middle, we have the half of the sum of these numbers. And this is indeed a very modest calculation, since if we add to the orbit of Saturn the distance from him to the zodiac, we shall have an infinite number of degrees'.

In the papyri pCtYBR 422 v°, preserved at Yale, and pLund 2058 v°, from the 1st to 2nd century AD, it is indicated that when a stone block fell from the temple of Heliopolis, a papyrus appeared which, on being studied by Petesis (a name from which could derive the Petosiris of other sources) and delivered to king Nechepsos ('Nekau the wise', *Njkꜣw pꜣ šš*, which must correspond to the Saite king of the 26th Dynasty Nekau II), turned out to be an astrological treatise written by the famous Imhotep (Ryholt, 2011: 62, 2013: 13). As the papyrus narrates: 'Here is a copy of the book of Imhotep the Great, son of Ptah, the great god' (*tw.s ẖt pꜣ ḏmꜥ Jy-m-ḥtp wr sꜣ Ptḥ pꜣ nṯr ꜥꜣ*).

Following Ryholt, the association of Nekau II with astrology stems from the fact that, shortly after being enthroned, while Psametik I was still being embalmed, an eclipse occurred in Egypt, as indicated in the demotic papyrus Berlin 13588. So the astrological work relating to Nechepsos was pseudepigraphically linked to king Nekau II. On the other hand, the mere fact of being an Egyptian king could also be a source of prestige for astrology, given the superhuman characteristics embodied in the figure of the pharaoh. Neither is it an accident that the papyrus was found in the ancient temple of Heliopolis, or that a person such as Imhotep is cited as the original author of the treatise. There is a clear intention to locate the origin of astrological knowledge in the ancient pharaonic world.

If the Nechepsos to which the sources refer clearly sought to establish the pharaoh Nekau II (610–595 BC) as the originator of astrology, then tracing the character that serves to project the figure of Petosiris is not so simple, since Petosiris ('the one whom Osiris has given') is a very common Egyptian name, especially during the Late Period. Despite these difficulties, many authors have attempted to identify Petosiris. Spiegelberg (1922) assumed that he could be the individual of the same name who was an important Hermopolitan priest of the late second half of the fourth century BC, whose monumental tomb is located, in excellent condition, in the necropolis of Tuna el-Gebel. However, it is difficult to corroborate this relationship, since, beyond the shared name, there are no indications in the tomb (Lefebvre, 1924) or solid elements that could serve to corroborate this relationship. Similarly, Neugebauer and Parker (*EAT* III: 216) looked for another option in a Petosiris buried in Atfih. In this case, it was the presence of an astronomical ceiling in the tomb that motivated the postulation of this identification.

An Egyptian sage called Peteesis is mentioned in a Greek papyrus from the 3rd century AD (pRylands 63; Johnson et al., 1915: 2–3) as the teacher of Plato who explained astrological questions. This seems to correspond with certainty to the Heliopolitan priest Petese of the demotic papyrus CtYBR 422 v° from the Tebtunis temple library. Petese, recognized as a great connoisseur of astrology and mainly a sage in the Egyptian tradition, in demotic and Greek, Ryholt (2006: 13–16) places him as an historical figure at the end of the 4th century BC, since a papyrus from that time (pPetese Saqqara) already mentions him. Even if he had really instructed Plato, who according to Mathieu (1987: 157) was able to visit Egypt during the reign of Achoris in 393 BC, it would have to be traced back to the beginning of the 4th century BC. According to Quack, he could even go back to the 8th–7th century BC if he is identified with a certain Petese of Heliopolis mentioned in an abnormal hieratic papyrus (pQueen's College) from the year 21 of Piye or Taharqa (Baines et al., 1998: 235–236). This Petese is designated as *pꜣ ḥrj*, not as a magician, but this reference could still be to a person *die besonderes Ansehen genoß und zumindest einen Gewissen 'Nachruhm' hatte*[18] (Quack, 2002b: 78).

Gysembergh (2018), on the other hand, considers that, even the astronomer Hipparchus of Nicaea (2nd century BC), in his *Commentary on the Phenomena of Eudoxus and Aratus*, written between 147 and 127 BC, already made mention of Nechepsos and Petosiris, if it is understood that when speaking of the 'ancients' (ἀρχαῖοι), as occurs in some astrological sources in Greek, it is to them that he refers. However, it is by no means certain that all references to the ancients or Egyptians that appear in various texts must necessarily refer to Nechepsos and Petosiris.

[18] 'Who enjoyed a particular reputation and had at least a degree of posthumous fame.'

Obviously, the inevitable discussion is whether these traditions, which transmit both Greek and demotic Egyptian texts, obey a literary *topos* or historical reality even with their deformations. For Aufrère and Marganne (2019: 505), 'from the first century BC, with Egyptians unwilling to be outclassed by Greek science, the growth of which was based on the integration of the Babylonian legacy of astronomy and astrology, there emerged cultural 'types' or legendary figures that seemed somehow to embody the primacy of Egyptian astrology: king Nechepsos or Nechepso and a man called Petosiris'.

Vettius Valens attributes to Petosiris the writing of a treatise called *Horoi* (*Definitions*) and in the *Suda* it is spoken about the *Astrologumena* (Ἀστρολογούμενα) or *Astrological Questions*. He is also credited with a letter addressed to Nechepsos (*Cod. Laur. Vict.* Pl. 38/24 fol 174v). And Nechepsos is credited with the authorship of at least 15 books (Riess, 1891), of which the 13th is titled *Lot of Fortune* (χλάρος τύχης), the 14th *Astrological Medicine* (Ἀστρομαθηματικά) and the 15th *Astral Magic* (ἀροτροπιασμοί), although nothing is known about the first twelve.

Although Nechepsos and Petosiris seem to share the authorship of some work (as Firmicus Maternus seems to indicate with respect to the *Thema mundi*, which has a clearly Egyptian origin, being based on the situation at the moment of the heliacal rising of Sirius), different works also attributed to them occasionally show a contrary opinion. This is manifested, for example, by Vettius Valens (VIII, 5) when he indicates, 'This procedure is from the King (in reference to Nechepsos), this other is from Petosiris, that one is from Critodemus, etc'.

The Louvre 2342bis papyrus (Neugebauer & Van Hoesen, 1987: 42), which is a horoscope for the year AD 137 made for a person named Anubion, indicates who were the 'teachers' of these two sages:

> After the examination of many books as it has been handed down to us from ancient wise men, that is the Chaldeans, and Petosiris and especially also King Necheus just as they themselves took counsel from our lord Hermes and Asclepius, that is Imouthes, son of Hephaestus.

Ryholt (2011: 71) thinks that the aforementioned Hermes must correspond to Amenhotep son of Hapu, the wise man of Amenhotep III, while Asclepius would correspond to Imhotep, thus responding with what was expressed by Clement of Alexandria (*Strom.* I. 21,134) when referring to two deified mortals that correspond to the Theban Hermes and the Memphite Asclepius. Bull (2018: 165–167), however, prefers to associate the ancestor of the two sages with Hermes Trismegistus, also linked to Thebes and superior and prior to Asclepius in the *traditio mystica*.

Asclepius/Imhotep could have played a relevant role in the beginning of the development of astrology, since it appears to be directly linked to several demotic astrological treatises and clearly predates Nechepsos and Petosiris. In pMichigan 149 it is indicated that it was he who devised the doctrine of the *octatropos*,[19] a reference that is also accredited by Vettius Valens (IX, 3, 5) in relation to the *octatropos*

[19]The *octatropos* ('eight-turning') is an eight-compartment system, while the *dodecatropos* is a system of twelve places.

but also to the *dodecatropos*, which does not appear in the Mesopotamian astrological context but in demotic astrological treatises (Quack, 2018b: 119). He is also credited with the *Book by Hermes to Asclepius on the Plants of the Seven Stars* and the *Sacred Book of Hermes to Asclepius* (in which the names of the decans of the Sethy IB family appear), with iatromathematical content, and other teachings taught to him by Hermes, among which the *Thema mundi* stands out.

In the *Summary of the Pinax* of Thrasyllus of Mendes or Alexandria (Schmidt, 1995: 57–60), one of the oldest Greek astrological texts, Petosiris and Nechepsos are again the only references mentioned, along with the so-called Hermes Trismegistus. And in the *Matheseos libri octo* ('Eight books of astrology'), work of Julius Firmicus Maternus (1975: 71–72) in the time of Constantine I (4th century AD), Petosiris and Nechepso appear as influential references of the *Thema mundi*:

> Those divine men, altogether worthy and admirable, Petosiris and Nechepso, who approached the very secrets of divinity, also handed down to us the birthchart of the universe in order to show us that man is made in the likeness of the universe according to those same principles by which the universe itself is ruled; and that he is sustained forever by those same everlasting fires.
>
> Petosiris and Nechepso in this doctrine followed Aesculapius and Hanubius. To them Most Powerful Mercury entrusted the secret. They setup the birthchart of the universe as follows: the Sun in the 15th degree of Leo, the Moon in the 15th degree of Cancer; Saturn in the 15th degree of Capricorn; Jupiter in the 15th degree of Sagittarius; Mars in the 15th degree of Scorpio; Venus in the 15th degree of Libra; Mercury in the 15th degree of Virgo, and the ascendant in the 15th degree of Cancer. By this they wished to prove that the fates of men are arranged in accordance with this birthchart, the conditions of the planets, and the influence they exert on the chart, just as is related in the book of Aesculapius which is called Myriogenesis.

On the other hand, the so-called *Anonymous of 379* (1993: 10), which locates the origin of astrology in Babylon, also comments on how, among the Egyptians, the first to develop this discipline with notoriety was Hermes, followed by others such as Nechao, Cerasphorus, and the already mentioned Petosiris and Nechepso. In this case, the differentiation between Nechao and Nechepsos is exceptional, since the *Anonymous of 379* text is the only one that, in reference to astrology, makes this double mention.

> The Babylonians and the Chaldeans, then, were just about the first to discover the knowledge of phenomena, as far as we knew from our progenitors. For Apollonius the Myndion and Artemidorus write accounts (…). And Berosus and those following wrote about them. And the ancestors of our Egyptians bore them in mind and made predictions concerning them, from whom Hermes first arose and wrote in his cosmic predictions concerning the annual rising of the Sirius, and Nechao and Cerasphorus, and Petosiris and Nechepso (…).

Just as in Nechepsos' sources, he is described as the king or sage, Petosiris appears sometimes identified as a priest. In the *Suda* (Riess, 1891: 330), he is described as an Egyptian philosopher versed in sacred books, astrological questions, and Egyptian mysteries. For Moyer (2011: 244), this association between the figure of a king (Nechepsos) and a wise priest (Petosiris) obeys a well-known motif in Egyptian literature, that of the priest who performs magic at court or presents a divine and hidden wisdom to the king. This was also shown by the well-known tale transmitted by the Westcar papyrus in which the magician Djedi displays his magic to King Khufu.

The Greek fragments linked to Nechepsos and Petosiris, collected in the first instance by Riess (1891), are dated with great caution by Heilen (2011: 23–24) in the second half of the second century BC by virtue of certain historical aspects or by Pingree (1997: 25–26) around the late 2nd or early 1st century BC. In fact, Petosiris and Nechepsos are believed to have done their astrological work in the mid-2nd century BC (Quack, 2018b: 113). Although their studies are not known to us from contemporary documents (they were first mentioned in the early first century AC by Thrasyllus), we do have references to the works of Nechepsos and Petosiris, so we know that they were linked to various types of astrology (iatromathematical or medical, universal or mundane, genethliacal or natal, and cathartic or elective).

2.4.2 *Astrology in Egypt*

In the final stage of the history of Egypt, as evidenced by the texts of the Tebtunis and Soknopaiou Nesos libraries and the astrological ostraca of Narmuthis, it becomes evident that in the temples astrology is practised with intensity, reaching its greatest moment of diffusion during the first half of the 1st century AD (Winkler, 2018: 298–299). Some temples, such as Narmuthis and Tebtunis, seem to be authentic astrological centres in Roman times (Jones, 1998: 211). However, there is also evidence that astrological consultations were available to many temple priests. In this sense, the best example is that of Narmuthis (Quack, 2002a: 169; Winkler, 2016: 275).

The representation of the zodiacal constellations in Egyptian temples, associated with the decans or other constellations of Egyptian tradition, is important at least in the late Ptolemaic and early Roman period in Egypt (see Chap. 4).

We have the outstanding case of the well-known circular zodiac of Dendara (*c.* 50 BC, the oldest surviving example of the representation of zodiacal constellations in Egypt; *EAT* III, pl. 35). To this we may add that of the now destroyed mammisi of the temple of Montu in Armant from Cleopatra and Ptolemy XV (*EAT* III, 71 fig. 17), a reused block from the ceiling of the temple of Min and Isis in Coptos from Ptolemaic to Roman period (*EAT* III, p. 82 fig. 18), that of the temple of Isis in Shanhur from the time of Augustus or Tiberius (*EAT* III, pl. 40 A), the rectangular zodiac of the pronaos of the temple of Hathor in Dendara of Tiberius (*EAT* III, pl. 42), that of Esna (B) from the time of Vespasian or Domitian (*EAT* III, pl. 43), that of the roof of the Temple of Khnum at Esna (A), originally dated to the time of Ptolemy III–V but probably from the 2nd century AD (*EAT* III, 62 and pl. 29; Quack, 2018b: 81–82), that of the missing temple of Triphis and Pan in Akhmin from the time of Trajan (*EAT* III, pl. 45), and that of the temple of Deir el-Haggar of the 2nd century AD (Kaper, 1995: 176 fig. 1). It is evident that the representation of these constellations in Egyptian temples is not only a confirmation of the introduction of these constellations of Mesopotamian origin into the Egyptian sky but, very probably, of the development and diffusion of astrology in the temples.

Already in Hellenistic times, Egypt had become the most important centre of astrological practice (Cramer, 1954: 18), and the temples had to develop an essential and preponderant role. This role played by astrology in the Egyptian temples during the last phase of Egyptian history transcends the information provided by the so-called *Book of the Temple*, partially and successfully reconstructed by Quack (2000a, 2003) from numerous fragments of papyri from the 1st and 2nd centuries AD, mainly from Tebtunis and Soknopaiou Nesos. It indicates how in the priestly hierarchy of the temple only the highest-level priests were familiar with astral divination and had to learn about omens linked to eclipses (Quack, 2002a: 168–169; Winkler, 2016: 270).

The origin of astrology has usually been attributed to Babylon. Sachs (1952: 53–54) postulated that between 600 BC and the beginning of the Hellenistic period a remarkable step was taken in Mesopotamia between the ancient astrology of the *Enuma Anu Enlil* and horoscopic astrology. As proof, he cites the invention of the zodiac, but also the cuneiform tablet AB 251, written in the time of Artaxerxes II, which provides the positions of the planets and the moon in 410 BC, with respect to the zodiacal signs at the time of birth of an individual. In total, 32 texts of this type, dated between 410 BC and the year 69 BC, are known (Rochberg, 1989: 102–103, 1998: 25–26), all of them predating the first known example in Egypt.

The documentary evidence clearly places the birth of astrology in Babylon, in the centuries after the fall of its last king, Nabonidus, a circumstance that may not be a coincidence. Contrary to the opinion of Pingree (1997: 20–24; Hunger & Pingree, 1999: 26–27), in some cases a greater connection between Babylonian astrology and the genethliacal horoscopic developed in Hellenistic times has been sought. Thus, for example, the cuneiform text of the tablet BM 36746 (dated in the fourth century BC), which follows the tradition of the *Mul-Apin*, shows, according to Rochberg (2010: 35–38), a series of concepts which, until that date, were known only in Hellenistic sources, thus establishing a connection between the omens of the *Enuma Anu Enlil* and horoscopic astrology. In fact, if we analyse the documentation, it seems that Hellenistic astrology combines various elements of Egyptian and Babylonian origin (Greenbaum & Ross, 2010: 153).

In spite of everything, on the question of origin there was a disparity of opinions among the Greek authors (Bouché-Leclercq, 1899: 51 n.1). Even Diodorus (I: 81) goes so far as to say that 'the Chaldaeans of Babylon, being colonists from Egypt, enjoy the fame which they have for their astrology because they learned that science from the priests of Egypt'.

The question we must address is when influences from Mesopotamian universal astrology or from the new natal astrology (developed at least from the middle of the first millennium BC) could have reached Egypt. In the 1st century BC Marcus Vitruvius (1874: 219, IX.7), the author of *De Architectura*, says of Chaldean astrology: 'The talent, the ingenuity, and reputation of those who come from the country of the Chaldeans, are manifest from the discoveries they have left us in writing. Berossus was the first of them. He settled in the island and state of Cos, and there established a school'. Hence, he points to Berossus, with his school on the island of Cos, as the transmitter of Babylonian astronomy/astrology to the Greek world. This island was for a time dominated by Ptolemaic Egypt, so there are those who have looked there for the beginning of this diffusion towards Egypt (Cooper, 2018: 387). However, there are authors who doubt the historicity of this version, in terms of both the stay in Cos and the very existence of the astrologer Berossus as the historian author of the *Babiloniaka* (Rochberg, 2009: 15 and n.9; Van der Spek, 2008: 288–289).

During the first Millennium BC Egypt was involved in major geopolitical changes that affected the Middle East. First, because of Assyrian pressure, which led to the taking of Memphis during the reign of Asarhaddon in 671 BC and Thebes during Ashurbanipal's reign in 664 BC at the end of the Nubian 25th Dynasty. During the invasion of Asarhaddon we know that the Assyrian monarch took seven astrologers (*tupshar*) to Egypt to accompany the king's entourage (Ulanowski, 2016: 77). On the other hand, cuneiform tablets from Nineveh, for example, indicate the presence of three Egyptian magicians (*hartibi*) and three Egyptian scribes on a list (ADD 851, K. 1276; Parpola, 2007: 456–457) of employees of the palace of Ashurbanipal which includes 45 specialists, among them were also Assyrian astrologers (Fincke, 2003/2004: 140 n.213). According to Zaccagnini (1983: 260), these scribes could have voluntarily marched to the Assyrian court and not necessarily have been simply prisoners of war, although in a text that refers to the transfer of people after the fall of Memphis, various Egyptian specialists are mentioned among the *hartibï* and scribes (Onasch, 1994: 32).

The presence of Egyptian scribes in the Assyrian court had been previously documented because we know that, in the time of Adadnirari III and Shalmanaser IV (first quarter of the 8th century BC), there were groups of Egyptian scribes there working with other Assyrian scribes and Arameans who, according to Wilson (1972: 62–63), were also scholars and specialists (possibly also dream interpreters) and not merely secretaries.

It is possible, therefore, that the arrival of Egyptian specialists at the Assyrian court and their subsequent return was a source of information transfer to consider. However, this transfer does not have to be seen only as one-way from Mesopotamia to Egypt.

The ivory prism fragment BM 123340 and the tablet fragments K 2077+3771+11044+ BM 54619 (dated 650 BC; Hunger & Pingree, 1999: 112–116), all dating from Neo-Assyrian times and found in Nineveh, provide information on the length of the day throughout the months, with the typical Egyptian 24-hour division of days and information for the 1st and 15th of the month, so they are proof, in this case, of that reverse influence, from Egypt to Mesopotamia.

The conflict with Assyria facilitated the installation of the Saite Dynasty in Egypt, initially indebted to the Assyrian power. Still on a literary and traditional level, let us remember how one of the kings of this Dynasty, Nechepsos (Nekau II), was later recognized as a great sage and astrologer. After the relationship with Assyria, it could undoubtedly have been during the first Persian rule that the transfer of astronomical knowledge and methods between Mesopotamia and Egypt was able to increase. The Persian world had since 525 BC (year of the Persian conquest of Egypt) unified the entire Near East under a single sceptre including, from 539 BC, Chaldean Babylon. Of course, the conquests of Alexander the Great would further facilitate these transfers.

The demotic papyrus Vienna D 6286 (Parker, 1959) of the 1st–2nd century AD (whose original Babylonian source could be dated between the Saite period and the beginning of the first Persian rule), seems to show the transfer of Mesopotamian astrological knowledge to Egypt. This is clear, given that it presents a relationship between Egyptian and Babylonian months. The text offers omens about solar and lunar eclipses that correspond to universal, judicial, or mundane astrology. In addition, a lacunar section of text A of the papyrus has been reconstructed as: 'Knowing the eclipses of the moon according to the new writings', which could precisely reflect the access to new foreign methods incorporated in Egypt. However, text B of the papyrus shows a whole series of peculiarities that seem to differentiate it from a clear Mesopotamian tradition. Omens made from the colour of the moon are described, with coloured vignettes (which obviously do not appear in the Mesopotamian tablet format), and other events related to the lunar disc (such as possible occultations), without being associated with specific months and in which Egypt is the protagonist. This format appears to show not a simple copy of Mesopotamian original but an adaptation to the Egyptian style and tradition (Altmann-Wendling, 2019: 233).

Also in demotic ostraca from the 1st century AD, such as some that are preserved in the Ashmolean Museum (where the calculation of an astronomical ephemeris referring to Mercury is explained), or in the well-known demotic Stobart Tablets (four tables written after 134 AD but with planetary ephemeris corresponding to the period between 71 AD and 132 AD: see Figure 2.22); *EAT* III, pl. 74), a Mesopotamian influence has been observed in the astronomical calculation procedure (Ossendrijver & Winkler, 2018: 382–419). Consider, for example, that the mere use of the sexagesimal system in Egyptian astronomical ephemerides of this time is a clear indication of Mesopotamian methods of calculation, since the traditional Egyptian number system was always decimal.

However, it must be borne in mind that in the astronomical/astrological documentation of the Graeco-Roman period there are not only clear indications of original Egyptian or Mesopotamian transfers dating back at least to Saite times. It should not be forgotten that the Carlsberg 1 papyrus (and other related papyri such as Carlsberg 1a, Carlsberg 228, Carlsberg 496, Carlsberg 497 and Oxford 79/105; Lange & Neugebauer, 1940; Quack, 2000b; von Lieven, 2007, 2010), being from the 2nd century AD, follows, even with commentaries, the text of the *Fundamentals of the Course of the Stars* that we know from three documents of a monumental nature, of which the oldest is located in the Osireion of Sethy I in Abydos, a document almost 1500 years older, which we know for certain partly provides astronomical information half a millennium older (Lull, 2004a: 125), perhaps even going back to the Old Kingdom (von Lieven, 2010: 140).

In contrast, the astronomical texts of the Ptolemaic or Roman period written in Greek do not necessarily have to derive in their entirety from Greek astronomy or be of Mesopotamian origin. It is evident that Greek was used by many Egyptian priests, and that this priestly elite could be bilingual, so even surviving texts in Greek may have been written by Egyptians and in an Egyptian cultural context. A possible example is offered by the pRylands IV 589 (formerly pRylands Inv. 666), from 180 BC, written in Greek. In it we read (Turner & Neugebauer, 1949: 94):

> Year 1 of Queen Cleopatra and King Ptolemy the son, gods Epiphaneis. Table of lunar new moons, showing how they are related to the days of the Egyptian twelvemonth. The period of the table is 25 years, 309 months (including intercalary months), 9125 days. It indicates the lunar months and which of them are full, which hollow, which intercalated; and in what sign of the Zodiac the sun will be during each month. When the sun has traversed the 25 years it will return to the same starting point and resolve in the same manner.

This papyrus provides the same information (the correspondence between 25 calendar years and 309 lunar months totalling 9125 days) as the demotic Carlsberg 9 papyrus (*EAT* III, 220–228), from the middle of the 2nd century AD (from Tebtunis), both based on the Egyptian civil calendar.

The lunar tablets on the Carlsberg 9 papyrus are the product of Egyptian mathematical methods and, in that sense, are inherited from the Egyptian tradition. The series of lunar cycles of this last papyrus seems to start around 357 BC according to Parker (1950: 23), although recently Depuydt (2016: 44–45) has denied this possibility on astronomical grounds and has proposed an origin of the cycle of AD 18 July 144, shortly before the making of the papyrus.

What this information was for has also been the subject of some disagreement. In the case of the Carlsberg 9 papyrus, the text indicates: 'Here is the procedure of enumerating the 25 years of the moon in order to make them known' (*EAT* III, 223). It could be considered as a useful tool for calculating, in the Egyptian civil calendar, when a lunation was going to begin and, based on this, organizing in advance religious festivals that had to do with certain lunar phases. All this would fall within the needs of the Egyptian priest. However, as in the Rylands IV 589 papyrus, there are also references to the zodiac, which falls within the new needs developed by Egyptian astronomical priests in Roman times. These astronomical tables could also have been used by Egyptian priests as a basis for astrological calculations.

Quack (2017: 199) warns that no Greek text in the Egyptian sphere referred to Mesopotamian models, while there are references to Egyptian models, such as the *lunarium* (Boudreaux, 1921: 105) composed from the one written by the hierogrammateus Melampous for king Nechepso and the one found in the temple of Heliopolis during the reign of Psametik. However, he also recognizes that the Egyptian material is not exempt from relationships with Mesopotamian because 'in some cases, even the possibility of adaptation or direct translation of cuneiform models must be seriously considered'.

It is precisely from these last phases of Egyptian civilization, when a much closer contact with Greek and Roman cultures is generated, that we receive a greater number of references from non-Egyptian authors about Egyptian astronomers or astrologers and about Greek sages who learned from the Egyptians during classical and Hellenistic times. These references serve the better to understand the importance of the work carried out by Egyptian astronomers and their true knowledge of the celestial vault, but also to verify how Egypt played a great role as a transmitter of knowledge, both its own and Mesopotamian.

The precedents of Greek sages studying or taking advantage of Egyptian knowledge, spending long periods in Egypt, are numerous. In these references it must be borne in mind that on many occasions the word 'astrology' actually refers to 'astronomy.'

Diodorus (I: 98), for example, describes how Democritus of Abdera (5th–4th century BC) 'spent five years among them (the Egyptians) and was instructed in many matters relating to astrology', how Oenopides of Chios (5th century BC) 'passed some time with the (Egyptian) priests and astrologers and learned among other things about the orbit of the sun, that it has an oblique course and moves in a direction opposite to that of the other stars' (that is, he learned from the Egyptians the obliquity of the ecliptic and the retrograde motion of the sun with respect to the stars), and how Eudoxus of Cnidus (4th century BC) 'studied astrology with them and acquired a remarkable fame for the great amount of useful knowledge which he disseminated among the Greeks'.

Later, the Cordovan Lucius Annaeus Seneca (*NḰ* VII: 3), who for a season lived in Alexandria, wrote in *Naturales Quaestiones* about the eclipses recorded by the Egyptians and the curious lack of information about comets:

> Eudoxus was, in fact, the first to import from Egypt into Greece the knowledge of these motions (of the planets), though he says nothing about comets. From this it becomes plain that, even among the Egyptians, the people that bestowed most care on observation of the sky, the portion of astronomy that relates to comets had not been worked out. Subsequently Conon, who was himself a careful investigator, made a record of the sun eclipses that had been observed by the Egyptians; but he made no mention of comets, though he would certainly not have omitted anything definite on the subject that he had learned in Egypt.

Of Eudoxus, on the other hand, we know that he used an astronomical observatory located in Kerkasore, located in Letopolis according to Goyon (1974: 135–147). According to Strabo (*Gr.* XVII: 1, 30), it was a kind of guard tower.

Democritus, Oenopides, and Eudoxus, renowned Greek mathematicians and astronomers, had to have taken advantage of the astronomical information (mainly solar, lunar, stellar, and planetary ephemerides) that the Egyptians had accumulated over many centuries. Even Plutarch (1st–2nd century BC), in *De Iside et Osiride* (10) indicates:

> Solon, Thales, Plato, Eudoxus, Pythagoras, who came to Egypt and consorted with the priests; and in this number some would include Lycurgus also. Eudoxus, they say, received instruction from Chonuphis of Memphis, Solon from Sonchis of Saïs, and Pythagoras from Oenuphis of Heliopolis.

Later, in the 3rd century AD, Diogenes Laërtius (I: 43) recalls a letter written by Thales (6th–5th century BC) to Pherecydes of Syros, in which he recounts how the Milesian met with Egyptian priests and astronomers to learn of their knowledge. Iamblichus (3rd–4th century AD), in his *Life of Pythagoras* (IV), also tells how Pythagoras of Samos (6th–5th century BC) 'spent therefore two and twenty years in

Egypt, in the adyta of temples, astronomizing and geometrizing, and was initiated, not in a superficial or casual manner, in all the mysteries of the Gods.'

The tradition, therefore, speaks of how renowned Greek mathematicians, astronomers and other wise men drank from the knowledge accumulated by the Egyptians. At the beginning of the Roman period in Egypt Strabo also visited the country of the Nile and in his *Geography* he expounded on how this knowledge continued to be transmitted thanks to the Greek translations of the (astronomical) treatises of the Egyptian priests:

> At that time the true year was unknown among the Greeks, as also many other things, until the later astrologers learned from the men who had translated into Greek the records of the priests; and even to this day they learn their teachings, and likewise those of the Chaldaeans. (*Gr.* XVII: 29)

This transmission is also detected in Graeco–Roman times in Greek astronomical texts. Thus, for example, the method of numerical calculation used by the Greeks in their planetary tables is the same as that used by the Mesopotamians, and is also present in demotic papyri. And the use of signs of demotic origin in Greek texts, such as zero, zodiacal signs, or signs to indicate the first and last visibility of a planet (Escolano-Poveda, 2018–2019) seem to show that the transference between Mesopotamian and Greek astronomy was developed through the Egyptians. Spiegelberg (1911; Ross, 2006: 153–155) already postulated more than a century ago the possible demotic origin of the signs used since the Middle Ages for zodiacal signs. Even the signs for Venus and Mars could have come from the demotic (Quack, 2018b: 78).

According to Neugebauer (1943: 121), the origin of Egyptian astrology could be traced in its first phase to the third century BC, in any case not before the Hellenistic period. The first proof of the presence of the zodiac in Egypt is the demotic ostracon Strasbourg D 521 (astronomically dated *c.* 250 BC although its composition could be from the 1st century BC), in which Egyptian months and zodiacal signs are associated (Ross, 2007: 13–14; Quack, 2018c: 66).

The first direct evidence of the use of the zodiac in Egypt, however, is provided by the Berlin papyrus 13146+13147 (Neugebauer et al., 1981), which contains an ephemeris for lunar eclipses between 84 and 73 BC with mention of the associated zodiacal constellations. For Holden (2006: 12), horoscopic astrology would not have evolved on its own while being subjected to external influences but would have been 'invented', during the 3rd–2nd century BC, by the Alexandrian circle of sages. Pingree (1997: 26), on the other hand, places Egyptian genethliacal astrology, also as an 'invention', between the end of the second century and the beginning of the first century BC. This definition cannot be correct, however, since it implies the denial of an even minimal evolution in which various cultural influences, traditions and methods had intervened. And, as we have already noted, in Egypt there are clear precedents for the use of the stars for divinatory purposes prior to the Graeco–Roman period, so the transition to astrology was not necessarily as abrupt as has usually been supposed.

On the other hand, within the various astrological practices one of the main groups were those linked to Sothiac omens (Jones, 1998; Hughes, 1951). In Greek works as late as the *Apotelesmatika* of Hephaestion of Thebes (1994: 53–58), 5th century AD, this astrologer indicates that it is a knowledge contributed by ancient Egyptian sages. An example of a Sothiac omen, undoubtedly the most properly Egyptian predictive technique, is shown in the demotic papyrus Cairo 31222 (2nd century AD, from Tebtunis), recently studied by Quack (2017: 200–201) after the original publication of Hughes (1951). Some excerpts are worth reproducing to see how these predictions were expressed:

> The prognoses of Sirius: (…)
>
> If she rises while Jupiter is in Sagittarius: The Pharaoh of Egypt will have power over his region. Enmity will occur against [him], and he will finally escape from it. Many people will rebel against the Pharaoh. Appropriate water is what will come to Egypt. Seed and grain will be high in money. An economy which is poor in money. The death of a god will occur in Egypt. A … will come up to Egypt, and it will go away again.
>
> If she rises while Mars is in Gemini: Some people will rebel [in] Egypt (and) in the country of the Assyrian. Pharaoh will set out against them with his army. He will destroy them. The sky will not be able to have much rain in the region of the Assyrian (…).

With the passage of time, and still maintaining various astrological methods and currents, it seems that three main aspects of astrology will be defined (Jones, 2007: 308), as transmitted by Hephaestion of Thebes, although all of them are already applied in the Egyptian papyri from the Graeco–Roman period. These were: (i) genethliacal astrology or horoscopic astrology (prediction of the character and destiny of a person through the interpretation of his horoscope, that is, of the positions of the sun, moon, and planets in relation to the zodiac and the horizon in the time of the person's birth), (ii) cathartic astrology, which indicates when a date is conducive or not for a general or specific activity depending on the position of the planets in the zodiac on that date, and (iii) universal or judicial astrology, which uses knowledge of the position of the planets in the zodiac, at the time astronomical events such as eclipses occur, to predict global situations.

In spite of the fact that, as we have seen, the origin of natal astrology comes from Babylon and clearly predates the Hellenistic period, it should be noted that in Egypt this type of astrology underwent a different development. The Mesopotamian documents linked to natal astrology do not show *cardines* (κέντρα *kentra* in Greek) in the information they provide. However, they already appear in the oldest of the Egyptian demotic documents linked to horoscopic astrology (oAshmolean DO 633, from the time of Cleopatra VII) and even in Tiberian Greek diagrams (as in pOxyrhynchus 235; (Neugebauer & Van Hoesen, 1987: 18 fig. 9).

The most important *cardines*, the ascendant and the *mesouranema*, have clear parallels with the ancient Egyptian astronomical tradition, in which the hourly decans showed their usefulness as such at the time of their rising on the eastern horizon, or in their transit across the central meridian (transit decans), as reflected in the diagonal star clocks, in the *Fundamentals of the Course of the Stars* and in the Ramesside star charts.

The word 'horoscopes' (ὡροσκόπος), which designates the ascending cardinal, literally means 'hourly marker'. Its association with the sense of the rising decan of the Egyptian tradition is obvious, so the latter is an antecedent. The Egyptian decans (δεκανός) thus found a new utility in this context of development, in Egypt, of horoscopic astrology.

The decans were not eliminated from Egyptian iconography during the Ptolemaic period, even in spite of the introduction of the zodiacal constellations. We see them, from the middle of the 1st century BC, in the Dendara zodiac configuring an outer circle of 36 decans, but also defining three parts of each zodiacal constellation, as in Esna A, configuring the division into 10° segments.

Greenbaum and Ross (2010: 157) recalled how the Egyptian astrologer Anubius (and other authors of the 1st and early 2nd century AD) refers to the 36 decans as the ὡρονόμοι or 'hour-regulators', not as the ὡροσκόποι or 'hour-markers'. Thus they deduced that this may be an indicator of a transition in the astrological development of the decans because 'the change in terminology reflects this transition from an hour-regulating decan to an hour-marking ascendant'.

The ὡροσκόποι or rising decans ('hour-markers') facilitated the appearance of the concept of ascendant or ὡροσκόπος so important in horoscopic astrology. In some papyri, moreover, the decan is completely assimilated into the horoscope. The most representative example is that of the London 98 papyrus, from the year 95 AD, written in Greek and with a part in Coptic. It speaks of two sets of decans: the decans and the '36 bright horoscopes' (36 λάμπροι ὡροσκόποι) that, according to Neugebauer and Parker (*EAT* III, 173; Greenbaum & Ross, 2010: 158–162), are associated with the transit decans (such as the so-called Sethy IB family) and the rising decans (Sethy IC decans family), respectively.

Another possible Egyptian contribution to astrology is the 'house' system (*oikoi*) (Quack, 2018b: 92), so related to the *Thema mundi*, since the houses define the place where the planets were at the time of the birth of the world and the *Thema mundi* associates a date with Cancer at the time of the heliacal rising of Sirius, the Egyptian Sopdet.

A final aspect of both astronomy and astrology that should be noted in Egypt, is the opacity that in many cases surrounds this knowledge. In fact, there are many classical authors who recall the secret nature of the knowledge of the Egyptian priests and astronomers. Thus, for example, Strabo (*Gr.* XVII: 29) wrote:

> In Heliopolis I also saw large houses in which the priests lived; for it is said that this place in particular was in ancient times a settlement of priests who studied philosophy and astronomy; but both this organisation and its pursuits have now disappeared. (...) However, at Heliopolis the houses of the priests and schools of Plato and Eudoxus were pointed out to us; for Eudoxus went up to that place with Plato, and they both passed thirteen years with the priests, as is stated by some writers; for since these priests excelled in their knowledge of the heavenly bodies, albeit secretive and slow to impart it, Plato and Eudoxus prevailed of them in time and by courting their favour to let them learn some of the principles of their doctrines; but the barbarians concealed most things.

There are also references to how complex it could be to understand the writings of some of the most famous astrologers who, let us not forget, in many cases cannot be dissociated from those we define as Egyptian astronomers. Vettius Valens of Antioch, an astrologer who lived a long time in Egypt to learn from the knowledge of Egyptian astronomers/astrologers, wrote in the third quarter of the 2nd century AD a ten-volume work in Greek, *Anthologiae*, one of the most complete surviving treatises on astrology from that time. In his seventh book (260P, 128) he refers to this obscurantism:

> I cannot tell whether the Ancients, although knowing the efficacy of forecasting, were driven by envy to hide this art because of their vainglory and because the human mind finds this art difficult; or whether they spoke in such riddles even though they had not, in fact, grasped what Nature had created, had prescribed, and had bestowed abundantly on mankind after sealing it with Fate. Of all the lovely elements of the numerous great creations in the world, none seems to me to have been begrudged by God for man's daily use. God would not have revealed it if He had not wished to provide it for use. In contrast, men have revealed this art only as they wished or as they were able. As a result, when I read over their chapter (on the following topic), I wonder at the crookedness and the obscurity of their thought. But I reveal whatever I have discovered by my experience, and in addition, I do not wish to conceal whatever I have gone on to discover (after writing my previous pages) this because of the remarkable quality of many forecasts, both good and bad, those happening in a short time or those remaining for a time at a steady state.

The evolution of the Egyptian astronomers, from their first documented beginnings in the Old Kingdom to the Graeco–Roman period, is of the greatest interest, despite the fact that the information on them available in the period covered by the Pharaonic period is certainly limited.

Furthermore, regardless of the development that astrology might have had outside the Egyptian clergy, we must not forget that many of the documents that we have mentioned come from Egyptian temple libraries or from the decoration in the temples themselves. This inevitably implies that, in the milieu of the Egyptian temples, and in the hands of priests, astrology found a focus of development and diffusion. As the numerous astronomical and astrological demotic texts, that have not yet been studied, are translated, the true importance of the role played by Egyptian astronomers and the originality of the contributions of the Egyptian astrologers that had at their disposal indigenous and external astronomical knowledge will possibly be verified.

Chapter 3
Measuring Time: Ancient Egyptian Star, Water, and Solar Clocks

In ancient Egypt, various methods were devised to measure the passage of time by day and by night. Most of the methods were astronomical, using either certain hourly stars at night or the sun during the day as a guide. But in addition to astronomical, stellar, and solar clocks, the Egyptians devised other mechanisms, such as water-clocks or clepsydras.

The clocks are based on the Egyptian premise that the 24-hour day is divided into two invariable halves, 12 hours for the night and 12 hours for the day, regardless of the season of the year. That is, although the Egyptian day is 24 hours long, its hours do not have the same duration, there will be some being more than 60 minutes in duration and others less, depending on whether the actual duration of the night or day is longer or shorter, according to the season of year. Egyptian clocks, therefore, had to adapt to their system of hours of variable duration.

3.1 Star Clocks

In Egypt several types of star clocks were devised to be able to measure the hours of the night. They all have in common the use of a series of reference stars, the so-called decans, which throughout the night and the year served to announce the hours. However, not all star clocks use the same method to measure hours. We can distinguish two main groups: that whose method consists in observing the rise of the hour star, as in the so-called diagonal star clocks, and those in which what guides the horologist is the transit of the hourly star through the central meridian, that is, its culmination, as in the Ramesside star clocks. This is essentially the idea that Neugebauer and Parker developed in their *Egyptian Astronomical Texts*. However, authors such as Depuydt (2010: 245) consider that these lists are only tables that represented the movement of the stars throughout the year and 'that the star tables never served as clocks'.

J. A. Belmonte, J. Lull, *Astronomy of Ancient Egypt*, Historical & Cultural Astronomy, https://doi.org/10.1007/978-3-031-11829-6_3

3.1.1 Diagonal Star Clocks

These are the oldest star clocks that we know of, the vast majority being embedded in the interior of decorated coffins lids, so that in terms of iconography, their stellar and celestial content is spatially opposed to the Netherworld represented by the texts and decoration of the coffin box (Willems, 1988: 237). Their chronology dates back mainly to the First Intermediate Period and the 12th Dynasty, and although they have been found in both Middle and Upper Egypt, many of them come from the Asyut area.

Since Neugebauer and Parker (*EAT* I: 1–32, pl. 1–23) catalogued the twelve known specimens in coffins in 1960, the number has gradually increased (Locher, 1998) to 27 specimens, although in many cases these are small fragments (Symons et al., 2013), and the full twenty are not actually all in terms of design. As an exception to the usual chronology of these clocks, there is part of one in the decoration (in this case from the King Merenptah period) of the cenotaph of Sethy I in Abydos (*EAT* I: 32–35), although it presents another format and, in addition, indicates the name of the hours (*rn n wnwt*), such data being ignored in the other examples.

The first to recognize the antiquity of these lists of decans was Daressy (1900), although Pogo (1932) was the first to observe the calendrical and hourly connotations (with the rise of the decans) of what he defined as 'diagonal calendars'. However, it would fall to Neugebauer and Parker to develop the hypothesis that these tables of decans are to be understood as star clocks. According to Parker (1974: 53) the use of stars as time markers could be traced 'soon after the introduction of the civil calendar'.

Doubt was expressed by both authors concerning whether the rising of a star indicates the beginning or the end of an hour, although it seemed more likely that 'the ends of hours were indicated since the first hour of night would begin with darkness after sunset and no star's rising would be necessary to bring it into being (…) and the heliacal rising would end the 12th hour' (*EAT* I: 106)—a hypothesis followed by other more recent authors (Gadré & Roques, 2008a). According to this hypothesis, 'The 1st hour on the star clock was then of indeterminate length since it began with darkness and ended only when a particular decanal star rose in the eastern horizon. This 1st hour would be longest at the beginning of a decade and would be a little bit shorter each night. At the other end the decanal star rising heliacally and so marking the end of the 12th hour and of night was on the 1st day closely followed by morning twilight. At the end of the decade the 12th hour would be followed by a period of darkness' (Parker, 1974: 55). On the other hand, assuming the importance of the heliacal rising of Sirius as the original example that led to the use of the stars as time indicators, 'years and decades were thus to begin in the morning (…) in line with the Egyptian counting of the days from sunrise' (*EAT* I: 107).

The diagonal star clock is shaped like a table in which series of twelve stars or groups of hourly stars, the decans, are ordered for each decade of the year and the epagomenal days. In its ideal format (see Figure 3.1), a diagonal stellar clock is made up of 40 columns, of which 36 correspond to the decades of the year, until

Epag.				IV shemu			III shemu			II shemu			I shemu			IV peret			III peret			II peret			I peret			IV akhet			III akhet			II akhet			I akhet			
40	39	38	37	36	35	34	33	32	31	30	29	28	27	26	25	24	23	22	21	20	19	18	17	16	15	14	13	12	11	10	9	8	7	6	5	4	3	2	1	Hours
Decans																						Decans																		
E1	25	13	1																									12											1	1
E2	26	14	2	E1																									12										2	2
E3	27	15	3		E1																									12									3	3
E4	28	16	4			E1																									12								4	4
E5	29	17	5				E1																									12							5	5
E6	30	18	6					E1																									12						6	6
E7	31	19	7						E1																									12					7	7
E8	32	20	8							E1																									12				8	8
E9	33	21	9								E1																									12			9	9
E10	34	22	10									E1																									12		10	10
E11	35	23	11										E1																									12	11	11
E12	36	24	12	E11	E10	E9	E8	E7	E6	E5	E4	E3	E2	E1	36	35	34	33	32	31	30	29	28	27	26	25	24	23	22	21	20	19	18	17	16	15	14	13	12	12

Fig. 3.1 Ideal design of a diagonal star clock. (Adapted from Lull, 2004a: 111)

completing a period of 360 days. Of the four remaining columns, the first three are simply a list of all the ordinary decans used in the clock, so these columns are not useful for measuring hours. The last of the columns contains the complete series of triangular decans, which are actually used by the horologist to determine the hours of the night during the epagomenal days that complete the Egyptian calendar year of 365 days.

In each column of the table there are twelve rows in which a list of twelve stars or groups of stars is shown, so it follows that the first in the list is the one that marks the first hour of the night and the last the twelfth hour of the night.

It has usually been considered that the hourly stars or decans of the diagonal star clocks marked the time of their rising, so that the first hour of the night would be indicated by the decan that appeared on the eastern horizon shortly after sunset, while the last hour of the night would be marked by the decan whose rising preceded dawn. Since the average solar day is $3^{m}56^{s}$ longer than the sidereal day, on each passing day, the stars rise approximately four minutes earlier than the day before, so that after a decade of ten days the decans marked their positions in a row denoting the previous hour; that is, the one that marked the twelfth hour goes on to mark the eleventh hour, and so on. For this reason each decan in the table seems to define a diagonal line, each decade marking a previous hour, which is why these types of clocks are known as diagonal star clocks.

The use of the rising of the decans on the eastern horizon for hourly purposes seems to be clearly defined in much more recent texts. Thus, for example, it can be understood from the reading of a text (text 406 in Sauneron, 1969: 9–10) present in an architrave of the temple of Esna (Quack, 1995: 100):

> *Heil euch, 'lebende Götter', Dekane in der Nekropole, die in der Dunkelheit leuchten, die die Stunden terminieren bei ihrem Kommen im Osten, die ankündigen, was geschieht, die nach ihrem Wunsch am Leben erhalten und Frevler töten, die am Himmel erscheinen, die die Länder mit Feuer beschießen, bei deren Hervorkommen jedermann zittert, deren Zeiten die Astronomen beobachten.*

> Hail, ye "living gods", decans of the necropolis', who shine forth in the darkness, who time the hours at their rising in the east, who announce what happens, who grant life and smite the wicked, who shine in heaven, who baste the land with fire, at whose sight all tremble, whose time astronomers mark.

However, Leitz (1995: 67) postulated that it must be the setting of the star that determined the counting of the passage of the hours of the night, the decans at the beginning of a decade showing the end of the respective hour of the night and not the beginning of it. As we shall see, in many coffins Sirius marks the 12th hour of the night in column 18, corresponding to II Peret 21–30. While for Neugebauer and Parker this would indicate the heliacal rising of Sirius and the immediate days, for Leitz it means that at that time Sirius had set before the new day began. In these clocks Sirius later appears as a marker of the first hour on II Shemu 11–20, on the supposition that, at the end of that decade, the star has had its heliacal setting (the 'moment', that is, when the star is seen for the last time in the western sky before entering into conjunction with the sun). At the end of that decade the star would enter into conjunction with the sun, according to the Egyptian belief reflected in the *Fundamentals of the Course of the Stars* or in the pCarlsberg I. After this, the star would remain for 70 days in the Duat; that is to say, in conjunction, following which the heliacal rising of Sirius would take place in I Akhet 1. Another option that has been considered is that there are two methods of operation, those clocks based on the rising of the stars and those based on their setting (Symons, 2007) or even culmination (Symons, 2015: 1498).

Leitz's hypothesis leads him to recognize that these coffins are based on an ideal civil calendar in which the heliacal rising of Sirius always occurs on I Akhet 1. It would indeed be a calendar that would only play its role in the religious and funerary sphere, thus not being affected by the cumulative lag to which the true civil calendar was subjected. This last consideration leads us to a problem that has been discussed concerning these clocks: their real usefulness. Authors such as Detoma (2014: 118) also consider that 'the stars could have been used to guide the deceased in his nocturnal journey in the Netherworld to the resurrection at dawn'.

On the other hand, Symons and Tasker's view is that the tables serve more as almanacs or charts than as practical clocks (2015: 36). In this sense, Depuydt considers that these tables 'are just records of celestial observation and not clocks, the hours can easily be interpreted as an integral and necessary part of the act of observation' (Depuydt, 2010: 245).

Taking into account that after forty years of use (owing to the lag between the Egyptian civil year and the solar year) an ordinary decan would mark an earlier hour, or that in half that time the series of triangle decans would have already been affected, this time interval would therefore determine the useful life of a diagonal star clock. In order to remain functional, the decans therefore needed to be rearranged or new ones introduced to adapt to the time lag. With a set of 73 decans the clocks could have been adapted for years.

Besides, these clocks have numerous errors in their manufacture because they were incomplete, they repeated series of decans, or some were omitted. Although Neugebauer and Parker proposed a grouping of coffins according to their characteristics (tradition and supposed archaeological antiquity), making the Coffin 1 group the oldest, other more modern analyses, such as those based on textual criticism, have offered different results, leading to the conclusion that the closest coffin to the original archetype would be S3P or EAT's coffin 9 (Kahl, 1993: 101).

However, we must never lose sight of the fact that the specimens that we know about all come from the funerary world, where what is paramount is not the precision of the clock (which would not be used by real horologists), but its inherent symbolism, as it would enable the deceased know the hours and the paths marked by the stars, knowledge that would evidently be useful in the context of the afterlife.

The complete and functional tables, of course, were to be used by horologists. They were probably on papyrus. Unfortunately, none of these tables of stars, useful for determining the hours of the night throughout the year, has come down us. The fact that the specimen of the cenotaph of Sethy I clearly indicates the name of the hours (Murray, 1904: pl. 12) must be taken into account when understanding these tables as clocks. In that example, the following hours are mentioned (*EAT* I: 32):

I Peret, the name of the hour
Her first hour of evening
Her second hour in early night
Her third hour in early night
Her fourth hour in early night
She creates middle night
Her beautiful hour in middle night

The first hour of the night was necessarily shortened in duration as the days of the decade passed, since its rising each day is a few minutes earlier. The first hour of the night ends when the decan of the second hour rises, and so on. On the other hand, the star marking the last hour of the night, at the beginning of the decade, marks its heliacal rising. Being a horizontal observation, the conditions of the atmosphere and the pollution of the horizon undoubtedly affected the horologist's ability to observe the outgoing decan at the expected time or later.

The total number of decans used in the table is 48 (although 68 are known; Symons, 2007), which have conventionally been classified as 36 ordinary decans (numbered 1–36) and twelve triangle decans (from A to M, not counting the I, or E1 to E12 or ED1 to ED12). However, although the tables do not show a different treatment between them, the appearance, separately, of ordinary decans in other types of astronomical lists does show this distinction.

The set of diagonal star clocks can be divided into two main groups based on the configuration of the decans presented in their tables. In the first group (called T) the first decan of the series is *ṯmꜣt ḥrt* or *ṯmꜣt ẖrt* while in the second group (called K) it is *Knmt* or *tpj-ꜥ Knmt*. Group T seems to be older than group K. For Symons (2015) the existence of two groups, T and K, with small variations, could be related to the use of two different observing methods.

As noted above, the majority of well-preserved diagonal star clocks do not reveal their full design, evident proof of their symbolic rather than functional meaning. One such example is that shown on the coffin of Idy, commander of the Egyptian fleet. This coffin is kept in the museum of the castle of Tübingen (Germany) (Pogo,

Fig. 3.2 Diagonal star clock belonging to Idy. ÄSUT Inv. Nr. 6. © Ägyptische Sammlung Universität Tübingen. (Photograph by courtesy of F. Kuyas)

1936a: 22–26, pl. 2; Brunner-Traut & Brunner, 1981: 209–226, pl. 47; Lull, 2004a: 111–120; Gautschy, 2020a: 43–53), and is the one that will serve as an example, since this model, in terms of the arrangement of several elements, is repeated in several copies.

The lid of Idy's coffin is 2 metres long and almost half as wide (see Figure 3.2). The clock was painted on a thin layer of stucco on the inside, although it is incomplete, since only eighteen of the 40 columns were represented. Following the longitudinal axis, the columns are divided by a line of text consisting of a formula for offerings in favour of the deceased:

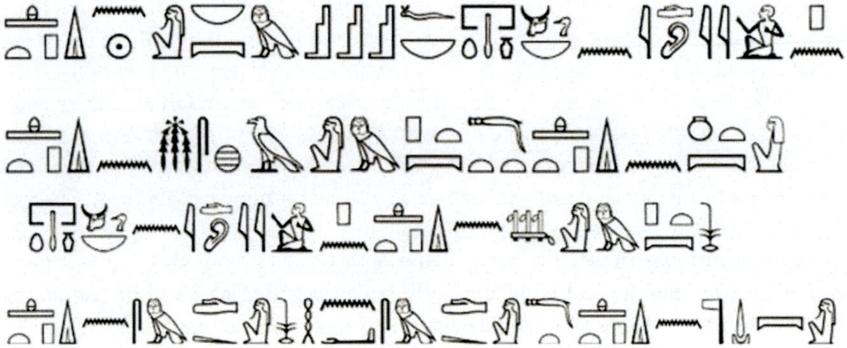

ḥtp dj.n Rʿw nb pt m swt.f nb(wt) prt-ḫrw t ḥnḳt k3w 3pdw n jdy pn
ḥtp dj.n msḫtjw m pt mḥtt ḥtp dj.n Nwt
prt-ḫrw t ḥnḳt jḥw 3pdw n jdy pn ḥtp dj.n s3ḥ m pt rsjt
ḥtp dj.n smd rsj ḥnʿ smd mḥt ḥtp dj.n nṯr d̠3w pt

Offering that *Re*, lord of heaven, has given in all its places, a funerary offering composed of bread, beer, bovids and birds for this Idy.
Offering that Meskhetyu has given in the northern sky. Offering that Nut has given,
Funerary offering consisting of bread, beer, bovids and birds for this Idy. Offering that *šʿh* has given in the southern sky.
Offering given by southern *Semed* and northern *Semed*. Offering given by the god who crosses the sky'.

This inscription reminds us that the document we are describing belongs to the funerary sphere, and that, in the celestial context, various deities had to be taken into account in the formula for offerings. Re, the 'Lord of the sky', but also Meskhetyu, an Egyptian constellation that grouped the main stars of the Big Dipper and Nut, the celestial goddess, are named for this reason. Also mentioned are Sah (partially equivalent to our constellation Orion), and the southern and northern Semed decans. The 'god who crosses the sky' is the epithet for a deity, possibly Sah.

At the beginning of the left half of the cover, after column 11 of the diagonal clock appears the only iconography. Four figures are shown accompanied by hieroglyphic inscriptions. The one on the left is Nut, who raises her arms holding the *pt* sign. It is just what the accompanying text describes: *Nwt f3j ʿwj.ṯ* ('Nut, raise your arms'). To the right of Nut is shown the bull's leg, Meskhetyu. In this case, the text reads: *Msḫtjw m pt mḥtt* ('Meskhetyu in the northern sky'). The following figure represents Sah, with his typical three-pointed crown reminiscent of the three stars of Orion's belt. Sah carries in his hands a *was* sceptre and the *ʿnkh* symbol. The inscription reads: *S3ḥ m pt rsjt S3ḥ pẖr ḥr.k r m33.k Wsjr* ('*sʿh* in the southern sky. *sʿh*, turn your head so that you can see Osiris'). Finally, next to Sah appears the goddess Sopdet, the star Sirius. The accompanying text says *Spdt dj.s ʿnḫ* ('Sopdet, she gives life'), in memory of Isis, with whom Sopdet is associated, because through her magic she brought the late Osiris (one of whose forms is Sah) back to life (see Figure 3.3).

All the other inscriptions that appear on the lid refer exclusively to the list of hourly stars, distributed by decades. Each column is headed by an inscription indicating the decade of the year for which the decans listed below are valid. Thus, the first column of the Idy clock, on the right edge of the inside of the lid, would be marked by the inscription *sw-mḏ tpj 3bd I 3ḫt* ('first decade of the first month of Akhet'). The next column is indicated by the inscription *sw-mḏ ḥrj-jb* ('middle decade'); that is, the second of the three decades of the first month of Akhet'. In the next column for that month, finally, the heading text says: *sw-mḏ pḥwj* ('final decade'). The same procedure is repeated in the following columns, so that only in the first decade of each month is the season and month number indicated (e.g. first decade of II Akhet', first decade of III Akhet', etc.). Only in the case of the Heqata and Iqer clocks (Detoma, 2014: 136; coffins 7 and 8 in *EAT* I: 26) the column corresponding to triangle decans is preceded by the inscription *rnpt 5 ḥrw*, 'five (days) upon the year'.

In Table 3.1 the reading direction is from right to left and from top to bottom. As we can see, each decan has a star on its left. The first decan of the right column, corresponding to the early evening in the first decade of I Akhet is *ṯm3t ḥrt*. In the following decade, it no longer worked as an hourly decan, since the one that continued to set the second hour of the night, *ṯm3t ẖrt*, started to mark the first hour in the second decade of the month. As explained earlier, from decade to decade, the decans start to mark an hour before, reflecting a diagonal in the table.

One of the hourly stars that appear in diagonal star clocks is Sirius, the Egyptian Sopdet. If the star was observed at its heliacal rising, as a marker of the twelfth hour of the night at the beginning of a decade, this data could then be used to calculate the

Fig. 3.3 Diagonal star clock belonging to Nakht, from Asyut. https://commons.wikimedia.org/wiki/File:Digaonalsternuhr2.jpg. (Author: NebMaatRa. Public domain, via Wikimedia Commons)

Table 3.1 The decans of the first three columns of the Idy clock

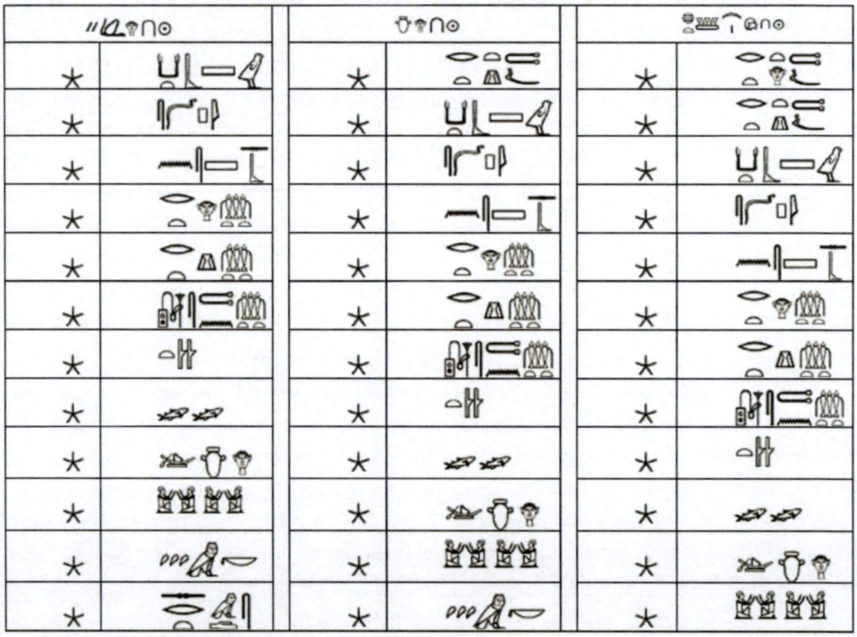

chronology of the coffin or, rather, of the design of the clock used. In Idy's coffin the star Sirius appears in column 18 as a marker of the 12th hour of the final decade (days 21–30) of the second month of the Peret season (II Peret). This implies a 140-day lag from the first Sirius *apokatastasis* on I Akhet 1, which is equivalent to about 680 years from the initial use of the Egyptian civil calendar sometime around 2782 BC (not correcting the Sothian cycle of 1460 years). That is to say, that position of Sirius in the coffin of Idy indicates that the design of the clock could be placed around 2102 BC. If the observation was made from Asyut and not from the Delta, this chronology could still be lowered by a few years, since for a latitude of 27.2° and an *arcus visionis* (β, see Glossary) of 9.4°, Sirius could have had its heliacal rising on II Peret 21 around 2083 BC, or around 2079 BC with an *arcus visionis* of 8.6°.

Sirius appears in the same position in the coffins catalogued by Neugebauer and Parker as coffins 1, 2, 3 (all of them from Asyut), 7 (from Gebelein), 8 (from Aswan), and in others catalogued later by Symons, such as A16 from Asyut (Symons, 1999: 18–26). In the Theban coffin A6, belonging to Ashayt, one of Mentuhotep II's wives (therefore a perfectly datable coffin from the historical point of view), Sirius does not appear as the 12th hour marker in column 18, but it does appear as the 11th hour marker in column 19, the 10th hour in column 20, and so on (except in two columns), until marking the first hour in column 29. Therefore, we may understand that also in this clock Sirius should have appeared pointing to the 12th hour in column 18. It is probable that both Idy's clock and that of Ashayt obey to the same reference model. But if Ashayt's was made taking into account the Theban sky, at latitude 25.7°, it could be dated to 2077 BC or 2073 BC, depending on whether it is β 9.4° or β 8.6°. Recall that, according to the chronology proposed in *Ancient Egyptian Chronology* (2006), the long reign of Montuhotep II would have started in 2009 BC, but if the Illahun Sothic date and the same latitude were taken, this heliacal rising of Sirius in II Peret 21 would indicate the early years of Intef II's reign, although we know that Ashayt died some 75 years later (Krauss, 2006c: 447).

In the coffins, Sirius appears as a marker of the 12th hour between II Peret 21 and IV Peret 21, which implies a chronology with a breadth of 240 years. However, it must be taken into account that these clocks show irregularities since, for example, in the A1–8 two decans are missing prior to the appearance of Sopdet (*EAT* I: 31), so if the table had been complete, it would have moved its position to III Peret 11, which would lead to a reduction of the previous chronology by 80 years, in any case still within the chronology of the 11th Dynasty. In the clock of coffin A9 Sirius appears in III Peret 1, so it could go back to the Heracleopolitan period (9th–10th Dyn.) and in A12 (coffin 12 of *EAT*) Neugebauer and Parker, with corrections, postulated a possible appearance of Sirius on IV Peret 21, which would take us to a 12th Dynasty chronology. These last two coffins would therefore set the chronological span of these clocks between the 9th and the 12th Dynasty. However, based on stylistic arguments, Zitman (2010) considers that the coffins must be dated to the early Middle Kingdom. In fact, Idy's coffin was dated by Brunner (1981: 209) to the middle of the Middle Kingdom. Another proposal is the one offered by Gadré and Roques (2009), who frame the chronology of the coffins (1–20) between 2101 and 1862 BC.

In any case, Neugebauer and Parker (see Table 3.2) suggested the possibility that the original clocks could reach as far back as the years 2830–2780 BC, considering

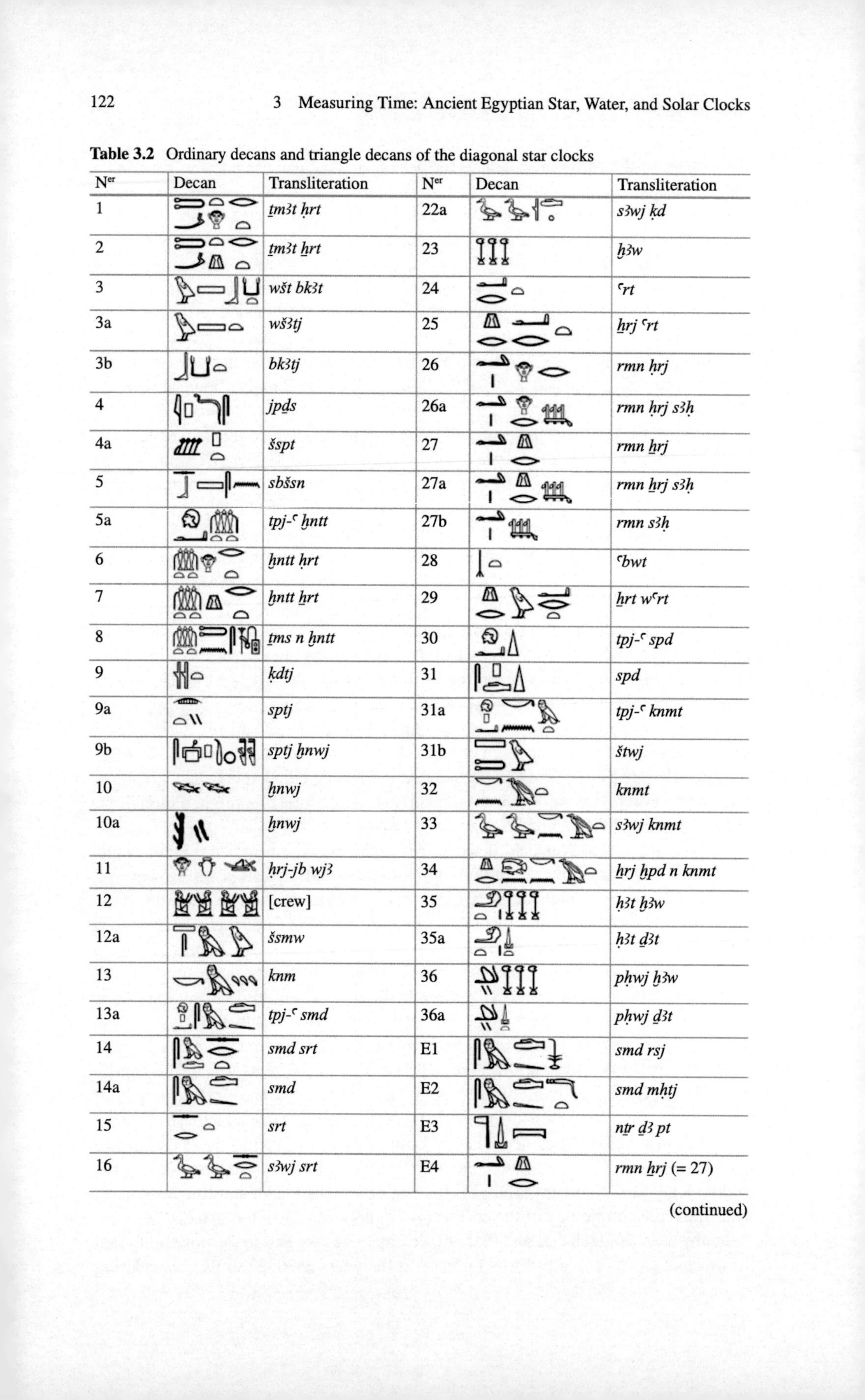

Table 3.2 Ordinary decans and triangle decans of the diagonal star clocks

N^er	Decan	Transliteration	N^er	Decan	Transliteration
1		*ṯmꜣt ḥrt*	22a		*sꜣwj ḳd*
2		*ṯmꜣt ẖrt*	23		*ḫꜣw*
3		*wšt bkꜣt*	24		*ꜥrt*
3a		*wšꜣtj*	25		*ẖrj ꜥrt*
3b		*bkꜣtj*	26		*rmn ḥrj*
4		*jpḏs*	26a		*rmn ḥrj sꜣḥ*
4a		*šspt*	27		*rmn ẖrj*
5		*sbšsn*	27a		*rmn ẖrj sꜣḥ*
5a		*tpj-ꜥ ḫntt*	27b		*rmn sꜣḥ*
6		*ḫntt ḥrt*	28		*ꜥbwt*
7		*ḫntt ẖrt*	29		*ẖrt wꜥrt*
8		*ṯms n ḫntt*	30		*tpj-ꜥ spd*
9		*ḳdtj*	31		*spd*
9a		*sptj*	31a		*tpj-ꜥ knmt*
9b		*sptj ḫnwj*	31b		*štwj*
10		*ḫnwj*	32		*knmt*
10a		*ḫnwj*	33		*sꜣwj knmt*
11		*ḥrj-jb wjꜣ*	34		*ẖrj ḫpd n knmt*
12		[crew]	35		*ḥꜣt ḫꜣw*
12a		*šsmw*	35a		*ḥꜣt ḏꜣt*
13		*knm*	36		*pḥwj ḫꜣw*
13a		*tpj-ꜥ smd*	36a		*pḥwj ḏꜣt*
14		*smd srt*	E1		*smd rsj*
14a		*smd*	E2		*smd mḥtj*
15		*srt*	E3		*nṯr ḏꜣ pt*
16		*sꜣwj srt*	E4		*rmn ẖrj* (= 27)

(continued)

Table 3.2 (continued)

N^er^	Decan	Transliteration	N^er^	Decan	Transliteration
17		*ẖrj ẖpd srt*	E5		*ḫꜣw*
18		*tpj-ꜥ ꜣḫwj*	E6		*tpj-ꜥ spd* (= 30)
19		*ꜣḫwj*	E7		*jmj-ḫt spd*
20		*jmj-ḫt ꜣḫwj*	E8		*ꜣḫwj* (= 19)
21		*bꜣwj*	E9		*ḫꜣw* (= 23)
21a		*ḫntw ḥrw*	E10		*nṯr ḏꜣ pt*
21b		*ḫntw ẖrw*	E11		*sꜣbw*
22		*ḳd*	E12	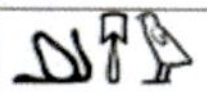	*pḥwj sꜣbw*

The numbering corresponds to Neugebauer and Parker

that they would have included Sirius, at its heliacal sunrise, at the end of the first decade of the year (*EAT* I: 111).

3.1.2 *Decanal Transit 'Clocks'*

Star clocks based on the rising (or setting) of the stars have the disadvantage of depending on the conditions of the atmosphere on the horizon. Dust in suspension, a common occurrence in Egyptian territory, could affect the accuracy of this type of observation. This must have been a probable reason why the Egyptians designed a new star clock system, this time focusing on observation of the culmination of the stars and, therefore, in more favourable conditions with the optimum position of the star in the sky.

Decanal transit 'clocks' possibly began to be developed during the Middle Kingdom and used practically the same decans as in diagonal star clocks. The oldest that have been preserved are those that were reflected as part of the *Fundamentals of the Course of the Stars* (or *Book of Nut*) in the cenotaph of Sethy I (Frankfort, 1933: pl. 81; *EAT* I, pls. 30–33) in Abydos (see Figure 3.4) and on the ceiling of the burial chamber of the tomb of Ramesses IV in the Valley of the Kings (Brugsch, 1968: 174–175; *EAT* I: pls. 34–35; Hornung, 1990: pl. 68–69).

However, some excerpts from the book also appear later in monumental contexts, such as in an abbreviated form in the tomb of Mutirdis (TT 410) (Assmann, 1977: 85–88, pl. 39), and also a small excerpt in the tombs (Régen, 2015) of Montuemhat (TT 34), Pedamenope (TT 33), and Pabasa (TT 279), all of them from the 26th Dynasty and located in the Asasif necropolis, and even in the Ptolemaic temple of Athribis. However, the most complete versions with commentaries (even without iconography) appear in the papyri of the Roman period Carlsberg I and Ia (*EAT* I, pls. 36–46). In the iconography of the *Fundamentals of the Course of the Stars*, the arched figure of the goddess Nut stands out, resting her feet and hand on the ground, as well as a figure of Shu on a smaller scale, holding the goddess.

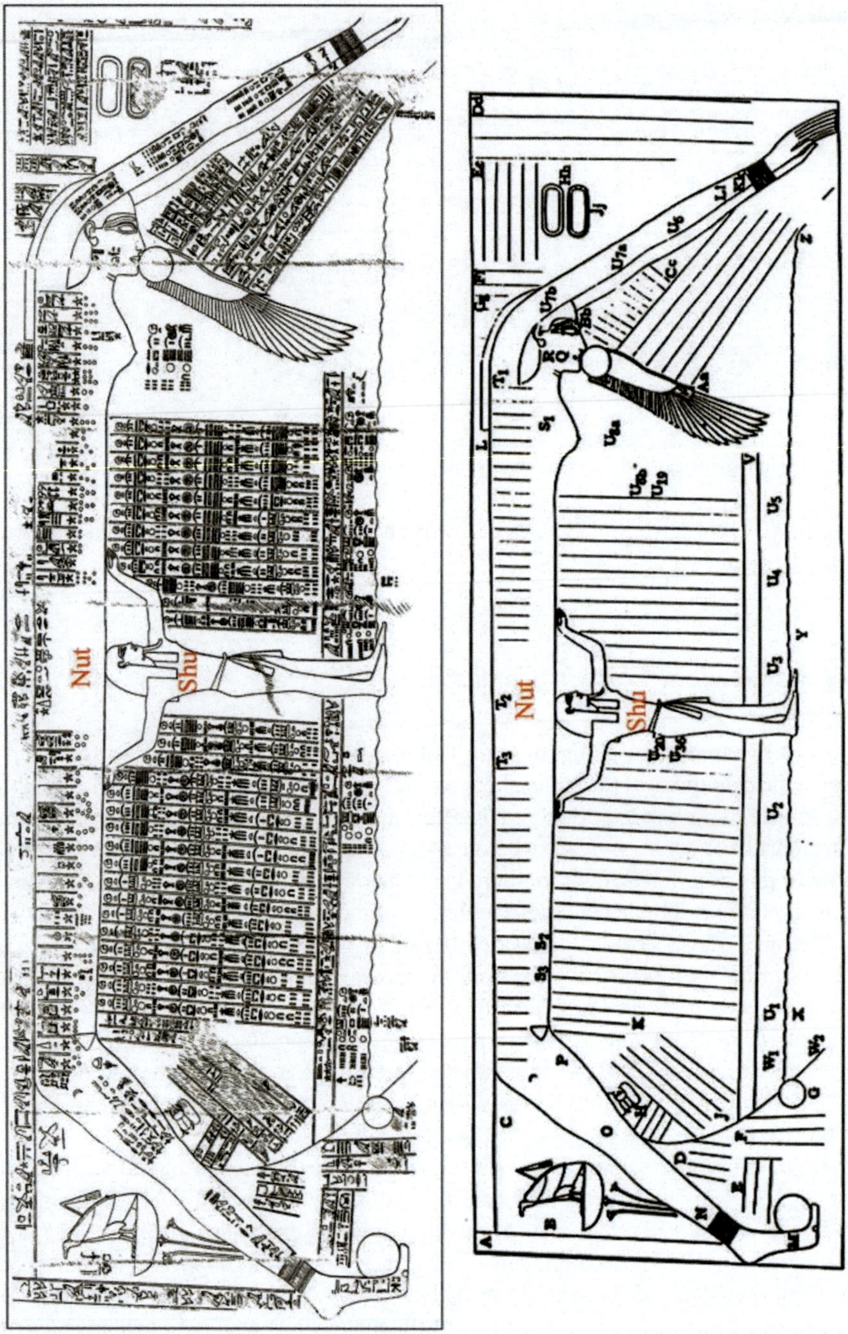

Fig. 3.4 *Fundamentals of the Course of the Stars* or *Book of Nut* in the cenotaph of Sethy I at Abydos (left), and distribution of texts (right). (Adapted from Frankfort (1933: pl. 81) and *EAT* I: 39)

Fig. 3.5 *Fundamentals of the Course of the Stars* or *Book of Nut* in the ceiling of the sepulchral chamber of Ramesses IV in the Valley of the Kings. (Image by courtesy of Guillermo Suay Montagud)

Actually, this compendium offers two lists of decans. One is included in columns inside Nut's body, from her armpits to her hips. In this list, taking into account the examples of Sethy I and Ramesses IV (see Figure 3.5), a total of 43 different decans appear, of which about seven are not documented among the decans used in diagonal star clocks (*sb3w mḥw, ḥrj-jb ḫnt, jwn sꜣḥ, msḏr sꜣḥ, sjꜣt, wꜥrt ḥrt sꜣḥ* y *wꜥrt ẖrt sꜣḥ*, although this could correspond to the decan *ẖrt wꜥrt* that appears in some diagonal star clocks). The second list is the one associated with the decanal transit 'clock', whose ephemerides are mainly under the body of Nut and on both sides of Shu.

Text V of the *Fundamentals of the Course of the Stars* offers us in the version of Sethy I (also in that of Ramesses IV and in pCarlsberg I) a description of how the decans are arranged throughout the night (*EAT* I: pl. 47; *EAT* I: 58):

> As for (what is) between the star which makes the first hour and the star which will be enclosed by the Duat, it is 9 (stars). Now as for (what is) between the star of birth to the star which makes the first hour, it is 20 stars; which gives 29, being those which live and work in the sky. One dies and another lives every decade (of days). Now as for (what is) between the star of birth and the star which will be enclosed by the Duat, it is 29 through that breadth of the sky as stars.

According to the scheme described in the previous text, a total of 29 decans are visible throughout the night. Of these, nine are inoperative in their time function, since they are to the west of the decan that indicates the first hour of the night on the central meridian as soon as night falls. The remaining 20 are those observed throughout the night, between that which indicates the first hour and that which has its heliacal rising at the time of the last hour, so that eight of these are to the east of the decan marking the twelfth hour of the night and do not yet intervene as hour markers. The remaining seven decans are invisible throughout the night, as they are in conjunction or, as the Egyptians would say, 'enclosed by the Duat'.

The Egyptians might have designated as *b3ktjw* the decans that were visible or only the twelve that actually 'worked' each night, while the invisible ones, in conjunction, might have been the demons *ẖ3tjw* (Von Lieven, 2007: 148–149), although this is not certain (Leitz, 2008: 13). Given that the stars rise about four minutes earlier each day than the day before, after ten days a decan stopped marking a given hour and indicated the preceding one so that every ten days the decan of the first hour also ceased to be operative. Thenceforth, a new one began to function as decan of the last hour of the night. For this reason, the text tells us, 'One dies and another lives every decade'.

The outline of the course of the stars is described in section E of the Carlsberg papyrus (Lange & Neugebauer, 1940: 30–31; *EAT* I: 58–60). Here we can extract additional information to the previous one. For example, 'the star which is at the door of the Duat' is mentioned in reference to the decan which is near its heliacal setting. Another expression used is 'the star which meets the lake', referring to one of the four κέντρα mentioned in the ostraca of demotic horoscopes (Neugebauer, 1943: 118). The 'lake' refers to the culmination point in the sky, thus distinguishing it from the 'lake of the Duat', which would consist of the opposite, or lower, point. The 'star which makes the first hour' is also known as the 'star of evening', and the 'star which meets the lake', the star culminating acronycally. It is precisely in the 'lake of the Duat' where the *Dramatic Text*, in the version of pCarlsberg I, says: 'The life of a star begins in the lake. They are the water-forms, which are in the form of fish, as (the book) of *j3t* makes it. When it begins in the water, it begins as a fish' (*EAT* I: 75–76).

This papyrus also refers to a book called *bnn*, in which it is stated: 'The (book) *bnn* makes them [the] 9 [stars] in the west of the sky, [saying]: "They have completed their work".' The nine stars located to the west of the sky are the nine stars that are to the west of the central meridian, and are therefore no longer in a position to mark the hours, so 'they have completed their work'. On the other hand, text E of the papyrus also says that between 'the day of setting' and 'the day of rising' the star spends 70 days in the Duat, that is, in conjunction. These 70 days of conjunction are also described in the so-called *Dramatic Text*. In reference to the star that enters into conjunction, it is said (versions of Sethy I and pCarlsberg I):

> The one which goes to the earth dies and [enters the Duat. It stops in the House of Geb] 70 [days. It looses] its [impurity] to the earth in 70 days. There is no speaking the name of the one loosening for 7(0) days. The name of 'living' is not said to the one loosened. It [sheds its impurities to the earth] (...) They are pure, they live, and they show their heads in the east. They rise, they shine, and they appear in the east (...). The life of a star begins in the lake'.

These 70 days of conjuction of the star suppose, from the religious point of view, a process of death and renewal perfectly assimilable to the course of the 70 days that, after the death of a person, embalming, etc., lead to his or her rebirth after the performing of the ritual of opening the mouth and entering the tomb. Obviously, the Egyptians must have equated this kind of limbo that supposes the time between the person's death and rebirth in the tomb to the 70 days of conjunction that they assign to the decans, and that, in origin, could well have been based on the direct observation of Sirius.

According to Leitz (1991: 51), to find a period of invisibility of Sirius of 70 days we must go back to the year 3324–3323 BC, considering an *arcus visionis* (β) of 9.3° to 9.1° or, according to Symons (1999: 110) to the year 3500 BC. Taking an *arcus visionis* of 9° for the heliacal rising and 7.5° for the heliacal setting of Sirius, from the latitude of Memphis (29.9°), we verify that (mainly owing to precession of the equinoxes) in 4000 BC Sirius was in conjunction for 73.62 days, while in 3000 BC that interval was 69.46 days (Ingham, 1969: 38). At the end of the 4th millennium BC Sirius therefore had an invisibility of 70 days.

Continuing with the description offered in the document, the text then mentions that 'it passes 80 days in the [east], before it does work', afterwards 'it passes 120 (days) doing work in the middle of the sky', and finally '[It passes three] months [in the west]', when the decan, still visible at sunset, has already passed the central meridian and is no longer operational as a time marker. This description adds up to 360 days, so the five epagomenal days are ignored. With these descriptions, it is clear how, in a general way, the hourly decans are distributed and when they will be operative in the transit clock. On the other hand, it follows that the Egyptian decans should be located in a celestial belt south of the ecliptic.

In the decanal transit 'clock' associated with the *Fundamentals of the Course of the Stars*, three ephemerides are indicated for each decan, headed by the words *tpt*, *šn dwꜣt* and *mst*, which we can translate as 'first' (or, according to Depuydt, 2010: 248, could come from the word *tp*, 'head', perhaps referring 'to what is high up or at the top'), 'enclosed by the Duat' and 'birth', respectively, and that would come to indicate the transit and culmination ephemerides by the meridian in the first hour of the night, of the heliacal setting and entry into conjunction, and of the heliacal rising of each decan, or as Von Lieven (2007: 144) pointed out *akronycher Kulmination, akronychem Untergang und heliakischem Frühaufgang* ('acronical culmination, acronical setting and heliacal rising').

However, the columns of text in which these ephemerides are included are not clearly associated with specific decans. This information, although essential, is missing, so Neugebauer and Parker proposed which decans should be linked by comparing the few decans (and their order) which here can be identified with the list of decans that appears on the ceiling of the pronaos of the temple of Hathor at Dendara. The only decans directly linked to the three ephemerides are: *Knmt*, *ẖrj ḫpd knmt*, *ḥꜣt ḏꜣt*, *pḥwj ḏꜣt* and *ṯmꜣt ḥrt ẖrt* (U1-5 texts) displayed on either side of Shu's feet above the horizon line (see Figure 3.6). It also appears to be the case for decan *sbššn* (text S1) and U8, on Nut's chest. Along the right arm of Nut, the texts U7a and U7b refer to *jpḏs* and (*wšꜣtj*) *bkꜣtj* (perhaps it should be associated with U6). At the other end of the body of Nut appear *ṯs ꜥrḳ* and *wꜥrt ẖrt* (texts S2 and S3), but without associated ephemeris (*EAT* I: 83–84).

Fig. 3.6 Texts U3-U5 an part of Text V of the *Fundamentals*. Tomb of Ramesses IV. (Image by courtesy of Félix Rodríguez)

If we look at the corrected values in the Table 3.3, the ephemerides are given for days 6, 16, and 26 of each month throughout the year, so that there are 90 days between the first two ephemerides, 70 days between the second and third (corresponding to the conjunction period), and 200 days between the date of the last ephemeris of a decan and the first of the next, thus totaling 360 days, again omitting the five epagomenal days. For all decans, therefore, the same amounts of time are offered between one ephemeris and another. However, no additional information is indicated about the stars throughout the night beyond the three ephemerides referring to a specific decan. For this reason, the widespread hypothesis of Neugebauer and Parker that contemplates in this list of ephemerides a star clock of transits, has been criticized in various ways by more recent authors, such as Symons (2002) and Depuydt (2010). In the opinion of Symons (2002: 445–446):

> The star calendar in the Book of Nut is an embellishment of a list of stars which displayed a common period of invisibility, not an improved diagonal star clock nor a 'transit star clock'. (…) The existence of a method for determining the hours of the night using the transits of decans across the meridian has been disputed and it has been shown that the Book of Nut star calendar is not an 'improvement' of the diagonal star clock but is closely related to the precursor of the diagonal star clock: a list of stars which disappeared for seventy days.

To be sure, in this case, the ephemerides offered are not enough to make known to a horologist the passage of twelve hours at night throughout the year. However, the main interest of this list lies in the very fact of defining two 'new' ephemerides in the apparent movement of the star in the celestial vault throughout the year. Whether one of them culminates at dusk is of great importance, since monitoring of the culmination of the decans will allow the development of the Ramesside star clocks.

In the text T2 we read: *pꜣ jrjt m ꜣbd* I *ꜣḫt ḫft prt spdt* ('The doing in I Akhet, when Sopdet rises'), a situation that certainly occurred during the

Table 3.3 Ephemerides set forth in text U of the *Fundamentals of the Course of the Stars*, as they appear in the versions of Sethy I and Ramesses IV, with corrected values and assignment of decans according to Neugebauer and Parker (*EAT* I: 84–86 and corrections by von Lieven, 2007: 64–67).

Text U Sethy I Ramesses IV	First (hour) *tpt*	Enclosed by the Duat *šn dw3t*	Birth *mst*	Related Decan	Number
1	**III Akhet 26**	**II Peret 26**	**I Shemu 6**	*knmt*	32
Sethy I	IV Akhet 16	II Peret 26	I Shemu 6		
Ramesses IV	IV Akhet 16	II Peret 26	[?] Shemu 6		
2	**IV Akhet 6**	**III Peret 6**	**I Shemu 16**	*ẖrj ḫpd knmt*	34
Sethy I	IV Akhet 6	III Peret 6	I Shemu 16		
Ramesses IV	IV Akhet 6	III Peret 6	I Shemu 16		
3	**IV Akhet 16**	**III Peret 16**	**I Shemu 26**	*ḥ3t ḏ3t*	35a
Sethy I	IV Akhet 16	IV Peret 16	I Shemu 26		
Ramesses IV	IV Akhet 16	IV Peret 16	I Shemu 26		
4	**IV Akhet 26**	**III Peret 26**	**II Shemu 6**	*pḥwj ḏ3t*	36a
Sethy I	IV Akhet 26	III Peret 16	II Shemu 16		
Ramesses IV	IV Akhet 26	III Peret 16	II Shemu 16		
5	**I Peret 6**	**IV Peret 6**	**II Shemu 16**	*ṯm3t ḥrt ẖrt*	1, 2
Sethy I	[?] Peret 6	[?] Peret 6	[?] Shemu 26		
Ramesses IV	I Peret 6	III Peret 6	II Shemu 26		
6	**I Peret 16**	**IV Peret 16**	**II Shemu 26**	*wš3tj bk3tj*	3
Sethy I	[?] Peret 26	III Peret 16	II Shemu 26		
Ramesses IV	I Peret 16	III Peret 16	II Shemu 26		
7a	**I Peret 26**	**IV Peret 26**	**III Shemu 6**	*jpḏs*	4
Sethy I	I Peret 26	III Peret 26	III Shemu 6		
Ramesses IV	I Peret 26	III Peret 26	III Shemu 6		
7b	**I Peret 26**	**IV Peret 26**	**III Shemu 6**	*jpḏs*	4
Sethy I	[?] Peret 26	III Peret 26	III Shemu 6		
Ramesses IV	I Peret 26	III Peret 26	III Shemu 6		
8a	**II Peret 6**	**I Shemu 6**	**III Shemu 16**	*sbššn*	5
Sethy I	II Peret 6	I Shemu 6	III Shemu 16		
Ramesses IV	II Peret 6	II Shemu 6	II Shemu 26		
8b	**II Peret 6**	**I Shemu 6**	**III Shemu 16**	*sbššn*	5
Sethy I	II Peret 6	I Shemu 6	IV Shemu 16		
Ramesses IV	II Peret 6	I Shemu 3	I Shemu 16		
9	**II Peret 16**	**I Shemu 16**	**III Shemu 26**	*tpj-ꜥ ẖntt*	5a
Sethy I	II Peret 16	I Shemu 16	IV Shemu 26		
Ramesses IV	II Peret 16	I Shemu 16	IV Shemu 26		
10	**II Peret 26**	**I Shemu 26**	**IV Shemu 6**	*ẖntt ḥrt*	6
Sethy I	II Peret 26	I Shemu 26	IV Shemu 6		
Ramesses IV	II Peret 20	I Shemu 17	IV Shemu 6		
11	**III Peret 6**	**II Shemu 6**	**IV Shemu 16**	*ẖntt ḥrt*	7
Sethy I	III Peret 6	II Shemu 6	II Shemu 26		
Ramesses IV	IV Peret 7	II Shemu 6	II Shemu 26		
12	**III Peret 16**	**II Shemu 16**	**IV Shemu 26**	*ṯms n ẖntt*	8
Sethy I	III Peret 16	II Shemu 16	IV Peret 26		
Ramesses IV	IV Peret 16	II Shemu 16	II Peret 26		

(continued)

Table 3.3 (continued)

Text U Sethy I Ramesses IV	First (hour) *tpt*	Enclosed by the Duat *šn dwꜣt*	Birth *mst*	Related Decan	Number
13	**III Peret 26**	**II Shemu 26**	**I Akhet 6**	*sptj ḫnwj*	9b
Sethy I	III Peret 26	II Shemu 26	I Akhet 26		
Ramesses IV	III Peret 26	II Shemu 20	II Akhet 6		
14	**IV Peret 6**	**III Shemu 6**	**I Akhet 16**	*ḥrj-jb wjꜣ*	11
Sethy I	IV Peret 6	III Shemu 6	II Akhet 6		
Ramesses IV	-omits-	-omits-	-omits-		
15	**IV Peret 16**	**III Shemu 16**	**I Akhet 26**	*sšmw*	12a
Sethy I	IV Peret 16	III Shemu 16	II Akhet 16		
Ramesses IV	IV Peret 16	III Shemu 16	II Akhet 26		
16	**IV Peret 26**	**III Shemu 26**	**II Akhet 6**	*knm*	13
Sethy I	IV Peret 26	III Shemu 26	II Akhet 26		
Ramesses IV	IV Peret 26	III Shemu 16	II Akhet 6-26		
17	**I Shemu 6**	**IV Shemu 6**	**II Akhet 16**	*tpj-ꜥ smd*	13a
Sethy I	I Shemu 6	IV Shemu 6	II Akhet 6		
Ramesses IV	I Peret 26	III Shemu 16	II Akhet 6		
18	**I Shemu 16**	**IV Shemu 16**	**II Akhet 26**	*smd*	14a
Sethy I	I Shemu [1]6	IV Shemu [?]	[?] Akhet 16		
Ramesses IV	I Shemu 16	IV Shemu 16	-omits-		
19	**I Shemu 26**	**IV Shemu 26**	**III Akhet 6**	*srt*	15
Sethy I	I Shemu 26	IV Shemu 26	II Akhet 26		
Ramesses IV	I Shemu 10	-omits-	II Akhet 26		
20	**II Shemu 6**	**I Akhet 6**	**III Akhet 16**	*sꜣwj srt*	16
Sethy I	II Shemu 6	[?] Akhet 6	III Akhet 16		
Ramesses IV	II Shemu 4	I Akhet 3	II Akhet 16		
21	**II Shemu 16**	**I Akhet 16**	**III Akhet 26**	*ẖrj ḫpd srt*	17
Sethy I	II Shemu 16	I Akhet 15	III Akhet 26		
Ramesses IV	II Shemu 16	IV Akhet 16	III Akhet 8		
22	**II Shemu 26**	**I Akhet 26**	**IV Akhet 6**	*tpj-ꜥ ꜣḫwj*	18
Sethy I	II Shemu 26	[?] Akhet 6	III Akhet 6		
Ramesses IV	II Shemu 26	II Akhet 26	III Akhet 6		
23	**III Shemu 6**	**II Akhet 6**	**IV Akhet 16**	*ꜣḫwj*	19
Sethy I	III Shemu 6	II Akhet 6	III Akhet 16		
Ramesses IV	III Shemu 6	II Akhet 16	III Akhet 16		
24	**III Shemu 16**	**II Akhet 16**	**IV Akhet 26**	*tpj-ꜥ bꜣwj*	
Sethy I	III Shemu 16	II Akhet 16	III Akhet 26		
Ramesses IV	III Shemu 16	II Akhet 6	III Akhet 26		
25	**III Shemu 26**	**II Akhet 26**	**I Peret 6**	*bꜣwj*	21
Sethy I	III Shemu 26	II Akhet 26	I Peret 6		
Ramesses IV	III Shemu 26	II Akhet 26	I Peret 6		
26	**IV Shemu 6**	**III Akhet 6**	**I Peret 16**	*ḫntw ḥrw*	21a
Sethy I	IV Shemu 6	III Akhet 6	I Peret 16		
Ramesses IV	IV Shemu 4	-omits-	I Peret 16		
27	**IV Shemu 16**	**III Akhet 16**	**I Peret 26**	*ḫntw ḥrw*	21b
Sethy I	IV Shemu 16	III Akhet 16	I Peret 26		
Ramesses IV	III Shemu 10	III Akhet 16	I Peret 26		

(continued)

Table 3.3 (continued)

Text U Sethy I Ramesses IV	First (hour) *tpt*	Enclosed by the Duat *šn dw3t*	Birth *mst*	Related Decan	Number
28 Sethy I Ramesses IV	**IV Shemu 26** III Shemu 26 III Shemu 26	**III Akhet 26** III Akhet 26 III Akhet [?]	**II Peret 6** II Peret 6 II Peret 6	*s3wj ḳd*	22a
29 Sethy I Ramesses IV	**I Akhet 6** I Akhet 6 I Akhet 6	**IV Akhet 6** IV Akhet 6 IV Akhet 6	**II Peret 16** II Peret 16 II Peret 16	*ẖ3w*	23
30 Sethy I Ramesses IV	**I Akhet 16** I Akhet 15 I Akhet 16	**IV Akhet 16** IV Akhet 16 IV Akhet 16	**II Peret 26** II Peret 26 II Peret 16	*ʿryt*	24
31 Sethy I Ramesses IV	**I Akhet 26** I Akhet 26 I Akhet 16	**IV Akhet 26** IV Akhet 26 IV Akhet 16	**III Peret 6** III Peret 6 III Peret 6	*rmn ḥrj*	26
32 Sethy I Ramesses IV	**II Akhet 6** [?] Akhet 6 II Akhet 6	**I Peret 6** [?] Peret 6 I Peret 16	**III Peret 16** III Peret 16 III Peret 16	*ṯs ʿrḳ*	26b
33 Sethy I Ramesses IV	**II Akhet 16** II Akhet 16 II Akhet 16	**I Peret 16** I Peret 16 I Peret 16	**III Peret 26** III Peret 26 III Peret 26	*wʿrt*	29
34 Sethy I Ramesses IV	**II Akhet 26** II Akhet 26 II Akhet 26	**I Peret 26** I Peret 26 I Peret 16	**IV Peret 6** IV Peret 6 IV Peret 16	*tpj-ʿ spdt*	30
35 Sethy I Ramesses IV	**III Akhet 6** III Akhet 6 II Akhet 16	**II Peret 6** II Peret 6 II Peret 16	**IV Peret 16** IV Peret 16 IV Peret 16	*spdt*	31
36 Sethy I Ramesses IV	**III Akhet 16** III Akhet 16 III Akhet 16	**II Peret 16** II Peret 16 II Peret 16	**IV Peret 26** IV Peret 26 III Peret 26	*štw*	31b

reign of Sethy I, when the heliacal rising of Sirius occurred in I Akhet. Either for this reason or because it refers to an ideal Sothiac calendar, the truth is that, as we have seen in Table 3.3, the ephemerides of text U35, assigned to decan 31 Sopdet, offer us other information:

> [Decan Sopdet]: First (hour), month III Akhet day 6. Enclosed by the Duat, month II Peret day 6. Birth, month IV Peret day 16'.

Taking into account that the 'birth' refers to the heliacal rising, this event provided by the *Fundamentals of the Course of the Stars* for Sopdet (IV Peret 16; see also Barta, 1981b) has important chronological consequences. Perhaps it is no coincidence that this same date is the one indicated in the Berlin 10012 papyrus referring to the heliacal rising of Sirius in the year 7 of Senuseret III. In any case, this makes it clear that this information did not have the slightest functional meaning in the time of Sethy I, since it would obviously have been a copy of a valid ephemeris from half a millennium earlier, in the time of Senuseret III.

The Naos of Saft el-Henna from the time of Nectanebo I (30th Dynasty) shows the importance of the three ephemerides referred to in the *Fundamentals of the Course of the Stars* by means of a remarkable iconography that must symbolize the various aspects that each decan passes through (see Figure 3.7). Habachi and Habachi (1952) have already suggested this relationship, although it was Leitz (1995: 8–12) who developed the hypothesis of associating the forms in which the decans are shown with their phases throughout the year. In the list of decans of this monument, each one is associated with five images that, from top to bottom, are a human-headed bird (like *Bꜥ*) on a boat, a falcon-headed sphinx, a lion-headed ram, a canine-headed standing mummy and a recumbent human mummy. All these images are accompanied by explanatory texts. In addition, in front of the bird (its head topped by a star included within a circle, similar to Gardiner's N15 sign) a small text with two or three columns indicates the month, season, and decade to which the decan mentioned is linked and an offering formula. Another two or three, longer, columns located in front of the remaining images have a mythological and astrological character.

According to Leitz, the figure in the first register (a human-headed bird) would represent the shape of the decan at the time of its heliacal rise; the second (a falcon-headed sphinx) would be linked to its 120-day period of work from its first completion as a marker of the 12th hour of the night. The figure of the third register (a lion-headed ram) would be associated with the period after its culmination as the first hour of the night. Finally, the last two registers are very descriptive, since they show the decan in his mummy form, standing in the fourth, surely related to his heliacal setting, and in the fifth, resting on a bed, undoubtedly in relation to the 70 days in which the decan is supposed to be in the Duat, in conjunction with the sun. However, in the opinion of von Bomhard (2008: 90–94), although the first and fifth images would have the meaning already indicated, the three centrals would be 'aspects of Shu related to the positions of the sun and the decanal stars as they are observed in the sky during the decade concerned'.

3.1.3 *The Ramesside Star Clocks*

On the ceilings of three tombs in the Valley of the Kings, from the 20th Dynasty, we find astronomical documentation for which there are no known precedents. Two specimens are found in the tomb of Ramesses VI (KV 9, one in room F and another in corridors B and C; see Figure 3.8), one especially disorganized in that of Ramesses VII (KV 1, corridor C, an improvised burial chamber), and another in that of Ramesses IX (KV 6, corridor C)—all of them, therefore, in a context of the second half of the twelfth century BC. It was Champollion (1845: pl. 272, 1868: 197–198), who discovered these 'star charts' in 1829 and was the first to make known the existence of these tables. However, he believed that they had astrological connotations, as he understood them to be tables of constellations and their influences for all hours of each month of the year.

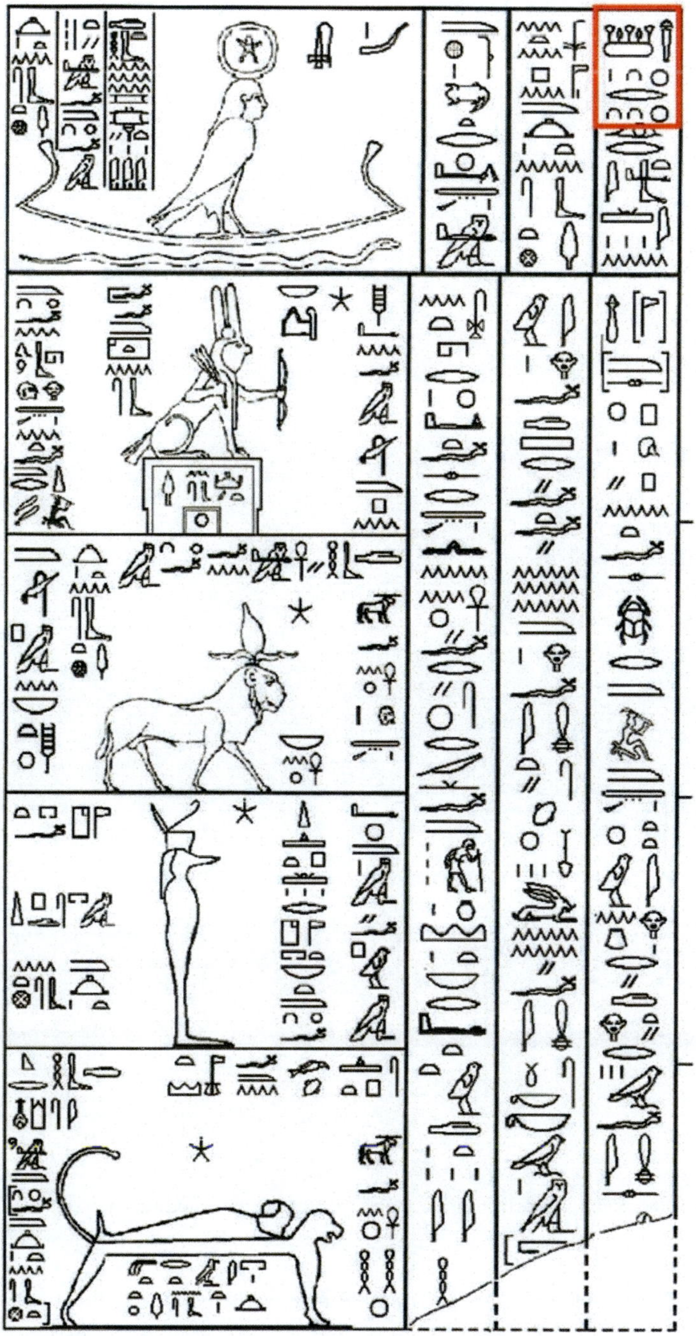

Fig. 3.7 The various aspects that each decan passes through. Naos of the decades, decade of I Akhet 11–20. (Adapted from von Bomhard, 2008: fig. 65)

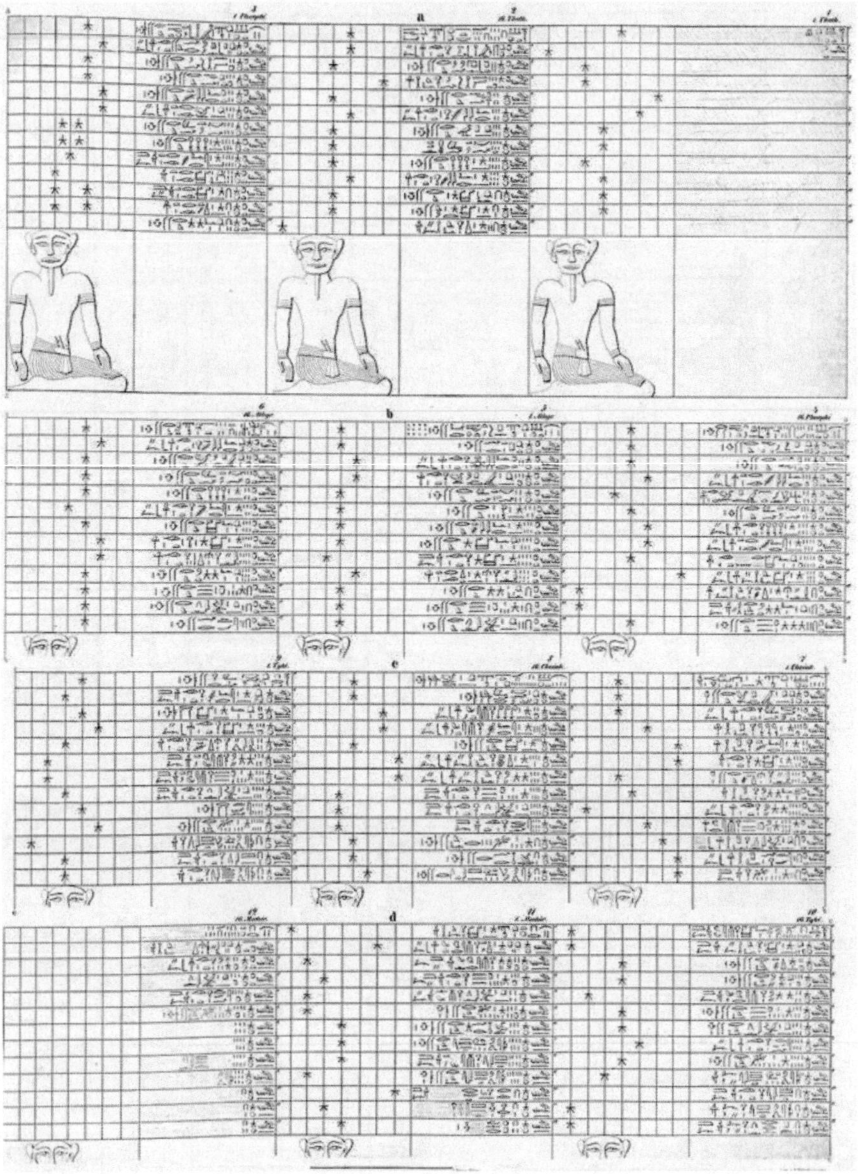

Fig. 3.8 Some Ramesside Star clock tables belonging to Ramesses VI. (After Lepsius D III B1 227)

Shortly after, Lepsius copied and published part of these documents in his *Dankmäler aus Aegypten und Aethiopien* (III B1 227–228bis). However, it was Le Page Renouf who, in an article published in 1874 (400–421), resolved the correct nature of these tables, when he deduced that they reflected the culminations of certain stars during the twelve hours of the night throughout the year.

Indeed, these tables constitute the so-called Ramesside star clocks that have the particularity of serving to count the hours of the night observations of the culmination (not the rising or setting) of certain stars (*EAT* II; Clagett, 1995: 405–465; Leitz, 1995: 117–264). This information, when presented in full, is organized into 24 tables, two for each month, so that the first is for days 1–15 and the second for days 16–30 of the month in question. In each table, on the other hand, there are lines of text with references to stars, from the one visible at dusk until twelve that will serve as markers for the twelve hours of the night. There is no information for the epagomenal days so it is possible that the last table also supplied this time.

A novel aspect of these clocks is the way in which the hourly stars are represented. Each of the tables could be accompanied by a frontal view of a seated man, with a network of seven vertical lines and twelve or thirteen horizontal lines serving as a coordinate system. The central vertical line, which could sometimes be made to coincide with the vertical of the figure's nose (as in some tables of the tomb of Ramesses IX; see Figure 3.9), represents the central meridian, that is, the imaginary north–south line in which the stars culminate, so that the other six vertical lines represent transit lines before or after the central meridian.

One might think, following this iconography, that for the making or use of these tables, two observers facing each other were needed, one as a mere reference to the vertical of the central meridian and the other, the observer, aligned with him in the N–S axis. In reality, a previously aligned fixed post would have been even more useful, so it cannot be ruled out that such instruments had been installed on the temple terraces, and that the position of the observer would also have been marked on the ground. To carry out these observations, the observer could have made use of a *merkhet* in order always to maintain verticality in correspondence with the meridian. Or the other way back, the stars with respect to their coordinate system, based

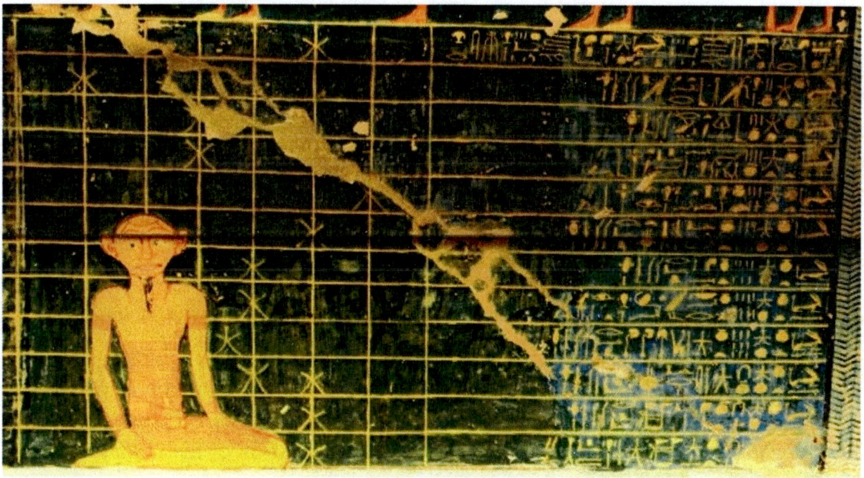

Fig. 3.9 One of the tables of the Ramesside Star clock of Ramesses IX. (Image by courtesy of Félix Rodríguez)

on observing the position of the star with respect to the head and the upper part of the observer's assistant (which, as we have indicated, could well have been a fixed post or even a statue).

However, there is no unanimous agreement on how the observations might have been performed. Symons (1999: 79–84), for example, proposes a new possibility, 'the reflection conjecture', which concludes that observations of the stars for the construction and use of the Ramesside star clock were carried out by looking into still water, making the temple's sacred lake an ideal place to perform this task.

The passage of the star through the meridian, its culmination, is referred to in the tables with the expression *r-ꜥḳꜣ jb* ('opposite the heart'). In relation to the grid, the star would appear in the centre of the seven vertical lines or, with respect to the figure of the assistant, if it were centred with respect to the said grid, which would be the moment when the star passed over his head. In the event that the star had not yet finished, but was close to doing so, three coordinates are offered: *ḥr kꜥḥ wnmj* ('on the right shoulder'), *ḥr msḏr wnmj* ('on the right ear'), and *ḥr jrt wnmjt* ('on the right eye'). If the passage through the meridian were equivalent to the 0 coordinate, the latter would be −3, −2, and −1, respectively, owing to its progressive proximity to the meridian. Similarly, once the culmination of the star has occurred, it would gradually move away from the central meridian and this would be referred to as como *ḥr jrt jꜣbjt* ('on the left eye'), *ḥr msḏr jꜣbj* ('on the left ear'), and *ḥr kꜥḥ jꜣbj* ('on the left shoulder'), that is, coordinates +1, +2, and +3.

The horizontal coordinates refer to the passage of the hours of the night. In total there are twelve (Ramesses IX) or thirteen (Ramesses VI–VII) rows indicated in the tables. The first refers to or *tp grḥ* ('beginning of the night') and the following to the hours starting with the first (*wnwt tpjt*) and ending with the twelfth (*wnwt* 12), except in the case of Ramesses IX, where the beginning of the night and the first hour are mentioned in the first line. In turn, each line offers complementary information: the name of the referenced star and its position with respect to the central meridian following the model explained above. Thus, for example, in the first line of the first table of the clock of Ramesses IX we read:

tpj ꜣḫt tp grḥ šwtj nt nḫt wnwt tpjt tp n nḫt ḥr msḏr (wnmj)

First of *ꜣkhet*, beginning of the night: the Two Feathers (*šwtj*) of the Giant (*nḫt*). First hour: Head (*tp*) of the Giant (*nḫt*), over the ear (right).

The Two Feathers refers to a star or group of stars in the constellation of the Giant, which, at nightfall, would be close to the central meridian. The first hour, however, would be marked by another star from the same constellation of the Giant, this time the Head.

The following table, corresponding to the second fortnight of I Akhet, shows us in the example of Ramesses VI the following information in its first two lines:

tpj ꜣḫt 16 smdt tp grḥ tp nḫt (ḥr jrt jꜣbjt)

First of Akhet, (second) half of the month, beginning of the night, Head of the Giant (over the left eye).

wnwt tpj nḥbt.f ḥr jrt jꜣbjt

First hour, His Neck (of the Giant), over the left eye.

While the first line has informed us of the fortnight to which the ephemerides correspond (I Akhet 16–30), and that at the beginning of the night the star 'Head of the Giant' was on his left eye (we do not know this last information from the text but from the drawing of the star in the adjoining grid), the second line tells us that in the first hour (at night) it is another star, His (the Giant's) Neck, which is over his left eye. That is, in both cases the star had already passed its culmination. The same scheme follows in the following lines, as we can see in Table 3.4, corresponding to table n° 2 of the example of the room of pillars of the tomb of Ramesses VI.

We can also extract chronological information, thanks to the ephemerides of these charts. In Table 11 of Ramesses VI, corresponding to II Peret 1–15, the Star of Sopdet *(sbꜣ n spdt)* is found in the position 'above the left ear' (+2) in the first hour of the night. So that in the following fortnight (table 12), corresponding to II Peret 16–30, that same must be the star that was observed in its culmination at the beginning of the night (the name of the star and its position are not preserved in the example of Ramesses VI, but partially in that of Ramesses VII):

wnwt tpj sbꜣ n spdt ḥr msḏr jꜣbj

'First hour, Star of Sopdet, on his left ear (position +2)'.

ꜣbd II *prt* 1[6 *smdt*] *tpj grḥ sbꜣ n* [*spdt*] [*r-ꜥḳꜣ jb*]

Second month of Peret, (second) [middle of the month], beginning of the night, Star of [Sopdet] opposite the heart (position 0)'.

Therefore, if Sopdet was in the central meridian in II Peret 16, we only have to follow the instructions of the *Fundamentals of the Course of the Stars*, which explains the work of the hourly stars in the sky, in order to conclude that the Star of Sopdet would have worked as an hourly star for 120 days. Before we would have to add

Table 3.4 Table n° 2 (I Akhet 16-30) from room F of the tomb of Ramesses VI (KV 9)

Hour – Decan – Position	-3	-2	-1	0	1	2	3
 First of Akhet, (second) middle of the month, beginning of the night, Giant's Head.					*		
 First hour, His Neck (of the Giant), over the left eye.					*		
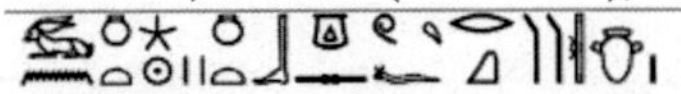 Second hour, His Hip (of the Giant), opposite the heart				*			
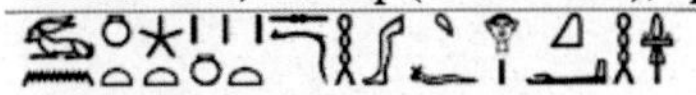 Third hour, His Knee (of the Giant), on the left shoulder							*
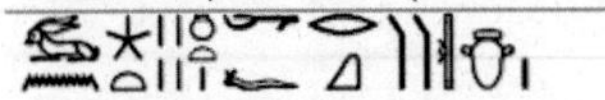 Fourth hour, His Pedestal (of the Giant), opposite the heart				*			
 Fifth hour, Aryt, over the left eye					*		
 Sixth hour, Head of the Bird, opposite the heart				*			
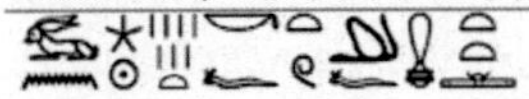 Seventh hour, His Tail (of the Bird), likewise				*			
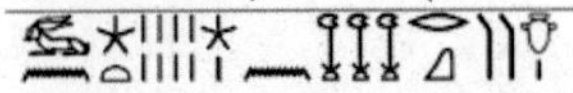 Eighth hour, Star of the Thousands, opposite the heart				*			
 Ninth hour, Star of Sar, over the left eye.					*		
 Tenth hour, Forerunner of Sah, opposite the heart				*			
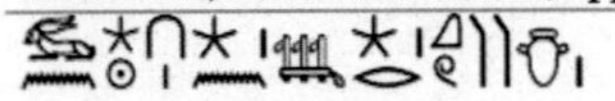 Eleventh hour, Star of Sah, opposite the heart				*			
 Twelfth hour, Star of Sopdet, on the right shoulder	*						

another 80 from its heliacal rising, which would take us to III Shemu 26 a date, that is, with a cumulative lag of 325 days since the last *apokatastasis* (the heliacal rising of Sirius coinciding with I Akhet 1). This would take us to a date close to the beginning of the reign of Thutmose III, far earlier than the reign of Ramesses VI (1140–1132 BC), the difference between the two being about 340 years.

According to Neugebauer and Parker, assuming the equivalence between the Star of Sopdet and Sopdet (Sirius) would place the chronology around the year 1470 BC, although they take it to be 1500 BC by taking into account other reasons (such as the variation in the length of the hours with the passing of the seasons). Le Page Renouf (1874), taking into account the culmination of the Star of Sopdet (= Sirius) at midnight (sixth hour) in the first fortnight of IV Akhet, came to the conclusion that the design of this clock must date to around 1450 BC.

A similar method is employed by Leitz (1997: 136), who, not recognizing the equivalence between the Star of Sopdet and Sirius, instead bases his conclusions on the culmination of Rigel (β Orionis, which he believes corresponds to *Sbꜣ n Sꜣḥ*) at midnight of III Akhet 23–24, which places the date of manufacture of the clock in the year 1463 BC. Whichever method is used, the truth is that the chronology of the clock dates to the first half of the fifteenth century BC.

At this point we must remember once again that most of the astronomical documents with which we are working come from the funerary sphere, so their function was merely symbolic. Those clocks were never going to be used by the living to calculate the hours but were intended symbolically to be used by the deceased, who would have need of them on their journey through the Duat. In any case, it seems incomprehensible that those in charge of the design of the tombs should use lists of decans so old when, without a doubt, they would have had at their disposal other updated ones. One reason may be the abandonment of star clocks in favour of clepsydras, which began to be developed with precision from the beginning of the 18th Dynasty.

There are 47 different hourly stars referenced in the Ramesside star clocks. The most remarkable thing is that there are almost no coincidences with respect to the decanal stars that we know from the diagonal star clocks, the lists of decans that appear on astronomical ceilings, and other associated sources. Possible coincidences are: the Star of Sah (*sbꜣ n sꜣḥ*) with some decan of Sah, the Star of Sopdet (*sbꜣ n spdt*) with Sopdet (*spdt*), and the Star of the Thousands/Myriads (*sbꜣ n ḫꜣw*) with the Thousands/Myriads (*ḫꜣw*). However, it should be noted that none of these three appears exactly with these names in other sources. In any case, that there are (almost) no coincidences is because the stars whose transits through the central meridian are used to mark the hours do not belong to the decanal belt of those stars whose risings were used for diagonal star clocks. This allows us to know another strip of the Egyptian sky. On the other hand, many of the stars mentioned in the Ramesside clocks belong to the same constellations, a fact which also serves in some cases to give us an idea of what their extension may be in right ascension (RA).

Following the nomenclature used by Neugebauer and Parker (gathering the stars into thirteen groups from E to R, omitting I), the hourly stars used in the Ramesside star clocks are summarized in Table 3.5.

As shown in Table 3.5, many of the hourly stars are selected from specific constellations that broadly coincide with the thirteen groups established by Neugebauer and Parker in their classification of these hourly stars. Group E, for example, corresponds to the constellation Nekhet, the Giant *nḫt*, which is the constellation that contributes the most hourly stars to the clock, 16, although it is not clear whether the stars that precede or follow the upper and lower parts of the figure of the Giant are part of this constellation. Through the name of the stars we can recompose the image that the Egyptians had of the Giant: a personage with two tall feathers, holding a mace and resting on a pedestal.

The next constellation with the most referenced stars is Reret *rrt* ( ), the 'She-Pig' (or 'female hippotamus'), with eight (group R). Also

Table 3.5 Hourly stars of the Ramesside star clocks

NP N^er	Name	Translation	NP N^er	Name	Translation
E1	tpj-a Swtj nt nxt *tpj-ʿ šwtj nt nḫt*	'predecessor of the Two Feathers of the Giant'	K1	tpj-a sAH *tpj-ʿ s3ḥ*	'predecessor of Sah'
E2	Swtj nt nxt *šwtj nt nḫt*	'the Two Feathers of the Giant'	K2	sbA n sAH *sb3 n s3ḥ*	'Star of Sah'
E3	*tpj-ʿ ḥḏ.f*	'predecessor of His Mace'	L1	*sb3 n spdt*	'Star of Sopdet'
E4	*tp nḫt*	'Head of the Giant'	L2	*jj-ḥr-s3 spdt*	'the one coming after Sopdet'
E5	*tp n ḥḏ nt nḫt*	'Head of the Mace of the Giant'	M1	*tpj-ʿ sb3wj*	'Predecessor of the Two Stars'
E6	HD nt nxt *ḥḏ nt nḫt*	'Mace of the Giant'	M2	*sb3wj*	'Two Stars'
E7	*nḥbt.f*	'His Neck'	N	*sb3w nw mw*	'Stars of the Water'
E8	*ḫ3b.f*	'Nape of his Neck'	O1	*tp n m3j*	'Head of the Lion'
E9	*mndt.f*	'His Breast'	O2	*sd.f*	'His Tail'
E10	*bgs.f*	'His Hip'	P	*sb3w ʿš3w*	'Many Stars'
E11	*sḏḥ.f*	'His Shank'	Q1	*[tpj-ʿ] mnjt*	'[Predecessor] of the Mooring Post'

E12	*Pd nḫt*	'Knee of the Giant'	Q2	*ṯ3 nfr*	*ṯ3 nfr*
E13	*sbk.f*	'His Foot'	Q3	*šmw n ḥ3t n mnjt*	'Follower of the front of the Mooring Post'
E14	*pt.f*	'His Pedestal'	Q4	*mnjt*	'Mooring Post'
E15	*jj-s3 pt.f*	'The one coming after His Pedestal'	Q5	*šmsw n mnjt*	'Follower of the Mooring Post'
E16	*sb3 n-s3 pt*	'Star of the back of the Pedestal'.	Q6	*šmsw jj-(ḥr)-s3 mnjt*	'Follower which comes after the Mooring Post'
F	*ʿryt*	Aryt	R1	*rdwj n rrt*	'Feet of Reret'
G1	*bʿnt nt 3pd*	*bant.f* Its Beak'	R2	*pd.s*	'Her Knee'
G2	*Tp n 3pd*	'Head of the Bird'	R3	*ḥrj-jb n mntj.sj*	'Middle of her Thighs'
G3	*ḥtyt nt 3pd*	'Throat of the Bird' *Htyt*	R4	*b3ḥ.s*	'Her private part'
G4	*kf3w.f*	'Its Rump'	R5	*ḫpd.s*	'Her Buttocks'
H	*sb3 n ḫ3w*	'Star of the Thousands'	R6	*mndt.s*	'Her Breast'
J	*sb3 n sʿr*	'Star of Sar'	R7	*ns.s*	'Her Tongue'
R8	*šwtj.sj*	'Her Two Feathers'			

in this case, the stars define the parts of the body of the figure of this constellation, again with two feathers but arranged in the sky in an inverted position to that of Nekhet. Another constellation, Menit, *mnjt* ('Mooring Post'), contributes six hourly stars (group Q), although many may not actually belong to the constellation (for example, both the stars that precede and those that follow Menit, as well as *ṯꜣ nfr*).

With four stars appears the constellation Aped *ꜣpd* ('Bird', group G), which in the sky would be orientated towards the west. Four other constellations each provide two hourly stars: *šꜥh* / *sꜣḥ* (group K), Sopdet *špdt* (group L), Sebawy *sbꜣwj,* 'two stars' (group M), and Mai *Mꜣj* (group O). Finally, five hourly stars apparently independent of the previous constellations are also indicated: Aryt *ꜥryt* (group F), *sbꜣ n sꜥr,* 'Star of Sar' (group J), and, forming part of open clusters, *sbꜣ n ẖꜣw,* 'Star of the Thousands' (group H), *sbꜣw nw Mw,* 'Stars of the Water' (group N), and *sbꜣw ꜥšꜣw,* 'Many Stars' (group P).

However, when we tabulate the total number of referenced stars for each fortnight of the year, we see how, while some have a notable follow-up, others are referenced almost anecdotally (see Table 3.6).

3.1.4 A Clock Based on the Movement Meskhetyu Around the North Celestial Pole

In a granite bull sarcophagus found in Kom Abu Yasin (Cairo JE 86723; *EAT* III: 49–52), from the reign of Nectanebo II, an exceptional table of positions of the constellation of Meskhetyu appears that undoubtedly show the use of this constellation for hourly and calendric purposes (see Figure 3.10).

The table presents two strips of diagrams, showing the changes in the position of Meskhetyu throughout the year. Three positions are indicated for each month so that a total of 36 Meskhetyu positions are displayed throughout the year. The first position of the month is accompanied by the text *wḫꜣ* ('beginning of the night'), the second position *ḥrj-jb grḥ* ('at the middle of the night'), and the third and last position of each month is accompanied by the text *ḥḏ tꜣ* ('dawn'). Therefore, the diagrams attempt to show the change in position of the constellation of Meskhetyu around the north celestial pole at three times of the night: the beginning of the night, midnight, and morning twilight.

The diagrams that appear on this sarcophagus are by no means exact, quite the contrary. The positions of Meskhetyu do not reflect the real movement of the constellation around the celestial north pole but have been presented without care or precision. However, we must bear in mind once again that this document belongs to the funerary sphere, where the overriding consideration is the symbolic value of what is represented, which would be known to all, and not the precision with which it is represented. Undoubtedly, there would be much more precise diagrams for use by horologists.

Table 3.6 Hourly stars for each fortnight of the year

Hour/ fortnight	0	1	2	3	4	5	6	7	8	9	10	11	12
I Akhet 1–15	*šwtj nt nḫt*	*tp n ḥḏ nt nḫt*	*nḥbt.f*	*bgs.f*	*sḏḥ.f*	*pt.f*	*ꜥryt*	*tp n ꜣpd*	*kft.f*	*sbꜣ n ḫꜣw*	*sbꜣ n sꜥr*	*tpj-ꜥ sꜣḥ*	*sbꜣ n sꜣḥ*
I Akhet 16–30	*tp nḫt*	*nḥbt.f*	*bgs.f*	*sḏḥ.f*	*pt.f*	*ꜥryt*	*tp n ꜣpd*	*kft.f*	*sbꜣ n ḫꜣw*	*sbꜣ n sꜥr*	*tpj-ꜥ sꜣḥ*	*sbꜣ n sꜣḥ*	*sbꜣ n spdt*
II Akhet 1–15	*nḥbt nḫt*	*bgs.f*	*sḏḥ.f*	*pt.f*	*ꜥryt*	*tp n ꜣpd*	*kft.f*	*sbꜣ n ḫꜣw*	*sbꜣ n sꜥr*	*tpj-ꜥ sꜣḥ*	*sbꜣ n sꜣḥ*	*sbꜣ n spdt*	*tpj-ꜥ sbꜣwj*
II Akhet 16–30	*bgs n nḫt*	*pd nḫt*	*pt.f*	*ꜥryt*	*bꜥnt n ꜣpd*	*kft.f*	*sbꜣ n ḫꜣw*	*sbꜣ n sꜥr*	*tpj-ꜥ sꜣḥ*	*sbꜣ n sꜣḥ*	*Jj ḥr sꜣ n spdt*	*tpj-ꜥ sbꜣwj*	*sbꜣw nw mw*
III Akhet 1–15	*pd nḫt*	*pt.f*	*ꜥryt*	*tp n ꜣpd*	*kft.f*	*sbꜣ n ḫꜣw*	*sbꜣ n sꜥr*	*tpj-ꜥ sꜣḥ*	*sbꜣ n sꜣḥ*	*Jj ḥr sꜣ n spdt*	*tpj-ꜥ sbꜣwj*	*sbꜣw nw mw*	*tp mꜣj*
III Akhet 16–30	*pt*	*ꜥryt*	*tp n ꜣpd*	*kft.f*	*sbꜣ n ḫꜣw*	*sbꜣ n sꜥr*	*tpj-ꜥ sꜣḥ*	*sbꜣ n sꜣḥ*	*Jj ḥr sꜣ n spdt*	*tpj-ꜥ sbꜣwj*	*sbꜣw nw mw*	*tp mꜣj*	*sd.f*
IV Akhet 1–15	*sbꜣ n sꜣ pt*	*tp n ꜣpd*	*kft.f*	*sbꜣ n ḫꜣw*	*sbꜣ n sꜥr*	*sbꜣ n sꜣḥ*	*Jj ḥr sꜣ n spdt*	*tpj-ꜥ sbꜣwj*	*sbꜣwj*	*sbꜣw nw mw*	*tp mꜣj*	*sd.f*	*sbꜣw ꜥšꜣw*
IV Akhet 16–30	*tp n ꜣpd*	*kft.f*	*sbꜣ n ḫꜣw*	*sbꜣ n sꜥr*	*sbꜣ n sꜣḥ*	*sbꜣ n spdt*	*sbꜣwj*	*sbꜣw nw mw*	*tp mꜣj*	*sd.f*	*sbꜣw ꜥšꜣw*	*ṯꜣ nfr*	*šmsw n ḥꜣt mnjt*
I Peret 1–15	*kft.f*	*sbꜣ n sꜥr*	*tpj-ꜥ sbꜣ n sꜣḥ*	*sbꜣ n sꜣḥ*	*Jj ḥr sꜣ n spdt*	*sbꜣwj*	*sbꜣw nw mw*	*tp mꜣj*	*sd.f*	*sbꜣw ꜥšꜣw*	*šmsw n ḥꜣt mnjt*	*mnjt*	*šmsw mnjt*
I Peret 16–30	*tpj-ꜥ sbꜣ n sꜣḥ*	*sbꜣ n sꜣḥ*	*sbꜣ n spdt*	*tpj-ꜥ sbꜣwj*	*sbꜣwj*	*sbꜣw nw mw*	*tp mꜣj*	*sd.f*	*sbꜣw ꜥšꜣw*	*šmsw n ḥꜣt mnjt*	*mnjt*	*šmsw mnjt*	*pd n rrt*
II Peret 1–15	*sbꜣ n sꜣḥ*	*sbꜣ n spdt*	*(tpj-ꜥ) sbꜣwj*	*sbꜣw nw mw*	*tp mꜣj*	*sd.f*	*sbꜣw ꜥšꜣw*	*ṯꜣ nfr*	*šmsw (n) ḥꜣt mnjt*	*mnjt*	*rdwj n rrt*	*pd.s*	*mntj.s*
II Peret 16–30	*sbꜣ n spdt*	*Jj ḥr sꜣ n spdt*	*(tpj-ꜥ) sbꜣwj*	*tp mꜣj*	*sd.f*	*sbꜣw ꜥšꜣw*	*ṯꜣ nfr*	*šmsw (n) ḥꜣt mnjt*	*mnjt*	*šmsw mnjt*	*(pd n) rrt*	*ḥrj-jb n mntj.s*	*ḫpd.s*
III Peret 1–15	*(tpj-ꜥ) sbꜣwj*	*sbꜣw nw mw*	*(tp n mꜣj)*	*s(d).f*	*sbꜣw ꜥšꜣw*	*ṯꜣ nfr*	*(šmsw n ḥꜣt mnjt)*	*šmsw mnjt*	*rdwj n rrt*	*pd.s*	*ḥrj-jb n mntj.s*	*ḫpd.s*	*mndt.s*

(continued)

Table 3.6 (continued)

Hour/ fortnight	0	1	2	3	4	5	6	7	8	9	10	11	12
III Peret 16–30	*sbꜣw nw mw*	*mꜣj*	*sd.f*	*sbꜣw ꜥšꜣw*	*ṯꜣ nfr*	*šmsw n ḥꜣt mnjt*	*mnjt*	*rdwj n rrt*	*pd.s*	*ḥrj-jb n mntj.s*	*ẖpd.s*	*mndt.s*	*šwtj.s*
IV Peret 1–15	*tp mꜣj*	*sd.f*	*sbꜣw ꜥšꜣw*	*ṯꜣ nfr*	*šmsw ḥꜣt mnjt*	*šmsw jj ḥr sꜣ mnjt*	*rdwj n rrt*	*pd.s*	*ḥrj-jb n mntj.s*	*ẖpd.s*	*mndt.s*	*ns.s*	*tpj-ꜥ šwtj nt nḫt*
IV Peret 16–30	*(sd n) mꜣj*	*sbꜣw ꜥšꜣw*	*ṯꜣ nfr*	*šmsw ḥꜣt mnjt*	*šmsw jj mnjt*	*rdwj n rrt*	*pd.s*	*bꜣḥ.s*	*ẖpd.s*	*mndt.s*	*ns.s*	*tpj-ꜥ šwtj nt nḫt*	*šwtj nt nḫt*
I Shemu 1–15	*tpj-ꜥ mnjt*	*šmsw ḥꜣt n mnjt*	*mnjt*	*šmsw mnjt*	*rdwj n rrt*	*pd.s*	*bꜣḥ.s*	*ẖpd.s*	*mndt.s*	*šwtj.s*	*tpj-ꜥ šwtj nt nḫt*	*šwtj nt nḫt*	*tp ḥḏ nt nḫt*
I Shemu 16–30	*šmsw n (ḥꜣt n mnjt)*	*mnjt*	*šmsw mnjt*	*rdwj n rrt*	*pd.s*	*bꜣḥ.s*	*ẖpd.s*	*mndt.s*	*šwtj.s*	*tpj-ꜥ šwtj nt nḫt*	*šwtj nt nḫt*	*tp ḥḏ nt nḫt*	*ẖꜣb.f*
II Shemu 1–15	*(šmsw) n mnjt*	*rdwj n rrt*	*pd.s*	*bꜣḥ.s*	*ẖpd.s*	*mndt.s*	*ns.s*	*šwtj.s*	*tpj-ꜥ šwtj nt nḫt*	*šwtj nt nḫt*	*ẖꜣb.f*	*mndt.f*	*bgs.f*
II Shemu 16–30	*rdwj n rrt*	*pd.s*	*bꜣḥ.s*	*ẖpd.s*	*mndt.s*	*ns.s*	*šwtj.s*	*(šwtj) nt nḫt*	*tp.f*	*ẖꜣb.f*	*mndt.f*	*bgs.f*	*pd.f*
III Shemu 1–15	*bꜣḥ n rrt*	*ẖpd.s*	*mndt.s*	*ns.s*	*šwtj.s*	*tpj-ꜥ šwtj nt nḫt*	*šwtj nt nḫt*	*nḥbt.f*	*mndt.f*	*bgs.f*	*pd.f*	*sbḳ.f*	*pt.f*
III Shemu 16–30	*ẖpd n rrt*	*mndt.s*	*ns.s*	*šwtj.s*	*tpj-ꜥ šwtj nt nḫt*	*šwtj nt nḫt*	*ẖꜣb.f*	*mndt.f*	*bgs.f*	*pd.f*	*sbḳ.f*	*Jj sꜣ pt.f*	*ꜥryt*
IV Shemu 1–15	*mndt nt rrt*	*tpj-ꜥ šwtj nt (nḫt)*	*ḥḏ nt nḫt*	*ẖꜣb.f*	*bgs.f*	*pd.f*	*sbḳ.f*	*pt.f*	*ꜥryt*	*ḥtyt nd ꜣpd*	*kft.f*	*sbꜣ n ẖꜣw*	*sbꜣ n sꜣḥ*
IV Shemu 16–30	*šwtj nt rrt*	*šwtj nt (nḫt)*	[…]	*nḥ(bt.f)*	*mn(dt.f)*	*bgs.f*	*s(bḳ.f)*	*(pt).f*	*ꜥryt*	*(tp n) ꜣpd*	*(kft).f*	*sbꜣ n ẖꜣw*	*sbꜣ (n) sꜥr*

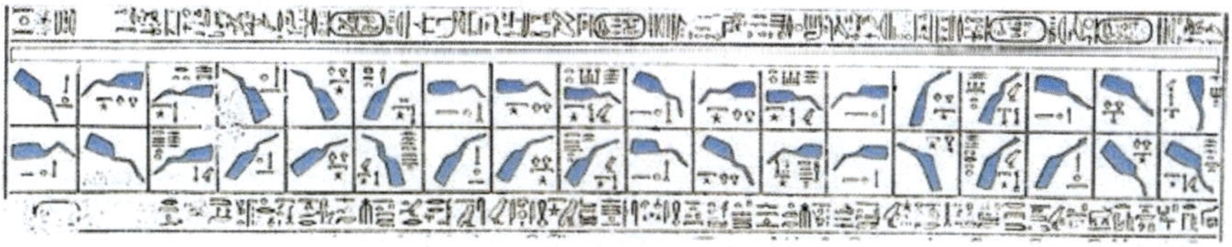

Fig. 3.10 Positions of Meskhetyu through the year depicted in a sarcophagus found at Kom Abu Yasin from the reign of Nectanebo II. (Adapted from *EAT* III: pl. 24)

At the time of Nectanebo II, the constellation was still circumpolar from the latitude of Memphis, although with Alkaid being only about 2° above the astronomical horizon at lower culmination. However, it had already ceased to be so from the latitude of Elephantine since the end of the Ramesside period because of the setting of Alkaid, although this star could still have been seen from Thebes, not without difficulty, at less than 2° above the horizon at lower culmination. Therefore, if we take into account that the constellation is represented as a bull's leg (and this includes the seven main stars of Ursa Major), we must assume that the use of Meskhetyu as a time marker must have been useful in the Delta region, especially if we refer to the time of Nectanebo II. Since when Meskhetyu may have been used in this sense is difficult to know. The constellation had been of great importance to the Egyptians for many centuries. Therefore, the observation of the different positions of Meskhetyu throughout the nights of the year could have been used not only for its calendrical interest but also for its hourly interest in periods of the history of Egypt long before Nectanebo II (Lull 2023).

The diagram, in its lower part, is accompanied by a text that does not offer relevant information:

> Hail Osiris, First Bull: Osiris, bull of the sky... The stars of the northern sky, they are your Foreleg (*msḫt*). , they never set in the western sky like decanal stars, they travel backwards at night as well as during the day. They are in the following of the Great Hippopotamus (*rrt wrt*) of the northern sky as your Foreleg, when it goes to the southern sky near the *bau* of the gods who are in Orion (*S3ḥ*).

On the one hand, it is striking that the bull's leg is identified with Osiris, when other, older, texts clearly identify it with Seth's leg lost in his fight against Horus. On the other hand, the text is also clear when it says that the Foreleg stars 'never set in the night'; that is, it recognizes the circumpolar character of Meskhetyu, as in fact it had been for all of Egypt (up to the latitude of the first cataract) until the middle of the 18th Dynasty (at the end of the 18th Dynasty, Alkaid was only 1.3° above the horizon at its lower culmination, so it would no longer be visible from Elephantine).

Owing to the possible similarities, it is worth remembering a system used by Spanish mariners in the times of Christopher Columbus: the *Reloj de las Guardas* ('Clock of the Guards'). In this case, the stars β UMi (Kochab) and γ UMi (Pherkad), called *Guardas* (the guardians of heaven) by Christopher Columbus, were used as a reference. For centuries Spanish sailors used these stars as a clock.

Table 3.7 Current movement of the Guardians around the north celestial pole, at the beginning of the night, middle of the night and dawn during the solstices and equinoxes

	Beginning of the night	Middle of the night	Dawn
Winter solstice	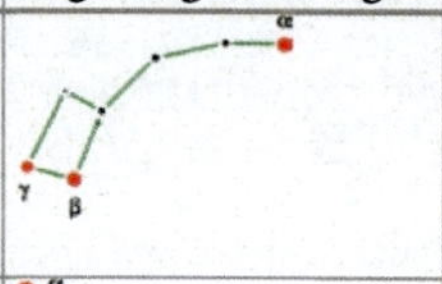	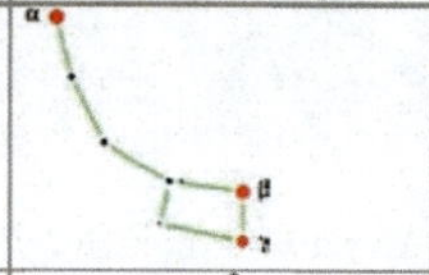	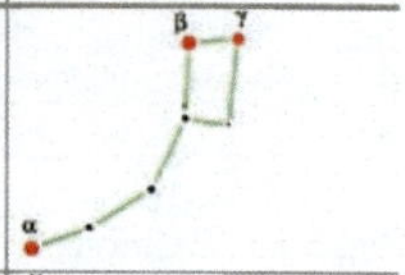
Vernal equinox	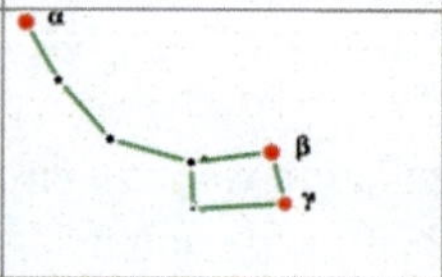	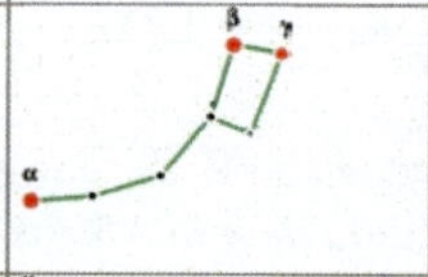	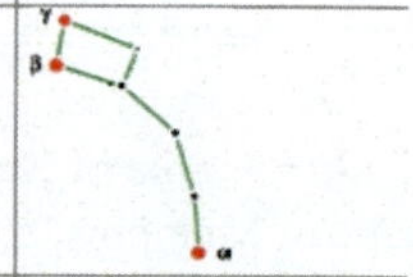
Summer solstice	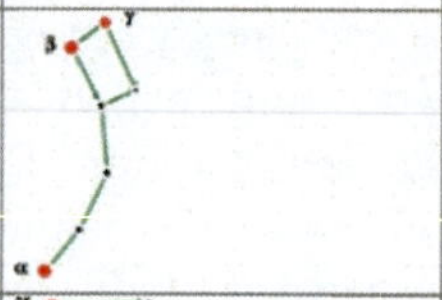	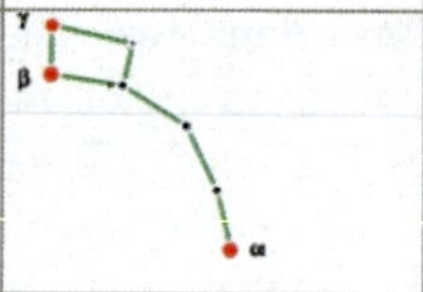	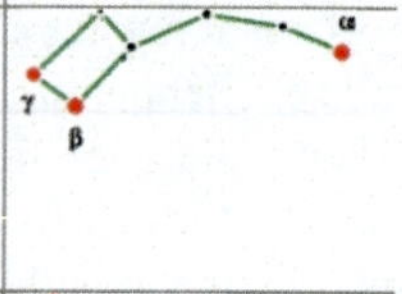
Autumnal equinox	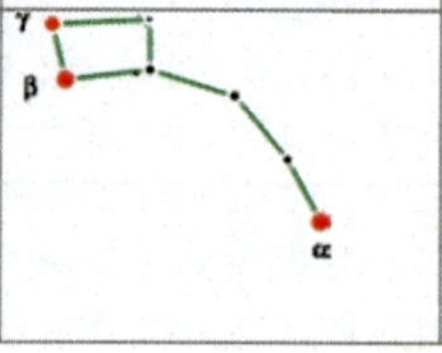	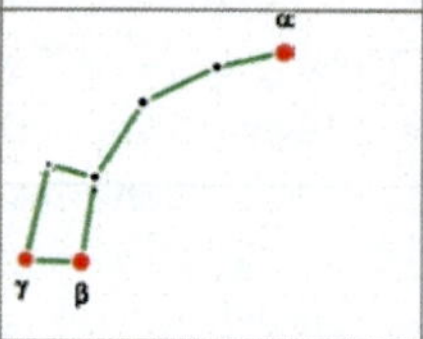	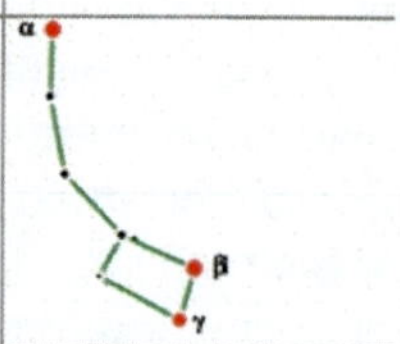

To make use of the clock, the sailors would memorize the positions of the Guards at three important moments of the night, taking advantage of their rotation around Polaris (north celestial pole): the beginning of the night, midnight, and dawn (Table 3.7) (Comellas, 2016: 50–53), just as we see in the Kom Abu Yasin sarcophagus for each month.

In turn, the sailors took an imaginary reference, the so-called *Hombre del Norte* ('Man of the North'), which consisted of a figure of a man, standing upright and on his back, with arms crossed and legs apart and with the pole star in the centre of his chest. The times were marked indicating when the Guardians were on his head, left shoulder, left arm, under the left arm, at his feet, etc.

Meskhetyu's positions that appear on the Kom Abu Yasin sarcophagus are erroneous, as the person in charge of its design paid no interest in showing even the slightest precision. He settled for the symbolic sense. A correct diagram, valid for the reign of Nectanebo II, is the one we see in Table 3.8. Probably, as a clock, the user would have had a complete month-by-month table, as shown in the sarcophagus, or memorized the three described positions of Meskhetyu at the beginning of each season, thus deducing the movement during the rest of the year, exactly the same that the Spanish sailors of past centuries did.

Table 3.8 Movement of Meskhetyu, at the beginning of the night, middle of the night, and dawn during the months of the Egyptian year under Nectanebo II (350 BC) as observed from Memphis

	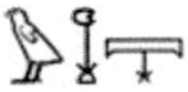 *wḫ3* beginning of the night	 *ḥrj-jb grḥ* middle of the night	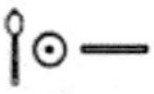 *ḥḏ t3* dawn
I Akhet 1	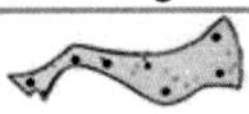		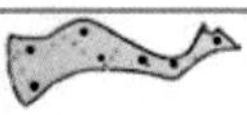
II Akhet 1	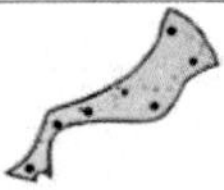		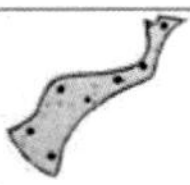
III Akhet 1		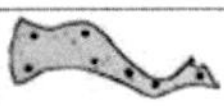	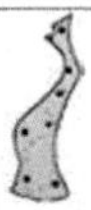
IV Akhet 1		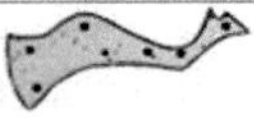	
I Peret 1	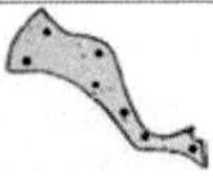	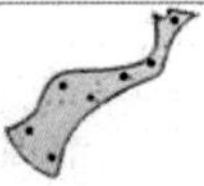	
II Peret 1	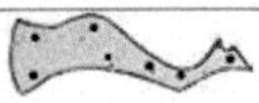		
III Peret 1	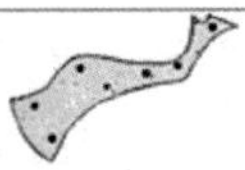		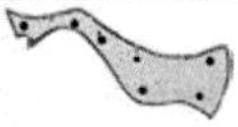
IV Peret 1	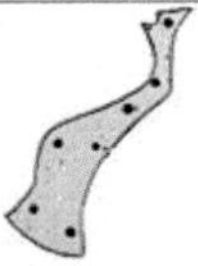		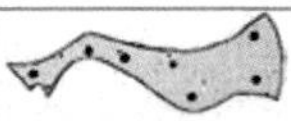
I Shemu 1		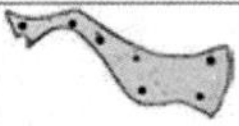	
II Shemu 1		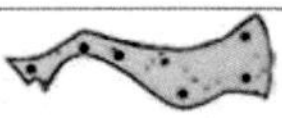	

(continued)

Table 3.8 (continued)

	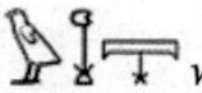 *wḫ3* beginning of the night	 *ḥrj-jb grḥ* middle of the night	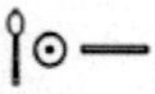 *ḥḏ t3* dawn
III Shemu 1			
IV Shemu 1	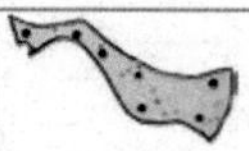		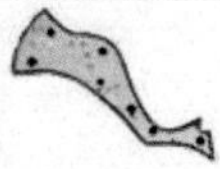

3.2 Sundials

In addition to the stars, the Egyptians also made use of the diurnal movement of the sun to be able to measure the hours of the day. They used both shadow clocks, in which the length of the shadow is measured to indicate the time, and sundials, in which it is the angular direction of the shadow that marks the hours.

In general, we can say that in ancient Egypt four main types of solar clocks were used (Borchardt, 1920), depending of the shape of the shadow-receiving area of the sundial. More recently, Symons (2020; Symons & Khurana, 2016) has proposed five categories. Typical of Ancient Egypt are the so-called L-shaped sundials or shadow clocks, where the shadow is projected on a horizontal surface in which the hours are indicated, and sloping sundials, where the shadow is projected on an inclined plane. Three other types of sundials categorized by Symons are semicircular sundials, made on a vertical plane, the concave sundials, where the shadow is projected on a spherical surface, and miscellaneous sundials.

3.2.1 *Shadow Clocks*

Although the oldest shadow clocks that have been preserved date from the 18th Dynasty, the possibility that they existed already in the Middle Kingdom is a useful working hypothesis. A proof could be drawn from a passage in the Neferty Prophecy (Papyrus Ermitage 1116B verso, l. 51–52; Helck, 1992: 45) in which the hour, noon, and their relationship to the shadow are discussed (Clagett, 1995: 84; Salmas, 2013: 355):

(jw) Rˁw jwd.f sw rmṯ [wb]n.f wn wnwt nn
rḫ.tw ḫpr mtrt nn tn(.tw) šwt.f

Re will separate from men. When he gets up, there will be the hour. We will not know (when) noon has come, we will not distinguish his shadow.

The oldest preserved complete clock from the 18th Dynasty is from the time of Thutmose III (Berlin, inv. 19744) (Borchardt, 1920: pl. XII) and measures 23.2 cm in length. There is another very similar one, from the Third Intermediate Period (Berlin, inv. 19743; Borchardt, 1920: pl. XII), which is 30.3 cm long. Both have five marks to tell the hours (see Figure 3.11). The head (almost 5 cm high) of a clock of Amenhotep III (Louvre N 781) is also preserved. There is another one from the same monarch's rule in Berlin (inv. 15573). L-shaped sundials are typical only of Egypt and were used until the Roman period.

Exceptionally, we also have references to these instruments both in a well-known example of the cenotaph of Sethy I in Abydos, where a relief shows an L-shaped shadow clock with instructions on how it should be built, and another, from the first century AD, preserved on a papyrus known as the Tanis Geographical Papyrus. Although, as we have mentioned, we have a number of physical copies of these clocks, the texts provide us with very interesting information.

The drawing and text referring to the shadow clock of Sethy I (Frankfort, 1933: pl. 83 II) are found on the west side of the ceiling of the sarcophagus chamber (see Figure 3.12). According to Frankfort (1933: 77 n1), who was the first to work on this text during his mission in the Osireion of Abydos, it must be a copy of another equally incomplete version, since, according to him, the first columns of the text

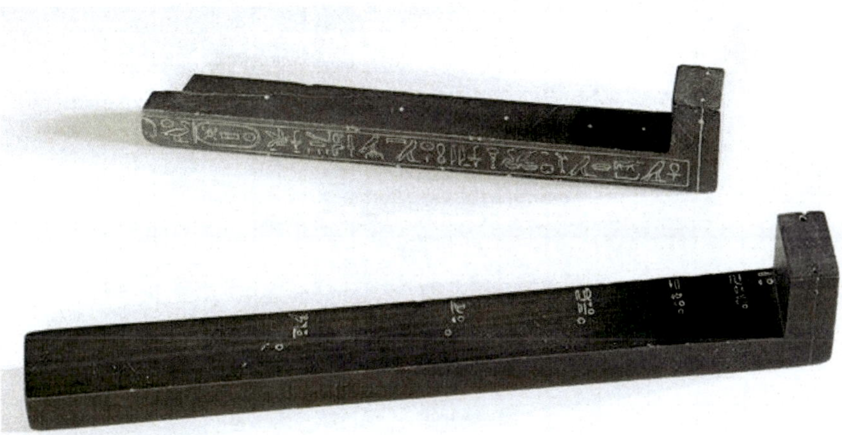

Fig. 3.11 Two original L-shaped sundials or shadow clocks from the time of Thutmose III and Third Intermediate Period. (Courtesy of the Staatliche Museen zu Berlin, Ägyptisches Museum und Papyrussammlung, Inv. No. ÄM 19744 (top: Thutmose III) and 19743 (bottom: TIP). Photograph: Margarete Büsing)

Fig. 3.12 The shadow clock of the cenotaph of Sethy I at Abydos. (Adapted from Frankfort, 1933: pl. 83 II)

show wide gaps that are not caused by deterioration or partial loss of the inscription; these parts were simply never written.

At the top of the text[1] there is a retrograde reading inscription in which, as a title, we read:

[knowledge] of the hours of the day and night. Example of setting noon.
[rḫ] wnwwt hrw grḥ tp n jrt mtrt

Another possible option for the final part of the title would be, according to Hoffmann (2016: 345). *dp n jrt mtrt Anweisung, eine richtige (Zeit-)Messung vorzunehmen* ('Instruction, a correct measurement to be carried out'). In the box under the title we can see the drawing of the shadow clock's profile, with five marks and, among them, the numbers 3, 6, 9, and 12. Below, there are thirteen columns of text, also with retrograde reading. This is where we find the most interesting part of this document. The first nine columns give information on the hours to be recognized, what the shadow clock should look like, and where to put the marks that serve to mark the hours:

(1) Hour of the day from the fixation of the place, [knowledge of the] hour
wnwt hrw šȝʿ m smnt dmj [rḫ] wnwt

(2) Hour […] after the first mooring post
wnwt […] m-ḫt mnjt tpjt

(3) Hour […] after the second mooring post
wnwt […] m-ḫt mnjt 2-nwt

(4) Hour […] after the third mooring post
wnwt […] m-ḫt mnjt 3-nwt

(5) Know the hour […] five fists in its length
rḫ wn[wt] […] šsp 5 m ȝw.s

(6) Height […] of two fingers in its height
ḳȝw […] nt ḏbʿ 2 m ḳȝw.s

(7) over the head[2] of the *sṯȝt* (*setjat*) […] these 5 fists in 4 parts […]
ḥr tp n sṯȝt n[…] sšp 5 pn m sp 4

(8) marked on this *setjat*. You must put 12 units there, to (mark) the first hour. You must put there 9 to (mark) the second hour. You must put there 6 to (mark) the hour
ȝb(w) ḥr sṯȝt tn rdj.ḫr.k 12 n hp jm n wnwt tpjt rdj.ḫr.k 9 jm n wnwt 2-nwt rdj.ḫr.k 6 jm n wnwt

(9) third. You must put 3 there to (mark) the fourth hour (…)
3-nwt rdj.ḫr.k 3 jm n wnwt 4-nwt (…)

In the first columns, a reference is made to several  *mnjt* or mooring posts that reflect the idea that the Egyptians had of the course of the sun through the celestial

[1] See the observations of Frankfort (1933: 77–78); Neugebauer and Parker (*EAT* I: 116–117); Clagett (1995: 465–470); Salmas (2013: 369–373).

[2] The translation of this part of the text can offer different interpretations, which means that it can change the meaning given to the word 'tp'. See Hoffmann (2016: 350).

vault. Let us remember that the Egyptians believed that the sun travelled through the celestial waters in a boat, called *mꜥnḏt* on its daytime route and *msktt* during the night, so that the hours, as fixed points, were associated with mooring posts where the sun, on its way, could dock at the end of every hour.

The measurements given in columns 5 and 6 are used to determine the dimensions of the shadow clock, called *sṯꜣt* (*setjat*) in Egyptian. The five fists (*šsp*) in length add up to 37.45 cm, since one fist is 7.49 cm long. That is the length of the horizontal and longitudinal base of the shadow clock, called a *mertut* (as we shall see later) *mrtwt*. Instead, the two fingers' height refers to the height of a piece that, depending on the translation, could be the *tp* placed at one end of the *mertut*, which will serve to project the shadow, or a piece over the end (*tp*) of the *mertut*. According to the text, this piece measures 3.74 cm, since each Egyptian finger (*ḏbꜣ*) has a value of 1.87 cm (one *šsp* 'fist' is divided into four *ḏbꜣ* 'fingers'). However, there are those who think (*EAT* I: 116; Clagett, 1995: 85) that this measure should not be associated with this piece but with another long and transversal one, a horizontal crossbar (supposedly the *merkhyt*) that we shall comment on later. However, bear in mind that the surviving head of Amenhotep III's clock (Louvre N 781) is even taller than the measurement given here.

Columns 7–9 give instructions on how the *mertut* should be divided into four parts to form the marks that will serve to tell the hours. The figures given are also shown in the drawing of the clock. As stated in the text, the explanation may seem somewhat confusing, because when it is indicated, 'you must put 12 units there', etc., the length measurement is not specified. In any case, we must understand twelve units, nine (units), etc., which add up to a total of 30 units. Obviously, the mark made 30 units from the head of the *setjat* should indicate the first of the hours of the day recognized by this instrument. If in the direction of the head we add twelve units, at eighteen units we would have the mark for the second hour, nine units later (at nine from the head) the mark of the beginning of the third hour, six units later (at three from the head) the mark of the fourth hour that would end at noon when the shadow stopped projecting.

This is the way in which we must understand the explanation that the text offers us. However, although the previous instructions assume the presence of four marks on the longitudinal piece, five marks appear erroneously in the drawing, and there is one left over. In any case, in other *setjats*, such as that of Thutmose III (Berlin ÄM 19744) or the TPI (Berlin ÄM 19743) (see Fig. 3.11), there are five marks. In the latter, in addition, the name of the hours was written next to the marks (Borchardt, 1910: 10): the first is *wbnt* ('the one that dawns'), the second *sšmt* ('the one that introduces'), the third *mk nb.s* ('the one who protects her lord'), the fourth *sštꜣt* ('the secret'), the fifth *nsrtt* ('the flame'), and, the sixth hour of the day, *ꜥḥꜥt* ('the standing'), the latter actually without an associated mark since it is found at the base of the head. Abydos's clock, according to the instructions, could mark the central eight hours of the day.

We should go back to the inscription. The following columns referring to the Sethy I shadow clock include instructions on how the clock should be used:

(9) (…) when you adjust this *setjat* with the Sun setting (with) its head to the east, where it is
jst ꜥḳꜣ.n.k sṯꜣt tn m ꜥḳꜣw rꜥ tp.s r jꜣbt ntj mrḫ(yt)

(10) this *(mr)ḫyt* (*merkhyt*), with which the shadow of the Sun is fixed on this *setjat*. But after the fourth hour is over you must turn
(mrḫ)yt tn ḥr.f r wnn šwt rꜥ ꜥḳꜣ m sṯꜣt tn jst m-ḫt pḥ.n wnwt 4-nwt spẖr.ḫr.k

(11) this *setjat* (with) its *mertut* to the east, after the Sun is on top (at its culmination) of this *merkhyt*. You must calculate these
sṯꜣt tn mrtwt.s r jꜣbt m-ḫt ꜥḥꜥ rꜥ m wpt nt mrḫyt tn ḥsb.ḫr.k grt nn

(12) hours until Re enters in the fourth hour according to the unit at the beginning, which makes the eighth hour, (with) 2 hours
wnwt r ꜥḳ Rꜥw ḫft wnwt 4 *mj hp jmj ḥꜣt dmḏ pꜣ ḏd?* 8-*nwt ḏr-ntt wntt wnwt* 2

(13) that have elapsed in the morning without Re getting up and two hours that will elapse after Re hides setting the place of the hours of the night.
ḫpr m dwꜣ n wbn Rꜥw jw grt kt wnwt 2 *ḫpr ḥr-sꜣ ꜥḳ Rꜥw n smnt dmj n wnwwt grḥ*

As indicated in column 9, the clock had to be orientated towards the east. In this way, the shadow cast by the clock's head on the horizontal plane could indicate the hours. Once the culmination of the sun arrived, at noon, the clock had to be turned 180° and orientated towards the west, since at that moment the instrument was no longer usable in its original position. This is just what is indicated in columns 10–11: 'after the fourth hour is over you will have to turn this *setjat* (with) its *mertut* to the east'. The  *mrtwt* (*mertut*), therefore, would be the name of the longitudinal piece where the hours were marked. This method, the E–W orientation, seems to be corroborated by modern calculations (Tupikova & Soffel, 2012). In relation to this there is a well-known passage from the Annals of Thutmose III (*Urk* IV: 655, 5–14) in which the reorientation of the clock seems to be evidenced, just as Nelson (1913: 39–40) assumed and Parker reaffirmed (1981):

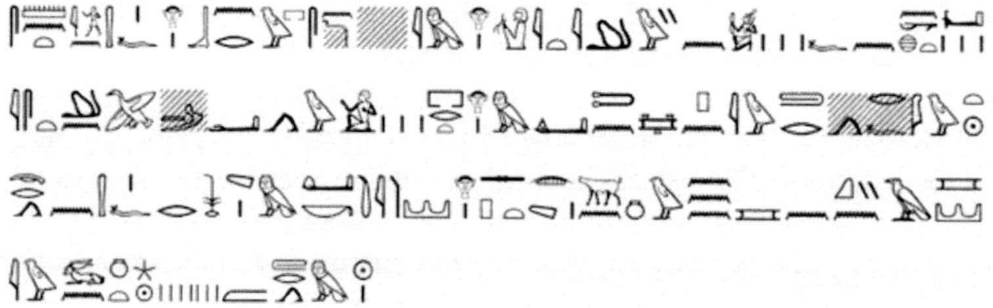

His majesty halted in the open. He sat (down) there to watch the rear guard of his valiant army. Now, when the last of the lead-troops had come out from this road, the shadow turned, his majesty arrived to the south of Megiddo. On the shore of the Qina brook (Lichtheim, 1976: 32), when the 7th hour was turning in the day. (Parker, 1981: 146–147)

The expression *jw pẖr.f šwt* 'the shadow turned' undoubtedly refers to the moment at which, once the sun reaches its culmination, the shadow will stop projecting towards the west and will do so towards the east; that is, the moment at which the clock must be reorientated. This is also clear, because after that it is indicated that Megiddo was reached at the seventh hour (*jw wnwt* 7 *m pẖr m hrw* 'when the 7th hour was turning in the day', since the sixth hour coincides with noon (as in clocks in which there are five dots marked on the *mertut*). If a clock with a head such as that of Sethy I was orientated on the E–W axis, the noon sun would not cast any shadow on the *mertut*, since it would be projected to north. For this reason, we must understand that the sixth hour of the day, *ʿḥʿt* ('the standing'), indicated on the Berlin clock ÄM 19743, is right at the base of the head.

However, authors such as Isler (1991) suggest that this would not be the method used by the clock wearer. They propose that the clock would be orientated towards the sun each moment the time was wanted, so that the shadow would always extend over the *mertut*. However, from reading the text of the cenotaph of Sethy I we cannot interpret this to be so.

At this point it is worth exploring the nature of the *mrḫyt,* whose word is split between columns 9 and 10, and which is mentioned again in column 11. The text indicates that the *mrḫyt* is located on the *tp*, at one end of the clock. Borchardt (1920: 33 fig. 9) assumed that on the head of the end of the clock there could be a long and horizontal crossbar, so this would be the *merkhyt*. Isler (1991: 177 fig. 28), on the other hand, speculates that this transversal piece would be placed directly on the longitudinal one.

Krauss (2018), who, like the previous author, also assumes a constant orientation of the clock towards the sun, recalls that neither in the representation of the clock in the cenotaph of Sethy I in Abydos nor in that of the Tanis papyrus, which we will comment on later, is a transversal piece observed that allows the annual projection of the shadow on the *mertut*. Symons and Khurana (2016: 378) consider that the instrument could not have contained a crossbar, and Hofmann (2016: 352) warns of the problems that this piece generates.

A transverse bar like the one proposed by Borchardt (see Figure 3.13) would certainly have been very useful since with the movement of the sun towards noon the projected shadow would tend to go north and leave the surface of the longitudinal bar. Bruins (1965), on the other hand, also proposes the existence of another piece that would allow the height of the crossbar to be varied.

However, none of these elements has been preserved in any of the specimens that have come down to us, nor has the plumb line that these instruments evidently had. If we look at the more complete specimens, for example, we can see how vertical marks appear on the sides of the head and small holes where the wire from which the plumb line hung should be inserted, which, following these vertical lines, verified the horizontality of the base. This is also suggested by the way in which some instruments were represented in hieroglyphics: . In the Ptolemaic and Roman periods, the profile of shadow clocks is used as a hieroglyph, as a determinative of the word 'hour' or as an ideogram (Borchardt, 1920: 52).

In reality, the basic problem is not only to know whether these clocks were orientated to the east and from noon to the west (with what could be considered the crossbar) or whether they were constantly orientated towards the sun, as many authors have postulated in recent years (so that the crossbar would not be necessary). We must also ask ourselves if by using L-shaped sundials great precision was really sought or if, as we may suspect, similar clocks were used in a given context so that they would give all their users similar results, so the reading of the hours would be equally useful despite its imprecision.

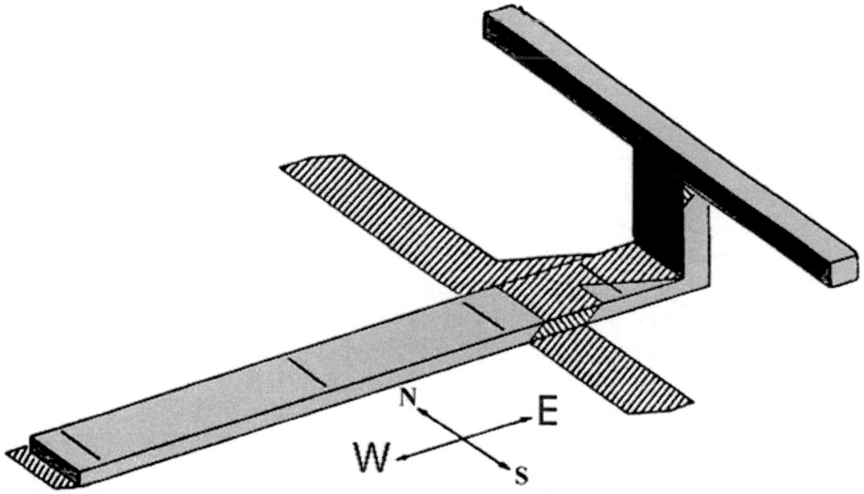

Fig. 3.13 Reconstruction of a shadow clock with long crossbar after the hypothesis proposed by Borchardt. (Adapted from *EAT* I fig. 36)

Columns 11 and 12 of the text give information about the hours that are not measured by the Abydos clock; that is, the initial two hours and the final two hours of the day. It is worth asking, then, what happens with these hours. For Neugebauer and Parker (*EAT* I: 118), the solution is in the text itself: 'The most plausible assumption is that sunrise and sunset, which are observable phenomena, mark the dividing points between the 2 extra hours before and after the central 8. This would mean that daylight would contain 10 hours and 2 more hours would be used for twilight before sunrise and after sunset in order to bridge the gap to the hours of night measured by the decans'.

However, the five marks of the Thutmose III clock imply the measurement of ten hours, not eight, so following the previous principle, in this case, as Spalinger (2012: 372) points out: '

> It could only have operated from sunrise to the end of each hour through the eleventh. At the conclusion of the twelfth hour it would have been sunset. To put it another way: the ten castings of a shadow would have covered the ten hours of daytime and the remaining two hours would have covered the extra time following sunrise and preceding sunset. (...) Thutmose's shadow clock implies that the night hours would also have sunrise and sunset as their boundary points. This means that a change had come to pass with regard to a regularized method of time keeping: the night hours no longer were based on the decanal system.

The shape of the *setjat* (without a hypothetical crossbar) is identical to that of the *merkhet* used by Egyptian astronomers, of which, as we also saw in a previous chapter, copies have been preserved. The only difference is that the *merkhet* has no hour marks. The similarity of the name in both cases, and in others that we can add (such as the *merkhyt* that is named in the Amenemhat inscription regarding the development of clepsydras) has led some scholars to think that its meaning would be *m-rḫ(y)t ce qui permet de connaître* ('that which permits knowledge'; Salmas, 2013: 359); that is, it would be a general word to designate measuring instruments, regardless of their type (shadow clocks, observing instruments, or clepsydras).

Part of a clock is preserved in the Louvre Museum (Louvre N 781; Figure 3.14), specifically the piece located at the end of the *mertut*. The interesting thing about this small piece is its decoration. On the reverse side appears the Pharaoh Amenhotep III offering the god Re-Horakhty a figure of Maat. And on other side Sopdet appears as 'lady of the year'.

As in the cenotaph of Sethy I, another sketch of a shadow clock comes from Tanis (Griffith & Petrie, 1889: pl. 15). In this case it is a papyrus from the Roman period, dated around 100 AD, in which the clock can be distinguished (see Figure 3.15) despite its poor condition. With what is preserved in three of the fragments of the papyrus, it can be seen that in the original design of the scheme a lateral focus was arranged at a certain height, from which six convergent lines were extended to various points of the horizontal base, where were marked the hours whose names were written at the bottom. If the model drawn by Sethy I had four marks (suitable for

Fig. 3.14 Original head of a reconstructed clock from the time of Amenhotep III. (Image kindly provided by Nacho Ares. Courtesy of Musée du Louvre)

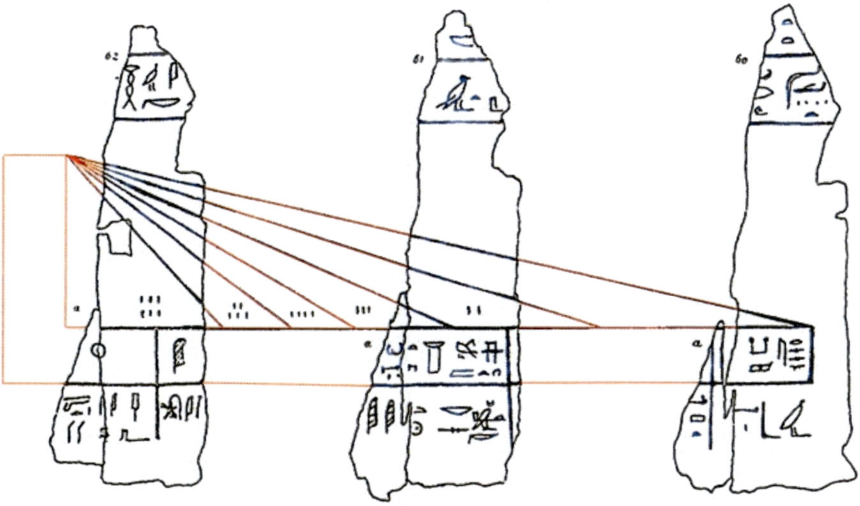

Fig. 3.15 Shadow clock from Tanis papyrus. (Adapted from Griffith & Petrie, 1889: pl. 15)

marking the central eight hours of the day) and those of the already mentioned examples of Thutmose III and TIP had five marks (capable of marking the central ten hours), the existence of six marks would indicate a more complete use throughout the day.

According to Krauss (2018: 175):

> *Misst die Tanis-Uhr mit einer zusammengesetzten Skala, bestehend aus einem Einzelwert und einer arithmetischen Reihe mit Gliedern proportional zum Einzelwert. Die Skala der Tanis-Uhr ist insofern eine Verbesserung als die durch die arithmetische Reihe gemessenen mittleren wnwt-Stunden anders als die entsprechenden wnwt-Stunden der älteren Uhren zu gleicher Länge tendieren. Es wäre bei beiden Skalenmodellen möglich gewesen die extrem lange letzte Vormittagsstunde (erste Nachmittagsstunde) zu unterteilen, aber zumindest in älterer Zeit gab es das Konzept von Stundenteilen nicht.*

> The Tanis clock lacks a composite scale and consists of a single value and an arithmetical row with fingers proportional to a single value. The scale of the Tanis clock is an improvement over the mean measured by the *wnwt*-hour arithmetic series or the corresponding *wnwt*-hours of older clocks of the same length. It would have been possible with both scale models to divide the extremely long last hour of the morning (first hour of the afternoon) into parts, but—at least in earlier times—the notion of subdividing the hour did not exist.

The precision of these types of shadow clocks is very relative. It must be taken into account that the *mertut* contains only one hourly scale valid for the whole year, so it is evident that something as important as the change in solar declination throughout the year is not considered (see Figure 3.16). Neither does it even seem to be taken into account that the clock should adapt to the latitude of the locality in which it was used.

In relation to these problems, a hypothesis that greatly simplifies all the complexity involved in calculating from where, how, and with what precision the time could be measured with these instruments is that proposed by Couprie (2013: 298–299).

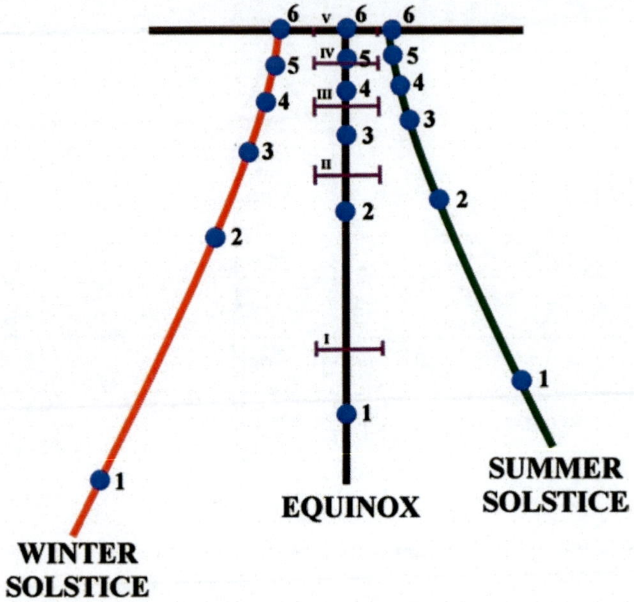

Fig. 3.16 Distribution of shadows throughout the year in a clock of the Sethy I type. (Adapted from *EAT* I fig. 37)

It is based on his interpretation of the use of the instrument and the value of the marks that appear in the *mertut*:

> The regular distribution of the marks means that the instrument was perfectly apt for making appointments or for fixing the moment of, say, a certain ceremony. When two or more persons had a copy of a *mrḫyt*, made according to identical instructions, they could easily agree to meet when the shadow had reached, e.g., the second mark in the morning, or start a celebration when it had reached the fourth mark in the afternoon, and so on. No matter the season of the year, they would all come at the appointed place at the same time. (...) (These instruments) are best equipped for making appointments in the relatively early morning and late afternoon. This is due to the way the marks were branded on the horizontal plank.

The problem inherent in the design of shadow clocks of the Sethy I type was improved over time. One way to improve the design was to make an instrument in which the shadow was not projected onto a horizontal plane but onto an inclined plane, as in sloping sundials, a type of sundial that is also characteristic of ancient Egypt. This meant, for the first and last hours of the day, that the clock did not to have to be very long. One of the best preserved specimens is one from Qantara (Borchardt, 1920: 45), a town in the eastern Delta.

The Qantara clock (see Figure 3.17) is made up of three pieces: a head with lateral vertical markings that indicate the original presence of a plumb line, an inclined plane, and a base on which the two previous pieces rest. Since the head had to project the shadow, the system that governed the use of this clock was the same one we have seen in previous examples. The clock had to be orientated towards the east and

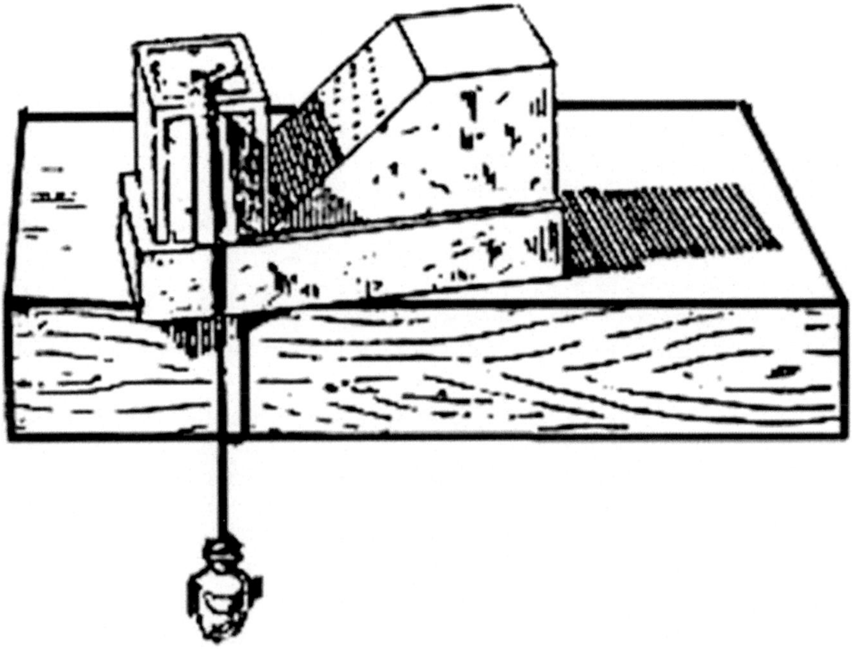

Fig. 3.17 Qantara shadow clock. (After Borchardt, 1920: 45)

at noon turned 180° to face the west to allow the shadow to continue projecting onto it. The inclined plane, as we have indicated, made it possible to shorten the length of the clock but, in addition, we see in this example that there is not one but seven scales with six dots marked on each one. So we have a clock that tries to adapt to the different declination of the sun throughout the year since the length and the hyperbolic arc of the projected shadows vary throughout the year, each month having its own scale that best approximates these changes in declination of the sun. On the other hand, the studies carried out by Borchardt (1920: 46) seem to show that this clock was designed to be used from a latitude close to 31°. However, see Henne (1991) regarding Borchardt's calculations.

In the upper horizontal part of the piece containing the inclined plane, the names of the months of the year associated with the clock scales were indicated in Greek. Thus, in relation to the first scale is indicated the month of Pharmouthi (IV Peret), in the second Phamenoth (III Peret) and Pachons (I Shemu), in the third Mecheir (II Peret) and Payni (II Shemu), in the fourth Tybi (I Peret) and Epeiph (III Shemu), in the fifth Choiak (IV Akhet) and Misore (IV Shemu), in the sixth Hathyr (III Akhet) and Thoth (I Akhet), and in the last scale Phaophi (II Akhet). The longest scale corresponds to the month Pharmouthi (IV Peret) and the shortest to Phaophi (II Akhet) since in these months the summer and winter solstices had to take place, respectively. This detail, on the other hand, helps us to calculate when this shadow clock was designed. If the summer solstice fell in the middle of IV Peret, then it must be dated around 320 BC; that is, at a date close to the beginning of the Ptolemaic Dynasty.

The diffusion and use of these clocks had spread in such a way in Graeco–Roman times that their shape was even introduced as a determinative hieroglyphic sign to designate the hour (Borchardt, 1920: 46). In fact, a good number of clocks from the Late period, and from the Ptolemaic and Roman periods have been preserved. Salmas (2014) has collected the examples that follow.

In the Louvre museum (E. 11558) there is a complete clock with an image of the god Thoth on the front. The piece corresponding to the head (6.4 cm high) of another clock on the side of which appears the figure of Re-Horakhty (Turin cat. 7353) could perhaps be from the New Kingdom or somewhat later.

Another, complete, but unfinished, clock 18.4 cm in length comes from Mit Rahine (Memphis), currently in the Cairo Museum (JE 42927). In the Petrie Museum (UC 16376) there is part of a clock made of steatite that is 13.7 cm long from the Ptolemaic period. It belonged to an Osiris priest called Esnun (Guermeur, 2003: 290–291). Only the block of the inclined plane is preserved, but the scales are distinguishable. Similar to this, of which the piece with the inclined plane has also been preserved, is another clock from the early Roman period (old Hoffmann collection) with seven scales, each with six dots to mark the hours, and the names of the associated months on each scale (Borchardt, 1920: 44, fig. 2).

From another clock from the Louvre (E 11737), of the Late Period, has only been preserved the piece (8.7 cm long) that contains the inclined plane. In this case, there is the added interest of an inscription in which the name of the astronomer Nimaatra appears. From that part of the clock there is also one that is 11.5 cm long in Cairo (JE 67342) and another, Ptolemaic, 9.3 cm high, 5.2 cm wide, and 14.7 cm long (weighing 1.3 kg) at the Metropolitan Museum of New York (Accession Number: 12.181.307). It was discovered by Howard Carter during his excavations at Dra Abu el-Nagah (see Figure 3.18). Another complete clock of the Ptolemaic period, made of wood, is kept in the Fitzwilliam Museum, Cambridge (E.GA.4596.1943).[3]

The design of this type of inclined-plane clock evolved with the appearance of composite clocks; that is, clocks in which the presence of two opposite inclined planes allowed the clock to be used throughout the day without need for reorientation after noon. One of the most interesting examples is a clock preserved in the Cairo Museum (CG 33401). It is made up of three double clocks in such a way that, by orientating the longest side of the instrument from east to west, it was no longer necessary to turn it to the west after noon (see Figure 3.19). The horizontal-plane shadow clock appears in the upper part, where a cubic piece, perhaps the *tep*, which this time is the cause of the shadow, is located right in the centre of the instrument. A series of parallel lines on both sides of this piece serve as time marks. The other two clocks are of the inclined-plane type, but while one of these is similar to that of Qantara the other has a stepped rather than a smooth plane. Symons and Khurana (2016: 380) do not believe this piece to be any type of sundial.

[3] Photograph in: https://collection.beta.fitz.ms/id/object/59048 (seen: 29/5/2021).

Fig. 3.18 Part of a Ptolemaic Period shadow clock with sloping face engraved with parallel and oblique lines. MET Accession Number 12.181.307. (Courtesy of The Metropolitan Museum of New York)

Fig. 3.19 Composite clock Cairo Museum CG 33401. (Image by Daniel Vives Menor. Courtesy of the National Museum of Egyptian Civilization in Cairo)

Another variant of this type of composite or multiple clock is the Tuna el-Gebel clock (Steinmann, 2006: pl. 22d–23a), discovered in 2004 and made of bone. The clock is from the Roman period. Curiously, the inclination of the gnomon of metal or wood that it originally possessed was 56°, instead of the 27° that its use in Tuna el-Gebel would imply.

3.2.2 *Vertical Sundials*

In 2013, a portable semicircular sundial from the Ramesside period, possibly from the late 19th Dynasty, was discovered in the Valley of the Kings (Bickel & Gautschy, 2014; Gautschy, 2020b; see Figure 3.20). The 17.5 cm × 13.5 cm clock was painted in black ink on a piece of local limestone. The upper part of its design is defined by two horizontal lines that are directed, in the central part, towards a hole 0.6 mm in

Fig. 3.20 Ramesside sundial from the Valley of the Kings. (© University of Basel, Photograph courtesy of M. Kacicnik)

width, in which the gnomon or metal or wooden rod that would project the shadow was originally inserted. Another eleven lines radiate from this hole, five to the left, a central one vertically, and another five to the right. Therefore, there is a total of 13 lines that start from the hole, all of them framed within a semicircle and with a point marked between each line in a position close to the inner edge of the semicircle.

The number of lines that radiate from the hole implies that the clock was used to measure twelve hours a day, from sunrise to sunset, the vertical line signalling noon. The gnomon projecting the shadows was probably inserted horizontally, in the manner of the so-called canonical sundials that were used during the Middle Ages to mark the canonical hours that followed the rhythm of monastic life. Otherwise, we would have to suppose a gnomon with an angle of inclination comparable to that of the latitude of the place, but the earliest-known examples of this type of sundial are from ancient Greece. However, according to a model elaborated by Vodolazhskaya (2014), applied to the clock of the Valley of the Kings, the model shows (according to this author) the 'reality of existence in ancient Egypt as a vertical sundial with a sloping gnomon, well as division of the day into 24 equal hour'.

The precision of this clock is very relative. It must be taken into account that, from the latitude of Thebes, the Egyptian hours varied in duration between 69 minutes during the summer solstice and 54 minutes during the winter solstice, owing to the changing azimuth and height of the sun in its annual movement between these two extremes. This actually implies that the design of the Valley of the Kings sundial could only be minimally accurate during the equinoxes. Well orientated towards the south (obviously), the only time of the day that would be accurately marked throughout the year would be noon, when the sun reached its culmination in the south. According to the calculations of Bickel and Gautschy, the clock presented a greater error in the indication of the hours of the afternoon, especially during the winter solstice. The clock could have been designed with 15° segments (which would make it perfect only for an equatorial observer), but the truth is that there are irregularities here as well, as this homogeneity in the segments is not observed in its design. This appears to be due simply to poor execution of the design.

An aspect of notable interest to bear in mind is what need the workers of the Valley of the Kings might have had in making use of a sundial. In the context of a temple we can see some logic in this need, given that ceremonies had to be carried out at specific times of the day. But in the context of a group of workers, the appearance of a sundial must imply a possible need to regulate working hours. This question has already been raised by Černý (1973: 53).

The Valley of the Kings sundial might be the oldest known in Egypt with the possible exception of a small Merenptah reign sundial, found at Gezer. The latter was discovered at the beginning of the twentieth century during Macalister's excavations at Gezer (1912: 331 fig. 456), and although it had initially been interpreted simply as a breastplate it was later explained as a portable sundial (Borchardt, 1920: 48; Pilcher, 1923; Sloley, 1931: 23). This last idea was maintained for much of the twentieth century (Clagett, 1995: 95–96), although more recently Symons (1999: 128 n.130) has questioned it.

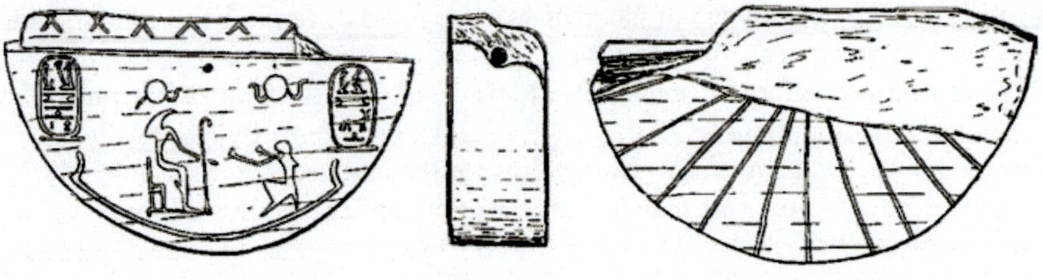

Fig. 3.21 Sundial found at Gezer. Reign of Merenptah. (After Macalister, 1912: fig. 456)

This ivory piece, currently missing, measures less than 6 cm in diameter (see Figure 3.21). On one side, under a frieze of stars, are shown the two cartouches of king Merenptah who appears on his knees worshipping the god Re-Horakhty, touched with a solar disc and *uraei* and placed on a boat. On the other side, which is where the clock should be present, there are ten radial lines. However, it is evident that, owing to the fragmented state of this part of the piece, other lines have possibly been lost. In fact, there would be enough space to trace the three that are missing and that would complete the thirteen lines on the Valley of the Kings sundial. It seems obvious that it is really a sundial, although its rudimentary nature and small size rather lead us to think that it is a model on a pectoral (Bickel & Gautschy, 2014: 10).

There is another 6 cm high specimen in the Louvre museum (E11738) (Salmas, 2014: 445 fig. 14). Its specific origin is unknown, as it comes from the antiques market. Above the gnomon hole there is a rectangular hole (between which there appear to be the two towers of a pylon) where a plumb line was probably placed to verify the verticality of the central line. Another specimen, 9.8 cm high, in this case made of wood and found in Meroë (Garstang Museum of Archeology Liverpool E.8501; Rondot & Török, 2010: 228 fig. 297), has above the clock a detailed decoration in the shape of a pylon (first century BC–second century AD). It is quite accurate.

Three other clocks show a similar design (Bickel & Gautschy, 2014: 9–11 fig. 7). One, made in faience and 7.4 cm long, is found in Berlin (Berlin ÄM 20322) (Borchardt, 1911: 67 fig. 1; Clagett, 1995: fig. III.56) and shows a design with 13 lines and a hole in the upper part that, according to Borchardt's reconstruction (1920: pl. 15), would have served for the insertion of a plumb line. Another, 9.1 cm long, is kept in Brussels (Musées Royaux d'Art et d'Histoire Bruxelles E.7330; Salmas, 2014: 445 fig. 15) and a third, from faience, in Leiden (Rijksmuseum van Oudheden, F 1987/2.2; Salmas, 2014: 445 fig. 16). These three are generally dated in the Graeco–Roman period.

A fragment of a stone sundial found in Dendara (Petrie, 1900: pl. XIX 18) and another from the temple of Amun in Naga are also known, as well as two fixed sundials found in the Ptolemaic quarry of Debod and in the Roman quarry of Kertassi (Bickel & Gautschy, 2014: 11–12). Some sundials of this type are very small, less than 10 cm in diameter, and show signs of being designed to be suspended and portable (Symons & Khurana, 2016: 379).

To cite a final example (different from the others and more interesting than Petrie UC 16500 and UC 36147; Salmas, 2014: 437 n.79), there is in the British Museum (EA 68475) a small ivory concave clock that was found in a late Ptolemaic house in Tanis during Petrie's excavations of 1884. It could possibly have belonged to an Egyptian priest whose home was located next to the temple. The clock was found incomplete and in multiple fragments that were restored in 2005 to facilitate its study. It belongs to a typology of Greek sundials (called πόλος or σκαφίς), based on Greek geometrical gnomonics, although the twelve-hour division of the day is Egyptian. It has an inscription (ΙΣΗΜΕΡΙΑ, equinox) indicating that the undercut front face signalled the equinox (Evans & Marée, 2008), since in this type of clock the undercut front face was designed to be parallel to the equator. In this case, the angle between the horizontal and the undercut front face is 57°, which corresponds to a latitude of 33°, close to the real value of the latitude of Tanis (31°). With this design, the undercut front face lies in the plane of the celestial equator, so that the sun leaves it in shadow during the spring or summer and begins to illuminate it from the autumn equinox and all winter, until with the vernal equinox when it is again in shadow. For this reason, the fact that the word ΙΣΗΜΕΡΙΑ ('equinox') appears here on this clock is proof that the illumination of this part was an excellent indicator of the equinoxes.

Finally, in relation to sundials—even if it is an example from a geographical area other than Egypt and Nubia (but not for that reason unrelated to Egyptian culture)—we must mention the so-called 'astronomical instrument of Meroë'.

In 1914 Garstang excavated the Building 950 at Meroë (Sudan, at 17° latitude), which he interpreted as an astronomical observatory. According to the excavator and other authors, some graffiti located on its west wall could represent an astronomical instrument. The ensemble appears to date from the late Ptolemaic period. Meroë was in ancient times one of the southernmost city known (mentioned in writings of Pliny, Strabo, etc.). During the Ptolemaic period there were contacts with this capital of the Meroitic kingdom, so it is also possible that the supposed astronomical instruments of the Meroë observatory were based on principles already known in Ptolemaic Egypt. Garstang (1916) discovered 'a number of graffiti representing both calculations or records of observations and actual sketches of two instruments which would seem, so far as they can be understood, to correspond to a transit instrument with a circle and an azimuth instrument' (see Figure 3.22).

According to Depuydt (1998a: 174–176), the purpose of the instrument that appears in graffito 1 would not have been to count the hours of the day, but would rather have served a purely astronomical function. The fact that two people appear to be needed to operate it distinguishes it from the sundials that we discussed earlier, since those, if we except the simple E-O turn, do not need any manipulation. In addition, this instrument seems to be more complex, since it might have consisted of a gnomon and a ring that could have had a similar function to that of a quadrant; that is, to measure angles.

Thus, if the instrument represented in graffito 1 of Meroë was used to measure the lengths of the shadow and the angular position of the sun, it is possible that there might have been some influence of trigonometry in its design, since that branch of

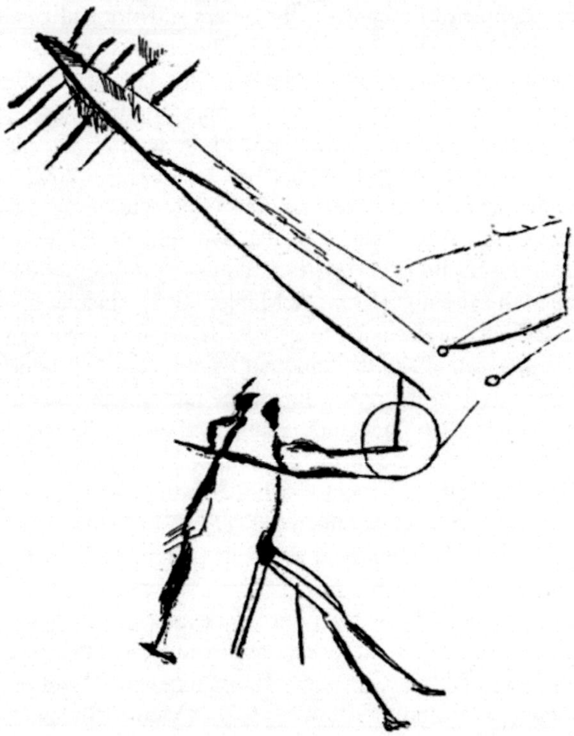

Fig. 3.22 Graffito number 1 from Building 950 at Meroë. (Adapted from Garstang, 1916: pl. VI)

mathematics had begun to be developed by Hipparchus in the second century BC. This author also recalls 'a passage in the Almagest (11.5) in which Ptolemy (second century AD) presents a proposition involving a drawing which combines a gnomon with a circle'. The circle may have been graduated and would have served to measure the angles of the shadows cast by the gnomon, whereby the Meroë instrument could have served to measure both the lengths and the angles of shadows cast from a gnomon.

Later on, Hollenback (2004) proposed an alternative interpretation of the instrument drawn on graffiti that was based on the design of the 'Qumran roundel', a 14.5 cm diameter piece found in 1954 at Qumran and similar in chronology to the graffiti from Meroë. The disc features four deeply incised concentric circles having at their centre a socket for a perpendicular gnomon, as well as three concentric, segmented, circular bands apparently intended to serve as graduated scales. The instrument would have worked as a solar compass to find the cardinal points, mark solstitial and equinoctial phenomena, and tell the time as a sundial. Therefore, comparing the Qumran roundel and the graffiti of Meroë, he considers that the latter could have had a similar function by making use of a disc with a gnomon inserted perpendicularly.

One of the graffiti from Building 950 at Meroë presents a series of lines, in groups, which evidently must constitute numerical notations in which each line symbolizes the passage of a day. Two series of strokes, that seem to add up to 45, could refer to the 45 days before and 45 days after the summer solstice in which (from Meroë) the shadow projected by a gnomon at noon would extend to the south and not to the north. This is a phenomenon that would have seemed striking to anyone who lived further north in lower Nubia. However, all in all, it is difficult to understand the meaning of this graffito.

According to Logan and Williams (2000: 77), however, the instrument drawn in Meroë would have served to determine the transits of stars. With the help of a mirror (which would be the disc shown in graffito 1, see Fig. 3.22), in which a cross would have been engraved, held in a horizontal position by the observer who remained seated and guided by the one behind him, the observer would have seen in this mirror the star that was located above it. At the same time, this operator would have used a block located in the observatory (as a fixed marker of transits) such that, when the star of the mirror was aligned with the tip of the triangle incised in the block, it would denote that the star had reached the meridian; that is, it would be in transit. In this way, this instrument would have served to count the hours of the night (Logan, 2000: 77).

3.3 Clepsydras

Clepsydras (from Greek Κλέπτω 'to steal' and τὸ ὕδωρ 'water', i.e. 'water thief') are water-clocks that could be used to measure the passage of the hours of both night and day, although quite possibly they were mainly employed for the former. They were therefore instruments for measuring time that could be used by Egyptian horologists and astronomers.

In Babylon there are references to this type of instrument (Neugebauer, 1947) mainly in textual sources such as the *Enuma Anu Enlil* and the *Mul Apin*. It seems that in the Old Babylonian period the name of the instrument was ^{giš}dib-dib, in Akkadian *dibdibbu* or *maltaktu*. Its shape is not really known, since no remains have survived, not even fragments, but it seems that it 'was a time-measuring device which could be filled and which involved a cylinder/prism out of which dripped water, the quantity of which may have been assessed by weight. It may also have included a part which filled slowly. Perhaps an alternative existed which used sand' (Brown et al., 1999–2000: 133).

Unlike in Mesopotamia, we have clear textual and iconographic evidence from ancient Egypt of the existence of clepsydras, in addition to numerous most valuable complete or fragmented copies from various periods. To be specific, the oldest reference comes from a text, absolutely exceptional in content, from the early 18th Dynasty. This text also provides the earliest evidence that the Egyptians tried to account for the seasonal differences in hours.

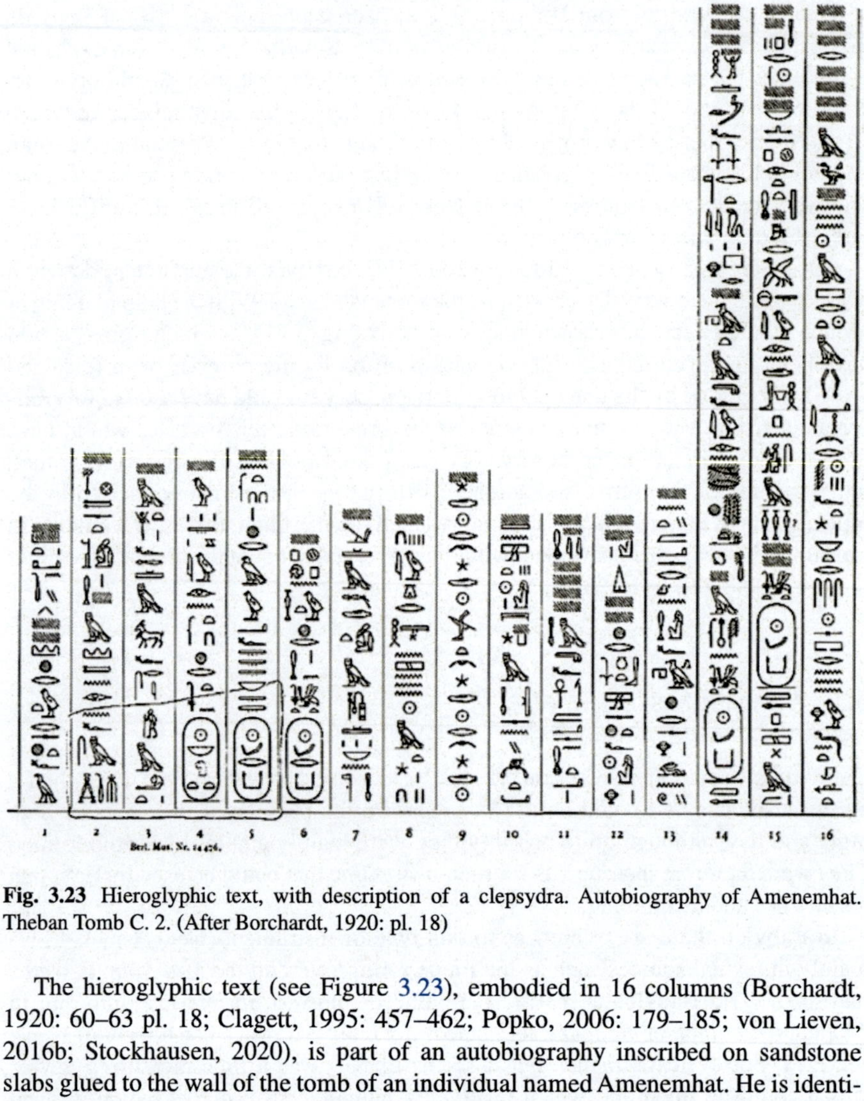

Fig. 3.23 Hieroglyphic text, with description of a clepsydra. Autobiography of Amenemhat. Theban Tomb C. 2. (After Borchardt, 1920: pl. 18)

The hieroglyphic text (see Figure 3.23), embodied in 16 columns (Borchardt, 1920: 60–63 pl. 18; Clagett, 1995: 457–462; Popko, 2006: 179–185; von Lieven, 2016b; Stockhausen, 2020), is part of an autobiography inscribed on sandstone slabs glued to the wall of the tomb of an individual named Amenemhat. He is identified with the titles *jrj-pʿt ḥꜣtj-ʿ ḫtmtj bjtj* '*iripat hatia*, bearer of the royal seal'; that is, a person of high rank. The tomb, listed as TT C. 2, in the Theban necropolis of Sheikh Abd el-Qurna, was discovered in 1885 but its exact location is currently unknown (Von Lieven, 2016b). It cannot be ruled out that, if rediscovered, still more remains of this interesting inscription could be found in the tomb.

We are faced with a document of incalculable, almost unique, historical value since it identifies a specific person as the author of a time-measuring instrument that would be prove its usefulness over many centuries. Unfortunately the text is incomplete. In general, only the final part of each column can be read, the rest is lost. The first column contains one of the oldest documented mentions of the name of one of the most important Near Eastern powers of the early 18th Dynasty, 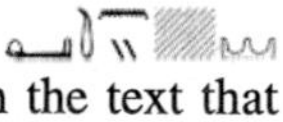 *Mj-tj-nj* (the Kingdom of Mitanni). However, there is a large gap in the text that prevents us from understanding the real purpose of this reference, although because of the little that remains legible both in this and in the second column it would seem that reference is being made to some conflict between the pharaoh and this enemy territory. However, this mention of Mitanni has been interpreted by authors such as Spalinger (1996a: 72–73) as a possible indication that knowledge of water-clock technology could have reached Egypt through Mitanni.

The third column already refers to the positions held by Amenemhat (or by his father): [...] *[wr] j3wt.f ˁ3 m s3ḥ.f sr m ḥ3t* ('[...] [great] in his office, important in his rank, an official at the forefront'), the autobiographical information that follows in the succeeding columns providing the reigns under which Amenemhat lived, which helps us to frame him quite precisely in the chronology.

Thus, in the fourth and fifth columns we read: [...]*w t3 jw jrj.n. (j) rnpt* 10 *ḫr nsw Nb-pḥtj-Rˁw* (5) [...] *[jrj].n. (j) rnpt* 21 *ḫr ḥrw wˁf t3w nb t3wj ḏsr-k3-Rˁw*, '[...] earth, I completed 10 years under king Nebpehtire (5) (I) [complet]ed 21 years under the Horus who curbs the lands, the lord of the two lands Djeserkare'. Therefore, Amenemhat lived for the last 10 years of the first king of the 18th Dynasty, Ahmose, and for the 21-years reign of his son and successor, King Amenhotep I. Since the reign of Amenhotep I was 21 years long, it is evident that Amenemhat continued to live at least during the reign of the next monarch, Thutmose I, although the huge gap in column 6 denies us that information. In any case, from the content of the rest of the text, it is evident that the 'invention', or 'improvement', of the water-clock whose authorship is due to Amenemhat took place during the reign of Amenhotep I. In fact, the sixth column again mentions Amenhotep I: [...] *[s]p tpj n ḥswt ḫr ḥm n nsw bjtj ḏsr-k3-Rˁw* ('[...] the first [t]ime of praise under the Majesty of the King of Upper- and Lower Egypt Djeserkare'), which probably indicates that, during that reign, Amenemhat accumulated more honours.

From the next column begins the part of the text that most interests us:

(7) [...] *m šdjt m sẖ3w nb*(w) *n mdw-nṯr*
[...] while reading in all writings of the god's words

(8) [...]14 *jw grḥ n šmw m wnwt* 12
14, while the night in *Shemu* is of 12 hours

(9) [...] *[3]bd r 3bd j3tt 3bd r 3bd*
[...] month by month and a reduction month by month

In column 7, after the gap, only the final part of a word is preserved: . Von Lieven (2016b: 221) suggests that it could be the verb *hbj* ('to enter'), understood here in the figurative sense of 'studying'. In any case, the meaning of what remains in this column suggests that Amenemhat instructed himself by reading texts, probably of astronomical content. Thanks to this, he was able to make certain deductions, which would be those shown in columns 8 and 9. Amenemhat indicates that the duration of the night in *Shemu* is 12 h, while in another season, coinciding with winter, it should be 14 h. That it refers to the variability of the duration of the hours throughout the months of the year, with their increase or decrease, is evident from what is indicated in column 9. The 14:12 relationship is the same as exists in the scale used in the clepsiydra of Amenhotep III found in Karnak that we shall refer to later. Therefore, '14' and '12' refer to the fourteen and twelve fingers in length of the scales used to measure twelve hours at night in specific months of the year. It does not refer to the use of equal hours. According to Fermor (1993: 364), however, the correct ratio for Upper Egypt would be 16:12, while for Lower Egypt it would be 17:12 or even 18–20:12 if considered from astronomical twilight.

(10) [...] *? šmwt Rꜥw ? m ?.snj ꜣwt*
[...] the movements of Re ? in their ?. The offering

Column 10 makes more difficult reading. Clagett (1995: 459) translated it as follows: '[I found them represented on the interior of the water-clock] and the movements of Re [and the moon-god?] With the utterances of both, and an offering'. Von Lieven (2016b) has recently proposed the following meaning: 'I represented the movements of Re, the leader of the hour priests, by their speech (ie dialogue)', while Stockhausen (2020) offers *[...] der Lauf des Re in ihrer Vereinigung. Das Überreichen.* ('[...] the course of Re in their association. The presentation.').

After Re, the word copied by Schiaparelli was , with no gaps indicated, while Borchardt copied . Doubtless, as Alexandra von Lieven has already observed, it is problematic, even taking into account errors in the copies, that the word *jaH* ('moon') could be impled here, since the appearance of the star would suppose an exceptional variant of the word.

Her suggested reading  *ḥrj wnwt* ('leader of the hour priests'), although intelligent, also suffers the disadvantage of not having examples of the use of this title in reference to Re. The group of signs that follows also presents problems. An alternative that Von Lieven gives to 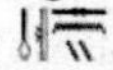 *ḫrw.snj* ('their speech') is 'their sound', which would lead us to think about the indication of the hours of the clepsydra by means of sound. Although this translation and interpretation cannot be certified, it is worth mentioning that possibility as well.

(11) [...] *mj[t]j* [...] *m-bꜣḥ.f ꜥnḫ wꜣs m-ꜥw.s mj nt-ꜥw*
'[...] a[s] [...]in front of him, *ankh* and *was* in her hand as it befits(?)'.

(12) [...] *[R]ꜥw dj* [...] *ḫr Nḫbt šm.s ḫft Rꜥw ḥr*
'[...] [R]e, [...] to Nekhbet. She goes in front of Re on'

(13) [...] *ntj m ꜥ.s r fnḏ n ḥm.f hꜣj.ḥr.s ḥr wnwj*
'[...] which is in her hand, to the nose of his majesty. She descends on two cords (?)'

(14) [...] *ḥꜥj mꜣꜣ.f nn nṯrywt ḥr prj[t] ḥꜥjt m-bꜣḥ.f*
[...] happy when he sees these goddesses going forth and descending in front of him (...)'.

The meaning of the content of these last three columns is not entirely clear but, once again, interesting possibilities present themselves. The text speaks of the movement of various goddesses (such as Nekhbet and, perhaps, Wadjet instead of the former as a representative goddess of Lower Egypt) around Re (and perhaps also the king). If it does not correspond to the description of a possible decoration of the clock, it could be a reflection of the description of a mechanical clock with moving figures, an idea initially suggested by Sethe and Quack, and supported by Von Lieven (2016b: 224). In this sense, the translation of *wnwj* as 'two cords (from *wnt*, 'cord') would be more interesting, since the cords would be part of the mechanism that would allow the movement. From this point on, column 14 continues with one of the most interesting parts of the text, since it is here that Amenemhat presents himself as the one who designed the clepsydra:

(14) (…)*jw jrj.n(.j) [mr]ḫ[yt] Hsbtj m rnpt nfr sj n nsw bjtj Dsr-kA-Raw mAa xrw*

'(…) (I) made a *merkhyt* calculated for the year. It was beautiful for the king of Upper and Lower Egypt, (Djeserkare), justified of voice'.

(15) *[jw.s] sjp.tj r tr.[s] nb n sp jrj.tw mjtt.[s] ḏr p3wt t3 jw jrj.n. (j) dbḥ pn šps m ḥswt [nt] nsw bjtj ḏsr-k3-Rʿw m3ʿ ḫrw psš m gs*

'[It was cou]nted in view of each of [its] seasons. Never before had something like it been made since the primeval time of the land. I made this august measuring vessel in honor [of] the king of Upper and Lower Egypt, Djeserkare, justified of voice, divided in half'.

The word  *mrḫyt merkhyt* is read in the copy made by Borchardt (1920: pl. 18); not so in Schiaparelli's, where the signs that he collects have nothing to do with . As we have seen, in the instructions for the clock of Sethy I in Abydos it is indicated that the  *mrḫyt merkhyt* is a piece located on the *tp* or head of the clock and, therefore, the piece in charge of projecting the shadow. The coincidence of words, taking into account that they refer to very different time-measuring instruments, clearly indicates, as we have already stated, that *merkhyt* would be a general word to designate measuring instruments regardless of their type. However, authors such as Quack[4] interpret the appearance of this word in Amenemhat's text as a demonstration that the inscription actually speaks not only about the clepsydra but also about other instruments: a *merkhet* or solar clock, a water-clock, and a mechanical clock with moving figures.

The text indicates that the instrument was calibrated. Stockhausen (2020: 56) gives the translation, at the beginning of column 15, *[…] geprüft für jede Jahreszeit* ('[…] calibrated for every season'), while von Lieven (2016b: 226), whom we have preferred to follow in this part, renders the text as '[It was cou]nted in view of each of [its] seasons'.

Another technical word used in the text in column 15 is  *dbḥ debeh*. It seems to refer to the clepsydra itself. Hannig (1997: 976) collects various meanings of this word, such as *Mass, Messgerät, Messgefäss* ('measure, measuring instrument, measuring vessel'), but also *Auslaufuhr* (that is, 'outflow clock').

(16) […] *r* […] *m ʿḳ [š]mw m prt m sḫn jʿḥ r tr.f wnwt nbt r sw.s prj mw ḥr wḏnt wʿt*

'[…] when entering Shemu, in Peret, embracing the moon at his specific time, every hour at its time. The water flows off through one spout'.

If, upon reading the text, there is still some doubt concerning the type of instrument that was being described by Amenemhat, this is completely dispelled at the end of the last column: *prj mw ḥr wḏnt wʿt* ('the water flows off through one spout'). Obviously this statement refers to the small hole through which the water exited the

[4] Reference seen in von Lieven (2016b: 228) from a work not available for the authors: J. F. Quack, *Beiträge zu den ägyptischen Dekanen und ihrer Rezeption in der griechisch-römischen Welt*, Habilitation Thesis, Berlin, 2002.

clepsydra. In this case, the word we translate as 'spout' is *wḏnt*. The determinative used is striking, the Gardiner N34 sign, denoting copper or bronze objects. This probably implies that a copper or bronze pipe or tube, the *wḏnt*, was inserted into the hole made in the clepsydra to evacuate the water. We may better understand the mechanism by describing the clepsydra of the Amenhotep III period. However, it should be noted that 'none of these plugs have so far been recognized in collections of Egyptian artefacts. If any have found their way into museum collections they have probably been identified as a large bead or similar item of jewelry' (Cotterell et al., 1986).

Amenemhat's autobiography might be much more interesting than it sounds. As indicated by Von Lieven (2016b: 229–30), 'surely, this is not the normal type of water-clock as is preserved in stone in several examples. Rather, it is in part a mechanical clock with divine figures moving on strings and lifting their arms to present symbols of life to another figure. Yet, there is also a shadow clock and a water-clock mentioned, and the description seems to imply a link between the moving figures and the water-clock. The likely explanation is that the outflowing water had a role to play in getting the figures to move'.

We have to take a very great leap in time—to the third century AD—to find another text in Egypt that refers to the characteristics of a clepsydra: pOxy. III 470 (Grenfell & Hunt, 1903: 141–146). Unfortunately, the Greek text is incomplete.

3.3.1 *Outflow Clepsydras*

In ancient Egypt we find two types of clepsydras, those of external and those of internal flow. The oldest preserved clepsydra, from the 18th Dynasty, could essentially correspond to the same type (outflow) as that apparently described in Amenemhat's text.

The calcite clepsydra from the time of Amenhotep III was found by G. Legrain in various fragments in the Karnak cachette in 1904 (Cairo JE 37525, CK 937). It has an outside height of 34.6 cm, a top diameter of 48.5 cm, and a bottom diameter of 26 cm (Figure 3.24).

For the clepsydra to work, it had to be filled with water inside, up to a higher level. The water exited through a small hole located a short distance from the base, so that the level of water inside went down as time passed. The container was therefore in the shape of a truncated cone because, as we shall see, this design is related to the compensation of the water pressure as it drops in level. For this reason, on the inner wall of the clepsydra there are twelve scales of different length that were used for each of the twelve months of the year. All the scales have a vertical series of eleven incised dots regardless of their length (except the one for the tenth month, which has twelve marks), proof that the clepsydra was designed not to measure equal hours but rather 'Egyptian' hours; that is, hours of variable duration according to the month of the year. Under each scale the ☥ *ankh* and *djed* signs appear alternately in relief (see Figures. 3.25 and 3.26). On the rim of this water-clock,

Fig. 3.24 Clepsydra of Amenhotep III found at Karnak. (Image kindly provided by Daniel Vives Menor. Courtesy of the National Museum of Egyptian Civilization in Cairo)

above each scale, are indicated the months in hieroglyph with no mention of the epagomenal days. The series of scales of the year continue approximately every 30° in an anticlockwise direction. However, there is no unanimous agreement on how many hours the clepsydra measured. Borchardt (1920) assumed it to mark twelve hours, and this idea has been followed by most authors. Dorner (1985) and Gautschy (2020c: 173) assumed that it should be ten hours, while Cotterell et al. (1986), estimate that this type of clepsydra was used to measure the civil night; that is, from sunset to sunrise. In the example of the Karnak clepsydra of Amenhotep III they estimate that the maximum error in one hour would be about 15 minutes (1986: 46).

The length of the scale corresponding to the month coinciding with the winter solstice must necessarily be different from that of the month coinciding with the summer solstice. In the clepsydra of Amenhotep III this maximum difference is observed for the months of III Akhet and I Shemu, according to new measurements by Schomberg (cited by Gautschy, 2020d: 89) or IV Akhet and II Shemu measured by Borchard (1920: 7; Clagett, 1995: 68), in which the length of the scale is fourteen and twelve fingers, respectively. The other scales vary with respect to the previous ones at an approximate rate of one-third of a finger per month, a progression that in

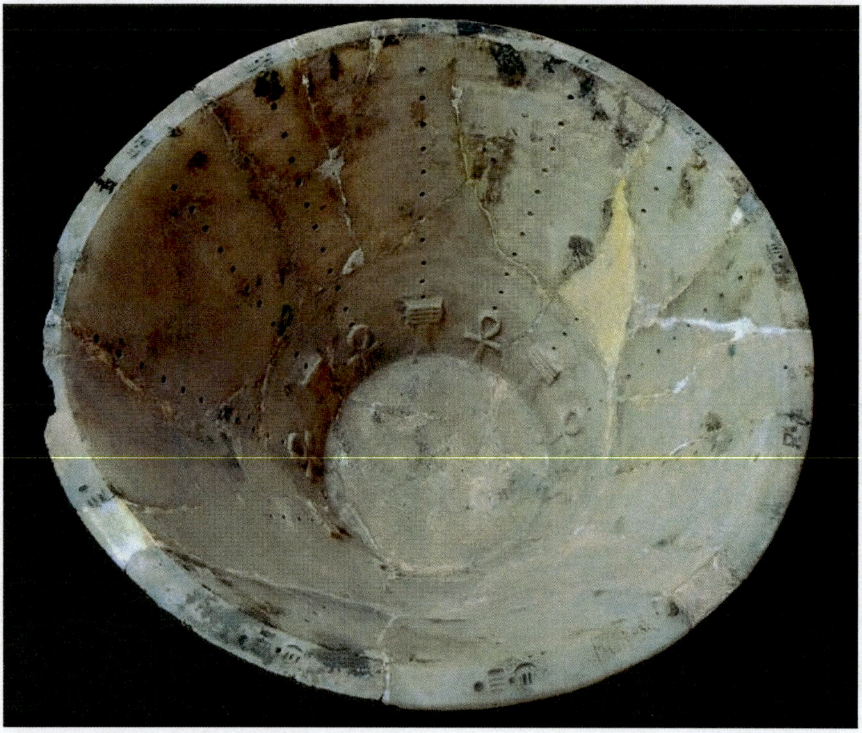

Fig. 3.25 Inside part of the clepsydra of Amenhotep III. Image kindly provided by Nacho Ares. (Courtesy of the Museum of Egyptian Antiquities in Cairo)

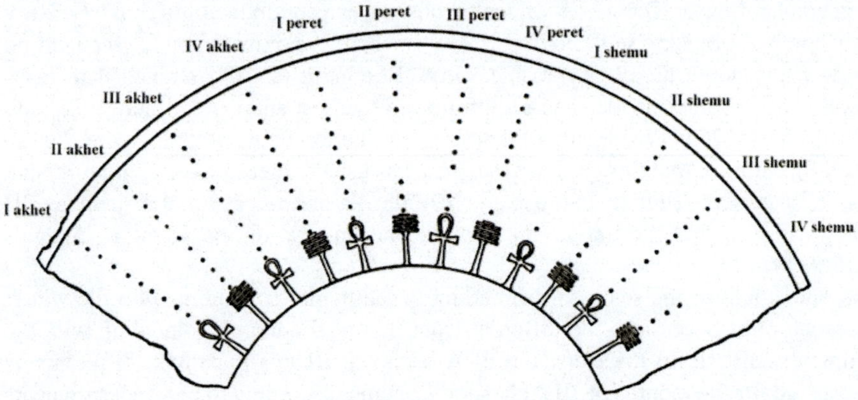

Fig. 3.26 Scales of the inside part of the clepsydra of Amenhotep III. (Adapted from Sloley, 1931: fig. 16)

no way corresponds to reality since the difference from month to month in the calendar year is not constant.

According to Clagett (1995: 68), the 14:12 ratio is not exactly suitable for Egypt; a 14:10 ratio would have been much more appropriate. Despite this, according to Naether and Ross (2008: 75), 'although this ratio differs the astronomically expected 14:10, the result is considerably better than the contemporary Babylonian approximations'. The internal inclination of the container walls is about 110°, but Sloley (1931: 175) was of the opinion that an inclination of 103° would have been more effective.

A problem with this and other clepsydras is that the first hours indicated are somewhat longer than the last hours, in such a way that only the central hours are those that are closest to our hour value of sixty minutes. There has been much discussion about the precision of these instruments. A good work in this regard, compiling and discussing various calculations made on the precision of these clepsydras, was published recently by Krauss (2015, 2016a). An obvious problem that affects our knowledge of the accuracy of the Amenhotep III clepsydra is that we do not know the size and shape of the clepsydra's outlet orifice. However, this part is known for other specimens, such as Cairo JE 67096, Museo Barracco n. 27, or St. Petersburg n. 2507a (Gautschy, 2020d: 91 n.12).

In the middle of the reign of Amenhotep III (1379–1342 BC) the summer solstice took place around July 3 (Julian calendar), which corresponded to IV Shemu 10 of the civil calendar. However, if the shortest scale presented by the clepsydra is in I Shemu, this indicates (understanding that this scale was used to count the hours of the night) that the interior design is based on a measurement valid for approximately 1768 BC (when the summer solstice was in I Shemu 1), or by the end of the eighteenth century BC, if the middle of the month was taken as a reference, in I Shemu 15. Clagett (1995: 68–69) took II Shemu as a reference to the summer solstice and calculated that this would rather make it valid for the times of Amenhotep I, making it coincide with the reference of Amenemhat in the reign of Amenhotep I. However, in this case our calculation takes us to the middle of the seventeenth century BC to II Shemu 1, early sixteenth century BC for II Shemu 15, and around 1528 BC for II Shemu 30, and therefore also before the reign of Amenhotep I.

This lag did not necessarily make it difficult to use the clepsydra, as long as the horologist took into account which scale was valid in his time. However, this is certainly a defect that could well have been corrected at the time of making this piece. Actually, there are more options that we may take into account in order to try to explain this defect. As Gautschy (2020d: 93) indicates, we may think that this clepsydra was never used as it might have been a votive object in the temple, or it could have been painted on each scale if the correct months had been used.

The iconography of the exterior decoration of the clepsydra of Amenhotep III, in three registers separated by blue bands, is similar to that of other known specimens. In the upper register there is a list of decans from the so-called Senenmut family and the planets (except Mars), and in the middle there is a representation of the so-called 'boreal constellations' with series of deities on each side, including Isis and the so-called four sons of Horus (Imset, Hapy, Duamutef, and Qebehsenuf) among others. The upper middle register is interrupted by a square in which Amenhotep III appears

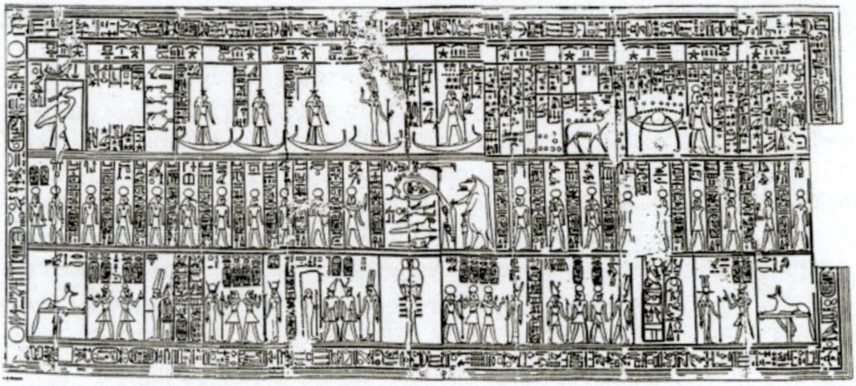

Fig. 3.27 Astronomical ceiling of the Ramesseum. (After *EAT* III: pl. 5)

accompanied by Thoth, making an offering to Re-Horakhty. In the lower register, however, we find six squares or vignettes in each of which the king is shown presenting an offering between two divinities (from the centre *tḫ - Ptḥ, Ḥwt-Ḥrw - Sḫmt, Jmnw-Rʿw nsw nṯrw - rkḥ wr* y *Rʿw-Ḥrw-ꜣḫtj - Jpt ḥmt, ḫnt* (*ḫt*)*jj - ḫnsw, Rnnwtt - rkḥ nḏs*). The outlet orifice and the highly probable metal tube are lost.

The most evident parallelism with clepsydras that we find of this decoration is to be found on the astronomical ceiling of the funerary temple of Ramesses II (see Figure 3.27), which could in fact be a copy of the previous one or have been based on the same model (Desroches-Noblecourt, 1976: 149). In the lower register of this ceiling, separating *tḫy* and *Rʿw-Ḥrw-ꜣḫtj* and below the boreal constellations, the frontal figure of a baboon appears on a *djed* pillar. Taking into account that the baboon also appears in other clepsydras, it is very likely that in the case of the clepsydra of Amenhotep III from Karnak, where this part is missing, the same motif was also repeated.

As is well known, the baboon is a hypostasis of Thoth who, in a way, we can present as the 'god of time'. At least from the 18th Dynasty the baboon was a reference to Thoth in the iconography (Larcher, 2016). As Bleeker (1973: 156) indicated, 'Thoth is not a god of the moon, but a moon-god'. Perhaps his association with the moon made it easier for him to be conceived as a god linked to time. For this reason, there are so many epithets that show this character of Thoth, such as that of *nb rnpwt* ('Lord of the years') or *wp trw ꜣbdw rnpwt* ('He who opens the seasons, months and years'). However, in late sources we find another explanation that is also directly linked to the operation of the clepsydra (albeit with internal flow). Horapollo (fifth century AD), in his *Hieroglyphica* (I: 16; Horapollo, 1840: 36–38), originally written in Egyptian but transmitted in Greek by Philippus, argues:

> Again, to signify the two Equinoxes they depict a sitting cynocephalus, for at the two equinoxes of the year it makes water twelve times in the day, once in each hour, and it does the same also during the two nights; wherefore not without reason do the Egyptians sculpture a sitting Cynocephalus on their Hydrologia (or waterclocks); and they cause the water to run

> from its member, because, as I said before, the animal thus indicates the twelve hours of the equinox (…). They also use this symbol, because it is the only animal that at the equinoxes utters its cries twelve times in the day, once in each hour.

It is understood, then, that it is in this position, through the phallus of the baboon, where the water outlet orifice of the outflow clepsydra is located, on account of its great symbolic meaning, since Thoth, as the god of time, would thus be controlling it. According to Gautschy (2020c: 173), 'the seated baboon was probably a representation of the thirteenth lunar month that was inserted into the calendar approximately every three years in order to keep lunar and solar years aligned'.

The Karnak clepsydra is the oldest that has survived. In fact, the next one that can be safely dated comes from the time of Nekau II in the 26th Dynasty (*EAT* III: 43 fig. 9), some 750 years later. Of this specimen (Cairo JE 67096) only three fragments are preserved. In the main fragment there are five text registers. In the upper one the months of the year are mentioned, of which only one (*wp rnpt*) remains complete that could be associated here with IV Shemu. The second record is a list of decans, the third one gods linked to the decans, the fourth one an address between Amun and King Nekau II, and the fifth one a list of hours of the night of which three are preserved. Coincidentally, in the main preserved fragment survives, between the fourth and fifth registers, the frontal figure of a baboon, with the drainage outlet of the clepsydra.

The particularity offered by this water-clock is that it associates groups of three decans with the months of the civil calendar. In the case of the month *wp rnpt*, the associate decans are *s3 ḳd*, *ḫ3w,* and *ʿrw*. Neugebauer and Parker (*EAT* III: 43–44) deduced that, since the beginning of the second fortnight of IV Shemu coincided around 600 BC with the beginning of the year, with a solar longitude of 275°, these decans should be located near the meridian after sunset, and therefore be able to mark the beginning or the end of the first hour of the night.

The majority of outflow clepsydra specimens that have come down to us, however, are from the Macedonian, Ptolemaic, and Roman periods. From the Macedonian period worth mentioning is the fragment BM 933, in the name of Alexander the Great, from the Tell el-Yahudiyah. Three other fragments of this clepsydra are documented, one in Berlin (ÄM 30508), another in Paris (Louvre E 30890), and a third in a private collection (Lodomez, 2007: 62). In the internal part of BM 933, several elaborate scales with dot-shaped incisions can be seen, ending in the signs *ankh*, *djed*, and *was*. At the top, above the scales and below the edge of the clepsydra, there is a horizontal line that was probably used as a reference at the time when the clepsydra was filled with water before counting time. The best-preserved scale seems to have originally had 12 marks.

We find this clepsydra (see also, Bosch-Puche, 2009: 65–69) especially interesting because of an inscription that has been preserved on the outside. It reads:

jw šsr wnwwt grḥ jm.f ẖft tm mꜣꜣ bꜣktjw r tm rdj th.tw nw nb n jr(t) jẖt-nṯr

The hours of the night are announced with it when the decans are not observed, so as not to allow each time of the offering ritual to be wrong […].

In this text, the word ‘decans’ has been written as ; that is, using the sign G31  instead of G29 , a probable scribe’s error in which, in addition, the two feathers of G31 would come from the hieratic *k*, allowing the reading *bꜣk*, and with the remaining signs *bꜣktjw*. The inscription makes it clear that the clepsydra was used at night, when the decans were not visible.

Another similar inscription appears on the Florence 2613 clepsydra (Borchardt, 1920: Taf. 4, n° 3): ‘[…] to announce the hours of the night if the decans are not visible […]’ (‘[…] *r šsr wnwwt grḥ ẖft tm dgꜣ b(ꜣ)k(tjw)* […])’; and on a fragment of a clepsydra of Ptolemy II discovered in the Tetragonos-Agora in Ephesos (now in Ephesus Archaeological Museum at Selçuk): *Die Stunden erkennen durch Thot, wenn sich der Himmel mit Wolken bezogen hat* (‘The hours are recognized by Thoth when the sky has become clouded’; Hölbl, 1984: 53). However, this does not mean that it was only used when the skies were cloudy. It is possible that the clepsydras were used from sunset or from the moment when the first stars became visible; that is, even when the decans were not visible. On the other hand, clepsydras could have been used in parallel with the observation of decans.

In the clepsydra of Brussels (E 4782) there is another interesting inscription: *Die Stunden des Tages zu erkennen, wenn Re nicht gesehen werden kann, die Stunden der Nacht kennen, [wenn die Sterne nicht erblickt werden können]* (‘The hours of the day are recognized, if Re cannot be seen, [when the stars cannot be seen]’; Schott, 1950: 909). In any case, the clepsydras could be used all day long. If the horologist used a month scale to know the hours of the night, he could also measure the hours of the day by observing the scale of a month located six months earlier.

Fragments of two other clepsydras are known from the reign of Alexander the Great (one in Brooklyn Museum 57.21.1 and another whose fragments are found between the Hermitage inv-2507a and the Naples Museo Nazionale, 2327). Also from Philippus Arrhidaeus, recognized as pharaoh in Egypt (323–316 BC) at the death of Alexander the Great, is preserved in the British Museum a fragment of a clepsydra (BM 938)[5] with a similar format to that of BM 933. On this water-clock there are two complete scales of twelve dots each and the upper horizontal line with a small vertical mark aligned with each scale. This is a detail that is not seen in the BM 933 example. On the other hand, next to the Egyptian inscription of the months were written the names of the months in Latin characters, as OCT and N[OV], because the clepsydra was reused in Roman times (Lodomez, 2007: 70).

Of the remaining known clepsydras, the following are worth mentioning, owing to their better state of preservation: one from the period of Ptolemy II found incomplete in several fragments in the Ilseo Campense in Rome but now kept at the Museo

[5] https://www.britishmuseum.org/collection/object/Y_EA938 (seen: 23/3/2021).

di Scultura Antica Giovanni Barracco (inv. MB 27).[6] This device has the particularity of being the only known example in which the outlet orifice is not located on the side close to the base but at the base of the clepsydra itself (Mengoli, 2019: 78). Another clepsydra (Roullet, 1972: pl. CXXIV) in good condition was found in 1859 in the temple of Fortuna in the Vigna Benolli, outside the Porta Portese, in Rome. From 1910 it was kept in Berlin (ÄM 19556), until its disappearance during World War II. Made of black granite, it was completely anepigraphic and without scales inside. Its iconography is similar to other well-known ones, with panels in which the king appears next to two divinities under the starry sign of the sky. This instrument could be dated to the Graeco–Roman period.

Finally, another exemplar (Ritner, 2016) that is kept in the Oriental Institute Museum in Chicago (OIM E16875). It was made of limestone, has a height of 52.5 cm (one Egyptian cubit) and a maximum width of 67 cm, although its manufacture was not completed, as evidenced by the fact that the exit orifice associated with the baboon relief was not made, or that part of the exterior decoration remained unfinished. The twelve scales inside have twelve drill holes in eight months and thirteen drill holes in the remaining four months (months 1, 3, 4 and 9). Ritner (2016: 389) dates this clepsydra to the Roman period, making it one of the last known examples in Egypt (see Figure 3.28).

A list of outflow water-clocks can be found in Symons et al. (2013)[7] and also on the website of the Berlin Waterclock Project.[8]

3.3.2 Inflow Clepsydrae

The operation of this kind of water-clock was different from the previous type, since in this case what was done was to fill the container with water, by dripping, through an external source. Undoubtedly, these clocks have a great advantage over outflow ones because, if the water supply is constant, the hours can be counted with greater precision, although following the Egyptian model, these are still variable in duration depending on the season.

The shape of inflow clepsydras is identified in greater numbers than for outflow clepsydras, in terms of its representation in reliefs, amulets, and votive objects. However, in contrast to outflow clepsydras, the number of specimens that has reached us is very small. From the known inflow clepsydra models, two main types

---

[6] http://www.museobarracco.it/es/percorsi/percorsi_per_sale/piano_primo/sala_ii_arte_egizia_e_mesopotamica/clessidra_ad_acqua (seen: 3/3/2021); a 3D model in http://repository.edition-topoi.org/collection/BWCP/single/0090/0 (seen: 24/3/2021). Anette Schomberg, Berlin Waterclock Project, 2019, Edition Topoi, DOI: 10.17171/2-10

[7] https://aea.physics.mcmaster.ca/index.php/en/database/water-clocks (seen: 23/3/2021).

[8] http://repository.edition-topoi.org/collection/BWCP/search (seen: 24/3/2021). Anette Schomberg, Berlin Waterclock Project, 2019, Edition Topoi, DOI: 10.17171/2-10

Fig. 3.28 A Roman period clepsydra kept in the Oriental Institute, Chicago (OIM E16875). (Public domain, via Wikimedia Commons. https://commons.wikimedia.org/wiki/File:Water_clock,_Egypt,_Ptolemaic_Period,_reign_of_Ptolemy_II,_285-246_BC,_limestone,_carnelian_beads_-_Oriental_Institute_Museum,_University_of_Chicago_-_DSC07912.JPG)

are distinguishable: prismatic inflow clocks and cylindrical inflow clocks, the former perhaps being of an older design than the cylindrical model.

We know that in the Ptolemaic period there was a greater development in the manufacture of more precise clepsydras. Ctesibius of Alexandria, for example, designed a model consisting of three vessels during the reign of Ptolemy II (Goodenow et al., 2015). One, higher up, was in charge of providing water to a second container, an overflow tank, which was kept at a constant water level thanks to an upper spillway and even to the use of a float in charge of gradually closing the opening of the first container while the second was filling with water. This second container would have in its lower part a water outlet in charge of filling the third and last container that would be in charge of signaling the hours as an inflow waterclock.

Therefore, this simple but ingenious system devised by Ctesibius ensured that the inflow clock received a constant and regular supply of water, thus facilitating greater precision. Ctesibius' writings have been lost, but fortunately we know of his improvement of the clepsydra and other inventions thanks to references from Hero of Alexandria (first century AD) and, especially, to those that appear in *De Architectura* by Marcus Vitruvius Pollio (first century BC), in which a chapter is dedicated to sundials and waterclocks (Schomberg, 2017: 305–306).

However, the only known specimen of an inflow clepsydra, found in 1901 by Maspero in Edfu (see Figure 3.29) is the cylindrical clepsydra kept in the Egyptian Museum Cairo (TR 2.12.26.16; Borchardt, 1920: pl. 9), which is dated around AD 100. In the first study that was published by Daressy (1902), it was concluded that

Fig. 3.29 Inner and outer section of Edfu's cylindrical clepsydra. (After Daressy, 1902: 239 fig. 2)

this object must have been originally used as a purification vessel, although it was later recognized as a clepsydra.

There is the relief of the baboon on the outside of this clepsydra. The device has inside twelve monthly scales (see Figure 3.30). The lower horizontal line of the net that decorates the inside of the cylinder marks the first hour of the night. The fact that this line is located at a variable height with respect to the bottom of the container made Petrie (1924: 50) think that it must be an outflow water-clock. The length of the various scales varies to accommodate changes in the length of the night throughout the months of the year, the shortest night being that of the tenth month and the longest that of the fourth (which would date its design to the second half of the Ptolemaic period). With respect to the Karnak clock, which is much older, the length of the scales referring to the solstices is kept at twelve and fourteen fingers for summer and winter.

According to Pogo (1936b: 412), the diagram used in this Edfu clock 'was copied either directly from an obsolete prismatic inflow clock, or, more likely, from a cylindrical inflow clock which was out-of-date'. Even more recently, Sambin (1986: 80) wrote: *La clepsydre d'Edfou étant seule de ce type, toutes les solutions proposées pour son fonctionnement restent des hypothèses* ('The Edfu clepsydra being the only one of this type, all the solutions proposed for its functioning remain hypotheses').

The other inflow clepsydrae that we know of consist of non-functional votive objects. In the Metropolitan Museum in New York we can see some beautiful examples. One of them (86.1.93) (see Figure 3.31), dated to the fourth century BC, was made in blue-green faience and is 8.8 cm high. Its shape is prismatic and between the baboon's legs it preserves the water outlet orifice. Another (17.194.2341; see Figure 3.32), only 5.7 cm high and 7.2 cm long, has been dated to the Late Period–Ptolemaic Period and was also made of green faience. In this case, the object consists of a model of a prismatic inflow clock on one side of which appears the figure of a baboon that has under its feet the water exit hole that leads to a little rectangular basin. A double row of steps below the baboon is marked on a vertical wall, and another double row of steps ascends to the rim of the basin. After Pogo (1936b:

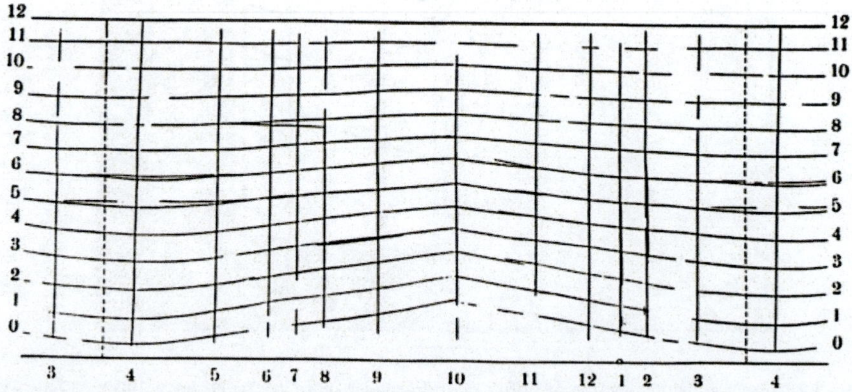

Fig. 3.30 Scales in the interior of the clepsidra of Edu. (After Borchardt, 1920)

Fig. 3.31 Votiv inflow clepsydra from the fourth century BC. Met 86.1.93. (Courtesy of the Metropolitan Museum of Art, New York)

417), these four rows may be considered as symbolizing the 24 hours of the day. This piece follows a model very similar to that of another kept in Leiden, 4.4 cm high (Leemans, 1843: 44, p. XIX no 42).

There are other types of votive objects that according to some researchers were intended to represent inflow clocks. They are called *shebet* offerings, consisting of a baboon on a basket and with a pillar (). Many were catalogued and analysed by Sambin (1987, 1988). A good example can be found in the Metropolitan Museum (10.176.45; see Figure 3.33). Dated in the Late Period, it is 11.8 cm tall and is also made of faience. In a scene in the room of the birth of the temple of Luxor, from the reign of Amenhotep III, the monarch seems to offer the goddess Mut this same type of object (Gayet, 1894: pl. LXXIV fig. 212; see Figure 3.34). The inscription mentions the offering of *šbt shebet*. For this reason, Pogo (1936b: 420–422) considered that inflow clepsydras must also date back to the 18th Dynasty. Indeed, one of the

Fig. 3.32 Votive inflow clepsydra from the Late Period-Ptolemaic Period. Met 17.194.2341. (Courtesy of the Metropolitan Museum of Art, New York)

oldest known mentions of the *shebet* appears in a temple in the fortress of Buhen (Caminos, 1974: pl. 71), from the time of Hatshepsut.

As for the iconography, there are numerous examples, from the 18th Dynasty to the Graeco–Roman period (Sambin, 1988), in which the offering of the object is usually made to a goddess. However, from the reign of Ramesses II we have an example from Buhen in which the recipient of the offering is the god Ptah (Handoussa, 1979: 67). Undoubtedly, the *shebet* in these inscriptions refers to the object that the monarch is giving to the goddess. Hannig (1997: 812) translates *šbt* as clepsydra, although this word does not appear in Erman and Grapow's *Wörterbuch*.

Handoussa (1979: 74) concluded:

> *La Sbt est le symbole du retour regulier du temps organise. Ce sont les heures du jour et de la nuit qui se succèdent. C'est le cycle lunaire qui s'ajoute au cycle solaire. C'est le temps regulier qui se suit, c'est l'ordre cosmique instaure.*
>
> 'The *šbt* is the symbol of the regulated return of organized time. It is the hours of the day and of the night that succeeds it. It is the lunar cycle that adjoins the solar cycle. It is the regulated time that follows, it is the established cosmic order'.

Fig. 3.33 Votive inflow clepsydra from the Late Period. Met 10.176.45. (Courtesy of the Metropolitan Museum of Art, New York)

However, Clagett (1995: 83) is of the opinion that what appears to be a clepsydra could be nothing more than a set of three hieroglyphic signs with the meaning 'Thoth, Lord of extended time'.

Sauneron (1963, 1969), in his study of the inscriptions in the temple of Esna, copied very interesting examples of hieroglyphics in which a seated baboon is observed, leaning against a container. In one case this object has two horizontal lines (II, n° 163, 25), in another a series of dots (II, n° 184, 22), and in a third example the truncated cone-shaped container presents a horizontal line in the middle height (IV, pl. XVII) (IV, n° 431, 2) with what appears to be the calibrated container of a clepsydra. The reading offered by this sign, *wnwt* (hour), also defines this association. However, given that the way in which these signs were represented is not exactly like those of the *shebet*, Sambin (1986: 82) concluded at first that, *il est bien peu probable que la «šbt» offerte par le roi soit la stylisation d'une horloge à eau* ('it is most improbable that the "*šbt*" offered by the king is a stylized representation of a water-clock'). In addition to the word *shebet*, two others also appear in the texts to refer to these offered objects: *wnšb* (*unsheb*) and *wtt* (*utet*), both with variants (Sambin, 1988: 236–238).

Fig. 3.34 Scene in the room of the birth of the temple of Luxor, from the reign of Amenhotep III. The monarch offers the goddess Mut a *shebet*. (After Gayet, 1894: pl. LXXIV fig. 212)

In the so-called *Book of Thoth*, Jasnow (2010: 242) has found the compound term *wnš-jb* and understands that it must be a demotic rendering of the *wnšb* (a term which in the Late Period is also written as *wšb*). Without being able to define with certainty whether or not it is a water-clock, this author considers that, 'the author of the Book of Thoth certainly brings it into association with Sothis, and the year, and the division of the year'.

Derchain (2006), based on the reinterpretation of the final episode of the ritual of IV Shemu 29 of the text dedicated to the annual development of the cult of the goddess Mut (Sauneron, 1983) at the door of the temple of Mut in Karnak, concludes that in this ritual what is delivered is a clepsydra (called here the *wnšb* object). Also to be taken into account is the latest analysis carried out by Kurth (2004b: 651), one of whose conclusions is: *Das Wenescheb wird Göttinnen dargebracht, deren spezifische Ka-Kraft die Fähigkeit zur zyklischen Erneuerung ist, und deren Ka daher durch die Opfergabe eines Zeitmessgerätes erfreut wird; insofern wird das Wenescheb, aufgefasst als Wasseruhr, den Aussagen der Texte gerecht* ('The *wnšb* object is offered to the goddesses, whose particular Ka power is their control over cyclic renewal, and whose Ka is therefore delighted by the offering of a timekeeper; thus does the *wnšb* object, understood as a water-clock, do justice to the declarations of the text'). It therefore seems, taking into account some of the contexts in which the object is offered, that it must refer to a clepsydra or, rather, to what it symbolizes, the passage of time and its renewal.

3.3.3 Clepsydras to Indicate Time- and Space-Distances

The *Heavens* (*Caelestia* in Latin or *Meteōra* in Greek) is the only surviving work by the Stoic philosopher Cleomedes, who possibly lived between the first and second century AD. This Greek author, speaking of the speed of the movement of the sky, says (110, *Heavens* II,1 184):

> The naïveté of this claim is also proved by water-clocks, since these are a means of showing that, if the Sun is 1 foot wide, then the greatest circle of the heavens will have to be 750 feet! For when the Sun's size is measured out by means of water-clocks, it is determined as 1/750th part of its own circle: that is, if, say, 1 kuathos (0.25 liters) of water flows out in the time it takes the Sun to rise completely above the horizon, then the water expelled in the whole daytime and nighttime is determined as 750 kuathoi. Such a procedure was reportedly first conceived of by Egyptians.

This idea, whose origin Cleomedes attributes to the Egyptians, is recorded in P. Berlin 23050, found during the German excavations at Elephantine between 1906 and 1908. The papyrus is dated between 650 and 550 BC. This papyrus 'indicates values (expressed in fractions) for the distance between different decans as well as within a decan constituted by several individual stars. While the numerical values as such are not easily understandable, the system seems to show similarities to the Mesopotamian system of *ziqpu*-stars where fractions of the volume of water in a water-clock are used to indicate time- and space-distances' (Quack, 2018b).

3.4 Hours of Constant Length

In all the types of clocks that we have seen so far, be they of stellar, solar or water typology, the Egyptian model of hours of unequal duration is repeated according to the season. The division of the day into 24 hours was in such a way that a half of these hours are assigned to the day and the other half to the night, with the result that, especially at the solstices, there is the maximum difference between the duration of the hours of the night with respect to those of the day. Obviously, the hours did not have fixed and equal values.

However, some documents show that the Egyptians knew the concept of hours of constant length. Of these, one of the oldest, from the time of Ramesses II (Bakir, 1966) or immediately before (Leitz, 1994b: 6), is the hieratic Cairo JE 86637 papyrus. The text that concerns us (Bakir, 1966: pl. 44A) is presented in twelve columns in such a way that each one gives information on the duration of the day and night for the twelve months of the year. Although the second half of the text presents some gaps, its restitution is simple:

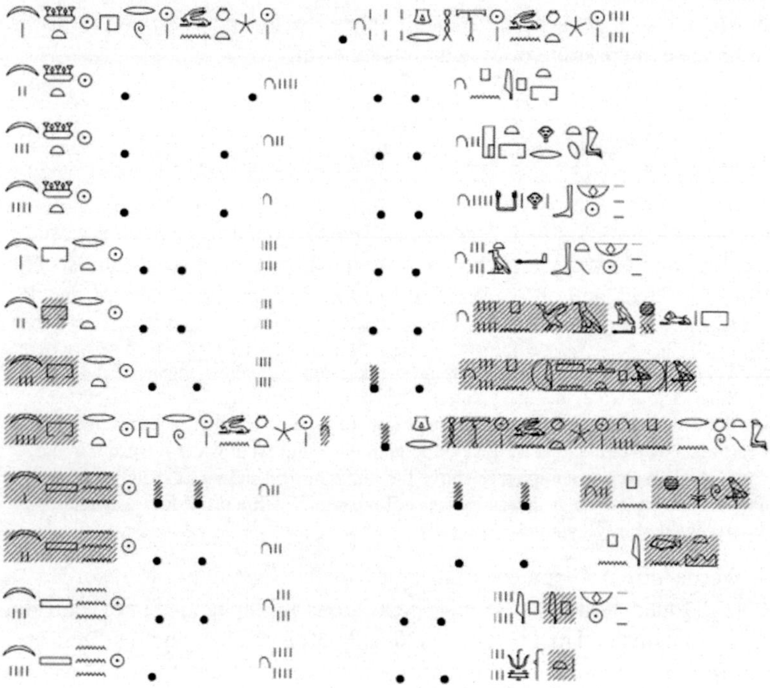

‘I Akhet,	the day (has)	:16 hours, the night:	8 hours	
II Akhet,	‘	:14 ‘ ‘	10 ‘	Pa-en-Ipet
III Akhet,	‘	:12 ‘ ‘	12 ‘	Hut-Heret
IV Akhet,	‘	:10 ‘ ‘	14 ‘	Ka-her-ka
I Peret,	‘	:8 ‘ ‘	16 [‘]	Ta-abet
II [Pe]ret,	‘	:6 ‘ ‘	1[8] ‘	Me[khi]r
[III Pe]ret,	‘	:8 ‘ ‘	1[6] ‘	[P(a)-en-Amenhotep]
[IV Pe]ret,	the day (has)	:[10] hours, the ni[ght:	14 hours	P(a)-en-]Renenutet
[I Shemu],	[‘]	:12 [‘] [‘	12] ‘	P(a)-en-[Khonsu]
[II Shemu],	‘	:12(sic) ‘ ‘	‘	P(a)-en-inet
III Shemu,	‘	:16 ‘ ‘	8 ‘	Ip[ip]
IV Shemu,	‘	:18 ‘	6 ‘	Wep-renep[et]’.

The monthly sequence is clear. The total of hours, of both day and night, although varying in proportion each month, adds up to 24 hours, which makes it possible to deduce that hours of equal duration are being considered. Month by month, the night and the day are shortened or lengthened regularly at the rate of two hours each month. Thus a regular monthly change in the length of daylight is assumed, although this does not adjust to reality because, as in dates near the solstices, the daily variation occurs more slowly; at the equinoxes the rate of variation is faster. Neither is it really the case for Egypt that the ratio of the longest to the shortest day is 3:1 (18-hour days or nights/6-hour days or nights, instead of 14–10 hours approximately), counting the day as the time elapsed from sunrise to sunset.

The longest day, as can be deduced from the data provided by the papyrus, occurred in IV Shemu, with eighteen hours of day and only six hours of night. The summer solstice took place in that month between 1400 and 1280 BC (*EAT* I: 119; see Table 3.9).

Another document, from which the same type of information can be extracted, although it presents certain differences from the previous one, was found in two

Table 3.9 Gregorian, Julian,– and Egyptian date (*EAT* I: 128)

	Gregorian Calendar			
	March 21		June 21	
Year	Julian Cal.	Civil Cal.	Julian Cal.	Civil Cal.
–2800	April 13	I Shemu 24	July 14	IV Shemu 26
–2700	April 12	II Shemu 18	July 13	I Akhet 15
–2600	April 11	III Shemu 12	July 12	II Akhet 9
–2500	April 10	IV Shemu 6	July 11	III Akhet 3
–2400	April 10	Epagomen 1	July 11	III Akhet 28
–2300	April 9	I Akhet 20	July 10	IV Akhet 22
–2200	April 8	II Akhet 14	July 9	I Peret 16
–2100	April 7	III Akhet 8	July 8	II Peret 10
–2000	April 7	IV Akhet 3	July 8	III Peret 5
–1900	April 6	IV Akhet 27	July 7	III Peret 29
–1800	April 5	I Peret 21	July 6	IV Peret 23
–1700	April 4	II Peret 15	July 5	I Shemu 17
–1600	April 4	III Peret 10	July 5	II Shemu 12
–1500	April 3	IV Peret 4	July 4	III Shemu 6
–1400	April 2	IV Peret 28	July 3	III Shemu 30
–1300	April 1	I Shemu 22	July 2	IV Shemu 24
–1200	April 1	II Shemu 17	July 2	I Akhet 14
–1100	March 31	III Shemu 11	July 1	II Akhet 8
–1000	March 30	IV Shemu 5	June 30	III Akhet 2
–900	March 29	IV Shemu 29	June 29	III Akhet 26
–800	March 29	I Akhet 19	June 29	IV Akhet 21
–700	March 28	II Akhet 13	June 28	I Peret 15
–600	March 27	III Akhet 7	June 27	II Peret 9
–500	March 26	IV Akhet 1	June 26	III Peret 3
–400	March 26	IV Akhet 26	June 26	III Peret 28

(continued)

Table 3.9 (continued)

	Gregorian Calendar			
	March 21		June 21	
Year	Julian Cal.	Civil Cal.	Julian Cal.	Civil Cal.
−300	March 25	I Peret 20	June 25	IV Peret 22
−200	March 24	II Peret 14	June 24	I Shemu 16
−100	March 23	III Peret 18	June 23	II Shemu 10
0	March 23	IV Peret 3	June 23	III Shemu 5
100	March 22	IV Peret 26	June 22	III Shemu 28
200	March 21	I Shemu 21	June 21	IV Shemu 23

pieces of a stone slab at Tanis in 1947 and 1948 (Clère, 1949). Neugebauer and Parker dated this inscription to the time of Nekau II (*EAT* III: 44), 26th Dynasty. Recently, however, taking into account the evolution of demotic writing and the presence in the text of some signs linked to this writing, Hoffmann (2017: 140) has proposed a dating between 650 and 300 BC.

Furthermore, the same author has recognized a Mesopotamian influence in the information of the data provided by the Tanite text (where a 3:2 relationship between the longest day and the shortest night could be deduced). This might also be useful in specifying the dating, since the 3:2 ratio appears there at the latest around 600 BC, and around 400 BC at the latest Egypt returns to the ratio 3:1, which in parallel with the ratio 2:1 will be documented until Roman times (Hoffmann, 2017: 153).

In text II of the document (Clère, 1949: 8 fig. 2), the following can be understood (*EAT* III: 45–46; Clagett, 1995: 101–102; corrected by Hoffmann (2017) from the reading of the ʒ and y signs as 1/3 and 2/3, respectively):

(1) Knowledge of the quantity of the Daylight in relationship to the Night

	I Akhet 1	daylight:	10 2/3 hours		nighttime: [13 1/3]
	[I Akhet 15	day] **(2)** light:	11		nighttime: 13
	II Akhet 1	daylight:	11 1/2		nighttime: [12 1/2]
	[II Akhet 15	daylight:	12]	**(3)**	nighttime: 12
	III Akhet 1	daylight:	12 2/3		nighttime: 11 1/3
	[III Akhet 15	daylight:	...		nighttime: ...]
(4)	IV Akhet 1	daylight:	13 1/3		nighttime: 10 2/3
	IV Akhet [15	daylight:	13...		nighttime: 10...]
(5)	[I Pere]t 1	daylight:	14		nighttime: 10
	I Peret 1[5	daylight:	14 (2/5 ?)		nighttime: 9 (3/5 ?)]
	[II Peret]1 **(6)**	daylight:	14		nighttime: 10
	II Peret 15	day[light:	13...		nighttime: 10...]
	[III Peret 1	daylight:	13]	**(7)**	nighttime: 11
	III Peret 15	daylight:	12 2/3 1/6		nighttime: [11 1/6]
	[IV Peret 1	daylight:	12...		nighttime: 11...]
	[IV Peret **(8)**]15	daylight:	12		nighttime: 12
	I Shemu 1	[daylight:	11...		nighttime: 12...]
	[I Shemu 15	daylight:	11 1/6	**(9)**	night]time: 12 2/3 1/6
	II Shemu 1	daylight:	22 (sic)		nighttime: [...]
	(II Shemu 15	daylight:	...		nighttime: ...)
	[III Shemu 1	daylight:	...		nighttime: ...]
	[III Shemu 15	daylight:	9 1/3	**(10)**	nighttime: 1]4 1/2 1/6
	IV Shemu 1	daylight:	[...		nighttime: ...]
	[IV Shemu 15	daylight:	...		nighttime: ...]

On this occasion, the time lapses occur at 15 days intervals, but we can see how the rate of growth or decrease of the hours, whether during the day or night, is not totally regular, so Clère (1949) suggested that the duration values of the text of Tanis certainly rest upon observations, an opinion shared neither by Neugebauer and Parker (*EAT* III: 47) nor by Clagett (1995: 104). The text, on the other hand, presents a series of errors linked to the mistakes of the copyist charged with passing the written information from an original papyrus to this stone slab. Thus, for example, in column 9 of the text, for II Shemu 1, it is indicated that the day lasts 22 h, which is obviously a glaring error. Likewise, we must see as an error the non-mention of the fortnight that begins in II Shemu 15.

According to the information provided by the Tanis text, it must be deduced that the longest day of the year took place in the second fortnight of I Peret, being more than 14 hours long (or $14\frac{2}{5}$ hours according to Hoffmann's proposal for a ratio 3:2 instead of the 3:1 of the Ramesside text). If we are looking for a summer solstice that coincides with that fortnight, we must go back from the time of the Nubian king Shabaka to the Saite king Psametik I; that is, during a good part of the seventh century BC (see Table 3.9). In keeping with this date for the summer solstice, the winter solstice would take place on III Shemu 15, and the equinoxes on II Akhet 15 and IV Peret 15.

Clère (1948: 18), in his study, made an interesting comparison between the values provided by the Tanis text and the values obtained by modern measurements from various cities (Tanis, Memphis, Thebes, and Aswan) located at different latitudes (see Table 3.10):

The biggest errors between the values offered by the Tanis text and the real values measured from Tanis appear in the following fortnights: III Peret 1 (−35′), III Peret 15 (−25′), IV Peret 15 (−20′) and III Shemu 15 (−45′). The rest of the data

Table 3.10 Comparison of the values of daylight length given by the Tanis text and the real value for four cities of Egypt (with corrected values for the Tanis text)

Text of Tanis				Real length of the day			
Date			Length day	Tanis	Memphis	Thebes	Aswan
Akhet	I	1	10 h 40 m	10 h 40 m	10 h 45 m	11 h 00 m	11 h 05 m
		15	11 h 00 m	11 h 05 m	11 h 10 m	11 h 20 m	11 h 20 m
	II	1	11 h 30 m	11 h 35 m	11 h 35 m	11 h 40 m	11 h 45 m
		15	12 h 00 m	12 h 00 m	12 h 00 m	12 h 00 m	12 h 00m
	III	1	12 h 40 m	12 h 30 m	12 h 30 m	12 h 25 m	12 h 25 m
	IV	1	13 h 20 m	13 h 25 m	13 h 20 m	13 h 10 m	13 h 05 m
Peret	I	1	14 h 00 m	14 h 00 m	13 h 55 m	13 h 35 m	13 h 30 m
	II	1	14 h 00 m	14 h 05 m	14 h 00 m	13 h 40 m	13 h 40 m
	III	1	13 h 00 m	13 h 35 m	13 h 30 m	13 h 20 m	13 h 10 m
		15	12 h 50 m	13 h 15 m	13 h 10 m	13 h 00 m	12 h 55 m
	IV	15	12 h 00 m	12 h 20 m	12 h 20 m	12 h 20 m	12 h 15 m
Shemu	I	15	11 h 10 m	11 h 25 m	11 h 30 m	11 h 35 m	11 h 35 m
	III	15	09 h 20 m	10 h 05 m	10 h 10 m	10 h 30 m	10 h 40 m

shows minor errors that can range from 5 to 15 minutes, or in their best precision, the null errors in I Akhet 1, II Akhet 15, and I Peret 1.

We have to wait until the end of the second century AD to find Egyptian texts (as in ostraca OMM 170, 796, and 844 from Medînet Madi) that provide information on changes in the duration of daylight from month to month with different fractions of change. This would go from 1/12 to 1/16, for each month of the year (Naether & Ross, 2008: 81–83).

Chapter 4
Mapping the Stars: The Skies of Ancient Egypt

From an early date, observation of the celestial vault enabled the Egyptian astronomers to distinguish the five visible planets from the background of stars; unfortunately, however, very little information about these planets or their movements and periods has reached us from the period before the Graeco–Roman period.

It is also rather depressing that the records that the Egyptians undoubtedly accumulated over thousands of years of observing comets, solar and lunar eclipses, and other regular phenomena, such as observations of storms or showers of shooting stars, or of unpredictable but equally striking events in the sky such as fireballs or bright meteors have been lost. Even so, it is possible to obtain information of interest from the interpretation of various texts.

As for the stars and constellations we have more information, although much of it comes from documents associated with the funerary world and with the limitations that this implies. Through the diagonal star clocks, the book of the *Fundamentals of the Course of the Stars*, lists of decanal stars, Ramesside star tables, and other similar sources we have lists of decans and hourly stars, so this specific group of stars is the best known in Pharaonic times. We also have the representation of certain Egyptian northern and southern constellations on astronomical ceilings and other similar documents, and of late but highly interesting representations, such as the remarkable Dendara zodiacs. However, the identification of these stars and constellations has been the subject of discussion, especially since Neugebauer and Parker described these identifications as practically impossible. Fortunately, this negative view has been superseded, at least partially, so the study of Egyptian decans, hourly stars, and constellations, in spite of the difficulties, has served to offer an approximate mapping of part of the Egyptian sky.

J. A. Belmonte, J. Lull, *Astronomy of Ancient Egypt*, Historical & Cultural Astronomy, https://doi.org/10.1007/978-3-031-11829-6_4

4.1 The Sun and the Moon

In the House of the Book of the temple of Horus in Edfu there is a list of works among which, owing to the astronomical theme that its title suggests, we may highlight the title *rḫ nmtt n ḥꜣytj* ('Knowledge of the periodic returns of the Sun and the Moon'; Alliot, 1949: 148). Monitoring of the apparent movements of the sun and the moon enabled the Egyptians to know the regularity of the displacements of both objects and their extreme positions on the celestial vault, and, from this knowledge, their calendrical connotations and the different duration of day and night throughout of the year. The duration of the tropical, or solar, year would have been established at 365 days from a very early time, at least at the beginning of the third millennium.

The sun, source of heat and life, was considered the most powerful heavenly object, the solar form of the Heliopolitan demiurge, so numerous hymns were dedicated to him as god. A set of words define the main characteristics offered by observing the sun, *rʿw* 're'. The solar disc was defined as *jtn*, 'iten' (von Lieven, 2001); its morning or evening reddish colour, being low above the horizon, was the *jmꜣw*, 'imau'; its glow the *jꜣḫw*, 'iakhu'; its light *šw* 'shu'; and its rays *stwt* 'setut', among other related words (Hannig & Vomberg, 1998: 255–256). As for the nature of the sun, the Egyptians related it to fire, but, so far as we are aware, they never went beyond that simple explanation. It is also interesting to observe how in the Graeco–Roman period many reliefs show the sphericity of the sun, especially if we look at the decoration of the lintels of the monumental doors (see Figure 4.1).

The observing conditions at the level of the horizon, which in Egypt are usually poor owing to air-borne dust, could facilitate the observation of sunspots, especially

Fig. 4.1 Winged sun disc at Philae. Chapel of Imhotep. Ptolemaic Period. (Image by the authors)

when they were large during solar maximum, the atmosphere acted as a filter for their observation. In fact, it seems that it was the morning mist conditions in some valleys of ancient China that facilitated the first observations of sunspots with the naked eye that appear to be found in the 9th–8th century BC (referenced in the *Chou Í* or book of *Changes of Zhou*, a divinatory and oracular text). However, no Egyptian documents are known where these observations can be verified.

Congdom (2000) came to interpret an unusual representation of the solar disc (Aten) as a possible representation of a solar halo, which appears in the tomb of the High Priest of Aten, Meryre, in the northern tombs of Amarna (TA 4; see Figure 4.2). In this iconography it is observed how between the solar disc and the rays that emanate from it a series of arches were drawn as necklaces which, according to this author, could correspond to the atmospheric phenomenon known as the 'solar halo', and which occurs when low temperatures in the upper layers of the atmosphere cause the formation of small ice crystals that modify the path of the Sun's rays, creating arcs that are easily visible to the observer. However, this hypothesis is difficult to corroborate since the inscription that accompanies this scene does not offer any clues in this regard.

Similarly, although it might be assumed that the ancient Egyptians observed the zodiacal light, a diffuse and faint white glow in the night sky that extends from the sun along the ecliptic caused by sunlight scattered by interplanetary dust, a phenomenon documented in modern times (Grusson, 1895). It is not easy to identify references to this phenomenon in Egyptian documentation.

Fig. 4.2 Curious image of Aten in Meryre's tomb, Amarna. (Image by the authors)

An astronomical phenomenon of great interest to the Mesopotamian priests were solar eclipses. Knowing the cycles of the sun and the moon, they were able to establish ephemerides in order to predict them, although the actual geographical path of a given solar eclipse could not be predicted in ancient times. The priests might know that a solar eclipse was due, but not where it would be visible from, or its degree of totality. Observation of both objects, if carried out continuously and accurately for dozens of years, can lead to a knowledge of the regularity with which eclipses take place. If the ancient Egyptians made observations of this kind, they could have come up with similar ephemerides. Seneca's reference (*Quaestiones Naturales* VII, 3) to the solar eclipses recorded by the Egyptians is well known, when describing the investigations of Conon (3rd century BC): 'Conon, himself a diligent inquirer, actually collected the eclipses of the sun recorded by the Egyptians'.

However, the sources that we preserve do not clearly mention observations of this type, although there are some references in documents that remind us of the observation of past eclipses or the prediction of these phenomena. One author has chosen to recognize references to total eclipses of the sun in texts of the New Kingdom, mainly from Deir el-Medina (Smith, 2007). According to Priskin (2019a: 167), although in a religious context as a mythological account, a total solar eclipse would also have been referred to in Spell 160 of the *Coffin Texts*.

Originally, the Berlin demotic papyrus 13588 (Möller, 1918; Erichsen, 1956), found at Abusir el-Meleq (1st century BC), was supposed to contain a possible reference to the partial eclipse of the sun on September 30, 610 BC, while Psametik I was being embalmed (Hornung, 1965): 'It is in Tjebne, my town, that I have heard that the sky swallowed up the sun-disk' (Caminos, 1958: 89).

The expression used to refer to the eclipse in pBerlin 13588 is 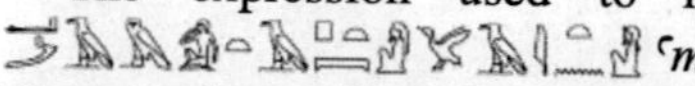 *ꜥm tꜣ pt pꜣ jtn* 'the sky swallowed up the disk'. There are several meanings of the verb 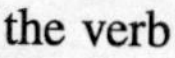 *ꜥm*: *verschlucken* ('swallow'), *trinken* ('drink'), *verschlingen* ('engulf'), *einatmen* ('inhale'), etc. (*Wb* I, 183–184), that we can understand from the context in which this verb is found in the sense of overshadowing, eclipsing.

This same sense is observed in other documents. Thus, for example, in an Egyptian mythological text it is said:  'let not the sun-disk grow dark',  'let not the sky swallow the moon' (Caminos, 1958: 89), referring, respectively, to eclipses of the sun and of the moon. Here we find the verb *verfinstern* ('darken'), *umnachten* ('cloud over'),  *knḥ*, whose noun is 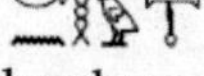 *ḳnḥw* (Hannig, 1997: 885), and the verb 'swallow',  *ꜥm*, already mentioned.

More recently, however, it has been concluded that the eclipse referred to in pBerlin 13588 would have been the lunar eclipse of March 22, 610 BC, visible from Daphnae (Smith, 1991; Ray & Gilmore, 2006; Krauss, 2006a: 377), and not the partial solar eclipse observed in September of that year. This hypothesis is partly based on the possible differential use of expressions to refer to eclipses of the moon and the sun (as we have seen in the previous paragraph). It is also based on a better translation and interpretation of the text. Thus, for example, it is understood that the eclipse would have occurred after the death of the king beyond the borders of Egypt

(Smith, 1991; Depauw, 1997: 170) and the term *wꜥbt* should be translated as 'death' rather than 'embalming place'.

An equally interesting aspect of pBerlin 13588 is the indication to the consultation of an astronomical text by the king: 'Let the book be brought (…) and let us find out whether Re has *jn-sꜣn* (writing for *jn-snj* pass by) the disk (that is, the moon)' (Erichsen, 1956: 73; Depuydt, 1995b: 54, n. 50).

Records of observations are very scarce (Jones, 1994: 37). One, from a much later period, is a Theban ostracon kept in the Turin museum and written in Coptic. It refers to an eclipse of the sun observed from Djeme (Medinet Habu) in March of the year AD 601 (Allen, 1947: 268): 'On the fourteenth of Phamenoth of the fourth indiction, the sun was eclipsed in the fourth hour of the day and in the year in which Peter, son of Palu, was made village official in Djeme'.

After Ray and Gilmore (2006), it is the only example that we have of a datable solar eclipse from the entire length of Egyptian history, before the coming of the Islamic era. On the other hand, it is curious to note that the text was surrounded with signs of the Cross, so that it can be deduced that these signs could have been used as amulets against the negative aspect of the eclipse.

It is possible that, already in the New Kingdom, the Egyptians had understood the relationship between the lunar conjunction and the possibility of an eclipse of the sun. In this sense, the sign N9 and N10 of Gardiner (𓇹𓇺) could be taken into account, because both could represent the sun and the moon at the same time (Depuydt, 1998c).

Eclipses of the sun were most probably signs of bad omen. According to 6th century AD author Joannes Laurentius Lydus: 'It must be understood that Manetho in his book *On Festivals* states that a solar eclipse exerts a baneful influence upon men in their head and stomach' (Waddell, 1964: 199, Fr. 84).

The moon was known by the name of *jꜥḥ*, 'iakh'. and its phases received the following names: new moon *psḏntjw*. 'pesedjentiu' (Depuydt, 1998c), first quarter *dnjt*. 'denit' (day 7), full moon *smdt,* 'semedet', and third quarter *dnjt,* 'denit' (day 23).

Knowledge of the lunar cycle undoubtedly dates back to very ancient times in Egyptian history, but the clearest picture of the whole lunar cycle is that which can be clearly seen in the recently restored pronaos of the Roman temple of Esna. Examples, chronologically earlier but especially from the Graeco–Roman era (Priskin, 2016), are limited to showing series of gods that symbolize each of the days of the lunar cycle. A well-known example is that seen in the pronaos of the Temple of Hathor at Dendara (see Figure 4.3), also in the Temple of Horus at Edfu (Barguet, 1977). Precisely in the Ptolemaic epoch an epithet used to refer to the moon is *wḥm ḳj.f* ('he who always renews his form'; see also Brugsch, 1891: 330).

As far as the observation of its details is concerned, we also know relatively little of the moon, but it is evident that its observation must have been very important, since many rituals took into account certain phases of the moon (Grimm, 1994), especially the new moon, but also the first crescent and full moon, and other

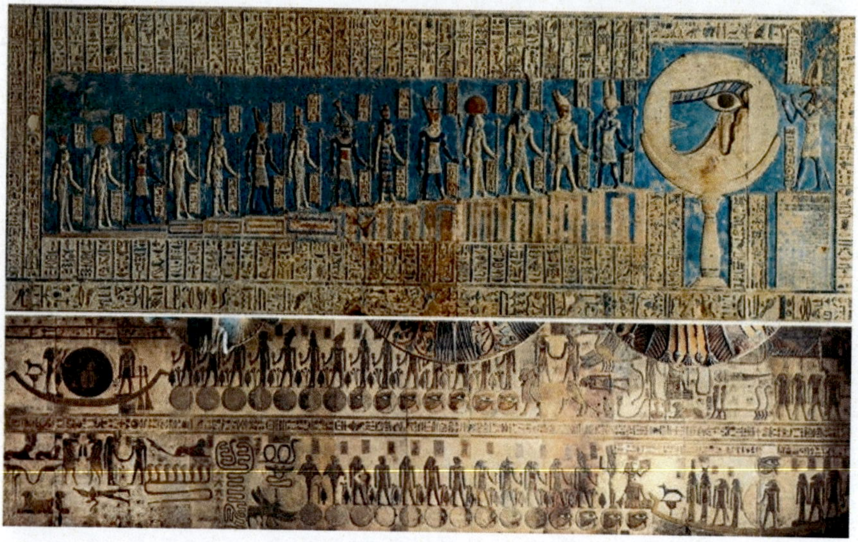

Fig. 4.3 Comparative lunar representations on the pronaos ceiling of the Dendara and Esna temples. (Images courtesy of Marc Mateos Comaposada)

apparently less significant phases such as the sixth lunation day (Altmann-Wendling, 2019: 214–215) and the fourth lunation day (Krauss, 2015). On the other hand, important annual festivals, even those linked to a specific month of the civil calendar, were set according to a specific lunar phase. In this sense, one of the most famous and important was *ḥ3b sẖn nfr* 'the 'Beautiful Embrace', in which the statue of Hathor joined the statue of Horus at Edfu in III Shemu, beginning with the new moon and ending with the full moon fifteen days later, although strictly speaking the full moon varies from 13.73 to 15.80 days after the conjunction. Also, the Festival of Min at Thebes began with the new moon of I Shemu, the Feast of the Valley was celebrated at new moon in II Shemu, and even the important Feast of Opet began on a new moon day, at least under Thutmose III (Fukaya, 2019) in II Akhet.

The synodic cycle of the moon was soon well established at 29.5 days and, for this reason, the lunar months are characterized by successions of months of 30 and 29 days. According to researcher Altmann-Wendling (2019: 235), who has dedicated several in-depth studies to the moon in ancient Egypt, designations as 'he who traverses the sky' (*ẖns nnt*), or 'Lord of the Path' (*nb mṯn*) might refer to its approximately 13° daily eastwards movement.

There are tables, especially from the Graeco–Roman period, with indications of when the new moons occurred (with respect to the civil calendar), in cycles of 25 calendar years (see also Chap. 5). Of particular note are the demotic Carlsberg 9 papyrus and the Rylands IV 589 papyrus (which although it is in Greek could perfectly well have been written by Egyptians), both from the 2nd century AD. Both papyri show that the Egyptians had arrived at the deduction that 25 calendar years correspond to 309 lunar months. Twenty-five civil years (365 × 25) amount to 9125 days, and 309 synodic lunar months of 29.53059 days five a total of 9124.95231 days; that is, only one hour less than the 25 civil years. This small lag would only produce an error of one day every 500 years. This 25-year cycle, fixed in the Egyptian civil calendar and therefore of Egyptian tradition, is also referred to in Ptolemy's *Almagest*, so it can be understood as a proof of knowledge transfer from the Egyptian world (Quack, 2016: 234).

The Carlsberg 638 papyrus (Hoffmann & Jones, 2009), written in demotic, gives the position of the moon in degrees inside the zodiacal constellation in which the moon will be observed, during part of the months of August and September of the year AD 13. It is the oldest document that presents the so-called 'Standard Lunar Scheme', which would last until the 4th century AD (Jones, 1997). Other texts, such as the Lund 35a papyrus (Jones & Perale, 2012: 2), records for the years AD 69–109, the moment when the moon is at its greatest distance from the earth and its position in the zodiacal band. The oBerlin 30539 is also of interest (Hoffmann, 2010), as it offers information on the new moon of the year AD 184 and 185 (reign of Commodus), specifying at what time this event took place.

It is interesting to note the possibility that the Egyptians had already assumed that the moonlight was a reflection of the sunlight on its surface. This has been deduced from the reading and interpretation of some texts. Thus, for example, in the *Great Hymn to the Aten* it is said, 'for the left eye receives the light of the right eye', or in Chapter 2 of the *Book of Coming Forth by Day* when it is said, 'O you Sole One who rises in the moon, O you Sole One who glistens in the moon' (reading *m* as a preposition 'in'; Beatty, 1998: 129–131).

Eclipses of the moon had necessarily to be observed in great numbers throughout the history of Egypt, but, as we have seen with the eclipses of the sun, here also the documentation is very scarce. An exception would be the lunar eclipse of March 22, 610 BC referred to, as explained above, in pBerlin 13588. According to Beatty (1998: 205–211); on the other hand, a symbolic reference to the lunar eclipse could be interpreted in chapter 112 of the *Book of Coming Forth by Day.*

In the *Chronicle of Prince Osorkon*, already mentioned in chapter 2, reference is made to the bad omen that the observation of a lunar eclipse in the time of Takelot II would have implied. But here it is also necessary to point out other possible interpretations of this text. Krauss (2007), for example, indicates the possibility that the text refers to a delayed appearance of the new crescent.

Solar and lunar omina, present in the Vienna demotic Papyrus (Parker, 1959), a text possibly written at the end of the 2nd century AD, seems to have contents of Mesopotamian influence from the first Persian period in Egypt.

In this sense, Quintus Curtius Rufus, in his *History of Alexander* (IV: 4–8), recalls an event associated with a lunar eclipse, prior to one of the encounters between the Macedonian army and the Persian army of Darius III. Part of the Macedonian army was scared, so Egyptian diviners with a good knowledge of the celestial vault had to intervene:

> Already the affair was approaching a mutiny, when Alexander, unterrified in the face of everything, ordered the generals and the higher officers of the soldiers to appear in full numbers at the king's tent, and the Egyptian soothsayers, whom he believed to be the most skilled in reading the heavens and the stars, to declare their opinion.

> But they, although they knew well enough that the heavenly bodies which determine the seasons have their destined changes, and that the moon suffers eclipse either when it goes behind the earth or is covered by the sun, do not teach the common people the knowledge which they themselves possess; but they declared that the sun represented the Greeks and the moon the Persians, and that whenever the moon suffered eclipse, defeat and slaughter was foretold for those nations, and they enumerated an eclipse of the moon showed to have fought without the favour of the gods. Nothing sways the common herd more effectively than superstition; generally uncontrolled, savage, fickle, when they are victims of vain superstition, they obey the soothsayers better than they do their leaders. Therefore the interpretation of the Egyptians, when they were made public, restored the downhearted to hope and confidence.

Aubourg (1995: 10), followed by Cauville (1997a: 12), considers that two eclipses were represented in the Dendara circular zodiac: on the one hand, a lunar eclipse that occurred in Pisces on September 25, 52 BC and, on the other, an eclipse of the sun that also took place in Pisces but on March 7, 51 BC (see Figure 4.4). According to these authors, the lunar eclipse would be indicated in the zodiac by a circle with an *udjat* inside it, located above Pisces. The solar eclipse, on the other hand, would be indicated by a larger circle in which a standing figure appears holding an animal that they identify with a baboon. Also, studying this same document, Trevisan (2011) considers that the circular zodiac of Dendara refers, in addition to the solar eclipse of March 7, 51 BC, to other solar eclipses visible from Dendara during those years: the eclipses of 1 October 62 BC, September 20, 61 BC, and August 21, 50 BC.

Fig. 4.4 Position in Pisces of the supposed eclipses of the moon and sun represented in the Dendara zodiac. Image by the authors. (Courtesy of the Louvre Museum)

Yet we have to go to a demotic treatise on lunar eclipses from the late Ptolemaic period in order to find one of the first pieces of direct evidence of Egyptian records of eclipses (Neugebauer et al., 1981). In the Graeco–Roman period the situation is certainly different, since we have received texts with astronomical ephemerides of these phenomena. Thus, as far as lunar eclipses are concerned, the demotic pBerlin 13146 + 47 is of great importance, since it indicates, with respect to the Egyptian civil calendar and the years 84–73 BC, the two days in the year in which a lunar eclipse could occur (when the moon crosses the lunar node) and the position of the moon in the sky (Neugebauer et al., 1981). The technical term for 'eclipse' is documented in demotic as *jrj ꜣbꜣ* ('to perform an eclipse') and *ꜣbꜣ jꜥḥ* ('eclipse of the moon'; Altmann-Wendling, 2019: 225).

4.2 The Planets

Observation of the celestial vault permitted the detection of the five planets visible to the naked eye (Mercury, Venus, Mars, Jupiter, and Saturn). Since the planets move with respect to the background of stars, regular observations of the sky facilitated their discovery and distinction with respect to the fixed stars. Krauss (1999) has suggested that the mentions made of Seth and Horus in the Cairo 86637 and Sallier IV papyri, from the time of Ramesses II and Merenptah, may conceal an astronomical reference to the planets Mercury and Venus. However, no planetary ephemerides have been preserved from the Pharaonic era that give us to any insight into any knowledge that the Egyptians had about the planetary cycles and the movements of the planets across the celestial vault. These kinds of documents are known dating only from the Roman period, with notable examples such as the pBerlin 8279, the Stobart Tablets, the demotic pCarlsberg 32 (*EAT* III, 240–241), and the Ashm. Dem. 483 and Ash. Dem. 525 + 732 + 763 (Ossendrijver & Winkler, 2018), among others.

There are extant documents from the Middle Kingdom onwards (Loprieno-Behlmer, 1986: col. 12), about 4000 years ago, in which the five planets observable with the naked eye are mentioned, but it is evident that this knowledge had already been reached centuries before. It is possible that they had all already appeared in Heny's coffin (Gunn, 1926; *EAT* III, 9 fig. 1), from the 11th Dynasty, although only remnants of the name of Jupiter remain. This coffin, now lost, was found very fragmented in Asyut, but even so it is evident that it showed a representational scheme relatively similar to that of the astronomical ceilings of the New Kingdom.

Although we know the names and epithets by which the five visible planets were designated, there is no agreement on what specific term the ancient Egyptians used (if indeed they ever had one) for 'planet'.

Brugsch (1891: 322) believed he had recognized in the *Pyramid Texts* (PT 245 § 251a W 156) the term planet and translated the following text in this way:

Du nimmst ein (?) deinen Sitz am Himmel mit den Planeten des Himmels
'Thou takest a (?) thy seat in heaven with the planets of heaven'

He therefore recognized the group  *abw* as 'planets'. However, this text should really be translated as, 'You shall part your place in the sky among the stars of the sky' (Allen, 2005: 40); that is, recognizing in *m-ꜥb* a compound preposition instead of part of a noun.

A term that could mean 'planet' comes from another passage in the *Pyramid Texts* (PT 273–74 § 393c W180a/T275a). Faulkner (1969: 80, 83 n.3) translated here as 'the planets (?) are stilled', while Allen (2005: 50) preferred the more literal translation, 'those who move have grown still'.[1] The word in question,  *gnmw*, refers to some kind of movement (Hannig, 1997: 901 'Art Bewegungen') and this is the reason, in a celestial context, why 'those who move' could be the planets. Also in the *Pyramid Texts*, reference is made to the *sḥdw*, which according to Briggs (1952: 59) could be the planets, given that they follow the path of the sun. However, Allen (1989: 4) considers that this term should refer, in general, to the starry night sky.

Krauss (1997: 157) has commented on another term that appears in the *Coffin Texts*: *tnm*. Thus, in CT II 147a it reads: *dass ich herabgestiegen bin ist, um zu sehen den Verirrten (tnm), der fünfte von euch Sternen ist er, die sich nähern dem sꜣḥ-Orion* ('that I am descended to see the lost (*tnm*), the fifth of ye stars, that is approaching *sꜣḥ*-Orion'), so the five stars may correspond to the five planets and the term *tnm* may be related to them. However, Krauss emphasizes: *Ich gehe nicht soweit, tnm/Verirrter mit unserem Terminus Planet, also Irr- oder Wandelstern, gleichzusetzen* ('I do not venture so far as to state that *tnm*/lost refers to our term "planet" or *Ꞽrr*- to "wandering star"').

Another option that has been discussed (Lieven, 2001: 277–278) is that the term *jtn* ('disc') may have been used to designate the planets, not just the sun and the moon. Thus, in the sarcophagus of Ankhnesneferibra appears the expression *wbn.ṯ ḫr jtn 7 m ḫnw pt* ('you shine under the seven discs in the sky'), which could refer to the sun, the moon, and the five visible planets.

On the other hand, at a later time, the expression 'the five living stars' (*pꜣ 5 sbꜣ (w) ꜥnḫw*) is used to refer to the visible planets, as we can see, for example, in the text of Dakka (Griffith, 1937: 27–8 Dak 30) belonging to Hornakhtyotef II (mid-3rd century AD). This expression is not known in earlier times.

The five planets are mentioned and represented (as forms of some gods) in astronomical ceilings of the New Kingdom. The order in which they appear is as follows: first Jupiter, Saturn and Mars, after the list of decans, and then Venus and Mercury with the triangle decans, although on some ceilings, such as those of the so-called

[1] This passage could also refer to the the standstill between prograde and retrograde motion, which must have been familiar to the Egyptians. We thank Terry Mahoney (IAC) for this comment.

Senenmut family, Mars is absent. This order was later altered owing to the influence of horoscopic astrology. For this reason, in the temples of the Graeco–Roman period the planets were ordered, in relation to the zodiac, according to their *hypsoma* (place of exaltation) or their houses. The latter has its logic, since this association corresponds to the *Thema mundi*, which implies the position of the planets at the time of the creation of the world. A later author, Macrobius Ambrosius Theodosius, who lived during the early 5th century, wrote in *Commentarii in Somnium Scipionis* ('Commentary on the Dream of Scipio') that there were two ways of ordering the planets: according to the Egyptians and according to the Chaldeans. The Egyptian order would be: moon, sun, Mercury, Venus, Mars, Jupiter, and Saturn, with eventual changes in the order between Mercury and Venus. The Chaldean or Babylonian order would be: Saturn, Jupiter, Mars, sun, Venus, Mercury, and moon (Guilhou, 2004: 27–28).

In Esna's zodiac, which Neugebauer and Parker (*EAT* III, 62, pl. 29) dated to the time of Ptolemy III–IV, but would be more suitably dated to a later time, considering that it marks the beginning of the year in Virgo, following the Alexandrian calendar, and has some royal cartouches belonging to Roman emperors of the 2nd century AD (Quack, 2018b: 81), Venus appears in Pisces, Jupiter in Cancer, Mars in Capricornus, and Mercury and Saturn (if preserved) would have been situated in Virgo and Libra (*EAT* III, 64); that is, in their places of exaltation. This same ordering seems to be repeated in the well-known Dendara zodiac, although authors such as Aubourg (1995) and Cauville (1997a: 12) (Aubourg & Cauville, 1998: 768) consider that the planets would have been represented in the zodiacal constellation where they had been observed at a certain time, specifically in 50 BC.

In private documents, such as coffins, however, the arrangement of the planets with respect to the constellations of the zodiac could be related to their situation at the time of the individual's birth. In astronomical tables from the Roman period in Egypt, the planets are ordered according to their apparent speed of displacement in the celestial vault: Saturn, Jupiter, Mars, Venus, and Mercury.

Some ancient authors made reference to knowledge that the Egyptians had of the planets. The opinion of Diodorus (I, 81), who visited Egypt in the 180th Olympiad (about 60 BC, possibly under the rule of Ptolemy XII) is known:

> For the positions and arrangements of the stars as well as their motions have always been the subject of careful observation among the Egyptians, if anywhere in the world; they have preserved to this day the records concerning each of these stars over an incredible number of years, this subject of study having been zealously preserved among them from ancient times, and they have also observed with the utmost avidity the motions and orbits and stoppings of the planets, as well as the influences of each one on the generation of all living things.

Even before the Ptolemaic period, Aristotle (*Meteorologica* I, 6), explaining questions related to comets, mentions the Egyptian knowledge about the planets:

> Objections which apply both to those who hold this theory and also to those who suppose comets are due to conjunction of two planets are that some of the fixed stars have tails. And for this we need not rely only on the evidence of the Egyptians who say they have observed it, we have observed it also ourselves. (…) And further, the Egyptians say that there are

> conjunctions both of planet with planet and of planets and fixed stars, and we ourselves have observed the planet Jupiter in conjunction with one of the stars in the Twins and hiding completely, but no comet resulted.

Similarly, Aristotle leaves interesting references in *De Caelo* II, 12 ('On the Heavens'):

> The movements of the sun and moon are fewer than those of some of the planets. Yet these planets are farther from the centre and thus nearer to the primary body than they, as observation has itself revealed. For we have seen the moon, half-full, pass beneath the planet Mars, which vanished on its shadow side and came forth by the bright and shining part. Similar accounts of other stars are given by the Egyptians and Babylonians, whose observations have been kept for very many years past, and from whom much of our evidence about particular stars is derived.

Representation of the planets underwent variations over time. In the examples of the New Kingdom it is usual that Saturn, Jupiter, and Mars are associated with the god Horus and are shown with the head of a falcon; Venus, related to Osiris, as a benu or heron bird; and Mercury as Seth with the head of an animal related to this god (see Figure 4.5). However, in later times more variants are observed (see Figure 4.6).

Although the farthest planet visible to the naked eye is Saturn, it is easily detectable, as it shines with a yellowish hue with an apparent magnitude of −0.5; that is, brighter than Tolimán (α Cen), the third brightest star. Saturn has a relatively slow apparent movement through the celestial vault, about 13° annually to the east. The Egyptians knew Saturn by various names and epithets. On the astronomical roof of Senenmut (TT 353) its name is: *Mwt kꜣ (n) pt rn.f* ('Mut, bull (of) heaven is his name'). However, the introduction of the Mut element is totally exceptional (Belmonte and Shaltout, 2007). Thus, in the astronomical ceiling of Sethy I it is called *ḥrw kꜣ (n) pt rn.f* ('Horus, bull (of) heaven is his name', his usual name). In fact, a deity called 'bull of the sky' that is mentioned several times in the *Pyramid Texts* and in the *Coffin Texts* could be Saturn (Brugsch, 1891: 322).

Fig. 4.5 Forms of Jupiter, Saturn, and Venus as seen in TT 353, tomb of Senenmut. (Image by the authors)

Fig. 4.6 The five planets as displayed in the rectangular zodiac of Dendara. (Images by the authors)

Its appearance in the *Pyramid Texts*, about 4350 years ago, leads us to think that the Egyptians already knew Saturn from much earlier dates, for which unfortunately we either do not have documents with references to the stars in the sky or, to put it simply, cannot interpret. Later we can find more abbreviated forms, such as *Ḥrw kꜣ* or *Ḥrw pꜣ kꜣ* ('Horus the bull', in demotic).

His name could appear accompanied by epithets such as *sbꜣ jmntj dꜣj pt* ('the western star which crosses the sky', in the tomb of Sethy I; *sbꜣ jꜣbtj dꜣj pt* ('the eastern star which crosses the sky'—actually, this name is a correction of *dꜣj pt jꜣbtj sbꜣ*, which is how it actually appears on the astronomical ceiling of the tomb of Senenmut); *dꜣ pt rn.f sbꜣ jmntj pt* ('crosser of the sky is his name, the western star of the sky'); and even in the sarcophagus of Nekhtnebef, grandnephew of Nectanebo I, it appears as *sbꜣ jmntj rsj* ('the southwestern star'). In later texts, the name of Saturn is not usually accompanied by any epithet, although an exception is found in Hermopolis, where it is once again called 'the eastern star' (*EAT* III, 178). The epithets of Saturn do not provide us with any data of astronomical interest. Taking into account that Saturn is an outer planet, it can appear in the sky in any part of the ecliptic and in any eastern or western elongation.

Saturn is represented as a falcon-headed god, as this planet was for the Egyptians one of the forms of Horus, the falcon god. He usually appears standing on a boat, usually with a star above his head. However, during the late period Saturn took other forms. Thus, in both Edfu and Dendara (B, C, E East, holding the *wꜣs* sceptre and the *ʿnḫ* symbol) and Nag Hamad B the planet is represented as a standing bull-headed god, and in other examples it appears as a bull-headed falcon, as in Dendara E West (*EAT* III, 179).

The largest of the planets in the Solar System, Jupiter, can reach an apparent magnitude of −2.9, so it is much brighter than Sirius, the brightest of the stars in the sky. Its outstanding brightness, therefore, made it easy to identify with respect to the fixed stars from very ancient times.

In the numerous variants of the names of Jupiter there is a real play on words, perhaps because of a metathesis. In Senenmut's tomb and in his family he is identified as *Ḥrw tꜣš tꜣwj* ('Horus, who bounds the Two Lands is his name'). At other times, as in the Ramesseum or in Tausert's tomb, for example, it appears as *Ḥrw tꜣš tꜣwj* ('mystery of the Two Lands is his name'); that is, simply changing *tꜣš* for *štꜣ*. This same wordplay is found in two other ways of designating Jupiter, such as *Ḥrw wpš tꜣwj* ('Horus, who illuminates the Two Lands') and *Ḥrw wp štꜣ* ('Horus, who opens mystery'); that is, modifying *wpš tꜣwj* for *wp štꜣ* (*EAT* III, 177), ('Horus the mystery') in demotic.

Regarding the epithets, Jupiter is sometimes described as (tomb of Senenmut and related group), which should be corrected to *sbꜣ rsj n pt*, 'southern star (of) the sky', or simply *sbꜣ rsj* ('southern star') in the ceiling of Sethy I. At other times he uses the epithet *šmsw n pt* ('follower of the sky'). In a fragment of Heny's coffin from the 11th Dynasty, [...] *rsj sbꜣ* can still be read, which should refer to the (corrected) epithet of Jupiter *sbꜣ rsj*. This is the oldest known mention of this planet in *Egyptian texts*. If Saturn is identified in the *Pyramid Texts* as 'bull of the sky', it is curious that the even brighter Jupiter has not been identified in these texts. Perhaps it simply had not yet been known in order to be identified. It is clear, however, that if Saturn was known in the Old Kingdom, Jupiter must also have been known with even more reason, both because of its greater brightness and because of its greater annual displacement over the background of fixed stars.

In the iconography, Jupiter is usually represented in a similar way to Saturn; that is, as a standing falcon-headed god with a star above his head, usually standing on a boat. However, in the late period Jupiter is represented in other ways: as a standing falcon-headed god with a star in a disc above his head (Edfu), a standing human-headed god with sun disc and uraeus (Dendara C), a standing falcon-headed god with atef-crown ( ) and *uraei* (Dendara E), a human-headed god with a Persian pointed hat and garmen, holding a sword in his right hand, and a bow and arrow in his left hand (Esna A), a standing lion -headed god (Nag Hamad), etc. (*EAT* III, 178).

Mars is not mentioned in the main group of the so-called family of Senenmut, although it does appear in the subgroups of this family and in the so-called families of Sethy IA and C. In the latter its name is *Ḥrw ꜣḫtj* ('Horus of the Horizon') in the tomb of Sethy I, also with variants of writing, as (for example, in the Ramesseum), etc. This name was documented until the end of the 30th Dynasty and early Ptolemaic times in the Nekhtnebef sarcophagus. It would later be known as *Ḥrw dšr* ('Horus the red'), in demotic. This name is very descriptive, as it defines the most outstanding characteristic of this planet (its colour). Just like Venus, Mercury and Jupiter have a white colour, but Mars is clearly identified in the sky by its reddish hue, relatively similar to stars like Antares (meaning'opponent to Ares/Mars', evidently for its reddish colour that rivals Mars).

Mars is designated by various epithets, mainly as *sbꜣ jꜣbtj (n) pt*, 'the eastern star (of) the sky', and also *sbꜣ jmntj (n) pt*, 'the western star (of) the sky'. However, the one of most astronomical interest is *sḳdd.f m ḫtḫt* ('he travels backwards'), an epithet that appears for the first time in the tomb of

Sethy I (*EAT* III, pl. 3). This epithet perfectly describes one of the characteristics of the apparent movement of the planet against the background of stars: the retrograde movement, a consequence of the different speed of translation and the different amplitude of the orbits of the earth and Mars, and of the different orbital inclination of Mars with respect to the plane of the ecliptic, so that an observer who follows this planet for several months will perceive how after a period of continuous advance towards the east, another of retreat will follow with variations in declination.

Mars takes the form of the god Horus represented standing on a boat, with a human body and the head of a falcon, usually with a star on his head. Other forms of Mars include a human-headed god (Edfu, Dendara C), a falcon-headed god wth a two-feather crown (Nag Hamad B), a falcon with a white crown (Dendara E), a falcon with serpent tail and long beak (Athribis, Zodiac B), a falcon with serpent tail and triple-serpent head (Athribis, Zodiac A), a Human-headed god in Egyptian dress with Greek helmet (Esna B), etc. (*EAT* III, 179).

Reaching its maximum apparent magnitude of –4.4 about five weeks before and after inferior conjunction, Venus is the brightest planet observable to the naked eye. Being an inner planet, it is always observed in the vicinity of the sun, with a maximum elongation of 47°. Its extreme brightness must have made it easily identified very early on, although it is difficult to know when the Egyptians first understood that both at its eastern and western elongations it was the same body and not two different objects.

In the astronomical ceilings of the New Kingdom, Venus is linked to Bah *Bʿḥ*, Osiris, and, in the case of the sarcophagus of Nekhtnebef, to *Bn*(*w*). Clark (1949–1950) believed that both Bah and the benu bird or the heron are all manifestations of Venus. For Neugebauer and Parker (*EAT* III, 182), the fact that Venus appears in relation to both Bah and Osiris in some tombs could mean that Bah is the evening aspect of the planet and Osiris its morning aspect.

On the ceiling of Senenmut Venus is called *ḏꜣ* (the 'crosser'), and on that of Sethy I it is called *sbꜣ ḏꜣ* (the 'star which crosses'). Later, in the Tanis family, in texts of Edfu () and in demotic, the name of the planet is (*Pꜣ*)-*nṯr-dwꜣw* ('The morning god'; *EAT* III, 181), a name that also appears in the *Texts of the Pyramids*, *nṯr dwꜣw* ('morning god'), and which Quack (2019: 17) considers to be also mentioned in the magical Leiden I 343 + 345 papyrus. Also in the *Pyramid Texts* we find *sbꜣ dwꜣw* ('morning star'), which also has been related to Venus (Krauss, 1997: 216–234), in demotic. Some passages in these texts indicate that the 'morning star' is the divine falcon, the Horus of the Duat (Graefe, 1982: col. 206). For Briggs (1952: 46), on the other hand, the 'morning star' is possibly the star that at any given moment is the brightest on the eastern horizon before sunrise, so that Venus could be so only occasionally. However, according to Faulkner (1966: 161) the 'morning star' can only correspond to the planet Venus.

More problematic is the identification of Venus, in its evening aspect, with *sbꜣ wʿtj* (the 'lone star'), as originally considered by Brugsch (1883: 73), followed also by Faulkner (1966: 161). A problem when making this comparison is that in the Tanis family of decans (documented since the 26th Dynasty, but with most of the examples in the Graeco–Roman period) the decan located between *tpj-ʿ*

smd and *smd* is *p*(ꜣ) *sbꜣ wꜥtj* ('the lone star'; *EAT* III, 144). In fact, Krauss (1997: 117) is of the opinion that the 'lone star' of the *Pyramid Texts* must be a star, not a planet, and must correspond to Capella. Despite this, Quack (2019: 17) considers the possibility that, at least in the Late Period, the 'lone star' should be considered as the counterpart of the 'morning god', these being the evening and morning aspects of Venus.

Venus is usually depicted as a heron on New Kingdom astronomical ceilings. However, in the Late Period its appearance takes other forms: a standing human-headed goddess (Edfu), a standing two-faced human-headed god (Dendara B, C), a falcon-headed god with a sun disc and uraeus (Dendara D), a standing two-headed god with a human head with the red crown and a falcon head with the white one (Dendara E), etc. (*EAT* III, 181). However, in the coffin of Heter (AD 125) Venus is also represented as a heron.

Mercury is the innermost of all the planets, and its maximum elongation never exceeds 27°, reaching an apparent magnitude close to −1 when it is in quadrature. Despite its brightness, its narrow elongation means that it is not always an easy planet to observe.

Unlike the other names of the planets, the meaning of the Egyptian name of Mercury is not known: *sbg* (TT 353), *sbgw* (KV 17). In Ptolemaic times the name of Mercury derived to *sbk* (in demotic) and in demotic there is an unusual variant: *swgꜣ*.

Usually, the name of the planet is accompanied by the figure of the god Seth, this being the only planet that is related to this deity. The god Seth, as the murderer of Osiris, usually has negative connotations, especially in the Late Period. It is foreseeable, therefore, that the observation of Mercury in specific positions would have been seen in a negative context. In this sense, Mercury carries an interesting epithet: , with variants for the name of the god as or ; that is, *Stš m wḫꜣ nṯr m dwꜣt* (*EAT* III, pl. 62), which we must translate as 'Seth in the evening twilight, a god in the morning twilight'. From this epithet we may draw two obvious conclusions.

We see how the possible negative aspect of Mercury, as a form of the god Seth, is associated only with his evening appearance, since at dawn Mercury is defined as a god, but not specifically as Seth. This means that if in its eastern appearance before dawn Mercury is a neutral entity, in its western appearance after sunset, Mercury transforms into a form of the god Seth and therefore into a negative object. As a second conclusion, this epithet shows that the Egyptians had recognized that Mercury was the same object that was observed sometimes at dawn and other times at dusk. Since the earliest date on which this inscription has been found is from the reign of Ramesses VI, we may say that we at least have evidence that since the end of the Ramesside Period the two appearances of Mercury had been identified as belonging to the same object and not two different bodies. However, while it is clear that Egyptian observers were able to reach this conclusion hundreds of years earlier, we cannot say precisely when it occurred.

The negative character of Mercury is observed in the treatment of the representation of the planet. In some cases, for example, the figure of Seth as the personification of Mercury is left incomplete, unfinished or with a mutilated head. This is a clear symptom of the existing superstition, since by mutilating it in some way the Egyptians believed that they could counteract or nullify its negative force (*EAT* III, 180). He can also appear as Horus, for the same reason. On the other hand, just as from the astrological point of view Mercury can change its negative or positive sense depending on its position, Mars and Saturn have clearer negative links while Venus and Jupiter, on the contrary, exert a positive influence (Quack, 2019: 16).

In the Late Period Mercury was represented as a standing human-headed god (Edfu, Dendara B, C, E, Shanhur), a frog-headed falcon (Dendara E), a falcon with a serpent's tail and head of Seth (Athribis zodiac A) (*EAT* III, 180). There is also an example in which he is shown with the head of an ape, which relates him to Thoth and his equivalent Hermes (Quack, 2019: 12). In fact, in the demotic papyrus Strasbourg D 521, Mercury is referred to as 'the star of Thoth'.

The way we currently refer to planets by means of symbols has, at least partially, its origin in ancient Egypt (Quack, 2019: 20). Thus, while the symbols used in Greek papyri for Saturn and Jupiter are simply monograms derived from the initial letter of the Greek names of these planets, in the case of Venus and probably Mars they come from a demotic form. As Quack (2018b: 78) has pointed out, the symbol of Venus might derive from the demotic form of the sign M9  (sometimes with the photenic value *dwꜣ*), whereas the symbol of Mars is possibly related with the demotic form of the sign T30 .

4.3 Comets

In the British Museum there are two Babylonian tablets that refer to a comet observed in October 675 BC. These are texts of an astrological nature. Thus, for example, one of them says: 'If a comet becomes visible in the path of the stars of Anu: there will be a fall of Elam in battle' (Hunger, 1992: 194). There is also a possible reference to a comet in a much older Babylonian text, from 1140 BC: 'A comet arose whose body was bright like the day, while from its luminous body a tail extended, like the sting of a scorpion' (Brandt & Chapman, 2004: 2).

However, it is the ancient Chinese who have bequeathed by far the greatest number of documented cometary observations. In most of their ancient texts the comets are referred to as 'broom star', 'sparkling star', 'bright star', 'long-tailed star', 'tangle star', or even because of their appearance, as a 'celestial magnolia tree' (Kronk, 1999). The Chinese were great observers and their documentation is abundant. In fact, during two thousand years they recorded more than 600 comets, many of which have served to verify previous passages of currently catalogued comets, such as Comet Halley.

If we take as valid a reference that appears in the *Huainanzi*, an ancient Chinese text from the court of Liu An, Prince of Huainan, during the reign of Emperor Wu of the early Western Han Dynasty sometime before 139 BC, a comet appeared at the end of 2nd millennium BC, during the campaign of King Wu against Zhou at the end of the Shang Dynasty (mid-11th century BC; Calder, 1985: 13):

> The king marches against Zhou, faces the East and welcomes Jupiter, reaches Qi, which floods, reaches Gongtou, which falls; a comet appears, with the handle towards the people of Yin.

Unfortunately, the information we have about comets in ancient Egypt is almost non-existent (Lull, 2022). In fact, from Pharaonic Egypt we do not have a single text in which the appearance of a comet is mentioned, although during the history of its dynasties many must have been observed. We are again faced with a problem with written sources. Seneca himself (*Quaestiones Naturales* VII, 3), referring to Conon's research (3rd century BC) and the astronomical information he obtained from the Egyptians, says, 'He made no mention of comets, and yet he would not likely have passed them by if he had learned anything definite among the Egyptians'.

In Antiquity comets were understood, in a very general way in all cultures, as messengers of catastrophes, diseases, wars, and other calamities. Comets had a clearly negative sense, and their appearance warned of the approach of an unpleasant event that had just happened or would be about to happen. The Egyptians possibly shared these common beliefs, especially since comets, being unexpected and unpredictable, and not obeying (known) regular cyclical movements, could be signs of the chaotic, of that not dominated by the order of ma'at.

In ancient times, each comet was an independent and occasional object from which a regular and predictable movement like that observed with the planets could by no means be expected. Moreover, even the status of astronomical body could be denied it; suffice it to remember that Aristotle's theory that comets were atmospheric phenomena lasted for more than two thousand years. It is possible, therefore, that the observation of comets or any celestial event considered out of the order of the known, was recorded but kept in the libraries of the temples, without further dissemination in other types of texts.

4.4 Shooting Stars and Meteorites

As for shooting stars, the references we have are meager. In the *Pyramid Texts* (PT 528 §1250a) there appears a deity named Suntju, 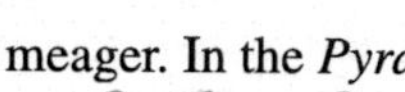 *swnṯw*, who is said to cross the sky at night. This made Sethe (1962: 150–151) think that the reference might be to a shooting star.

A text from the 18th Dynasty might be the recollection of an observation of a bright shooting star. The inscription (Reisner, Reisner, 1933: 35; *Urk* IV, 1238–1239; Klug, 2002: 202) comes from the stela of Thutmose III at Gebel Barkal (Boston MFA 23733), found in the first courtyard of the Temple of Amun at Gebel Barkal:

Jw wn wnwtj sp 2 jjt sbꜣ m jjt n rsj.sn n ḫpr mjtt wdj.n.f r.s m ꜥḳꜣ.f n ꜥḥꜥ wꜥ jm smꜣ.n.j sn mj ntjw n ḫpr ḥdb(w) ḫr snf.sn ḫrw m gbgbjwt
jsṯ rf nsrt m sꜣw.sn m sḏt r ḥrw.sn

It was the second hour when a star came from their south. Never had happened something like that. He (the star) threw himself towards him in opposition. No one stood there. I massacred them like the ones that don't exist, being they lying in their blood fallen in a heap.
Then, there was the uraeus behind them with the fire towards their faces.

If the text refers to the observation of a shooting star, its magnitude was such that we should rather classify it as a fireball; that is, a shooting star of great luminosity which is sometimes even associated with a sound similar to that of thunder.

However, the interpretation of this text is not absolutely clear either. Among the astronomical hypotheses, it is worth mentioning that of Sauneron (1960: 156), who suggested that the star described in the Gebel Barkal stele of Thutmose III could be a comet, specifically Halley's Comet. More recently Henriksson (2020: 38–39), has assumed that it could be Comet Encke. However, as understood by reading the text, the star appeared suddenly, from the south, in the second hour. This description cannot correspond to that of a comet. Another, non-astronomical, interpretation is that offered by Galán (2002: 123 n. 13), who interprets the appearance of the star to be a metaphorical description of the pharaoh's attack on the enemy. In this sense, it is worth highlighting the way in which the word 'star' was written with the determinant of god, along with the fact that at the beginning of the stele (*Urk* IV, 1229: 4) the pharaoh is described as a star that crosses the sky:

sšd.f r jmjtw pḏtj 2 mj sbꜣ ḏꜣj.f ḥrt

He shines between the two bows like a star that crosses the sky.

We have no further historical references of this type but there is a text, that of the well-known *Tale of the Shipwrecked Sailor* (pHermitage 1115), which records the observation of a meteor that manages to overcome the friction of the atmosphere and impacts causing the death of many beings. It is believed that the main serpent in the story must be identified with a form of Re, in such a way that the remaining 75 serpents would be the 75 manifestations of Re in the *Litany of Re*, while the youngest daughter would be Maat (Derchain-Urtel, 1974: 27–31).

The story occurs in an Egyptian literaturary work and is not meant to be historical. However, it should not be ruled out that its narrative might have been based on a real background that had occurred some time previously and had been retained in the collective memory. We reproduce here the part of the *Tale of the Shipwrecked sailor* that we are interested in highlighting (Blackman, 1972: 45):

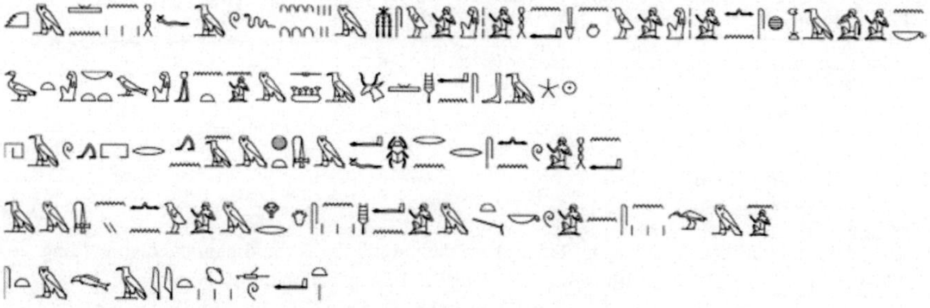

Km.n.n. ḥfꜣw 75 m msw.j ḥnꜥ snw.j nn sḫꜣ.j n.k
sꜣt ktt jnj.tw n.j m sšꜣ ꜥḥꜥ.n sbꜣ
hꜣw pr.n nꜣ m ḫt m-ꜥ.f ḫpr.n r.s nn wj ḥnꜥ.(sn)
ꜣm.nj nn wj m ḥrj-jb.sn ꜥḥꜥ.n.j mt.kwj n.sn gm.n.j
st m ẖꜣyt wꜥt

We were in total 75 snakes with my children and my brothers without mentioning you a little daughter who was brought to me by prayer. So a star came and they went out with fire because of him. It happened when I was not with (them).
They were burned and I was not among them. I was shattered for them when I found them like a single pile of corpses'.

Simpson (1984: col. 620) supposes that this text can be understood as an astronomical metaphor in which the protagonist serpent can be interpreted as the sun, a falling star appears, and its stay in the place is four months (which corresponds to a season), etc.

Somewhat later, references to stars falling from the sky and causing harm to people, as they appear in oracular texts, have also been interpreted as possible meteorites (Edwards, 1960: I, 2 n.11); according to Meeks, however, many of these references might rather be linked to lightning (Meeks, 1982: col. 117).

In both the Gebel Barkal stela and the *Tale of the Shipwrecked sailor*, the term used to designate what appears to be a shooting star is simply *sbꜣ* ('star'), which does not imply a specific concept. However, there is a more specific term used for a shooting star:  *sšd* (Hannig, 1997: 778), although this could serve to designate both shooting stars and lightning bolts (Meeks, 1982: col. 117).

In another text from Thutmose III, in the so-called Poetic Stele of Karnak, a comparison is made between the king and the term *sšd* (*Urk* IV, 615: 13), which we can translate as shooting star:

dj.j mꜣꜣ.sn ḥm.k mj sšd stj bsw.f m sḏt

I make them see your majesty like a shooting star, throwing its fire with flame.

Another text, from the year 8 of Ramesses III, provides a possible variant of the same term and, due to the context, it seems appropriate to translate it as a shooting star as well. It refers again to the king (Epigraphic Survey, 1930: pl. 46, 3rd col.):

pḫrr mj sb3w ḥr sšdw jmj ḥrt

Running as the shooting stars which are in the sky.

In *The Chronicle of Prince Osorkon* we find a similar expression,  *mj sb3 ḥr sšd*, that Caminos (1958: 81) translated as 'like a star darting up', although we could assimilate it to the text seen previously and translate rather as 'like a shooting star'. In this text, the way in which the word 'star'  has been written stands out. As Caminos indicates, the numeral 6 (*sjs*) placed before of the determinative of star indicates the way in which the word 'star' was probably pronounced at this time. In the time of Osorkon III the number six was pronounced *sow*, as in coptic ⲥⲟⲟⲩ. Taking into account that in the Graeco–Roman period the word 'star' is also found with variants such as 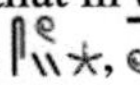, 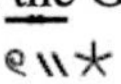 and 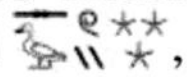, and in coptic ⲥⲓⲟⲩ, it is clear that the term mentioned must be read as 'star'.

Another example from the year 11 of Ramesses III, like the previous one from Medinet Habu, seems even clearer:

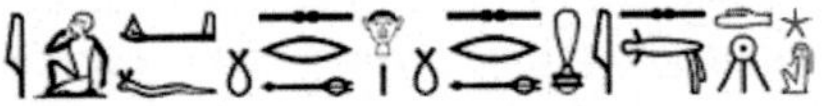

j.dj.f šsr ḥr šsr mj sšdw

He sends arrow after arrow like the shooting stars.

The comparison between continuous arrow shooting and *sšdw* is reminiscent of a shower of shooting stars. It is difficult to know whether the ancient Egyptians perceived a regularity, from year to year, of at least some of the most important showers of falling stars, just as it is difficult to identify in the texts an observation that we can relate to this phenomenon.

When an object of extraterrestrial origin overcomes the friction of the atmosphere and hits the earth, we call what remains a meteorite. Did the ancient Egyptians know where meteorites came from? Did they know how to identify them? The truth is that the ancient Egyptians used meteoric iron to make valuable utensils. Some of the oldest documented examples come from two intact tombs, tomb 67 and tomb 133 (Petrie et al., 1912: 16–17) found in 1911 in the El-Gerzeh necropolis.

Specifically, these are nine small beads made with meteoric iron (Rehren et al., 2013; Johnson et al., 2013) that come from two graves securely dated to Naqada IIC–IIIA (*c.* 3400–3100 BC). A later example, but of great quality, is the iron dagger blade (Carter 256 K) found wrapped with the mummy of king Tutankhamun (Comelli et al., 2016).

Generally, it has been said that the way to differentiate meteoritic from telluric iron is that meteoric iron has more nickel. In fact, meteoric iron contains between 5% and 26% nickel. In contrast, in telluric iron, except in very specific veins, it is not usual to find traceable nickel levels. However, it should be noted that nickel-containing iron can derive from some smelted ores and long buried meteoric iron objects can have most of the nickel leached from them (Ogden, 2000: 167).

This type of iron was known, at least since the 19th Dynasty (Harris, 1961: 59), according to one of the earliest examples on the Ani papyrus (Budge, 1913: III pl. XV, Chap. 23 l. 4) as  *bjꜣ n pt* ('iron from heaven'), although it is clear that this name does not necessarily imply that under that name only meteoric iron is included, or that the Egyptians suspected an extraterrestrial origin of meteoric material. In fact, observations of the fall and impact of meteorites are an extraordinarily rare and unusual event. Wainwright (1932a: 5) documented that in AD 856 'five stones fell from the air, one of which burned up a Bedawy tent. Four of them were taken to Fostat and the other to Tennis'. And in more modern times four have been observed to fall during the 39 years between AD 1877 and 1916.

It is estimated that the yearly impact rate of meteorites weighing more than 10 grams is one per 2590 km^2, and meteorites larger than one metre in diameter are estimated to have an impact rate on the planet of one per year. According to a recent study (Bland & Artemieva, 2006: 607), iron meteorites 2.5 m in diameter arrive at the earth's surface approximately once every 50 years, iron meteorites a few metres larger in diameter which form impact craters *c.* 100 m in diameter will impact the earth's land area every 500 years, larger bodies able to form craters 0.5 km in diameter every 20,000 years, and craters 1 km in diameter will be formed on the earth's land area every 50,000 years. Although the impact rate is low, the undoubted use of meteoric iron by Egyptians and other peoples makes it equally clear that the location of specimens of iron meteorites should not be as unlikely as the observation of their fall.

East of Gebel Uweinat, and less than a kilometre from the current border with Sudan, a 45-metre wide and 16-metre deep meteorite impact crater with an ejecta ray structure has been identified that seems to have occurred 5000 years ago; that is, at the beginning of the dynastic period in Egypt. This crater, known as Kamil crater (22° 01′N, 26° 05′E), appears to have been caused by an iron meteorite 1.3 metres in diameter. After its discovery in 2008 by Vincenzo de Michele using the Google Earth program (see Figure 4.7), the crater was visited in 2010 by an Italian–Egyptian expedition that recovered 5178 iron meteorite specimens totalling *c.* 1.71 tons of material in the crater and surrounding area (Folco et al., 2010; D'Orazio et al., 2011). It is known that the meteorite approached the earth from the north-west, travelling along a moderately oblique trajectory. Whether it entered the atmosphere during the day or at night, the bolide must have been extremely bright, so there is no doubt that it is highly probable that the iron bolide was observed by the Egyptians. On the other hand, although the region where the crater is located is today a total desert and at that time the process of hyperaridization of the Sahara had already begun, it is not ruled out that the impact was witnessed by nomadic groups that were still circulating in the region.

Fig. 4.7 Kamil Crater, a meteoric crater located east of Gebel Uweinat. (Courtesy of Google Earth)

Ferric material from heaven could have been of great importance, for example in the making of ritual instruments. Wainwright (1932a: 11) had already mentioned the fact that the adzes used during the ritual of opening the mouth could have been made of this material. If we look at the representation of the ritual embodied in the tomb of Tutankhamun, we can see how the end of the adze that the successor Ay brings closer to the mummiform Tutankhamun, is a dark piece tied to the adze. There is no doubt that this piece is iron, perhaps meteoric. This element, therefore, served to give the deceased the breath of life. In relation to this, it has also been commented that the emblem of the city of Letopolis could symbolize lightning, and it must be taken into account that generally in the past meteorites and shooting stars were believed to be thunderbolts, and this could also have been the case in ancient Egypt. And it is precisely in Letopolis where we find a high priest with the title 'Opener of the Mouth'.

The importance of the metallic element *bjꜣ* is already mentioned in the *Pyramid Texts*. In PT 21 §13 (N 100) (Allen, 2013) it is said:

Your mouth has been parted with Anubis's adze, the metal Striker that parted the mouth of the gods'. (Allen, 2005: 252)

> I open your mouth for you with the adze of Wepwawet, I split open your mouth for you with the adze of iron which split open the mouths of the gods'. (Faulkner, 1969: 3)

In PT 21 it is also indicated that the mouth opens *m bjꜣ pr m Stš* ('with the metal that came from Seth'; Allen, 2005: 252). As we shall see later, Seth is represented in the Egyptian sky by the constellation Meskhetyu, since the bull's foot with which this constellation was recognized (in addition to the adze shape) is his. Plutarch's comment in his well-known work *De Iside et Osiride* (62) is no accident: 'They still call the loadstone "the bone of Horus", and iron "the bone of Typhon", as Manetho records'. Typhon is the Greek version of Seth.

In the centre of the disc of the Dendara zodiac, a constellation appears whose image corresponds to a jackal mounted on a hoe. Undoubtedly, this jackal is none other than Upuaut, and this constellation must correspond to our Ursa Minor (see Figure 4.8) because, like Ursa Major, this constellation also has the shape of the adze used during the ritual of the opening of the mouth, an adze known by the name of *nṯrj*. In addition, the word 'adze', as seen in the text, is written *msḫtjw*, a name also given to the Egyptian constellation that corresponds to our Big Dipper, and that we can see written both *msḫtjw* (with the determinative of adze) as well, with the bull's foot clearly reflecting the two forms with which the Egyptians saw this constellation.

During the time of the pyramids the two adzes (*nṯrtj*) of the Egyptian sky corresponded with the two bears. Such was its importance that the two holes of the *serdab* of the Pharaoh Netjerkhet in Saqqara, are orientated (in its time)

Fig. 4.8 Pole constellations in the round zodiac of Dendara. (Image by the authors)

precisely towards the declination in which the stars Dubhe and Kochab are located, which would represent the meteoritic iron tip of the instrument in the form of an adze (Belmonte et al., 2009b: 250–253; see Chap. 6). Roth (1993: 78) believes that the shape of the adze has its origin in the way in which midwives placed their fingers in the mouth of newborns to help them breathe.

Apart from the use of iron meteorites to make beads or other types of small objects, we should not ignore their possible presence in temples as pieces fallen from the sky linked to a celestial divinity. Even today there are some temples in which meteorites are preserved as part of the elements that surround the cult of the sacred. The best-known example is undoubtedly the Black Stone (al-Ḥajaru al-Aswad) located in the eastern corner of the Kaaba in Mecca (Burke, 1991: 221–223).

We have similar other examples in Antiquity, although these pieces have long since disappeared, being sometimes replaced by omphaloi that, to a certain extent, might recall the shape of the so-called orientated ferrous meteorites, with an approximately conical appearance. Wainwright (1935: 42–43) cites some of these possible sacred meteorites: the 'star fallen from the sky' kept in the temple of Astarte on Tyre, the old Stone of Cronus at Delphi, which had fallen to earth when vomited forth by Cronus, the obscure stone at Emesa which arrived as 'a ball of fire running down from above', and at Ephesus the sacred object belonging to Artemis that 'fell down from Jupiter'.

In Egypt itself, in dynastic times, meteorites could have existed that would have reached the rank of the sacred. One of these is said to have been the Theban fetish that appears in some representations; separated from its throne and human head, it has a conical shape. Related to the god Amun, as a divinity of celestial character, this fetish could have been copied and exported to other temples, such as Siwah or Napata, where the conical omphaloi, just as happened in Ephesus, etc., were manufactured forms that were intended to recall the authentic stone fallen from heaven (Wainwright, 1935: 44).

Another sacred object for which a meteoric origin has been suggested is the benben stone itself, which was kept in the temple of Heliopolis (Bauval, 1989b: 8–9) and which, as we know, is also represented in a conical shape.

4.5 The Milky Way

In places that enjoy dark skies, the Milky Way is a striking element of the celestial vault. Its diffuse appearance meant that its nature was not understood in ancient times, so that cultures devised many different explanations and mythological relationships to make sense of this milky band. In fact, Aristotle thought that it could have an atmospheric origin. The name we use today, Milky Way, has its origins in Greek mythology. Hermes placed Heracles on Hera's breast, to suck the divine milk, and when Hera realized that the one who suckled was an offspring of a love affair of Zeus, her husband, she abruptly separated him from the breast and the milk spilled over the sky, thus forming the Milky Way. In Spain, the Milky Way has been called the Camino de Santiago, as it served to guide the pilgrims on their way to Santiago de Compostela. Other cultures have seen heavenly roads and rivers in the Milky Way.

In ancient Egypt there are several terms that have been proposed as equivalents of the Milky Way. One, which appears in the *Pyramid Texts*, is *msḳt sḥdw* (Sethe, 1962: II 20), translated by Faulkner (1969: 72 n.11) as 'Street of Stars (?)', Indicating that it must undoubtedly refer to the Milky Way, or translated as 'beaten path of stars' by Allen (2005: 78) or 'Path of Sailing Stars' (Allen, 2000: 21), a term that, in contrast, is understood by Conman (2003: 39) to be the ecliptic. The term *msḳt* (Hannig, 1997: 365) by itself has also been translated as 'Milky Way'.

Another option that has been postulated is (Davis, 1985: 102), a term that is mentioned on numerous occasions in the Pyramid Texts. *Mr-nḫꜣ* translates as 'Winding canal', a name that could recall the same shape of the Milky Way, but this equivalence has been criticized and it is proposed instead that this term, as *mr n ḫꜣ*, 'Kha canal', could designate the band of the ecliptic (Krauss, 1997: 63–64).

It also seems possible the Milky Way could be seen as a manifestation of the goddess Nut, the goddess of the sky. If we observe the Milky Way over the months, we see how, when it reaches the constellation Cygnus, it splits in two so that we could imagine the two legs of the goddess, while at Gemini we would find the goddess's head. This suggestive hypothesis, formulated by Wells (1992: 309), may have other interesting implications. During the predynastic era, around 3500 BC, the winter solstice sun rose at dawn, when, before the sunlight made the stars disappear from the sky, the constellation of Cygnus was seen above the eastern horizon. Thus, the birth of Re or *mswt Rꜥw*, on that important event, occurred after observing the part of the Milky Way equivalent to Nut's reproductive organ, located in Cygnus, where the Milky Way bifurcates, where the two legs begin to part. On the other hand, at that same time the setting of the sun at the spring equinox occurred when the head of Nut (Gemini) was observed on the western horizon, so that, curiously, between this and the previous event, nine months had passed, a period reminiscent of human gestation.

Chatley (1940: 124) considered the possibility that the Milky Way might be indicated by the set of stars on the astronomical ceiling of the tomb of Senenmut that appear to the right of the constellation of the Sheep. However, this correlation is by no means certain.

4.6 Stars and Constellations

The ancient Egyptians had a complete set of constellations spanning the entire sky visible from their territory throughout the year. The most common term to refer to the sky is *pt* ('sky'), but this is in reality just one of the many words that the Egyptians had for designating the celestial region. Thiers (2009: 53) recalls that one of the papyri in the Tebtunis temple library (pBerlin 7810 + pLouvre AF 11112) refers to 39 different terms to refer to heaven.

In the Graeco–Roman period, the northern sky was called *ghrt/khrt*, while the southern sky seems to be associated with the term *gbt* (Thiers, 2009; Von Bomhard, 2012). But throughout their history we see that the Egyptians refer to the northern stars and the southern stars that therefore form two groups of differentiable stars in two sectors of the celestial vault.

Krauss (in the press) has recently studied, especially from some references from the *Pyramid Texts* and the *Coffin Texts*, two other terms that imply a division of the sky into two sectors: the upper sky *pt ḥrt* and the lower part of the sky *ẖrj-ẖt pt* or lower sky *pt ẖrt*, which would be linked to the vision of the Duat in the sky. The separation of both parts of the sky would occur, according to the German researcher, by the ecliptic strip into a northern and southern sky. Also, in the *Coffin Texts* (*EAT* I, 28–29) reference is made to *pt mḥtjt* ('northern sky') and to *pt rsjt* ('southern sky').

Another characteristic of the night sky is that it is divided between two main groups of stars: the *jḫmw skjw* *die das Untergehen nicht kennen* ('those who do not know death'; Hannig, 1997: 99), those who do not know the twilight, located in the *mḥtjw pt* (the northern part of the sky) as indicated, for example, in PT 519 §1220a, and the *jḫmw-wrḏ* *die das Ermüden nicht kennen* ('those who do not know tiredness'; Hannig, 1997: 99).

Brady (2015: 84) understands that the *jḫmw-wrḏ* would be equivalent to those that Ptolemy describes as the Arising and Lying Hidden types, but adds a third type, which would correspond to those that Ptolemy classifies as belonging to the phase of Curtailed Passage, 'the ones that could defy the Duat, for they appeared to die by setting with the sun but unlike other stars they rose out of the Netherworld that very same night and then eventually could take a leap into the heavens to join the Imperishable stars'.

In the *Pyramid Texts*, there is talk of the late pharaoh's desire to go to the place of the *jḫmw skjw* stars ('those who do not know the setting'), which are northern stars. The desire for this further intensifies the symbolic value of this region. On the other hand, it is also deduced from the *Pyramid Texts* that the northern stars or some of them and the *ꜣḫw* ('*akhu*') are closely related, such that these stars could be the same *Axw*. Therefore, it is evident that the orientation of a pyramid to the N–S axis, especially the placement of its entrance on the north side, follows some guidelines that must be related to the importance that the polar region of the celestial vault would have in the funerary world.

For a long time, the majority opinion was that such stars should be identified with circumpolar stars. But, according to Krauss (1997: 104–109, 117–120) this belief is incorrect, since in the *Pyramid Texts* it is also indicated that some of these stars rise and set, so it supposes that the *jḫmw skjw* could be stars north of the ecliptic plane. The heliacal rising of such a star precedes the heliacal setting, when the sun passes nearby on its annual trajectory. This means that these stars are visible every night and do not have an annual period of invisibility. It is possible that the Egyptians took the fact that these stars are always visible, and that they lack an annual period of invisibility as proof of their 'eternity'. The imperishableness of the fixed northern stars is contrasted with the annual period of invisibility of the southern stars (Krauss, 2006b: 191). The circumpolar stars would be, according to Krauss's proposal, only a part of the *jḫmw skjw*.

Fig. 4.9 Astronomical ceiling of Sethy I in KV17. (Image courtesy of Guillermo Suay Montagud)

A division of the representation of the sky into two sectors is observed on the astronomical ceilings, as in the first example preserved in the tomb of Senunmut (TT 353) in Deir el-Bahari (Pogo, 1930: pls. AK; Dorman, 1991) and later examples such as those of the tombs of Sethy I (KV 17; see Figure 4.9), Tausert (KV 14), Ramesses VI (KV 9), and Ramesses IX (KV 6) in the Valley of the Kings, that of the funerary temple of Ramesses II (Ramesseum), and in the Amenhotep III's clepsydra from the Karnak cachette. A sector considered boreal, in which a celestial diagram appears, in which the constellation of Meskhetyu stands out in diagonal stellar clocks, is expressly described as associated with the boreal sky. The southern sky, with which the decans, planets, and other constellations are associated, is represented by Sah (part of our Orion), Sopdet (part of CMa) *sjt/srt* (Sit or Siret, the Sheep), and *wjꜣ* (Wia, the Boat). As the decans had already apparently ceased to be useful as time-keepers in the New Kingdom, a new system was developed (the so-called Ramesside star clocks or star tables), which used only the meridian transit of some stars belonging to constellations and certain significant asterisms that mostly do not coincide with the stars or asterisms that appear mentioned in the decanal lists. These tables were found in the tombs of various pharaohs of the 20th Dynasty and provide a little more information on other regions of the Egyptian sky.

After the New Kingdom, there are several useful, but scarce, sources for examining knowledge of the Egyptian celestial vault. Among these, we may mention the celestial diagrams on the astronomical ceilings of the tombs of Pedamenope and Montuemhat in the necropolis of el-Asasif (immediately east of Deir el-Bahari), dating from the Late Period, and a number of clepsydras or coffins from this same period. Finally, during the Graeco–Roman period, the documentation is again abundant, including new astronomical ceilings, some of them quite spectacular, such as those in the temple of the goddess Hathor in Dendara, and also various astronomical representations in the decoration of a certain number of clepsydras, coffins, and sarcophagi.

Most of these sources were brought together by Otto Neugebauer and Richard Parker, in their remarkable four-volume collection: the *Egyptian Astronomical Texts* (*EAT*), which still remain a basic source for the study of the ancient Egyptian sky. However, neither author considered it to be possible to use such information to map the sky of the ancient Egyptians for more than a few examples. In fact, they wrote (*EAT* II, x):

> Just as in the case of the decans the crudeness of the underlying procedures is so great that only under severely restrictive assumptions could numerical conclusions be abstracted from the given lists. If we add the fact of obvious errors and carelessness in details in the execution of the texts as we have them one would do best to avoid all hypothetical structures designed to identify Egyptian constellations from the analysis of the Ramesside star clocks.

Despite these unfortunate words, it should be noted that some interesting (and controversial) approaches have been made in recent years, and recognition of the Egyptian sky is an aspect of Egyptian astronomy in which our research, we believe, has contributed a fruitful advance (Lull & Belmonte, 2006, 2009).

While accepting that these hypotheses should be treated with caution, we believe that, given the existence today of extremely powerful computers that did not exist in the 1960s, it is possible today to use both diagonal star clocks and Ramesside star clocks, with the support of the so-called celestial diagrams of the astronomical ceilings, to identify at least some of the stars or asterisms included in these diagrams and their respective constellations. The same approach could be advocated for the Dendara zodiac.

Two masterpieces of ancient Egyptian art, the astronomical ceiling of the tomb of Senenmut, in Deir el-Bahari and the circular zodiac of the temple of Hathor in Dendara, will be the frames of reference with respect to which the analysis of the Egyptian sky will be circumscribed that we now summarize. Before continuing with the identification of the Egyptian stars and constellations, however, it is convenient that we go on to describe an astronomical ceiling of the New Kingdom (although it must be remembered that different families have their own peculiarities).

4.6.1 An Example of an Astronomical Ceiling: The Celestial Diagram of Senenmut

The oldest known astronomical ceiling appears in a private tomb from the 18th Dynasty, on the ceiling of chamber A[2] of Senenmut's tomb TT 353 (Pogo, 1930; Dorman, 1991), from which the Senenmut family of decans acquired their prototype. However, it should be noted that the origins of the design of this ceiling must go back at least six centuries before, since part of a coffin from the 11th Dynasty, now lost, included a list of decans and the representation of the boreal constellations (Gunn, 1926).

Following Senenmut's exceptional example, astronomical ceilings will appear in many royal tombs in the Valley of the Kings from Sethy I onwards and in other contexts. The astronomical ceiling tradition continued in the 22nd Dynasty at Tanis (Lull, 2002: 96–98) but a few later examples are also observed, such as the newly discovered in the tomb of Karakhamun in South Asasif (TT 223, 25th Dynasty; Molinero Polo, 2014) and the later ones from the antechamber of the pyramid P 26 of Gebel Barkal (Berenguer, 1997: 131), perhaps belonging to a king of Napata and that of the southern Begrawiyah pyramid 503 of the queen Hennuwa, 3rd century BC, in Meroë (Yellin, 1984: fig. 1), among others.

The ceiling of Senenmut, discovered during the 1925–1927 excavations of the Metropolitan Museum at Thebes, is divided into two clearly distinguishable parts: that dedicated to the southern sky (see Figure 4.10) and the other dedicated to the northern sky (see Figure 4.11), respectively. The ceiling covers an area of $3 \times 3.6\ m^2$, the greatest length corresponding to the N–S axis. The two parts are placed in their correct position with respect to the cardinal axes. Thus, the section of the ceiling that is dedicated to the southern sky is in the southern half of chamber A, while that of the northern sky is in the northern half. Both parts, framed by a set of five-pointed stars, are separated by several lines of hieroglyphics, of which the central one, with larger hieroglyphs, includes only the royal titles of Hatshepsut and the reference to two titles of Senenmut and his filiation. The other lines are religious in nature and are made up of passages from the *Pyramid Texts*.

Because of its astronomical nature, we shall explain in detail what appears in the southern and northern halves of this ceiling, starting with the southern half. To understand this record we have to start reading the hieroglyphs that remain on the right side of it. In the first of the columns is inscribed the name of a decan called 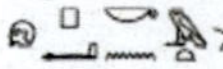 *tpj-ʿ knmt* ('Predecessor of Kenmet', Senenmut family of decans, number 1), with the determinative of star. This is the name of the first of the decans on the Senenmut family list. After the name of the decan, the inscription in the first

[2]In many works carried out by specialists, such as Pogo (1930: 305), Clagett (1995: 112), Leitz (1995: 35), etc., a recurring error persists in placing the astronomical ceiling in chamber C of TT 353, a chamber with a vaulted ceiling that is located at the end of the tomb, and that would correspond to the burial chamber. However, the truth is that Senenmut's chamber C does not have any decoration on its ceiling, and it is in chamber A, the first chamber, that we find this example.

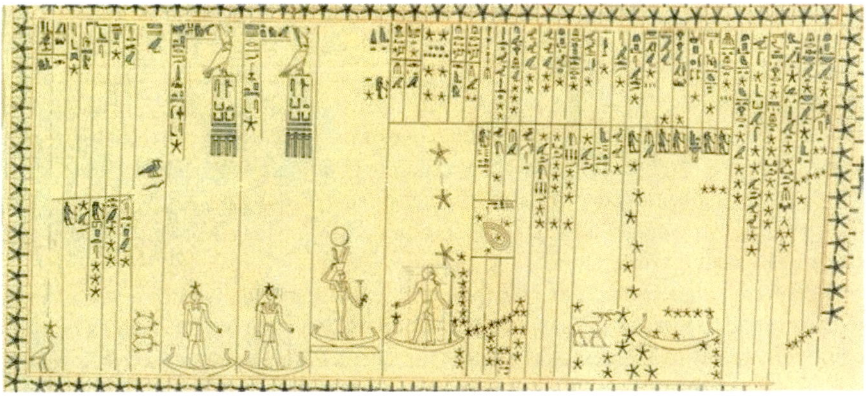

Fig. 4.10 Astronomical ceiling of Senenmut. South section. Facsimile. (Courtesy of The Metropolitan Museum of New York (Rogers Fund, 1948 Accession Number: 48.105.52))

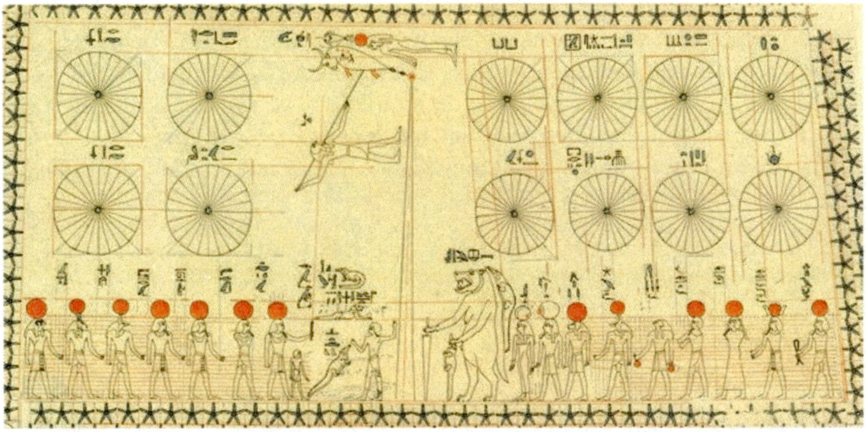

Fig. 4.11 Astronomical ceiling of Senenmut. North section. Facsimile. (Courtesy of The Metropolitan Museum of New York (Rogers Fund, 1948 Accession Number: 48.105.52))

column continues with the names of two gods related to the first decan, Hapy and Imset, with two other stars. The second column contains the third decan, *ẖrj ẖpd Knmt* (the 'One that is under the final part of Kenmet', decan n. 3) and the second, *knmt* ('Kenmet', decan n. 2), with the goddess Isis as a related divinity. Two descending lines of five stars follow the names of related deities and the bottom of the first two columns. This must indicate that the first three decans are linked to the same constellation of stars, which must correspond to the Egyptian constellation of Kenmet.

The third column includes two other decans, the fourth, *ḥ3t ḏ3t* ('Beginning of the Djat', decan n. 4) and the fifth, *pḥwj ḏ3t* ('End of the Djat', decan n. 5), to which the god Duamutef and the Sons of Horus are related. The latter, as there are four, are determined by four stars. The fourth column continues with the sixth decans, *ṯmꜣt ḥrt* ('Upper Tjemat', decan no. 6) and seventh *ṯmꜣt ẖrt* ('Lower Tjemat', decan no. 7), associated with Duamutef and the Sons of Horus. The next column contains the eighth and ninth decans, *wšꜣtj* (the 'Two wesha Birds', decan no. 8) and *bkꜣtj* (the 'Two pregnant women' or 'Two Ladies', decan no. 9), respectively, related to the gods Duamutef and Hapy. The sixth column continues with the decans called *tpj-ʿ ḫntt* ('Predecessor of Khentet', decan n. 10) and *ḫntt ḥrt* ('Upper Khentet', decan n. 11), both determined by a star and related to the god Horus, followed by three stars.

The decans that are included between the seventh and the twelfth column seem to correspond to parts of the same constellation, which must have covered a considerable extension of the celestial vault. This constellation is represented under its decans. It is shaped like a boat, so it is known as the Egyptian constellation of the Boat (see Figure 4.12). A number of stars are drawn next to it. In the seventh column is inscribed the decan *ḫntt ẖrt* ('Lower Khentet', decan n. 12), in the eighth *ṯms n ḫntt* (the 'Red of Khentet', decan n. 13), in the ninth *sptj ḫnwj* ('Sapty of Khenwy', (decan n. 14), in the tenth *ḥrj-jb wjꜣ* (the 'One who is in the middle of the Boat', decan n. 15), in the eleventh *sšmw* (the 'Guides', decan n. 16), and in the twelfth column the decan named *knmw* ('Kenmu', decan n. 17). The associated divinities are Seth, Horus, Isis and Nephthys, Seth, and the Sons of Horus, respectively.

The decans of the thirteenth and fourteenth columns refer to another constellation called Semed. Decan 18 is called *tpj-ʿ smd* ('Predecessor of

Fig. 4.12 The constellations of the Boat and the Sheep. Astronomical ceiling of Senenmut. (Image by the authors)

Semed'), and decan 19 *smd* ('Semed'). These are associated with the gods Horus and Hapy, under which a group of stars develops and must somehow be linked to this constellation, although it has been proposed that they could indicate the presence of the Milky Way (Chatley, 1940: 124).

A new Egyptian constellation is glimpsed between the following columns, since a Sheep is drawn under them and this reference also appears in the name of the decans. Decan n. 20 is called *sjt* ('Sheep'), which corresponds to the constellation of the same name (see Fig. 4.12). After the horizontal line we see the sign that represents the goddess Isis and then a very interesting indication that reads *3 nwt ẖt* (the 'third cluster'). In the sixteenth column, the next decan is called *sꜣwj sjt* (the 'Two sons of the Sheep', decan n. 21), determined by a single star and associated with the god Duamutef.

Inscribed in the seventeenth column, we have a decan which could be independent of the Egyptian constellation of the Sheep. It receives the name of *ẖrj ḫpd srt* (the 'One that is under the final part of the Sheep', decan n. 22). It is related to the god Qebehsenuf, who is followed by four stars in an upright position.

Another possible constellation or group of stars, although not a figurative one, could be understood from the name of the decans inscribed in the next two columns. The first is called *tpj-ꜥ ꜣḫwj* ('Predecessor of the two Akhu', decan n. 23) and the second *ꜣḫwj* (the 'Two Akhu', decan n. 24). These decans are associated with Duamutef, followed vertically by four stars, and with Duamutef and Qebehsenuf, determined by two stars.

The twentieth column includes the decan who is called *bꜣwj* (the 'Two Bau', decan 25). It is determined by two stars and related to Hapy and Imset, two of the four sons of Horus. At the end of the column, a group formed by seven stars seems to show that this decan is not constituted by a single star but by a group. In fact, at the end of the following column there is, between a group of stars, the 4 *nwt ẖt* ('fourth cluster').

The next three columns are linked to an oval-shaped constellation (see Figure 4.13), also known from other representations in tombs in the Valley of the Kings. Here it is called *mw nwt ẖt*. The twenty-first column has inscribed two decans, called *ḫntw ḥrw* ('Upper Khentu', decan n. 26), and [ ] *[ḫntw] ẖrw* ('Lower [Khentu]', decan n. 27), associated with the (Sons of Horus).

The next column contains two more, *ḳd* ('Qed', decan n. 28), and *sꜣwy Ḳd* (the 'Sons of Qed', decan n. 29), associated with the gods Hapy and Qebehsenuf, respectively. At the end of this column there are three stars. We continue with the twenty-third column, in which a single decan is inscribed, with the very descriptive name of *ḫꜣw* (plural of 'thousand', that is, 'Thousands' or 'Myriads', decan n. 30). The associated divinity is, in this case, the Sons of Horus, after whose inscription the hieroglyphic text warns us that it is the 'fifth cluster', 5 *nwt ẖt*.

One of the best-known constellations of ancient Egypt is represented in the space occupied by the next five columns. Here we see a boat on which we have the standing figure of a man (the god) with his head turned to the right (see Figure 4.14). The god, dressed in a *shendjut*, holds a large *wꜣs* sceptre in his left hand and the *ꜥnḫ*

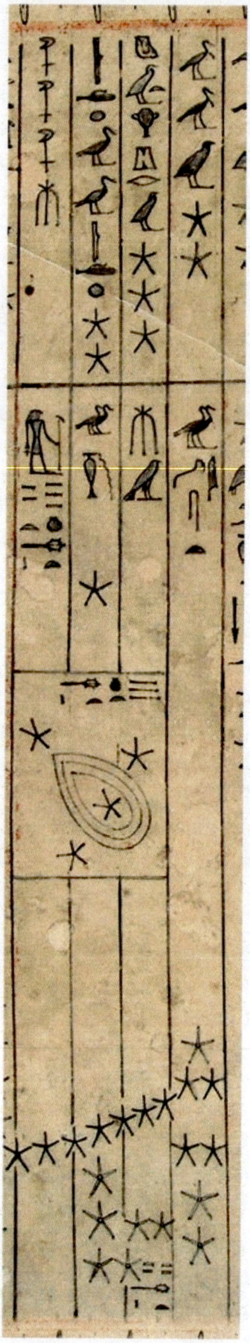

Fig. 4.13 Columns 20–23 and clusters 4, 5, and ‘6’ (or *ẖt nt mw* ?). Astronomical ceiling of Senenmut. (Image by the authors)

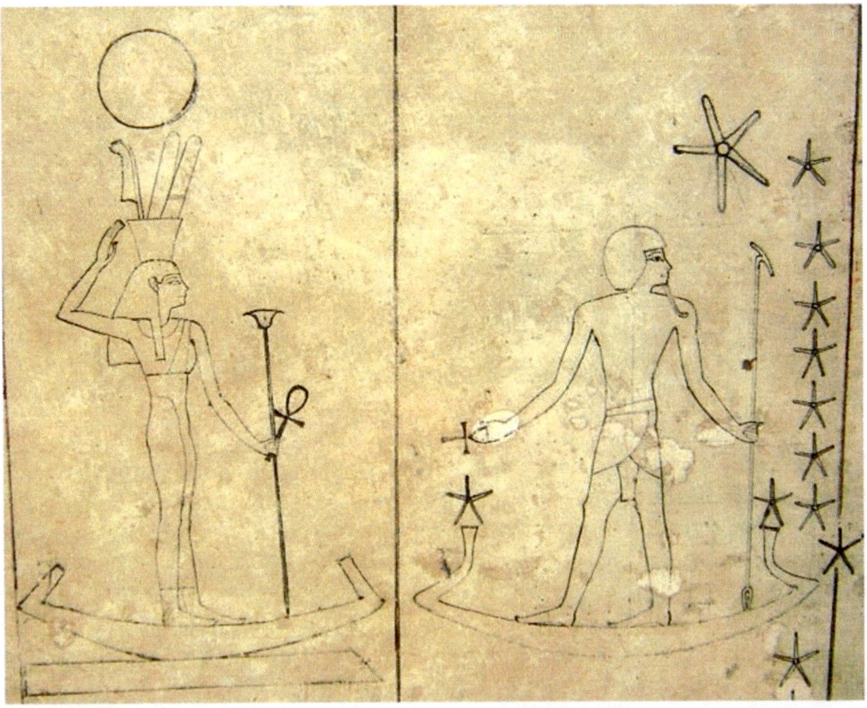

Fig. 4.14 The constellations of the Sah and Sopdet. Astronomical ceiling of Senenmut. (Image by the authors)

symbol in his right. It is *Sꜣḥ* (Sah), a god personified in a constellation that is essentially equivalent to our well-known Orion. This god is a form of the god Osiris. The set of stars that surround the figure of Sah, three large stars above him, nine stars to his right, and one above the prow and stern of the ship, must somehow symbolize the stars that make up this constellation. The three big stars, presumably, may represent Orion's belt.

The decan that appears in the twenty-fourth column is called *ḥrj rmn Sꜣḥ* (the 'One who is under the arm of Sah', decan n. 33). The next column repeats the same name, although in the first case the Sons of Horus were associated with two determinative stars, and in this one with Osiris, with no star. The twenty-sixth column, wider than the previous ones, contains only a set of stars grouped symmetrically (it may correspond to decan number 34, which in other lists is called *rmn Sꜣḥ* (the 'Arm of Sah'). The twenty-seventh column contains the decan *ʿrt* ('Aret', decan n. 31), determined by a star and related to the Eye of Horus. In the next column is *ẖrj rmn sꜣḥ* (the 'One on the arm of Sah', decan n. 32), with the Sons of Horus as associated divinities. Decan number 35, absent as an independent decan on Senenmut's list, is known on other lists as *Sꜣḥ* ('Sah', our Orion). According to Neugebauer and Parker (*EAT* III, 114), in the Senenmut

Table 4.1 The Orion decans in Middle Kingdom coffins, Senenmut, and Senenmut family soubgroups A–B, and Sethy A–C families (*EAT* III, 112–113)

M.K. Coffins	Senenmut	Senenmut subgroup A – Ramesses II A	Senenmut subgroup A – Ramesses III A	Senenmut subgroup B	Sethy A – Sethy C
ꜥrt	*ẖrj rmn sꜣḥ*	*ꜥrt*	*rmn sꜣḥ*	*ḥrj rmn sꜣḥ*	*ꜥrt*
ḥrj ꜥrt	*ẖrj rmn sꜣḥ*	*ẖrj rmn sꜣḥ*	*ḥrj rmn sꜣḥ*	*ꜥrt*	*jwn sꜣḥ*
rmn ḥrj	Constellation	Constellation	*Ḳd, sꜣwj ḳd, ḫꜣw*	*sꜣḥ*	*rmn ḥrj*
rmn ẖrj	*ꜥrt*	*rmn sꜣḥ*	*ꜥrt*		*msḏr sꜣḥ*
ꜥbwt	*ḥrj rmn sꜣḥ*	*ḥrj rmn sꜣḥ*	*ẖrj rmn sꜣḥ*	*ẖrj rmn sꜣḥ*	*rmn ẖrj sꜣḥ*
ẖrt wꜥrt					*ꜥ sꜣḥ*
					sꜣḥ

list, the decans of the Orion area were not listed in the correct order simply owing to a small change in the representation of their figure, which instead of holding the *wꜣs* sceptre and the *ꜥnḫ* symbol with the right and left hand, respectively, does the opposite. In the main models that we can compare, from the decan *ḫꜣw* the order is as given in Table 4.1.

However, it is understood that the correct order would be: *ꜥrt, ḥrj rmn sꜣḥ, ẖrj rmn sꜣḥ, rmn sꜣḥ* constellation, and *sꜣḥ* (*EAT* III, 114).

The twenty-ninth column is much wider than the previous ones and includes another of the better-known Egyptian constellations, that of Sopdet, which should roughly occupy the space of our Canis Major with Sirius (see Figure 4.15). This represents the last of the decans, the 36th, called here *ꜣst-Spdt* (Isis-Sopdet). The constellation is represented under the column as a female divinity standing on a boat. In the same way that the Greeks brought the images of their divinities to the celestial vault, grouping them together to be able to follow through them a narrative thread of their myths and legends, the Egyptians had long before also distributed the stars of the sky into recognizable groups identifiable with their gods, although in this case we have difficulties not only in accurately identifying their constellations with modern equivalents but also with their origin and meaning. One exception, however, is the series of constellations related to Osiris and Isis that are called Sah and Isis-Sopdet (in the position related to Orion and Canis Major). Osiris is personified in the sky with the figure of Sah in the same way that the goddess Isis is with Sopdet. That the Egyptians placed both constellations together in the celestial vault make sense in the same way that the Greeks saw Cassiopeia and Cepheus together, since Osiris was the brother and husband of Isis, the great magician. In the sky, Sah-Osiris always precedes Sopdet-Isis, so it is as if Isis followed Osiris, which is directly linked to the legend according to which the goddess was a long time looking for the body of her murdered brother.

Columns 30 and 31 include two of the three outer planets known to the ancient Egyptians, both represented here as hieracocephalic deities standing on a boat. In the upper right part of both columns we have a serekh, on which perches a falcon

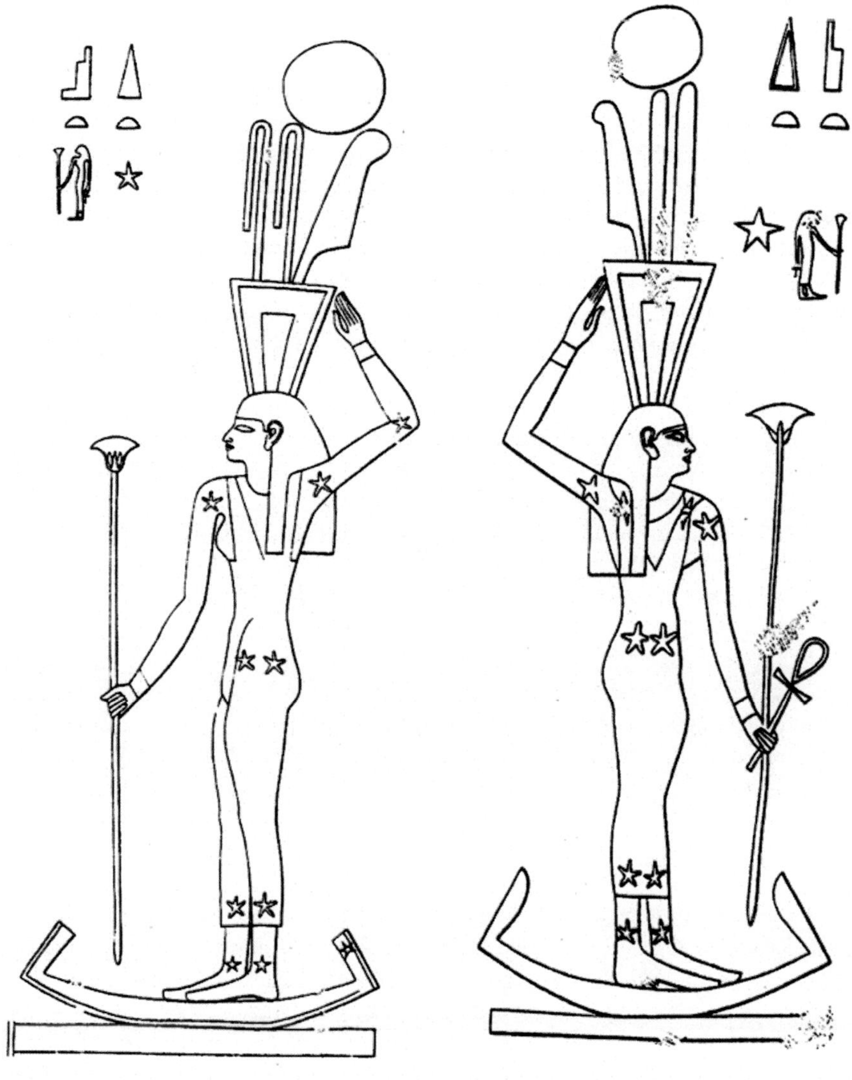

Fig. 4.15 Sopdet as seen on the astronomical ceilings of Montuemhat (left) and Pedamenope (right), 26th Dynasty. (Adapted from *EAT* III, pls. 18 and 20)

with the double crown that represents Upper and Lower Egypt. Inside the serekh is inscribed Horus name of Queen Hatshepsut, *wsrt k3w* ('She who is mighty of kau'). Next, in column 30 we have the name of the planet Jupiter *Ḥrw t3š t3wj rn.f rsj pt* (*n*) *sb3* ('Horus the one that unites the Two Lands is his name, the southern star of the sky'), and in the next column Saturn, identified here with a singular name: *Mwt k3* (*n*) *pt rn.f ḏ3 pt j3bt sb3* ('Mut, the bull of the sky is his name, the eastern star that crosses the sky').

Generally, Jupiter and Saturn are followed by the planet Mars, but in Senenmut's list Mars is not mentioned. Several hypotheses have been proposed in response to this absence. Some researchers have tried to show that the Senenmut ceiling is more than an ideal and symbolic outline of the Egyptian firmament, and that it actually represents a certain and precise view of the starry sky on a certain night. For example, Von Spaeth (2000) has argued that the ceiling represents the night of May 22, 1534 BC and identifies Mars with another celestial object mentioned in the diagram, decan *ṯms n ẖntt*, the 'Red (star) of the Bow' or 'Red (star) of Khentet'. But this hypothesis is untenable, since the identification of this star with Antares seems clear. Another hypothesis is that proposed by Leitz (1991: 44–47), criticized with sound judgment by Krauss (1992b, 1995). According to Leitz, various sections of Senenmut's astronomical ceiling actually represent the sky as it was observable on the night of November 14, 1463 BC, when Mars was invisible. Therefore, the celestial diagram (and consequently the tomb TT353) would have been designed within the accepted dates for the usurpation of power by Queen Hatshepsut. However, Leitz's idea is incorrect. If the calculations obtained for November 14, 1463 BC are repeated in the Julian calendar it can be seen that, on that night, not only Mars but also Mercury was invisible in the sky (as Krauss had already shown), although this second planet certainly appears on the astronomical ceiling. Even if we accept a definition error and consider the night November 14, 1463 BC, but in the proleptic Gregorian calendar, as the referred to by Leitz (which apparently is not the case), it would have to be explained why Venus was visible in the sky ahead of Mercury, a situation that is not reflected in the ceiling. Therefore, we may infer as true that no part of Senenmut's astronomical ceiling represents the real sky on any given night.

Significantly, the absence of Mars is not the only striking 'irregularity' concerning the sector of the celestial diagram relating to the outer planets. In all other celestial diagrams, Saturn is defined, among other titles, as 'Horus, the Bull of Heaven'. However, in the astronomical ceiling of Senenmut, the epithet of this planet is , which could be translated as 'Mut, bull of the sky is his name' or 'Mother of the Bull of the sky is his name', possibly because of the royalty of Hatshepsut. (see e.g. Belmonte and Shaltout 2007), Von Lieven (2021: 210) argues against this hypothesis and thinks that it must instead be a scribe's error.

If the decoration of the Karnak clepsydra is analysed, there are several important facts relating to its exterior decoration. Mercury and Venus are represented in the standard order of most representations; there is a scene where the king is making an offering to the god Re-Horakhty in the presence of the lunar god Thoth (see Figure 4.16); and, most importantly, only Jupiter and Saturn are represented. Furthermore, the king is represented alongside each planet in his boat. Thus, Mars is again absent, but (oddly) Horakhty is not and is represented in another scene in his form of solar divinity. It is therefore proposed that Mars may have been a nocturnal aspect of Re-Horakhty so that it was not necessary for him to be represented a second time in the celestial diagram of the clepsydra.

Another more prosaic option is that the astronomical ceiling of Senenmut's tomb is actually a copy of a papyrus representation of a celestial diagram that may have existed, and that was used for the decoration of clepsydras. Owing to lack of space

Fig. 4.16 Re-Horakhty, Nebmaatre, and Thoth on the clepsydra of Amenhotep III found at Karnak. (Image by Nacho Ares, courtesy of The Cairo Museum, and adapted drawn from Desroches-Noblecourt et al., 1976: 144)

when converting the design prepared from a conical shape to a flat surface, some of the decoration might have been lost. Consequently, only two of the outer planets would have remained in the final representation.

It must be concluded that the astronomical ceiling of Senenmut's tomb at Deir el-Bahari does not represent an actual astronomical event but rather a schematic celestial diagram that may have previously been used to decorate water clocks, where Mars was not present among the outer planets because Horakhty was already represented elsewhere as a manifestation of the solar god Re (Belmonte & Shaltout, 2007). This supposed master copy was copied on the ceiling of the first room of TT353, while incorporating a new element that was relevant considering the particular situation of Senenmut (who lived under the rule of the enthroned Hatshepsut): the 'feminine' name of the planet Saturn.

Returning to the astronomical ceiling of Senenmut, after Jupiter and Saturn, columns 32 to 36 contain the so-called epagomenal decans. Six of the twelve used by the Egyptians are named in Senenmut's list. In column 32 we have the epagomenal decan 1 of the Senenmut list, called *štwj* (the 'Two turtles'), associated with the gods Hapy and Duamutef. Under this column are shown two turtles, an image from which a small group of stars should be recognized. Column 33 has the epagomenal decan 2, which is called *Nsrw* ('Neseru'). After a horizontal line that divides this and the following columns in two, it reads *ẖt pw*, which in Egyptian means 'it is a cluster'.

The next column has the epagomenal decan 3, *šspt* ('Shespet'), determined by a star and related to the (Eyes of Horus), in turn determined by three stars. The epagomenal decan 5 is in the next column and is listed here as *ꜥbšs* ('Abshes'). After the horizontal line we find the figure of Horus, which we must relate to the previous decan, and after this is the epagomenal decan 4, called *ḥpds* ('Hepedes'), followed by three stars. The last of the epagomenal decans, 6, was entered in column 36, under the name *nṯr wꜣš* and is associated with Duamutef.

In the last two columns, as is usual in this type of list, the inner planets are mentioned. In column 37 we read Mercury's name (*Sbg*). It is determined by a star and linked to the god Seth. Column 38, the last one, is dedicated to the planet Venus. The inscription begins with *ḏꜣ* (the 'One who crosses'), related to Bah and Osiris. A heron topped by a star is represented at the end of this column.

Of the other half of the ceiling, the north, we are especially interested in the representation of the boreal constellations. At the top, we see an ovoid-shaped animal with a bull's head and a tail made up of three stars. This is a constellation that in other representations appears in the shape of a bull or, in the Middle Kingdom, as a bull's foreleg, being known by the name of *msḫtjw* (Meskhetyu). It corresponds to our Big Dipper. Of the three stars that make up the tail of this bull, the last one is indicated in red, thus differentiating it from the others (see Figure 4.17). In turn, we can see how two divergent lines start from the star reaching the bottom of the panel and forming a triangle. This must be related to the importance of this star, perhaps used for the orientation of buildings or, in its upper or lower culmination, to verify the central meridian. Pogo (1930: 309–310) postulated that the three stars of

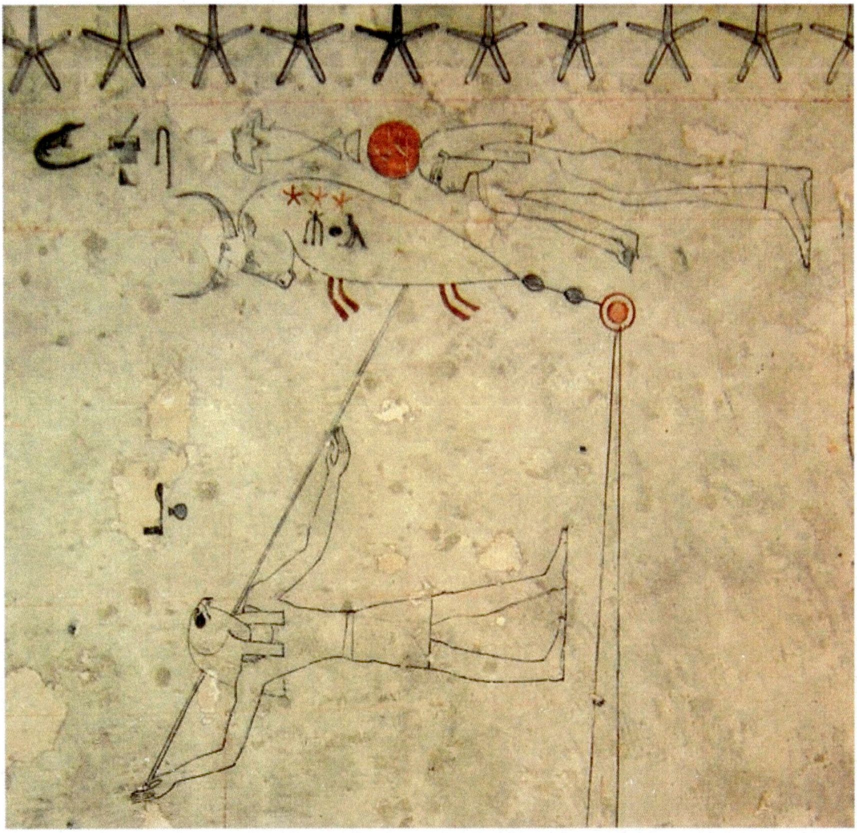

Fig. 4.17 Meskhetyu, Serqet, Anu, and Saq, among the northern constellations on the astronomical ceiling of Senenmut. (Image by the authors)

the bull's tail should correspond to δ, ε, and ζ UMa (Megrez, Alioth, and Mizar, respectively); if so, remember that ζ UMa would be the only 2nd magnitude star less than 10° from the polar axis around 3000 BC.

The boreal constellations are similarly recognized on the astronomical ceilings of other tombs. Above Meskhetyu (Ursa Major) we have another constellation that appears in the form of a woman in the tomb of Senenmut. An inscription identifies her as *Srḳt* (Serqet), the goddess who sometimes appears in the form of a scorpion. Another boreal constellation is called *S3ḳ* (Saq) and is shaped like a crocodile with its tail tucked up. It is just to the left of Serqet. Below Meskhetyu a hieracocephalic deity holds a long spear with which it seems to be harpooning the figure of the bull. This new constellation is called *ʿnw* (Anu).

Further down we find the remaining the boreal constellations (see Figure 4.18). To the right of the triangle that ends in the third star of Meskhetyu's tail, we have one of the best-known constellations in the northern Egyptian sky. It is *3st ḏ3mt ḥb pt* ('Isis-Djamet, feast of heaven'). The figure with which this constellation is represented consists of a female hippopotamus on whose back there

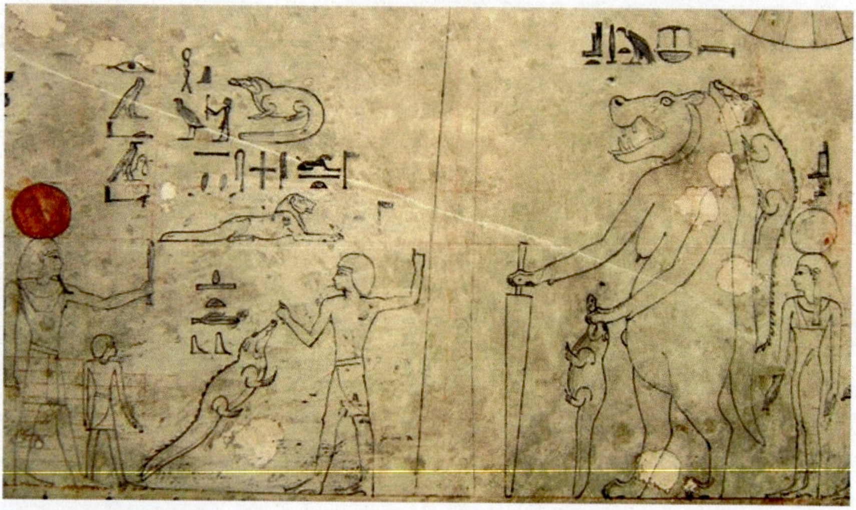

Fig. 4.18 Isis-Djamet and other northern constellations on the astronomical ceiling of Senenmut. (Image by the authors)

is a crocodile. Because the name of the constellation is compound, it might be understood that Isis refers to the hippopotamus and Djamet to the crocodile. The hippopotamus leans with her left hand on a crocodile placed vertically as a cane, while with her right hand she holds what appears to be a kind of mooring post.

To the left of the triangle, we have another constellation in the shape of a crocodile with a tucked tail. It is called *Ḥ3kw* (the 'Looter'). Under this crocodile is a figure in the shape of a lion with a crocodile's tail. It is called *rw nṯrj jmj.snwj* (the "Divine Lion that is between them'), a name reminiscent of the two crocodiles that accompany him, because the figure under this lion is another crocodile, this time called *Ḥtp rdwj* ('Lying on His Feet'), an epithet of the god Sobek (although the text that appears in the tomb of Senenmut is). To the right of this last figure we have a man with raised arms, almost in the attitude of harpooning the animal that precedes him, although neither in this nor almost in any other representation, such as in the tomb of Ramesses VI (*EAT* III, 186, fig. 29), do we see that he wields any throwing weapon.

We may compare the set of boreal constellations exhibited in the tomb of Senenmut with those depicted on the ceiling of the burial chamber of Sethy I in the Valley of the Kings, or in other later tombs (see Figure 4.19). Although there are small variations in the position and size of the figures, we can see that they are basically the same. In any case, the final problem is that of identifying the northern Egyptian constellations with our own. The most obvious and sure equivalence is that of the Egyptian constellation Meskhetyu with Ursa Major. When it appears in the shape of a bull or bull's leg, we recognize without the slightest doubt the shape of a cart by which the Big Dipper is also known.

Fig. 4.19 Northern constellations. Tomb of Ramesses VII, KV 1. (Image by the authors)

On either side of the northern constellations we have two rows of divinities that seem to be heading towards the centre of the scene. Seven are located on the left side. They are called (Irmaua), (Teknu), (Shedheru), (Nehes), (Aaner), (Imysehnecher), and (Heruhekenu); and there are nine to the right. These are called Isis, (Imset), (Hapy), (Duamutef), (Qebehsenuf), (Maanitef), (Irendjetef), (Irrenefdjesef), and (Haqu). These are sixteen protective deities, eleven of which Parker (1950: 43) identified as patrons of the festivities on the days of the lunar month. All are crowned by a red disc, except the first two on the right side.

One of the most outstanding details of this north panel of the ceiling of chamber A of TT353 are twelve circles, eight on the right side and four on the left. The circles are inscribed above the names of the twelve months of the Egyptian calendar. The four in the upper right correspond to the months of the Akhet season, from right to left: *Tḫy* (Tehy), *Mnḫt* (Menhet), *pt ḥnʿ 3ḫ3ḫw ḥt-ḥrw* ('the sky together with the stars: Hathor'), and *K3*-[*ḥr*]-*k3* (Ka-[her]-ka). The circles on the left side correspond to the four months of the Peret season, the second of the three seasons of the Egyptian civil calendar. Their names are (top row from right to left) *šf bdt* (Shefbedet) and *rkḥ* (Rekeh). In the bottom row, from left to right, we read the names of *rkḥ* (Rekeh) and *rnwtt* (Renutet). Finally, the circles corresponding to the months of the season of shemu, the third, are in the lower right row. These are also inscribed with the names from left to right: *ḫnsw* (Khonsu), *ḫnt-ḫty prtj* (Khent-Khety-Perty), *jpt ḥmt* (Ipet Hemet), and *wp rnpt* (Wepet renpet).

The 24 sections with which each circle is divided have been associated with the 24 h of the day and night, although it has not been understood how they might be marked. Clagett (1995: 120) believes that the sectors were made with the intention of indicating two groups of 12 meridian hours indicated by the transit of bright stars, in the same way as happens in the Ramesside transit clocks whose origin is at least contemporaneous of Senenmut. If so, each circle should contain the transit stars responsible for marking twelve hours at night on the first day of the month in one half of the circle and the transit stars responsible for marking those same hours on the 16th of the month in the other half of the circle. Von Bomhard (2012: 98–100), based on an idea by Manuel Mª. Valdés, indicates that the circles would be radial star clocks that would relate the positions of the boreal constellations with the hours of each month. For von Lieven (2021: 209) it is likely that the gods of the month in question received offerings, perhaps once an hour, which according to her would explain the division into 24 parts.

The astronomical ceiling (Lull, 2002: 157–158) constitutes a subject that is mainly directly related to the immediate funerary environment of the body of the deceased pharaoh. The lists of decans and the representations of the boreal constellations should offer the deceased a knowledge of the celestial realm and the passage of time that would doubtless be of great use to him for his new life in the Duat.

Since the end of the 5th Dynasty, the ceiling of the king's sepulchral chamber, although gabled, is decorated with a uniform background of five-pointed stars, in an attempt to provide the pharaoh with that celestial sphere. On the ceiling of the royal burial chamber, an evolution is observed that tends to assimilate its shape with the celestial vault. As we saw in chapter 1, King Menkaure, during the 4th Dynasty, was the first to adopt a vaulted roof for his burial chamber, followed by his successor Shepseskaf, although during the 5th and 6th Dynasties the gabled roof would continue being the most used.

During the Middle Kingdom the vaulted ceiling was recovered again, as we have examples such as those of Senuseret II (see Figure 4.20) and Amenemhat III. However, it is in the royal tombs of Biban el-Moluk, beginning with Sethy I, that this form is adopted in most cases. This form of construction had been adopted since

Fig. 4.20 Vaulted sepulchral chamber of Senuseret II at Lahun. (Image courtesy of Mónica Moritz)

the 19th Dynasty by the pharaohs Sethy I, Ramesses II, Merenptah, Siptah, Tausert/Setnakht, Ramesses III, Ramesses V/VI, Ramesses VII, and Ramesses XI. The exceptions that we find from the time of Sethy I are related not to a change in the construction model but to unfinished tombs. Among these exceptions we have the KV 10 of Amenmesse whose sarcophagus was located in the second corridor of the lower level of the tomb; that is, in an unfinished tomb. KV 15 of Sethy II was also not completed and the first corridor on the lower level had to be used as a burial chamber. KV 2 of Ramesses IV, KV 19 perhaps originally belonging to Ramesses VIII and tombs KV 9 and KV 18, of Ramesses IX and Ramesses X, respectively, do not have a vaulted ceiling either owing to the prompt interruption of the work that had to be carried out for the complete excavation of the tomb.

Thus, the vaulted ceiling decorated either by an astronomical ceiling or by the *Books of Heaven* only intensifies the relationship between the celestial world and the burial chamber, the last abode of the late pharaoh. Since Sethy I, only the unfinished tombs are those that do not have a vaulted ceiling, but this was also done in KV 4, the unused tomb of Ramesses XI, the last royal tomb in the Valley.

This important feature of the Ramesside tombs was abandoned in the royal Tanite necropolis of the 21st and 22nd Dynasties. At Tanis, the covering of the burial chamber was made by means of large slabs with smooth faces, which meant that this vaulted form was not used. Despite this, it is necessary to remember that in the only case where a representation on the ceiling has been preserved (that is, with Shoshenq III) it is of an astronomical type (Lull, 2002: 155).

The vaulted ceiling tradition, abandoned at Tanis, continues in the private tombs of the late Third Intermediate Period and Saite Period (Eigner, 1984: 139–144). Equally interesting is to see how during the Nubian Dynasty the Kushite royal tombs followed this practice, extending in time even until the Meroitic Period. In the burial chamber of Tanutamun (a.k.a. Tanwetamani), the last monarch of the 25th Dynasty, the vaulted ceiling was decorated with a field of stars, just as in the tomb of his mother Qalhata. In these tombs, however, the astronomical ceiling is missing.

4.7 Hypotheses Concerning the Identification of Stars and Constellations

Neugebauer and Parker (*EAT* II, x; *EAT* III, 3), considered that it was not feasible to proceed to identify the Egyptian constellations, since the variants that occur in the lists and representations would make this task, in their view, a game of conjecture. Even more recently, Depuydt (2010: 59) wrote: 'The identities of the stars in the star tables are, in my opinion, difficult if not impossible to know except for *spd(t)*, which is Sirius, and perhaps also *sꜣḥ*, which may be or overlap with Orion'.

In our opinion, and with great respect for these authors of such a fundamental reference work for the study of astronomy in ancient Egypt, this statement needs to be qualified. With the help of lists of decans, of iconography, of the zodiacs related to decans, and of Egyptian star clocks and star tables, and with the help of modern astronomical programs, valuable conclusions may be drawn that, in some cases, allow us establish with a high degree of confidence equivalences between Egyptian constellations and stars, and their current counterparts. Obviously, many other proposals will be recognized as mere hypotheses.

4.7.1 Northern Constellations

During the Pharaonic period only a small group of constellations (Lull & Belmonte, 2015: 1477–1497) were represented by the ancient Egyptians. In the oldest representations we know from some coffins of the late First Intermediate Period, along with a long list of decans (Belmonte, 2002), we generally find only such representations of Egyptian constellations as Sah, Sopdet, and Meskhetyu, perfectly located in Orion–Lepus, Canis Major (with Sirius as reference), and Ursa Major, respectively (see Figure 4.21).

New Kingdom astronomical ceilings, however, offer a wider space for figures and inscriptions, so a good set of constellations and other figures are depicted along with decans lists. For the northern sky the number of constellations represented in an astronomical ceiling is relatively high: Meskhetyu, Anu, Serqet, Isis-Djamet, identifiable with the Reret of the Ramesside star clocks, the crocodile Saq, the crocodile Hetep-redwy (the 'Divine lion who is between them'), the anonimous harpooner (perhaps a figuration of the vast constellation of Nekhet), and a few more constellations (Lull & Belmonte, 2006, 2009).

The ancient Egyptians spoke of a type of stars that they called *jḫmw-skjw* (*Ikhemu-sekiu*), which literally means 'those who do not know the setting'; that is, the 'immortals'or 'imperishable stars'). It could reasonably be suspected that these are the circumpolar stars, those that never dip below the horizon.

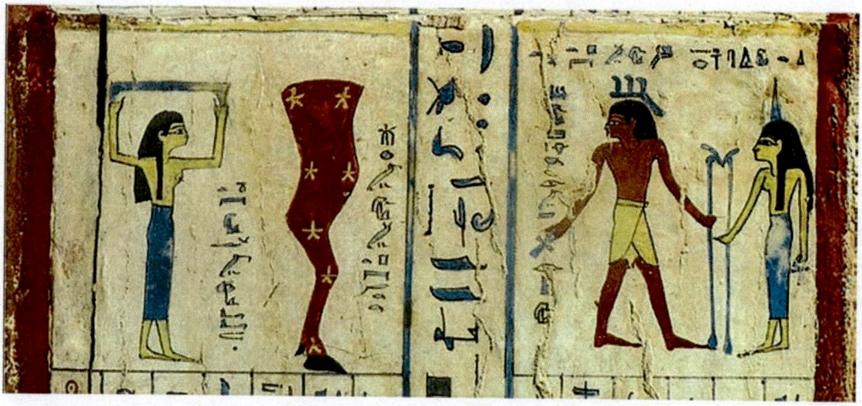

Fig. 4.21 Nut, Meskhetyu, Sah, and Sopdet in the coffin of Idy. (ÄSUT Inv. Nr. 6 (© Ägyptische Sammlung Universität Tübingen. Photograph: F. Kuyas))

In the *Pyramid Texts*, these stars are mentioned on numerous occasions. Thus, the formula PT 302 § 458 a–c (W 207) reads:

The sky has been bled and Sothis lives, for Unis is the living one, Sothis's son, for whom the Dual Ennead have cleaned the imperishable Striker (*jḫmw-skj*). (Allen, 2005: 56)

However, religious texts are open to various interpretations. In fact, this type of text is always abstract, and to this we must add grammatical and terminological difficulties.

Mainly in astronomical ceilings, but also in the decoration of many coffins, sarcophagi, and reliefs found in temples, a series of constellations are represented that we usually classify as 'northern', and that belong without doubt to the region of the sky dominated by the *jḫmw-skjw* stars. The most characteristic of these constellations is Meskhetyu.

In PT 302 §458c we find a relationship between Meskhetyu and the (stars) that do not know sunset. We must conclude that the stars that make up the Egyptian Meskhetyu are of the *jḫmw-skjw* type. On the other hand, we also know that the Egyptians considered these stars to represent gods. Thus, for example, in PT 441 § 818c (P 37) we read:

You might go to yonder gods, the northern Imperishable Stars. (Allen, 2005: 107)

However, not all stars of the *jḫmw-skjw* type are circumpolar. In fact, from some passages in the *Pyramid Texts* it might be deduced that they comprise all the stars north of the ecliptic, including the circumpolar ones. If we consider the formula PT 520 § 1222 c–d (P 468), it is indicated that one has to cross one side of the sky to reach the stars of the *jḫmw-skjw* type:

He might proceed to the side where the Imperishable Stars are and be among them. (Allen, 2005: 161)

In the *Pyramid Texts* a 'channel' called 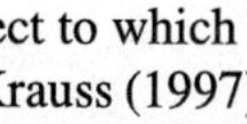 *mr n ḫꜣ* ('channel of Kha') is cited on numerous occasions, with respect to which it is said that it can be crossed on its north or south side. According to Krauss (1997), this channel must refer to the ecliptic, so, supposing that in PT 520 (§ 1222 cd) the text also refers to the crossing of this channel to the north, we would only have to assume, according to him, that the *jḫmw-skjw* ('immortal') stars include not only circumpolar stars but also those located north of the ecliptic. However, there are different opinions regarding this because the literal translation of the term *jḫmw-skjw* could be understood to refer to circumpolar stars and, as we have seen previously,  has also been interpreted as the Milky Way. In any case, Meskhetyu, according to the texts, is within the stars located in the northern part of the sky.

The earliest surviving mentions of Meskhetyu date, in the *Pyramid Texts*, to the reign of Pharaoh Wenis. Recall that the oldest mentions of Meskhetyu refer to this constellation as , with the adze determinative, so it is possibly the way in which they originally saw this constellation. Later, Meskhetyu is written as , with the bull's leg determinant, thus recognizing the way in which this constellation was also observed.

The earliest representations of this constellation date back to the late First Intermediate Period and the Middle Kingdom (Lull, 2008c). Thus, in some characteristic coffins of this time, we have very interesting examples. In these, Meskhetyu is represented in the shape of a bull's leg, and the seven stars indicated in it lend it a shape that reminds us of the Wagon. In Figure 4.22 we can see the examples of Itib, Idy, Aashyt, and Heqat from the 9th–11th Dynasties.

On the other hand, in a text from Idy's coffin we read:

 (...)

An offering of Meskhetyu in the northern sky; An offering of Nut (...).

Meskhetyu is in the northern sky. We know that the *jxmw-skjw* stars, including those of Meskhetyu, symbolize gods. That is why, in addition, it is a wish of the late pharaoh to be able to travel to that side of the sky, since it is there that the immortals, the gods, gather.

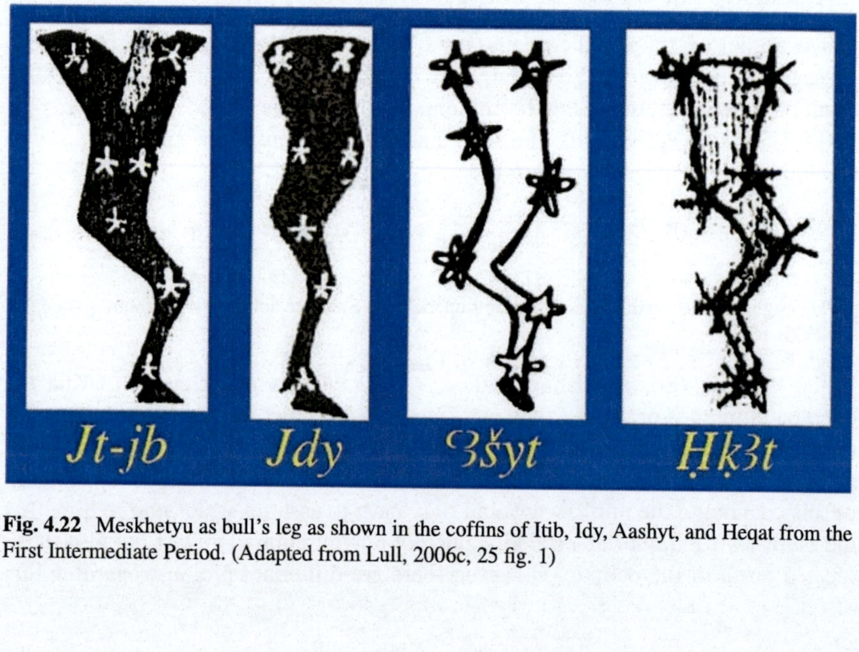

Fig. 4.22 Meskhetyu as bull's leg as shown in the coffins of Itib, Idy, Aashyt, and Heqat from the First Intermediate Period. (Adapted from Lull, 2006c, 25 fig. 1)

What did the bull's leg in the form of Meskhetyu mean to the Egyptians? A first explanation can be found in the *Book of the Day*, which appears illustrated in two versions in the tomb of Ramesses VI (Piankoff, 1942: 24, 95):

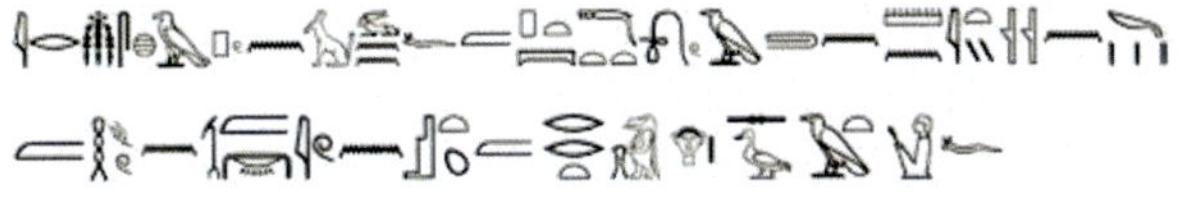

> *Quant a cette Cuisse de Seth, elle est dans le ciel du nord, attachée a deux piquets de silex (dsw) par une chaîne d'or. Elle est corifiée a Isis sous la forme d'un hippopotame femelle qui la garde.*

> 'With regard to this Thigh of Seth, it is in the northern sky, attached to two pegs of flint (*dsw*)' by a golden chain. It represents Isis in the form of a female hippopotamus that guards it.'

Meskhetyu, then, represents the leg of Seth, a god who, as the murderer of his brother Osiris, presents an evil aspect. It is his sister Isis, in her hippo form, who has the mission of guarding that leg in the northern sky. To do this, she has it tied by gold chains to two mooring posts.

This image may be better appreciated on some astronomical ceilings, such as, to name a few examples, that of Sethy I (see Figure 4.23), Pedamenope (see Fig. 2.6), the roof of the pronaos of the temple of Hathor at Dendara (see Figure 4.24), and in the later and recently restored temple of Khnum at Esna (see Figure 4.25). In these ceilings boreal constellations were represented, among them Meskhetyu (this time in the form of a bull or oval bull, instead of in the form of a leg, as was usual in the Middle Kingdom), and the hippopotamus who holds in her hands the two poles of mooring (well outlined in the example of Pedamenope), towards which the chains are directed towards Meskhetyu.

Fig. 4.23 Northern constellations in the astronomical ceiling of Sethy I. (Taken from Lull, 2019a, b, fig.1, adapted from *EAT* III)

Fig. 4.24 Meskhetyu in the pronaos ceiling of the temple of Hathor at Dendara. (Image by the authors)

Fig. 4.25 Meskhetyu and Weret (Reret) in the temple of Khnum at Esna. (Adapted from europa-press.es. Courtesy of The University of Tübingen)

In the Jumilhac papyrus, from the Ptolemaic period, we find further complementary explanations about the origin of Meskhetyu in the northern sky (Vandier, 1961: 129):

> *Après avoir fait cesser le combat et avoir taillé en pièces les rebelles, il anéantit Seth, extermina ses alliés, détruisit ses villes et ses nomes, effaça son nom dans ce pays; après avoir mis en pièces! ses statues dans tous les nomes et avoir coupé son ḫpš, il l'emporta au milieu du ciel, des génies étant là pour le garder; c'est la Grande Ourse (msḫt) du ciel septentrional, et la Grande Truie (rrt wrt) le tient, de telle sorte qu'il ne puisse plus naviguer parmi les dieux.*

> 'After bringing the combat to a close and having hewn the rebels, he annihilated Seth, exterminated his allies, destroyed their towns and nomes, effaced their name in that country. Once having shattered their statues in all the nomes and cut off his [Seth's] *ḫpš*, he transported it to the sky, having spirits there to watch over it; it is the Great Bear (*msḫt*) of the northern sky, and the Great Sow (*rrt wrt*) retains it in such a manner that it can no longer navigate among the gods'.

In both this and other documents from the Ptolemaic era, it is indicated that Seth had been defeated by Horus and dismembered. After this, Horus placed Seth's leg in the northern sky, leaving the great hippo to keep him tied, confined to that area of the sky.

We also have information, with some variation, in a Roman period text from the Temple of Khnum at Esna:

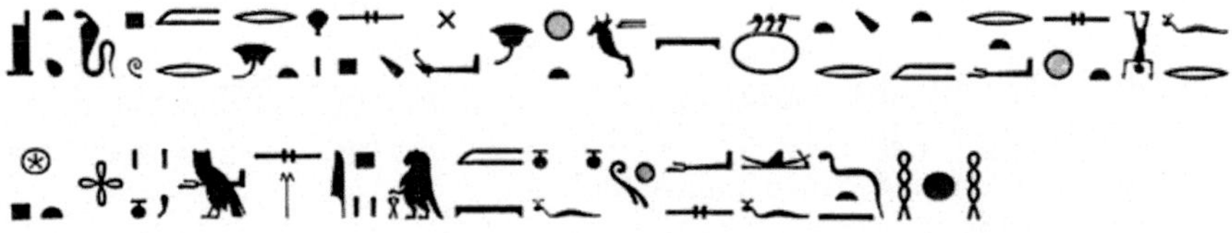

(Sauneron, 2012: 3)

> It is Isis as a hippopotamus (*rrt*) who tethers the Foreleg (*msḫt*) in the northern sky in order not to let it go outside down into the Duat. It shall be with her, being (the goddess) Ipy, in the sky and she shall not release it forever and ever (*EAT* III, 191).

However, although these texts are late, there is no doubt that the origin of this story goes back to the Old Kingdom, since in the *Pyramid Texts* (PT 61 § 42) it is said: 'Accept the foreleg of Seth, which Horus has torn off' (Allen, 2005: 318). The god Horus, in his Harendotes form, receives the epithet  *hrw-nḏ-jt.f* ('Horus, the avenger of his father'). It was the god Horus, then, who was responsible for this leg going to the northern sky, and thus dismembering his uncle Seth is how he avenged his father (who will also appear in the sky in his form of *Sꜣḥ*, which corresponds to Orion).

Egyptian mythology is an important source for understanding how the Egyptians explained the existence of this constellation. However, Egyptian mythological texts sometimes seem contradictory. Thus, in the sarcophagus of Kom Abu Yasin (30th Dynasty), the inscription associates Osiris, bull of the sky, with the stars of the northern sky and the Foreleg, Meskhetyu (*EAT* III, 191). According to this last text, then, the bull's leg would be a form of Osiris, in complete contradiction with the previous texts. But, as we have already commented, even Plutarch, in *De Iside et*

Osiride, tells us that Typhon's soul is the Great Bear, Typhon being, of course, the Greek god corresponding to the Egyptian Seth.

However, Meskhetyu was not always circumpolar (see Chap. 6). Although the Egyptians did not know the nature of the phenomenon of precession, they were aware of the change that, over the centuries, their sky underwent. This observation is evident in some texts. Waitkus (2010: 179) mentions one in which reference is made to the loosening of the chains that hold Meskhetyu to allow it to hide in the west; that is, verifying how this constellation had already ceased to be circumpolar, which could not fit the mythological explanation in ancient texts of the confinement of Seth's leg.

The importance of Meskhetyu is evident not only in that it had already been mentioned in the *Pyramid Texts* since the 5th Dynasty but also in the way it was represented and in the specific indications concerning its stars. A clear example is provided by the astronomical ceiling of Senenmut (Fig. 4.17). Following a typical model of the New Kingdom, the bull's leg has been transformed into a bull in the shape of a pointed ovoid, with four small legs. However, what is interesting in this representation is how the bull's tail is made up of three stars, the last of which was coloured red, circled in red, and even indicated by a pointed object in the shape of an isosceles triangle. This can be explained only if we understand that this Meskhetyu star had a special significance for the Egyptians. In our opinion, this star is Alkaid (85-η UMa). As we shall see later, this star could have been used as a reference to establish the axes of various temples, such as Dendara, where Alkaid would probably be the star mentioned in the hieroglyphic texts of the temple of Hathor as the *ꜣḫ* of Meskhetyu, the celestial object to which the pharaoh directs his gaze during the ceremony of tensioning the rope in the founding rites.

In relation to this, it is possible that the triangle-shaped object that reaches Alkaid on the astronomical ceiling of Senenmut represents an observing instrument (Lull, 2006–2009). This is most clearly seen in the astronomical ceiling of Pedamenope, from the 26th Dynasty (see Fig. 2.6), where we see how this pointed object has been transformed into a kind of pole; it should be no coincidence that the first of the divinities on its left raises its right arm and holds what could be a *bay*; that is, a sighting instrument. If we associate the *bay*, the pole, and the star marked in red in the tomb of Senenmut, it seems even more evident that this star, Alkaid, aided the Egyptians in establishing orientations, measurements of transit through the central meridian, etc. In the example of the Ramesseum (see Fig. 2.14), it is also interesting to see how to the left of this triangle there is a perfect vertical line that reaches Serqet's hands. In the Petosiris example (see Figure 4.26), on the other hand, it is the object (now with the upper end not pointed but rounded) that reaches the hands of this female figure, thus suggesting the possibility that it is indeed an instrument dedicated to establishing the central meridian and possibly observing the culminations of stars such as Alkaid. The example of the Ankhhapy sarcophagus is also interesting, since on the one hand it shows the hippopotamus holding the two mooring posts, but also the triangular object attached to Meskhetyu and one of the posts, and Serqet in relation to it (Fig. 4.26).

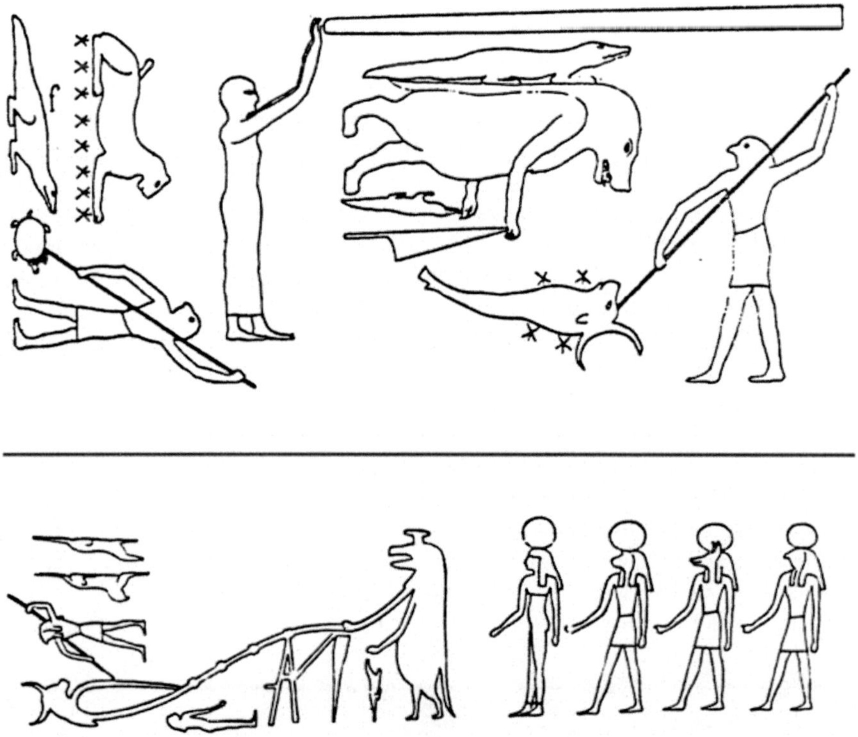

Fig. 4.26 Northern constellations in the ceiling of Petosiris (top) and in the sarcophagus of Ankhhapy (bottom). (Adapted from *EAT* III, 65 fig. 15 and *EAT* III, 38 fig. 13)

Le Page Renouf (1874) believed the Big Dipper to be the thigh and Cassiopeia the leg of Meskhetyu. However, most authors agree on recognizing the Big Dipper with Meskhetyu. Perhaps it is in the coffins of the Middle Kingdom where, better than in the astronomical ceilings, the shape of this constellation is recognized. It is presented there as a bull's leg within which are included the seven stars that give it its shape.

Thus, Alkaid (85-η UMa), the star that at the present time is at the end of the Ursa's tail, would correspond to the hoof of the bull's foot represented in the coffins of the Middle Kingdom or, in the New Kingdom, at the end of the bull's tail. Pogo (1930: 309), however, estimated that the star at the end of the leg or tail would be Mizar (79-ζ UMa), thus not taking into account the seven stars that appear drawn on the leg during the Middle Kingdom. Finally, Etz (1997: 155) proposes Muscida (1-o UMa), at the head of the Big Dipper, as the end of the bull's tail.

We have seen which are the seven stars that make up the bull's leg, but the way of representing this constellation changed during the New Kingdom. This could indicate that other stars were added to give the shape of a bull to what was initially only one leg of the torus. Given that one of the main characteristics of this animal are its horns, it is probable that these prominences were added to the bull's leg, which could result in the ovoid shape of the bull's body, a product of the iconographic fusion of the leg with what was intended to be the body of the animal. In any case, on some ceilings, such as that of Pedamenope, the bull's horns appear marked by a series of stars that give it its shape. This could be coincidental, because in this sense the Egyptians were not very fortunate in the representation of the stars in their constellations, as we see now, almost unanimously, these are nothing more than decorative elements.

However, assuming that the bull's horn has a real value in the sky, and given the orientation of the bull's leg, it is most likely that it would have to be formed with the stars 1, 23, 29, 30, 25 and 9, 12 UMa. In this way we can imagine how the Egyptians viewed Meskhetyu in his bull's leg form and in his bull form (see Figure 4.27).

From chapter 17 of the *Book of Going Forth by Day*, it is understood that the children of Horus are part of the constellation Meskhetyu (Mathieu, 2008: 12-13): 'As for the Council around Osiris, these are Imset, Hapy, Duamutef, and Qebehsenuf, who are back of the Great Bear in the northern sky' (Allen, 1974: 29). According to Mathieu, these four gods would have their correspondence in four of the stars of the Great Bear: Megrez, Phecda, Merak, and Dubhe; that is, those that form the square or upper part of the bull's leg.

If the Egyptians saw two adzes in the boreal sky and one corresponded, in its original form, to what would later become the bull's leg, Meskhetyu (that is, the Big Dipper), it is evident that the second adze must correspond to the Ursa Minor, since

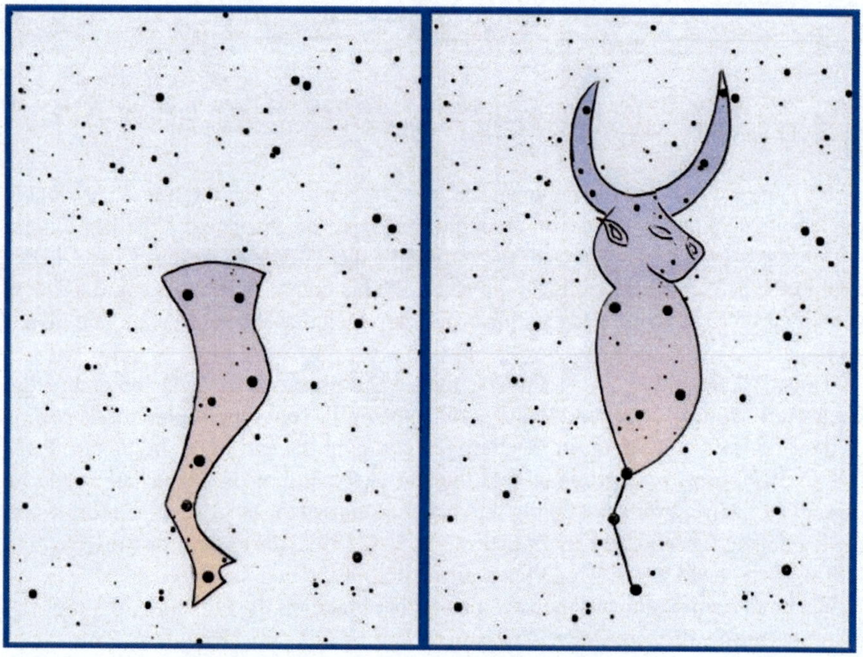

Fig. 4.27 Transformation of Meskhetyu englobing more stars for its horns belonging to UMa. (Adapted from Lull, 2006c, 29 fig. 6)

this constellation, even fainter in brightness, is almost a mirror image of Ursa Major. However, it is difficult to recognize precisely which Egyptian constellation among those represented on the astronomical ceilings may correspond to Ursa Minor. In the Dendara zodiac there is a hoe on which a canid stands (Fig. 4.8), while Meskhetyu is represented as a bull's leg. We may therefore suppose that this hoe with the canid must correspond to Ursa Minor, although it is difficult to define this constellation among those represented in the boreal sky of the astronomical ceilings.

Another well-known constellation from astronomical ceilings, also present in the Dendara zodiac and other documents from Roman times, is identified on the Senenmut ceiling as *ꜣst-ḏꜣmt* ('Isis-Djamet'), in the shape of a hippopotamus to whose back is attached a crocodile. Pogo (1930: 311) believes that it may be the constellation Draco. Biegel (1921: fig. 7) recognized this Egyptian constellation mainly in the stars of Boötes, with the crocodile spreading through other stars in Corona Borealis and Hercules, and the mooring posts in Coma Berenices (see Figure 4.28). In the opinion of Locher (1985: 152 fig. 1) it would be formed by stars of Draco (hippopotamus), Cepheus, and Cygnus (crocodile), while the stars of Ursa Minor and some from Draco would form the mooring posts and the chain that Isis-Djamet holds in the representations of the New Kingdom (see Figure 4.29).

Davis (1985: 104), on the other hand, identifies Isis-Djamet with Ophiuchus, Libra, and Scorpius. Chatley (1940: 123) offered another option, assimilating the hippopotamus with stars from Boötes, Hercules, Lyra, and Draco. As we can see, contrary to what happens with the comparison between Meskhetyu and the Ursa Major, here there is no longer unanimity. Finally, to cite another proposed equivalence, Etz (1997: 158) suggests the constellation of Auriga and stars of the neighbouring constellations.

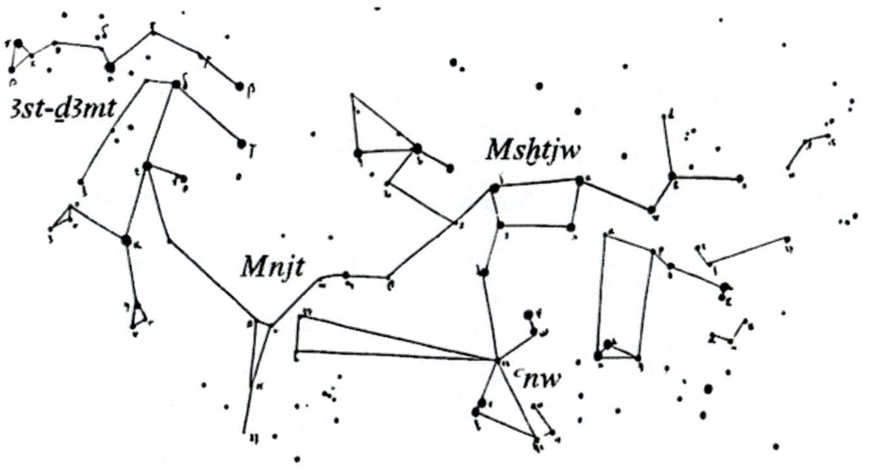

Fig. 4.28 Northern constellations after Biegel, 1921: fig. 7. The reconstruction is based on a very simple exercise

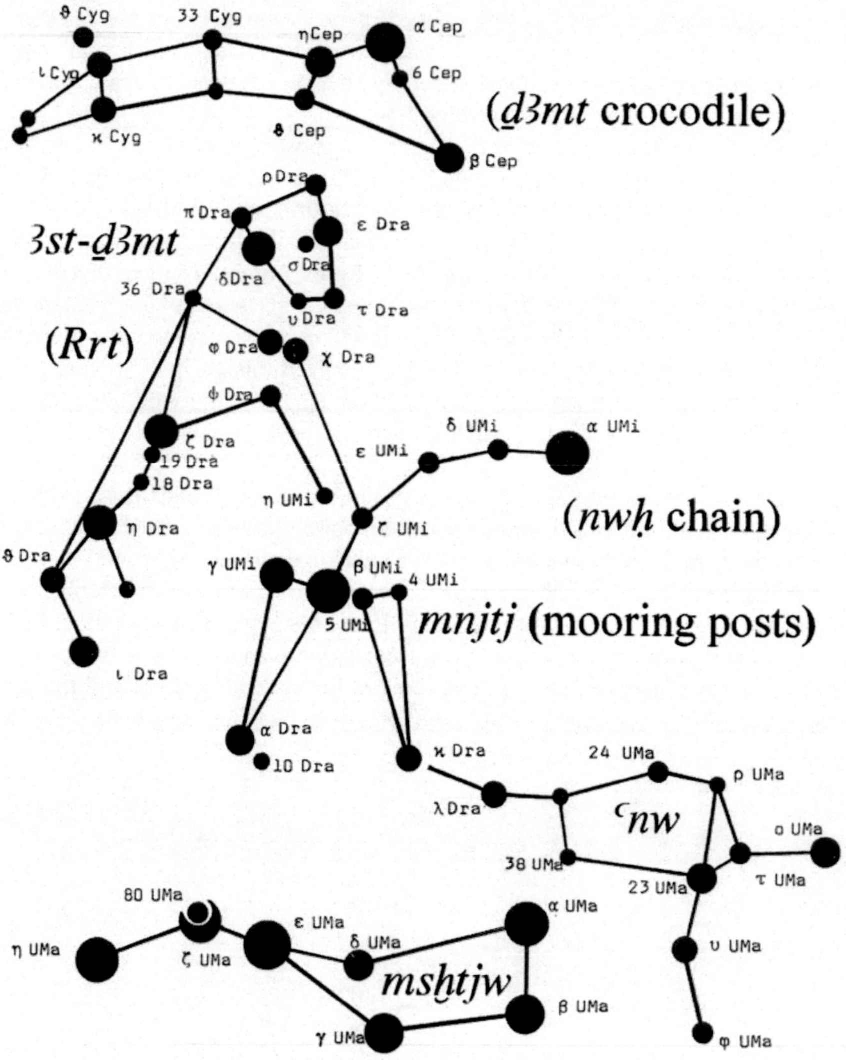

Fig. 4.29 Northern constellations after Locher. Adapted from Locher, 1985, fig. 1. Locher settled most northern constellation in a very small area of the sky

Our suggestion is that the area occupied by the hippopotamus should be much larger, located between the constellations of Boötes and Lyra, with the crocodile on its back reaching Serpens Caput. The head would be located roughly on Draco's head.

In the *Book of the Day*, of the Ramesside period, Isis, as hippopotamus, is called *rrt*. In the Kom Abu Yasin bull sarcophagus, from the Nectanebo II period, she is called *rrt wrt*, on the Jumilhac papyrus, from the Ptolemaic period, *rrt wrt*, and on Esna's ceiling she appears as *wrt*. With this, there is textual evidence that the hippopotamus located in the northern hemisphere is often called Reret or Great Reret; that is, the great hippopotamus.

The name of *rrt wrt* never appears in any of the versions of the northern constellations of the ancient astronomical ceilings, but it does appear among the constellations that served to mark the hours on the Ramesside star clocks and star tables. In fact, the second constellation with the most hourly stars mentioned in the tables of the Ramesside clocks is that of *rrt* ('the pig', or hippopotamus), with eight stars. The determinatives of pig and hippopotamus can be confusing at times. In any case, the hippopotamus of the Jumilhac papyrus and other texts is known as *rrt*, in the same way as the constellation of the pig (or hippopotamus) mentioned in the Ramesside tables. On the other hand, another constellation called *mnjt* ('mooring post') also appears in the Ramesside star clocks.

We recall that Isis, in her hippopotamus form, as indicated in the *Book of the Day*, fastened to two mooring posts ( ) with gold chains that were tied to the Meskhetyu bull's leg. However, if we look at the representations of astronomical ceilings, the chains (or the linear or dotted shapes that attempt to symbolize them) are rather attached only to one of the posts that holds the hippopotamus, not to the one that has the shape of a crocodile but the other, the one that most closely resembles the sign P11 ( ), the sign *mnjt*.

Besides, in all the examples in which the hippopotamus holds these two posts (and the chains appear), such as the Ramesseum, that of Ramesses III A (KV11), the block of Ramesses III of Medinet Habu, the case of Ramesses VI C (KV9), Ankhhapy (Saqqara, 30th Dynasty or early Ptolemaic), Hermopolis A (Tuna el-Gebel, reign of Ptolemy I), and Harendotes (BM 6678, reign of Ptolemy III), the chains only touch the *mnjt*-shaped pole.

Similarly, in those cases where only a pole held by the hippopotamus appears, as seems to be the case in the example of the clepsydra of Amenhotep III from Karnak (see Figure 4.30), from the astronomical ceiling of Sethy I (KV 17), that of Tausert A (KV 14), and that of Esna (Fig. 4.25), this post also corresponds to the one in the shape of a *mnjt*. An exception would be the case of the representation existing in the Dendara pronaos (Fig. 4.24), although here the chain does not go to the crocodile but to the only object that the hippopotamus supports, in this case the sign V17 *s3* ( ).

Furthermore, the fact that in the Ramesside clocks the hours marked by the stars of the *mnjt* (mooring post) constellation follow those of *rrt* (in the same way as in the circular Dendara zodiac the mooring posts precede the boreal hippopotamus, see Fig. 4.8) seems to confirm a possible equivalence between the constellation *3st-d̠3mt* of the astronomical ceilings of the New Kingdom, and other later documents, with the *rrt* of the Ramesside clocks and late texts. It also seems clear that it is only a pole that is used to chain Meskhetyu, the one that has the shape of *mnjt* in the representations and the one that as a constellation *mnjt* is mentioned in the Ramesside tables.

However, one of the authors has expressed the possibility that two mooring posts were supported by the hippopotamus in many representations, and a third one (which in the representations would correspond to what we have previously defined as an astronomical instrument) was the one referenced in the Ramesside star tables (Belmonte, 2012: 66). Similarly, he hypothesized that the posts handled by the

Fig. 4.30 Northern constellations in the clepsydra of Amenhotep III. (Image kindly provided by Nacho Ares, courtesy of The Cairo Museum, and adapted drawn from Desroches-Noblecourt et al., 1976)

Hippopotamus should be located in Ursa Minor and Draco, and might represent the celestial and ecliptic poles (2009: 163). This is difficult to prove with the evidence at hand.

In the opinion of one of the authors (Lull, 2003), the pointed object or post that appears between the boreal constellations cannot be a mooring post, as it would be repeating it unnecessarily. On the other hand, the mooring posts are always characterized by being wider at the top so that the ropes do not escape, with a more cylindrical shape at the bottom so that they fit better. This definition is well suited to the mooring post and even to the crocodile that, placed vertically, is held by Isis-Djamet. Furthermore, the pointed object would be useless as a mooring post, since it is wider at its base than at its upper end.

Fig. 4.31 Goddess Ipet (for the eponymous Paophi) in a procession of gods associated with the twelve months of the civil year. Temple of Amun at Deir el-Haggar, Dakhla oasis. (Image by the authors)

In spite of the comments just made, Neugebauer and Parker (*EAT* III, 191) do not agree with the equivalence between the hippopotamus of the astronomical ceilings and the Reret of the Ramesside star tables, because they think that the 'two feathers' (the name of one of the stars of the constellation of the *rrt wrt* in the Ramesside star tables) are never represented on the head of Isis-Djamet on astronomical ceilings. However, the hippo goddess Ipet is frequently depicted with a crocodile tail on her back and, more importantly, with two feathers on her head. The celestial diagram of the temple of Deir el-Haggar in the Dakhla Oasis is also relevant in this particular context (see Figure 4.31).

The goddess Taweret (Figure 4.32) is shown in a very similar way, and she does not always wear feathers. Moreover, she can also be compared to representations of the goddess Ipet. In fact, on the ceiling of the temple of Isis at Shanhûr, dating from the early Roman period, the hippopotamus is identified as Ipet (see Figure 4.33). Thus, for example, the representation of the constellation on a block that apparently comes from the ceiling of the second hypostyle hall of the temple of Ramesses III in Medinet Habu (see Figure 4.34) offers us a similar aspect to that of these goddesses when they do not wear feathers.

Neugebauer and Parker also believed that if *rrt wrt* is used as a time marker it is possible that it was between or near the decanal constellations (that is, in the ecliptic area or further south), which is why it could not correspond to Isis-Djamet, which lies between the boreal constellations. These reasons and the fact that the hippopotamus Isis-Djamet is never called *rrt* in its representations mean that, according

Fig. 4.32 Statue of Taweret (Courtesy of Science Museum, London) and relief coming from the Horus temple at Edfu. (Composite diagram by the authors, based on: https://commons.wikimedia.org/wiki/File:Statue_of_Taweret,_Cairo,_Egypt,_1925-1933_Wellcome_L0058412.jpg. Wellcome Collection gallery. Public domain, via Wikimedia Commons)

Fig. 4.33 Ipet as hippo constellation in the astronomical ceiling of the temple of Isis at Shanhûr. (Adapted from *EAT* III, pl. 40)

Fig. 4.34 Northern constellations. Temple of Ramesses III at Medinet Habu. (Adapted from *EAT* III, 29 fig. 5)

to them, both hippos must, in fact, be different constellations. Furthermore, among the Egyptological community, there has been a strong tradition, never questioned since it was first proposed by Sloley (1931), that the observations described on the Ramesside clocks were made by an observer looking south towards a second priest, who acted as a frame of reference. If this was so, Reret must be located in the southern part of the sky and could not be identical to the hippopotamus in the northern skies.

Any star in this constellation that had a declination greater than 26° would culminate in the zenith or north of the zenith and, since *ꜣst-ḏꜣmt* is assumed to be a boreal constellation, with a high probability that all, or almost all, of its component stars would exceed this declination. Usually it has been said that in a Ramesside star clock the observer directs his gaze towards the south, which is why the stars that culminate north of the zenith would not be used. However, we do not agree with this assertion, since Reret had to be observed, like Menit, regardless of its position north of the zenith, since many of its stars are mentioned in the Ramesside clocks.

In the Dendara zodiac *ꜣst-ḏꜣmt* has her feet at the level of Libra and Virgo. At about that right ascension we should look for the mooring posts that she is holding. Isis-Djamet had to be constituted by the stars of our Boötes, Corona Borealis, Hercules, and Draco, while the crocodile *ḏꜣmt* could coincide with stars of Serpens Caput (δ, β, γ Serpentis) and others of Hercules (γ, β, δ, λ, μ, ξ, o Herculis). The Menit mooring post could therefore be made up mainly of Boötes stars (specifically, Arcturus and Muphrid), with its extreme to the north of Boötes marked by Alkaid in Ursa Major (Belmonte, 2012: 67).

The Dendara Planisphere shows that the orientation and position of the Isis-Djamet hippopotamus and the Meskhetyu bull's leg is, on most astronomical ceilings, reversed from what is observed in the sky. In the planisphere the bull's leg is orientated in correspondence to what would be expected for the seven main stars of the Big Dipper, and in relation to it, Isis-Djamet.

Above Meskhetyu, according to the ceiling of Senenmut, the clepsydra of Karnak, the ceiling of the Ramesseum, that of Pedamenope, and the sarcophagus of Nekhtnebef, for example, we have another constellation in the shape of a woman called *Srḳt* (Serqet), the goddess who sometimes appears in the form of a scorpion. On the ceilings of Sethy I, Tausert, and Ramesses IX, she also appears face down, horizontally, but before Meskhetyu. Chatley (1940: 123) believes that it could be Coma Berenices, and Davis (1985: 104) that it is Virgo, at a lower declination but in the same right ascension. For his part, Biegel (1921: figs. 6a-by 7) reconstructed this Egyptian constellation from stars such as 23 and 1-o Ursae Majoris, and, according to Etz (1997: 158), by α, ι, and θ Draconis.

Serqet is not mentioned in the star clocks, so it is very difficult to place it more specifically in one area of the sky. If we compare all the representations of the Egyptian boreal sky we see how, in addition, there are important variants. Meskhetyu seldom looks at Isis-Djamet, as on the ceiling of Room 25 of the Osiris sanctuary at Medinet Habu of Ramesses III. In the other examples, both figures look in the same direction, although on some occasions it is to the left (ceilings of Senenmut, Sethy I, Ramesseum, Ramesses VI, Ramesses IX, Pedamenope, etc.) and less often to the right (ceiling of Tausert, sarcophagus of Psusennes I, zodiac of Dendara, zodiac of Esna), which suggests that Isis-Djamet and her mooring posts are tied to Meskhetyu by the hoof or tail of the bull through the star Alkaid. This is the reason that leads us to place Isis-Djamet between Boötes, Hercules, and Draco. From here we can venture to locate the constellation Serqet, which almost always appears above the bull's leg or to one side of it.

From the Jumilhac papyrus we understand how Meskhetyu, being tied to the hippopotamus by gold chains, according to the *Book of the Day*, does not escape from the circumpolar sky. In some representations, such as the ceiling of Pedamenope (26th Dynasty), a ceiling of the underground gallery C of Tuna el-Gebel (Ptolemy I, Hermopolis), and, apparently, the clepsydra of Karnak (Amenhotep III), Serqet holds a rope that joins with that of Meskhetyu to the base of Isis-Djamet's mooring posts. This perhaps indicates that Serqet was also a circumpolar constellation, trapped in the same area of the sky as Meskhetyu. Likewise, it should be noted that on the ceiling of Senenmut, where the boreal constellations are divided into two groups (leaving aside considerations of space), only Serqet and Anu were represented together with Meskhetyu, perhaps because of their greater declination.

Nemes's (2020: 28) observation regarding the interpretation of an inscription from the *Pyramid Texts* is interesting. In PT 571 §1469a (P 511), the late king is said to be a *jḫmw-skj*: 'Pepi is an Imperishable Star, they are of the great sky that is in the midst of Selket's enclosure' (Allen, 2005: 179). The Great Sky being also a personified form of the northern sky, it is clear that the place of Serqet is also there.

At this point, the authors hold differing opinions, one believing that Serqet could be closer to the celestial north pole, perhaps sharing stars from Draco and Ursa Minor, especially Polaris (Lull, 2004a: 241).

Fig. 4.35 Northern constellations in the Ramesseum. (Image by the authors)

In this sense, there are a number of possible interpretations concerning the arrangement of Serqet's hands that lead Lull to conclude that the hands could coincide with Polaris. These were the position of Serqet's hands on the other side of the star marked in red (which we assume to be Alkaid) on the ceiling of Senenmut, or, in the same position, receiving a vertical line (as a central meridian) in the Ramesseum (see Figure 4.35), or, as also happens in the sarcophagus of King Psusennes that shows a vertical line that goes to Serqet's hands, or picking up a rope that passes through Alkaid in Pedamenope, or picking up a rope that, like Alkaid's, is part of one of the mooring posts held by the hippopotamus in Hermopolis A, or an unusual variant such as that of the ceiling of Petosiris, from the Ptolemaic period, in which Serqet raises his arms, reaching the elongated object with a rounded point (the instrument represented on the astronomical ceilings and which should have been used to observe the passage of the stars through the central meridian). In the boreal sky the culmination of Alkaid coincided, with some approximation, with that of Polaris, so that the relatively simultaneous culmination of both defined the central meridian and at the same time symbolized the imprisoned stars of the circumpolar world.

However, in other representations present in the astronomical ceilings, such as in the tomb of Sethy I or in Tausert, for example, Serqet seems to be more related to the constellation of the Lion, for which Belmonte (2012: 91) has considered that Serqet could occupy a sector of the constellation Virgo. As we have indicated, the identification of this constellation is difficult because of the variations that exist in the position of Serqet in the astronomical ceilings and because none of its stars appears among the hourly stars.

Another constellation associated with Meskhetyu on astronomical ceilings is ꜥnw ('Anu'), shown as a falcon-headed deity holding a long spear with which he appears to be harpooning the figure of the bull. According to Wainwright (1932b), it could correspond to Cygnus, as also claimed by Volten (1958), although this has been questioned with respect to the distance that separates Cygnus from the polar constellations (*EAT* III, 183 n. 2; Clagett, 1995: 119). Davis (1985: 103) placed it in Ursa Minor. Biegel (1921: figs. 6a and 7) identified it with the stars of Ursa Major and Coma Berenices (the head of this figure would be the stars ω and ψ UMa, the arms and spear from ξ to ν UMa, and the base of his feet in 37 and 31 Com) and, according to Locher (1985: 152 fig. 1), in the part of the head of the Big Dipper. For his part, Etz (1997: 161) indicates that Anu is perhaps not a constellation but rather a divinity with astronomical connotations. On the other hand, Waitkus (2003) considers that one of the instruments used in the Ritual of the Opening of the Mouth may be paralleled with the constellation Anu, or more appropriately *dwn-ꜥn.wj* or *dwn-ꜥ.wj*. Meskhetyu and Anu (Seth and Horus) were in turn mirrored in the two ritual adzes, *dwꜣ*, *dwn-ꜥ*, or *dwn-ꜥn.wj*, and *msẖtjw* (Nemes, 2020).

Anu almost always appears below Meskhetyu, so we think it should be found in a wide area south of Meskhetyu that included parts of Canes Venatici, the Big Dipper, and Lynx, with Cor Caroli (α CVn) as the brightest star in the constellation.

The following boreal constellations form, from our point of view, an independent group from the previous one. In this group, the constellation of reference would be the Lion, which we believe corresponds to the zodiacal constellation of Leo. From here, we can look up the position of the other boreal constellations in this group.

The constellation *Sꜣḳ* (Saq), shaped like a crocodile with a tucked tail, is present among the boreal stars. On the ceiling of Sethy I it is situated between the Lion and Anu, in the Ramesseum above Meskhetyu preceding Serqet, as on the ceilings of Senenmut and Pedamenope. Above the Lion it appears in a fragment of a ceiling from Abydos, of Ramesses II, associated with Serqet, and on Tausert's ceiling in front of the Lion. In other examples, more variants appear, making their relationship with a specific group of stars difficult to ascertain. In fact, on many occasions, it does not even appear.

Another constellation in the shape of a crocodile with a tucked tail, *Ḥꜣkw* ('the looter'), may be linked to the previous one in such a way that in the sarcophagus of Nekhtebef it appears under the name of *Ḥꜣkw n Sꜣḳ*. However, on the ceiling of Senenmut they are clearly differentiated, although this is really the only example in which *Ḥꜣkw* is present.

On the ceiling of Senenmut, under *Ḥꜣkw* there is a figure in the form of a lion with a crocodile's tail called *rw-nṯrj jmj.snj* (the 'Divine Lion who is between them'). The earliest mention and representation of this constellation appears on Heny's coffin (11th Dynasty), where it is called (the 'Divine Lion'; see Figure 4.36). The indication 'between them' must refer to the fact that the lion is situated between two constellations that look like a crocodile (*EAT* III: 193), since

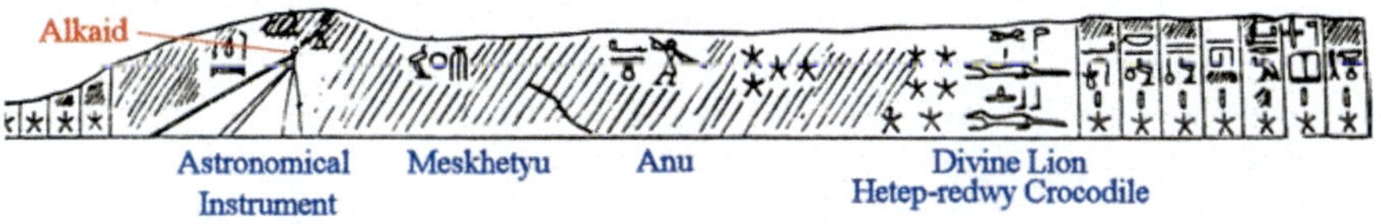

Fig. 4.36 Part of an astronomical diagram, precedent of that of the astronomical ceilings, in the now lost coffin of Heny, 11th Dynasty. (Adapted from *EAT* III, 9 fig. 1)

the figure under this lion is another crocodile, this time called *Ḥtp rdwj* ('Lying on his feet'). This name is used as an epithet by the god Sobek, whose best-known form of manifestation is precisely that of a crocodile. On the ceiling of Sethy I and that of Tausert, a crocodile with its tail tucked in, identified as , does not appear to be another constellation but a confusion of the final part of the name of the constellation of the Lion. In any case, the crocodile appears next to the Lion, under which another larger crocodile stands with its tail extended (obviously, *Ḥtp rdwj*). So, here we have a group of three constellations in which the central one is 'the divine lion that is between them (the crocodiles)'.

According to our hypothesis, the Lion may correspond to the zodiacal Leo. Actually, Leo is one of the few constellations in which an observer can easily recognize the shape of the animal that it symbolizes. It should not be ruled out that the Egyptians had also recreated the figure of a lion in this group of stars. On the other hand, in the ceilings of Sethy I, Tausert (*EAT* III, pl. 3 and 9), and Ramesses VI (*EAT* III, 187 fig. 30), the Lion is associated with a bird, which is not accompanied by hieroglyphics that could identify it. In these cases it looks in the opposite direction and the bird is in front of the lion or on its front legs. Although the bird is only found in three astronomical ceilings, in the Dendara zodiac we find it again, not in front of but behind the figure of the zodiacal constellation Leo, and also looking in the opposite direction (see Figure 4.37). We cannot be sure that the bird on the astronomical ceilings matches the Dendara bird (most probably equivalent to Corvus), although they both look in the opposite direction from the lion and are asociated with it. The crocodile *S3ḳ* could correspond to the Leo Minor region, and the crocodile *Ḥtp rdwj*, which always appears on a larger scale, would perhaps cover a greater area south of Leo, extending through Hydra.

In the star-charts of the Ramesside clocks there is a constellation called *m3j* ('Lion'), of which two stars work as hourly stars: *tp-n-m3j* ('Head of the Lion') and *sd.f,* 'His (the lion's) tail'. Since in the Ramesside star tables the star 'Head of the lion' culminates before the one called 'His tail', it is evident that the Ramesside lion looks towards the west, in the same way the zodiacal Leo. The divine Lion, Mai, and Leo must be the same constellation (Lull & Belmonte, 2006; Lull, 2019a).

Taking into account that the references to the Egyptian hourly stars can be transformed, to within a certain approximation, into equatorial coordinates (although only in right ascension), we can check in what approximate coordinates the two hourly stars of the Lion are located, and for this we can start with the equatorial coordinates of ('Star of Sopdet'), ('Star of Sah'), and ('Star of

Fig. 4.37 Detail of the rectangular zodiac of Dendara (up) and the round zodiac of Dendara (down), showing the following constellations and figures: A, Leo; B, Woman holding the tail of the Leo; C, Serpent (Hydra); D, Bird (Corvus); E, Woman with child (Virgo, Spica); F, woman with an ear of grain (Virgo); G, small king (Regulus). (Images by the authors)

Khau') in the Ramesside star tables, assuming that these correspond to Sirius, Rigel, and Alcyone (with the Pleiades), respectively. Taking this into account, we may postulate that the 'Head of the Lion' could be Regulus and 'His tail' Denebola.

It is also worth mentioning that one of the constellations with the most hourly stars referenced in the tables, (Reret), has as its first star ('Feet of Reret') six Egyptian hours from the 'Head of the Lion', precisely in a position where we might suppose the constellation in the shape of a hippopotamus (Reret) that appeared on the astronomical ceilings with which it must definitely be equated.

In the rectangular zodiac of Esna, Leo is associated (in RA from west to east) with three hourly stars (Hery Kheped Kenmet, Hat Djat, and Pehwy Djat), preceded in Cancer by Sopdet, Shetu, and Kenmet. If we compare it with the oldest decan lists, those of the diagonal star clocks, the decans that follow Sirius to the east are (Kenmet), (Sawy Kenmet), and (Hery Kheped) in Kenmet. Therefore, in broad strokes, we see how the beginning of Leo was located 3 hours from Sepedet (the same distance as in Tables 8, 11, and 12 of the Ramesside star clocks), and the 'Star of Sepedet' separates from the 'Head of the lion', from which it is possible to suppose that Leo occupied the position of Mai, the Egyptian lion in the sky.

In Heter's coffin (see Figure 4.38), just above the register representing the group of figures in which the 'Divine Lion' appears, several zodiacal constellations were captured in such a way that Leo was represented just above it. Far from indicating that they are different constellations, it could well be an allegory to their relationship, in the same way, as we shall see, that could happen between the Egyptian constellation of Wia and Sagittarius.

On the other hand, in the Dendara zodiac we see how next to the constellation of Leo there is a small figure that represents a king (G in Fig. 4.37). It is possible that this does not personify a constellation but, rather, a star. The best candidate would be Regulus, whose name in Latin, perhaps not coincidentally, means 'little king'. This important star has the particularity of being on the ecliptic and of having been located for many centuries very close to the position of the sun at the summer solstice. As we have deduced, it must be equivalent to the star Tep-en-Mai ('Head of the Lion' from the Ramesside tables, while the star Sedef ('His tail', in reference to the lion's tail), could be an equivalent of β Leonis (Denebola, a name that comes from the Arabic Danab al-Asad which literally means the 'lion's tail'). Leitz (2006: 296–297) also agrees with this identification.

If Leo overlapped the Egyptian Mai (lion), and Mai corresponds to the 'Divine Lion' that appears in the oldest Egyptian astronomical iconography, it is also reasonable to speculate that some of the constellations associated with the 'Divine Lion' could be better syncretized with the new groups of constellations that arrived to Egypt in Ptolemaic times.

Fig. 4.38 Detail of the coffin of Heter, AD 125. (Adapted from Brugsch, 1862)

Such is the case of Hydra, a long serpent-shaped constellation (C in Fig. 4.37), and Corvus (D in Fig. 4.37) which already appear among the 48 constellations mentioned by Claudius Ptolemy in the *Almagest*. Although both constellations were integrated into the Greek mythological legends, their origin is Babylonian, like that of the zodiacal constellations (Leo is the Babylonian Mul Ur-Gu-La—'lion'—that appears in the *Mul-Apin*). Hydra seems to correspond to Mul-Dingir-Mush and Corvus to Mul-Uga-Mushen. The shapes of these constellations and their location, still approximate, could have facilitated their comparison with the crocodile Hetep-redwy ('Lying on his feet') and the bird that is sometimes associated with the Divine Lion. In the case of the former, Hetep-redwy, it could have been simpler, since as Hydra we have placed it below the Divine Lion (this corresponding to Leo). However, the correspondence of the bird associated with the Divine Lion with Corvus would be more complex to verify, especially if this bird corresponds to the constellation of *ꜣpd* referred to in the Ramesside tables, since the latter would be further west.

According to Davis (1985: 103), the boreal lion should correspond to Leo, the bird to Leo Minor, the large crocodile to Hydra, and the small one to Cancer, while the man who aims to harpoon the large crocodile would be Gemini. As we can see, our reconstruction of this part of the sky is partially similar. Biegel (1921: figs. 6a and 7) saw the lion and the bird in the stars above Ursa Major, in the same way as the large crocodile and the harpooner. Chatley (1940: 23) placed the crocodile in Cassiopeia and Perseus, and the lion in Auriga. Locher (1990: 50 fig. 1), finally, recognizes the small crocodile in stars of Aries, Triangulum, Andromeda, and Perseus, the lion in Perseus and Auriga, the bird in the Pleiades, and the large crocodile in Taurus and Auriga (see Figure 4.39).

Etz (1997: 158) proposes similar counterparts for the Egyptian boreal constellations, although with notable differences. The main basis for his reconstruction is to suppose that the tip of Meskhetyu's tail is at 1-o UMa, since in this way the bull would be orientated towards the east, as in most of the ceilings. From there, he respects the positions and orientations of the figures represented on the astronomical ceilings, so that he identifies Isis-Djamet with Auriga, the lion with Leo, the big crocodile with Hydra, the harpoon man with Gemini, the little crocodile with Leo Minor, Serqet between Ursa Minor and Draco, and the harpoon hieracocephalus figure with Lynx and part of the Ursa Major.

However, it is much more difficult to reach a consensus on the identification of the anonymous human figure that appears in the celestial diagrams. To get closer to a possible answer, two constellations that are mentioned in the stellar lists of the Ramesside star tables must first be discussed: *nḫt*, the Giant (Nekht), and *ꜣpd*, the Bird (Aped).

Nekht (as its name suggests) is a gigantic constellation spanning several hours of right ascension. This extensive constellation would be located between the constellation Aquila, in which its head and the feathers that are located on it should be found, and the Square of Pegasus, a singular asterism belonging to this constellation, where its pedestal could be found. The Bird follows the Giant in the star tables and therefore has to be located between the stars of Triangulum and Perseus. The

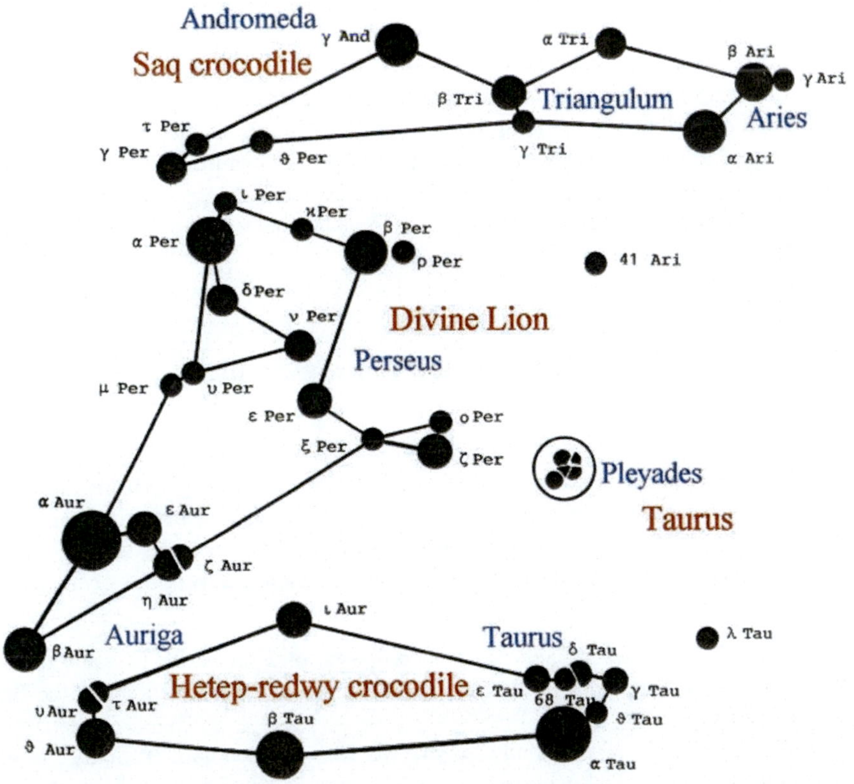

Fig. 4.39 Identification, after Locher, of the Divine Lion, the Bird, and two Crocodiles. (Adapted from Locher, 1990: 49 fig. 1. This complements Fig. 4.29, and poorly compares to our own hypotheses for the northern constellations (see Table 4.7))

problem arises when trying to combine these data with the iconographic representations on astronomical ceilings.

It was precisely a star of Perseus that has been the protagonist of a series of recent works (Porceddu et al., 2008, 2018) mainly focusing on the study of pCairo 86637 (mostly known as the Cairo Calendar), which Hardy (2002–2003) had previously interpreted thus as a stellar almanac. In these studies, it was postulated that the Cairo Calendar, a calendar of lucky and unlucky days dated at the reign of Ramesses II in the 19th Dynasty, assigns luck with the period of 2.85 days, concluding that this was the period of the eclipsing binary Algol three millennia ago. However, this attractive hypothesis has recently been criticized, correctly in our opinion, by Krauss and Reijs (2021). Unfortunately, though, Algol is not among the stars that we have identified from Egyptian documentation.

The anonymous human figure and the bird in the celestial diagrams could be identified with the Giant and the Bird (Belmonte, 2012: 70), respectively. However, this is difficult to verify, as both figures are unnamed and celestial diagrams should certainly not be viewed as actual maps of the sky, but rather as an artistic and symbolic representations (in accordance with the well-known *horror vacui* of the Egyptians) in which all the northern constellations, including those used in the Ramesside clocks, would be duly represented.

Furthermore, it may be proposed that the later addition of the standing human figure on the ceiling of Senenmut could be related to the development of the star tables of the Ramesside clocks as a new device for the nocturnal calculation of the hours. This would replace the Decanal clocks, now out of date, precisely at the time of this exceptional character, who would combine his many virtues with an accurate knowledge of the sky.

4.7.2 Hourly Stars and Southern Constellations: From Diagonal Star Clocks to Ramesside Star Tables

Thanks to the compilation of lists of decans (more than fifty ranging from complete to fragmentary) carried out by Parker and Neugebauer, and collected in their well-known Bible of the field, the *Egyptian Astronomical Texts*, these have been organized into six groups (called 'families'), each of which is composed of the same series of decans or stars (*EAT* III, 1). These are:

1. The family of decans of Senenmut is composed of 20 lists, the first and oldest being that of Senenmut (15th century BC) and the most recent from the 3rd century BC;
2. The Sethy IA family, made up of eight lists from the cenotaph of Sethy I (13th century BC) to the 2nd century BC, introducing two new decans (*sb3w mḥw* and *ḥrj-jb ẖntw*) as well as variants in the group of decans belonging to the constellation of *S3ḥ* (part of our Orion);
3. The family of Sethy I C, of which seven lists are part, from Sethy I to Tausert;
4. The family of Sethy I B, with 11 lists, from the time of Sethy I to that of Trajan (2nd century AD), with new decans *tm3*, *tpj-ʿ b3wj* and *ṯs ʿrḳ*;
5. The Tanis family, from the 26th Dynasty to the 1st century AD, with new decans such as *ḏ3t*, *p3 sb3 wʿtj* and *pḥwj ḥrj*; and
6. A series of four incomplete leftover lists that could not be related to the five families of decans (see also von Lieven, 2021: 208).

To these families (*EAT* III, 158–159), we must also add the lists of hourly decans that appear on the lids of coffins from the Middle Kingdom. According to Parker and Neugebauer, the decans were located in a band approximately parallel to the southern part of the ecliptic (*EAT* I, 100 fig. 27). A compilation of the decans can be also consulted in Gadré and Roques (2008b).

A good starting point to find the position in the sky where the different decans are located is to take into consideration not only the indications that we can find about their risings, settings, or transits in star clocks of the New Kingdom but also in relation to the zodiacal constellations during the Ptolemaic Period. In fact, the zodiacs in which decans are related are a helpful source to consider because, although the decan may remain unidentified, at least we can place it with certain guarantees in a specific position on the sky in right ascension.

Apparently, the decans changed their role during the Ptolemaic era. Even if during the Middle and New Kingdom these had served to indicate the hours of the night, they could still serve now, regardless of their fitness for that purpose, as dividers of the zodiacal constellations. The 36 main decans divided the celestial circle into 36 10-degree sectors, totalling 360°. The 10° sections represented one-third of a zodiacal constellation, so that for astrological considerations the new 'zodiacal decans' could be seen as increasing the complexity and refinement of the zodiac. Each zodiacal constellation was divided into three sectors corresponding to three decans. Regarding the latter, it is interesting to note that this division already seemed to be intuited in older Egyptian documents. Thus, for example, in a fragment of a clepsydra that has been dated to the reign of Nekau II of the 26th Dynasty (see Figure 4.40), there should appear a list of months of the year, each one related to three decans. The only month that remains complete, *wp rnpt* (IV Shemu), is

Fig. 4.40 Fragment of a clepsydra with months divided into three decans. (Adapted from *EAT* III, 43 fig. 9)

associated with the decans *s3 ḳd*, *ḫ3w* and *ʿrw* (*ʿrt*); that is, the same series that is seen in the zodiac of Esna A2 (Tanis family) for Taurus (*EAT* III, 169).

If we compare the lists of decans of the diagonal star clocks of the Middle Kingdom and those of the astronomical ceilings of the New Kingdom, we observe that, although there is great similarity, it is difficult to find a complete concordance between the positions of the decans that in one list or another bear the same name. This lack of coherence could reside in the way in which observation was carried out in each case. While in the former it was the rising or setting of the decan that marked the time, in the latter it might have been the moment of culmination, as it was doubtless going to be in the Ramesside star clocks.

Neugebauer and Parker, followed by other authors, believe that the decans mark the time at their rising. However, as we indicated in Chap. 3, other hypotheses have been made about the operation of diagonal star clocks. Leitz, for example, has proposed that the decans marked the end of the hour of setting, and that the preserved clocks followed an ideal civil calendar in which the heliacal rising of Sirius would take place on I Akhet 1. Hence, these clocks cannot be used to determine the position of Sirius or to calculate the absolute chronological moment in which they were made.

Let us share some hypotheses regarding the observation of the decans. It should not matter too much if they were observed at rising or setting, since the visibility conditions at the horizon level would be the same.

However, both methods would differ in a notable way from each other in consisting of the observation of the culminations of the decans, as we propose for those in the decan lists of the New Kingdom: when at culmination the right ascension of the object is an important parameter that we must know, with the risings or settings. The declination also plays an indisputable role, since at the same right ascension a star of greater declination is hidden by the horizon after another of lesser declination. This means that factors such as the luminosity of the star must be taken into account much more in order to calculate at what height above the horizon it is no longer visible and, in the same way, the latitude of the point of observation.

For the culmination of a given star, latitude is not so important, as it will only affect whether it is seen higher or lower above the southern horizon. On the other hand, latitude does irremediably affect a system based on the observation of risings or settings, since the visibility of the star will be modified by variations in its height and the change of azimuth.

Thus, several problems arise in identifying the decans of diagonal star clocks. We must start from the reconstruction of the celestial vault for a specific date in the Middle Kingdom, assume a specific latitude of observation, and finally consider the possible variability of the hours of the night and the unknown of determining whether the beginning of the night occurs after the astronomical, nautical, or civil twilight. Neugebauer and Parker dispensed with any attempt to search for equivalences, but not other authors. Leitz (1995: 96), for example, in his proposal starts from the date 2100 BC from a latitude of 30° (see Table 4.2).

Table 4.2 Equivalences proposed for the decans by Leitz (1995: 96)

Decan	Star	Decan	Star
1. *tm3t ḥrt*	α Crucis	19. *3ḫwj*	?
2. *tm3t ḥrt*	β Crucis	20. *jmj-ḫt ʿḫwj*	?
3. *wšbt bk3t*	β Centauri	21. *b3wj*	ζ,θ,χ Ceti γ Pegasi
4. *jpḏs*	α Centauri	22. *ḳd*	α Andromedae
5. *sbššn*	θ Centauri	23. *ḫ3w*	M 31 Galaxy?
6. *ḫntt ḥrt*	γ Lupi	24. *ʿryt*	α Arietis
7. *ḫntt ẖrt*	ζ Scorpii	25. *ẖrj ʿryt*	δ Ari/o Tau ?
8. *ṯms n ḫntt*	ε Scorpii	26. *rmn ḥrj*	γ Tauri
9. *ḳdtj*	η Sagittarii	27. *rmn ẖrj*	?
10. *ḫnwj*	δ Sgr – ε Oph	28. *ʿbwt*	β Eridani
11. *ḥrj-jb wj3*	σ Sagittarii	29. *ẖrt wʿrt*	ζ Orionis
12. *šsmw*	ρ Sagittarii	30. *tpj-ʿ spdt*	β Canis Maioris
13. *knm*	λ Aquilae	31. *spdt*	α Canis Maioris
14. *smd srt*	α Capricorni	32. *knmt*	?
15. *srt*	δ Capricorni	33. *s3wj knmt*	?
16. *s3wj srt*	β Aquarii	34. *ẖrj ḫpd knmt*	?
17. *ẖrj ḫpd srt*	α Aquarii	35. *ḥ3t ḫ3w*	?
18. *tpj-ʿ 3ḫwj*	?	36. *pḥwj ḫ3w*	?

In the attempt to identify the decans or, at least, place them to within a certain approximate position in the sky, we can also make use of late documents, such as the rectangular zodiac of Esna, which has the peculiarity of dividing the zodiacal constellations into three sectors, each of which is occupied by a decan. In this way, since the zodiacal constellations are recognized in the sky, we are able to know what approximate right ascension the decans mentioned occupied. Esna's lists are related to the list of decans of the Sethy IB family and that of Tanis. In the table that we reproduce below (see Table 4.3) we include only the decans from the Seti I B list, starting from Cancer with *špdt* (Sirius) heading east.

Regarding the more complex identification of epagomenal decans, Leitz proposes π Orionis for *nṯr d3 pt* (ED3), γ Orionis for *rmn ẖrj* (ED4), β CMa for *tpj-ʿ spdt* (ED6), α CMi for *jmj-ḫt spdt* (ED7), η CMa for *3ḫwj* (ED8), and α Hydrae for *nṯr d3 pt* (ED10), leaving ED1, ED2, ED5, ED9, ED11, and ED12[3] unidentified.

špdt-Sirius is preceded by a series of decans belonging to the constellation of *S3ḥ* (Orion). The decan *tpj-ʿ Spdt* (the 'predecessor of Sopdet' of the diagonal clocks) we suspect could be the star Mirzam (β CMa). The next four decans of diagonal clocks, *ẖrt wʿrt,* 'the one under (Sah's) leg'; *ʿbwt* 'the sceptre (of Sah)'; *rmn ẖrj,* the 'lower arm (of Sah)'; and *rmn ḥrj* the 'upper arm (of Sah)', are all parts of the constellation of Sah (Orion) and in some way describe how the Egyptians saw it.

Between Sirius (*špdt*) and the Pleiades (with high probability *ḫ3w*), even though there is only a difference of about 3h 40m (epoch –2100) in right ascension, the

[3] ED1, ED2, ED3, etc. were catalogued by Parker and Neugebauer as A, B, C, etc.

Table 4.3 The decans as dividers of the zodiacal constellations in Esna's zodiac

Zodiacal const.	Decan	Zodiacal const.	Decan
Cancer	*spdt*	Capricornus	*tpj-ꜥ smd*
	štw		*smd*
	knmt		*srt*
Leo	*ẖrj ḫpd Knmt*	Aquarius	*sꜣ srt*
	ḥꜣt ḏꜣt		*ẖrj ḫpd Srt*
	pḥwj ḏꜣt		*tpj-ꜥ ꜣḫwj*
Virgo	*ṯmꜣt*	Pisces	*ꜣḫwj*
	wšꜣtj bkꜣtj		*tpj-ꜥ bꜣwj*
	jpsd		*bꜣwj*
Libra	*sbẖs*	Aries	*ḫntw ḥrw*
	tpj-ꜥ ḫntt		*ḫntw ẖrw*
	ḫntt ḥrt		*sꜣ Ḳd*
Scorpius	*ḫntt ẖrt*	Taurus	*ḫꜣw*
	ṯms n ḫntt		*ꜥrt*
	sptj ḫnwj		*rmn ḥrj*
Sagittarius	*ḥrj-jb wjꜣ*	Gemini	*ṯs ꜥrḳ*
	ššmw		*wꜥrt*
	knmw		*tpj-ꜥ Spdt*

diagonal clocks include seven decans, counting ten-day intervals in the culmination would far exceed that difference. This could show that in the diagonal clocks the culmination of the decan is not being observed, but rather its rising or setting, which for a small difference in right ascension the different declination of the stars could be reflected in the high number of entries that the Diagonal clocks show between Sirius and the Pleiades. In fact, if we see the division of the zodiacal constellations into decans, according to the zodiac of Esna, there would be only five decans between *Spdt* and *xAw*, a number that is repeated in the list of decans of Senenmut and in the list of the text U of the cenotaph of Sethy I.

The decan 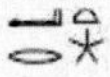 *art*, which precedes the decans associated with *sꜣḥ*, seems to correspond to the star Aldebaran (α Tau). Decan 30, *ḫꜣw* ('Myriads' or 'Thousands'), is commonly identified with the Pleiades. The relationship of 'Thousands' with the Pleiades star cluster seems clear but, in addition, another proof of its position in the constellation of Taurus (in the same way as *ꜥrt*) can be found in the rectangular zodiac of Esna. However, we can see in the equivalence table proposed by Leitz (1995: 93) for the Middle Kingdom how he indicates that *ḫꜣw* could correspond to M31, the Andromeda galaxy. We cannot share this hypothesis in any way. Very close to them, *ḳd*, the Circle or the Fold, could perhaps be located in the head of Cetus.

In another area of the sky characterized by the presence of the Milky Way, the bright β Cen (Hadar) and α Cen (Toliman), and the stars of the Southern Cross stand out, which, according to our results, should be identified with the decanal stars *jpḏs* and *sbšsn* (Arcturus for Böker, 2007) and with the double constellation *wꜣštj-bkꜣtj*, respectively. In the same area, it is proposed to locate the Ferry, *ḏꜣt*, in the area of

Argo Navis, a classical constellation now broken up into smaller constellations. The group of decans of the constellation *Knmt* must also be found between The Stern and the southern areas of the Canis Major. A much more complicated matter is the identification of individual stars or asterisms with individualized decanal stars within the framework of these constellations.

For the southern sky, there are very few constellations represented in astronomical ceilings or related documents. In this sense, wee can list: Sah and Sopdet (also present in the coffins of the First Intermediate Period); the 'Two turtles' *štwj* (constellation formed possibly by the stars Procyon and Gomeisa of Canis Minor) (Lull, 2004a: 258); the asterism known as , perhaps the Hyades (Lull, 2006b: 205), an open star cluster around Aldebaran (the star that we identify with *ꜥrt*), or even a figuration of the Pleiades (Lull & Belmonte, 2009: 175–176); *sjt* 'Sheep' , and finally the constellation of *WjA* the 'Boat' (Fig. 4.12). The fact that very few southern constellations were represented could be an indicator of the importance they may have had for their symbolism or easy visualization by the ancient Egyptians in the sky. Such must be the case of *Wjꜣ*, the constellation of the 'Boat' (Lull, 2018a, b).

In the astronomical ceiling of Senenmut (Dorman, 1991: pl. 85) the constellation of the Boat is the first that is represented in the southern sky. The Boat is placed under six columns (columns 7–12) in which some decans are listed, so they can serve to identify the approximate position of the constellation (Table 4.4).

The tenth decan announces a star located west of Khentet, but the eleventh marks the beginning of Khentet, which continues in the following columns. The word *ḫnt* is translated as *Stirn, Vorderseite* ('prow'; Hannig, 1997: 607). As preposition *ḫnt* means *vorn an, an der Spitze von* ('at the head of'; Hannig, 1997: 607), etc. Given that the constellation we are dealing with is a boat, and that the regular movement of the sky is from east to west, it is opportune to find the prow of the boat facing the west, that is, following the movement of the sky. Therefore, we may assume that *ḫntt* (written or) could refer to the front of the boat, its

Table 4.4 Decans 10–18 in the astronomical ceiling of Senenmut

Decan	Column	Name		
10	6		*tpj-ꜥ ḫntt*	'Predecessor of Khentet'
11	6		*ḫntt ḥrt*	'upper Khentet'
12	**7**		*ḫntt ẖrt*	'lower Khentet'
13	**8**		*ṯms n ḫntt*	'the red (star) of Khentet'
14	**9**		*sptj ḫnwj*	'sapty of the two khenwy fishes'
15	**10**		*ḥrj-jb wjꜣ*	'the (star) which is in the middle of the boat'
16	**11**		*sšmw*	'the guides'
17	**12**		*knmw*	'(stars of) Kenmu'
18	13		*tpj-ꜥ smd*	'Predecessor of Semed'

prow. Jones (1988: 180), in his comprehensive study of Egyptian nautical terminology, also proposed that possibility, although in Egyptian texts we do not known with certainty any case in which the prow of a boat is called *ḫntt*.

As shown before in Table 4.2, Leitz (1995: 96), studying the decans list coffins of the First Intermediate and Middle Kingdom Period, concluded that these three decans of Khentet (*ḫntt ḥrt* 'upper Khentet', *ḫntt ẖrt* 'lower Khentet', and *ṯms n ḫntt* 'the red (star) of Khentet') should be γ Lupi, ζ Scorpii, and ε Scorpii, respectively. However, only one of the decans of Khentet is identifiable with its current counterpart with complete certainty. There is not much room for doubt that the 'red (star)'  'of Khentet' , *ṯms n ḫntt*, should be equivalent to Antares (α Sco; Lull, 2004b: 73–77), a red giant star distinctive for its brightest and reddish colour in that region of the sky, and not to the star Wei (ε Sco), as indicated by Leitz as his first option, or to a star located in Sagittarius (Al Kiladah - Psi Sagittarii, 43 Sagittarii, or 52 Sagittarii), as supposed by Gadré (2008: 186). Furthermore, such recent proposals as matching the prow of the Boat with the stars γ Pyx, α Pyx, β Pyx, and ζ Pup (Priskin, 2019b: 143–146) is totally unsustainable and incomprehensible.[4]

In fact, the colour index[5] $B-V$ of Wei is 1.14, and its apparent magnitude 2.28, while Antares' colour index is 1.86 (redder) and its apparent magnitude 1.03 (brighter).[6]

The number of stars that we can see as red to the naked eye are very few. Under visual magnitude 3.5, only Betelgeuse and 34-μ UMa (Tania Australis) have a $B-V$ index higher than Antares. Moreover, in Egyptian texts there are very few cases in which a star is distinguished by its colour. Another star that is also known as red is 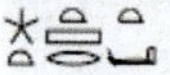 ('the red star'), which similarly appears as a representative of the 34th decade in an inscription from the naos of Saft el-Henna from the period of Nectanebo I (30th Dynasty). This has been identified with Aldebaran (Leitz, 1995: 33–35; von Bomhard, 2008: 177).

The accurate identification of the 'Red (star) of Khentet' with Antares is essential as a starting point for positioning the constellation of the Boat with better guarantees, although the correspondence of most of their decans with current stars can only be approximate. Both upper Khentet and lower Khentet must necessarily be situated west of Antares, so both decans must be located most probably somewhere in the strip occupied by Libra.

The fact that a constellation shaped like a boat could be split up into smaller ones representing the parts of the craft has a curious parallel in our present-day sky: Argo Navis (the ship of Jason and the Argonauts in Greek mythology) was a constellation listed among the 48 cited by Claudius Ptolemy in the *Almagest*. This constellation

[4] We fear that Priskin (2019b) is an example of how the programme Stellarium must not be used to identify ancient Egyptian constellations without making use of further astronomical (e.g. intervals in RA) and textual information, and, indeed, a deep knowledge of what the sky really looks like.

[5] Colour index is the difference between the apparent magnitudes of a star at two different wavelengths. We are citing colour indexes calculated from B and V magnitudes. Higher magnitude values refer to fainter stars.

[6] Consulted in the database of the *Starry Night Pro* astronomical program.

covered so great an expanse of sky that in the eighteenth century it was divided by French astronomer Lacaille into three smaller constellations: Carina (the Keel), Puppis (the Stern), and Vela (the sail). In this case, the stern of Argo Navis is situated to the west and hence contrary to the direction of rotation of the sky. However, regarding Egyptian *wjꜣ* there is a trait that should not be neglected. Argo Navis is located in the Milky Way, which acts as a celestial river, as was the Egyptian boat.

An Egyptian boat in the most prominent part of the Milky Way is a very significant matter from a symbolic point of view, because we must not forget that in the Egyptian conception is quite possible that the celestial goddess Nut could be found in the Milky Way. Consider, as seen in religious compendia such as the *Books of the Hereafter*, that the solar boat sails with its entourage along the body of Nut. Therefore, the Milky Way could also be seen by the ancient Egyptians as a celestial river, so nothing would have seemed to them to be more appropriate than to place the constellation of the Boat where the Milky Way is most prominent.

Three main clues may be used for the identification of *sptj ḫnwj* ('Sapty Khenwy', the 'Sapty of the two khenwy-fishes') in the sky. On the one hand, we know that it is east of Antares, in the eastern part of Scorpius or at an equivalent RA; on the other hand, the '(two) khenwy-fishes' may be sought in the celestial river of the Milky Way; and finally, the fact that two fishes are concerned suggests two stars close together.

With this in mind, we may reasonably suggest the very possible identification of Khenwy with Shaula (λ Sco) and Lesath (υ Sco), two prominent stars very close together (36′), that form the sting of Scorpius. In the zodiac of Esna, the 'Red (star) of Khentet' appears as the central decan of Scorpius. Obviously, recognizing the equivalence of Antares with the decan, the place where we expect to find this decan is precisely the abdomen of the scorpion in the centre of the constellation, just as we identified the two fishes with Shaula and Lesath, the stars in the eastern third of Scorpius that form the sting of the scorpion.

According to Leitz (1995: 90), the '(star) which is in the middle of the boat' *ḥrj-jb wjꜣ* (decan 15) must match the star Nunki (σ Sagittarii). In our opinion, however, a better option would be Kaus Australis (ε Sgr), the brightest star in the constellation of Sagittarius and one that is closer, compared with Nunki, to the brightest part of the Milky Way in this region. Moreover, this identification would fit better with the fact, as we see later in the zodiac of Esna, that *ḥrj-jb wjꜣ* is linked to the western third of Sagittarius, and Kaus Australis lies at that end.

The next decan is *sšmw*, which can be translated as 'Guides'. The 'Guides' could refer to the pilots of the Boat. Recall, for example, that while the fifth hour of the night is called '(The one) which is in the middle of his boat' in the *Book of Nut* (Roulin 1996: 53; Lull, 2002: 140), or in the first division of the *Book of the Imiduat*, the fifth hour of the *Book of the Imiduat* receives the full name of *sšmyt-ḥrt-jb-wjꜣ.s*, '(She who) guides, in the middle of his boat' (Hornung 1992: 387). Therefore, and since this decan is over the figure of the Boat, it is clear that it is also part of it. Since the 'Guides' should be formed by a group of three or more stars, the equatorial coordinates lead us, east of Kaus Australis, to the group composed of 27-φ Sagitarii, Nunki (34-σ Sagittarii), and Ascella (38-θ Sagittarii), although Böker (2007) insists on Vega.

Fig. 4.41 Possible identification of the stars of the Boat. (Adapted from Lull, 2018b)

Finally, above the figure of the Boat a final column appears in the astronomical ceiling of Senenmut with the decan *Knmw* (decan 17), which may correspond to the group of stars around Albaldah (41-π Sagittarii) like 39-o Sgr, 37- ξ Sgr, and 44-ρ Sgr further east (see Figure 4.41).

The decans that were part of the constellation of the Boat are numbered 11–17 (i.e. from *ḫntt ḥrt* to *Knmw*) with *Tpj-ʿ ḫntt* as the decan preceding the prow of the boat and *tpj-ʿ smd* the decan following its stern. The order of decans between *tpj-a xntt* and *knmw*, as we have seen in the astronomical ceiling of Senenmut, is repeated, regardless of writing variants, not only in other cases of the so called 'family of Senenmut' but also in other families of decans (*EAT* III), with tens of known examples. Therefore, we are able to deduce that this is a well-defined, error-free list (Table 4.5).

Chatley (1940: 124) suggested that the stars represented in *Knmw* and the next decan, *tpj-ʿ smd* ('Predecessor of Semed'), might symbolize the Milky Way although this hypothesis does not seem to correspond with what we would see in the sky, because the showiest part of the Milky Way is precisely that which runs through the constellation of the Boat, between Sagittarius and Scorpius.

Neither the Boat nor any of the decans that were part of it mentioned in diagonals star clocks and lists of decans of the astronomical ceilings was included in the Ramesside star tables (Belmonte, 2003c), where the vast constellation of Nekhet served as a new reference at this right ascension.

However, during the Ptolemaic era a notable change occurs in the Egyptian heavens when zodiacal constellations of Mesopotamian origin are introduced. With this change the ancient Egyptian constellations occupying the zodiacal band were

Table 4.5 Proposed equivalents of the decans of the constellation of the Boat

Decan	Name		Likely equivalence	Magnitude	Angular Distance in RA
11		*ḫntt ḥrt*	α Lib Zuben Elgenubi	2,6	
12		*ḫntt ẖrt*	γ Lib Zuben Elakrab	3,9	40′
13	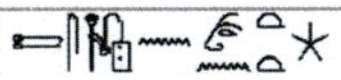	*ṯms n ḫntt*	α Sco Antares	1	41′
14	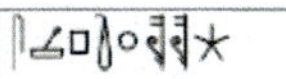	*s3ptj ḫnwj*	λ Sco Shaula/υ Sco Lesath	1,6/2,7	40′
15		*ḥrj-jb wj3*	ε Sgr Kaus Australis	1,8	48′
16		*sšmw*	σ Sgr Nunki	2	40
17		*Knmw*	ρ Sgr – 44 Sgr	3,9	37′

eliminated, or at best minimized, in favour of the new constellations. This last option is what must have happened in the case of the constellation of the Boat. At the beginning of the Ptolemaic period the coffin of Harendotes (BM 6678) is one of the latest examples to show the list of decans with the old representation of the Boat. In fact, it was catalogued by Parker and Neugebauer as the last exponent of the 'family of Senenmut' (*EAT* III, 61).

With the new zodiacal constellations the sky of the Egyptians underwent significant changes. The former Boat constellation was associated—in a minimized form—with the constellation of Sagittarius. This can be checked even in the zodiac of Dendara, where we see that at the foot of Sagittarius a small boat was represented, just as in the rectangular zodiac of the western part of the roof of the pronaos of the same temple, or in the rectangular zodiac of Esna (where the boat appears under the hind legs and not the forelegs of Sagittarius; see Figure 4.42).

It could be argued, following the *Mul-Apin* (Mesopotamian astronomical tables from the seventh century BC) that the boat associated with Sagittarius corresponds to Mul-Ma-Gur or Makurru (Hunger, Pingree, 1989), the 'load boat', which could correspond to the star ε Sagittarii, or to the constellation of Corona Australis. In fact, in the zodiac of Dendara we find other parazodiacal constellations of probable Mesopotamian origin (Lull & Belmonte, 2009: 182).

Aubourg (1995: 4 fig. 2) and Cauville (1997a, b: 35), however, identify this boat with the constellation of Corona Australis (CrA), named Στεφάνος νότιος ('Southern Wreath') by Claudius Ptolemy. However, keep in mind that the stars of this constellation are not at all bright or prominent. Both α CrA (Alphecca Meridiana) and β CrA are stars of apparent magnitude 4.1.

Even though, in ancient times, the stars of CrA might have been associated with Sagittarius (the Mesopotamian Mul Pa-Bil-Sag), as well as a boat, crown, or garland according to Mesopotamian or Graeco–Roman tradition, our hypothesis is that the small boat that is associated with Sagittarius in ancient Egyptian documents is *wj3* (the Boat), a constellation that, far from being eliminated by the introduction of Sagittarius, was associated with it in the iconography. The introduction of the

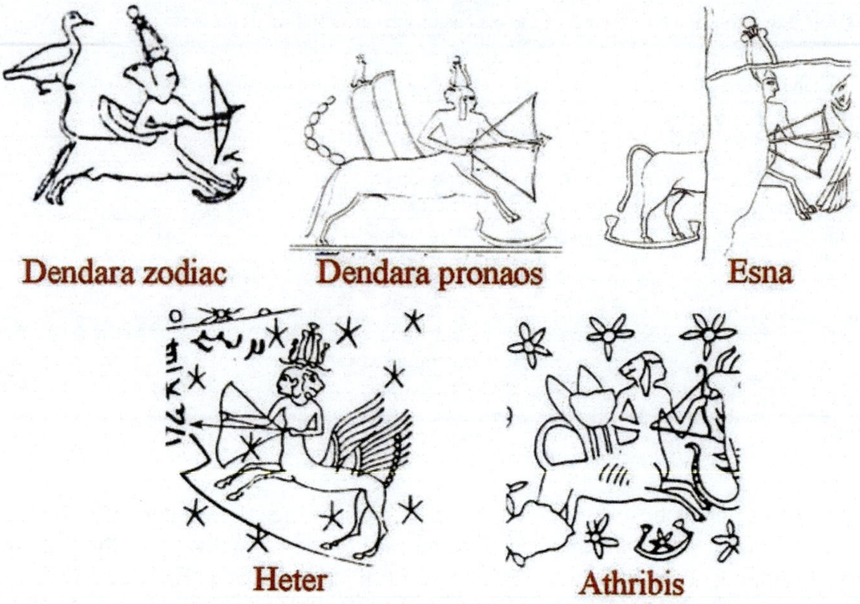

Fig. 4.42 Some examples of Sagittarius with a related Boat. (Adapted from Lull, 2018b)

Mesopotamian 'load boat' in the zodiac of Dendara could be perfectly subsumed in the Egyptian conception, being transmitted iconographically as *wjA* associated with Sagittarius.

Another of the few constellations represented in the southern sector of the astronomical ceilings and associated documents is the Sheep (see Fig. 4.12). The constellation of the Sheep would be developed into the present-day Aquarius and Capricornus. Decan 22, *ẖry ḥpd Srt* ('the one under the back of the Sheep') may refer to α PsA (Fomalhaut), while decan 21, *S3wj Sjt* ('the two sons of the ram') could correspond to δ and γ Capricorni. Decan 20, 'the star of) *Sjt*' (perhaps Albali, 2-ε Aquarii), is another part of the Egyptian constellation of the Sheep.

Locher (1981: 74 fig. 1; see Figure 4.43) had reconstructed this constellation with stars from Capricornus and to a lesser extent from Aquarius, a hypothesis that we follow to a large extent, although we understand that the figure could extend through the southernmost stars of Aquarius to form its hind legs. However, the most notable difference from Locher is the position of the Sheep in the sky. According to this author, the head would be among the stars around 48-γ Aquarii, with the Sheep looking towards the east. In all representations of the sheep, however, it always faces west.

The fact that the Sheep was found between Aquarius and Capricornus, has led some authors to suggest that the origin of the current Capricornus (the goat which, according to Greek mythology, originated when Pan, god of the forests, was transformed into such an animal to sneak out of Typhon) could go back to ancient Egypt (Gundel, 1969: 333–334). However, this constellation also appears in Babylonian astronomy as the 'goat-fish'. The Egyptian Sheep is unlikely to be related to its

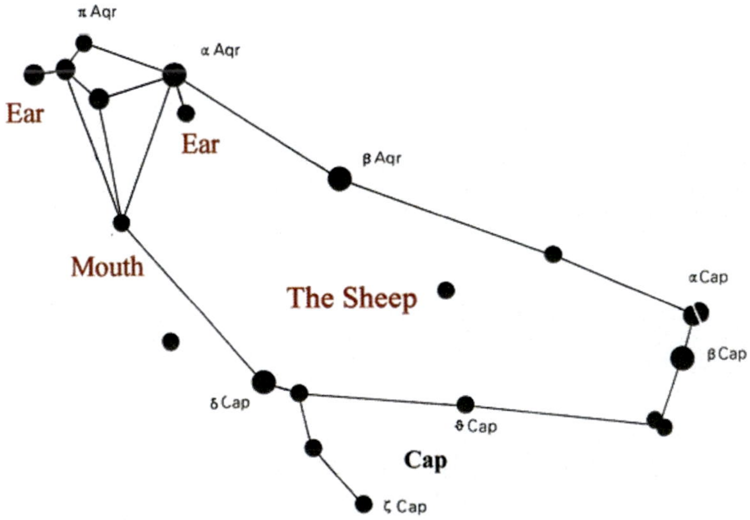

Fig. 4.43 The Sheep after Locher. (Adapted from Locher, 1981: 74 fig. 1)

Babylonian counterpart, but its Egyptian form could have facilitated its adaptation to the new zodiacal constellation.

The next constellations represented on the astronomical ceilings are Sah and Sepedet. In the coffins of the Middle Kingdom, Sah is usually shown standing facing backwards, with a long *was* sceptre on his left arm and an *ankh* symbol on his right. Sah's head is topped with his typical three-pointed crown, Gardiner's D61 hieroglyph ( ).

Furthermore, in the lists of decans of the astronomical ceilings of the New Kingdom, Sah always appears as standing on a boat, in a pose similar to that of the Middle Kingdom, although never with the three-pointed crown, even though it does appear in the sarcophagus of King Psusennes I, from the 21st Dynasty (Montet, 1951: pl. 79). Only in the tomb of Ramesses VI (*EAT* III, pl. 13) was Sah depicted kneeling on the boat. In both the Middle and the New Kingdom, he always directs his gaze in the opposite direction to his direction of departure.

The last changes to be highlighted took place during the Ptolemaic Period. Sah is still, as from the New Kingdom, on a boat looking back (although many exceptions already appear, such as Nag Hamad A, Dendara, Kom Ombo, etc.) and holding one or two sceptres, but now he is generally distinguished by wearing the royal crown of Upper Egypt.

The figure of the Egyptian constellation of Sah is not well defined in the celestial vault, even though we know that he is identified with stars of Orion. A very suggestive proposal is that of Locher (1993), the three points of Sah's crown would correspond to the three stars of Orion's belt, while the *was* sceptre would have its head in Saiph and its base in a position so far south in declination as the star Phact (α Columbae). Thus, Sah would develop between the belt of Orion, and further south,

Fig. 4.44 Sopdet and Sah in the sky (a) and in the coffin lids of the First Intermediate Period and the Middle Kingdom: Idy (b) and Ashayt (c). The former is compared with the identifications of Kurt Locher. In Ashayt's coffin, Sopdet is crownedd by a peculiar poligonal sign which could eventually be glimpsed in the sky (see panel a, above the name Sopdet). (Diagram by the authors, with panel (c) adapted from von Bomhard, 1999, fig. 17)

Lepus, and Columba. This same author reconstructs the figure of Sepedet with a *was* sceptre parallel to that of Sah, between 3 Monocerotis and Wazn (β Columbae), and with the body of the divinity among the stars of Canis Major, the head being marked by the luminous Sirius (see Figure 4.44).

Another of the constellations represented in the southern sky is located among the so-called triangular decans. It is a constellation formed by a pair of turtles, *štwj* (the 'Two Turtles'), which could be identified with Procyon and Gomeisa, the two brightest stars of Canis Minor, (Lull, 2004b: 258).

From some correspondences that we consider to be certain within the frame of our hypothesis (Antares, Sirius, etc.), it is possible to try to propose further identifications, even with a certain degree of approximation, for the remaining decans. It must be recognized, however, that many identifications are difficult to contrast and thus remain uncertain and are a matter of conjecture (Lull, 2004b: 261–262 tab.12).

As we have indicated elsewhere, the Ramesside tables are also an important source allowing us to propose new identifications of stars. In fact, we have already advanced some identification proposals, such as those of the great constellations of the Hippopotamus, the Lion, the Mooring Post, the Giant, and the Bird. With the other hourly stars and constellations used in the Ramesside tables, we are able to propose new identifications. Forty-seven stars in total are mentioned in the Ramesside clocks.

However, if we compare the list of decans of the diagonal star clocks of the Middle Kingdom, or the decanal lists of the New Kingdom, with the stars mentioned in the Ramesside tables, we will see how, surprisingly, there are only three similar entries: *sbꜣ n sꜣḥ* (the 'Star of Sah'), *sbꜣ n spdt* (the 'Star of Sopdet'), and *sbꜣ n ḫꜣw* (the 'Star of Thousands').

Le Page Renouf (1874) was the first to carry out a serious study of the Ramesside star clocks and the equivalences of their stars, and came to suggest identifications, starting from the basis of the accepted correspondence between *špdt* and *Sꜣḥ* with Sirius and Orion, respectively. These were as follows: *ꜥryt* with Cassiopeia, *Tp n ꜣpd* (the 'Head of the bird') with α Arietis-Hamal, *ḫꜣw* (the 'Thousands') with the Pleiades, *sbꜣ n sꜥr* (the 'Star of Sar' or 'Star of Fire') with α Tauri-Aldebaran, *sbꜣwj* (the 'Two stars') with Castor and Pollux (α y β Geminorum), *sbꜣw ꜥšꜣw* ('Multitude of stars') with Coma Berenices, the stars of *mꜣj* 'Lion' with Leo, *ṯꜣ nfr*, which he translates as 'Bearer of the lute', with α Virginis-Spica, and the constellation of *Mnjt* as part of a group where α Bootis (Arcturus) and α Scorpii (Antares) would be.

Also, Petrie (1940: pls. III–IV) published his hypothesis identifying the constellations and stars of the Ramesside star clocks (see Figure 4.45). From Sirius to the Pleiades, this coincides with Renouf's proposal. However, he places the 'Head of the bird' in Mirach, the beginning of its neck in Hamal, and its lower end in the variable star Mira (o Ceti). He located the legs of *Nḫt* in the Square of Pegasus, the head of his sceptre in Aquila, his chest in Cygnus, and the upper end of his feathers in Draco's head, extending the constellation over about 7 hours of right ascension. Also for Petrie, *Rrt* and *ꜣst-ḏꜣmt*, the hippopotamus, seem to be the same constellation. He places the head in Hercules, that of the crocodile that covers it in Lyra, the feet of the hippopotamus between Libra and Scorpio, and the end of its tail in Antares, thus occupying about 5 hours of right ascension. As for *Mnjt* ('Mooring post'), Petrie reconstructs its ends between Arcturus and Spica (as the base of the post), so that the constellation does not extend over more than one hour of RA.

By the way, Leitz (1995: 159–192) carried out a complete study of the stars of the Ramesside star tables and arrived at the following equivalences (see Table 4.6).

There are several points in Leitz's deductions that attract our attention. First, he identifies the Ramesside constellations in regions where he had previously recognized possible equivalences to the decanal constellations. If so, we would have to admit that the Egyptians, at some time in the beginning of the New Kingdom or earlier, replaced their old constellations with new ones. Furthermore, some of the identifications that he establishes, such as *sbꜣ n Spdt* with Saiph (53-κ Orionis) instead of Sirius, or *sbꜣ n ḫꜣw* with 89-π Ceti instead of Alcyone or any star of M45, particularly call our attention for their unexpectedness and illogicality.

As for the first, it is actually very likely that *sbꜣ n Spdt* refers to the star *špdt* (Sirius). *špdt* was both the name of the constellation and the name of the star Sirius, so *sbꜣ n Spdt* must refer to the latter. More doubtful is the identification of his Follower Star, *jy sbꜣ ḥr-sbꜣ n Spdt*. Within the group of stars associated with Sah, on the other hand, π Ori or one of its neighbours in the Orion arc could correspond to *tpj-ꜥ sꜣḥ*, and Rigel could coincide with *sbꜣ n sꜣḥ* (the 'Star of Sah').

As for *sbꜣ n ḫꜣw*, this would be nothing more than the most luminous star in *ḫꜣw*. As we shall see, Senenmut's list indicates that *ḫꜣw* is a star cluster, possibly M45. Thus, *sbꜣ n ḫꜣw* should be Alcyone (25-η Tauri), of magnitude 2.85, the main star of the Pleiades. The relationship with 89-π Ceti, of apparent visual magnitude 4.24 and located in a poor stellar region with dim stars, therefore seems very inappropriate.

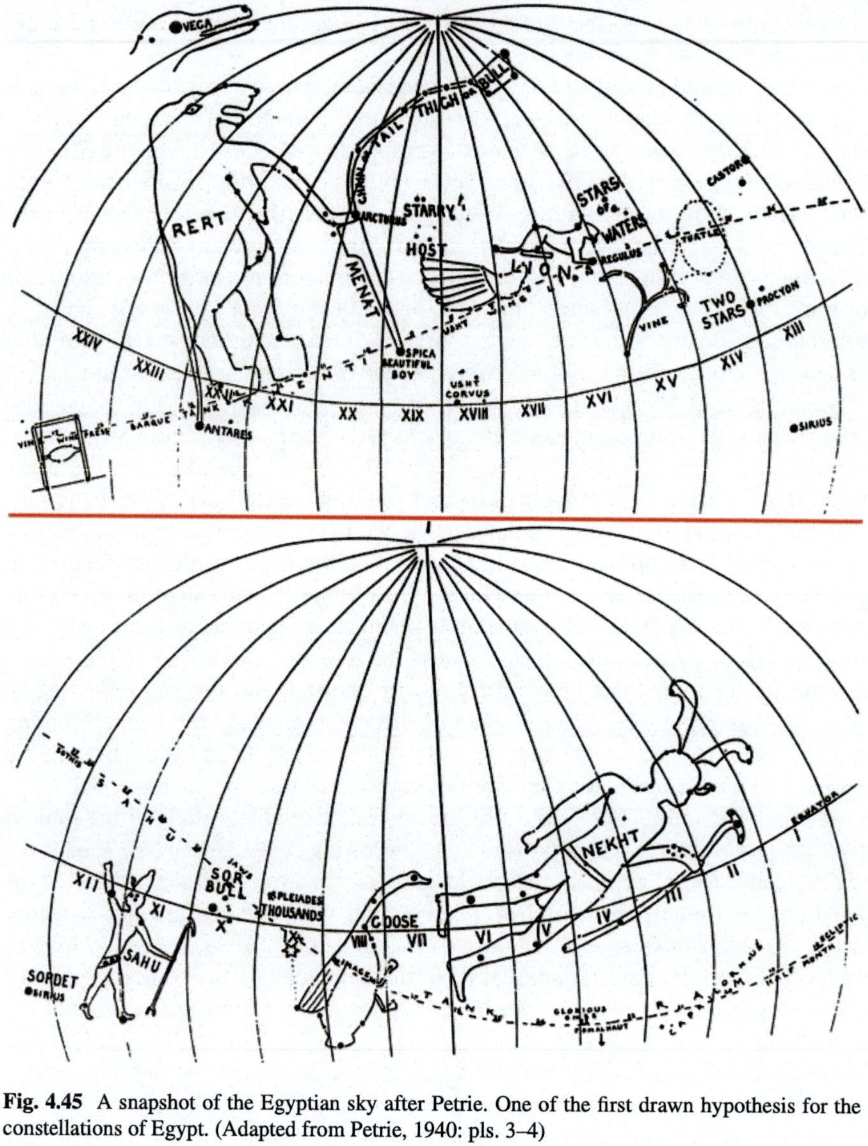

Fig. 4.45 A snapshot of the Egyptian sky after Petrie. One of the first drawn hypothesis for the constellations of Egypt. (Adapted from Petrie, 1940: pls. 3–4)

Cassiopeia may correspond to *ʿryt* (not to be confused with the similarly named constellation of the diagonal star clocks), although it is certainly a difficult identification to verify.

Another interesting star may be *sbꜣ n sʿr* ('Star of Fire'), which could correspond to Capella (α Aur), one of the brightest stars in the sky and with an orange colour that could be reminiscent of fire itself. Belmonte (2003c), however, earlier suggested Aldebaran as an alternative.

Identifying the asterisms associated with *sbꜣwj* (the 'Pair of Stars') and *sbꜣw nw mw* (the 'Stars of the Water') is also difficult. However, it is probable that the 'Pair

Table 4.6 Leitz's controversial proposal of Equivalencies for the stars of the Ramesside star clocks

Ref.	Egyptian Star	Star	Ref.	Egyptian Star	Star
E1	*tpj-ꜥ šwtj nt nḫt*	ρ Sgr	K2	*sbꜣ n sꜣḥ*	Rigel
E2	*šwtj nt nḫt*	β Cap	L1	*sbꜣ n spdt*	Saiph
E3	*tpj-ꜥ ḥḏt.f*	ε Aqr	L2	*jy ḥr-sꜣ sbꜣ n spdt*	η Lep
E4	*tp nḫt*	μ Aqr	M1	*tpj-ꜥ sbꜣwj*	o CMa
E5	*tp n ḥḏt nt nḫt*	ν Aqr	M2	*sbꜣwj*	σ CMa
E6	*ḥḏt nt nḫt*	ε Aqr	N	*sbꜣw nw mw*	ξ Pup
E7	*nḥbt nt nḫt*	β Aqr	O1	*tp n mꜣj*	9 Pup
E8	*ẖꜣb.f*	ι Cap	O2	*sd n mꜣj*	λ Vel
E9	*mndt.f*	δ Cap	P	*sbꜣw ꜥšꜣw*	φ Vel
E10	*bgs nḫt*	α Aqr	Q1	*[tpj-ꜥ] mnjt*	*Non-existent*
E11	*sḏḥ.f*	δ Aqr	Q2	*ṯꜣ nfr*	η Car
E12	*pd nḫt*	λ Aqr	Q3	*šmsw n ḥꜣt n mnjt*	δ Cen
E13	*sbḳ.f*	98 Aqr	Q4	*mnjt*	ξ Cen
E14	*pt.f*	ω Aqr	Q5	*šmsw n mnjt*	ε Cen
E15	*jy sꜣ pt.f*	–	Q6	*šmsw jy ḥrsꜣ mnjt*	ζ Cen
E16	*sbꜣ n sꜣ pt*	–	R1	*rdwj n rrt*	η Cen
F	*ꜥryt*	ι Ceti	R2	*pd n rrt*	γ Lup
G1	*bꜥnt nt ꜣpd*	θ Ceti	R3	*ḥrj-jb n mnty.s*	H Sco
G2	*tp n ꜣpd*	η Ceti	R4	*bꜣḥ n rrt*	N Sco
G3	*ḥtyt nt ꜣpd*	–	R5	*xpd n rrt*	θ Sco
G4	*ktf.f*	α Psc	R6	*mndt nt rrt*	γ Sgr
H	*sbꜣ n ḫꜣw*	π Ceti	R7	*ns.s*	λ Sgr
J	*sbꜣ n sꜥr*	τ Eri	R8	*šwtj nt rrt*	σ Sgr
K1	*tpj-ꜥ sbꜣ n sꜣḥ*	o Eri			

of Stars' must be identified with the celestial pair formed by Castor and Pollux, and the 'Stars of the Water' must be found in the nebular, almost watery aspect of M44 (the Praesepe Cluster, also known as the Beehive) in the constellation Cancer. The logic of the identification is due, not only to the RA at which we suppose that we should look for the 'Stars of the Water', but also to the surprising parallelism between some of the Egyptian names and the celestial reality.

The final identifications can be proposed by taking a jump in right ascension after the constellation of the Lion; for example, the open star cluster of Coma Berenices with *sbꜣw ꜥšꜣw* ('Many Stars'), a very appropriate name for an object of this nature. Similarly, the hourly star *ṯꜣ nfr* (the 'Beautiful Child') has as a possible correspondence with α Virginis, the luminous Spica. As we shall see, this star could be represented twice in the Dendara zodiac, where it would appear as an ear of grain in the hands of Virgo, certainly from an influence external to Egypt, and perhaps as a child in the hands of a goddess, located just below Virgo, possibly following the local Egyptian tradition. In Table 4.7 is a summary of the main identifications, exact or approximate, that we have commented on in the previous paragraphs, including our own most cherished hypotheses.

Table 4.7 Summary of the main exact or approximate identifications of ancient Egyptian stars, asterisms and constellations according to the authors' hypotheses (see also Belmonte, 2003b)

	Name	Possible translation	Exact identification or approximate location
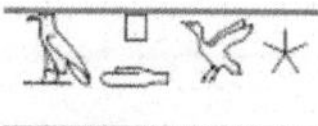	*3pd*	The Bird	Area of Triangulum & Perseus
	*ꜣn(w)*	Anu (an avatar of Horus)	Lynx to Canes Venatici
	*3ḫwj*	The Two Spirits	Area of Pegasus, Sculptor or Pisces
	*jpḏs*	Bright Star	βCen (Hadar)
	*ʿryt*	(The 2) Jaw(s) ?	Cassiopeia
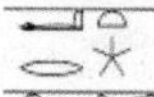	*ʿrt*	Jaw ?	α Tau (Aldebaran)
	*wj3*	The Boat	Sagittarius
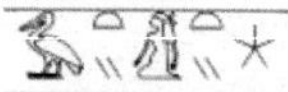	*wš3tj bk3tj*	Two Wesha and Two Ladies	Southern Cross
	*b3wj*	The Two Souls	A couple of stars in Pegasus or Cetus
	*m3j*	The Lion	Leo
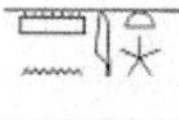	*mnjt*	Mooring Post	Area of Boötes, including α Boo (Arcturus)
	*msḫtjw*	The Bull's Foreleg	The Plough, UMa
	*nḫt*	The Giant	From Aquila to the Square of Pegasus
	*rw-nṯrj*	The Divine Lion	Leo
	*rrt (3st ḏ3mt)*	The Female Hippopotamus	Large area near the Pole covering from Lyra to Boötes
	*ḥrj-jb wjꜣ*	The (star) which is in the middle of the boat	ε Sgr (Kaus Australis)
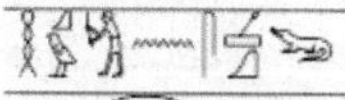	*ḥkw n sʿḳ*	The Plunderer	Leo Minor
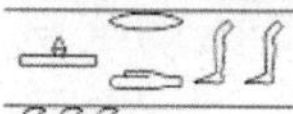	*ḥtp rdwj*	Lying on His Feet	Hydra
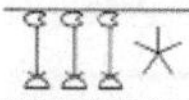	*ḫ3w*	Myriad or Flock	The Pleiades Cluster (M45)
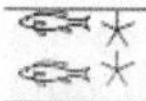	*ḫnwj*	Khenwy Fishes	Area of λ Sco
	*ḫntt ḥrt*	Upper Khentet	α Lib (Zuben Elgenubi)
	*ḫntt ẖrt*	Lower Khentet	γ Lib (Zuben Elakrab)
	*s3ḥ*	Sah	Parts of Orion. Head at the Belt
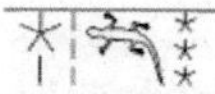	*sb3w ʿš3w*	Many Stars	Coma Berenices
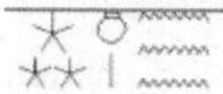	*sb3w nw mw*	Stars of the Water	The Praesepe Cluster (M44)
	*sb3wj*	Pair of Stars	Castor and Pollux

(continued)

Table 4.7 (continued)

	Name	Possible translation	Exact identification or approximate location
	*sb3 n sʿr*	Star of Fire	α Aur (Capella)
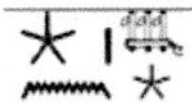	*sb3 n s3ḥ*	Star of Sah	Rigel
	*sb3 n spdt*	Star of Sopdet	Sirius
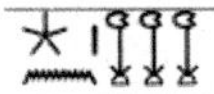	*sb3 n ḫ3w*	Star of the Myriad	Alcyone
	*sbšsn*	Sage's Star	α Cen (Rigil Kent or Toliman)
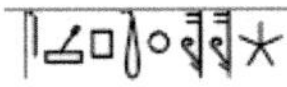	*sptj ḫnwj*	Sapty of the two khenwy fishes	λ Sco (Shaula)/υ Sco (Lesath)
	*srḳt*	Selkis Goddess	Area of Ursa Minor or Virgo
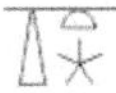	*spdt*	Triangle	Sirius & its companions
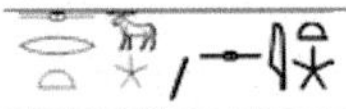	*srt or sjt*	The Sheep or Goat	Capricornus – Aquarius
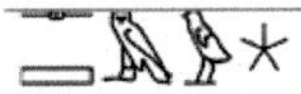	*sšmw*	The Guides ?	σ Sgr (Nunki)
	*sd.f*	Its tail (of the Lion)	Denebola
	*štwj*	The Two Tortoises	Procyon and Gomeisa
	*ḳd*	The Circle or Sheepfold	Head of Cetus
	*ḳdty*	The 2 Nets	One of them would be Corona Australis
	*Knmw*	(Stars of) Kenmu	ρ Sgr – 44 Sgr
	*knmt*	Cow?	Canis Maior & Puppis
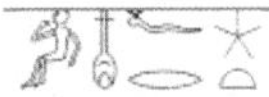	*ṯ3 nfr*	Beautiful Child	Spica
	*ṯm3t*	The Wings	In the area of Argo Navis
	*tpj-ʿ sb3wj*	Predecessor of the Two Stars	Alhena, in Gemini
	*tp-n-m3j*	Head of the Lion	Regulus
	*ṯms n ḫntt*	The Red of the Prow	Antares
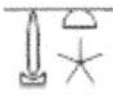	*ḏ3t*	The Ferryboat/Crosser	Area of Argo Navis.
Anonymous figures			
		Crocodile on back of Reret	Area of Serpens Caput
	*nwt*	(The Goddess) Nut	The Milky Way
		Standing man of celestial diagram	It would be identical to the Giant
		Triangular shape of celestial diagram	An astronomical instrument similar to a gnomon

4.7.3 Star Clusters and Fuzzy Deep-Sky Objects

Among the stellar objects indicated by the Egyptians in the lists of hourly decans there are some designated as 𓐍𓏏 *ẖt* (see Figure 4.46). This term (*Wb* III, 356–358; Hannig, 1997: 629) is usually translated as 'body', but also 'troop', in relation to the army and 'group', which in our case must refer to a group of stars; that is, a star cluster. Thus, translating *ẖt* as '(stellar) cluster' in the astronomical context, we find a type of object that the Egyptians wanted to differentiate from other stars. The identification of these star clusters is very attractive, since being in relation to the decans, they can also serve to improve the identifications of the latter.

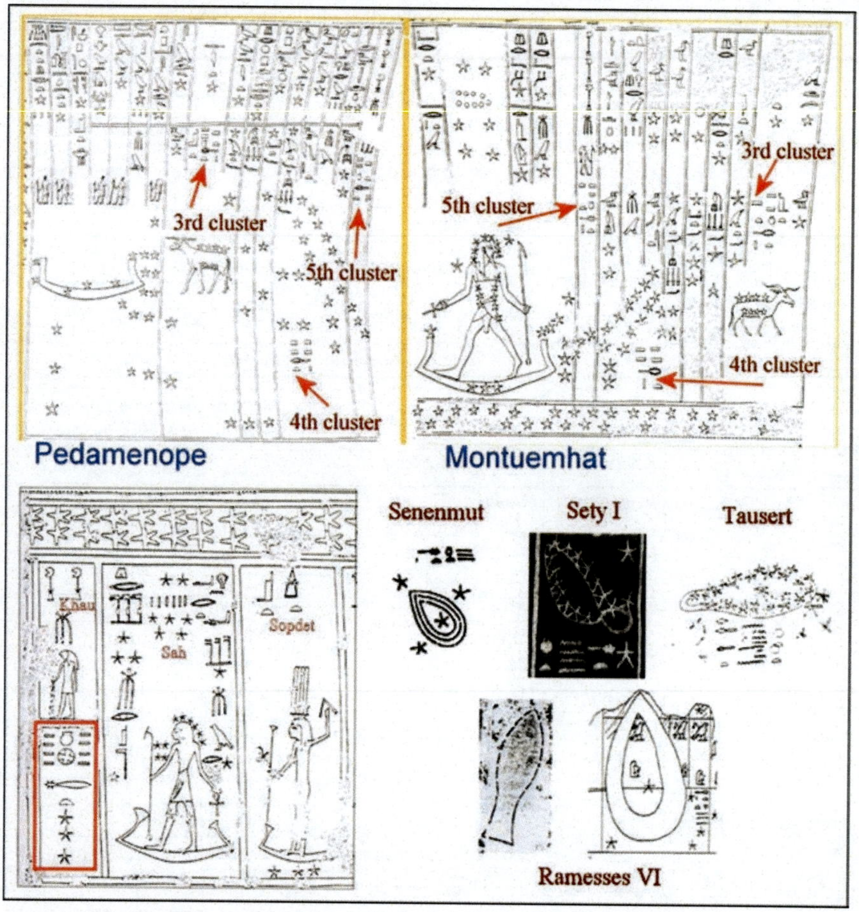

Fig. 4.46 Top: clusters 3–5 in the ceilings of Pedamenope and Montuemhat. 26th Dynasty. Botton left: the 5th cluster as *ḫmnw ẖt* on the ceiling of an underground chapel at Tuna el-Gebel (Hermopolis), from the reign of Ptolemy I. Botton left: The '6th' cluster or cluster 'of water' from different examples. (Diagram by the authors, with images adapted from *EAT* III, pls. 18, 20, and fig. 14 and Lull, 2006b: fig. 4)

In the decanal list of the astronomical ceiling of Senenmut a cluster called 3-*nwt ẖt*, 'third (star) cluster', appears to be related to the decan *sjt* ('Sheep'). The fact that this cluster is called 'third cluster' evidently means that, being numbered in increasing order, the first and second stellar clusters had to be located, although they were not cited, prior to its position in the celestial vault.

The fourth cluster, 4-*nwt ẖt*, appears on the Senenmut list in a position where it can remain unnoticed, with the inscription surrounded by small stars. In this list it is placed at the level of the 26th and 27th decans (*ḫntw ḥrt* and *ḫntw ẖrt*, respectively), which are the ones that follow *bꜣwj*, number 25. In other later lists this is clearly indicated.

The next one is already the fifth. This, 5-*nwt ẖt*, 'fifth (star) cluster', appears linked to decan *ḫꜣw* ('Myriads', decan 30), whose name alone should already indicate the presence of a large group of stars. Curiously, on the ceiling of an underground chapel at Tuna el-Gebel (Hermopolis), from the time of Ptolemy I, the name of 5-*nwt ẖt* has been transformed into *ḫmnw ẖt*, no doubt because Hermopolis is the City of Eight (Fig. 4.46).

Another cluster appears on the ceiling of Senenmut as *mw nwt ẖt*, although Neugebauer and Parker (*EAT* III: 115) proposed that the term ('water') should be replaced by *sjs* ('six'), so instead of the above there should be been written 6-*nwt ẖt*, 'sixth (star) cluster'. Regarding this cluster, it should be noted that it is not only named but also represented, usually in the shape of an oval object pointed on one of its sides (see Fig. 4.46).

A last cluster, although not numbered, appears on the roof of Senenmut within the so-called epagomenal or triangular decans. This one is called *ẖt pw* ('it is a cluster'), in reference to the decan *nsrw* (epagomenal decan 5), which on this ceiling occupies an intermediate position between the decans *štw* ('Turtle') or *štwj* the ('Two turtles') and *šspt* (epagomanal decans 4 and 6, respectively).

A compilation of the main documents in which these clusters are mentioned can be seen in Table 4.8. In none of them can we make any mention of the first and second clusters, as there are only references to clusters 3–'6' and that of Neseru. On the other hand, it is truly remarkable that the writing offers sensitive variants that sometimes make us think that the person who copied these texts did not really know what he was writing about. Thus, for the third cluster we see, for example, the variant , where the three strokes of the numeral 3 have been replaced by three Egyptian N's, perhaps because the copyist was using a hieratic document. In other documents, the cardinal number is not indicated for the third cluster, as in the ceilings of Ramesses VI, VII, and IX. Confusion is also seen in the naming of the clusters. For example, in the cases of Pedamenope and Montuemhat the fifth cluster bears the same name as the third.

The third cluster is present in all the lists in relation to the constellation of *sjt* or *srt* ('Sheep'), specifically in relation to decan 20, which in the Esna list appears in the eastern third of the constellation Capricornus. Locher (1981: 74 fig. 1) placed this constellation between Capricornus and Aquarius, which corresponds to the same region that we propose. The Star of *sjt* could be 2-ε Aqr (Albali). In any case,

Table 4.8 Clusters in the astronomical diagrams, whenever present

	3rd cluster	4th cluster	5th cluster	'6th' cluster	Neseru
Senenmut	*sjt* (20)	*ḫntw ḥrw/ẖrw* (26-27)	*ḫ3w* (30)	26-30 [drawn]	5th ep.decan
Clep. A. III	*sjt* (20)	*B3wj* (25)			5th ep.decan
Sethy I	–	–	–	26-28 [drawn]	
Ramesses II	*sjt* (20)	25-27	28-30	–	5th ep.decan
Tausert	–	–	–	23-24 [drawn]	–
Ramesses III	*sjt* (20)				
Ramesses VI	*srt* (20)		*ḫ3w* (30)	26-27 [drawn]	
Ramesses VI	*sjt* (20)		*ḫ3w* (30)	24-25 [drawn]	
Ramesses VII	*srt* (20)		*Kd* (28)	*ḫntw ḥrw* (26)	
Ramesses IX	*srt* (20)		*ḫ3w* (30)	26-28	
Pedamenope	*sjt* (20)	*B3wj* (25)	*ḫ3w* (30)		5th ep.decan
Montuemhat	*sjt* (20)	*B3wj* (25)	*ḫ3w* (30)		5th ep.decan
Abu Yasin	*sjt* (20)		*ḫ3w* (30)		5th ep.decan
Harendotes	*sjt* (20)				

this right ascension means that the third cluster may correspond to the constellation of Delphinus, which in just five degrees gathers a group of stars with brightnesses close to the 4th magnitude.

Most of the lists place the fourth cluster in relation to the 25th decan, *b3wj*, which according to the Esna division of the zodiac would correspond in RA to the easternmost third of the constellation Pisces. The Andromeda galaxy, M31, which is easily observed as a nebular object on a dark night, is at a similar right ascension and could therefore be the aforementioned 'cluster'. On the other hand, this region of the celestial vault does not contain any notable open clusters, so this fact may lead us to propose M31 with more guarantees (Lull, 2004: 253).

The fifth cluster, clearly named in the Senenmut and Ramesses VI examples, is associated with the decan 30 *ḫꜣw* 'Myriads' and obviously coincides with M45, the Pleiades. This same identification has been affirmed by other authors since Renouf (1874), although Leitz (1995: 93) thinks that it should be M31, the Andromeda galaxy.

Accepting that the clusters are denominated correlatively in numerical order in relation to their position in the celestial vault, we should inevitably understand that the 'sixth' cluster must be sought at a higher right ascension than that of the fifth cluster, *ḫꜣw*. Having identified the fifth cluster in M45, we must find the sixth more to the east, although in that case it must be recognized that its position was not well marked in any case. Since decan 31 corresponds to *ꜥrt*, probably Aldebaran, it is possible that the 'sixth' cluster is related to this decan and corresponds to the Hyades. The solution, for Neugebauer and Parker (*EAT* III, 115), lies in recognizing the 'sixth' cluster as one or all the decans of Sah (Orion) and therefore to suppose it was represented out of place. Although the concentration of stars in the constellation Orion is high, we cannot know if their density was noticeable enough for the Egyptians to designate this area as a cluster. Leitz (1991: 38) identifies the Hyades with this group of stars, a group that Locher (1993) believes could correspond to stars in the constellation Cetus.

The cluster of *Nsrw* (Neseru), served as an hour marker east of Sirius, after the epagomenal decan of the 'Two Turtles', as shown on the ceiling of Senenmut. If Neseru were really behind the 'Two Turtles', we could conclude that it was the most outstanding cluster in that region, M44, located in the centre of Cancer, a cluster that we have already designated the *sb3w nw mw* ('Stars of the Water').

However, being aware that Senenmut's list is incomplete, we rely on that of Sethy IA's family to reconstruct it. In this case, the twelve epagomenal decans would be ordered as follows: *štwj, Sjꜣtw,* [...]*bn, hnhn, kꜣkꜣ, nbj wr, Nsrw, šspt, Nhs, Jpds, šbšsn,* and *nṯr wꜣš,* so *Nsrw* would not be the second but the seventh. In that case, an excellent candidate would be IC 2602 with θ Carinae. This cluster, which can be seen with the naked eye with about seven stars packed together in a region about 70′ in diameter, is also known as the 'Southern Pleiades'.

In summary, the clusters mentioned in the lists we have discussed could correspond to the equivalences shown in Table 4.9, as a working hypothesis.

Table 4.9 Identification of the clusters mentioned in the astronomical diagrams, an hypothesis

ḫt	Cluster	Related decan	Identification
	3rd cluster	20 – *sjt*	Delphinus
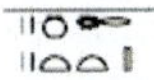	4th cluster	25 – *bꜣwj*	M 31 galaxy
	5th cluster	30 – *ḫꜣw*	M 45 star cluster
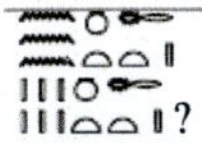	'6th' cluster	31 – *ꜥrt*	Hyades star cluster
	Neseru	*Nsrw*	IC 2602 star cluster

One last question remains regarding the first and second 'clusters' that do not appear in the celestial diagrams and lists but whose existence can be explicitly deduced. In this regard, it would be interesting to consider a couple of asterisms from the Ramesside star clocks that have been reasonably identified with star clusters, such as *sb3w nw mw* ('Stars of the Water') with the Manger cluster (M44), or *sb3w ˁš3w* ('Many stars') with the Coma Berenices cluster. Very suggestive is the idea that these were really the first and second clusters, and that they are not mentioned in the lists of the celestial diagrams because they belong to a totally different set of constellations.

An important corollary to this hypothesis would be that ancient Egyptian sky watchers would have recognized several small but notable abnormal objects in the sky vault, listed them, and may even have used them as suggestive temporal markers (the Pleiades are a practically universal 'time marker'). On the other hand, with respect to their nature, the name of some of the objects, such as those given to some clusters, would suggest that they were related to the supposed presence of water in the sky. Given its nebular or hazy appearance, perhaps this idea could be extended to the other members of this strange but extremely interesting family of celestial objects.

4.7.4 The Dendara Zodiac and Its Different Traditions

The so-called Dendara zodiac is found on the ceiling of Chapel No. 2 on the terrace of the Temple of Hathor in Dendara, on the east side. When the sages who accompanied Napoleon to Egypt stopped with General Desaix's troops near the ruins of Dendara, Viscount Vivant Denon took the opportunity to visit the temple and discovered the magnificent monument. Once Denon's drawings, published in 1802 in his *Voyage dans la Basse et la Haute-Égypte*: *pendant les campagnes du général Bonaparte*, were seen in France this ceiling aroused great interest (Buchwald & Greco Josefowicz, 2010). Thus, in 1822, a trip was organized to try to take that piece to France and, with the agreement of Mohamed Ali, the zodiac was torn from its place and transported to France, arriving in Paris in January 1823, having been since 1919 at the Louvre.

The zodiac, 2.55 metres on a side, occupies almost half the ceiling of Chapel No. 2, in the middle of which appears the figure of the goddess Nut, the goddess of the sky (see Figure 4.47). The first hypotheses about the meaning of the zodiac related it to a vision of the creation of the biblical world or to the announcement of future catastrophes. Even the first researchers who dealt with it, before the decipherment of hieroglyphic writing, understood that since the zodiac must be from the Pharaonic period, the zodiacal constellations represented in it must date back to an ancient time and be original to the Egyptians.

The 'planisphere' is held by four pairs of kneeling hieracocephalic deities and four other standing female deities. The latter are the goddesses of the four cardinal points and their position on the ceiling was not placed at random but they were well orientated to the cardinal points they represent.

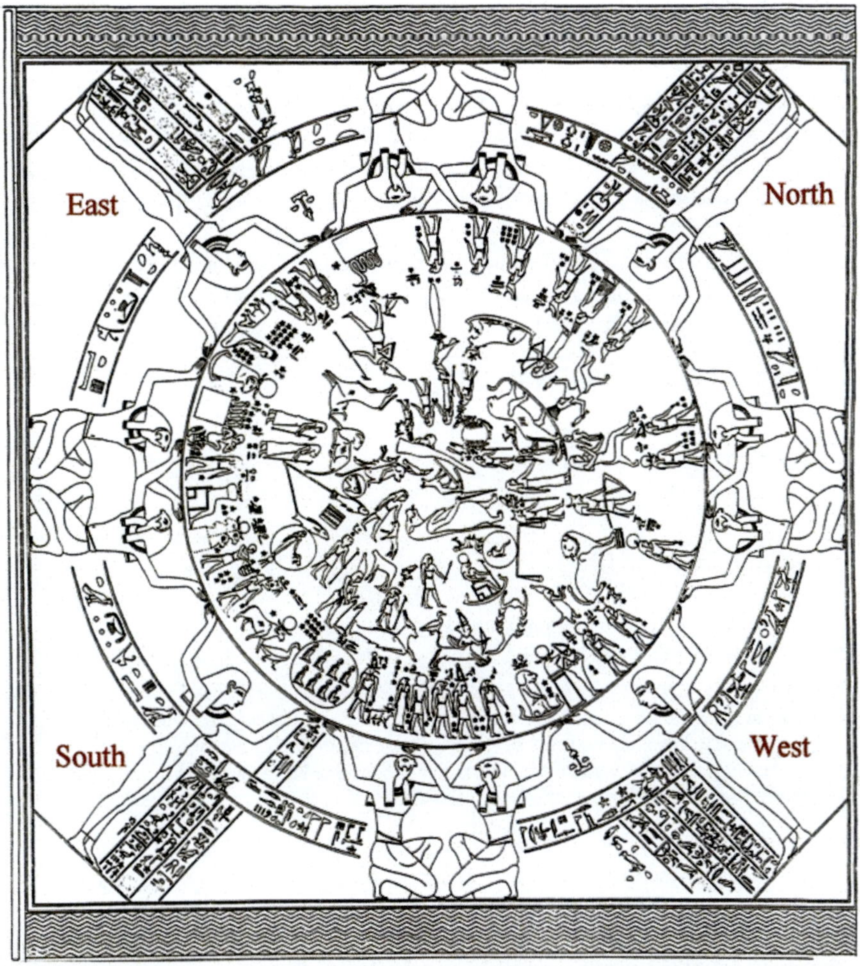

Fig. 4.47 Round zodiac of Dendara. (Adapted from Cauville, 1997a, b: pl. X 60)

The goddess of the east, called *fꜣyt* ('the one that carries'), is located near the feet of the great goddess Nut in the centre of the ceiling. Next to it appears the word *jꜣbt* ('East'). To the left of it is the goddess of the north *mḥt*, called *ḫyt* ('the one who raises'), accompanied by four hieroglyphic columns in which, as an example, can be read (Cauville, 1997b: 175):

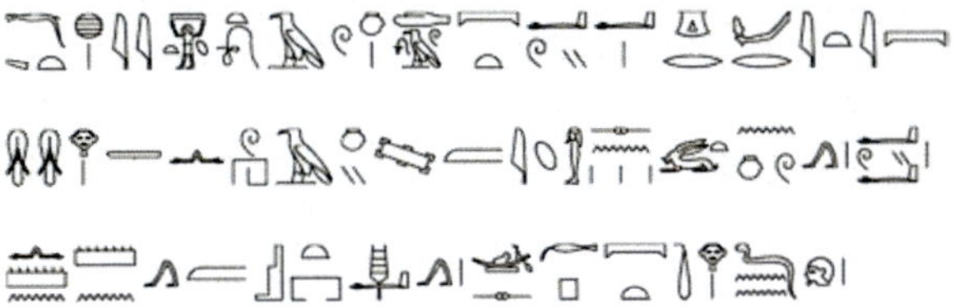

> North, Khyt: I carry the sky, my arms carry the sky, (my) feet on the ground do not miss their task, I extend my arms
> without moving from the place where I am. The sky is high over my head.

Then follows the western goddess *Jmnt* who received the name of *dwꜣyt* ('the one who carries') and finally the southern *rsj* goddess, called *ʿḥʿyt* ('the one who stands'). These goddesses had the function of supporting the sky, for this reason the texts to which they are associated refer to their duty as bearers of it.

Surrounding the zodiac and passing behind the waist of the divinities that support the sky, there is a hieroglyphic text that develops describing a circle. It is a kind of hymn (Cauville, 1997b: 175):

> The sky of gold, the sky of gold, Isis the great, mother of god, lady of Iatdi who lives in Iunt.
> The golden sky. The great gods are her[7] stars. Harsiesis her god of the morning.
> Sokar her light, Ihy her visible star, Osiris her Moon, Sah her god, Sepedet her goddess.
> They enter (and leave) through the *kyw* of the valley.

The 'constellations' are represented in high relief in the centre of a disc surrounded by the 36 decans (see Figure 4.48). These in turn, identified by their names, are symbolized by a series of divinities that march in an anticlockwise direction. In principle, as we have seen, the decans served the Egyptians to set the hours of the night throughout the year. However, it is not clear whether in Graeco–Roman times the decans were still employed for this purpose. At that time they were used as dividers of the zodiacal constellations, dividing each constellation into three parts of 10°. In that role they appear in the zodiac of Esna A, now destroyed.

Each of the decans (with the head turned to the right) has its name in front of it. Thus, for example, the first decan is called *Knm* and the second, located on the left hand of the goddess of the north, is called *ẖrj Knm*. Both the third and fourth decans receive the same name, *ḏꜣt*, the fifth *pḥwj ḏꜣt*, etc. For Neugebauer and Parker (*EAT* III, 72), however, the first decan on Dendara's list could be *wʿr* (35) or *pḥwj-ḥrj* (36), although this is not at all clear, as *Knm* is placed just below the cow that represents the star Sirius and could indicate its role as the first decan of the series.

[7] 'Her', taking into account that 'Sky' is a feminine noun in Egyptian.

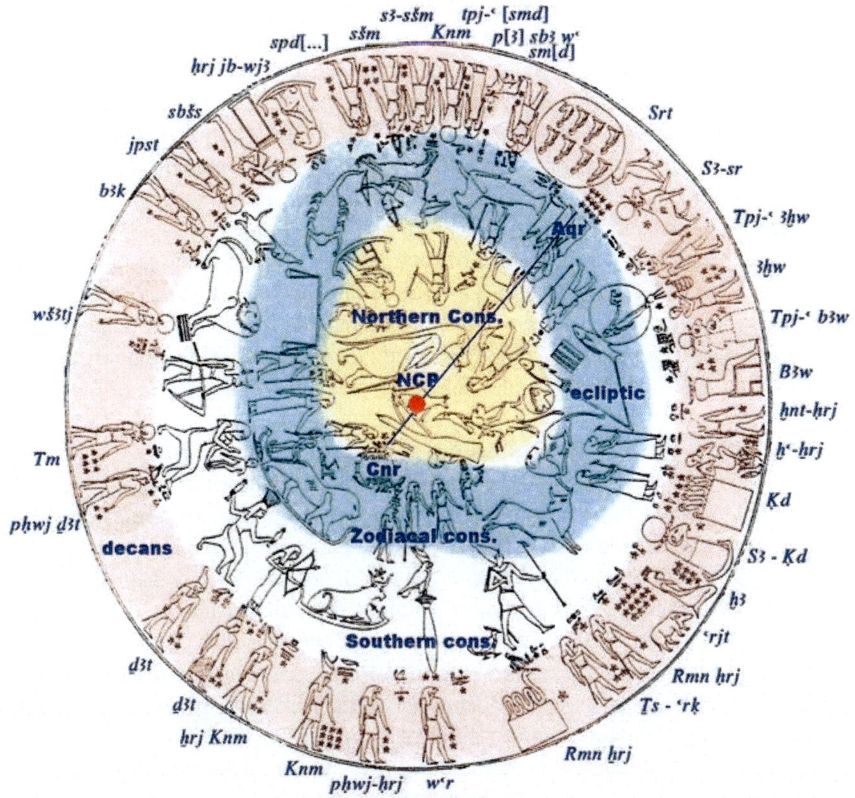

Fig. 4.48 An idealized representation of the celestial diagram of the circular zodiac where four clear regions have been identified. The decans are represented in a row of divine beings marching anticlockwise in the outer ring. Their names in hieroglyphics, typical of the Graeco–Roman period and slightly different from earlier lists, have being transliterated. NCP for north celestial pole. (Adapted from Lull, 2008b: 79 fig. 4)

The Dendara zodiac is the only circular-shaped celestial diagram found in Egypt[8]. To some researchers this fact has suggested its identification with a celestial planisphere where the stars and constellations would be located more or less realistically. However, this claim must be qualified by the particularities it offers, although it is not properly a zodiac since the constellations corresponding to the classical zodiac occupy only a small, but significant, part of it.

If we focus on the zodiacal constellations, we can see that they were not all represented at the same distance from the centre of the 'planisphere'; that is, they are not at the same declination in the 'planisphere'. In the celestial vault, the maximum declination of the ecliptic currently occurs at the eastern end of Taurus, while the minimum declination is located in Ophiuchus, between Sagittarius and Scorpius. This is

[8] There is also a small late example at the Greek National Archaeological Museum, Athens.

due to the inclination of the earth's axis with respect to the celestial equator. In the Dendara 'planisphere' this was taken into account, which leads us to think that the constellations were represented following a model that, along general lines and despite the originality of the representation (where symbolic elements also appear and many constellations are displaced owing to lack of space), should approximate to what they were observing. The lowest declination in the Dendara planisphere is between Aquarius and Sagittarius, and the highest in Cancer and Gemini, on the opposite side. In the middle of the 1st century BC, when this 'planisphere' was made, the lowest and highest declination of the ecliptic was located precisely and, not by chance, between Capricornus and Sagittarius, and between Cancer and Gemini, respectively. In the pronaos zodiac, on the other hand, the sun is born in Cancer (summer solstice) and approaches the mouth of Nut (winter solstice) in Capricornus.

The circular zodiac can be divided into three sections by the ecliptic ring implicit in the diagram: the zodiacal, northern, and southern constellations, in parallel to what is found in the astronomical ceiling of Senenmut (where there are obviously no zodiacal constellations but there is that north–south division). Typically, each study of the diagram that has been carried out has followed an independent analysis of these three preselected regions. However, we shall use a different approach.

This approach is presented in Figure 4.49 and will consist of a hypothetical series of constructive phases in the making of the diagram, which will take into account the well-accepted fact that the zodiac is actually a mixture of constellations of purely Egyptian origin, and of others that have their roots in the astronomical tradition of Mesopotamia, expressed possibly through a Greek nuance during the Hellenistic period of the Middle East, of which the Dendara zodiac is undoubtedly one of its masterpieces.

But the zodiac, which has been the subject of much debate since its discovery until today (Ross, 2019), could also be related to mythological matters. Thus, in the words of Priskin (2015a, b: 137): 'The understanding of the circular zodiac leads onto the realization that the linear zodiac in the pronaos is a variant of the same theme and was also intended to narrate celestial phenomena that were seen as the pivotal moments of the story about the family of Osiris'. In that sense, this researcher believes that the astral myth about the birth of Horus can be found even in the 26th Dynasty tomb of Benaty in the Bahariya oasis (Priskin, 2017).

4.7.4.1 The Zodiacal and Parazodiacal Constellations and the Ancient Egyptian Tradition

The zodiacal constellations would have been among the first to be represented in the diagram (see Panel I of Figure 4.49), in the form of an ellipse undoubtedly symbolizing the ecliptic, with its centre located more or less at Reret's head (see Panel II), where the pole of the ecliptic is actually located and consequently far from the centre of the 'planisphere' itself, a position equivalent to that of the north celestial pole. This phenomenology is obviously associated with the inclination of the axis of rotation of the earth with respect to the axis of the orbit of translation of our planet around the sun. This circumstance could be used to obtain a very approximate dating of the zodiac to the middle of the 1st century BC. This dating would be fairly in

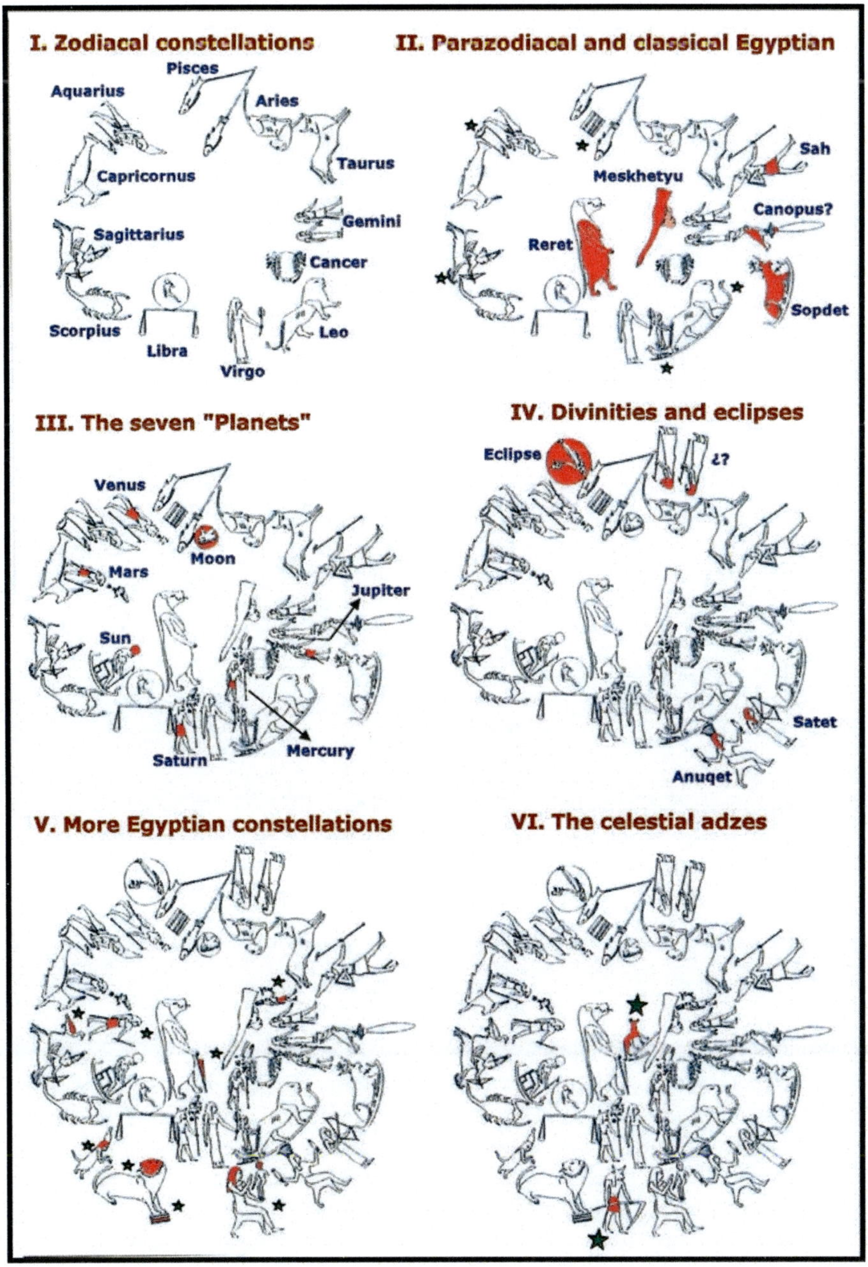

Fig. 4.49 A hypothetical exercise demonstrating how the zodiac was perhaps designed by successive levels of astronomical traditions and perceptions: (I) zodiacal constellations; (II) parazodiacal constellations (marked with a star) and ancient southern and northern Egyptian constellations (plus Ursa Maior); (III) the 'seven' planets; (IV) 'symbolic' drafts including a solar eclipse; (V) Egyptian constellations of the Ramesside star tables twice repeated; (VI) the celestial 'adzes' (VII and VIII) filling gaps; (IX) bright-star representations; and (X) the complete circular zodiac

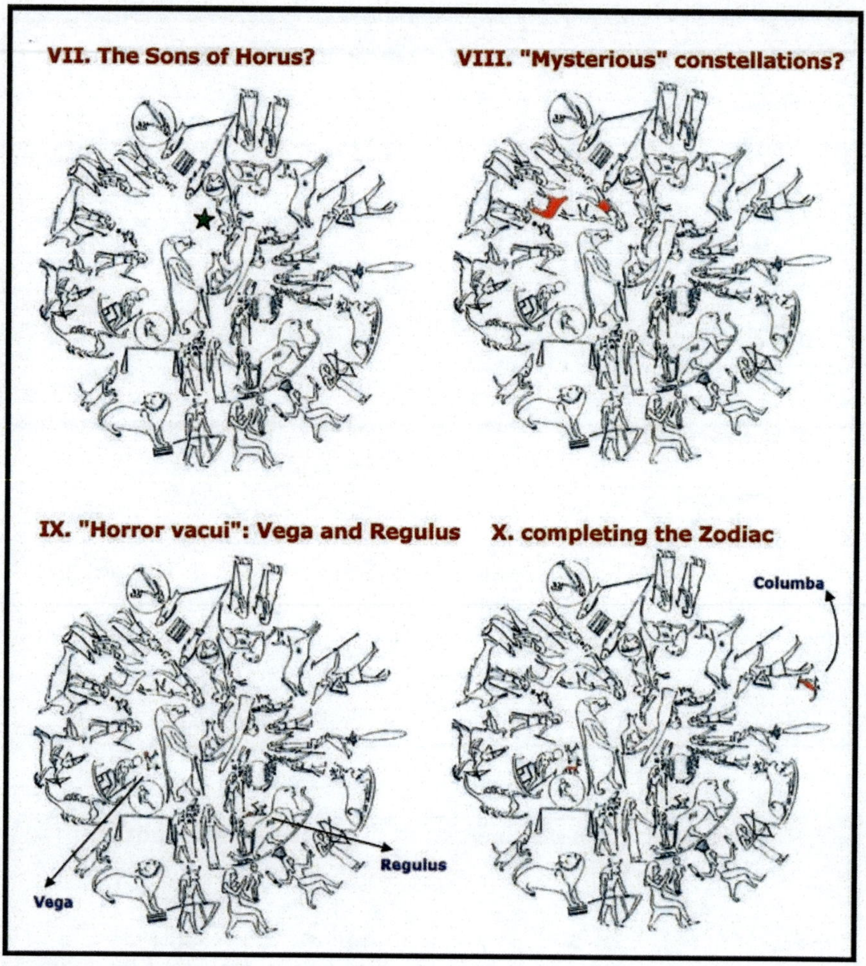

Fig. 4.49 (continued)

accordance with historical facts as defended by Cauville since the upper chapels of the Dendara temple would have been built during the reign of Ptolemy XII Auletes (who ended his reign in 51 BC), whereas the decoration of this part of the temple would be dated during the reign of his daughter and successor Cleopatra VII. However, the pronaos, where another version of the zodiac is found, would have apparently been finished and decorated in Roman times.

The zodiacal constellations are easily distinguished. These star associations were unknown in Egypt prior to the Persian or Ptolemaic periods, when they would have been imported from Greece or perhaps directly from Mesopotamia. However, some of the images show typical Egyptian features, such as Gemini (*n3 ḥtrw* 'the Twins'), represented by the two primordial deities Shu and Tefnut rather than by the twins Castor and Pollux, as seen in both the circular and pronaos zodiac (see Fig. 4.50).

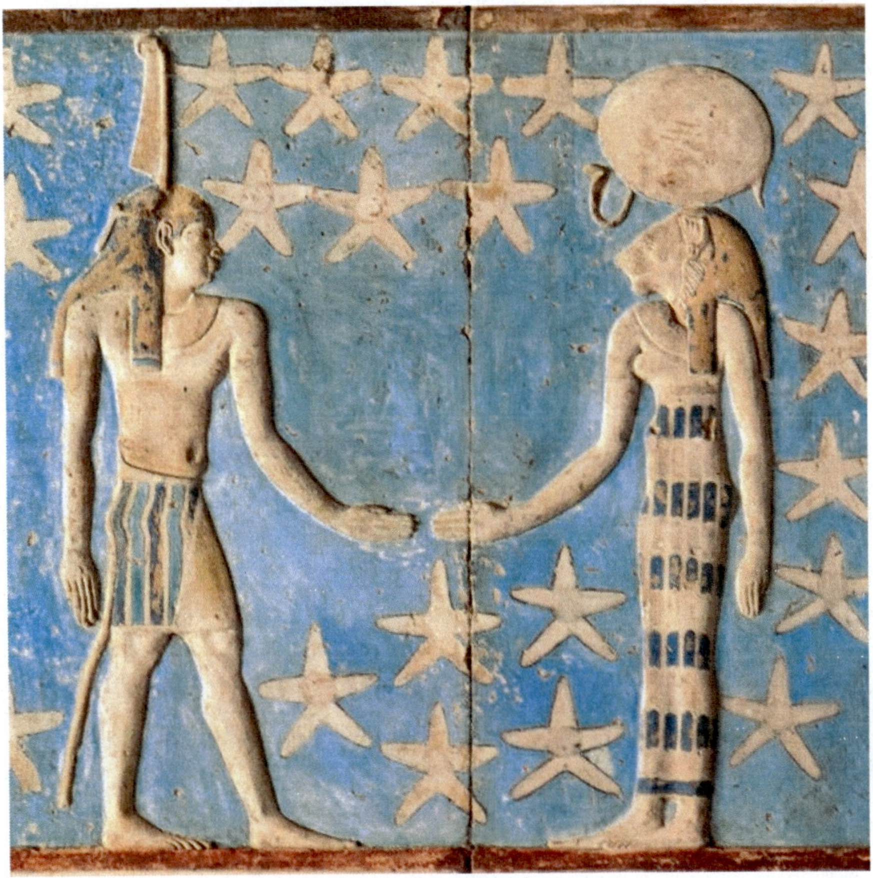

Fig. 4.50 Gemini, as Shu and Tefnut, shown in the pronaos of Dendara. (Image by the authors)

We must remember that the Egyptians already saw a pair of stars (*sb3wj*) where the Graeco–Roman world saw the Dioscuri. Cancer (*p3 gnḥḏ*) is represented as a crab, although this is often replaced by the scarab of the god Khepri in other diagrams.

Next comes the lion representing Leo (*m3j* for ancient Egyptians, *p3 m3j* 'The Lion' in Graeco-Roman period astronomical and astrological texts) followed by Virgo (*t3 rpjt*) with an ear of grain in his hand (a typical Mesopotamian shape) representing the bright star Spica. Libra, however, is represented by the balance for the first time in Egypt, the place in the sky where the autumnal equinoctial point was located in late Antiquity. The figure of a child inside a disc can be seen on the scale. This image is also represented on the horizon sign (*3ḫt*) in the zodiac of the roof of the pronaos, as shown in Figure 4.51. In fact, in Graeco-Roman period astronomical and astrological texts Libra is normally called Akhet (*3ḫt*). For Priskin (2015a: 142) it would represent the first occasion of *psḏntjw*, the new moon, after the autumnal equinox, considering that the Egyptians customized the representation of Libra to

Fig. 4.51 Libra as shown in the pronaos of Dendara. (Image by the authors)

mark a momentous cosmic event that in their mythology corresponded to the conception of Horus by his mother, Isis.

After Libra, Scorpius (*tꜣ ḏlt*), Sagittarius (*pꜣ ntj ꜣtḥ*), and Capricornus (*pꜣ ḥr n ꜥnḫt*) have an identical appearance to other representations from Mesopotamia, such as those found in various Babylonian kudurrus. Aquarius (*pꜣ mw*), however, is again represented as a typically Egyptian deity: the god of the Nile, Hapy (see Figure 4.52). Pisces (*nꜣ tbṯw*) appears in its usual form, followed by Aries (*pꜣ jsw* 'the Ram') and Taurus (*pꜣ kꜣ* 'The Bull'), which, curiously, is represented in a very atypical way looking backwards (Fig. 4.4). None of these last images has a parallel with the Egyptian constellations of the previous period, although the ram is reminiscent of certain representations of the god Amun.

The zodiacal constellations would have been sculpted simultaneously or perhaps shortly after some older and especially important Egyptian ones (Panel II of Fig. 4.49), such as the boreal constellations of Meskhetyu and Reret.

Sah is easily identified as the man who, wearing the crown of Upper Egypt, holds a *was* sceptre and a flagellum, and walks with great strides, undoubtedly being a personification of Osiris. Sopdet (Sirius – α CMa) appears as a cow reclining on a typically Egyptian boat with a five-pointed star between its horns, which is almost certainly Sirius. The cow, a symbol of the goddess Hathor, the mistress of the temple, suggests the assimilation of Hathor with Isis and therefore with Sopdet herself.

More difficult, however, is the interpretation of the prominent figure of a falcon on a pillar (actually a *wꜣḏ* symbol) that appears represented between Sah and Sopdet in a position where there is actually little space in the sky for more constellations (see Figure 4.53). In a typically Egyptian setting, this is the well-documented image of Horus-on-his-pillar (*ḥrw ḥrj wꜣḏ.f*) that would complete the Osirian triad of Osiris, Isis, and Horus. The image may represent Horus as a third member of the triad, which already appears in the *Pyramid Texts* as the Sah-Sopdet-Sopdu

Fig. 4.52 Aquarius, as Hapy, shown in the pronaos of Dendara. (Image by the authors)

grouping, and which in this case could have only a symbolic character as a representation of 'Horus found in Sothis', or perhaps as an archaic image of the morning star, a form of the planet Venus, since both images constitute possible interpretations of certain recitations of the *Pyramid Texts*. In this sense, it is curious that, if the hieroglyphic text that surrounds the zodiac is taken into account, it mentions 'Horus son of Isis (Harsieses), god of the morning' as one of the 'stars' represented in the stellar diagram. This would mean that the planet Venus would be represented on two occasions (Belmonte, 2012: 88–89).

Another option is that this curious image of Horus represents another celestial object neighbouring Sirius and Orion that could well be mentioned in the circular inscription; we refer to the god Ihy, the third member of Dendara's own triad of divinities. In this case, an obvious candidate would be the bright star Canopus, a hypothesis supported by Cauville (1997a: 35–36). In another chapter, the importance of Canopus as one of the stars to which the Egyptian temples were orientated is examined and we are reminded of the idea, mentioned by Allen (1963), that Canopus could well have some relationship with the figure of the king himself and

Fig. 4.53 Sah, a falcon upon a *wAD* symbol and Sopdet. Round zodiac of Dendara. (Image kindly provided by Ángel Ferrer. Courtesy of Louvre Museum, Paris)

therefore to be a personification of Horus, the divinity with which the king was identified. However, this hypothesis poses a minor problem in that it would be contrary to physical reality, since the rising of Canopus always follows that of Sirius and not the other way around, although it is also true that if the passage through the meridian of the 'triad' is observed, Canopus would be seen between Sah (southern sector of Orion) and Sopdet.

Both hypotheses, (i.e. whether merely a symbolic character or an image of Canopus) are attractive, and perhaps the solution could be a mixture of both, a very common and appropriate solution for the Egyptian mentality in ancient times.

The area around the celestial pole is occupied by two traditional Egyptian constellations. The giant crocodile-tailed hippopotamus is surely Reret, the constellation often depicted in astronomical ceiling diagrams and used to measure time on the Ramesside star clocks.

The bull's leg in the centre of the diagram is undoubtedly Meskhetyu. However, in this case it is not isolated, as there is a small figure of a reclining animal attached to the leg itself.

Surprisingly, the zodiac is the only case among the dozens of celestial diagrams found in Egypt where this happens, so it has been a matter of debate among specialists who have seen a variety of animals in it ranging from a ram, without major connotations, to a reclining lion (*EAT* III, 200), which, according to some authors who doubt the identification of the Divine Lion with Leo, could be identical to the former and in reality be a small constellation of the northern skies, as had been suggested, for example, by Locher.

However, there could be another solution. The circular zodiac is a diagram composed of traditions from Egypt and Mesopotamia with an indisputable influence of Greek origin and was probably designed to be correctly interpreted by people related to the tradition of all these civilizations. Consequently, it is suggested that this little animal is a bear, and that it represented the constellation Ursa Major, to which the constellation of Meskhetyu is undoubtedly equivalent. This Bear would act as a mnemonic reinforcement. Therefore, what we are really dealing with is a duplication of a celestial object in a peculiar mixture of properly Egyptian knowledge and an alien tradition that overlaps with it. We shall see that this idea can be supported by several more examples in the zodiac itself where a similar phenomenology has been discovered.

Leitz (2006: 295) also thinks that it could be the bear, since it also seems to be looking towards Orion, as stated in Homer's *Iliad* (18: 485–490): 'And therein all the constellations wherewith heaven is crowned—the Pleiades, and the Hyades and the mighty Orion, and the Bear, that men call also the Wain, that circleth ever in her place, and watcheth Orion, and alone hath no part in the baths of Ocean.'

In this regard, Panel II itself shows the presence of a series of what Rogers (1998) has called parazodiacal constellations. The series is a set of 'constellations' that were frequently associated with the original zodiacal constellations in Mesopotamian astronomical texts, particularly the *Mul-Apin* tables. Among them, we may mention Ku, the Fish, identical to our Piscis Austrinus (PsA), with Fomalhaut as the brightest star; Ash-Iku, the Field, which is located in the Square of Pegasus; Uga, the raven (Corvus); Mush, the Hydra, normally related to Leo; and, finally, Ma-Gur, the Boat (possibly Corona Australis), associated with the archer, Mul Pa-Bil-Sag, an alter ego of Sagittarius. Surprisingly (or perhaps not), all of these star groups can be easily identified in the Dendara zodiac, such as a fish located under Hapy on which Hapy pours water, a rectangular flooded field in Pisces, a bird and a snake near Leo and, finally, a small boat located under Sagittarius' front hooves. The parallelism is so striking that the solution is, in our opinion, self-evident.

However, special mention should be made in the case of the small boat. In previous sections it has been shown that the ancient Egyptians recognized a Boat (*wjꜣ*) in the area of Sagittarius. Therefore, we could identify this boat with *wjꜣ* and not with Ma-Gur. However, another extremely interesting possibility is that, once again, we are faced with an image with multiple meanings that could be interpreted differently according to the background of the person looking at the image.

Closely linked to the previous analysis is that of the last of the parazodiacal constellations in the diagram which, in this case, apparently does not have a Mesopotamian origin. It is a small figure of a woman clinging to the Lion's tail (Fig. 4.37), situated above Corvus (the bird that appears at the end of the tail of the serpent on which Leo walks). Because of this relationship with the bird identified with Corvus, Gundel (1936a, b: 259) also identified it with stars in Corvus. In later diagrams, it is obvious that it was intended to represent Virgo, an identification

Fig. 4.54 Zodiacal constellations in the astronomical ceiling of the tomb of the two brothers in Athribis. Virgo (marked in green) was drawn as a woman holding the tail of Leo in the upper register, and as a woman holding an ear of grain. (Adapted from *EAT* III, pl. 51)

suggested by Hübner (1995: II, 50), and perhaps in this case we are faced with a new double representation in the zodiac. Note, for example, that in the astronomical ceiling of the so-called tomb of the two brothers (2nd century AD), at Athribis (see Figure 4.54), the twelve zodiacal constellations are twice represented. And while in one register Virgo is identified as the woman who wears the ear of grain, in the other Virgo is identified with the woman who clings to Leo's tail. It is clear, then, that both figures are associated with Virgo. However, the hypothesis of Leitz (2006: 310) that sees in this woman not the representation of the constellation Virgo but of its main star, Spica, is also appropriate.

Another option, considering the arrangement of the boreal constellations on some of the astronomical ceilings of the New Kingdom, is that the constellation representing the goddess Serqet (*srḳt*), who was often depicted above or near the divine Lion, has moved towards the tail of Leo and therefore appears in a close relationship with Leo. The identification of Serqet in the sky is very complicated, as we already discussed elsewhere, so, taking into account the particularities of the zodiac, we may also venture to make this suggestion.

The truth is that the various figures associated with Virgo could refer to the constellation, to the most important of its stars, or even to symbolic figures. Piperakis (2017: 146), regarding the figure of the woman sitting on a throne holding a child (Fig. 4.37) below Leo, makes some interesting observations:

> Isis-Thermouthis was assimilated to Demeter, and so the goddess rules over a segment of the constellation of Virgo. Ancient astronomers represented Virgo as a winged woman holding an ear of grain, which refers to the constellation's brightest star, Spica. As ears of grain were the symbol of Demeter and later of her Egyptian equivalent Isis, Virgo was equated with either Demeter or Isis. By way of analogy, the first decan of the sign soon was drawn into this equation. The Egyptian astrologer Teucer of Babylon (who lived between the end of the second century BC and the middle of the first century BC) remarks that with the first decan of Virgo rises the paranatellon, 'a goddess seated on a throne and feeding a young child', which is construed by Teucer as Isis feeding Horus in the entrance to a temple.

On the other hand, it should be remembered that the term paranatellon signified stars either rising together with the sun or being in other conspicuous positions with respect to it, or a constellation that was co-rising within certain degrees of a zodiacal sign or with a decan. According to Quack (1999), this concept could have its antecedents in ancient Egypt, at least from the 13th century BC. On the other hand, the astrologer Antiochus, who lived between the 1st and 2nd centuries AD, describes this group simply as 'a woman carrying a young child' (Boll, 1903).

4.7.4.2 The Planets, Sun, and Moon

As shown in Panel III of Fig. 4.49, the five naked-eye planets known to the ancient Egyptians are represented in the zodiac and identified by one of their names in hieroglyphics. Mercury is in Virgo (its place of exaltation) and his name was evidently inserted after both Cancer and Meskhetyu had already been sculpted. Saturn, with the head of a bull, is located between Virgo and Libra (in Libra if it is in exaltation). Mars, here called Horus the Red, is on the back of Capricornus (its place of exaltation). Two-faced Venus is between Aquarius and Pisces (in Pisces if it is in its exaltation). Lastly, Jupiter is between Gemini and Cancer (in Cancer if it is in its exaltation).

There have been long debates as to whether these positions of the planets between the zodiacal constellations are merely random, symbolic or whether, on the contrary, they are intentional. Of the various proposals submitted, there are two that appear to have the most substance. The first, proposed by Neugebauer and Parker (*EAT* III, 203–204), and recently defended by Conman (2013: 213) and other authors, suggests that the planets are shown in their places of exaltation or *hypsomata*. Their positions would therefore have only a symbolic character and not an astronomical meaning. The second, defended by the French astronomer Aubourg, suggests that the planets are found in the constellations where they were located at a time close to the construction of the temple in the 1st century BC. In fact, the two hypotheses need not be at odds and are actually compatible with each other within certain limits, so both could be considered as reasonable and valid explanations of the position occupied by the planets in the diagram.

For the cultures of the Middle East, including those of classical Greece, the sun and the moon were also considered planets, although of a different category. In Egypt, however, the planets were considered at a lower level than the great gods of the sun and moon. However, the Egyptians may have tried to represent them in the zodiac. Our preliminary proposal is that the moon can be identified as the Eye of Horus, the *udjat*, inserted in a disc, a frequent representation of the moon in Graeco–Roman times that appears sculpted above Pisces, while the sun can be identical with the imposing falcon-headed figure that is above the Scorpion, as this is a familiar representation of the god Re-Horakhty when he is identified with Sokar, travelling in the Duat (that is, in the night sky), a divinity that we know is found among those represented in the circular zodiac as recorded in the inscriptions themselves.

Continuing with the analysis, in Panel IV a series of images relevant to the debate have been introduced. Just below Pisces is found what seems a goddess, situated inside a circle with a small pig or baboon (see Fig. 4.4). Cauville and Auburg have proposed that it is the symbolic representation of a solar eclipse that occurred in Pisces on March 7, 51 BC, visible from Dendara, a date close to the death of King Ptolemy XII. In the rectangular zodiac, in the pronaos, this same figure also appears and the figure of a god (not a goddess) is perfectly recognizable, and it is clear that the animal being held is a pig (see Figure 4.55). We should therefore think that this same animal is the one that appears in the circular zodiac, although Aubourg (1995: 10) and Cauville (1997a, b: 22) think that it is a baboon, which, because of its lunar character, would give more meaning to the interpretation of the eclipse. In fact, Priskin (2015b: 32) concludes that the baboon could symbolize the lunar conjunction and therefore the possibility of a solar eclipse. This relationship between the baboon and the moment of the moon's conjunction was already expressed by Horapollo in his *Hieroglyphica*.

On the other hand, Aubourg and Cauville estimate that the *udjat* over Pisces could also represent an eclipse, lunar on this occasion, which occurred, again in Pisces, on September 25, 52 BC. Also, Priskin (2015a: 138) accepts the interpretation of both eclipses and estimates that these show that the zodiac wants to make references to a period that only includes September 52 BC and March 51 BC. However, not all authors agree with this interpretation. Von Lieven (2000: 157), for example, considers that the *udjat* within the disc represents the full moon. In fact, that any *udjat* inscribed inside a disc could be interpreted as a lunar eclipse

Fig. 4.55 A man holding a pig inside a disc as shown in the pronaos of the temple of Dendara. A similar image can be observed in the circular zodiac. (Image of the authors)

would be more of a problem, since as we well know the *udjat* inside a disc is used to represent the full moon.

The dating of the zodiac design also shows no unanimity, although the different dates are based mainly on astronomical motifs that have been interpreted as eclipses or planetary positions. Neugebauer and Parker (*EAT* III) dated it in the Late Ptolemaic period, Aubourg and Cauville between June and August in 50 BC, Trevisan (2011) also in 50 BC (but considering the importance for the configuration of the zodiac of four solar eclipses from 62 to 50 BC), and Park and Eccles (2012) in the latter part of the reign of Augustus.

Another problem is that of the two standing figures sculpted below Aries (see Fig. 4.4), one with a lion's head and a female figure. According to Conman (2002), they represent the two remaining planets of Antiquity that apparently are not in the zodiac, that is, the sun and the moon, respectively. To make this curious and controversial claim, Conman relies on the fact that they carry sceptres, like the 'other' five planets, and that the lion is indeed a very ancient Egyptian solar symbol. Additionally, in this way the sun would be located in its sign of exaltation (Aries) and the moon near his, that is, Taurus; therefore, the 'seven' planets would be in their corresponding *hyposomata*, a fact that is certainly suggestive. The main problem with this hypothesis, already recognized by Conman herself, is that the moon is always a male divinity in ancient Egypt. Actually, it is very difficult to find a reasonable alternative for the identification of these two figures (see below); therefore, Conman's hypothesis could be accepted as another external influence present in the Zodiac, presumably of Greek origin, since the moon was also a dominant male deity in the Middle East during ancient times.

Fig. 4.56 Sah, the falcon upon the pillar and Sopdet as a cow. Pronaos ceiling of the temple of Hathor at Dendara. (Image by courtesy of Txema Sanz)

Additionally, in Panel IV, we have also included two extra personifications that are controversial. On the astronomical ceiling of the pronaos (see Figure 4.56) these personifications appear together in a boat behind Sopdet and are identified, by their appearance, with the goddesses of the First Cataract of the Nile, Satet and Anuket, whose temples have been found on the islands of Elephantine and Sehel, respectively. In the circular zodiac (see Fig. 4.55), Anuket is easily identifiable, but in the case of Satet, who was often identified with Sopdet as one of her avatars, the issue is not so clear. But in the pronaos, the inscription that appears in front of the goddess who has the appearance of Satet indicates the name of Sopdet.

The supposed image of Satet in the circular zodiac does not wear the typical tiara of the goddess (she does, however, in the pronaos, see Fig. 4.56), but nevertheless she is armed with a bow and an arrow ready for use. In Mesopotamia, Sirius was known as Kak-Si-Sa, the 'Arrow Star', while large areas of Canis Major were interpreted as the bow ready to be used. Perhaps we are faced with the same situation as before, where the typical Egyptian constellation (Sopdet) is followed by an allogeneic interpretation (bow and arrow) which, moreover, would fit well with an alternative Egyptian interpretation of the same image (the goddess Satet). Therefore, the constellation Canis Major and the star Sirius would also be doubly represented in the zodiac.

As for Anuket. This goddess is represented with a pair of libation glasses in her hands, as if she were prepared to pour out the waters of the flood, which occurs just after the heliacal rise of Sirius, and perhaps her presence here is merely symbolic.

It has often been argued that the zodiac should also have represented a series of heavenly bodies from the earliest Egyptian tradition. This is evident in the case of the list of decans in the outer circular area of the 'planisphere'. It is now going to be shown that the second set of stars and constellations, those of the tradition represented by the Ramesside star clocks, is also present in the zodiac.

By right ascension, a list of the stars and constellations of these devices, beginning with Sirius (*sbꜣ n spdt*, the Star of Sopdet), would be the following: *sbꜣwj* (the Pair of Stars), *sbꜣw nw mw* (the Stars of the Water), *mꜣj* (the Lion), *sbꜣw ꜥšꜣw* (the

Multitude of Stars or Many Stars), *ṯꜣ nfr* (the Beautiful Child), *mnjt* (the Mooring Post), *rrt* (Reret, the Hippopotamus), *nḫt* (Nekhet, the Giant), *ꜥryt* (Aryt) , *ꜣpd* (the Bird), *sbꜣ n ḫꜣw* (the Star of Myriad), *sbꜣ n sꜥr* (the Star of Fire), and finally *sbꜣ n sꜣḥ* (the Star of Sah). Reasonable identifications have been provided for all of them in Table 4.7. Panel V (see Fig. 4.49) incorporates a new series of representations, which, adding some already analysed, would almost complete the previous list.

For the Star of Sopdet and the Pair of Stars we have their representation in the star located between the horns of the goddess Hathor, showed in the form of a reclining cow, and with the stars that would be located on the head of Shu and Tefnut, respectively.

The circular zodiac twice shows the figure of a lion: that representing Leo and the lion located below Libra, which according to Neugebauer and Parker would be Mai, the lion of the Ramesside tables (*EAT* III, 202), an opinion shared by Leitz (1995: 251). This second lion has under its front legs the possible representation of the Stars of the Water in the form of a small pond. However, in our opinion, if it were Mai, he would have simply been unfolded from Leo, duplicating a figure that, in reality, would correspond to the same constellation.

The Beautiful Child, probably Spica, can certainly be found in the hands of the female figure sitting just below Virgo. Since we have also identified it with the ear of grain that Virgo carries, Spica would also be represented twice in the diagram. The lady who carries the Child could be identical to Virgo and once again we would have duplication. Teucer of Babylon (Boll, 1903: 210), a 1st–2nd Century AD Egyptian astrologer, mentions a constellation that corresponds to this figure, and that he identifies with Isis and her son Horus in relation to Virgo.

The mooring post, Menit, is now in the hands of the great hippopotamus, with the star Arcturus at its apex. This pair of constellations was surely from the beginning represented as a single group. Leitz (1995: 251) has proposed that the Reret of the Ramesside clocks may be behind what he considers Mai (the second lion shown in the cicular zodiac), but this is unlikely. The Giant, Nekhet, would have spread from Sagittarius to Pisces; however, the presence of the hieroglyphic text that identifies the planet Mars as 'Horus the Red', the name with which it was known at that time, already filled much of this space. Consequently, the Giant was depicted with his mace in his hand (this is the name of one of his stars) but in the empty space above Capricornus, followed by the Bird (*ꜣpd*). In the Ramesside tables, the stars of the Bird, ordered in RA, are: *bꜥnt nt ꜣpdm* ('Beak of the Bird'), *tp n ꜣpd* ('Head of the Bird'), *ḥtyt nt ꜣpd* ('Throat of the Bird'), and *ktf.f,* 'Its rump (of the Bird)', so it is evident that the bird is orientated facing west. This is the same orientation as the bird below the standing human figure with mace that we believe may be the Giant (Nekhet), so perhaps this also supports our proposal to recognize Nekhet and the Bird here.

The Star of Fire (Capella) could be identified with the small human figure wearing a headdress with feathers, holding a ram-headed staff in one of his hands and standing to the right of Gemini. In the pronaos this figure is located next to Taurus (see Figure 4.57). However, for Priskin (2015a: 150) the waxing crescent of the

Fig. 4.57 A god holding a ram-headed staff, next to Taurus. Pronaos, temple of Hathor at Dendara. (Image by the authors)

moon first seen after the spring equinox in Taurus was represented in the zodiac by this figure. For him, the ram-headed object in his grasp may be an allusion to the setting sun, because it was often depicted as a cryocephalus being.

The Star of Sah, probably Rigel, does not appear to be shown individually, as it does with the figures that represent Regulus or Capella, or the star that on Sopdet's horns represents the Star of Sepedet. It would be in the place occupied by Sah's own head.

The Star of Myriads is undoubtedly a member of the Pleiades, predictably Alcyone, which are already mentioned in the ring of stars corresponding to the decans, as well as perhaps implicitly included in the constellation of Taurus. However, it is strange that, considering the large number of duplications in the diagram, this very important constellation has not been explicitly represented.

Aryt should have been represented in the sector near Pisces. The translation of his name and the identification of this constellation are still problematic and we do not have an accurate answer to this absence, unless Aryt is in the strange figure located just above Aries and the disc with the *udjat*.

Finally, the Multitude of Stars or Many Stars should be behind Leo, but their identification is very complex. It could be postulated that it corresponds to the strange figure that is located behind the lion just below Libra, understanding that both constellations appear misplaced, but we do not find clear elements to aid identification.

If any case, we see how the zodiac appears to include many of the individual stars from the Ramesside star tables, clearly showing that this unique masterpiece might in fact well be considered a classical Egyptian celestial diagram despite numerous foreign influences.

4.7.4.3 Problematic 'Constellations'

Panel VI (see Fig. 4.49) includes two figures that we assign to a very peculiar group, because both are associated with the image of an adze. One of them is located precisely in the centre of the diagram. Consequently, it should be located in the vicinity of the celestial pole at maximum declinations. One of the celestial adzes, used in various religious ceremonies, in particular that of the Opening of the Mouth, could be identical to Ursa Minor (UMi), while the other is undoubtedly, as has been shown, an alternative representation of Meskhetyu itself. Although the shape of the two adzes is not the same in the zodiac as it is in religious ceremonies or in older hieroglyphic texts (versus), we dare to suggest that the symbolism behind these representations may actually be identical.

The Jackal depicted on the adze, with which it has normally been associated since the discovery of the zodiac, is interpreted as an image of the god Upuaut, the 'Opener of Paths'. However, it should be noted that this image is totally absent in earlier celestial diagrams. Interestingly, the falcon-headed image of Anu, previously depicted in this region of the sky on astronomical ceilings, is absent from the circular zodiac. However, it is present in close connection with Meskhetyu on the ceiling of the pronaos of the temple (see Fig. 4.24).

The second figure with an adze in his hand is that of a bull-headed god situated just below Saturn. In reality, it is extremely difficult to find an adequate explanation for this 'constellation' and, although we could make other proposals, all of them highly speculative, we choose to consider it as one of the unsolved images of the zodiac. For Park (2008), the god with the bull's head and hoe could represent Boötes/Arcturus.

Panels VII, VIII, IX, and X (representing the inner sector of the zodiac in its entirety) (see Fig. 4.49) are the last stages necessary to complete this very particular analysis that has been carried out. In Panel VII we have included, in the empty space between the moon and Meskhetyu, a figure composed of a baboon, a hawk and a peculiar-looking animal. The strange animal has the appearance of an ass and is similar to some later representations of the god Seth. The interpretation of this group is very difficult. This composite figure, which is difficult to interpret, is also present, between Aries and Taurus, on the ceiling of the pronaos. Moreover, to confuse matters even more, a baboon and a hawk are represented in some less complex celestial diagrams, such as the one represented in Heter's coffin (Fig. 4.38), where they are related to some unmistakable representations of classical Egyptian constellations. From its position it could be Cassiopeia, but the problem is not easy to solve. Leitz (2006: 302) comments on the possibility that it is, at least in part, a constellation in the shape of a baboon that is mentioned at three degrees from Aries in the *Liber Hermetis*.

The situation is also quite complicated in relation to the images that complete the great gap above Reret's head (see Panel VIII in Fig. 4.49): a man holding an animal by the head would correspond, according to Boll (1903: 239), with Cepheus and an animal. On the other hand, a headless human being on all fours is also difficult to identify. For Cauville (1997a: 32) he would correspond with Equuleus.

Both seem to be related since they are represented together, near Aquarius, in the ceiling of the hypostyle hall. Unfortunately, the explanations offered to date by various researchers, and as a consequence the identification of these images with constellations in the 'corresponding' region of the sky, are unprovable as they are impossible to verify, there being no parallels or elements of comparison.

However, there is an alternative explanation for this troublesome set of figures. On the astronomical ceiling of Senenmut the four sons of Horus (Imset with a human head, Hapy with the head of a baboon, Kamutef with that of a jackal, and Qebehsenuf with a falcon head) are represented in a row behind the goddess Isis and the constellation Reret. Without entering into the controversy about their meaning in the astronomical ceilings themselves (this is the worst understood sector of the classical star diagrams), there are cases in which the children of Horus could be interpreted as starry beings in the area of the celestial north pole. In this sense, it is at least surprising to realize that we have a jackal, a hawk, a baboon, and a human figure all together, near Reret's head, and in an obviously northern location of the sky.

4.7.4.4 Completing the Zodiac

Panel VIII (see Fig. 4.49) already shows a circular zodiac almost completely filled with images at this stage. However, there are still some small empty spaces where the artist (or designer of the work, if it comes from a papyrus copy) still felt able to include some small figures. One of these spaces is just above Leo where it was chosen to represent the image of a seated Egyptian king. It is possibly a representation of the star Regulus (α Leonis), called Lugal, (King) in the *Mul-Apin* tables. Following this reasoning in which it has been proposed to identify some small figures in the diagram with individual bright stars, such as Capella or Regulus, we tend to do the same in the case of a small seated female figure that is located just behind Reret (see the Panel IX in Fig. 4.49). In various celestial diagrams, Reret is followed by a representation of the goddess Isis which, in this case, is not unequivocally equivalent to Sopdet (Sirius). However, there is another very bright pale white celestial object in that region of the sky that would meet our expectations. Therefore, we propose that the seated Lady is Vega, the brightest star in the northern celestial hemisphere.

To complete the diagram (see Panel X), we still need two additional figures: a miniature jackal located just below the Lady and a strange looking bird marching just behind Sah. Regarding the first, it is suggested that this Jackal is another atypical parazodiacal constellation. The *Mul-Apin* tables include two possible alternatives: it could be either Ur-Dim, the Mad Dog, located in Mesopotamian texts to the left of the Scorpion, which is often identified with Lupus, or it could be Ur-Ku, the Dog, located south of Hercules, a position that would produce a reasonable fit with the position of our figure in the diagram.

Regarding the Bird that walks behind Sah, we speculate with the idea that it is the first representation of the constellation Columba, the Dove. According to Allen (1963), the first planispheres with the image of Columba, always close to Orion, could be dated to the 17th century. However, Allen himself informs us that in the writings of Clement of Alexandria a Dove is mentioned in connection with the Ship (Argo). This could suggest that a constellation of the Dove had already been created in the centuries before this Christian apologist, and that this could be the bird represented in the zodiac.

Figure 4.58 summarizes visually our proposal in which the various groups, into which the images in the diagram have been divided for their correct identification as stars and constellations, are clearly differentiated by the use of various colours.

Fig. 4.58 A colour-coded summary of our analysis of the zodiac of Dendara, which may be considered as a working hypothesis. Zodiacal and parazodiacal constellations are represented in dark and light green, respectively. The most important traditional Egyptian constellations are represented in yellow. The planets are scattered among these and painted orange. The Bear of the Greek tradition (Ursa Major) is represented adjacent to Meskhetyu in pale yellow. The constellations of the Ramesside star charts are depicted in deep blue, including some of the zodiacal ones (e.g. *sbꜣwỉ* in Gemini) and the yellow-shaded Egyptian constellations (e.g *sbꜣ n sꜣḥ* and *sbꜣ n spdt*). The solar eclipse occurring in Pisces on March 7th 51 B.C. is depicted in dark brown. Satet and Anuket are painted in light and dark grey, respectively. The Small Dipper (Ursa Minor) can be found in deep purple, either including or not excluding the image of the jackal. Regulus and Vega are the seated king and lady painted in red, respectively. Finally, Columba, behind Sah, is painted fuschia. This proposal would explain nearly 90% of the images of the Zodiac. However, some problematic representations, painted in pink and light blue, purple and brown, respectively, are still quite controversial. These are extensively discussed in the manuscript. See the text for further details.

Our main conclusion is that the Dendara circular zodiac is not a precise description of the sky sculpted in stone but rather a highly symbolic representation of the sky where the zodiacal constellations would have been represented more or less realistically, followed by the also more or less realistic images of some traditional local constellations, in particular Meskhetyu and Reret. Around this basic template, the planets, other constellations, both parazodiacal and traditional Egyptian, fundamentally composed of the individual stars, asterisms, and constellations of the Ramesside clocks, would have completed the image of the 'firmament' in such a way that almost all of them filled empty spaces in order to avoid the dreaded *horror vacui*. Other images, sometimes difficult to ascribe, would fill in the remaining gaps.

This unique task, delivering a sound working hypothesis, is the culmination of our previous success in identifying classical Egyptian constellations (see Table 4.7), following a series of basic approaches such as: (a) the presence of mixed but convergent foreign traditions (Mesopotamia and Greece) along with Egyptian traditions; (b) strong symbolism; (c) the duplication of various celestial bodies; and (d) altering, often drastic, the position of some images in order to fill gaps.

With these basic premises, it has been possible to identify an important part of the internal area of the circular diagram of the zodiac. We hope that new discoveries, other intelligent alternative proposals or the more precise interpretations of the existing texts will help to unravel the remaining part.

Chapter 5
The Calendar of Ancient Egypt: A Gift of the Nile

After three decades of research in the field, the authors are often astonished of how difficult it is for some scholars of Egyptology, not to say to the numerous *aficionados*, novelists, and outreachers to comprehend the nature of ancient Egyptian calendrics. One often reads that the year started with the rising of Sothis in July (see below), and that the seasons followed the behaviour of the Nile. This, we imagine, could be related to the fact that during two of the most important periods of ancient Egyptian history, the Old and New Kingdoms, this was roughly, although not precisely, the situation. However, during such important periods of Egyptian history as the Middle Kingdom or the Low Epoch it was just the opposite. For example, such an important event as the heliacal rising of Sirius, or Peret Sopdet (*prt spdt*) happened in the middle of the year, nearly 6 months after New Year's Eve.

This misunderstanding is probably due to the difficulty in comprehending the wandering nature of the Egyptian civil year and, perhaps, that Egyptologists working on the astronomy of ancient Egypt, including the authors, have not taken all the necessary pains to explain this fascinating, but indeed troublesome, issue in an easy and clear manner.

Hence, our intention in this chapter is to show that the ancient Egyptians had just one calendar in operation, the so-called civil one, during most of their history and before the overwhelming influence of Hellenic culture (Belmonte, 2003a). This calendar may have been developed for a specific purpose in the first half of the third millennium BC, when hypothetical previous local Nile-based lunar calendars were rendered useless as the result of the unification of the country and the emergence of new social, economic, and administrative requirements. The civil calendar started from the onset and to the end of independent Egypt at the feast of *wpt rnpt* in the first day of the first month of the Inundation season (I *ꜣḫt* 1). We suggest that its peculiar length (of 365 days), might have been established from simple astronomical (presumably solar) observations and simple arithmetic.

A series of very important lunar festivals and cult activities related to the lunation, or synodic month, were articulated within the framework of the civil calendar. This means that ancient Egyptians followed a sort of lunar timekeeping system or

J. A. Belmonte, J. Lull, *Astronomy of Ancient Egypt*, Historical & Cultural Astronomy, https://doi.org/10.1007/978-3-031-11829-6_5

lunar computus, a parallelism to the Christian tradition of establishing the date of Easter. However, this does not necessarily imply the presence of an independent lunar calendar, such that existing in countries with a Muslim tradition. This civil calendar had a well-documented set of month names, numbering twelve, at least from the beginning of the New Kingdom, if not earlier, but several of these months altered their names, mostly in the Ramesside period or a little later, presumably for social or religious reasons.

This early contention is needed because, on certain occasions, Egyptology has been slow in reacting to advances in research, and calendrics has perhaps been one of the worst examples of this. The whole story started 70 years ago when the seminal work of Richard Parker, *The Calendars of ancient Egypt,* was published (Parker, 1950). This fascinating volume is a must read for any scholar wishing to enter this slippery field. However, it unfortunately includes the original sin, which has complicated the matter since then. This can be summarized in the following paragraph:

> Exactly when the second lunar year was introduced remains uncertain, but it was probably not too long after the divergence between the two forms of the year (civil and lunar) became apparent. A good guess might be to put it in the neighbourhood of 2500 BC From that date the Egyptians had three calendar years, all of which continued in use to the very end of pagan Egypt (Parker, 1950: 56).

We had to wait 45 years before Anthony Spalinger felt secure enough to make the following statement:

> With the removal of the lunar calendar from all of these scenes [he refers to diverse aspects where the lunar calendar had been identified], the importance of the civil calendar is stressed. At the risk of anticipating myself, I do not see the lunar calendar playing an important role in Egypt outside of some feasts. It is my contention that the Egyptian calendric system is simpler than usually maintained. ... The implication for a complete revision of Egyptian calendrics is evident (Spalinger, 1995a: 122).

Following in the wake of Spalinger's assertion, Belmonte (2003a) performed just such a revision and concluded that the civil calendar was probably the only calendar in full operation in Pharaonic Egypt from its inception in the early Dynastic period. This chapter is basically an updating of that still uncontested work.[1]

Unfortunately, the two paragraphs just cited are still quite representative of the state of studies of the ancient Egyptian calendar in recent decades, actual discussions being reduced to a handful of specialists seldom reaching the general public or the community of Egyptologists. Richard Parker's *Calendars* has reigned supreme in the view of a majority of scholars as the last word on Egyptian calendrical matters. This was so in spite of the fact that several of the ideas and hypotheses expressed in that book had been questioned. Hence, it is not uncommon to read in various manuals, either of Egyptology or of the history of astronomy and archaeoastronomy, that the Egyptians had three calendars working at once, with no

[1] Interestingly, Spalinger (2018: 88) has recently written, 'Lunar sightings still occurred. They were, however, now embedded within the new "rational" Civil calendar', thus confirming, apparently without knowing, Belmonte (2003).

attempt being made to enter into further discussion or to challenge that hypothesis (as happened to Belmonte, 1999: 146–154, 2000: 109).

This situation predominated until the mid-1990s when several scholars decided to enter this slippery field of research and started to produce new and interesting approaches to ancient information by proposing new interpretations in several cases, as well as publishing new material that had been previously unknown or never discussed in an appropriate way. The names of several of these scholars that should be mentioned in this context include Ulrich Luft (1992a), Leo Depuydt (1997, model qualified in 2009), Rolf Krauss (2002), and eventually Christian Leitz (1991), Ronald Wells (1994), and indeed Anthony Spalinger (1994a, 2002a, b), all of whom have been pivotal in recent decades and will be fundamental to our discussion.

We must also take into account the important review volumes by Marshall Clagett (1995) and Anne-Sophie von Bomhard (1999), whose works defend two different aspects. Clagett wonders how ancient Egypt timekeeping system would have worked (a sound position), and von Bonhard suggests that at least a second fixed calendar, connected with Peret Sopdet, should have been operating in the Nile Valley. As we now show, this double calendar is at least questionable.

After a detailed examination of a large percentage of the bibliography generated in the last decades (and several of the relevant earlier works), our main conclusion is that the study of the ancient Egyptian calendar suffers from what may be called the 'Ebers syndrome'. This may be said because the calendar written on the verso of the Ebers Medical Papyrus, discovered in Thebes in 1862 and first published a few years later, has contributed little, if anything, to the solution of any of the open questions on Egyptian calendrics in the last two centuries. It is important to note that from the moment of its discovery to the present, dozens of papers or book chapters have been published on it. The vast majority of these papers present attempts at interpretation in open contradiction with one another (see, Depuydt, 1996, for a review). There is even a recent article, forthcoming in *Orientalia* by Rolf Krauss, with an easy explanation for the calendar layout. This is a reaction to Spalinger (2018: 69), who interprets it as 'nothing more than a native Egyptian schematic survey of the star's passage [referring to Sopdet] through time'. Krauss tries to save the Sothic date value of the papyrus for chronologists, Spalinger does not.

The confusion that the Ebers calendar represents was first analysed by Belmonte (2003a), and the situation has not changed much since then. In that paper, a working hypothesis was proposed advocating reconsidering the situation by evaluating the idea of what might have happened if the Ebers Medical Papyrus had never been discovered—or if the calendar had never been written on the verso of the papyrus. This was a hazardous, problematic, and indeed questionable approach since the Ebers Calendar was the first in a series of important documental discoveries associated with Egyptian calendrics and chronology, and had hence always been one of the pivotal strands of evidence in every discussion on the subject. However, it is our contention still that, by ignoring it, the discussion of the Egyptian calendar becomes unexpectedly simplified, and that several apparent problems may easily find a fairly reasonable solution. This chapter is an updated, simplified, and illustrated version of that work.

5.1 One or Several Calendars?

Considering what has been early discuss, how many calendars were simultaneously in operation in ancient Egypt? This is the basic question to be answered and the most difficult one. Without finding a reasonable solution for this, it would be impossible to go any further in the analysis or solve other important problems and questions. In this chapter, we will defend the hypothesis that throughout Egyptian history, from the very creation of the civil calendar—in the 2nd and 3rd Dynasties—to the Roman conquest, the Egyptians had only one calendar. This would have been the civil one, in spite the fact that some of their feasts and the cultic activities in the temples could be established according to the moon. Indeed, under Persian and, above all, Macedonian rule (sixth to first centuries BC), the lunar calendars of these peoples might have produced a certain influence in local timekeeping systems (e.g. Bennett, 2011). These were also used to date documents or monumental inscriptions in either Aramaic (Depuydt, 1995a) or Greek (Spalinger, 1992), respectively.

The civil calendar of ancient Egypt consisted of 12 months of 30 days each, grouped into three seasons (see Figure 5.1), which from the very beginning received the names of *ꜣḫt* (Akhet, translated as Inundation in Greek texts), *prt* (Peret, translated as Winter) and *šmw* (Shemu, translated as Summer). Every season had 4 months. Each divided into 3 decades (*mdw*) of 10 days, at least from the Old Kingdom. This 12-month year amounted to a total of 360 days.

Normally, a civil date is expressed by a number for the regnal year (each reign represented a new era in ancient Egypt), followed by a set comprising a Roman numeral for the ordinal month within a season (which are similar to the actual Egyptian strokes for the numbers), then the season proper and finally the day of the

Fig. 5.1 The three seasons of the Egyptian calendar, *Akhet*, *Peret* and *Shemu*—or *ꜣḫt*, *prt* and *šmw*—as beautifully represented in a door-jamb at the gate of the tomb of the vizier Mereruka in Saqqara. The '5 above the year' are absent. (Photograph by the authors)

month. One example is III *3ḫt* 26, which reads as the 26th day of the 3rd month of the Inundation. However, we shall argue later on, following Griffith (1909: 185), that we should probably read 26th of *ḥwt-ḥr* (Hathor, or Athyr in Greek papyri), exactly as we write 29/02/2020 but we read 29th of February of 2020, for an additional day in leap years (something ancient Egyptians never developed before Macedonian rule).

Five extra days known as the 'Five above the Year' (5 *ḥryw rnpt*), or epagomenals, were added to the end of this year of 360 days throughout most of Egyptian history, although there is some evidence that they might have been placed at the beginning in earlier times (Clagett, 1995: fig. III.81a). However, this discussion is probably nonsense since, as Spalinger (1993a, 1995b) has lengthily argued, they were considered as outside the year and were not taken into account at all on most occasions (see Fig. 5.1). Furthermore, these 5 days have often been assigned a nefarious character, despite the fact that, thanks to the god Thoth who managed to get those extra days, the 'Birth of the Gods' (these were Osiris, Horus the Elder, Seth, Isis, and Nephthys) took place during them. With the 5 epagomenals, the civil year completed 365 days, a number very close to the length of the tropical year (the year of the seasons) which in 3000 BC lasted for some 365.2425 days.[2] Consequently, the civil calendar has frequently been quoted as the Egyptian 'solar' calendar, which, to a first approximation and in one generation (*c.* 25 years) would be a very appropriate term, as Herodotus (II, 4) recognized.

One peculiarity of the civil calendar, however, is that no days were added to the standard 365 (as we do in leap years). Hence, the civil dates of the calendar wandered through the seasons, completing a circuit in nearly 1507 years. As we have previously argued, this is certainly an exceptional aspect, which has confused generations of students and scholars. Egyptian skywatchers, or *imy-wnwt* (see Chap. 2) most probably knew that the tropical or seasonal year was almost a quarter of a day longer but they never did anything to resolve this, apparently, annoying situation.

Interestingly, according to Nigidius Figulus, a first century BC Roman grammarian, it was reported in the Aratea (Lull, 2004; Aveni, 1989; Krauss, 1985, 2011; Allen, 2005),[3] that the Egyptian kings were obliged to swear at coronation that they would never tamper with the calendar, thus keeping the year (*rnpt*) and cosmic order (*m3ʿt*) in perfect agreement (see Figure 5.2). Consequently, important astronomical events, such as the stations of the sun (solstices, certainly known by ancient Egyptians, and equinoxes) and the heliacal rising or settings of stars would also wander throughout the civil year, there being roughly a one day's delay every four years and one month's delay every 123 years.

It is one of the immutable bases of Egyptian chronology that no reform was performed (although it was attempted notwithstanding the prohibition) so that the

[2] The tropical year varies with time. Today it has a value of 365.2422 days.

[3] *... et jurejuranddo adigitur neque mensem neque diem intercalatum, quem in fesstum diem immutarent, sed CCCLXV dies peracturos siccut institutum est ab abtiquitis* ('and it is justly obligatory that neither a month nor intercalated day, that on a feast day be changed, but that 365 days be completed, as long ago decreed'), Germanicus (Aratea, 285).

Fig. 5.2 Images of the twin goddesses Ma'at and Renpet together in a beautifully carved relief of the temple of Sethy I in Abydos. The Egyptian kings swore at the moment of their accession not to alter the year (*rnpt*) as one of their duties to keep *mꜣꜥt* on earth. (Photograph by the authors)

seasons wandered throughout the civil year, from the creation of the calendar, in the early third millennium BC to Augustus' reform and the creation of the Alexandrian calendar in 23 BC. The consistency of this wandering up to the New Kingdom has been proved by Depuydt (1995b) and is widely discussed by Krauss in many of his works (e.g. Krauss, 1985, 2002).

In this respect, a special case is that of Sirius (Egyptian *spdt,* Sopdet, Sothis in the Graeco–Egyptian context), the brightest star in the sky then and now. The high proper motion of this star, owing to its proximity to the earth (*c.* 9 light-years), caused the Sothic year (i.e. the period of time between, for example, two successive identical position of the star in the earth–sun reference frame) to be almost exactly 365.25 days throughout most of the history of ancient Egypt. This meant that the important phenomenon of the heliacal rising of Sirius moved forward by one day in each four-year period, a quadrennium (e.g. from II *3ḫt* 3 to II *3ḫt* 4). This phenomenon was known as *prt spdt,* certainly from the Middle Kingdom and perhaps earlier in the late Old Kingdom (Habicht et al., 2015). Peret Sopdet wandered through the entire civil year in 1460 years (a Sothic cycle or Great Year). Actually, Ingham (1969) has shown that two whole slightly shorter Sothic cycles elapsed, the first of 1454–1456 years (from *c.* 1315 to 2770 BC) and the second of 1452–1453 years (from *c.* AD 139 to 1315 BC), according to modern calculations, i.e. during most of the Pharaonic period.

As the opening paragraphs of this chapter have shown, Parker (1950: 56) strongly supported the idea that there were three calendars in operation at the same time in ancient Egypt. The first and simplest would be the civil calendar, as already discussed. There is no argument about this. However, problems begin to arise when the other two calendars are considered.

First, there was the 'old' lunar calendar. This, according to Parker, would have been the original calendar prior to the development and inception of the civil one. This timekeeping system would have worked like a lunistellar calendar, consisting of 12, or occasionally 13, lunar months of 29 or 30 days, heralded by the heliacal rising of Sirius. This calendar is often referred to in literature even today. According to Parker (1950: 24–30), *prt spdt* had to be named *wpt rnpt,* a long-standing error (Clagett, 1995). This lunar calendar would start by the following conjunction or new moon (*psḏntjw*) after Sirius' heliacal rising. This New Years's Eve at *neomenia* would be called *tpj rnpt.*

Then there was the 'new' lunar calendar. This would have been invented when the divergence between the civil and the former lunar calendar had become evident. A new set of 12 or 13 lunar months was attached to the civil calendar, heralded by the 1st day of the civil year, I *3ḫt* 1, also called *wpt rnpt* in a secondary use of the term, according to Parker. However, this is almost certainly its actual significance.

These ideas—or rather hypotheses—had several weak points. However, they were rarely questioned for decades and we had to wait until the 1990s for a serious threat to most of Parker's arguments from various scholars such as Luft (1992b), followed by Clagett (1995: 31), Grimm (1994: 439), and Spalinger (1995c). The last of these scholars had even reached the point of questioning the independent operation of any sort of lunar calendar in ancient Egypt, which as he wrote, would make

Egyptian calendrics *simpler than usually maintained*, (Spalinger, 1995a, b, c, d: 122). But it was only recently that he fully rejected their existence (Spalinger, 2018: 88).

However, Parker's views (especially regarding the existence of the 'new' or civil-based lunar calendar) received a breath of life from the work of the Flemish scholar Leo Depuydt, now at Brown University, where Parker spent much of his career. His erudite book *Civil Calendar and Lunar Calendar in Ancient Egypt* (Depuydt, 1997) is a stupendous reference for an understanding of what has been going on in the study of the Egyptian calendar since the nineteenth century, in spite of the fact that we may disagree with some of his conclusions (still subscribed to in Depuydt, 2017). It is worth noting that, in his most recent works, this scholar has introduced the term lunar timekeeping system as an alternative to the lunar calendar (Depuydt, 2007). We are perfectly happy with the use of the term lunar timekeeping system or *lunar computus*, while still rejecting the existence of an articulated independent lunar calendar.

The simultaneous use of more than one calendar by a certain culture is not an unusual phenomenon. Furthermore, it is also a well-established fact that, once a good calendar has been developed, it is fairly complicated to reform it and still more difficult, if not impossible, to abandon it. Hence, we would not in principle be against the simultaneous use by the ancient Egyptians of more than one calendar, although in our context and with the evidence at hand, we consider it to be unlikely.

In contrast, the use of two calendars simultaneously is very common in the Muslim world today. In fact, we shall illustrate the situation by reference to two countries, each with a long history, facing each other across the Mediterranean Sea: modern Spain, the homeland of the authors, and Islamic Egypt. In Spain, Roman tradition reigned supreme for several centuries and hence all aspects of daily life were and still are governed by the Gregorian (or, prior to the sixteenth century, Julian) calendar. However, Christian tradition is also dominant and several religious feasts move according to a lunar *computus*. The most important is Easter, which is fixed as the first Sunday after the first full moon (which must not be a Sunday) after the Spring Equinox. Once Easter is fixed, the dates of Ash Wednesday (with Carnival), Palm Sunday, Good Thursday, Good Friday, Pentecost (Pinkster) and Corpus Christi are accordingly established.

Obviously, there are what we may called moon-controlled feasts, but there is no lunar calendar in parallel operation with the Gregorian; indeed most Spaniards—and Westerners in general—do not concern themselves with the behaviour of the moon. Can the reader, for example, confirm what the phase of the moon is today while reading this paragraph?

However, the situation in traditional Islamic Egypt was quite different. The Muslim calendar governs the religious and social life of the majority of the people but, as a purely lunar calendar, it is quite unsuitable for governing the economic life of the country. The reason for this is that the Muslim calendar is a purely lunar calendar of 12 lunar months and is thus 11 days shorter than the tropical year, rendering it quite unsuitable for an agrarian culture. Consequently, in every country converted to Islam, it has been almost mandatory to keep previous older calendars in use for practical purposes.

In traditional Islamic Egypt, this task was the responsibility of the Coptic calendar, the direct descendent of the Alexandrian one, the corrected version—to include leap years of the civil calendar (Depuydt, 1995b). It is worth noting that the Gregorian calendar has mostly substituted the Coptic calendar today. Indeed, in Egyptian society, whether modern or traditional, one calendar cannot work without the other—although in our experience the Muslim calendar is followed only for festivals such as Aïd el Adha or Ramadan. In this case, the obvious result is the need for two completely independent calendars, one solar and the other lunar, with different dates, different beginnings, and different month names operating at the same time. The point of view we defend is that the timekeeping system of the Egypt of the Pharaohs was closer to that found in modern Spain than to that found in traditional or even modern Egypt.

Not surprisingly, a similar argument for the necessity of operating two calendars in Islamic Egypt has frequently been claimed as the reason for the existence of more than one calendar in ancient Egypt, especially the 'earliest' ones developed, the old Sirius-based lunar and the civil one. This idea rests on the supposed inability of the civil calendar to follow the seasons because it follows a vague (wandering) year. For example, von Bomhard (1999: 8) associates the term *rnpt gbt* (upset year) in the text 'Come to me, oh Amon! Save me of this upset year. It happened that the sun did not rise, that winter arrived in summer, month follows month in the wrong order, the hours are disrupted' with the 365-day year, thus leaving the impression that the ancient Egyptians considered their civil calendar to be unsuitable.

However, this argument is aprioristic and fallacious. Firstly, this text does not apply at all to the civil calendar, since only the sentence 'winter arrived in summer' might be applied to it—and certainly not in the context of the time when this paragraph was written, in the early Ramesside period. Besides, it is completely unfair to argue that the civil calendar was useless in daily life, and that the Sothic heralded lunar calendar was necessary to keep the correct track of the seasons.

In fact, the civil calendar was ideal for daily life. It diverged by only 10 days from the tropical year over a period of 40 years (an average human life-time in Antiquity). Hence, considering that the most important natural seasonal phenomenon in Egypt, the rising of the Nile, can vary by as much as 70 days (see Figure 5.3), the civil calendar would have been perfectly capable of handling local agriculture and even state administrative policy, over very long periods of time.

This was also clear to the father of history, who wrote:

> Their calendar is in my opinion better than that of the Greeks, because these introduce an intercalary month every two years, in consideration of the seasons, whereas the Egyptians, with their twelve (12) months of thirty (30) days, add five (5) supernumerary days to each year; so that the cycle of the seasons always appears at the same date for them. (Herodotus II, 4)

That the harvest occurs in the season named *ꜣḫt,* rather than in the season named *šmw* (often translated by inference as 'Harvest'; later we will discuss this particular problem of the season names), as occurred in the early Middle Kingdom. So what? We have a month with a name that means '7th' and is our 9th month and nobody

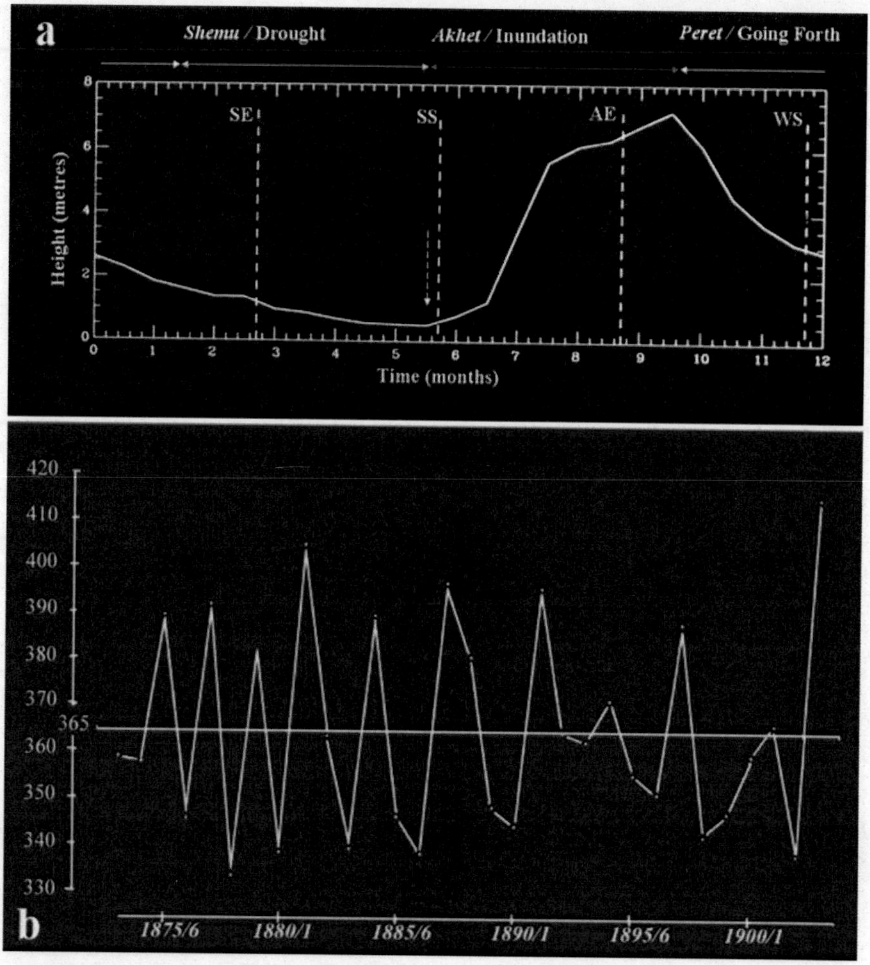

Fig. 5.3 'Egypt is a gift of the Nile'. Without the river and its inundations there would have been no Egyptian civilization. The river controlled the economy of the country with its different ecological phases: the flooding or inundation often starting close to the summer solstice (SS), followed by the lowering of the waters four months later, allowing the land to be cultivated, and finally the almost four-month drought period, when the level of the waters was very low. The time-period assigned for the seasons in the upper panel of the diagram would work for the period of the creation of the calendar (**a**). In principle, the average period between successive arrivals of the flooding season should be that of a tropical year. However, periods as large as 410 days or as short as 320 days were reported in the historical record during the Nineteenth Century (**b**). This was certainly the case in ancient times. Bottom graphic adapted from Neugebauer (1938)

seems unduly perturbed about that. Besides, as we argued before, this problem did not happen in the period of maximum prosperity during the Old and New Kingdoms, when festivals (and the calendar itself) were either created or substituted by new ones, respectively.

There is a further argument to raise against the existence of a well-structured original lunar calendar as proposed by Parker. It is probable that, if the Egyptians had ever developed such a good timekeeping device as an 'old' lunar calendar (once the site of observation of Peret Sopdet had been agreed upon), they would never have needed (and thus invented) the civil one. The lunistellar nature of this old calendar, having dates in harmony with the seasons, would have made the implementation of yet another calendar unnecessary, and even undesirable according to *mꜣꜥt*.

However, it is worth mentioning that several pages in *Calendars* (Parker, 1950: 30–50) are devoted to explaining the operation of this 'old' (Sirius-based) lunar calendar and defending its existence. Today, it is quite clear that most of his arguments were inaccurate. Clagett's (1995: 31) concerns are categorical, when he argues:

> In brief, it appears to me that Parker's opinion that the old lunar calendar was intercalary may be correct (though not certainly so) but that (1) the use of the Sothic heliacal rising as the mechanism of intercalation, that (2) the intercalary month (if it existed) was named Thoth, and that (3) the lunar calendar in schematised form is that given in the Ebers calendar and in the astronomical ceiling of Senenmut's tomb and the Ramesseum are all unproved and indeed untenable.

To this, we could add that the identification of *prt spdt* with *wpt rnpt* is nowadays completely dismissed (see below). Actually, the only support today for the old lunar calendar seems to come from the Ebers Calendar, a sad example of the syndrome (see Epilogue).

Moreover, another of the classical alleged proofs, the calendar of supposed lunar months of Papyrus Berlin 10056A briefly following the heliacal rising of Sirius found in the Illahun archive, has been recently challenged by Luft (1992a, b: 198) and Krauss (2003). These scholars have demonstrated that the document shows periods not from lunar day (hereafter LD) 1 (*psḏntjw*) to LD1 (i.e. a lunar month in Egyptian mentality as suggested by the name of LD6), but rather from LD2 (and even LD3) of the Egyptian lunar month; they might thus be better described by the term *wrš* (moon period service). More recently, Krauss (private communication), defending its lunar character, has argued that 'the Illahun lunar year may be interpreted as a standard calendar, used and modified for a temple roster', but, as we see the problem, a modified lunar timekeeping system does not necessarily mean a fully operative, monthly expressed, lunar calendar.

When all this evidence is taken into consideration, we could even argue against the use of the term 'lunar month' in ancient Egypt (with the exception of the archaic and perhaps Ptolemaic periods; Bennett, 2011). We could instead, following Belmonte (2003a), restrict the use of the term lunation to any period of time between two consecutive identical phases of the moon, be it conjunction, first crescent, full moon, LD1, LD2 and LD15, respectively, or any other aspect. This will leave the term 'month' only for the well-attested months of the civil year. This policy will be followed in the rest of this chapter. However, it is worth noting that the Egyptians had a certain sense of order in the days of the lunation since, as will be shown later, they had clear numerals to name LD6 and seldom LD15.

Indeed, to be able to continue with this proposal, we must first show that the other alleged lunar calendar, the civil-based one, was not expressed as an actual calendar, and that the term lunar month is consequently not needed. The reason for justifying the later invention of the new lunar calendar is less clear than in the case of the Sirius-based lunar calendar (which theoretically even predated the civil one), and those defending its existence apparently argue for the correct placing of lunar feasts within the framework of the civil calendar (e.g. Depuydt, 1997 and 2009). However, to have lunar-governed festivals within a calendar does in itself not imply the need of an articulated lunar calendar.

We described above how in Western Christian countries there is a large collection of festivals (religious and profane) dictated by a lunar computus but nobody in this cultural tradition would claim that there are two calendars(a solar and a lunar one) in simultaneous operation. This is so even if the Metonic cycle is used to correlate the phases of the moon and the tropical year, but again a lunisolar cycle does not make a calendar. A similar argument could be applied in the case of those arguing that the Egyptians' knowledge of the 25-year cycle (as found in Papyrus Carlsberg 9) in the Graeco–Roman period supports the idea of the civil-based lunar calendar. In Papyrus Carlsberg 9, the dates of the beginning of 300 lunations in 25 civil years are presented (out of the 309 possible lunations—those starting in the epagomenals and those belonging to 'blue' months are not considered), according to Depuydt's (1998b) clever and simple interpretation of the Papyrus. 'Blue' months are those civil months where two synodic months can start at the same time within a certain year, one in day 1 and the other in day 30.

Only in the case that the civil-based lunar calendar had left unmistakable signs of a complete operative system might we be confident of its existence. This is in fact what the otherwise excellent volume of Depuydt (1997) had tried to show, but was challenged in a long discussion on the topic by Belmonte (2003a: 14–17). There, the first author presented some hypotheses, offering an alternative, simple and plausible explanation for the controversial 'undisputed' double 'lunar-civil' dates reported so far. These were civil II *šmw* 13, equivalent to the full moon of I *šmw,* and civil IV *šmw* 18, equivalent to the second lunar quarter of III *šmw*, respectively (Parker, 1957a; Brugsch, 1883). This would avoid the necessity of resorting to the development of a new civil-based lunar calendar.

In considering the presumed 'design' of this civil-based lunar calendar there are in principle two possibilities for these dates (Belmonte, 2003a): yearly or monthly pairing. The former means that 'lunar' I *ꜣḫt* would start at the first *psḏntjw* (conjunction or new moon) of civil I *ꜣḫt* and then the 'lunar' months would continue their counting consecutively until the end, even in the presence of a 'blue' month. The second means that, if a blue month is present, the second lunation within that civil month is not counted and it is the following lunation, starting in the appropriate month, that receives the name. This poses the problem of how this unnamed synodic month would be counted; hence, this second possibility is completely abnormal and does not define a functioning calendar anywhere on earth. Interestingly, this is exactly the situation of the two dates under discussion, since the 'lunar' months are named according to monthly pairing. If yearly pairing had been applied, the months would have been II *šmw* and IV *šmw*, being exactly the same as their civil counterparts, and thus further complicating the situation.

It is worth mentioning that there would have been a series of actual and definitive proofs of the existence of the civil-based lunar calendar. For example, one determinant proof would have been established if we had found two 'lunar' months mentioned consecutively, both with the same name and simply distinguished by a numeral (e.g. IV *prt* and IV *prt sn.nw*), as in the Babylonian calendar, when a full (13 lunar month) year is present. Although, in Babylonia, this is always found at the middle (Elulu II) or at the end (Adaru II) of the year and not in a random position within the year. Further unmistakable evidence would have been a report of a proper 'lunar' month starting in one of the *ḥryw rnpt* (the straddle month in the five cases out of 309 when this happens) and the specification of this peculiar lunar month's name. However, neither of these two cases has ever been reported in the sources!

As we have seen, the proposed civil-based lunar calendar leaves itself open to a number of serious challenges such as the use of the illogical monthly pairing and the open question of what would have been the name of this second lunation in a blue month, or on an epagomenal day. To summarize, the existence of this calendar is very unlikely.

To close the question of how many calendars were operating at once in Pharaonic Egypt, we shall deal with an argument that is often invoked as a proof of the existence of two calendars operating simultaneously in the Nile valley (e.g. Neugebauer, 1942). This is the mention of a great (*ˁ3t*) and a small (*nḏst*) year in one of the lists of feasts in the tomb of the provincial governor Khumhotep II at Beni Hassan (dated in the 12th Dynasty). The list is presented in Figure 5.4 and the two 'years' are presented in the second row of the text.

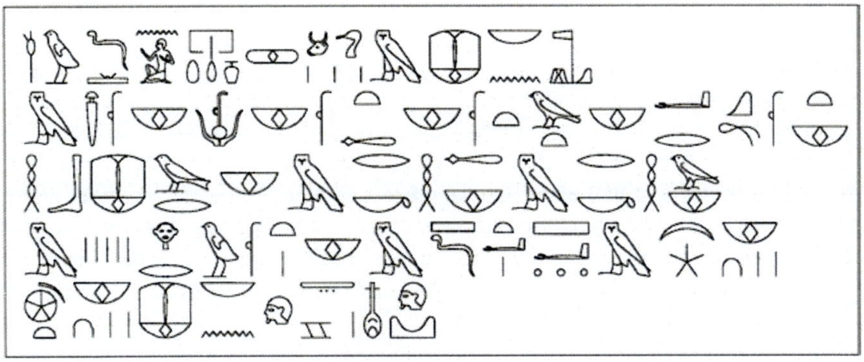

Fig. 5.4 Hieroglyphic inscription of one of the festival lists of the tomb of Khnumhotep II at Beni Hassan. In the second row, the two 'beginnings' of the year (*tpj rnpt* and *wpt rnpt*) and the large and small year feasts (*rnpt ˁ3t* and *rnpt nḏst*) are reported. Note also the existence of the two Burning festivals (*rkḥ ˁ3* and *rkḥ nḏs*) in the following row, the five epagomenal days (5 *ḥryw rnpt*) and the 12 first crescent (*3bd*) and full moon (*smdt*) lunar festivals associated with the civil calendar of 12 months. All these items play an important role in our discussion. Interestingly, *prt spdt* and *psḏntjw* are absent of the list

The hypothesis is that the 'great year' would be associated with the civil (or solar) year of 365 days or, maybe, with a full lunar year of 13 lunations of 384 days, and the 'small year' with a hollow lunar year of 12 lunations of 354 days. This would be supported, as first noticed by Neugebauer (1942), by the fact that 16 great years and nine small years are mentioned in Papyrus Carlsberg 9. This is exactly the correct ratio of 12- and 13-month years in the 25-year cycle presented in the papyrus.

The papyrus was produced in Tebtunis sometime after Antoninus' year 7 (AD 144/145). The main issue is that it possibly reproduces a simple arithmetic rule applied to the 25-year cycle and not a calendar according to Depuydt (1996b). Besides, there is a whole 2000-year gap between it and Khnumhotep's list, without any other parallel case being documented in that long period of time. Actually, Spalinger (1996b: 38) completely rejected that assumption and once more proposed a 'simpler' explanation: the 'small year' and 'great year' rather refer to the short civil year of 360 days, frequently used in many aspects of Egyptian society, and to the full year of 365 days, including the 5 *ḥryw rnpt*.

However, an overriding issue is that both *rnpt ꜥ3t* and *rnpt nḏst* are in a list of festivals and both have the determinative of festival close to them (see Fig. 5.4), so they must refer to specific feasts (either one day or a short-period term) and not to a period of one year considered an entire feast. This might already be expressed by the closing formula *ḥb nb*, 'each feast' in the last row of the inscription.

All in all, after considering several options, we must agree with Anthony Spalinger and a few others that Egyptian calendrics must have been much simpler that is usually maintained. Hence, we endorse Belmonte's (2003a) hypothesis, and accept that one—and only one—calendar would have been in operation in Egypt from the creation of the civil calendar in the first half of the third millennium BC to the reform of Augustus and the imposition of the Alexandrian calendar after 23 BC. This important device was, of course, the civil calendar. Obviously, lunar feasts and a lunar *computus* were taken into account for various purposes (mostly ritual and religious), but they were often, if not always, connected to the civil calendar, exactly as they are in present Gregorian calendar, and never as independent features of any alternative calendrical system.

For obvious reasons, this hypothesis poses new questions regarding the 'various' lists of month names, some of which had been assigned to one or other lunar calendar and to the various 'beginnings' of the Egyptian calendar 'year'. We deal with these open questions in later subsections. However, prior to that, we have to deal with perhaps the most mysterious and fascinating of all the problems that we must consider: why, how, and when the Egyptians developed such an unusual 365-day calendar.

5.2 The Solar Origin of the 365-day Calendar: A Working Hypothesis

There must have been a powerful reason for the archaic Egyptian society to invent a calendar that, in the words of Otto Neugebauer, was 'indeed the only intelligent calendar which ever existed in human history'. However, other clever systems of timekeeping were invented in other places, as for example the Maya long count and the associated circular calendar (e.g. Aveni, 1989). What is certain is that it was one of the most efficient that had ever been developed, surviving in human society since its implementation in the first half of the third millennium BC (see below), until the Julian Date was invented for continuous time keeping and astronomical calculations in the Sixteenth Century by the astronomer and chronologist Scagliero.

However, Neugebauer (1942) also pointed out that 'every theory of the origin of the Egyptian calendar which assumes an astronomical foundation is doomed to failure'. As trained astronomers and Egyptologists, we regret such an affirmation because the vast majority of cultures worldwide and throughout history have used observation of the heavens as the most reliable way of controlling and keeping track of time, leading eventually to the creation of a calendar. To our knowledge, there is no calendar on earth that would have not have had a certain connection to the sky, be it lunar, solar, stellar or a combination of the previous ones. The origin of the 365-day year is a highly controversial issue that is far from reaching consensus. Can we put forward a reasonable solution? If so, would it be unique?

Since the beginning of Egyptology, several theories have been proposed, discussed, and established on a certain basis, with more or less success, and almost every specialist in the topic has tried to postulate his own hypothesis. In this context, the most reasonable have been:

i) A Sothic origin, 365 being the average value of days between two successive heliacal risings of Sirius. This has been defended by several scholars since the very beginning of Egyptology (even before the decipherment of hieroglyphs), because several classical sources associated the rising of Sirius with the beginning of the Egyptian year (e.g. Censorinus XXI, 10). Considering the length of the Sothic cycle to be 1460 years, the inauguration of the civil calendar would have been around 4241 BC according to nineteenth century scholars, or around 2781 BC once the origin of Egyptian history was brought by Meyer (1904) to nearly 3000 BC in the early twentieth century. Spalinger (2010 and 2018: 68), following Berlev (1999), has recently championed this hypothesis.
ii) A solar origin, based on the determination of the period of time between two successive repetitions of the same station of the sun, either a solstice, a zemith pass or, less likely, an equinox. This was never seriously considered (the previous hypothesis was more popular), but it was defended by Sethe (1919–1920: 38) and Gardiner (1906) and has more recently been revived by Wells (1994).
iii) A lunar origin, based on an average lunar month and an average lunar year. Parker (1950: 51) was the champion of this hypothesis, based on his defence of the original lunar calendar heralded by Sirius.
iv) A Nilothic origin, based on the average value of the interval of days between successive Nile risings. Neugebauer (1938 and 1942) was the pioneer of this idea.

The first two were astronomical and both were discounted by Neugebauer (1942), because the civil year was almost a quarter of a day shorter than the Sothic and tropical years, and it would have been very clear from the very beginning—after just 8 or 12 years had passed—that the phenomenon would not be repeated within a period of 365 days. This is not completely true! When the Julian reform was applied to the Roman calendar, leap years were to be included 'every four years'. However, because of the Roman inclusive way of counting numerals, this was made effective every 3 years. The problem lasted and, in fact, was not corrected until 8 BC by Augustus [all leap years between 8 BC and AD 8 were suppressed], some 36 years after the reform had been introduced (González-García and Belmonte, 2006).

As a matter of fact, the actual problem with the Sothic origin (i), as often stated by Krauss in lectures and private conversations, is that it is difficult to establish clearly that 365 is the number of days between two successive heliacal risings of Sirius. On the one hand, one thing that is frequently forgotten is that this celestial event is highly dependent on latitude (with differences of nearly 8½ days for 2800 BC and nearly 6 days for 200 BC between the First Cataract and the Delta, respectively (see Leitz, 1992) and, notwithstanding, on atmospheric conditions. These effects could have introduced variations of several days in observations of the phenomenon from different locations along the Nile valley.

On the other hand, even if we admit that we are faced with a well-defined place of observation (e.g. a temple in the country's capital, Memphis, or a very important sanctuary in a relevant location such as Buto or Elephantine, or even Thebes in the New Kingdom) and a perfect atmosphere, we have yet another problem. Imagine that you are observing the rising for three consecutive years, there would be four possible sequences of observations:

365	365	365			
	365	365	366		
		365	366	365	
			366	365	365

So, only in 1 out of 4 possibilities (25%), would we have reached the unequivocal value of 365 days. As a consequence, it is highly improbable that the heliacal rising of Sirius was the phenomenon observed to establish the length of the civil year as 365 days.

However, based on the rare mention of a '*wpt rnpt* of the ancients' in I *ꜣḫt* 9 in a feast calendar at Esna, Berlev (1999) has suggested that this date (a lunar one in origin) would represent the first observation of Sirius' heliacal rising in 2767 BC, the one inaugurating the civil calendar in year 18 of King Djoser. Spalinger (2010, 2018: 68) seems to endorse this hypothesis. Leaving aside chronological problems for the reign of Horus Netjerkhet (the dates are too high, see Chap. 7), there is no indication of how observations might have been performed. There is also a huge gap of *c.* 28 centuries between the Esna calendar (Roman Era) and the epoch of the creation of the calendar. Could a tradition have survived for so long without intermediate references—leaving aside the Ebers calendar? It could have done, but we doubt it.

The first author has had an interesting epistolary interchange with Rolf Krauss about this topic. The idea was that if I Akhet 9 served as the New Year's Eve of the ancients we would not simply be confronted with confusion where an alternative simplified writing choice between *psḏ* ('nine') and *psḏntjw* is produced. Hence, what Esna would mention as the New Year's Eve of the ancient was the original location of *wpt rnpt* at day I *ꜣḫt* 1 of an ancient (pre-civil) lunar calendar starting with conjunction at I *ꜣḫt* (see below). Krauss completely rejected this possibility.

Regarding the stations of the sun, 40 years of observation must have made it obvious that the year fell short by 10 days. But, did the Egyptians ever observe the solstices or the equinoxes? We shall return to this question later on (see also Chap. 6). For now, it is worth mentioning that it was proposed almost a century ago that the Egyptian year once started at the summer, or even the winter solstice—Gardiner's (1906) so-called Misore year. More recently, Wells (1992) produced a new, and in certain aspects too elaborate, theory. This relates to the mythology of the birth of Re by the sky goddess Nut and the development in Lower Egypt of a lunisolar calendar heralded by the rising sun at the winter solstice. There are, however, no traces of such a calendar in any historical register. Simultaneously, a lunistellar calendar, heralded by the heliacal rising of Sirius (Parker's old lunar calendar version, see below) would have evolved in Upper Egypt. The civil calendar would have been the outcome of an amalgamation of both calendars around 3250 BC. As we later show, the Palermo Stone (Wilkinson, 2000b; Nuzzolo, 2020) demonstrates that there was no organized calendar in Egypt at such an early epoch.

To avoid Neugebauer's (1942) categorical judgement, Parker (1950: 30–50) proposed instead an astronomical approach, but one that would not suffer from the gradual shift within the year or, if indeed it did suffer such a shift, would not be immediately apparent and thus easily detected. He then arrived at hypothesis (iii).

The bases of his theory were threefold: (1) the presence of an already well-regulated and working lunar calendar (the lunistellar one related to Sirius); (2) the dependence of the civil calendar not on Sirius but rather on some event that was variable in itself or that in the forward shift of the year would not be immediately apparent; and (3) the requirement that the theory include an explanation of why the later lunar calendar tied to the civil one was inaugurated.

Parker then argued that there were economic disadvantages of a lunar year of now 12 months, and now 13. We wonder why, however, becaus even today, billions of people in Eastern Asia do not seem to identify this disadvantage. He then proposed that a lunar schematic year of 12 months of 30 days was developed and that the extra 5 days would have been easily worked out by averaging the excess between the actual lunar year and the new 360 day year in a period of some 25 years. According to his proposal, he finally argued for a date of introduction of the calendar between 2937 and 2821 BC, presumably during the First Dynasty.

The idea is worth considering and it has received further support in Depuydt's works (Depuydt, 2007, 2009). However, we have argued before that if a culture had developed a precise and useful calendar such as the Sirius-heralded lunar calendar, they would probably never have felt the need to create yet another calendar. Another relevant fact is that there is no real hint, with the unfortunate exception of the controversial Ebers papyrus (*testis unus, testis nullus*), that this calendar was in operation at any stage of Egyptian history. Hence, premise (1) is unproven. Additionally, we have shown that there are serious doubts that the civil-based lunar calendar ever existed (premise 3). Hence, this left us only with point (2), the necessity of a variable event, as the remaining workable contour condition for any reasonable theory.

Neugebauer (1942) sought that event in the inundation of the Nile. Each year, the heavy monsoon rains on the Ethiopian Plateau produce the rising of the waters of the Blue Nile. By early spring the waters reach Khartoum and the White Nile, moving north at such a rate that they reach the first cataract and the traditional frontier of Egypt at Elephantine at the latest for the time of the summer solstice (Gregorian 21st of June) and Memphis some 10 days later. Actually, the phenomenon, although cyclic, is highly non-periodic and relatively unpredictable, with reported risings of the waters arriving at Cairo (or Memphis) as early as April 25 and as late as July 3. This means that, within two consecutive Nile risings, a period as short as 10 lunations or as long as 14 might have elapsed (see Fig. 5.3).

For the inhabitants of the Nile valley, the most important fact of their year would undoubtedly have been the arrival of the new waters, which would control the vegetative cycle and, consequently, the economic life of local societies. Indeed, the arrival of the Inundation (as the name of the first season might show, see below) may have acted as the herald of that calendar and as the point from which to start counting the cycle of moons. When the unified state was formed, it was perhaps necessary to unify the criteria for the entire country and a new calendar would have been developed.

For Neugebauer (1942), the simplicity of the Egyptian calendar 'was a sign of its primitivity; it was the remainder of prehistoric crudeness, preserved without changes'. He alleged this primitive astronomy also to be responsible for the 30-day length of the civil month. Hence, he proposed that an average Nile year could be established through observations of successive Nile risings at a certain place. As a first step, a crude average of 360 days (30 × 12) would have been selected and then the five remaining days would have been calculated by averaging the differences with 360 during a period (as with Parker's lunar) of nearly 25 years.

He pointed out that the civil calendar was developed for the necessity, in private and public economy, of determining future dates regardless of the irregularity of the moon (and of the Nile risings we may add), following a quite materialistic point of view. However, it is surprising that he had, at the same time, proposed a highly developed theoretical astronomy for the Babylonians, who implemented a working lunisolar system, arguing that the differences with Egypt can perhaps be found in the differences in social and economic structures of the two countries. These points can be criticized as aprioristic and gratuitous.

One suggestive fact concerning this hypothesis is that, considering the high variability of the arrival of the inundation (see Fig. 5.3), only after two or three centuries would this Nile calendar be considered as no longer accurate. This might happen when I *ꜣḫt* 1 was systematically ahead of the arrival of the inundation anywhere in Egypt. Consequently, he argued, the Egyptians were forced to adopt a new criterion for heralding the flood, which happened to be the reappearance (aka heliacal rising) of the star Sirius (i.e. *prt spdt*).

In the light of this last piece of evidence, the last point is quite feasible, as we discuss further in the following sections. However, Krauss (1985) has gone even further when dealing with the wandering problem when he states: 'a way out of this dilemma is the presumption that the 365-day calendar was intentionally planned and inaugurated as a calendar which moved forward through the seasons'. Although the seasonal displacement of the calendar should not be a problem under this clever ad hoc hypothesis, this solution is quite improbable and impossible to falsify, at least for the Old Kingdom.

Coming back to the Nile hypothesis, there are doubts about the possibilities of the averaging method over long periods of time (Belmonte, 2003a). To perform such a task requires not only a well-organized society deciding where and when the systematic measurements should be done (which may be probable), but also somebody, who would need to have had a long life, with the ability to look ahead and a capacity for prediction, and an ability to decide when the average was finally suitable. In this sense, curiously enough, the data used by Neugebauer (1938: fig. 3) between years 1875 and 1905 to establish his theory contradict his own reasoning. A person in charge of selecting the appropriate average would probably have chosen 366 instead of 365, since, for five years, between 1892 and 1896, i.e. only 20 years after starting the experiment (saving time is always important) an average value of an additional 6-day period to 360 would have been yielded.

Consequently, the discussion has left us with no uncontroversial theory for the origin of the civil calendar so far. However, there must be one at least! To support his theory, Otto Neugebauer felt obliged to ignore any other options, especially the most challenging for his purposes, the astronomical ones. Indeed, to produce a new reasonable theory is not an easy task and for this purpose, Belmonte (2003a) selected the best of the previous theories, used new data, and allowed himself a bit of speculation.

5.2.1 *The Role of the Nile*

A few years ago, when visiting the University of La Laguna for a graduate course, Rolf Krauss offered the attendants a completely new perspective for the Palermo Stone, the largest fragments of a stela with the annals of the first five Egyptian dynasties. Belmonte (2003a) had already noticed another relevant aspect of the verso of the stone related to calendrics. One such piece of evidence was the 100% probability that the 365-day civil year was fully implemented in the transition between the reigns of Sahure and Neferirkare (*c.* 2480 BC). Even so in the transition of Menkaure and Shepseskaf, three decades before (see Figure 5.5).

However, the most important lesson learnt was that Palermo contained a clue to the importance of the Nile year in ancient Egypt. Firstly, and most important, every year in these annals was identified not only by a special event occurring in it but also by the height reached by the waters of the Nile somewhere in the country—unfortunately we do not know where. However, it is in the first row of the annals of the

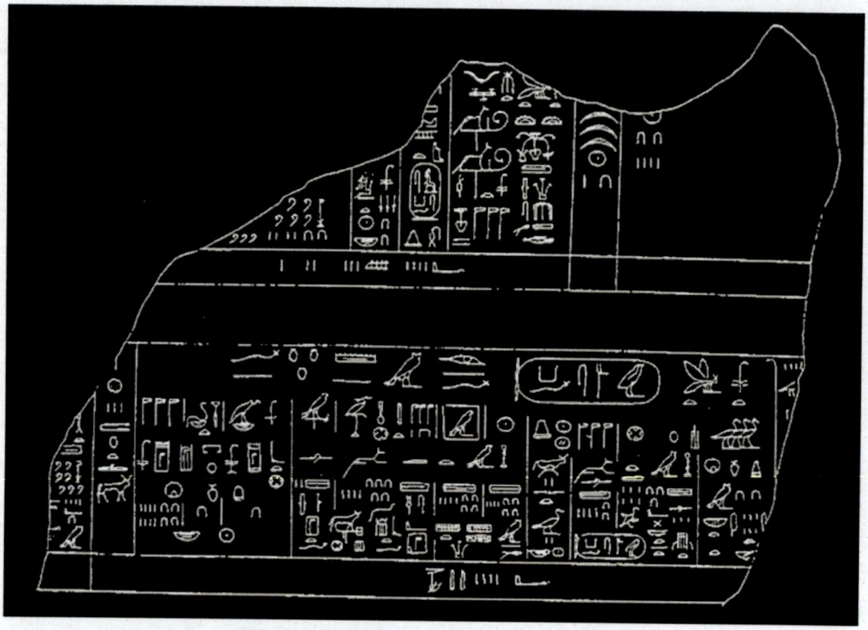

Fig. 5.5 A fragment of the annals presented in the verso of the Palermo Stone, representing the beginning of the reign of Shepseskaf (upper row) and a year of the reign of Userkaf (lower row). The upper row shows that Menkaure reigned x months plus 24 days in his last year, which was shared by $x + 3$ months plus 11 days of the reign of Shepseskaf. This is an indirect proof of the existence of the 30 × 12 + 5 (365) day civil calendar in that period. (Image adapted from Naville, 1903)

kings of the 1st Dynasty where the most relevant datum is found (see Figure 5.6). There, we have four consecutive 'yearly' registers, with the two central ones (2 and 3) including a certain number of months and days (6 months and 7 days, and 4 months and 3 days, respectively), but just one notation of the flood level below the third register. Moreover, between registers 2 and 3, there is the typical vertical stroke, separating two consecutive reigns.

Krauss interprets this fact (and the authors are inclined to agree with him) as reflecting a single year, straddling both reigns (perhaps those of Djer and Den) and amounting to a total of only 10 months and 20 days. Of course, this 'year' is too short for either a civil year of 365 days or even for the shortest possibility for any kind of lunisolar or lunistellar year (354 days). However, a year in which a period between 315 and 320 days has elapsed between two successive measurements of the lowest (or highest) level of the inundation may fit a Nile calendar perfectly. This is perhaps the best proof, not known to Neugebauer, of the possible existence of a Nile-governed calendar before the invention and implementation of the civil one.

Another fascinating piece of evidence lies in the names of the seasons. The three seasons: 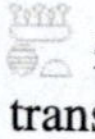 *3ḫt*, *prt*, and *šmw* (i.e. Akhet, Peret, and Shemu in the best accepted transcription) are often translated as Inundation, Winter or Growing, and

Fig. 5.6 A very peculiar and interesting fragment of the annals in the second row of the Palermo stone showing a year of reign change between two kings of the 1st Dynasty. The length of this year was only 10 months and 20 days, too short for a civil year and even for any kind of lunisolar or lunistellar year, but still within a reasonable limit for a very short Nile-year. See the text for further discussion. (Photograph by J. A. Belmonte by courtesy of the Palermo Museum of Arcaheology)

Summer or Harvest, respectively, on the grounds that this set of names would be clearly related to an agricultural year. However, as recently argued by Belmonte (2003a) and later by Depuydt (2007), these names could be considered as controversial; hence, this paradigm is far from secure. Actually, when the names were first discovered by Champollion (1842), the French scholar translated *ꜣḫt* as 'Vegetation' and *šmw* as 'Inundation' (see also Champollion, 1868). However, in the late nineteenth century, Heinrich Brugsch (1883) defended the theory, with several sound arguments, reasoning that *ꜣḫt* was the season that ought to be identified with the period of thc Inundation (and so translated), and since then any other readings have been considered almost a heresy.

But, as Depuydt (2007: 35) has stated, there are still some old epigraphic hints that would better relate *ꜣḫt* with the roots 'to be verdant' or 'to flourish' instead of the root 'to flood'.

Interestingly, we can deal with this dilemma by studying the ecology of the process followed by the Nile. On the one hand, when the waters started to rise, more or less at the period of the summer solstice (see Fig. 5.3), a formidable biological process occurred. Millions of micro-organisms, whose parents had been dormant for months in the waters of the river, blossomed as the level of the waters started to rise, changing the water colour from a pale blue-grey to a dark green. This would last for several weeks until the typical red waters, full of sediments from the Ethiopian Plateau, would become dominant. This idea would perfectly fit both the epigraphy and nature, and *ꜣḫt* could be translated as the arrival of the Inundation in the sense of becoming verdant. On the other hand, the other two seasons may refer to other important periods in the Nile's ecology: going forth (*prt*) of the land, after the period in which it had been covered by the waters and, finally, drought, with the river at its lowest level and with most of the useful water (*mw*) stored in the basins (*š*). Hence, in a calendar year, we would have a complete Nile year.

Afterwards, the original meaning of the seasons was probably forgotten as time passed, especially when, owing to the wandering nature of the civil calendar, they no longer adjusted the actual behaviour of the river. This happened twice in the Middle Kingdom and during the Third Intermediate Period and the Low Epoch. The seasons' meaning do not seem to have been forgotten in the Middle Kingdom because in the Ramessside period they were fully acquainted with them (at least Akhet for the inundation). However, in later periods, other meanings had to be developed when the names were translated into foreign languages. According to this line of argument, the current Coptic meaning, which has been translated into other languages as 'Inundation', 'Winter', and 'Summer', is perfectly explained by the fact that the Coptic year starts on August 29 Julian (September 11 by today's Gregorian calendar). This implies that Inundation covers the period from late August to December (a period when the waters are still high; see Fig. 5.3). Furthermore, Winter extends from December to April and Summer from late April to August. The parallelism with the traditional Mediterranean winter and summer seasons is appealing and the translation is thus obvious.

All in all, we would be tempted to believe that, prior to the invention of the civil year, the Egyptians did have a year connected with the Nile. The logical way of operating it would have been to wait for the arrival of the rising waters and then start to count the months, in this case obviously lunar months, with the first subsequent conjunction or the first crescent visibility of the waxing moon. This is not obvious for such an early epoch, especially because *psḏntjw* is absent from Old Kingdom festival lists while *ꜣbd* is recurrent (see below). This Nile year would then run until the next rising of the waters or perhaps until the end of the harvest epoch. Such a calendar would have been efficient for a small local community, but would have given rise to several problems.

Although average years would normally have been 12 or 13 lunar months long, they might occasionally have included only 10 or 11 lunations or as many as 14 lunar cycles (see Fig. 5.3). Since the flood lasted some 12 days from Elephantine to the Mediterranean, different communities along the river's course (to the north), would certainly have started their year one month after other groups to the south of the country, whenever a new lunar month had started in any of those 12 days.

Once the country became politically and economically unified, such a situation would have been unacceptable. Actually, we do not know of any procedure that the Egyptians might have used to pass information quickly from the extreme south of the country to the north. Hence, it is difficult to imagine what might have been the easiest solution to choosing a single reference place, and then passing the information across the rest of the country, so that every community would start their year at the next conjunction or, perhaps for this early epoch, the incoming new crescent.

A reform would certainly have been necessary. In any case, if such a place ever existed, it would probably have been Elephantine, regarded during most of Egyptian history as the place of origin of the Nile floods and the site of an important cult (Bard, 1998: 283).

5.2.2 A Brand New Clever Calendar … and a Date

For the new calendar, two solutions could have been adopted: first, the creation of a standard month of 30 days, very probably taking as a model the synodic month of alternative 29 and 30 days (29.5306 days on average); then a year of 365 days would have been inaugurated whose origin that, *pace* Neugebauer, we believe was certainly astronomical. Both processes are intimately related to each other and, at the same time, conform to the Egyptian way of understanding the cosmos.

Apparently, from the very beginning, the Egyptians used a base ten counting system (Ifrah, 2005: 316), which was also applied to timekeeping, as it is the origin of the decades, an extremely important period of time throughout Egyptian history (see Chap. 3; Spalinger, 1996a, b, c). A lunar synodic month is on average 29½ days long. In many societies, the best way to approach 29½ is by the alternation of 30- and 29-day periods, as in Hebrew or Muslim calendars. However, 29 is not divisible by 10 and when creating a new average month, the Egyptians might have preferred to choose a single value that, at the same time, could easily be counted in three decades.

The name chosen for that period was  (*ꜣbd*), which is equivalent, at least in the consonants, to the traditional name of LD2 of the Egyptian lunation (the first being the conjunction, *psḏntjw*, see Sect. 5.6). This might reflect the hypothesis that, at the very beginning of Egyptian history, the lunar month of the Nile year might have started at first crescent visibility (Belmonte, 2003a). However, this relation might have been completely forgotten very early on in daily life and the months of the civil year might have completely lost any connection with the actual moon, as happened in many other languages and societies (e.g. English with the terms 'month' and 'moon' is an obvious case).

For the origin of the 365-day period, Belmonte (2003a, 2009) introduced a new astronomical concept, the zenith pass, that has been hardly taken into account in Egyptian historiography. In the second half of the Third Century BC, the Alexandrian scholar Eratosthenes, born in Cyrene (in today's Libya), made a revolutionary measurement of the circumference of the earth (Cleomedes, *De motu circulari corporum caelestium* I, 10). To accomplish this, he made use of what must have been a well-known fact to contemporaneous Egyptian society, that the noonday sun at the

Fig. 5.7 (**a**) Archaic sacred enclosure of the goddess Satet in Elephantine. This sanctuary was enclosed by three natural large granite boulders. The gate opened approximately towards the southeast horizon, where the heliacal rising of Sirius happened in 3200 BC. At least four temples dedicated to the goddess would be erected successively on this sacred spot, where the primordial waters of the flood played an important role. (**b**) A well in the ruins of ancient Syene in the island of Elephantine. This old well was in use until very recent times. Although it has suffered many

summer solstice was able to illuminate the water at the bottom of a very deep well (see Figure 5.7) in the city of Syene (Aswan). This happened because at that exact moment of the year and at that latitude (*c.* 24°) the sun (in fact only part of it, see below) passed exactly overhead. This is a phenomenon known in astronomy as the zenith pass of a celestial body.

For the sun, this phenomenon happens on two occasions each year only in those places located between the tropics of Cancer and Capricorn, which of course receive the name of tropical zones. The limits are located exactly at the tropics, where the sun has a zenith pass only once on the day of the local summer solstice, when it reaches its maximum declination (actually in a series of days because the solar declination varies very little at the solstices). Curiously, in 3000 BC, the maximum declination of the sun was 24° 5', exactly the latitude of central Aswan and, of course, of one of its most important areas, the island of Elephantine.

The island of Elephantine was already at that time a very important settlement and archaeologists have shown that a sanctuary (and perhaps also a Nilometer) was already in operation on the site of the later temples of Satet and Khnum (see Fig. 5.7), the gods of the first cataract and of the inundation, at least from 3200 BC (Bard, 1998: 283). Was the zenith pass observed at Elephantine in that epoch? Following Belmonte (2009), we suggest that it was. Besides, we may endorse the first author's hypothesis: this observation could be one of the possible, perhaps the most sensible, ways of determining the length of the (solar) year as 365 days.

We have not yet considered the interval between two sunrises or sunsets at consecutive summer, or winter, solstices as a good candidate for the determination of the period of 365 days. This is because it was argued (Neugebauer, 1942) that, within a short time, it would be obvious that the exact moment of the solstices was moving backwards in relation to the civil year. However, this is not exactly true. During the solstices, the sun stands at almost the same declinations for several days and, consequently, its rising or setting positions arrived at a standstill (hence the name). Even if the proto-dynastic period Egyptian hour-watchers were able to determine precisely these positions on the horizon (which we believe they were, as will be shown in Chap. 6), they would hardly have reached a precision better than 2–3 arc minutes. This is equivalent to a variation of 8–12 days in the Sun's orbit and, considering the wandering of the civil calendar of 4 years per day, it would have lasted at least 32–48 years before the displacement became obvious.

However, the effect becomes much more dramatic when zenith pass is considered. If the non-shadow effect on a gnomon (an obelisk or little pyramid for example, since both are known as extant monuments in early dynastic Elephantine; Wilkinson, 2000a, b: 96) or the illumination of a deep well, were the actual

Fig. 5.7 (continued) reconstructions, this could perhaps be the original, or a predecessor, of the one mentioned by the Hellenistic sage Eratosthenes. At low waters, with the river some 9–10 m below the rim of the construction, the summer solstice zenith sun was still able to illuminate the water at the bottom of the well. A similar device—perhaps a rudimentary Nilometer—could be tentatively used in proto-dynastic Egypt to establish the length of the solar (seasonal) year at 365 days. (Photographs by the authors) phenomenon observed, the length between two consecutive zenith passes would have been easily established as 365 days. At the same time, however, once the civil year had been in operation, it would have been extremely difficult to detect the displacement of zenith pass events during the months and years to come.

The reason for this is that the sun is not a point source of light, but has a well-defined circular shape with a mean diameter of some 36′. That is also the reason why the illumination phenomenon was still observed in the well in Aswan in Eratosthenes' times. The extreme declination of the sun at that moment was 23°47′ owing to the decreasing value of the inclination of the axis of the earth with respect to the plane of its orbit (the obliquity of the Ecliptic). This is now decreasing at a rate of 0.46″ per year. Hence, one third of the solar disc was still able to shed light on the bottom of the well.

With this fact in mind, we can even give an estimate of the interval for the inception of the civil calendar, provided the hypotheses we are endorsing are correct.

Imagine as a hypothesis, that the civil year was inaugurated at the beginning of a lunar month following the summer solstice and the moment of zenith pass at Elephantine. This could also be the latest average date of the arrival of the flood at this particular spot and we may thus consider it as the beginning of the corresponding lunar Nile year. We can easily calculate that the earliest time in Egyptian history when I *3ḫt* 1 coincided with the summer solstice was in the quadrennium centred more or less on 2760 BC.

However, considering that the new lunar month might have started as much as 29½ days later, this could have happened 118 years earlier. Also, using reasonable numbers, we can estimate an error of roughly 72 civil years (equivalent to 16′, half a solar diameter) before it was obvious that the sun was not clearly crossing the zenith in I *3ḫt* 1.

Summing up all these numbers, we reach to an interval roughly between 2950 BC and 2690 BC for the inauguration of the civil calendar, although slightly later dates could also be acceptable if a period of trial and error is considered. This is an interval of time more or less between the beginning and the end of the proto-dynastic period (see Table 7.1), when several relevant aspects of the Egyptian civilization would have been developing. Indeed, the calendar could be one of them.

We can place the earliest presumed mentions of the seasons in this context. The first is a most controversial piece of evidence from the reign of Djer (see Figure 5.8), whose reign may fit with the upper limit of the interval, but the civil calendar was not yet operative, as has been suggested before (see Fig. 5.6), even if the name of the seasons already existed at this early period.

The second is an inscription undoubtedly mentioning the month III *šmw* in the Step Pyramid complex in Saqqara, which is perhaps the first evidence of the civil calendar after its invention (Berlev, 1999). This inscription is logically placed in the reign of Djoser, located by most chronologies immediately after our lower time limit. However, this lower limit fits different chronologies for the reign of his father Khasekhemuy (Belmonte & Shaltout, 2009, Appendix 1; see also Table 7.1), who unified the country after a convulsive period.

Fig. 5.8 A facsimile of an ivory tablet of the reign of Djer, 4th king of the 1st Dynasty (original at the Pennsylvania Museum). According to Altenmüller (1965), it may represent the first mention in Egyptian history of *wpt rnpt*, the 'Opening of the Year', in I *3ḫt,* and to the contemporary heliacal rising of *Sopdet*. However, owing to the difficulty in interpreting the hieroglyphs at these early stages, this hypothesis has been seriously contested by Clagett (1995: Fig. III.3a), who recognizes the goddess Sekhet-hor, instead of Sopdet, in the stylized image of a recumbent cow. The interpretation is hence controversial

In a brief but most elucidating essay, Kahl (2007) has shown that the solar cult possibly developed at the beginning of the 2nd Dynasty, during the reigns of Hotepsekhemuy and Raneb (*c.* 2750 BC). Krauss (2011) later argued that the calendar was created in this precise epoch, in total agreement with the central date of the above-postulated quadrennium (2760 BC).

Any of these candidates sounds reasonable but in our present state of knowledge, it is impossible to make a decisive choice.

Interestingly, it is also worth noting that if 'Inundation' had been the original meaning of *šmw*, as proposed by Champollion and other scholars in the beginnings of Egyptology, the inauguration of the calendar would move to a date *c.* 3300 BC, when it is highly probable that there was neither a centralized state nor a common calendar in ancient Egypt. This argument also contradicts any kind of support for this outdated and abandoned translation.

Another important feature of this hypothesis is that a zenith pass event did not occur at any other important city to the north of Elephantine and thus its people would not have been cognizant of the phenomenon. This is relevant to Neugebauer's (1942) point about the impossibility of an astronomical origin of the calendar because with the nucleus of the kingdom being well established in the Memphis region since the beginning of the 2nd Dynasty, nobody would have cared about the displacement of the civil year from an unobservable—by them—celestial event.

Once the time periods of 30 and 365 were established, the internal distribution of the calendar was forced by the simplest arithmetic: 36 decades of ten days, grouped in 12 months of 30 days, plus 5 extra troublesome days located 'above the year'. As a heritage of the old Nile year, the set of three seasons was kept, with four months as their standard length (this might not have been necessary the case in Nile years).

To conclude the arguments, we may quote in evidence a cultural parallelism that has never been considered in the discussion of Egyptian calendrics. This is the Mesoamerican calendar of 365 days. Apparently, like the Egyptian civil calendar, it worked with a permanent period of 365 days (i.e. without leap years) during well-established historical periods such as during the classical Maya. In this epoch, the Haab (i.e. the 365-day year) was an integrated and dynamical part of a very elaborated system of time-reckoning, together with the long count, the 260-day ritual calendar, the lunar series and other time cycles, all of them well attested by archaeology and ethnography (e.g. Aveni, 1990: 135; Šprajc, 2018). Nobody in Mesoamerican studies doubts that the origin of the Mayan and Aztec 365-day year was solar. Indeed, when confronted by similar situations, humans often develop similar intelligent solutions to certain problems.

In this same sense, presumably nobody in late Antiquity doubted that the origin of the Egyptian civil calendar was also solar. The following texts are extremely illustrative:

> … the Egyptians organise months and years in a very special way. Relating the days not to the movement of the moon but of the sun, … [Diodorus Siculus I, L]

and

> It is said that the priests in Thebes are mostly astronomers and scholars: to the priests we owe the habit of calculating the days, not according to the moon but to the sun, and to add each year 5 days to the 12 months of 30 days each [Strabo, *Egypt* XVII, I, 46].

One question that we might ask is what would have happened once it became obvious that the civil calendar and the climatic (ecological) seasons were no longer in agreement; i.e. when I *ꜣḫt* 1 (aka *wpt rnpt*) was systematically ahead of the arrival of the inundation anywhere in Egypt. This may have happened some 120–200 years after the creation of the calendar, or nearly 500 years for a complete seasonal displacement (the complete flooding period occurring in *prt*). Further systematic astronomical observations would have been made in connection with the solstices or other cyclic annual phenomena in order to test the accurate behaviour of nature.

Belmonte's (2009) contention is that the relevance of solstitial observations during the Pyramid Age (Belmonte & Zedda, 2007) and the introduction of the heliacal rising of Sirius (*prt spdt*) as the harbinger of the Flooding, attested perhaps from the Old Kingdom onwards (see below), were the collateral effects of this necessity.

5.3 The Names of the Months: A Tale of Mystery and Imagination

Not surprisingly, historically speaking, there has always been a 'problem' with the names of the Egyptian months (Gardiner, 1955; Parker, 1957a, b), considering the ongoing discussion on the number of calendars and the likelihood of which set of month names belonged to which calendar. The problem has reached the level that, in Depuydt's (1999) words, 'after many decades of debate, each single item has been tirelessly discussed'.

The key to the problem is that, in the well-known and widely used *Concise Dictionary of Middle Egyptian*, Faulkner (1962) identifies *ḥwt-ḥr* as the fourth month of the Egyptian calendar or *ẖnsw* as the tenth, among others. However, all later documents demonstrate that Athyr (Coptic and Greek sources equivalent to Hathor) and Pachons (equivalent to That of Khonsu) were clearly the third and the ninth months of the civil year (see Table 5.1). The main reason for that is apparently Faulkner's loyalty to his former master, Gardiner, who, basing his argument primarily on the Ebers Calendar, had proposed that Misore (a later name for *wpt rnpt*) had

Table 5.1 The months of the Egyptian calendar in different denominations, including the standard transcription of civil months, Coptic and Greek sources, Egyptian modern Arabic (in Latin alphabet transcription) and, finally, the transcription of month at Deir el Medina and Edfu (see Table 5.2 for transliterations)

	Civil month	Coptic	Arab Misri	Deir el Medina	Edfu Temple
1	I Akhet	Thot	Tot	Djehuty	Tehy
2	II Akhet	Paophi	Babe	Pa en Ipet	Menkhet
3	III Akhet	Athyr	Hatur	Hathor	[Hathor]
4	IV Akhet	Cohiac	Kyak	Kaherka	Kaherka
5	I Peret	Tybi	Tobiyi	Ta Abet	Shef Bedet
6	II Peret	Mesir	Mashir	Pa Mehir	Rekeh Ur
7	III Peret	Phamenoth	Buramar	Pa en Amenhotep	Rekeh Nedjes
8	IV Peret	Pharmouti	Baramunde	Renenutet	Renenutet
9	I Shemu	Pachons	Bashans	Pa en Khonsu	Khonsu
10	II Shemu	Payni	Bauniyi	Pa en Inet	Herti khedet
11	III Shemu	Epiphi	Abib	Ipet	Ipet Heb
12	IV Shemu	Misore	Meshri	Meswt Ra	Ra Horakhty
	5 hrwy rnpt	Epagomenoi	Nisi		

Note the striking similarity between the data of Deir el Medina and the original Coptic names

Table 5.2 Extant lists of month names from the tomb of Senenmut (18th Dyn.),[a] the Karnak Water Clock (18th Dyn.), the Ramesseum at Thebes (19th Dyn.) and the astronomical frieze of the Horus temple at Edfu (Ptolemaic Period)

Senenmut Tomb	Karnak Clepsydra Months	Divinities	Ramesseum Months	Divinities	Edfu Freeze Months	Divinities	Data from Deir el Medina
ṯḫ	*tḫ* (I *ꜣḫt*)	*tḫ*	I *ꜣḫt*	*tḫj*	*tḫ* I *ꜣḫt*	*tḫ*	*ḏhwty*
mnḫt	II *ꜣḫt*	*ptḥ*	II *ꜣḫt*	*ptḥ rsy.f*	*mnḫ*(*t*) II *ꜣḫt*	*ptḥ*	*p n jpt*
ḥwt ḥr	III *ꜣḫt*	*ḥwt ḥr*	III *ꜣḫt*	*ḥwt ḥr*		*ḥwt ḥr*	*ḥwt ḥr*
kꜣ [*ḥr*] *kꜣ*	IV *ꜣḫt*	*sḫmt*	IV *ꜣḫt*	*sḫmt*	*kꜣ ḥr kꜣ* IV *ꜣḫt*	*kꜣ ḥr kꜣ*	*kꜣ ḥr b*
šf bdt	I *prt*	*jmn-rꜥ*	I *prt*	*mn*	*šf bt* I *prt*	*šf bt*	*ṯꜣ ꜥbt*
rkḥ	II *prt*	*rkḥ wr*	II *prt*	*rkḥ wr*	*rkḥ wr* II *prt*	*hippq or pig*	*p n pꜣ mḫyr*
rkḥ	III *prt*	*rkḥ nḏs*	III *prt*	*rkḥ nḏs*	*rkḥ nḏst* III *prt*	*hippo or pig*	*p n jmnḥtp*
rnwtt	IV *prt*	*rnnwtt*	IV *prt*	*rnnwtt*	*rnn* (*wtt*) IV *prt*	*rnwtt*	*p n rnnwtt*
ẖnsw	I *šmw*	[*ḫnsw*]	I *šmw*	*ḫnsw*	*ḫnsw ḥb* I *šmw*	*ḫnsw*	*p n ḫnsw*
ẖnt ẖii prty	II *šmw*	[*ḫnt-ḫty*]	II *šmw*	*ḫnt tiy*	*ḥr yt ẖdtt* II *šmw*	*ḥrt y ẖdtt*	*p n jnt*
ỉpt ḥmt	III *šmw*	*jpt-ḥm*(*t*)	III *šmw*	*jpt ḥmt*	*jpt hb* III *šmw*	?	*jp jp*
wpt rnpt	IV *šmw*	*rꜥ ḥr ꜣḫty*	IV *šmw*	*rꜥ ḥr ꜣḫty*	*rꜥ ḥr ꜣḫty* IV *šmw*	*rꜥ ḥr ꜣḫty*	*wp rnpt*

In the case of the last three examples, the table also presents the list of accompanying deities. The last column shows, for comparison, the set of month names that can be obtained from data of Deir el Medina such as the Cairo Calendar (86637, verso), and several ostraca (BM29560, DM1088, DM1265)

[a]Most recently, Spalinger (2020b) has definitely endorsed this list of months as belonging to the civil calendar

originally been the first month of the year (Gardiner, 1906), yet another consequence of the Ebers syndrome.

Thus, for our discussion, we will rely only on the four extant lists of months that have reached us in an almost complete state of preservation, excluding the Ebers. These are presented in Table 5.2. The youngest of the four is the list of months on the astronomical frieze of the temple of Edfu. This list, by itself, would have been sufficient to prove the names of 11 of the 12 months of the civil calendar (Belmonte, 2003a), since these names are clearly associated with the seasonal names of the months (e.g. the two *rkḥ* to II *prt* and III *prt*, respectively). Other inscriptions of the temple further emphasized that parallelism for month IV *ꜣḫt*, I *prt*, II *prt,* and III *šmw*, and also provide an alternative name (*wpt rnpt*) for IV *šmw*. However, because of the historical debate, they had been seen alternatively as the names of the civil-based lunar calendar or as the names of Parker's original lunar calendar that had been borrowed by the civil calendar owing to the archaic style of the inscriptions of Edfu (Depuydt, 1997: 116, 1999).

This is surprising, especially when one compares these names with those in the oldest month list, that of the astronomical ceiling in the tomb of Senenmut (18th Dynasty, *c.* 1470 BC). Nine names are virtually identical, and those which are not (two) are strikingly similar. The remaining one, *ḥwt-ḥr*, can be found in many other references, including the papyri and ostraca of Deir el Medina, as the name of the third month of the civil year (see e.g., Clagett, 1995: Fig. II.7).

Thus, from any logical point of view, we might think that in Senenmut we have the oldest list with the proper names of the months of the civil calendar. However, once more, Senenmut's list has too often been identified as the twelve names of the lunar months of the Sirius-based lunar calendar or as a list of festivals associated with those months. Its obvious civil arrangement as proposed in Belmonte (2003a, and reference therein) has seldom been neglected.

Fortunately, there is an almost contemporaneous proof that this set of names is certainly civil and not lunar. This is to be found in the upper rim of the Karnak water clock, dating from the reign of Amenhotep III but presumably following much earlier models of the sage Amenemhat, dating to the beginning of the 18th Dynasty. There, the original designer of the clock identified the hour marks for each month of the civil year, starting surely with I *ꜣḫt*, followed certainly by II *ꜣḫt,* and ending with IIII *šmw*, with the months clearly identified by their seasonal names. However, at some stage, the border of the clepsydra was broken in the area where I *ꜣḫt* was written and subsequently restored. Curiously, the person who made the restoration did not again write I *ꜣḫt* but *tḫ* instead (Spalinger, 1995a, b, c, d; see Fig. 3.25). This is stressed in Table 5.2. Thus, I *ꜣḫt* is obviously identical to *tḫ* and, if this is so, why not the rest of the whole list?

If any doubt remained, we might still draw the attention of the reader to the oldest reference to some of the month names, appearing in the papers of Hekanakhte (James, 1962), a set of hieratical documents dating from the reign of Mentuhotep II (*c.* 2000 BC). There, the names *šf bdt*, one of the *rkḥ* and *ḫnt ḫtii prty* are clearly mentioned in a civil calendar context for the first time in Egyptian history.

The list of names of the civil months is presented in Table 5.3. This list is basically that of the tomb of Senenmut with some minor modifications and simplifications suggested from the other lists. These names, as Edfu demonstrates, were used as the proper names of the months, at least in monumental inscriptions, until the Graeco–Roman period (e.g. Černý, 1943). However, at some point during the New Kingdom, a new set of names was developed for less official documents, such as papyri and ostraca, which would finally be the origin of the month names known from later Aramaic, Greek, and Coptic sources (see Table 5.1). This new set of names is also presented in Table 5.3. The corresponding Greek forms of this new set are presented in the first column of the table.

As Table 5.3 shows, two names (3 and 4) are identical and another two can be considered as mere New Kingdom variants of previous forms (8 and 9), where the article *pꜣ* and the preposition *n* has been added to the original name. One name (11) has suffered a little variance. In the case of the 12th month, *wpt rnpt*, this name is kept for hieroglyphic and hieratic sources, whereas it never appears in the later

Table 5.3 Proper names of the civil month

	Classic sources	Civil months	Old names	Translation	New names	Translation
1	Thoth	I *ꜣḫt*	*tḫ(j)*	Drunkenness	*ḏḥwtyt*	Thoth's
2	Phaophi	II *ꜣḫt*	*mnḫt*	Clothing	*pꜣ n jpt*	That of Opet[d]
3 G	Athyr	III *ꜣḫt*	*ḥwt-ḥr*	Hathor	*ḥwt-ḥr*	Hathor
4 G	Choiak[a]	IV *ꜣḫt*	*kꜣ-ḥr-kꜣ*	Ka on ka	*kꜣ-ḥr-kꜣ*	Ka on ka
5 G	Tybi[b]	I *prt*	*šf-bdt*	Emmer Spring	*tꜣ ꜥbt*	The Offerings
6 G	Mechir	II *prt*	*rkḥ (ꜥꜣ/wr)*	(Great) Burning	*(pꜣ n pꜣ) mḫyr*	Low Lands
7 G	Phamenoth	III *prt*	*rkḥ (nḏs)*	(Small) Burning	*p n imn-ḥtp*	That of Amenhotep[d]
8 G	Phasmouthi	IV *prt*	*rnwtt*	Renutet's	*p n rnwtt*	That of Renutet[d]
9	Pachon(s)	I *šmw*	*ḫnsw*	Khonsu's	*p n ḫnsw*	That of Khonsu[d, e]
10	Payni	II *šmw*	*ḫnt-ḫty*	First of …	*p n int*	That of the Vꜥlley[d]
11 G	Epiphi	III *šmw*	*jpt-ḥmt.s*	Her majesty Ipet	*jp jp*	(Feast of) Ipet
12 G	Misore[c]	IV *šmw*	*wpt rnpt*	Year Opener	*wpt rnpt mswt rꜥ-ḥr-ꜣḫty*	Yeꜥr Opener Birth of Re[f] (Hor-Akhty)

The old (traditional) and new proper Egyptian are presented, together with their possible translation. G stands for those months suffering from the Gardiner phenomenon: a feast of the same or similar name in the first day of the following month (Belmonte, 2003a). In bold type, those months whose name changes completely in the Ramesside period. See text for further discussion

Notes: [a]Feast of Nehebkau; [b]Only a possibility. It was celebrated in I *prt* 20 at Dendara; [c]This needs the obliteration of the 5 epagomenal days; [d]Reflects the domination of Thebes, her gods and her festivals; [e]Same name with New Kingdom grammar; [f]Later form of IV *Smw*

Aramaic, Greek, demotic, or Coptic documents where the name Misore is preferred (Quack, 1996). Interestingly, the hieroglyphic counterpart of Misore, *mswt-rꜥ-ḥr-ꜣḫty* is attested in hieroglyphs as a feast name but never as a month name. For the first month, it might be possible that it was known as *ḏḥwtyt* in earlier times (see below) as an alternative name.

Five months changed their names completely from the old to the new set. We have no clear reason for this, neither is it possible to find an explanation in the extant sources. However, one fact is clear, months 2 (*p n jpt*), 7 (*p n jmnḥtp*), 10 (*p n jnt*), and perhaps also 9 (*p n ḫnsw*) reflect the new dominant political, social, and religious situation of Egypt in the New Kingdom. This is because those names clearly derived from three of the most important festivals celebrated in Thebes from that time onwards (Clagett, fig. II.6, and references therein). Hence, at some moment during the New Kingdom, the Egyptians changed the names of several of their months—as was the case in our calendar twice. This was the situation in some of their documents and probably in the speech of daily life, so that this new set of names was the one that passed to later sources and survived to our time in the names of the months of the Coptic liturgical calendar.

It is worth mentioning that Depuydt (1997 and 1999, qualified in 2017) has defended the existence of these two parallel sets of names because one belonged to the lunar calendar (almost our old set) and the other to the civil calendar (the new set), and because the second derived from the first. However, as Belmonte (2003a) showed, we do not have any irrefutable proof of the existence of a whole set of 12 (or 13) well-articulated lunar months at any moment in Egyptian history.

Consequently, applying the principle of economy, we should assume that in Egypt there was just one single set of 12 names for the 12 months of the civil calendar, at least from the beginning of the New Kingdom, and perhaps even during the Middle Kingdom. For unknown reasons, some of these names were either altered or slightly changed during the New Kingdom or later (this would be the case of Misore), but still kept the same internal coherency. However, the old set of names was certainly maintained for some monumental inscriptions, as the data of Edfu suggest, in parallel to the new set of typical seasonal names.

Griffith (1909: 185) had suggested that the seasonal set of names was written on the inscription but that the proper names of the months were the ones usually read. This is indeed a sound hypothesis. Would it therefore be possible to go further in time, to earlier periods of Egyptian history, and try to see what the names of the months of the civil calendar were during its inception?

5.3.1 Problems, What Problems?

To finish the 'problem' of the month names and try to understand how some of the months earned their names, we must devote some time to discussing the Brugsch and Gardiner phenomena, which has been addressed in detail by Depuydt (1997) and Belmonte (2003a). Here we will merely summarize the problem:

i) The Brusgch phenomenon comes from the fact that the last month of the Egyptian calendar, *wpt rnpt*, the Opener of the Year, has a name that apparently refers to a beginning.
ii) The Gardiner phenomenon is connected to the well-known fact that several important festivals, which bore the same name as the months, were not celebrated in their eponymous months but rather in the first (or second) day of the following month. Six unmistakable cases of this phenomenon are attested (see Table 5.3), with two more doubtful cases.

Depuydt (1997) found an explanation for both phenomena in the mutual influences and derivations produced first between the Sirius-based lunar calendar and the civil calendar, and later between this and the civil-based lunar one. However, as we have accepted in this chapter, all the names attested so far would have belonged to the civil calendar probably since the Middle Kingdom onwards. What would the original names (if they ever existed) of the Nile year lunar months have been prior to the invention of the civil calendar?

Fig. 5.9 Part of a list of festivals in the walls of the tomb of Mereruka in Saqqara. *wpt rnpt*, or Wepet Renpet ('New Year's Eve'), is followed by the Feast of Thoth, the controversial *tpj rnpt* and the *wꜣgj* Feast. This order of feasts, typical of the 4th and 5th Dynasties, may suggest a lunar computus with the celebration of *wꜣgj* in III *ꜣḫt*. The precise nature of this lunar timekeeping system, perhaps connected with the arrival of the actual Inundation, is still a matter of debate. (Photograph by the authors)

The Old Kingdom inscriptions do not offer much information on this particular issue. However, there is a large number of festival lists from various sites (especially the necropolises at Giza and Saqqara), which offer a good set of festival names (see Figure 5.9). From a compilation of all the sources, it is clear that the most complete list was that formed by eleven festivals plus the closing formula *ḥb nb* (*rʿ nb*), 'each festival, each day', completing twelve entrances (Spalinger, 1996b: 110).

However, on several occasions, a twelfth feast, called *wꜣḥ ʿḫ*, was added frequently between the original position 7 and 8, seldom between 8 and 9, and rarely in other positions, perhaps indicating the moveable nature of the feast. The standard list, including *wꜣḥ ʿḫ* in its most common eighth position, would be as follows:

(*1*) *wpt rnpt*	(*5*) *ḥb skr*	(*9*) *prt mn*
(*2*) *ḏḥwtyt*	(*6*) *ḥb wr*	(*10*) (*ꜣbd n*) *sꜣḏ*
(*3*) *tpj rnpt*	(*7*) *rkḥ*	(*11*) (*tp*) *ꜣbd*
(*4*) *wꜣgj*	(*8*) *wꜣḥ ʿḫ*	(*12*) (*tp*) *smdt ḥb nb* (*rʿ nb*)

It is suggestive that the standard list included a total of 12 feasts, exactly the same as the numbers of months in the civil calendar. Belmonte (2003a) has even argued that each of these feasts might somehow represent each one of the 12 months of the civil calendar, with the closing formula at the position of the epagomenal days. This working hypothesis is schematically presented in Figure 5.10, where the feasts are related to the months.

First, we have the obvious parallelism of *wpt rnpt* with day I *ꜣḫt* 1 and with the last month of the year, also named *wpt rnpt*—we shall come back to the Brugsch phenomenon later. Then we have the Feast of Thoth (*ḏḥwtyt*), well attested in I *ꜣḫt* 19 in later sources, and even the later name of the first month a posteriori. We now jump forward and leave *tpj rnpt* and defer discussion until later. The celebration of the Feast of *wꜣgj* is well attested in I *ꜣḫt* 17 and 18 from the Middle Kingdom onwards. However, in the Old and Middle Kingdom, there was a second, moveable *wꜣgj* Feast (Luft, 1994; Krauss, 1998b, 2021; Depuydt, 2000). There are firm hints

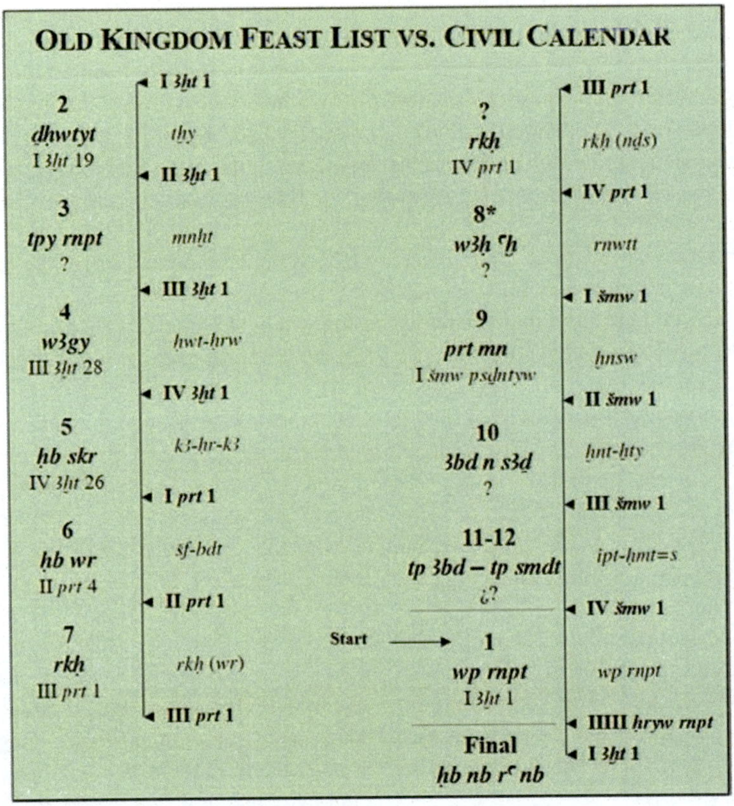

Fig. 5.10 Tentative comparative table between the 12 festivals usually mentioned in the standard feast list of the Old Kingdom and the 12 months of the civil calendar. Belmonte (2003a) has speculated that, generally speaking, each of these 12 festivals might have been vaguely associated with each of the months within a civil year. Doubts indeed remain for several fits, notably those of the 12 lunar dates. See text for further discussion. (Adapted from Belmonte, 2003a)

that during the Old Kingdom, it might have been celebrated once in III *ꜣḫt* 28, on lunar day 18 (or 17) of the lunation following the rising of the Nile waters, the summer solstice, or *prt spdt*. For IV *ꜣḫt*, we have the well-attested Feast of Sokar (*skr*) in IV *ꜣḫt* 26 (Sauneron, 1962; El-Sabban, 2000).

The list follows on the Great Festival (*ḥb wr*). There is extremely little information on this feast, but in the very late temple of Edfu there is a 'Great big' festival (*ꜥꜣ ḥb wr*) celebrated on II *prt* 4 and thus close to the beginning of the station *prt* (El-Sabban, 2000: 173). Then comes *rkḥ*. This is one of the long-lasting festivals, certainly lending its name to a pair of months of the calendar, II *prt* and III *prt*. Middle Kingdom sources (the Illahun archives) indicate that the *rkḥ* festival was celebrated twice, in III *prt* 1 and IV *prt* 1 (Luft, 1992a, b: 215), so we might perhaps relate the *rkḥ* of the Old Kingdom list to both of them and, consequently, we may associate it with two months. Eventually, we might assign *wꜣḥ ꜥḫ* to IV *prt* and, by doing that, we have another useful clue. In later sources such as the festival list of the temple of Ramesses III at Medinet Habu (El-Sabban, 2000: 127), the Procession of Min (*prt mn*) was established in connection with the *psḏntjw* of I *šmw*.

This parallelism might also tell us that, in the case of the Old Kingdom, the feast list might reflect a much older situation, one in which we are in fact dealing with an older lunar calendric system, which, as demonstrated by *wꜣgj*, was increasingly overshadowed by the civil calendar as Egyptian civilization evolved. Belmonte's (2003a) hypothesis (endorsed by the authors) is that this older system was the original Nile year.

The list ends with three events that are clearly lunar. One is poorly known, (*ꜣbd n*) *sꜣḏ*, and it might account for month 10. However, (*tp*) *ꜣbd* and (*tp*) *smdt* are clearly the first crescent day feast and the full moon feast and might consequently have been celebrated twelve times per year. However, the uncommon adjective *tp* might mean that these were two special events.

Belmonte (2003a) speculated that these events could be related to the last month in a peculiar way. In the Nile year, the last lunar month—and even the previous one—might not be present in the case of a very short year (like the one in the Palermo stone) and thus might not have had a proper name assigned and have been simply ignored, as in the primitive Roman calendar, or merely numbered like our September or October. However, this hypothesis lacks any proof that might support it.

Of the original nine feasts of the Old Kingdom hypothetically related to months, only three survived as month names in the Egyptian calendar in later epochs (*wpt rnpt*, *ḏḥwtyt,* and *rkḥ*). Consequently, we could imagine that the names of the months may have been properly assigned later, presumably in the Middle Kingdom, when three of them are attested. However, as the late renowned astronomer Carl Sagan used to say, 'absence of evidence is not evidence of absence' and, following Belmonte (2003a), we may argue that the months of the civil year already had a proper name from the very beginning.

In fact, the only way to solve the problem of the Brugsch and Gardiner phenomena is to assume that the vast majority of the months (especially those suffering from the Gardiner phenomenon) must have taken their names from another source, from the very moment of the inauguration of the civil calendar. This situation is perfectly illustrated in Figure 5.11.

5.3.2 *Calendric Mechanics: A Model*

To find a solution, Belmonte (2003a) imagining the moment of creation of the civil calendar, assumed the following hypotheses:

(a) The calendar was inaugurated at the same time as a Nile lunar year began; i.e. the first I *ꜣḫt* 1 was the first day of a lunar month (a *psḏntjw*). From that point to the very end of Egyptian history I *ꜣḫt* 1 would be the date to celebrate the Feast of *wpt rnpt*.

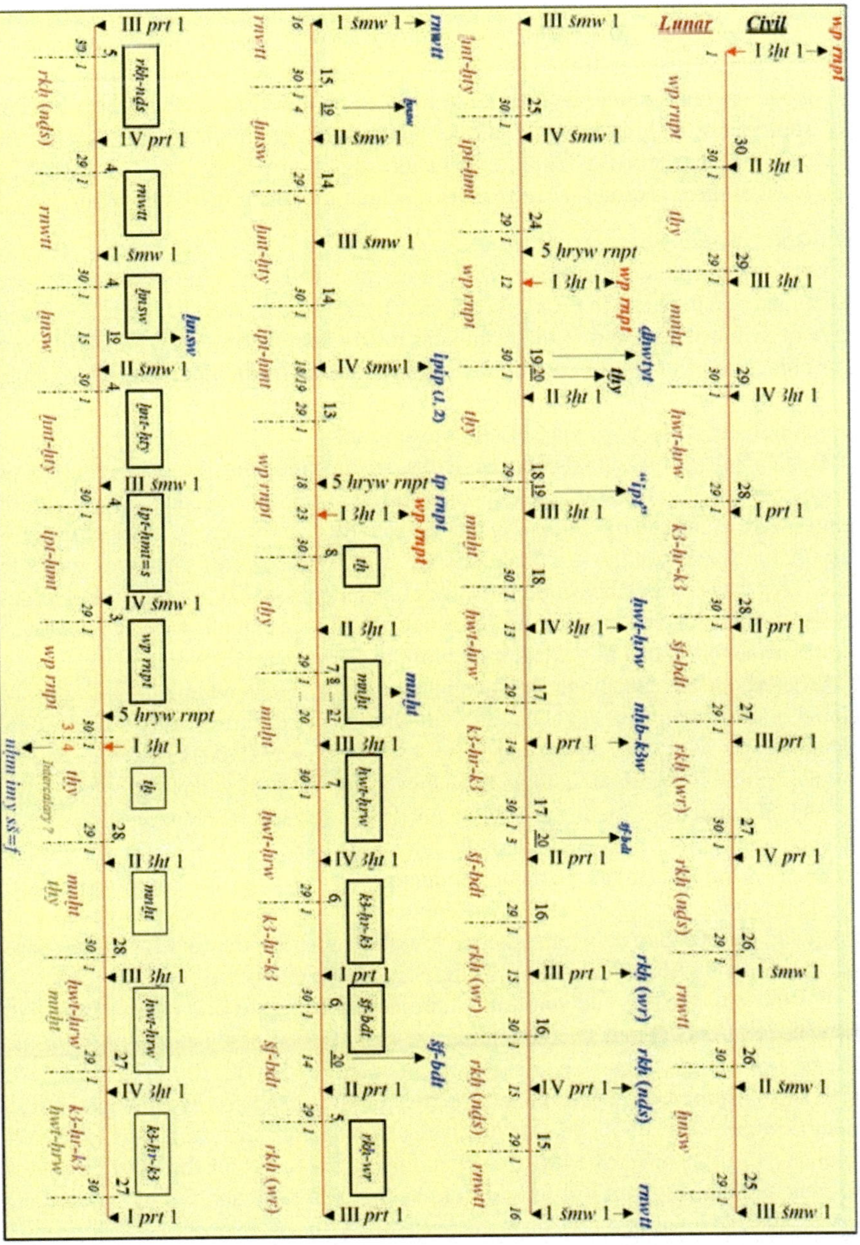

Fig. 5.11 Diagram showing a hypothetical moment in which the civil (right line of time) calendar was created from an original lunar calendar (left line), starting with the *wpt rnpt* feast as the first day of both the new civil calendar and the old cycle of lunar months. In three years of parallel but wandering running, the civil calendar would have run completely independently, keeping within its structure old features of its lunar counterpart that are now controversial; the Brugsch and Gardiner phenomena. See text for further discussion. (Adapted and qualified from Belmonte, 2003a)

(b) The Nile year had a set of month names, starting with the Opener of the Year (*wpt rnpt*), eponymous with its corresponding feast, followed by *tḫ*(y), and so consecutively. Not all the months would necessarily have had a name and for those having a name, this does not need to be the same as that given in later sources (18th Dynasty, see Table 5.3). However, for simplicity, we assume the old set of 12 names in Table 5.3 to be already in operation.
(c) A set of three 'normal' Nile years of 12 lunar months follow each other.

With these working hypotheses, we might obtain the following conclusions. Already in year 1 of the series, some part of the lunar months of the Nile year would have fallen one month ahead in the civil calendar. In the second year, the time lag would have been substantial and in the third year nearly complete, so that lunar *wpt rnpt* was coincident with civil IV *šmw*, *tḫ* with I *ꜣḫt,* and so forth. Since no name for the intercalary month is reported (if it ever existed), the situation would have been still more dramatic at the beginning of the fourth year.

Hence, Belmonte (2003a) speculated that, at some moment, either in year two, or much better in year three after the inauguration of the civil calendar, several month names were borrowed by the civil calendar from the simultaneous names of the corresponding lunar months—or lunations we might already call them—of the Nile year. This was presumably in the process of abandonment. This explanation might easily account for the Brugsch phenomenon and might explain why *wpt rnpt* was the contradictory name of the last month of the civil year.

What about the puzzling Gardiner phenomenon? Here we might find a more global solution that would explain, not only the six months affected by the phenomenon but also why other important feasts such as *ḏḥwtyt* (I *ꜣḫt* 19), *tḫj* (I *ꜣḫt* 20), *šf-bdt* (I prt 20 in Dendara), and *ḫnsw* (I *šmw* 19) were fixed where they were.

As first noticed by Spalinger (1994b; 45, 1995), the special location of the Feast of Thoth in the civil calendar could have something to do with the importance of the god as a timekeeper (in his character of lunar god) since day I *ꜣḫt* 19 had a peculiarity. This would correspond to the difference in days between 13 lunations (384 days) and the 365-day civil calendar, and might have been relevant for lunar phase calculations or, perhaps, may simply have a symbolic character.

For the other feasts, the solution is simpler still. Figure 5.11 shows how five instances of the Gardiner phenomenon (*ḥwt-ḥr*, *nḥb kꜣw*, associated with *kꜣ ḥr kꜣ*, *rkḥ wr*, *rkḥ nḏs,* and *rnwtt*), which were celebrated on the first day of months 4, 5, 7, 8, and 9, respectively, are related to days 13–16 of the corresponding synodic months in the second Nile year of the cycle. Considering that the moon looks full from day 13 to 15 or from 14 to 16, depending on the length of the lunation (29 or 30 days, respectively), all five feasts could be the *smdt* festivals (full moon) of their eponymous old lunar months. Since the feast falls at the same time on a very special day within the new civil calendar (the first day of the following month), it might have been decided that this peculiar situation should be thenceforward frozen.

Another curious situation is the case of the Feast of *tḫj* in I *ꜣḫt* 20 (or *tḫ* 20), well reported at least from the Middle Kingdom and perhaps earlier. In this case, the feast is located within its eponymous month. Parker suggested that it should be also associated with a *smdt*. However, as Parker (1950: 57) and Spalinger (1993b) had suggested and Fig. 5.11 illustrates, it fits much better if the Feast of *tḫj* was originally the *psḏntjw* (LD1) of its eponymous lunation. Thus, *tḫj* could be fixed in day 20 of I *ꜣḫt* as soon as the second year of implementation of the civil calendar, when the corresponding *psḏntjw* occurred exactly on that day.

This model, although certainly hypothetical because there is no evidence of things happening this way, would have the advantage of offering a sound explanation for the controversial, and somehow deterrent, Gardiner phenomenon and beyond.

Interestingly, a similar line of reasoning might have dictated the decision, taken some 12 centuries later, to fix the first day of the Feast of Opet (*jpt*) in II *ꜣḫt* 19, since as we can see, it might also have corresponded to the *psḏntjw* of *mnḫt*, the month to be known later as Phaophi (*p n jpt*). However, a stronger possibility would have been the special relevance taken by days 18 and, overall, 19 within the civil months as important dates for the celebration of festivals. This could explain the last case of the Gardiner phenomenon, that of *jpjp*, the eponymous feast of *jpt-ḥmt.s* (III *šmw*, later Epiphi), but that was celebrated on days 1 and 2 of IV *šmw*, which might have corresponded to lunar days 18 and 19 of the 'almost' eponymous lunation.

Summarizing, we have endorsed Belmonte's hypothesis that there was just one set of month names, pertaining to the civil calendar, that changed some of their names later in Egyptian history. The majority of these names could be taken from a set of original lunar months of the previous Nile-based lunar year, thanks to a process of time lag, from the very moment of creation of the civil calendars in the first half of the third millennium BC.

This rapid assimilation might explain certain problems, later identified as the Brugsch and Gardiner phenomena. These were inherent to the Egyptian calendar from the very beginning, and not the result of the side effect provoked by the interaction throughout history between the civil calendar and the alleged Sirius-based and civil-based lunar calendars.

5.4 The Solstices as Time-Markers in the Framework of the Egyptian Calendar

Since the beginnings of Egyptology, there has always been an acrid debate between those scholars supporting knowledge of the stations of the sun by the ancient Egyptians and those attacking that possibility (e.g. Sethe's theory for the beginning of the year at the winter solstice). Leitz (1994a) has defended the solar roots of the Egyptian calendar, while Wells (1992) has related the mythology of the birth of the god Re from the goddess Nut (identified as the Milky Way, a reasonable possibility, see Chap. 4) to the relative positions of both the sun and the Milky Way at the vernal equinox and at the winter solstice. However, the model for the calendar (Wells, 1994), proposed as a corollary of that hypothesis, is unreliable, as we have previously mentioned.

What is almost a certainty, and difficult to deny any longer, is the fact that several Egyptian temples were orientated towards the solstices. This will be lengthy discussed in Chap. 6 and could be related to both calendric and symbolic aspects. However, Egyptian knowledge of the equinoxes is not so clear since the orientation of temples facing east, and thus the equinoxes, could be a by-product of the northern orientation of Old and Middle Kingdom pyramid complexes (Belmonte, 2022).

It has often been claimed that there are two features of the Egyptian calendar that might have a solstitial origin. First, the name of the 12th month (Misore) of later periods, the Birth of Re, and the eponymous feast at I *ꜣḫt* 1 have been related to the birth of the sun at the winter solstice. This could be a common link with many other cultures throughout the Mediterranean but—apart from Wells's (1994) hypothesis—never convincingly postulated in the case of ancient Egypt. On the other hand, the name of the sixth and seventh months, *rkḥ*, i.e. 'Burning' (see Table 5.3), has been related to the heat of the sun at the summer solstice. This is a hypothesis defended by Sethe (1919–1920: 38). Alternatively, it has been related to a much more prosaic solution of the artificial heat needed in Egypt at the time of the winter solstice, a hypothesis proposed and defended by Parker (1950).

The name Misore, instead of *wpt rnpt,* for the 12th month has been reported only in the Late Period and, especially in the Aramaic, Greek, and Coptic papyri written after the conquest of Egypt by the Persians in 525 BC. However, there is an early mention, in a necropolis report from Deir el Medina, of a feast under the name of *mswt rꜥ-ḥr-ꜣḫty*, celebrated in I *ꜣḫt* 1 as early as the 20th Dynasty (Gardiner, 1906). This feast could have a precursor in Amarna, where a festival *mswt.jtn* took place in I *ꜣḫt* (Krauss, 1998a, b), so the association between the feast of Misore and the first day of the civil year must be much older, but, by how much?

Belmonte (2003a) proposed the promising hypothesis that this link could effectively be associated with a moment when I *ꜣḫt* 1 fell at the time of the winter solstice, but not in prehistoric or proto-historic times, as generally argued, but well into the historic period. To be precise, because of the wandering of the civil calendar across the seasons, there have been two occasions when I *ꜣḫt* 1 has fallen at the moment of the winter solstice: in a four-year period centred on 2004 BC and in 500 BC, respectively, as shown in Figure 5.12. Considering the 20th Dynasty mention of the feast, we may exclude 500 BC. This leaves us with the year 2004 BC.

This was a fascinating and most relevant moment in Egyptian history. Mentuhotep II from Thebes had just re-unified the country and new buildings on a large monumental scale, were constructed for the first time in the very south of the country. The most significant of all was his mortuary temple at Deir el Bahari (see Chap. 6). Also, a few years later, the temple of a new aspect of the solar god, Amun, was re-erected by Senuseret I on the other side of the river, in Karnak, also on a larger monumental scale (Gabolde, 1998; Belmonte, 2006). Not surprisingly, both monuments were orientated towards the rising of the sun at the winter solstice and thus, for a few years, to the rising of the sun at the first day of the civil year, I *ꜣḫt* 1, or New Year's Eve (see Chap. 6). We may suggest that at this precise moment, when the actual Birth of Re at the winter solstice occurred in I *ꜣḫt* 1, the feast of *mswt-rꜥ* could be frozen at *wpt rnpt* for the rest of Egyptian history.

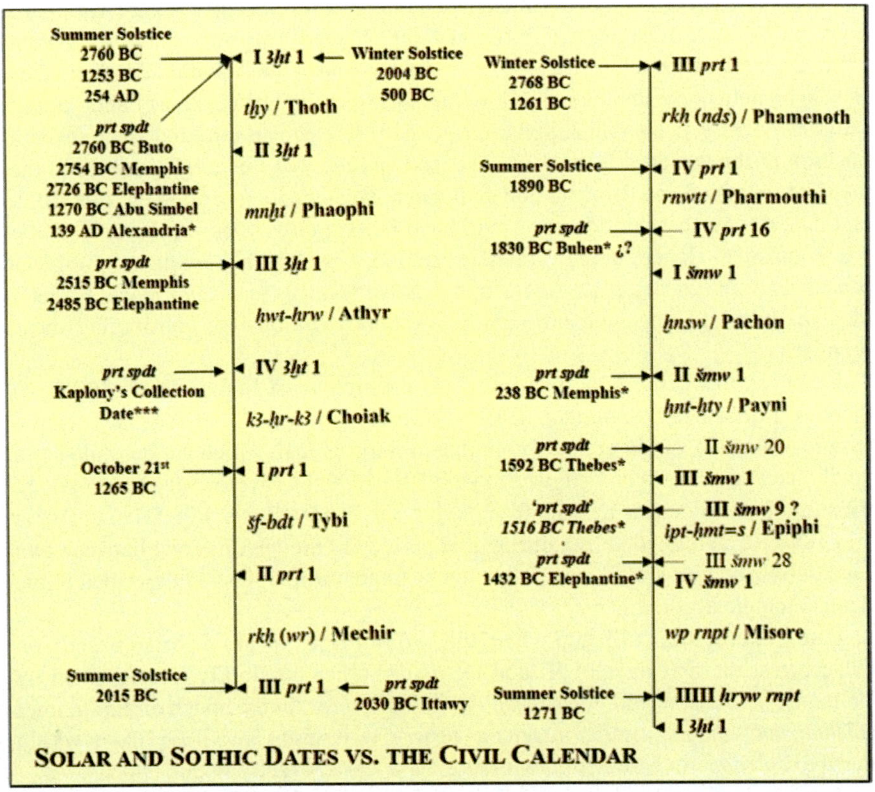

Fig. 5.12 Civil dates of the solstices as a function of time and the heliacal rising of Sirius (*prt spdt*) as a function of both time and location. Asterisks marks those reported by ancient sources. The much-discussed Ebers Sothic date is printed in italics. See text for further discussion. The recently discovered (Habicht et al., 2015) controversial Sothic date in the Kaplony's personal collection (***) is tentatively fixed in IV *ꜣḫt* 1. See the text for further discussions. (Adapted and updated from Belmonte, 2009)

A further reflection of this ancient tradition could be inherited by two of the most important festivals of Amun, celebrated during the Ramesside period when the winter solstice was not centred on I *ꜣḫt* 1, but rather occurred at the time of the months II *prt*, III *prt,* and IV *prt*. These were the festival of 'Lifting Up the Sky' (*ʿḥ pt*), celebrated from II *prt* 29 to 30, ending when the god 'Enters the sky' (*ʿḳ pt*) in III *prt* 1, followed by the festival of 'Entering the Sky' from III *prt* 29 to 30, and finishing when the god 'Re-enters the sky' (*sʿḳ pt*) in IV *prt* (El-Sabban, 2000: 60). In 1261 BC (see Fig. 5.12), in year 18 of Ramesses II, the winter solstice occurred more or less at III *prt* 1, and we hypothesize that this is what was commemorated at the very moment when the god 'enters' the sky. Obviously, this would have occurred for a period of just a few years. In any case, the winter solstice would have fallen in II *prt* or III *prt* during most of the New Kingdom.

This fact may have had some connections with the problem of the two *rkḥ*. This was the name of two months of the civil calendar (II *prt* and III *prt,* months 6 and 7 respectively), according to the model defended in this chapter. There were two eponymous festivals celebrated in III *prt* 1 and IV *prt* 1, respectively, as the Illahun archive clearly showed (Luft, 1992a, b: 215). Since the Middle Kingdom (see Fig. 5.4), both were distinguished by the adjectives 'great' (*ꜥꜣ* or *wr*) and 'small' (*nḏs*). Originally, in the Old Kingdom, just one *rkH* feast was reported in the seventh position of the standard festival list. We have argued that this might have been the name of months 6 and 7, within the civil calendar, since its inception (*c.* 2760 BC), and thus during that period, *rkḥ* would have fallen more or less at the time of the winter solstice. Hence, in the Ramesside period, when the *rkḥ* feasts are not mentioned at all, the equivalent Amun winter solstice festivals took their place. Parker's idea of 'burning', meaning artificial heat, sounds reasonable, considering these premises.

However, there is another possibility. In the archive of Illahun (Luft, 1992a, b) there is a document (Spalinger, 1992: 8) which shows a list of revenues for the temple of Anubis, at the pyramid complex of Senuseret II, which are delimited to a yearly interval, from year 1 III *prt* 1 to year 2 II *prt* 30. During the Middle Kingdom, the summer solstice was moving from III *prt* 1 in 2015 BC (see Fig. 5.12) to IV *prt* 1 in 1890 BC and beyond. So, during this period, a Nile year interval between two successive flooding would have fallen, on average, more or less as described in the Anubis temple account.

Furthermore, it is at Illahun where the two *rkḥ* festivals are reported exactly on those dates, the first days of III and IV *prt*. Another possibility, as suggested by Belmonte (2003a), would be that *rkḥ* was in fact a term for the epoch of the summer solstice, and that, in a similar manner as *mswt-rꜥw*, it might have been frozen within the civil calendar, in association with months 6 and 7, exactly in that epoch.

Against this idea, we have the striking evidence presented before that *rkḥ* might have been associated in some way with months 6 and/or 7 already in the Old Kingdom (see Fig. 5.10). Hence, we accept either *rkḥ* as a name to describe the period close to the summer solstice or that this was already the name of the sixth civil month in the Old Kingdom. Both alternatives are mutually exclusive. Belmonte (2003a) was unable to give preference to either hypothesis. However, we must admit that *rkḥ* as equivalent to the hot and waterless period embracing the summer solstice is a more sound and more appealing hypothesis. This would situate the relationship of the two solstices to important festivals precisely at the beginning of the Middle Kingdom.

Interestingly, following a similar kind of reasoning, Belmonte (2003a) proposed a hypothesis that would explain why the name Misore supplanted that of *wpt rnpt* as the name of month IV *šmw* only in the Late Period. This was actually not the case in documents at Deir el Medina, where other new names, like *p n jpt* or *p n jnt*, are already present (see Table 5.3).

This hypothesis relies in a promising piece of evidence in one of the fragments of a geographical papyrus from Tanis of the Late Period where the following inscription is recorded (Clagett, 1995: Fig. III.5):

.

This reads *4 šmw wpt rnpt ḫd r'w*, i.e. IV Shemu, 'Wepet Renpet, travel north of Re'. Between *c.* 645 BC and 520 BC, or within the reigns of Psametik I (664–610 BC) and Darius I (521–486 BC), the winter solstice fell in IV Shemu, suggesting that the ancient Egyptians were aware, during this epoch, that the sun started to move back north at the horizon in this month, as it happens after the winter solstice. If we interpret the winter solstice as the Birth of Re', then the adoption of the name Misore for this month could have occurred at any moment within this period.

In 525 BC, the Persians conquered Egypt. The association of the winter solstice with the birth of the sun god Mithra could have been obvious to the invaders. Finally, in 500 BC, I *3ḫt* 1 and the winter solstice were once more coincident. This might have been the trigger for the definitive association of Misore with month IV *šmw* and of the Feast (of the Birth) of the sun (*ḥb rʿ*) with New Year's Day.

5.5 The Beginning(s) of the Egyptian Year: An Egyptologist's Nightmare

From any logical point of view, if there was just one calendar in ancient Egypt, there must have been just one beginning of the year. However, the perennial discussion on the existence of other calendars has tried to find 'beginnings' for the different proposals. Moreover, it is an established fact that, in ancient Egyptian sources, there are mentions of more than one term that might be interpreted as the beginning of the year. In accordance with this line of argument, from the feast lists of the Old Kingdom and of the Middle Kingdom (see Figs. 5.11 and 5.4, respectively), we have at least two terms that could be interpreted as the beginning of the year. These would be:

1. , *wpt rnpt* (*ḥb*), the Opener of the Year (Festival), and
2. , *tpj rnpt* (*ḥb*), the First of the Year (Festival).

The second one must not be confused with the also very common term: , *tp rnpt* (*ḥb*), either Head, At the Front or Beginning (also First) of the Year (Festival). Fortunately, there is a consensus that this is an alternative term for the first of the epagomenal days (also known as *mswt-wsir*), which could also be considered to be at the 'beginning of the year', because the epagomenal days were 'above the year' (Spalinger, 1990).

This must be distinguished from , *tp tr*, the Beginning of the Season. Also, to this sample we might add (see Fig. 5.4):

i) , *rnpt ʿ3t* (*ḥb*), the Great Year (Festival), and

ii) , *rnpt nḏst* (*ḥb*), the Small Year (Festival), terms which, as we have seen, are very difficult to interpret. Fortunately, for the End of the Year we found just a single term:

, *ꜥrḳ rnpt* (*ḥb*), the Closing of the Year (Festival),

which is always undoubtedly associated with the last day of the civil year, excluding the 5 *ḥryw rnpt*, IV *šmw* 30 (see Fig. 5.4).

The situation, however, is further complicated by the appearance, perhaps in the Old Kingdom or with certainty in the Middle Kingdom (Krauss, 1998a, and reference therein) of:

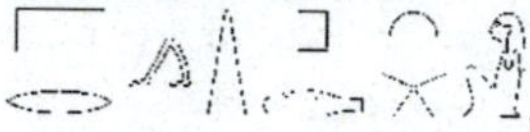,

prt spdt, Peret Sopdet, the Going Forth of Sothis or the Heliacal Rising of Sirius, either in transliteration, transcription, translation or astronomical terminology, respectively.

From the very beginning of Egyptian calendrical studies, this term was associated with *wpt rnpt* for two reasons especially, based on two particular documents: the Ebers calendar (remember the Ebers syndrome) and the Tanis version of the Decree of Canopus (Parker, 1950, and references therein).

In the hieroglyph section of this extremely important bilingual (tri-writing) inscription we read:

On the day of the Going Forth of Sothis, called Wepet Renpet in name in the writings of the House of Life.

From this and other lesser documents, such as the astronomical ceiling of the Temple of Dendara, Parker (1950: 33) argued the validity of the equation *wpt rnpt* = *prt spdt* from the very beginning of Egyptian history. As a consequence, *tpj rnpt* was proposed as the first day of the Sirius-based lunar calendar, a hypothesis that has not been seriously contested (Belmonte, 2003a).

However, this is not the case for the equation *wpt rnpt* = *prt spdt*, which has been severely challenged. The vast majority of the documents from the Old Kingdom (Abusir Archive; Posener-Kriéger, 1976: 571 with note 3), passing through the Middle Kingdom (Illahun Archive; Luft, 1992a, b), the New Kingdom (e.g. the Buto Stela; Bedier, 1994; Spalinger, 1996c, see also below), and beyond clearly indicate the inequality of both terms.

Hence, *wpt rnpt* was from the invention of the civil calendar the name of the feast associated with the first day of the civil year I *ꜣḫt* 1. What, then, can scholars do with the clear statement of the Canopus Decree?

Spalinger (1992: 31) has cleverly argued that the most reliable version of the decree is the one written in Greek, and that the demotic and hieroglyphic version are mere translations of the former. The Greek version states: 'on the day of which the star of Isis heliacally rises which is regarded in the sacred writings to be a new year', which gives in the demotic version the term *ḥ3t rnpt*, the 'beginning' of the year, not the 'opener'.

So what the Decree clearly says, as Spalinger has pointed out, is that *prt spdt* is associated with the beginning of a new era and not necessarily with the precise date of *wpt rnpt*. However, it is obvious that Sothis was somehow associated with the beginning of the year, and that a certain kind of year begins with its rising (see below Sect. 5.7), as several later-period texts demonstrate. For instance, one example from the Graeco-Roman period at Dendara explains that:

Years are reckoned from her shining-forth.

It is worth mentioned that the special relation between Sothis and *wpt rnpt* might have been generated in the New Kingdom. During the 18th Dynasty, the star actually had its heliacal rising during the month *wpt rnpt* (IV *šmw*) and, depending on the latitude, exactly at I *3ḫt* 1 in the decades around the beginning of Ramesses II reign, as the famous inscription in the Ramesseum apparently demonstrates (see also Figure 5.13):

May he let you shine as Isis-Sothis on the sky in the morning of the Opening of the Year.

However, it is important to notice that we lack evidence of any other earlier connection between Sopdet and *wpt rnpt*. In any case, it is clear that Sirius was in one way or another related to the calendar even in the Old Kingdom, since in the utterance 965 of the Pyramid Texts (Faulkner, 1969; Allen, 2015) we read:

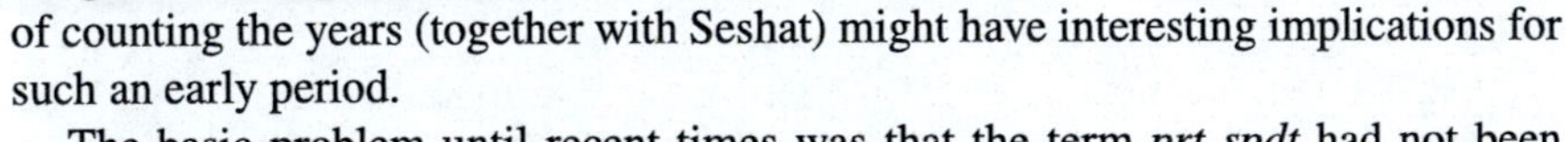

It is Sothis, your daughter, your beloved, who has made your year-offerings in this her name of Year.

The direct calendrical meaning is not obvious, but the assimilation of Sopdet with the goddess Renpet, companion of the god Min and one of the goddesses in charge of counting the years (together with Seshat) might have interesting implications for such an early period.

The basic problem until recent times was that the term *prt spdt* had not been reported in documents earlier than the Middle Kingdom, when it frequently appears in festival lists and is mentioned twice in the Illahun Archive—the famous and controversial Sothic date of IV *prt* 16 (see below).

5.5.1 *The Enigma of Tepj Renpet vs. Peret Sopdet*

Does this mean that the heliacal rising of Sirius was not observed (or was of lesser importance) in previous periods of Egyptian history, including the Old Kingdom?

Fig. 5.13 Close up of the astronomical ceiling of the Ramesseum in Thebes, where this suggestive text is inscribed to the right in the image … 'may he let you shine like Isis-Sopdet in the sky in the morning of Wepet Renpet' … (see text). This should actually have happened in Thebes in the years around 1285 BC, during the reign of Ramesses II's father, Sethy I, owing to the wandering nature of the Egyptian calendar. (Photograph by the authors)

Not at all! Krauss (2002) has suggested that knowledge of the progressive delay of the heliacal rising of Sirius had been a planned part of the firm establishment of the civil calendar of 365 days since its very beginning. Also, Nolan (2003) has published a very interesting study where he unmistakably demonstrates that the 'cattle count' was not at all regular during most of the Old Kingdom and the number of 'years after' the count (see Chap. 7) should be dramatically reduced to one third. This must have strong implications for the chronology of the age of the pyramids, still not fully understood and even less applied by modern studies.

In that work, Nolan also suggested that 'cattle counts' were made on a semi-regular basis, in agreement with Parker's lunisolar calendar, heralded by *prt spdt*. However, an in-depth analysis of his own data demonstrates that this possibility is unrealistic and aprioristic—there is a data selection effect (see Chap. 7). So, the reality of the observation of the heliacal rising of Sirius before the Middle Kingdom was a very controversial topic.

To save the situation, Spalinger (1996b: 36), after disqualifying Parker's equation (i.e. *wpt rnpt* = *prt spdt*), re-enacted a previous proposal. The Hungarian scholar V. Wessetzky (1954) argued that the Old Kingdom term for *prt spdt* was none other than *tpj rnpt,* so conspicuous in festival lists and located, in the standard arrangement, after *wpt rnpt* and *ḏḥwtyt.*

In our discussion of these lists (see Fig. 5.10), we have associated *tpj rnpt* with month II *ꜣḫt* and the transition to III *ꜣḫt* (with *wꜣgj* related to the following month). In effect, as shown in Fig. 5.12, *prt spdt* would have taken place in III *ꜣḫt* 1 in the decades around 2500 BC, in the middle of the Fourth Dynasty (according to most accepted chronologies) and at the precise moment when the festival lists started to appear in the private necropolises at Giza and Saqqara.

Another fact has been claimed to support this hypothesis: that *tpj rnpt* is sometimes present in typical Middle Kingdom festival lists (where *prt spdt* is not explicitly mentioned) just after *rkḥ*. Considering that at the beginning of the Middle Kingdom, *prt spdt* took place, in the decades around 2030 BC, at III *prt* 1, the classical date of *rkḥ*, it would have occurred after that date during most of the Middle Kingdom, exactly as the feast lists suggest (Belmonte, 2003a).

Doubt, however, can unfortunately arise from Middle Kingdom data as well. In a statue of a certain Amenemhet, found at Abydos and dated during the reign of Senuseret I or Amenemhat II, the following list of festivals is reported (Spalinger, 1996b: 47): *ꜣbd/smdt tp/wꜣgj/ḏḥwtyt/ḥb wr/ḥb skr/prt mn/prt spdt/tpj rnpt/wpt rnpt/ḥb rʿ nb.*

Although the order does not seem chronological (the location of *wꜣgj* before *ḏḥwtyt* is an innovation of the end of the Old Kingdom, when the civil-based *wꜣgj* was resettled in I *ꜣḫt* 18 just before *ḏḥwtyt* in I *ꜣḫt* 19), there is one important objection: *prt spdt* is mentioned near both *tpj rnpt* and *wpt rnpt*. So, unless we accept the correction that *tp rnpt* (standing for the first of the epagomenal days, see above) should have been written instead of *tpj rnpt*, we would be confronted with a serious challenge to the above-proposed hypothesis. Actually, the recent 'discovery' of a new Sothic date, presumably of the Old Kingdom, in a previously unpublished small jar in the personal collection of Swiss Egyptologist Peter Kaplony, has turned the situation upside-down (Habicht et al., 2015), even though the reading has not yet been clarified, as we shall discuss later on.

Consequently, the equation *prt spdt* = *tpj rnpt* is not as obvious as many scholars would like and we might therefore need to search for an alternative explanation for the term *tpj rnpt* as the beginning of something.

Belmonte's (2003a) answer was that this alternative can be found in the Nile year that we have discussed exhaustively early on. Once the civil calendar was inaugurated, the Nile year probably ceased to be operative. However, the arrival of the rising waters must still have been a primordial event, even within the frame of the wandering civil calendar, as the archaeological proofs found at the temple of Satet in Elephantine and others certainly suggest (Bard, 1998: 283). A feast is unlikely to be celebrated haphazardly. Hence, when the arrival of the waters was detected by the Nilometers, it is possible that the celebration of the arrival of the flood would have been moved to *psḏntjw*, *ꜣbd*, *smdt,* or any other important day of the following synodic month. This would have been the first (*tpj*) lunar event of that class in the new Nile year, which might also be called *rnpt*. We find this solution still defensible. It is worth mentioning once more, that this does not mean that a whole set of lunar months would have been counted after that event, although there might perhaps have been some special lunar computus with a set of lunar festivals whose dates could have been estimated from that point onwards, as in the Christian ecclesiastic calendar.

In the same sense, another important point to discuss is whether *prt spdt* could be considered as the beginning of some of these periods of time, either a certain set of lunations (not properly a year) or a set of lunar festivals, as has frequently been claimed (Depuydt, 2000). In the following sections, we deal once more with this particular point, regarding either *prt spdt* or a first lunar event after the arrival of the flood.

We need to clarify, for this is really important, the point that particular problem of these two 'beginnings' (Tepy Renpet and Peret Sopdet) is that both have a markedly local character. The inundation would arrive, on average, twelve days earlier at Elephantine than at Buto in the Delta, and, in a similar fashion, the heliacal rising of Sirius would have been observable, under good atmospheric conditions, 8½ (*c.* 2800 BC) to 6½ (*c.* 200 AD) days before at Elephantine (latitude 24°) than at Buto (latitude 31°).

This means that any lunar event calculated from those 'beginnings' could have a difference of a whole lunar month from Elephantine to some important city further to the north. There is a parallel to the standing problem of Catholic and Orthodox Easter because each church uses a different calendar (Gregorian and Julian, respectively). In this case it is a cultural issue, whereas in the Egyptian case it is a problem of different latitudes and hence timing. This will be relevant for the argument that follows.

To finish this discussion on the 'beginnings' of the year, we should briefly discuss the festivals known as *rnpt ꜥꜣt ḥb* and *rnpt nḏst ḥb* that can be found in a few feast lists of the Middle Kingdom (see Fig. 5.4). We can hardly imagine another festival, with a different name, celebrated either on the first day of the civil calendar (*wpt rnpt*), on its last days (*ʿrḳ rnpt*), or on any of the epagomenal days (5 *ḥryw rnpt*) since all of these are already present in the list.

Should we go back then to a lunar interpretation? We have seen that this was Neugebauer's idea in connection with Papyrus Carlsberg 9. Also, in a demotic papyrus edited by Parker (1959), there is a mention of 'small' and 'great' years is associated with the key word *wrš*, which could be translated as a lunation period of service (Krauss, private communication). So it is probable that the Festivals of the Great and of the Small Year might have been celebrated at some particular lunar events. However, with currently available data it is impossible to take these hypotheses any further.

5.6 Lunar Days, Lunar Festivals, and the Ancient Egyptian Lunar Computus

It is evident that the ancient Egyptians were especially keen on the phases of the moon and that several of their major festivals were celebrated at key moments in the lunar cycle. This was either for every lunation within the civil calendar or at some important festivals, such as Peret Min, which was at once the name of LD30 and the name of a Feast celebrated at the *psḏntjw* of I *šmw* (see Fig. 5.10).

However, it is also quite clear that these lunar feasts were expressed within the framework of the civil calendar, and that there are no unmistakable proofs that these lunar festivals were ever articulated in the form of a calendar once the civil one was in operation (Belmonte, 2003a). This is illustrated in the feast lists of the tomb of Khnumhotep II (see Fig. 5.4), where there is a mention of 12 *ꜣbd* and 12 *smdt*, even though there could have been an extra *ꜣbd* or *smdt* in several years either in one of the *ḥryw rnpt* or as an additional one in a civil 'blue' month. Apparently, these extra lunar events were not taken into account.

Spalinger (1995d) has gone one step further in proposing that the 12 *ꜣbd* and the 12 *smdt* of the lists might actually refer to days 2 and 15 of the 12 civil months. However, this hypothesis probably lacks solidity. There are in fact several occasions when a clear lunar event is reported with an independent civil date, such as the famous occasion of a *psḏntjw*—whether predicted or observed, nobody knows—at the battle of Megiddo, which is reported to be exactly (sic) on day I *šmw* 11 in the 23rd year of Thutmose III. The hieroglyphic texts reads:

We will return to this important lunar date in Chap. 7. Here, *psḏntjw* (often transcribed psedjentiu) refers to the day of new moon or conjunction, the first day of the Egyptian lunation (synodic month) according to most authors. This feast day is already mentioned in the *Pyramid Texts* where, however, it is written without the preposition *n*:  , this could be relevant to the early discussion on the calendar of Esna.

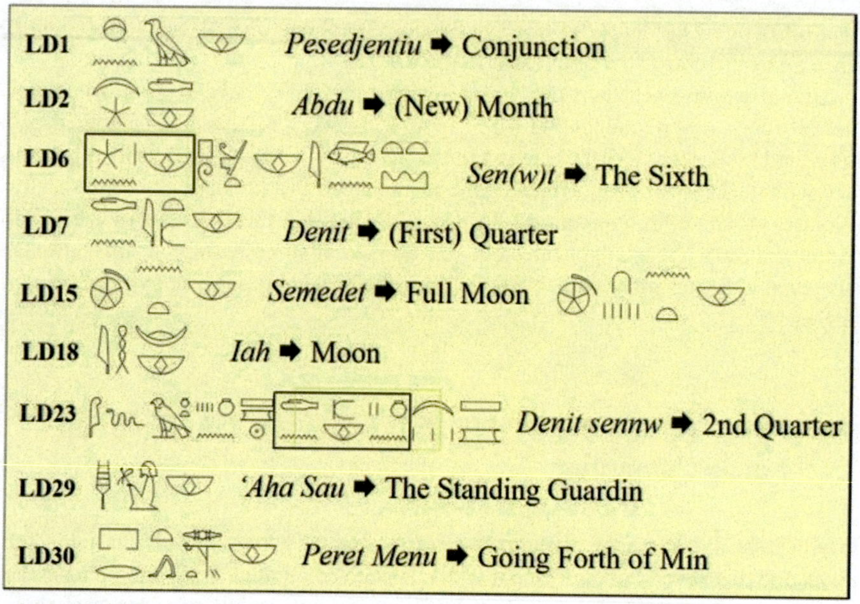

Fig. 5.14 Some of the most important days of the lunar cycle throughout Pharaonic history. In particular, LD29 was of special relevance because on that day, as its name suggests, the duration of the corresponding lunation was fixed by naked eye observations at 29 or 30 days (Belmonte, 2003a). LD30 was present only on a slightly more than half of the occasions. Lunar dates at LD6 and LD23 are squared within longer reported inscriptions. LD6 and LD15 were often written with numbers, indicating their position within the lunation which started at ‘conception’ at LD1. The quarter days (LD7 and LD23) are also straightforward because they are represented by half a cartouche, suggesting that only half of the moon was visible in those days. See the text for further discussions

Other common lunar days (see Figure 5.14) from the Old Kingdom onwards were , or simply , *ȝbd* (First Crescent, often transcribed abd(u)), for LD2, and , *smdt* (or semedet, full moon, read also *mḏ-ḏi-nt*, according to Luft, 1992a, b), for LD15 of the cycle (also reported as). In the Middle Kingdom, *snwt*, the LD6, is also found. From the early New Kingdom, a standard list including of the order of half a dozen of these festivals was often reported.

However, in the sources of the Graeco–Roman period, such as the astronomical frieze of the temple of Edfu (for the temple of Dendara after a fascinating restauration, see Figure 5.15), a whole set of days of the month can be found. Table 5.4 reports this set from the eclectic list reported by Parker in his *Calendars* (Parker, 1950: table 2), with minor modifications:

Given that the list offers 30 days, and that it is located within a civil calendar context, it has been argued that these are actually the proper names of the thirty days of the month of the civil calendar. In fact, many of them have never been reported in either a lunar or in a civil context and some others, such as LD18, *jꜥḥ* (see Fig. 5.14) have only ever been found in a civil context, representing day 18 of a certain civil month (Depuydt, 1997: 84).

Fig. 5.15 As in many other sectors of the sky diagram, the cleaning and restoration of the astronomical ceiling of the hypostyle hall at the Temple of Hathor in Dendara has uncovered the beautifully coloured long-hidden inscriptions, including the list of the 30 days of the (lunar) month. The top panel shows the seven first days of the lunation, from *psḏnty*(*w*) to *dnjt* (including LD6 by number), heralded by an introductory formula. The bottom panel shows LD28 to LD30, the latter called *ḥb nw pt* instead of *prt mn*. Note the human figure in LD29 with a sort of device on his hands (perhaps a *bꜥi* or a similar astronomical instrument), suggesting an *ỉmy-wnwt* waiting for his duties. Adapted from the forthcoming *The Sky over Dendara* by José María Barrera. We really appreciate Barrera's willingness to allow us to use some of his extraordinary photographic material prior to publication

It is therefore apparently not easy to deny this hypothesis. However, it is quite clear that several of these names (for days 1, 2, 4, 5, 6, 7, 10, 15, 16, 27, 29, and 30) have been reported in a quite obviously lunar context, and so it is highly probable that the whole set would have been developed to offer suitable names for each day of a lunation. This might have been particularly important since it would have not been clear exactly when the lunation commenced. Recently, Spalinger (2020a, b)

Table 5.4 List of the days of the lunar month

LD	Names	Glyph	LD	Name	Glyph
1	*psḏntjw*	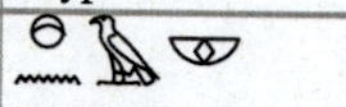	16	*mspr sn-nw*	
2	*3bd*	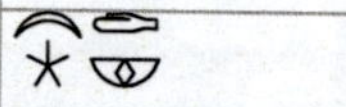	17	*sj3w*	
3	*mspr*	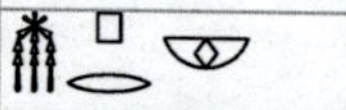	18	*jʿḥ*	
4	*prt sm*	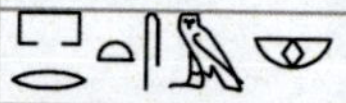	19	*sḏm mdw.f*	
5	*jḫt ḥr ḫ3wt*	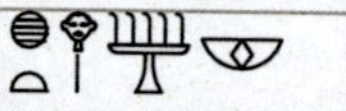	20	*stp*	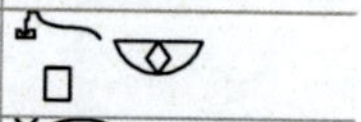
6	*snwt*	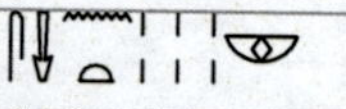	21	*ʿprw*	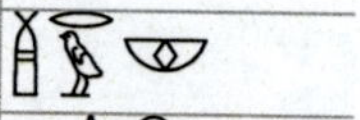
7	*dnjt*	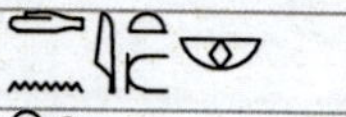	22	*pḥ spdt*	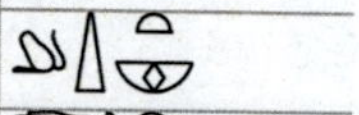
8	*tp*	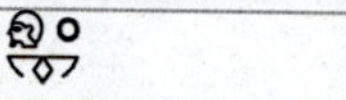	23	*dnjt*	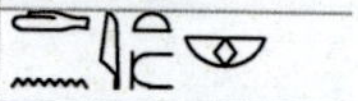
9	*k3p*	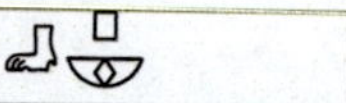	24	*knḥw*	
10	*sjf*	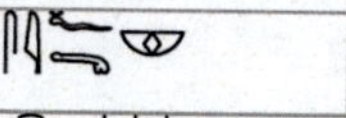	25	*stt*	
11	*stt*	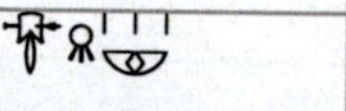	26	*prt*	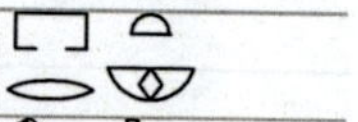
12		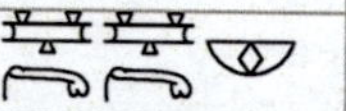	27	*wšb*	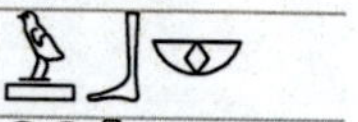
13	*m33 sty*		28	*ḥb-sd nwt*	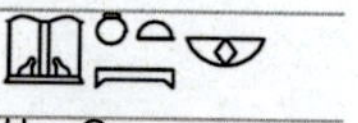
14	*sj3w*		29	*ʿḥʿ s3w*	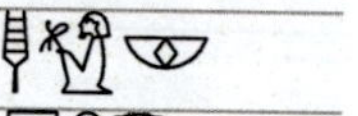
15	*smdt*	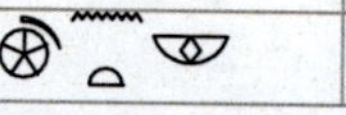	30	*prt mn*	

Adapted from Belmonte (2003a). See the text for further discussion

has argued, 'many names are visual descriptions of the moon at a specific point of time while some days are decidedly cultic', and suggests that they are related to the actual behaviour of the moon and very ancient Egyptian religious traditions. We suggest that some, like LD29, can even go further.

This implies that these names are very old and possibly identified the lunar days without any need to an articulated monthly cycle within a lunar calendar, as certainly were the dates of the civil calendar. The LD count would hence be an arbitrary modern way of naming the days for the sake of simplicity. A good example of this is LD 28 *ḥb-sd nwt*, the Jubilee of Nut, which might relate to the possibility of the sky goddess being ready to be fecundated on this day so that the lunar god would be conceived in LD1, *psḏntjw*. Actually, Spalinger (2020a, b) proposes a series of suggestive hypotheses for the interpretation of several names within this context.

A dramatic confirmation that the lunar days actually represent the real behaviour of the moon in the sky comes from the also astonishing restoration (see Figure 5.16) of the astronomical ceiling of the hypostyle hall of the Temple of Khnum at Esna.

Fig. 5.16 Snapshots of the section representing *smdt*, the full moon, of the Esna astronomical ceiling before and after restoration. The reappearance of long by-gone names and sections of the painted decoration, including colours, is astonishing. As an example, just note the Wadjet Eyes on the moon discs in the lower-left corner. (Images by courtesy of Jose Ricardo Belmonte, taken in 2004, and Artiom Gizun, taken in 2021)

Priskin (2016) had made an interesting analysis, prior to restauration, of the organization of the lunar section of this ceiling, suggesting the presence a full 30-day lunar 'month' from two periods of lunar invisibility. The images would be read in an anti-clockwise direction starting from the lower register. This is beautifully confirmed now with the newly cleaned decoration and register (see Figure 5.17).

This figure illustrates the moon section of the astronomical ceiling in the north of the hypostyle hall. A whole 30-day synodic month is represented. It starts in the

Fig. 5.17 (**a**) Marvellously restored north section (light enters from the east) of the astronomical ceiling of the hypostyle hall of the temple of the god Khnum at Esna. The section represents the moon cycle, mixed with other astronomical images (presumably constellations) and winds, including texts. A synodic month, or lunation, of 30 days is shown in panels b, c, and d: (**b**) *psḏntjw* represented by the conjunction of the dark moon disc and the solar god Khepri; (**c**) *smdt* or full moon, represented by the conjunction of the complete Wadjet Eye on the moon disc in opposition to the sun god, presumably Khnum-Re, the temple owner. The last panel (**d**) shows the period of waxing (below) and waning (top) moon as beautifully illustrated by the increasing and decreasing visibility of the Wadjet Eye in the 14+14 lunar discs. Each of these is associated with deities of the lunar days. We thank the Belorussian scholar Artiom Gizun for kindly permitting the use of these extraordinary images

lower section, where the days of the waxing moon are represented, including perhaps the eventual one or two days when the moon is still invisible. Afterwards, the Wadjet Eye grows day by day for 14 days until the disc is completed and presented by Thoth. Hence *smdt* or the full-moon is nicely represented in opposition (they face different directions) to the solar (nightly) god, presumably identified here by the god Khnum-Re, owner of the temple.

The scene continues in an anticlockwise direction in the top section represented by the 14 days of the waning moon. The elegant diagram ends with *psḏntjw* (see below) represented by the conjunction of the dark disc of the moon and the solar god Khepri, the sun at dawn, escorted by two lunar deities Thoth and a falcon-headed god, perhaps Khonsu. The images are so nicely expressed that there is hardly any doubt that the diagram symbolically represents a real observable lunar cycle as it happens in the sky. The only concern might be that it is a very late work of the Roman period, but we do not see a problem with that.

It is worth noting that, since Parker's *Calendars*, it has usually been accepted that the Egyptian lunation started at *psḏntjw*, on the day following that in which the last crescent of the moon could be seen at dawn. Frequently, an illustrative text at the Ptolemaic propylon of the temple of Khonsu at Karnak is advocated, supporting this idea (Brugsch, 1883):

He (Khonsu) is conceived (*bḳꜣ*) on *psḏntjw*, he is born in *ꜣbd*, he grows old after *smdt*.

The rendering of *smdt* as LD15 (different to more frequent ones, see Fig. 5.14) is fascinating: the standard glyph for number 10, plus the star standing symbolically for number 5.

However, there is scant mention of *psḏntjw* in the Old Kingdom (whereas *ꜣbd* and *smdt* are very frequent). The same can be said for the Middle Kingdom. For example, in the Illahun Archives, *psḏntjw* is mentioned three times, while *ꜣbd* and *smdt* have more than 15 entries each. Most importantly, it has been demonstrated that it frequently occurs that there are not 12 full days between one and the other (as would be expected if they were always LD2 and LD15), calculating the interval between various *ꜣbd* and their corresponding *smdt*. Sometimes there are instead 11 or 13, indicating that we might be faced with a real (and not a calculated), observed synodic month, starting on the evening visibility of the first crescent, which then grows until full moon (Luft, 1992a, b: 116 and 198; see also Depuydt, 1997: 183). This idea might be reinforced by the following text, also of Middle Kingdom provenance):

I know, Oh souls of Khmenu (Hermopolis), what is small on *ꜣbd* and what is great on *smdt*, this is Thoth.

This issue would also speak against an explicitly expressed lunar month (at least in the Middle Kingdom) and thus would not favour any possible well-structured lunar calendar (Belmonte, 2003a). Clagett (1995: 22) is also very explicit on this point.

However, the most important day in each Egyptian lunation was probably LD29. This day was called *ꜥḥꜥ sꜣw* (Belmonte, 2003a) which we may translate as the ‘Standing Guardian’. The most likely significance is that somebody (probably an *imy wnwt* or ‘hour-priest’, who would maintain vigil all night long, see Chap. 2) was observing at the end of the night on that day (see Fig. 5.15). It should be remembered that, according to Krauss and other scholars (e.g. Krauss, 2003), the Egyptian day started at dawn, at the moment of *ḥḏ-tꜣ*. The duty of this standing observer would be to discern the last crescent of the waning moon. If it was clearly visible, then the next day would still be LD30 of the current lunation, but if, on the contrary, the last crescent was not visible, then the next day would be *psḏntjw* of the following lunation. Furthermore, that LD29 was a most prominent day can also be inferred from the list of Feasts of Sky (*ḥbw nw pt*) in the Festival Calendar of the funerary temple of Ramesses III at Medinet Habu (see Figure 5.18). There lunar days 29, 30, 1, 2, 4, 6, 10, and 15 are mentioned and exactly in that order (El-Sabban, 2000: 60; Belmonte, 2003a).

Nevertheless, it is highly probable that the ancient Egyptians considered their lunations to start by conjunction, *psḏntjw*, called *neomenia kata selene* (‘new moon according to the moon’) in line 125 of Papyrus Carlsberg 9, at least from the New Kingdom onwards, and indeed in later periods (see Fig. 5.15).

Apart from the lunar day dates, there are frequent references in Egyptian documents and monumental inscriptions that refer to important festivals associated with a particular day of the moon. This particular day of the moon (often a *psḏntjw*) usually fell at the same time associated with a certain month of the civil calendar. This is obvious for most of the ‘lunar’ feast of the New Kingdom (Opet, the Procession

Fig. 5.18 The southern wall of the majestic Million Year Mansion of Ramesses III at Medinet Habu, in Thebes, on the west bank, where the feast calendar was written (**a**). This includes a list of earthly and heavenly festivals. The Feats of the Sky included a series of important lunar days. The Feasts of the earth, curiously started by Peret Sopdet in her day (*sic*) during the first month of the year, I Akhet (**b**). This was indeed the case in the time of Ramesses III epoch. (Photographs by the authors)

of Min, and various Amun festivals), as clearly stated in the Medinet Habu calendar (El Sabban, 2000: 60; Spalinger, 2020b). However, several doubts arise when the documentation on the Old and Middle Kingdoms and of the Beautiful Feast of the Valley in the New Kingdom are considered. Discussion of the Feast of the Valley is frankly controversial since the information reported is often contradictory in nature and would need a chapter in itself (e.g. Krauss, 2015). We therefore limit our discussion to the earlier periods.

In the Illahun Archive, apart from the two mentions of *prt spdt* on day IV *prt* 16 (or 17) in the year 7 of Senuseret III (see below), there are several mentions of some other feasts that have been classified as 'lunar' by Luft (1992a, b: 141) in his analysis of the data. The most important ones are the Proceeding of the Land (*ẖnt nt tꜣ*), Jubilation (*jhhj*), the Measurement of the Cord of the Nile-mile (*šspt jtrw*), and that of Wagi (*wꜣgj*).

There are more than seven entries for the feast of the Proceeding of the Land and, for the dates that are reported, it was always celebrated between III *prt* 1 and III *prt* 17. This means that it was perhaps located at *psḏntjw* of the civil month III *prt* and was thus probably expressed within the framework of the civil calendar. Regarding *jhhj*, there were apparently two feasts of the same name but the most common and most important was always celebrated on I *ꜣḫt* and probably at the first *psḏntjw* of the civil year.

Finally, we have six dates for the feast of *šspt jtrw*. This has normally been associated with *prt spdt*, on the basis of the argument that it might have been established with the latter as a reference point. However, the extant data show that the feast was celebrated between III *ꜣḫt* 25 and IV *ꜣḫt* 14, so we might then relate it to a feast celebrated at the *smdt* (indeed another important lunar day) of the lunation started in III *ꜣḫt*, which would always fall exactly between III *ꜣḫt* 15 and IV *ꜣḫt* 14. Consequently, these three feasts were apparently well articulated within the framework of the civil year, in the same fashion as Easter is articulated within the Gregorian calendar. Hence, there is no need to claim any sort of lunar calendar to explain their behaviour.

The situation of the important 'beautiful feast of *wꜣgj*', however, is further complicated. Unlike the others, which never had a fixed civil-based twin, there are news of a fixed *wꜣgj* feast in I *ꜣḫt* 18, with its corresponding Eve on day 17, at least from the Middle Kingdom onwards. This situation probably originated during the last years of the Old Kingdom (6th Dynasty), when a new order is imposed on the festival lists, with *wꜣgj* falling before *ḏḥwtyt*. However, there was its moveable counterpart whose first mention must be the oldest feast list of the 4th Dynasty, and which lasted at least until the end of the Middle Kingdom.

The *wꜣgj* feast was probably an important festival associated with wine, as shown by Merenre's *Pyramid Texts* (utterance 442; Faulkner, 1969: 819c–820b). Consequently, it has been suggested that it may have been a vintage feast, associated with the cult of Osiris (and its celestial counterpart, *sꜣḥ*). Curiously, when the other *wꜣgj* was fixed within the civil calendar, it was located close to *tḫj* ('Drunkenness') and *ḏḥwtyt* ('Thoth'), both festivals clearly associated with wine and inebriation (Luft, 1994; Spalinger, 1996b: 163; Krauss, 1998a, b). Another connection between *w3gj* and the vintage might come from Papyrus Berlin 10007 (recto). Here, an entrance of the moveable *w3gj* (in II *šmw* 10+X), is followed two months later (in IV *šmw* 1) by the performance of *ḥtp nṯr ḥ3t jrp*, where perhaps the 'first divine offerings of wine' of the new vintage—after nearly two months of maceration—would be mentioned (Spalinger, 1992: 8).

Unlike the custom in other Mediterranean countries, where vintage is normally in September, in Egypt grape picking starts much earlier because of the special climatic regime of the country, so that the grape harvest normally lasts from mid-July to mid-September (Robinson, 1999: 253). This means that a feast celebrating the vintage could be performed at any point within that period of time. In 2620 BC, nearly at the beginning of the 4th Dynasty (when the first festival lists were recorded), mid-July Gregorian would have corresponded to III *3ḫt* 1 in the civil calendar and mid-September to the beginning of I *prt*. In Fig. 5.10, we had provisionally assigned *w3gj* to month III *3ḫt*.

In the archive of Abusir, there apparently was a *w3gj* feast reported (besides the fixed one) at a date III ? 28, where '?' might stand for either *3ḫt* or *prt*. The chronological relevance, or otherwise, of this date will be analysed in Chap. 7. If *w3gj* were III *3ḫt* 28, it would correspond to mid-July around 2500 BC and thus we might be faced with a feast associated with the beginning of the grape harvest, before the maximum rising of the Nile flooding. If on the contrary it were III *prt* 28, we would need to move further forward by more than 250 years in Egyptian history (*c.* 2250 BC, at the end of the Old Kingdom) for it to be located in mid-September. Consequently, we might be dealing with a feast at the end of the vintage, when the river was already very high. A date nearer to this second possibility sounds more promising because a feast would be better celebrated at the end of an agricultural event and, especially, because a date a little earlier of 2250 BC is much more probable than 2500 BC for the dating of the Abusir archive. How, then, was the date of the feast calculated?

In the case of the Illahun archive, we clearly have two cases of the moveable *w3gj* feast in days II *šmw* 29 of year 9 of a king and in II *šmw* 17 of year 18 of yet another king. The kings are probably Senuseret III and Amenemhat III. These dates have been assigned by Krauss (2003), following Luft (1992a, b: 232), to lunar day 17 (or 18) of the second full 'lunar month' after the heliacal rising of Sirius (occurring in IV *prt*), considering that *prt spdt* might have acted as a sort of zero point for the determination of at least this lunar event. Curiously, when Luft proposes the same experiment for the Abusir archive, the first full 'lunar month' after *prt spdt* is the one claimed. Moreover, in the Old Kingdom, when the fixed *w3gj* was located within the civil calendar, days 17 and 18 of the first month of the year, *tḫ*, and not of the second, were chosen accordingly. In the present state of our knowledge, we cannot offer an easy solution for this apparent dichotomy. Belmonte (2003a) has doubted that *prt spdt*, was the harbinger of *w3gj* in the Old Kingdom, and opted for *tpj rnpt*. We have seen that this festival could be either the older name of *prt spdt* (unlikely now) or the name of the first lunation after the arrival of the flooding.

The determination of the date of the moveable *w3gj* feast through a Sirius-heralded lunar computus—associated with either the first or the second lunation after the heliacal rising—has not been proven. Hence, Depuydt's (2000) asseveration that 'the link of certain lunar dates (such as *w3gj*) to *prt spdt*, whatever it was, suffices to claim the existence of an original lunar calendar', is far from certain. This is as if, in the Gregorian calendar, the link between Easter and the full moon after the vernal equinox guaranteed the existence of an equinox-heralded lunar calendar.

Belmonte (2003a) had doubts about the use of *prt spdt* to articulate a complete series of synodic months (i.e. of lunar months within a Sirius-based lunar calendar) during most of the existence of ancient Egyptian civilization. However, there is indeed some striking evidence that it may have been the beginning of something, at least in the Ptolemaic period, when, as we have mentioned, many inscriptions related Sothis to the 'beginning' of a certain 'year'. This evidence, although scarce, would point to the possibility that in the Ptolemaic period *prt spdt* was considered as the beginning—or herald—of a certain cycle of lunar festivals just as the spring equinox is the herald of Easter and other Christian feasts determined by the moon.

All in all, our wish is to have shown that the vast majority of the lunar feasts in the Egyptian year were clearly expressed within the framework of the civil calendar throughout most of Egyptian history. Often, they were heralded by civil dates. There are possible exception such as the moveable *w3gj* in the Old and Middle Kingdom and some doubts concerning *ḥb jnt*, the important New Kingdom Feast of the Valley (see Belmonte, 2003a: 46–47, for a discussion on *ḥb jnt* and the eponymous name Payni; most relevant is Krauss, 2015). Consequently, our contention is that there is no need to claim for any sort of articulated lunar calendar—even in the case of *w3gj*—to explain the *modus operandi* of the lunar festivals in the calendrics of ancient Egypt, thus confirming Belmonte (2003a) original hypothesis.

5.7 The Going Forth of Sopdet: The Magic of Egyptian Sothic Dates

It is not the intention of this chapter to undertake a complete review of the role played by astronomical calculations in the determination of the chronology of ancient Egypt. This will be modestly attempted in Chap. 7. Currently, there is a large set of 'Sothic dates' (mentions of *prt spdt* in a civil calendar context), although only some of them would be useful for chronological purposes (Krauss, 2006c). It could be interesting to make a summary of all 14 of them in a rough chronological order:

(1) Interestingly, what could be the oldest Sothic date has been the most recently discovered (Habicht et al., 2015). The text was found in the decoration of a small jar identified in the collection of the Swiss Egyptologist Peter Kaplony shortly after his death. Stylistically, this has been dated in the late Old Kinddom (Habicht

et al., 2015), although a Middle Kingdom chronology cannot be discarded. The hieroglyph text certainly reads as:

However, the problems comes when the translation is considered. Habicht and his collaborators (2015) have read it as: 'Ointment made for the protection of the year, month 4, Peret-season, for the forthcoming of Sothis, month 4, Akhet season it is, made for the first day of the month'. This certainly is a very free translation of the text because two unmentioned uses of 'made' are assumed and the determinative *pw* is assigned to IV *3ḫt*. This is certainly a strange position for this determinative and offers other alternative translations, as we shall argue.

Under these premises, they have calculated an interval between 2419 and 2406 BC, depending on the *arcus visionis*, and assigned the jar to the 5th Dynasty, a sound possibility, perhaps to Niuserre's reign.

However, there are a couple of perfect alternative translations. One possibility would be: 'This is the ointment for the protection of the year [at] IV Peret according to Peret Sopdet [at] IV Akhet, at the time of the beginning of the month'. Here, the determinative *pw* makes reference to the ointment which would seem more logical, grammatically speaking, and no additional actions are needed. Then, the 'beginning of the month' does not necessarily refer to *prt spdt* in IV *3ḫt* and a much wider interval at the end of the Old Kingdom, up to 6th Dynasty would be perfectly manageable. This will keep this text as the oldest reference to *prt spdt* in the Old Kingdom, when it was perhaps assumed as the Herald of the Flooding (see above).

However, another, more uncertain, reading is feasible if a peculiar position for *pw* is assumed: 'This is the ointment for the protection of the year [at] IV Peret at the moment (*ḫtf*) of Peret Sopdet, IV Akhet at the time of the beginning of the month'. In this case, *prt spdt* would occur in IV *prt* during the Middle Kingdom in a close parallelism to the next Sothic date. Kaplony's collection date would then not be exceptional at all. This is a clear example of the difficulty of interpretating ancient Egyptian sources related to astronomy.

(2) The two mentions of *prt spdt* in the Illahun archive, dated in the reign of Senuseret III (Luft, 1992a, b: 224, and references therein). As shown in Figure 5.19 (to be discussed in Chap. 7) these are exceptional because one of them is a prediction and is controversial.

(3) The observation, apparently real (verb *m33* is mentioned), in *c.* 1592 BC of the heliacal rising of Sirius at Jebel Tjauti, near Thebes. The text reads (Darnell and Darnell, 2002):

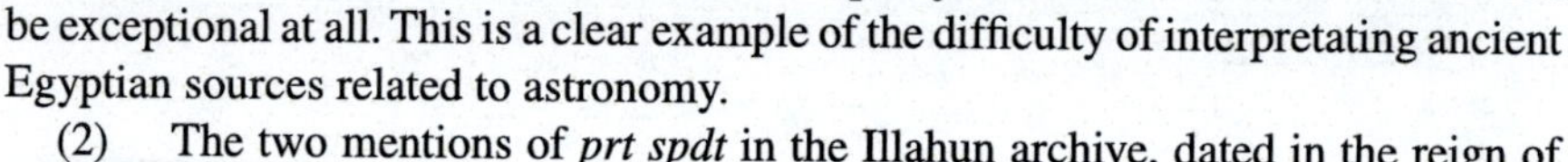

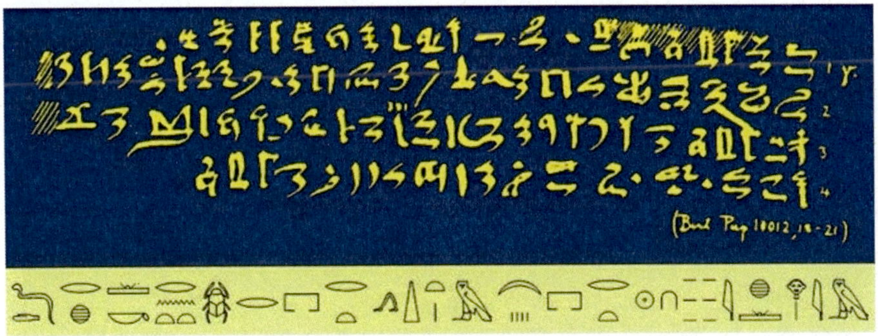

Fig. 5.19 Transcription in hieroglyphic writing of the second line of a text of Papyrus Berlin 10012, as seen in hieratic writing in the original (Möller, 1923: 19). The text reads: … 'I inform you that the Going Forth of Sirius (Peret Sopdet) will take place there in IV Peret 16' … The text indicates that the astronomer priests of ancient Egypt were able to make predictions about astronomical events. This in turn complicates the correct interpretation because we do not know if the heliacal rising occurred on the estimated date or not, nor the place for which the prediction was carried out. Controversy is indeed served

which dates it in II *šmw* 20 of year 11 of an, unfortunately, unknown king, presumably of the 17th Dynasty. If the king's name had been recorded, this would have been an excellent time-mark.

(4) The too-long disputed mention of *prt spdt* in the controversial calendar of the Ebers medical papyrus. We return to this later on.

(5) A block at the Open Air Museum in Karnak, presumably dated to the reign of Amenhotep I. Apparently, *prt spdt* is located between two feasts celebrated in I prt 3 and I *prt* 20 (Spalinger, 1992: plate II). This could be another reference to *prt spdt* in the Old Kingdom if the fragment is conceived as part of a very old Feast Calendar, because the interval of the event would be *c.* 2301–2182 BC. However, date (6) and (7) offer a different possible scenario.

(6) A second block, related to the previously mentioned one but not so easily dated, where *prt spdt* is located between IV *prt* 1 and an unknown day of I *šmw* (see Figure 5.20; also Spalinger, 1992: plate III). If the interval were correct, this will date this out of context block in the Middle Kingdom. However, see Sothic date (7).

(7) The *prt spdt* entrance on the Buto Stele, undoubtedly made in the reign of Thutmose III. The day of crossing of Sothis is unexpectedly placed between two feasts celebrated in I *šmw* 4 and I *šmw* 30. This will be discussed at length in the following paragraphs.

(8) The 1432 BC heliacal rising of Sirius at Elephantine, in III *šmw* 28, in an unknown year of a king who is almost certainly Thutmose III (El-Sabban, 2000: 31). This is from a festival calendar of the temple of Khnum at Elephantine (see Figure 5.21).

(9) The monumental version of the Fundamentals of the Course of the Stars (von Lieven, 2007; previously known as The Cosmology of Nut) at the Osireion in Abydos, built in the reign of Sethy I (Clagett, 1995: 357). The heliacal rising of

Fig. 5.20 A fragment of a calendar of festivals in an out-of-context block in the Open Air Museum in Karnak. Found inside one of the temple pylons, it has been tentatively ascribed to Amenhotep I. However, the list includes a rare and controversial mention of *prt spdt* (fourth preserved line) between IV *prt* 1 and an undetermined day of I *šmw* (perhaps 1); too early a date for the heliacal rising or Sirius in the reign of this king. See the text for further discussion. (Photograph by J. A. Belmonte)

Translation

Third month of Shemu, day 28,
day of the feast of the Going Forth of Sopdet

What is offered in this day in the name of …

Flock bull	1
Common birds	5
White fig bread, loafs	10
White bread for the offerings, baskets	35
Incense, bowls	10

Fig. 5.21 A fragment of a list of offerings and festivals from the Temple of Khnum in Elephantine (the original in now almost hidden on top of a wall in the Louvre Museum). It includes the earliest uncontroversial New Kingdom date of the heliacal rising of Sirius (*prt spdt*) in III *šmw* 28 of an unknown year of a king who is almost certainly Thutmose III. The figure also illustrates the translation of the text, including the related offerings. (Photograph by J. A. Belmonte)

Sirius can be implicitly dated in IV *prt* 16, exactly the same date as the prediction in Sothic date (2). There is no comprehensive explanation for this dichotomy (see Chap. 3).

(10) The Ramesseum astronomical ceiling where *prt spdt* is not explicitly mentioned but the simultaneity of this with *wpt rnpt* can be implicitly understood (see Fig. 5.13).

(11) The festival calendar of Ramesses III at Medinet Habu. According to this calendar, *prt spdt* occurred in I *ꜣḫt* in this period (see Fig. 5.18).

(12) The Canopus Decree, where there is an attempt to fix *prt spdt* in II *šmw* 1 (First of Payni), as occurred in 238/237 BC, probably in Memphis. This frustrated attempt of reform of Ptolemy III had no major consequences. However, the inappropriate translation to hieroglyphs of the original decree in Greek has been responsible of more than one Egyptologist's headaches, as previously argued.

(13) The date of instauration of the Alexandrian Calendar implicitly indicates that in 25 BC, *prt spdt* took place in III *šmw* 25.

(14) The accredited and famous report of the Roman scholar Censorinus in his De Die Natali Liber (XXI, 10). The heliacal rising of Sirius and the First of Thoth (I *ꜣḫt* 1) coincided in AD 139, probably in Alexandria. This is the pivotal date for the Sothic approach to Egyptian chronology as the end of a Sothic cycle of, theoretically, 1461 wandering civil years (Belmonte, 2012: 34).

Of these, only the last three options are completely reliable but are from a period when Sothic dating is unnecessary. In contrast, none of the dates for the period when the chronology is not fixed by independent sources is completely reliable (i.e. before Thutmose III). As we shall argue in Chap. 7, the case is especially dramatic for case (2), which has been considered to be the most important of all Sothic dates in Egyptian historiography, since the discovery of the Illahun archives more than a century ago.

To discuss other problematic Sothic dates, we now move forward to the ancient city of Buto, in the marshes of the western Delta. Buto was a very old city already existing in pre-dynastic times and was well known as the city of the cobra goddess Wadjet. The archaeological site is dominated by two huge mounds, the remains of the ancient quarters of Pe and Dep. Between the two, the ruins of the temple of Wadjet has been discovered, but the excavation has only reached the foundations of New Kingdom buildings and most of the stonework dates from the Late Period.

Indeed, one of the most interesting discoveries on site is the so-called Stela of Buto, dating from the reign of Thutmose III (*c.* 1479–1425 BC). This man-sized slab contains a royal decree regarding the restoration of the temple under Thutmose and the calendar of feasts established accordingly. The relevant sketch of the calendar is presented in Figure 5.22.

This feast calendar, which start at *wpt rnpt* or New Year's Eve in I *ꜣḫt* 1, has been the cause of a major controversy. This is because the heliacal rising of Sirius—Sothic date (7) aka *prt spdt*—is located between two dates of I *šmw* (indeed not in *wpt rnpt*). This is in apparent contradiction with what one would expect from the widely accepted chronology of the New Kingdom (see Table 7.1). This would

Fig. 5.22 A close-up of the Festival Calendar of the Stele of Buto (*wpt rnpt* at the first row), dating to the reign of Thutmose III (*c.* 1450 BC). The second line from the bottom contains a reference to the heliacal rising of Sirius (*prt spdt*) between two feasts celebrated in I *šmw* 4 and I *šmw* 30. This has created a profound controversy because Peret Sopdet presumably occurred in III or IV *šmw* during the reign of Thutmose III. This problem could be solved if the text actually reflected a heliacal rising prediction related to the heliacal setting of Sirius in I *šmw* at that particular period. See the text for further discussions. (Photograph by J. A. Belmonte)

instead situate *prt spdt* at the end of this calendar season, during the reign of Thutmose III (see Fig. 5.20). As a consequence, von Bomhard (1999: 45) urged for a complete revision of the chronology of the beginning of the New Kingdom, which is unreasonable, while Spalinger (1996c), when analysing the calendar of festivals, avoided this slippery issue.

However, Krauss (2006 private communication) argues:

> If one concludes that the rising of Sothis in the time of Thutmose III happened in the 1st month of Shemu, because the entry is positioned between two 1st month of Shemu entries, one gets into severe chronological problems. Between year 7 of Sesostris (or Senuseret) III and year one of Thutmose III at least 302 years are accounted for. 302 years move the rising of Sothis from IV Peret 16 to III Shemu 1, which means that in the time of Thutmose III, the rising of Sothis took place after III Shemu 1. An equivalent conclusion follows from the Elephantine block (see Fig. 5.20), which gives III Shemu 28 as rising date of Sothis [*c.* 1432 BC, in the second half of the reign of Thutmose III]. There is no Egyptological doubt that the block originated in a building of Thutmose III. Under these circumstances one has to conclude that the position of the Buto date does not mean that the rise of Sothis took place in the 1st month of Shemu in the time of Thutmose III. The text itself says in any case only that Sothis rises on her calendar day(s). Similar problems are raised by the two Sothic dates which are mentioned on unpublished blocks in Karnak from the time of Amenhotep I. In

one case, the line with the entry 'rising of Sothis on …' is positioned between entries which concern the 1st month of Peret. In the other case 'rising of Sothis on her day(s)' comes after an entry in the 4th month of Peret and before an entry in the 1st month of Shemu. The problem, as I see it, is not one of calendars and astronomy, but of organizing the texts of steles and wall inscriptions.

We show here, following Belmonte (2009), that these dates are perhaps a 'problem' of calendars and astronomy, but without the need for any change in the accepted chronology, by the use of Egyptian grammar, archaeoastronomy, and a working hypothesis.

The Egyptian–Spanish Mission of Archaeoastronomy (Belmonte et al., 2009b, and reference therein) has found that the temple of Wadjet at Buto was orientated towards the western half of the horizon to a declination of −18½°±¾° (see Belmonte & Shaltout, 2009, Appendix II). Interestingly, during the reign of Thutmose III, who is known to have restored the temple, the declination of Sirius was of the order of −18¼°. Considering that there is certain evidence that Wadjet—along with several other goddesses such as Isis, Hathor, Satet, and Seshat—was related to Sopdet, at least from the New Kingdom onwards, the Mission members suggested that the temple at Buto would actually have been orientated to the setting of Sirius at the time of Thutmose III. And this is the nub of the question. The corollary is that the temple could be related to the heliacal setting (not rising) of Sothis on the western horizon during the reign of this king.

The phenomenon of the heliacal setting of stars was known in ancient Egypt by the term *šnj dw3t*, 'Encircled by the Duat' (see Chaps. 3 and 4), at least from the New Kingdom onwards, if not earlier. From Fig. 5.22, we extract the text associated with *prt spdt*, which reads as follows:

No date is mentioned explicitly and the sentence contains the particle *r*, used for a future action or to describe a future task. In contrast, in the mention of *prt spdt* at the festival calendar of the south wall of the Million Year Temple of Ramesses III at Medinet Habu (Fig. 5.18), the text reads

On this occasion, the particle used to describe the action is *m*, typical of the present tense, and a date is explicitly mentioned, First month of the Inundation. Hence, while the sentence at Medinet Habu can be read precisely as 'First Month of the Inundation, the Heliacal Rising of Sopdet happens in her day', that of the Buto Estela ought to be read as 'The Heliacal Rising of Sopdet will occur in her days' (note the plural). The former is an explicit statement. The latter, however, seems to be once more a statement of a predictive character. The hypothesis is that the best moment to make the 'prediction' that the heliacal rising of Sopdet will occur in the near future would have been at the instant of her heliacal setting nearly seven Egyptian decades before.

We have shown above that a heliacal rising of Sirius took place in III *šmw* 28 in 1432 BC at Elephantine. From simple astronomical considerations, this means that the heliacal setting of the star at Buto occurred, with one or two days' margin of error, given varying atmospheric conditions, at I *šmw* 25 for this year. Hence, for the entire reign of Thutmose III, the helical setting of Sopdet at Buto happened roughly between I *šmw* 13 and I *šmw* 27. This is perfectly compatible within the interval between I *šmw* 4 and I *šmw* 30, as described by the festival list of the Buto Estela. Hence, the astronomical event actually emphasized at this singular monumental inscription was the heliacal setting of Sopdet: an astronomical event.

Now, it is the turn of the two dates at the isolated, and out of context, blocks of Karnak – (5) and (6) in the list. One of them (see Fig. 5.20) contains a reference to *prt spdt* between two feasts celebrated at IV *prt* (1) and an uncertain date of I *Smw*. It reads as follows:

Again, no date is mentioned explicitly and once more the particle *r*, is used in this context. The second mentions *prt spdt* between I *prt* 3 and I *prt* 20. The inscription is broken at the end but the evidence suggests that it was almost identical to the previous one. As in the case of Buto, the inscription of the two blocks ought to be read as a prediction that the 'Heliacal Rising of Sopdet will happen in her day'. Once more, Belmonte (2009) played with the hypothesis that this prediction be related to the instant of the heliacal setting of the star.

Astronomical analysis permits the helical setting of Sirius in Thebes to be settled between I *prt* 3 and 20 for an interval between *c.* 1976 BC and 1908 BC, with a margin of a couple of years. This interval contains most of the reign of Senuseret I, whichever chronology we consider. Senuseret I was the founder of the Temple of Karnak and it would be logical to assume that this festival calendar (like his contemporaneous White Chapel) was prepared for that original temple. Once more, the festival list would include a reference to the interval of the heliacal setting of *Sopdet*, and, by inference, of her heliacal rising seven decades after (between *c.* III *prt* 13 and 30).

The former *prt spdt* mentioned is contained in a much wider interval. This allows for dates of the heliacal setting of Sirius, according to the same astronomical considerations, to fall within a longer uncertain temporal spectrum of not less than 120 years and a maximum of 240, including part of the reign of Amenhotep I. Hence, the corresponding festival lists could be ascribed to the reign of Amenhotep I.

Interestingly, under simple epigraphic and astronomical premises, a reasonable explanation for a serious challenge to the standard chronology of ancient Egypt has been reached by simply performing an alternative reading of three controversial inscriptions. These take into account that the setting of Sopdet on the horizon, as suggested by the temple orientation, and her heliacal setting, could have been important land-and time-marks for ancient Egyptians.

5.8 Epilogue: More on the Ebers Calendar

In the previous sections, we have discussed simple solutions and proposed working hypotheses for a series of endlessly discussed problems of Egyptian calendrics (more will be dealt with in Chap. 7). Belmonte (2003a, 2009) was convinced, and we endorse this, that these sound solutions were reached because research had been completely freed from the Ebers syndrome.

The Ebers Medical Papyrus is, from the palaeographical point of view, a New Kingdom copy of an earlier document with medical prescriptions, which had been ascribed by the Egyptians to the legendary king Athothis of the First Dynasty. In the verso of the papyrus, another hand later wrote the famous calendar in a hieratical script that has also been considered typical of the New Kingdom, although this datum is controversial. It was discovered somewhere in Thebes in the 1860s, and has been available to the researchers since 1862. From its discovery, the calendar was a key point in any discussion on Egyptian calendrics, receiving almost as many interpretations as scholars who have investigated it (for a detailed story of the Ebers, see Clagett, 1995: 193: also relevant is Depuydt, 1996).

The structure of the 'calendar' can be seen in Figure 5.23, where a facsimile of the original (in Leipzig University) is presented. From right to left, and from top to bottom, it consists of:

1. An indented horizontal row (D), where a typical dating formula is encountered (year 9). This does not start at the upper-right corner of the text, as would have been expected, because Egyptian manuscripts very rarely, if ever, present an indentation. The reading of part of the name of the king (G) has been the subject of controversy in the past. However, most scholars now agree to read it as *ḏsr-kꜣ-rꜥ*, i.e. Djoserkare, the throne name of Amenhotep I, which would place the calendar at the beginning of the New Kingdom rather than in the Middle Kingdom.
2. A vertical column (A), with 12 names. These are, under the hypothesis defended in this book, those of the months of the civil calendar but in a strange order, starting with the twelfth month, *wpt rnpt*.
3. A column (B), with dates in the civil calendar, using seasonal names and starting with the month III *šmw* (equivalent to *jpt-ḥmt.s*). The digit symbol C, following the symbol for day (*sw*) is normally read as 9, but it has been suggested that it could also be read as *psḏ-[ntjw]* (Borchardt, 1917: 56; Krauss, personal communication, considers this hypothesis absolutely unrealistic).
4. An entrance (E), where the heliacal rising of Sirius (*prt spdt*) or the equivalent festival is mentioned in apparent association with *wpt rnpt* in column A and with III *šmw* 9 in column B.
5. A column of dots (F), just below the heliacal rising entrance.

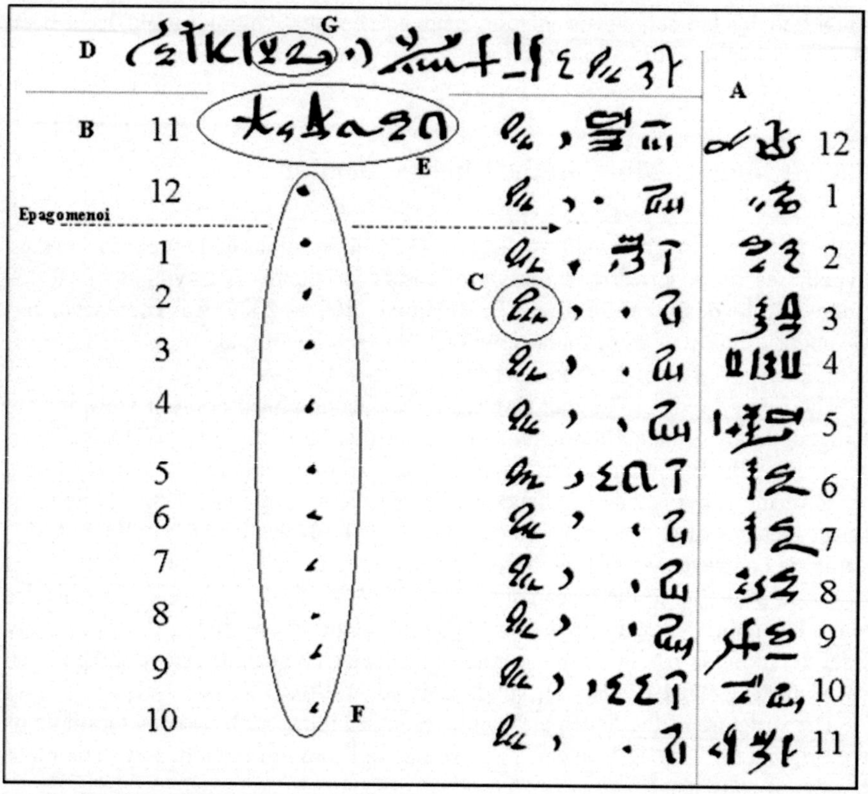

Fig. 5.23 Facsimile of the hieratic original of the calendar in the Ebers Medical Papyrus, defining the different areas that are relevant to any attempt at interpretation. These are: the month name column (A), the civil date column (B), the number nine (9) column (C), the regnal year row (D), the rising of Sirius (E) and the associated dot column (F), and finally the problematic signs of the king's name (G). See the text for further discussion. (Adapted from Belmonte, 2003a)

As can be seen, the structure is puzzling as it relates seasonal and proper names of the months of the civil calendar in a way that is contradictory. However, at the same time, it apparently offers one of those rare jewels for chronologists, a Sothic date. One of the most important problems with the calendar is the complete absence of the epagomenal days, that should have been present between IV *šmw* 9 and I *3ḫt* 9. Actually, the second date should have been I *3ḫt* 4, provided that the first column is a set of hypothetical feasts controlled by the moon or of lunar months. The solution has seldom been to blame the scribe for an inexcusable fault (e.g. Edgerton, 1937). As far as we are concerned, of the fourteen or so traditional proposals to explain the Ebers calendar (see Belmonte, 2003a, for an analysis of the different, often contradictory, hypotheses), that of Depuydt (1996), of days 9 as points of time and not intervals, seemed the most reasonable. In this case, we might then even retain the chronological importance of the Sothic date.

Rolf Krauss has recently argued with the authors, that the Ebers calendar 'can be explained as the Sothis-based lunar year of the Illahun type, adapted to a regnal year in order to make known and commemorate the very rare coincidence of a royal accession day (that of Amenhotep I) with the beginning of an intercalary lunar month in which Peret Sopdet occurred'. He thus suggests the existence of an operative Sirius-heralded lunar calendar that is, to say the least, problematic, after the invention of the civil calendar, whose names are identical to the ceiling of Senenmut's tomb and the freeze of Edfu (see Table 5.2). However, this hypothesis will save, at face value, the Sothic date fixing *prt spdt* in III *šmw* 9 of year 9 of Amenhotep I. Both things, however, could be independent of each other.

Ultimately, Krauss (2022b), in a most recent paper forthcoming in *Orientalia,* has proposed an easy explanation for the calendar layout in parallel and vertical lines of certain amplitudes, where dot-column F would simply be a stylistic solution of the scribe to fill an empty space (*horror vacui*). This paper is a reaction to Spalinger (2018: 69), who interprets it as nothing more than a native Egyptian schematic survey of the star Sirius' passage through time. Hence, it would represent a succession of *prt spdt* events through time. This would be an example of the 'Phoenix Year', a concept of which the Egyptians of the New Kingdom had probably no idea at all. This will put aside the value of the Sothic date. This is, as we see it, a battle of intellectual giants still pursuing a solution for a puzzle. A solution that perhaps does not exist.

Significantly, our impression is that there is no satisfactory and simple answer for the question of what the Ebers calendar represents. Recently, Depuydt (2017) has written that the Ebers calendar is 'in relation to its diminutive size, the most discussed historical source of ancient Egypt', but no solution has been found.

In conclusion, we will have to agree with Edward Meyer who, a dozen decades ago in his *Ägyptische Chronologie* (Meyer, 1904), seemed to have given up all hope that the enigma of the Ebers calendar might ever be solved. Hence, it would be better if we assume that the Ebers calendar was never discovered (*testis unus, testis nullus*). Without it, the evidence of the existence of an independent set of lunar months (or were they feasts?) deferring from the civil ones and wandering among these, is questionable. A lunar feast such as *w3gj* and a lunar computus such as Middle Kingdom temple *wrš* do not make a calendar.

Under this simple assumption, the functioning of Egyptian calendrics can easily be explained.

There was a unique calendar, the civil one, created in the first half of the third millennium BC with a duration of 365 days derived from detailed solar observations, an internal structure dictated by basic mathematical and astronomical concerns, and a seasonal and a monthly structure inherited from an older Nile-based lunar calendar. A series of lunar festivals existed, but they were articulated within the civil calendar or within a lunar time-reckoning system strongly related to it. This lunar computus could be heralded either by Peret Sopdet or by the arrival of the Flooding, or even of both because since the late Old Kingdom, Peret Sopdet could be chosen as the herald of the rising waters once the civil seasons were clearly ahead of the climatic ones.

In the Late Period, when the Egyptians were familiarized with alien calendars, such as the Babylonian calendar of Assyrian and Persian invaders, they even developed a system to equate these alien months with his own lunar time-reckoning system. The term *pꜣ wrš* (indeed coming from temple services) followed by a civil month numerical name was developed to identify the equivalent periods of time to this purely synodic months (Krauss, 2015: Excursus 3). However, this does not mean that they were using them in their daily life. Perhaps this was developed for administrative purposes related to their foreign rulers.

As a matter of fact, our working hypothesis is that Egyptian timekeeping systems were simple and governed by the civil calendar, a gift of the Nile and of detailed astronomical observations. This is what we have shown within the paragraphs of this chapter. Pending much-desired new discoveries, the reader now has the last word.

Chapter 6
Ancient Egyptian Land- and Skyscapes

The walls of the temples of the goddess Hathor at Dendara and Horus at Edfu offered many representations of the stretching of the cord ceremony in which the northern, imperishable, ancient Egyptian constellation of Meskhetyu is advocated as the target the Pharaoh was looking at when fixing the orientation of the temples. However, it is often asserted that temples built along the Nile were most commonly orientated along an axis according to local topography as determined by the river, a dominant component of Egyptian landscapes, notably during the Inundation (see Figure 6.1). Our two-decades-long work in Egypt has proven that both possibilities are not only true but also complementary.

From the birth of their civilization, ancient Egyptians selected certain sectors of their geography, blessed by peculiar orographic and celestial conditions, with the aim of creating cosmic landscapes. This tradition can be traced from the dawn to the dusk of their civilization in the proto-Dynastic and Roman periods, respectively. Throughout this chapter, which is actually an essay review, several of the most sacred places of the Black Land will be visited mostly following a diachronic perspective. This will show how Abydos, Memphis, Thebes, Heracleopolis, and Dendara, among other sites, were transformed into real cosmic landscapes, reflecting the Egyptian world-view of each epoch: from the astral eschatology of the *Pyramid Texts* to the highlights of the solar cult as represented by Karnak.

In this chapter, we deal specifically with the relationship between astronomy and geography, land- and skyscape, and the role that observation of the sky can play not only in the orientation of a particular monument, but also in the choice of its location, if not in a combination of both. This phenomenology will be related to certain other aspects of the Pharaonic civilization such as the calendar (Belmonte, 2003a, b, c; Depuydt, 2007; Spalinger, 2018; see Chap. 5), the mapping of the sky (Lull & Belmonte, 2009; see Chap. 4), and the celestial aspects of the religion (Brugsch, 1883; Krauss, 1997; see Chap. 1).

Landscape archaeology—the discipline within which we are going to move—is a subject which has not in the past enjoyed great prestige in Egyptological studies. Only recently has an approach to the discipline begun timidly to develop, as in the

J. A. Belmonte, J. Lull, *Astronomy of Ancient Egypt*, Historical & Cultural Astronomy, https://doi.org/10.1007/978-3-031-11829-6_6

Fig. 6.1 The pyramids of Giza at high Nile. The relationship between the Nile and ancient architecture is clearly emphasized. (Adapted from original old images by Lehnert and Landrock)

study of ritual landscape in ancient Thebes and our own work throughout the country (Ullmann, 2007; Belmonte, 2012: 215–250; Belmonte, 2016). Earlier, most such studies had been limited to a very prosaic relationship with the Nile (Wilkinson, 2000a, b: 36–37), other elements of the landscape, such as the horizon, and of course, with very rare exceptions (Gabolde, 1998, 2009: 145–157), the celestial phenomena occurring on it, the rest of the celestial vault being ignored.

However, the authors and their collaborators—mostly but not all within the context of the Egyptian–Spanish Mission on Archaeoastronomy of ancient Egypt (Belmonte et al., 2009b)—have made a series of crucial analyses that have largely demonstrated the important role played by the heavens in the alignment of sacred buildings in ancient Egypt. In this chapter, we briefly analyse some of the most striking cases, in particular those in which astronomy was combined with local topography, on a large scale, to fight chaos (*isfet*) and help maintain cosmic order (*ma'at*), as described by Hornung (1999).

We focus on the most famous archaeological sites in Egypt, such as the archaic royal cemetery at Abydos, the pyramid fields, notably the 4th Dynasty behemoths at Dahshur and Giza, the complex of Karnak in Luxor, the temples of Deir el Bahari, also in Thebes, and the temple complex of the goddess Hathor at Dendara, among others. For a detailed archaeological description of a majority of these sites, see: Aufrère and Golvin (1997), Aufrère et al. (1991). The chapter closes with an analysis of the monuments erected by the Kings of Kush, under strong Egyptian influence (Belmonte et al., 2010).

6.1 Introduction: A Story of Discovery

For the last two decades, the Egyptian–Spanish Mission for the Archaeoastronomy of ancient Egypt has been hard at work in the country of the Pharaohs. In a series of seven campaigns, their members have visited nearly every corner of the land of the Nile and measured more than 500 alignments, including more than 350 in temples, in nearly 100 archaeological sites. Campaigns were initially planned following a geographical criterion (Upper, Middle, and Lower Egypt, the Western and Eastern Desert Oasis, etc.). However, in subsequent campaigns, we performed selective experiments, in an attempt to falsify or test earlier hypotheses, often revisiting sites for new measurements, or in order to get direct observations of predicted astronomical phenomena (see Figure 6.2).

The project developed campaign by campaign, showing how some preliminary hypotheses had to be abandoned in the process and how new ideas became more and more substantiated as time passed. The journey started in the splendid temples of Upper Egypt and finished in the domains of the lunar god, Thoth, a clear demonstration that astronomy did play a fundamental role in the search of the ancient Egyptians for cosmic order. This work was completed over the past decade within the framework of the Spanish Archaeological Mission at Heracleopolis Magna (Belmonte et al., 2015, 2016) and the European Project STARS4ALL, when certain temples at Thebes were revisited (Belmonte et al., 2020).

Were the temples of the ancient Egyptian civilization astronomically orientated? This was an important question that was far from resolved at the turn of the century. Wilkinson (2000a, b) clearly stated that temples built along the Nile were most commonly orientated along an east–west axis, according to local cardinal directions as determined by the river, so local topography would be the determining reason for temple orientation. This is demonstrated beyond doubt in Figures 6.3 and 6.4 (Belmonte et al., 2009b). However, Wilkinson also pointed out that, on occasions, orientation towards the sun or important stars was definitely the priority. We have shown (for a first approach see, Belmonte, 2000), and will demonstrate in this

Fig. 6.2 A splendid light hierophany at sunrise at the winter solstice (December 21st 2006) at the temple of Sobek-Re in Qsar Qarun (Dionysas) in the El-Fayum Oasis, when our team discovered the phenomenon. Notice the illumination of the inner sanctuary (left), where the statue of the god would have been located, and of the various successive portals. The winged-disc symbol of Horus Behedety blossoms above the first gate in the pale yellow light of the first rays of his physical counterpart, the solar disc. As the sun rises, its light moves to the north illuminating the small shrine to the right of the sanctuary, while the niche to its left, where the mummy (the underworld aspect) of the god was presumably located, always remains in darkness. The parallelism with the illumination phenomenon at Abu Simbel is striking. (Photographs by the authors)

chapter, that this principle is much more important than is often recognized. The Nile as the main sources is far too restrictive, and solar (see Figure 6.5) and stellar orientations were much more common in ancient Egypt than had previously been suspected.

The first campaign of the Egyptian–Spanish Mission was performed in the late winter of 2004 and was concentrated on the temples of Upper Egypt and Northern Nubia. Our main task was to measure all the temples, giving similar weight to those magnificently preserved and to temples where not more than a few walls remain. One of our first intentions was to test the Nile hypothesis. Figure 6.4 shows an extraordinary outcome of our data since it demonstrated, beyond any reasonable doubt that ancient Egyptian temples were orientated in such a way that the main gate of the building directly opened onto a direction perpendicular to the Nile, as did the fields. This had been frequently argued in specialized circles but had never been proven statistically as we were able to do. The relevance of the Nile as the most important reference for temple orientation is clearly stressed. The c.170 temples analysed in Fig. 6.4 give an average difference of only 5 ± 7 min of arc between their orientation and the direction either perpendicular or parallel to the river.

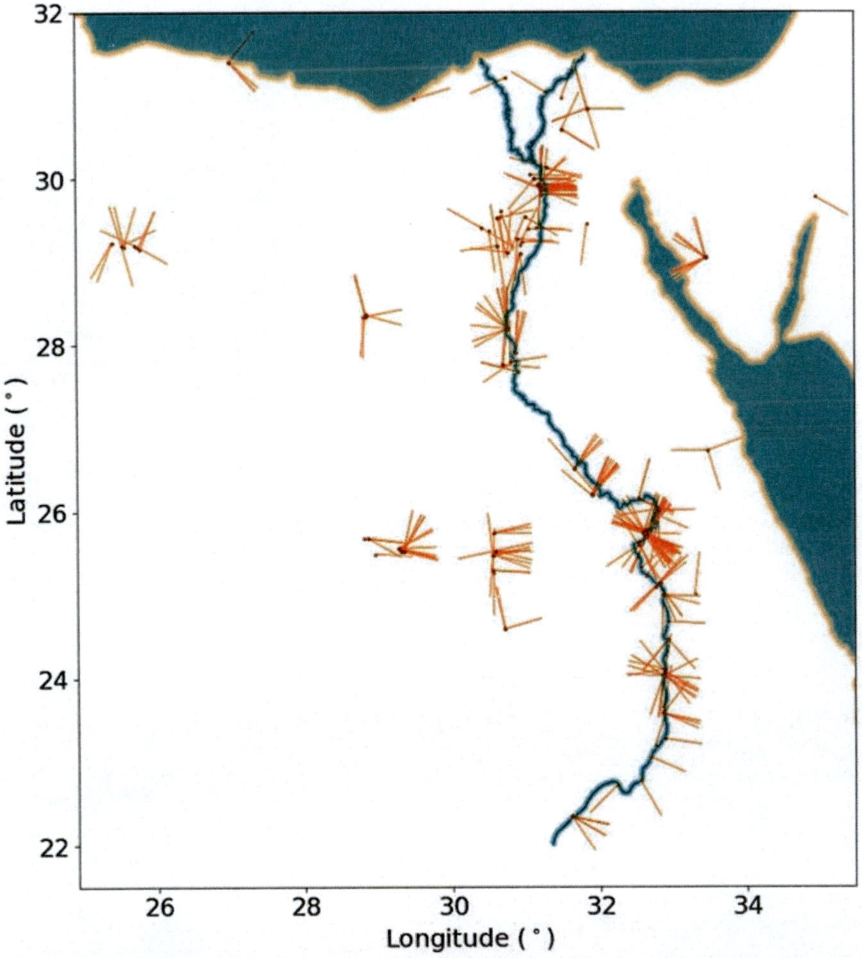

Fig. 6.3 Azimuth map of Egyptian temples. Each orange line represents the orientation of a temple (*c.* 350 are represented). The diagram clearly shows how temples along the Nile are orientated mostly either perpendicular or parallel to the river. Note the concentrations near Memphis and Thebes. However, when temples in the oases are considered, meridian and solar orientation are certainly preferred. (Diagram by courtesy of Maitane Urrutia-Aparicio, adapted from Belmonte & Urrutia-Aparicio, 2022)

What may we infer from the data presented in Fig. 6.3? On the one hand, the data of Middle Egypt seem to follow the pattern for Upper Egypt. On the other hand, we have the data of the valley, pyramid and sun temples of the area surrounding the ancient capital of Memphis. The vast majority have azimuths close to 90° or due east. Despite the fact that the river has frequently changed its route in that area, it is widely accepted that it has always had a more or less south–north course. Obviously, this means that this entire group of temples was at the same time perpendicular to the Nile.

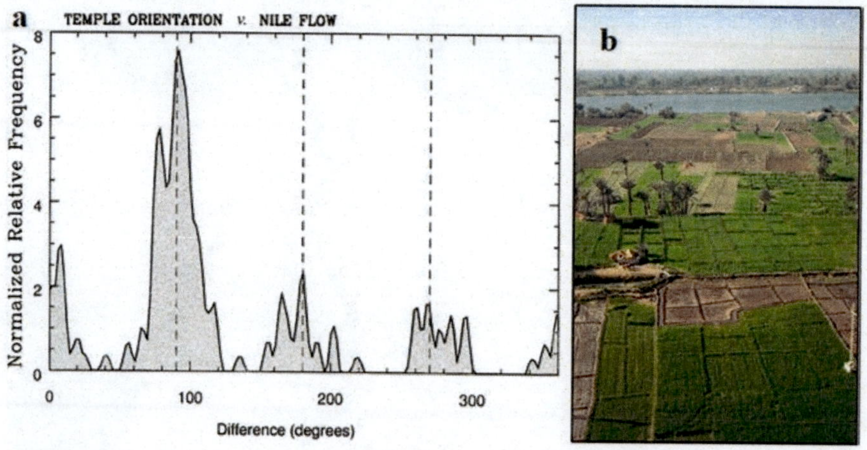

Fig. 6.4 Testing the Nile hypothesis. (**a**) This histogram represents the difference in orientation between the main axes of 170 temples in the Nile Valley, and the average course of the river at their corresponding locations. This figure demonstrates beyond any reasonable doubt that local topography (the course of the Nile) was very important when laying the foundations of the temples, although it was not the only factor to be considered. (**b**) The course of the Nile in Middle Egypt with the fields nearly perpendicular to the river. Panel (**a**) adapted from Belmonte et al., (2009b). Panel (**b**), photograph by the authors

Fig. 6.5 Solstitial orientations in western Thebes. Today, just as 3350 years ago, the colossi of Mennon still face sunrise at winter solstice, as did the Million Year Mansion of King Amenhotep III behind them. (Photograph courtesy of M. A. Molinero and N. Delgado)

However, the need for an orientation perpendicular to the river cannot explain the precision of this group of alignments to due east, as will be shown later on. Consequently, we believe that the importance in this particular group of cardinal directions in the area of Memphis could be partially explained by the peculiar course of the river, but that they would finally have been actually manifested through celestial mythology and obtained through astronomical procedures.

For the Delta, the situation is a bit more complicated. In this region today the Nile is divided into two major branches, those of Damietta and Rosetta (see Fig. 6.3). However, in ancient times, the river had as many as seven main branches and innumerable secondary ones. The branches continuously changed their course forcing the abandonment of important cities, such as Pi-Ramesses, and the foundation of new ones, such as Tanis. Although it is difficult to know the exact courses of these ancient river branches, approximate reconstructions could be attempted.

The 18 Delta temples analysed (Shalthout et al., 2007), have revealed an average difference of $-5° \pm 1°$ between their orientation and the direction either perpendicular or parallel to the corresponding Nile branch. This result is far from the extremely accurate results obtained in the Nile Valley but is still close to 0°. Hence, we may conclude that even in the Delta the orientation of the temples according to the course of the river was a common practice. The case of Tanis is especially attractive and will be further discussed in one of the study cases because it is an extraordinary example of the combination of land- and skyscape.

However, for an archaeoastronomical mission, the most outstanding early result was clearly in determining that astronomical orientations were indeed present in our extensive, but still incomplete, cluster of temples (Shaltout & Belmonte, 2005). This is well illustrated in Figure 6.6, where the histogram of declination of the temples is shown. Declination is an astronomical coordinate that defines a particular sky position and is independent of the observer's latitude, as opposed to azimuth. For a star, it is fixed for long periods of time, and varies slowly with the precession of the equinoxes. The Sun's declination varies throughout the year between its extreme positions at the solstices and its midpoint at the equinoxes when it is equal to 0°. The extremes of the Sun's declination today are about ±23½° and have been progressively decreasing since ancient times at a rate of *c.* 0.46″ per year (the so-called ecliptic variability). It was closer to ±24° in the times of the Pharaohs.

There was a dominant peak, with a certainty of more than 99%, which could be associated with winter solstice orientations, ubiquitous in the area of Thebes, although this assumption will be qualified in later sections. There are a couple of significant peaks representing an attempt to orientate temples perpendicular to the Nile (see Fig. 6.3). These were mostly related to temples in Abydos, Thebes, and the area close Edfu.

The second campaign of our mission, in the summer of 2004, was a very brief one, straddling a study journey into Egypt of the pupils of the hieroglyph course of La Laguna University, which included the first author. This was a very festive event but we took the opportunity to check previous ideas and to measure several temples and tombs in Middle Egypt, including the monuments at Amarna and the suggestive necropolis of Beni Hassan. Once more, the presence of the Nile was determinant

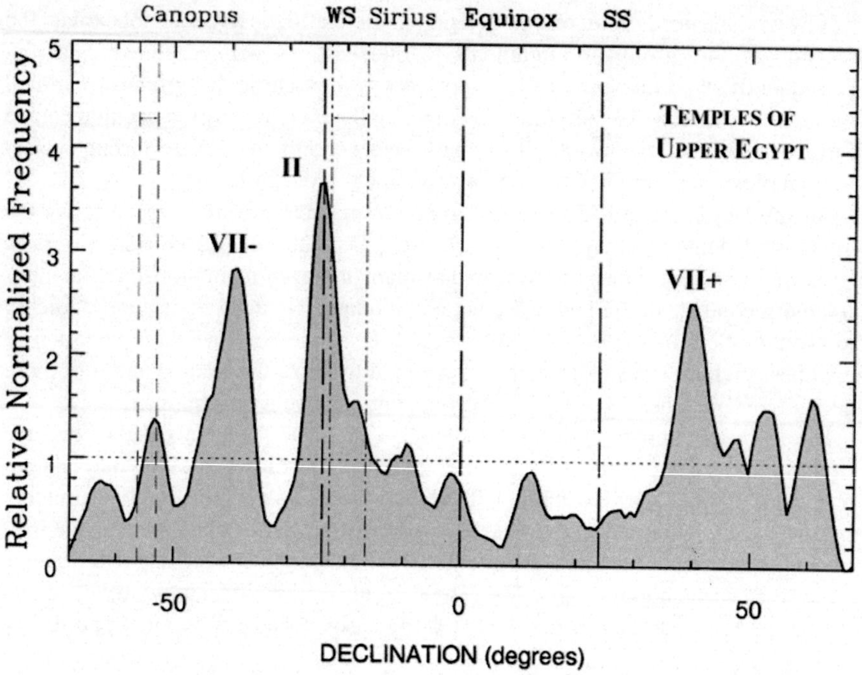

Fig. 6.6 Declination histogram of more than a hundred temples in Upper Egypt and Lower Nubia (Uauat). This figure illustrates that the observation of heavenly bodies must also have played a role in temple orientation. (Adapted, updated and upgraded from Shaltout & Belmonte, 2005)

and we decided that a new, and more specific and clever campaign ought to be organized.

In private conversations, some Egyptologists, discussing our first outcomes, have argued that local topography would have been the predominant rule, the heavens playing only a secondary role, the astronomical orientations in some cases even occurring by chance, as in that of the winter solstice alignments in Thebes. Thus, we were dealing with a frustrating dichotomy.

Consequently, our team decided that the situation deserved a new campaign in Egypt (a third one); but not of the usual kind. The idea was to falsify the Nile hypothesis by doing fieldwork in areas where the river could play a null role in the orientation of sacred structures. To achieve this, we went to the Oases of the Western Desert, where the presence of the ancient Egyptian culture, well attested since the Old Kingdom to Roman times, guaranteed the utility and relevance of the data, while the absence of a 'Nile' would provide an interesting new perspective (Belmonte & Shaltout, 2006).

The campaign was performed in the early spring of 2005, visiting the oases of Fayum, Bahariya, Farafra, Dakhla, and Kharga (Siwa was also visited but in a later campaign). Data were collected at all the oases but Farafra where the ancient remains were too scant for our purposes. This was the first systematic fieldwork campaign in search for a relationship between land- and skyscape ever performed at the oases.

Our results showed that when the river does not dominate the landscape, ancient Egyptian temples tend to be astronomically orientated according to two dominant customs, one of a presumably solar character—with several temples orientated within the lunisolar arc—and the other following the meridian line (see Figure 6.7). We suggested for the first time that this could be the case for the vast majority of Egyptian temples, and that Nile orientations were actually part of a wider general plan where both local topography and astronomy (i.e. land- and skyscape) played a significant role simultaneously. However, it is worth noting that orography, and even meteorology, can sometimes dictate the orientation in a peculiar way. This is well illustrated by the case of the Kharga Oasis, where Roman forts and the related *centuriatio* were organized according to the movements of the dunes, which followed the north dominant winds (Rossi & Magli, 2019).

As an additional result of the third campaign, a new experiment was designed to account for other solar orientations, a subgroup of our temples, in cases where these are evidently not dictated by the Nile. This exercise consisted of finding possible alignments to either sunrise or sunset on the wandering New Year's Eve (Wepet Renpet) of the Egyptian civil calendar (Belmonte & Shaltout, 2006). The preliminary results of the experiment showed that, under some special historical and

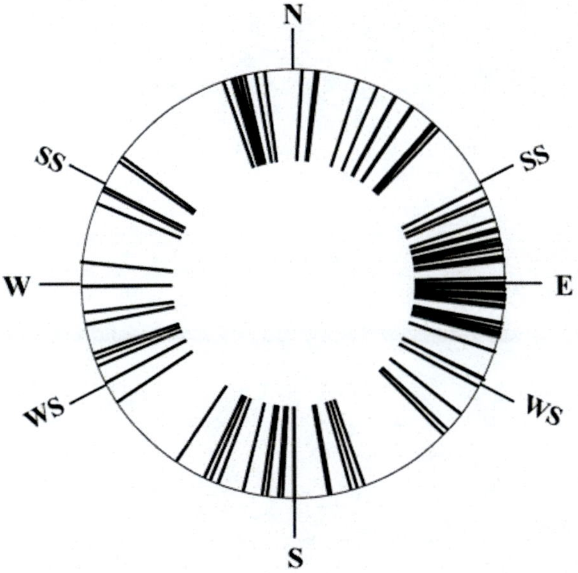

Fig. 6.7 Falsifying the Nile hypothesis. Orientation diagram of the data assembled for temples measured outside the Nile Valley, where the river influence would be absent. Notice the Maltese Cross form of the diagram typical of astronomical orientations with a preference for cardinal and solar orientations. WS and SS stand out for winter and summer solstices, respectively. (Adapted, updated and upgraded from Belmonte & Shaltout, 2006)

archaeological assumptions, some of these temples might have been orientated according to this quite simple hypothesis. We suggested that, provided the Nile alone does not necessarily govern eastern and western orientations in the valley, this hypothesis could be further tested in other temples across Egypt. This would become evident in some of our most suggestive results, as we shall later show.

Another suggestive outcome of this campaign was the idea that most of the temples in our sample, aligned nearer to the meridian line, and open to either north or south, were actually orientated by observations of the northern 'imperishable' constellation of Meskhetyu (the Plough, see Figure 6.8). These would perhaps respond to certain configurations of the asterism such as meridian transits, extreme polar distances or the rising or setting of certain stars (notably Alkaid). This hypothesis could also be further investigated in several sacred buildings in the Nile Valley, where we were certain, according to the inscriptions, that some of the temples were orientated towards this constellation.

Finally, the timing of this campaign was selected with another special objective in mind. The observation, in situ and at the appointed time, of an important astronomical event in locations where it had been suggested that equinoctial alignments occur, notably the Giza Plateau and the solar temples of the 5th Dynasty at Abu Ghurob. Figure 6.9 shows one of these impressive alignments in Giza, illustrating the celestial relationship between the Sphinx and the pyramids.

Fig. 6.8 The seven bright stars of the ancient Egyptian constellation of Meskhetyu, the Bull's Foreleg, definitively identified with the asterism of the Plough (Thuault, 2020). We defend in this chapter the importance of the observation of different configurations of this celestial object in temple orientation. (Image adapted from a photograph by courtesy of A. López)

Fig. 6.9 The equinoctial sun at 21st March 2005 is setting at the bottom-left corner of the second pyramid as seen from the main axis of the temple of the Sphinx of Giza. This image suggests a relationship between the two monuments. (Photographs by the authors)

The fourth campaign of the Egyptian–Spanish Mission was performed in June 2006 in the north of Egypt, including the Cairo region, the Delta, the Mediterranean coast, the Oasis of Siwa, and the Sinai. The vast majority of the relevant archaeological sites in those areas were visited, including the impressive and evocative sanctuary at Serabit el Khadim (Belmonte et al., 2008). On this occasion, more than 50% of the data belonged to the important sacred precincts related to the pyramids of the Old and Middle Kingdoms, which provided extremely interesting results. To our knowledge, this was the first systematic archaeoastronomical campaign ever undertaken at a complete sample of pyramid complexes. We also took the opportunity to observe relevant astronomical phenomena related to the summer solstice, notably in Giza (see below, Fig. 6.22).

The data for Lower Egypt included the vast majority of, if not all, the measurable monuments in the Delta. We visited temples that are seldom studied or analysed (for example, the temple of Isis at Behedyt el Haggar and the temple of the Ka of Pepi I at Bubastis, see Figure 6.10), with some most interesting and pleasant surprises (Shalthout et al., 2007).

As on previous missions, but certainly with a rather more emphasis, the analysis of the data obtained in the third campaign proved extremely fruitful and allowed us to obtain excellent results (Shalthout et al., 2007). The most outstanding outcome was the proposal of the existence of seven different families of astronomical orientations for the temples of Lower Egypt and neighbouring regions in the light of the archaeological evidence and knowledge we had accumulated on the ancient Egyptian skies. Later on, this hypothesis was easily extended to the rest of the country after subsequent campaigns and even farther beyond, to the domains of the King of Kush, as we later demonstrate (see Figure 6.11). We have recently updated, upgraded, and qualified these results as shown in Figure 6.12 (see also Belmonte & Urrutia-Aparicio, 2022).

Two of the families (II and III, see Fig. 6.11) have a marked solar character with temples orientated to conspicuous landmarks of the annual cycle (the tropical year),

Fig. 6.10 Remains of the temple of the Ka of Pepi I at Bubastis (left) and the temple of Isis at Behedyt el Haggar (right). These are among the few preserved temples in the Delta. However, some of them present interesting and suggestive astronomical alignments. In this particular case both are seemingly orientated to sunrise at the winter solstice, although for the poorly preserved temple of Isis, a western orientation to summer solstice sunset is also possible. (Photographs by the authors)

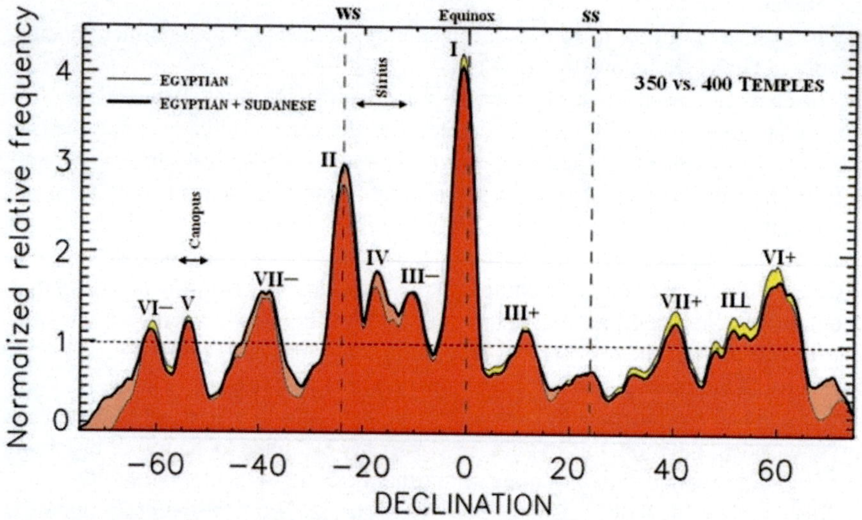

Fig. 6.11 The core of the astronomical hypothesis. Declination histogram of *c.* 350 temples of ancient Egypt (red and yellow areas), obtained from the data measured in our field campaigns across Egypt, plus 50 temples in Sudan (red and pink areas). Every peak is identified by a Roman numeral referring to each of the seven families of astronomical orientations as first defined in Shalthout et al. (2007). Only peaks I and II are statistically significant beyond any doubt. (Adapted from Belmonte, 2012)

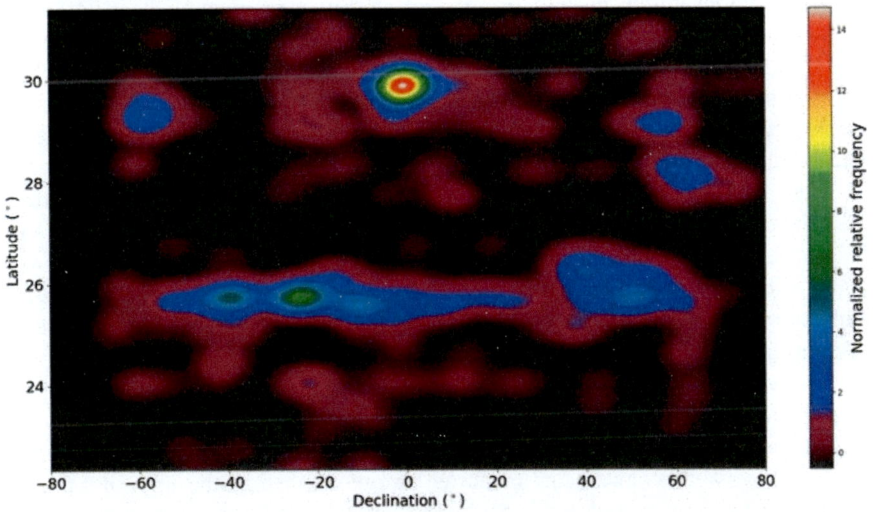

Fig. 6.12 Declination frequency map of 350 Egyptian temples. This three-coordinate diagram (latitude, declination, and normalized frequency) additionally shows how the significant peaks at the 'equinox' and the winter solstice are concentrated at the latitudes of Memphis and Thebes, respectively. Other peaks in turquoise also become statistically significant, notably the inter-cardinal family in Upper Egypt. The *c.* 50° concentration at Theban latitudes has several explanations, as discussed in the text. (Image by courtesy of Maitane Urrrutia, adapted from Belmonte & Urrutia-Aparicio, 2022)

such as the solstices but also to other important time markers related to the real moments of sowing and harvesting in ancient Egypt. Peak I is problematic because, as shown in Fig. 6.12, it is mostly related to the pyramid fields in the area of ancient Memphis. The peak is close to declination 0°, and hence to the 'equinox' (see Belmonte, 2021, for an updated discussion of the term). However, it is worth mentioning that, being associated with the funerary and valley temples of the pyramid complexes, it is not clear whether the dominant orientation should be considered to be due east, facing sunrise, or due north, where pyramid entrances are often open, and to the imperishable stars (see the Introduction).

Data additionally showed that selected solar orientations evolved, at certain historical periods, with buildings being orientated to the beginning of the seasons of the civil calendar, thus confirming earlier hypotheses. This was particularly the case for Wepet Renpet—the First Day of the first month of the season of the Inundation, I Akhet 1, or New Year's Eve—but also I Peret 1, as will be discussed later on. The coincidence of these dates with these important points of the tropical year (i.e. summer or winter solstices and spring equinox) could have acted as a mutual reinforcement in the interest of ancient Egyptians for these special days of both the tropical and the civil year.

As a corollary of this, we proposed as the result of the summing up of high-quality accurate data, that a brilliant combination of astronomy and landscape was achieved in many areas of ancient Egypt. There, a combination of astronomy and local topography produced an extremely interesting phenomenology in different regions of the country. One of the most fascinating case studies was that of the pyramid complexes of the Old Kingdom (see below). According to this hypothesis (first stressed in Shalthout et al., 2007), the location and orientation of several monuments were deliberately chosen in order to fit a general pattern of topographical and astronomical alignments.

After 3 years of intensive fieldwork, and four campaigns devoted to measuring the orientation and studying the spatial location of ancient monuments across the Nile Valley and beyond we were achieving exceptional results (Belmonte et al., 2008). More than 500 pyramids, hypogea, chapels, sanctuaries, and small and large temples had been measured, and the vast majority of the country had been scanned in detail from north to south and west to east.

However, in earlier campaigns, we had not measured some temples that were located either beyond the limits of standard circuits (such as Aïn Labkha in Kharga Oasis), at difficult locations (such as Mons Claudianus), or even temples whose existence we were unaware off (such as various shrines in the area of Luxor). Besides, some important places in Middle Egypt, such as Heracleopolis and southern Athribis had not been visited yet. A new campaign (the fifth) was indeed necessary to complete the sample.

This campaign was performed in December 2006 with the important aim in mind of carrying out solar observations at the precise moment of the winter solstice in several selected locations across the country, notably at Karnak (see Figure 6.13), Western Thebes, Qsar Qarun (see Fig. 6.2) and Dahshur.

Some of the places we visited in our fifth mission were not easy to find and were completely off the standard routes, leading to abnormal, sometimes curious situations. For example, in Kom Mir, south of Esna, we received complaints and demands from local authorities, since no one from the Supreme Council of Antiquities had gone there for years. In Naqada, the chief inspector of Qena province was happy to help us find, with the help of satellite images and old plans, the foundations of the temple of Seth, whose precise location had been forgotten since it had been excavated a century ago. Other places, such as the fascinating ruins of Mons Claudianus, in the Eastern Desert, deserved a dedicated programme involving many people, including desert tribes and the director of the antiquities service of the province of Bahr el Ahmar (Red Sea).

During the fifth campaign, we also took the opportunity of completing our sample of minor step pyramids, a group of fascinating, very ancient monuments (see Figure 6.14). Our analysis of these forgotten pyramids has proven extremely fruitful and we were able to demonstrate that a combination of land- and skyscape could help easily to explain their purpose, location and orientation.

The December 2006 mission was also centred on a study of the iconography of Seshat, the Egyptian goddess of time reckoning and calendrics, and her related epigraphy, including the stretching of the cord ceremony (see Figure 6.15), a most important aspect of interest for us (Belmonte et al., 2009a). Figure 6.16 illustrates some of the outcomes resulting from a most intriguing hypothesis that developed as a separate exercise in parallel to the main results of our mission.

Fig. 6.13 Sunrise at the winter solstice in December 2006 on the main axis of the temple of Karnak as seen from the quay. The phenomenon would have been more accurate 4000 years ago when the temple was first aligned, since at that precise period the disc of the sun would have been seen within the square frame of the distant eastern gate (the present one is from the Ptolemaic period). (Photograph by the authors)

With the fifth field campaign performed in December 2006, the Egyptian–Spanish Mission had accomplished some of the most important objectives we had in mind for this archaeoastronomical investigation into ancient Egyptian culture. Thc principal dilcmma wc wantcd to solvc was whcthcr or not thc tcmplcs of this civilization were astronomically orientated. Epigraphic sources clearly mentioned solar and stellar targets as references for temple orientations (see Fig. 6.15). However, the scientific community at the time agreed only on the planning of orientations strictly with reference to the Nile, the relevant inscriptions being sometimes considered as a mere remembrance of long forgotten practices. At that moment, we had measured 330 temples and shrines throughout the geography of Egypt belonging to all periods of her history, representing approximately 95% or all the temples in any state of preservation still existing in the country.

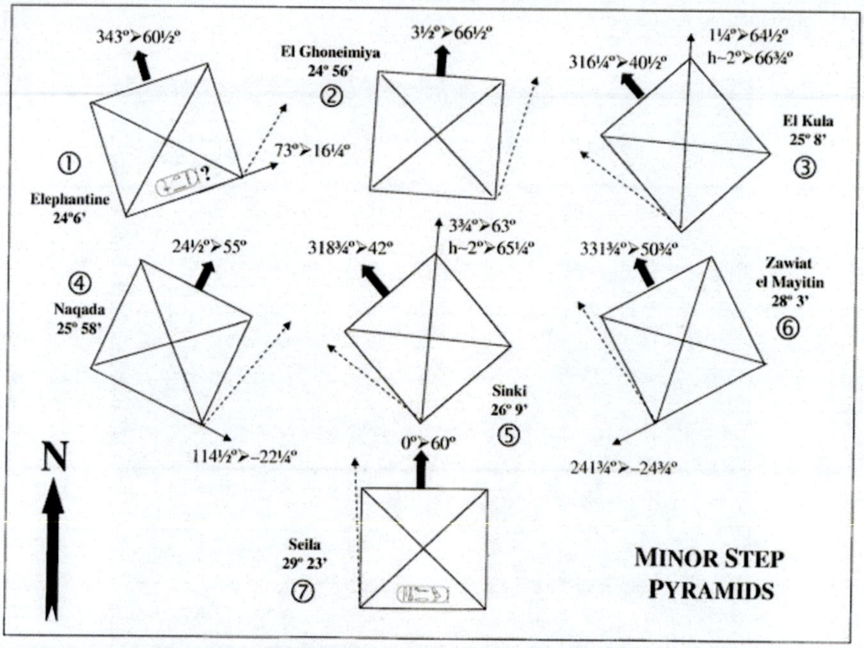

Fig. 6.14 Schematic diagram of the orientation of the seven minor step pyramids, numbered from south to north. Each pyramid shows a characteristic orientation that might be explained within an astronomical and topographical context. (Adapted from Belmonte & Shaltout, 2009)

Fig. 6.15 The stretching of the cord ceremony in the southwest wall of the pronaos of the Ptolemaic temple of the Falcon God Horus at Edfu. These scenes include long hieroglyph inscriptions with detailed information on the ceremony, including astronomical quotations. These indicate that Meskhet(yu) was the relevant asterism for these purposes. (Photograph by M. Sanz de Lara)

Fig. 6.16 This sequence of images illustrates the hypothesis of the use of the sign of Seshat as a topographical instrument similar to the Roman groma (see Chap. 2). Panel (**a**) shows a relief from the solar temple of Niuserre at Abu Ghurob, where the sign appears as a standard or portable object. Panel (**b**) stands for the nucleus of the idea with the sign transformed into a real object, when changing the two-dimensional version of the representation into a three-dimensional image. The uppermost elements of the sign would define a sighting device, or eyepiece, in the style of the 'merkhet', as shown in Panel (**c**). Once the alignment had been obtained, the eight radii of the device would directly offer the four cardinal and four inter-cardinal directions, as illustrated in Panel (**d**). Diagrams courtesy of SMM/IAC. (Adapted from Belmonte et al., 2009a)

At this stage, there were three particular results, which were real highlights of the analysis of the complete series of data (Belmonte et al., 2008). The first was that temples of the Nile Valley and the Delta were orientated according to the Nile as the data had clearly illustrated (see Figs. 6.3 and 6.4). However, as a second outcome, the temples were also astronomically orientated beyond any reasonable doubt as all the successive analyses we had done to the data fully demonstrated. This means that the ancient Egyptians had to deal with special situations to accomplish both needs.

This problem was solved in two ways. One alternative was the selection of appropriate orientations of one or the other class at different sites, so that they would be more or less compatible with the Nile course (inter-cardinal directions are a good example of this). The other was deliberately to choose places in Egypt where the Nile prescription and a conspicuous astronomical orientation were simultaneously achieved. In the following sections we analyse several examples of this phenomenon of combined land- and skyscape.

Finally, there were three, and only three, kinds of targets used for astronomical orientations. One was probably related to different and conspicuous celestial configurations of the stars of Meskhetyu in order to get an approximate, or even accurate, Meridian orientation. This primary axis could have been rotated later by an eighth, a quarter or half a circumference to obtain any possible cardinal or inter-cardinal direction (families I, VI and VII; see Fig. 6.11). The second kind of target had a markedly solar character and was fundamentally related to important time-marks of the annual cycle and/or the civil calendar (families I, II, II⊥, and III). A third group of targets would be formed by the two brightest stars of the ancient Egyptian skies. These were Sirius and Canopus, families IV and V, respectively), although these peaks had less statistical significance. These customs were in use during most of Egyptian history and in the different areas of the country, although some areas presented substantial concentrations (see Fig. 6.12).

A complete summary of the main results of the Egyptian–Spanish Mission in its first 5 years of existence can be found in Belmonte and Shaltout (2009, see also Belmonte, 2012). Indeed, we considered our sample of temple alignments to be statistically representative beyond any doubt and we were convinced that new data would only serve to reinforce or slightly qualify our results. It was thus the moment to falsify such statement. Two experiments were selected for this purpose.

On the one hand, a new campaign, the sixth and last of the Mission, was conducted in Egypt in December 2008 at a small number of sparse temples in Middle Egypt and the Oasis of Fayum, which could not be measured in previous campaigns for various reasons, mainly related to questions of security. The idea was to complete the sample of Egyptian temples to the point of exhaustion. Little spaces in the Eastern Desert, such as Mons Porphyrites and Sukait (Sidebotham et al., 2008), remained finally unexplored. On the other hand, a new complete experiment was designed for temples in Sudan (Belmonte et al., 2010). This is dealt with at the end of the chapter.

A most interesting Egyptian case is that of the temple of Thoth at Seikh Abada (see Figure 6.17). This New Kingdom temple is the best-preserved shrine of Thoth

Fig. 6.17 The New Kingdom temple of Thoth at Sheikh Abada (on the site of classic Antinoopolis). This building is aligned to the direction of moonrise at the epoch of the major northern lunastice. Either by chance or by deliberate selection of the site, the corresponding moon also passed across the local zenith. (Photograph by the authors)

in the area of ancient Hermopolis, the city under his patronage. Actually, it is located on the opposite side of the river, where an important wadi connects the Nile valley and the Red Sea and the precise spot where Emperor Hadrian founded the city of Antinoopolis centuries later. The temple axis is not perpendicular to the Nile and is directed nearly, but not precisely, towards Hermopolis on the other side of the river so local topography does not seem mandatory.

However, a very interesting situation arises when the direction from outside looking in is considered. In this case, the northernmost moonrise is produced over the hills of the Eastern Desert. Besides, considering the apparent diameter of the lunar disc, this moon would also pass across the zenith of the temple. It is fascinating to find a temple of the lunar god perhaps orientated to the northernmost lunar rising position (the lunastice) in the precise geographical area in Egypt where the same moon crosses the zenith. In previous studies, our team was surprised by the lesser importance of lunar alignments in ancient Egypt. This case indeed was the exception to the rule.

At the turn of the century, our team envisaged a project to answer a quite simple question. Now, near two decades later, our team is proud of the quantity and variety of the results obtained so far. These were the gateway for new requirements and questions, and further inquiries, as will be shown in this chapter.

All in all, this story of discovery in ancient Egyptian archaeoastronomy clearly illustrates what we could only have imagined at the very beginning of our project: ancient Egyptians undoubtedly scrutinized the sky in a permanent search for their correct orientation not only in time but also in space. Beyond any reasonable doubt, astronomy certainly did play a fundamental role in Egyptian culture, in terms of religion, the architectural design, and the sacred geography of the Nile Valley civilizations. This was undoubtedly reflected in some of their most sacred places.

The authors and their collaborators, on several occasions in the context of the Egyptian–Spanish Mission of Archaeoastronomy of Ancient Egypt (Belmonte et al., 2009b), have addressed a number of fundamental analyses that have convincingly demonstrated the important role played by the heavens in the alignment of sacred buildings in the ancient Nile civilizations. In contrast to China, however, there was no implicit tradition, such as Feng Sui (Baratta & Magli, 2021), relating architecture and the environment reflected in Egyptian texts, so we have to look for it in the monumental remains. We focus briefly here on a series of the most striking cases in which astronomy, combined with local topography on a large scale, has made it possible to curb chaos (*isfet*) and help maintain cosmic order, the *ma'at* (Hornung, 1999).

6.2 History Begins at Abydos

Our research team has proposed that structure HK29A at Kom Al-Ahmar (Hierakonpolis or ancient Nekhen, Friedman, 1996) was one of the first buildings erected in Egypt to be orientated astronomically (Belmonte et al., 2008), and it is of course the oldest that has been found. Hierakonpolis was presumably the city where the cult of the Falcon God Horus was developed and it is certain that the first kings of Egypt had close ties with the town. A peculiar configuration of Meskhetyu would have been used in Kom Al-Ahmar to find directions, perhaps because the constellation was not circumpolar on site during the Predynastic period, since Merak (β UMa) dipped slightly below the local horizon at that time.

However, the Egyptian imperishable constellation of Meskhetyu became entirely circumpolar a few hundred kilometres to the north of Nekhen at a site, Abydos, that would be transformed into one of the most important sacred sites of Egypt (see Figure 6.18). Umm al Qab, in the desert area of Abydos, is the site of a huge Pre- and Proto-Dynastic cemetery, which also includes the tombs of the first Kings of a unified Egypt (0th and first Dynasties), at the moment when the Egyptian state was founded. In this epoch, certain metaphysical aspects relating to the figure of the King were first developed, perhaps including the stellar eschatology of the *Pyramid Texts* (for the PTs, see Faulkner, 1969; Allen, 2005, Morrow, 2015).

Abydos is located at a very striking site of Middle Egypt. The hieroglyph for mountain in the ancient writing was *dju*. Actually, there are few *natural* mountains in the Nile valley. Most of the orographic elements are the bluffs of the desert ridges, crossed by seasonal water streams or wadis. This orographic accident probably is the origin of this sign, which is extremely ancient, as it already appears on the seals found in the pre-dynastic tomb U-j at Umm el Qab, where a 'King Elephant mountain' is mentioned (Dreyer, 1998). Its connection with the afterlife and, in particular, with Abydos is clear. To the southern horizon of the necropolis a huge *wadi* opens into the Umm el-Qab bay, functioning as a sort of 'mouth to the afterworld' (see Fig. 6.18). The sign appears explicitly in the name of Abydos, which was:

 *Abdju*

Giulio Magli (2013) has presented a detailed analysis of the site arguing the existence of a direct connection between the topographical feature, the necropolis at Umm el Qab and the funerary precincts at the limit of the cultivable lands (Fig. 6.18). All of them are aligned within a meridian axis that suggests a certain relationship to the northern skies, as we show below. The Royal Tombs or the associated funerary complexes, including the largest surviving early structure at Abydos, the precinct of King Khasekhemuy at Shunet el-Zebib (Fig. 6.18), are distinct members of the inter-cardinal family of orientations (developed in Upper Egypt to comply with restrictions imposed by the Nile course). This was possibly a particular case of a general cardinal pattern related to the northern stars (Belmonte et al., 2009a, b).

However, Magli probably fails in his interpretation of inter-cardinal directions in Upper Egypt, notably in proto-Dynastic Abydos (but also in New Kingdom Thebes). He argues that the idea that the Egyptian temples were mainly orientated towards the Nile tends to be overplayed in Egyptological literature. According to him, the hypothesis defended by our research group and demonstrated in Figs. 6.3 and 6.4 is also a workaround to reconcile Nile orientation with the astronomical one, an explanation which, as Magli argues, would be too far from 'what we know about the

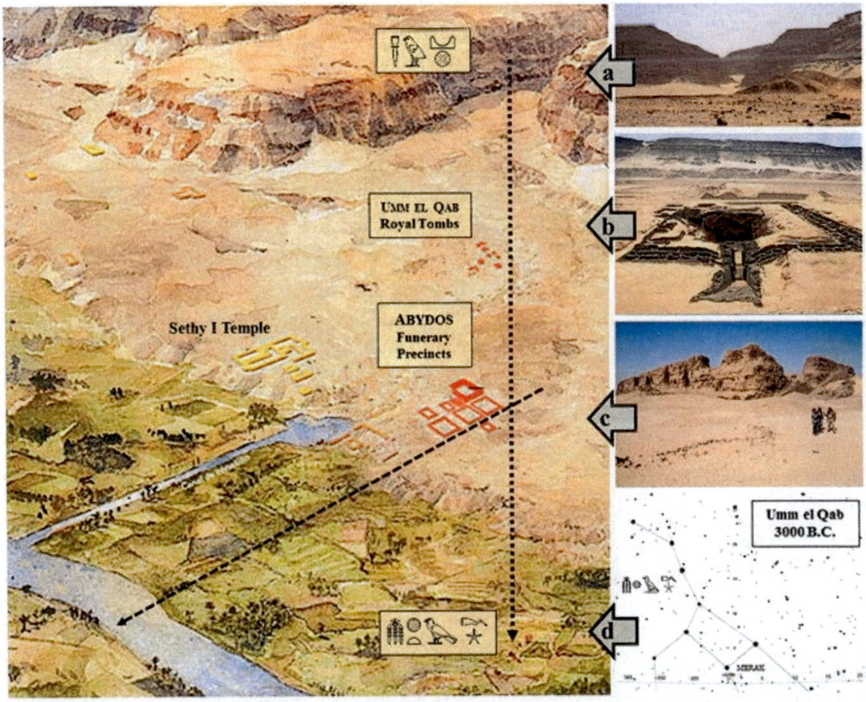

Fig. 6.18 A cosmic landscape at Abydos (ancient Egyptian *Abdju*), the site of the burial monuments of the first kings of a unified Egypt. A distinct orographical feature on the local landscape (**a**), the royal tombs at Umm el Qab (**b**, tomb of Den), and the funerary precincts (**c**, Khasekhemuy's 'Shunet el-Zebib') were aligned along a south-north axis (dotted line). This perhaps pointed to the lower culmination of Merak (β UMa)—the lowest for any star of Meskhetyu (**d**)—as seen from the Royal Cemetery, where the 'imperishable' asterism was circumpolar during the First Dynasty. This might have justified the sacredness of the place. The tombs and the precincts themselves were inter-cardinally orientated (dashed line). See the text for further details. (Diagram of the authors, adapted from Belmonte, 2016)

Egyptian architect's way of thinking'. He finds a far more natural astronomical interpretation, possibly inspired by his earlier work on the Inka: the Milky Way. However, there is a lack of ancient texts supporting his ideas (Magli, 2013: 14–24) and Fig. 6.3 seems to prove otherwise.

Middle and especially New Kingdom Thebes replicates Abydos, the place of the ancestor burial. In Abydos, NE orientations are dominant (Fig. 6.18), while SE is the preferred orientation in Thebes. While the Galaxy could explain the orientation in Abydos, only a custom to place temple façades perpendicular to the Nile valley can explain both orientations simultaneously. Inter-cardinal directions are present, with minor exceptions, only in Upper Egypt, where the corresponding alignments would be nearly perpendicular to the Nile, and they are absent elsewhere in Egypt. This would not happen if the reference were a relevant celestial body. Similarly, the orientation family peaks are rather sharp in the related histograms, which raises difficulties when attempting to explain by means of an orientation with respect to a large and rambling object covering a wide range of declinations, such as the Milky Way (González-García & Belmonte, 2014).

The question now is why the northern skies should be relevant at Abydos. Two scenarios are possible. First, it could be argued that Umm the Qab was selected as the site for the early dynastic Royal Cemetery because it was the first site, following the northward flow of the Nile from Nekhen, where Meskhetyu would have been a circumpolar asterism in *c.* 3100 BC (see Fig. 6.18, i.e. it behaved as an imperishable sky entity). This would plainly justify the sacred character of Abydos and suggests that the site was deliberately selected. In contrast, the alternative hypothesis would be that Meskhetyu was important as the quintessential imperishable constellation because it was circumpolar at the site of the royal necropolis. Both solutions may be possible and certainly indistinguishable from the astronomical and topographic point of view.

However, the first option seems the more likely because it would represent a closer relationship between astronomy and local geography with the creation of a true cosmic landscape. In addition, Toby Wilkinson (1999) has suggested that the founder Kings of Egypt (0th and 1st Dynasties) were originally from Nekhen, but that they established their Royal Necropolis a few hundred miles further to the north, at Umm the Qab and Abydos for reasons unknown. Our suggestion could support the idea that the place of the royal necropolis (and by extension Abydos) was deliberately chosen for religious reasons related to astral eschatology. This would convert Abydos into the primaeval Egyptian Cosmic Landscape later to be imitated at other royal sites of Egypt under similar but not necessarily identical frameworks.

6.3 Astronomy, Architecture, and Symbolism in the 4th Dynasty

The two pyramids built during the Old Kingdom by the 4th Dynasty King Sneferu at Dahshur are usually considered as two consecutive projects, the second—that of the Red Pyramid—presumably being generated by a failure of the first, the Bent Pyramid. In the paragraphs that follow we show that the archaeological proofs of such a scenario are far from obvious and that, on the contrary, a series of architectural, topographical, epigraphic and astronomical hints point to a unitary project probably conceived from the very beginning in terms of the two pyramids and their annexes. Similar conclusions could be reached for other royal tombs of the 4th Dynasty, notably at Giza, Abu Rowash and Zawiyet el Aryan. The pyramids of the period taken all together are thus shown to form a conceptual, sacred landscape related to the power of the Pharaoh and his afterlife.

In this section, we discuss the role that sky observation might have played not only in the orientation of the pyramids of Dynasty IV, but also in the choice of their location, thereby producing a synergy between both astronomy and landscape realities. This phenomenology is closely related to other aspects of Pharaonic civilization such as the calendar, celestial mapping, and the astral aspects of religion (Brugsch, 1883; Krauss, 1997). We therefore concentrate on one of the most famous archaeological areas in Egypt: the pyramid fields, and in particular on the giants of the 4th Dynasty at Dahshur, Giza, and elsewhere (Aufrère & Golvin, 1997; Lehner, 1997).

6.3.1 Kha(wy) Sneferu: Symbolism and Power

Generations of Egyptologists have postulated as a fact, almost as a dogma, that during the reign of Sneferu (*c.* 2575 BC) there was an evolutionary process of trial and error in the construction of pyramids, of which the so-called Bent Pyramid at Dahshur (*ḫʿ-snfrw-rsy*) would be a magnificent example. This eventually led to the first perfect pyramid, the so-called Red Pyramid (*ḫʿ-snfrw*) built later on the same site by the same king (see, for example, Edwards, 1993; Lehner, 1997). However, recent ideas based on different, including architectural (Magli, 2013) and astronomical (Belmonte & Zedda, 2007) criteria postulate that the pair of pyramids of Sneferu in Dahshur would actually form a single project conceived as such from its inception (Belmonte & Magli, 2015).

Given this suggestive possibility, a detailed study of the available information was carried out, both from the chronological point of view (Krauss, 1996), including epigraphy (in a new reading of the Palermo Stone, Wilkinson, 2000a, 2000b; see also Figure 6.19) and from a symbolic viewpoint (in a new analysis of the *Pyramid Texts*, Goebs, 2008), including astronomical investigation. New ideas were analysed, including a reasonable and pioneering explanation of the reason for the anomalous slope of 43°23′ of the faces of the Red Pyramid (Belmonte & Magli, 2015; see Figure 6.20).

The set of ideas expressed by Belmonte and Magli (2015) seems to confirm the hypothesis that the two great pyramids of Sneferu in Dahshur were conceived as a unitary project in which astronomy was mixed with architecture to capture in stone, on a gigantic scale, a tangible reality laden with symbolism. This could support the idea that the two great pyramids of Giza had originally been part of a single unitary project on an even larger scale during the reign of Sneferu's son, Khufu (*c.* 2550 BC): *ꜣḫt ḫfw*.

The leitmotif of the ideas defended by Belmonte and Magli (2015, 2017) was the independent research on the subject carried out by both scholars but using different approaches (symbolism versus architectural design). There was, however, a common interest in the role of cultural astronomy. Thanks to this multidisciplinary approach, they were able to reach a similar conclusion: the pair of gigantic pyramids erected in Dahshur were completed, and perhaps even conceived, as a unique project. Thanks to his enterprise, apart from the strong symbolic aspect and the obligation to comply with the needs of the sovereign in the afterlife, King Sneferu was

Fig. 6.19 Astronomy, architecture, and symbolism in Dahshur: the Red (left) and Bent (right) Pyramids as a symbolic representation of the red and white crowns, of Lower and Upper Egypt, respectively, which would appear mentioned as such in the Palermo Stone. In this guise, they could be interpreted as the petrified counterpart of pertinent celestial phenomena, of respective reddish or whitish hues, such as the northern lights (or the aurora) or the zodiacal light (to which Venus would be added). See text for details. (Diagram by the authors courtesy of the IAC Multimedia Service, based on images courtesy of Graham Parkin and Daniel López, and on a tracing of a photograph by Margarita Sanz de Lara obtained courtesy of the Palermo Archaeological Museum)

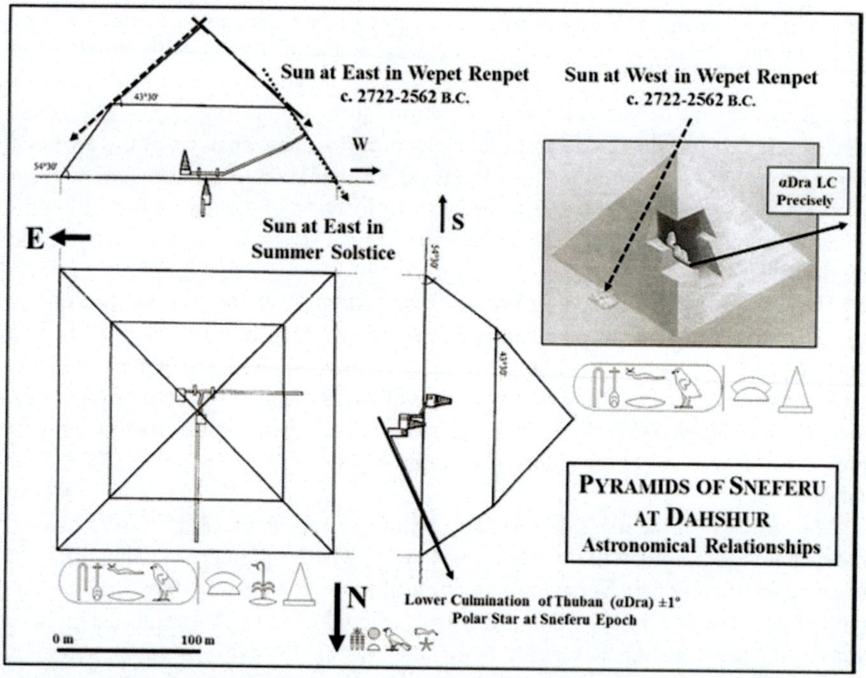

Fig. 6.20 Astronomical relationships in the Sneferu pyramids at Dahshur. Both pyramids are properly orientated to the north so that the slope of the corresponding access corridors would point to the lower culmination of Thuban, the Pole Star during the Old Kingdom. Uniquely, the slopes of the two pyramids could be related to a couple of important time-markers in the Egyptian time-reckoning system: the summer solstice and Wepet Renpet, the New Year's Festival. (Adapted from Belmonte & Magli, 2015)

able to show his absolute control of the territory by simply contemplating his fascinating monuments.

From Sneferu onwards, if not before, each royal funerary complex, with a pyramid as its main element, would receive a name that has come down to us in the reliefs of the tombs of the officials and priests in charge of the complexes (Quirke, 2003: 145). The name of the Red Pyramid was , which includes the hieroglyph ḫʿ, and should be read as Kha-Sneferu, which is usually translated as 'Sneferu is brilliant'. However, the glyph ḫʿ describes both the rise (and therefore its first glow) of a celestial object, as well as the assumption by the king of royal paraphernalia (particularly crowns) and his appearances at festivals and ritual processions (Goebs, 2008: 24–31; Borrego Gallardo, 2014).

Interestingly, the ancient name of the Bent Pyramid was the same as that of the Red Pyramid: , but with the addition of a sign that denotes the 'south' or also 'Upper Egypt'. This could be read as Kha-Sneferu-Resy and has been standardly translated as 'Sneferu is brilliant in the South'. However, taking into account the previous premise, the name of the Bent Pyramid could be translated as 'Sneferu appears wearing the crown of the south' or, directly, as the 'Sneferu's Southern Crown'.

The similarity, or even complementarity, of the names of the two pyramids is undoubtedly evidence that supports the idea of a common project. In fact, if we look for new clues, a reference to Sneferu's mortuary complex as , Kha(wy) Sneferu, could be recognized in a decree issued by the 6th Dynasty King Pepi I (*c.* 2250 BC), regulating the administration from the pyramid city at Dahshur, and found on an inscribed stele not far from the Red Pyramid (Moret, 1917).

This name could be translated as 'the two pyramids: Sneferu is brilliant', or even, in the new scenario defended here, as 'the two crowns of Sneferu'. This again would support the hypothesis that the two great pyramids of Sneferu at Dahshur were conceived as a unitary project. In astronomy, religion, politics, and architecture combined to produce a tangible reality, charged with symbolism on a truly gigantic scale.

The rising idea is that the two pyramids of Sneferu at Dahshur could perhaps, reasonably, be identified as a huge symbolic representation of the white crown of Upper Egypt, in the case of the Bent Pyramid , and of the red crown of Lower Egypt for the Red Pyramid . This hypothesis rests on arguments previously defended by Cintron (2003) in his analysis of the Palermo Stone. These would at the same time be the materialization in white and red stone, respectively, of cosmic realities that could well be relevant in the process of catasterization or astralization of the king in the afterlife, as would be beautifully expressed in the *Pyramid Texts* two hundred years later.

If this were the case, then the pyramids of Sneferu at Dahshur would be physically real manifestations on a gigantic scale, symbolizing the eternal power of the king, in the form of a huge and impressive cosmic landscape (see Fig. 6.19). Once again, this fact would speak of the actual existence of a unified project that would allow the king to be manifested as determining celestial aspects of royalty and divinity, both during his lifetime and in the afterlife.

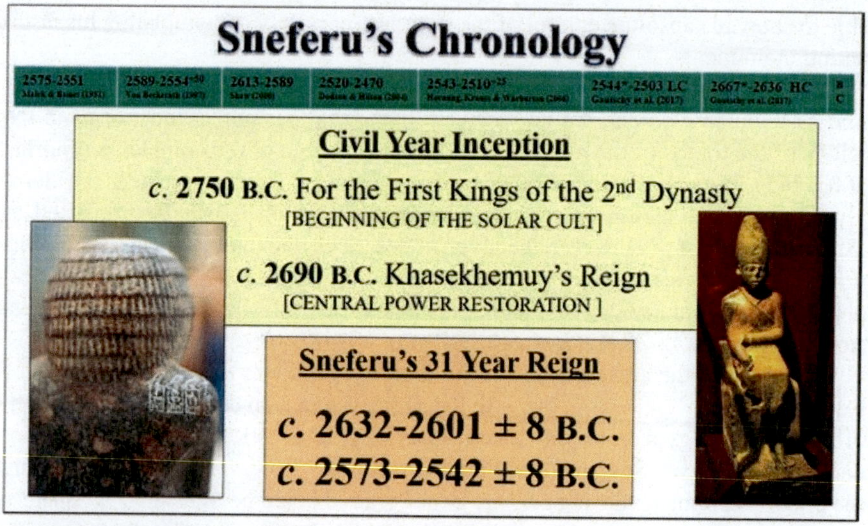

Fig. 6.21 The different chronologies of Sneferu's reign according to different authors (green box) and the two possible proposals based on the dates analysed and raised by the authors in agreement to the founding of the Egyptian Civil Calendar (see Chap. 5)

According to this hypothesis, the story could be summarized as follows: in the year 8 or 9 of his reign (for Sneferu's chronology, see Figure 6.21, also Chap. 7), and for unknown reasons, the king decided to abandon his necropolis at Meidum, where a huge step pyramid had already been built. Sneferu then created a new residential area on the border between Upper and Lower Egypt. In this area, a new, unique, dual-character project would begin to develop, which would serve as a guarantor for Pharaoh's eternal life, but also as a true image of his overwhelming power in life. This project would be conceived as a funerary complex made up of two gigantic pyramids (with their respective mortuary temples), a satellite pyramid (perhaps for the King's Ka), and a 'Valley' temple possibly associated with the celebration of the Sed Festival (Staedelmann, 2011).

Therefore, Belmonte and Magli (2015) have argued, the Bent and the Red pyramids at Dahshur would be the manifestation of the sovereign's power as Dual King of Upper and Lower Egypt by symbolic imitation of the white and red crowns of Upper and Lower Egypt, respectively (see Fig. 6.19). This would include colour and location, perhaps also in the shape of the monuments themselves: pseudo-conical and flattened, respectively. The slopes of the pyramids (interchangeable, as they appear in both) would have been defined with a specific astronomical symbolism (summer solstice and New Year's festival or Wepet Renpet) that might be related to the creation, inception and subsequent development of the Egyptian civil calendar (Belmonte & Magli, 2015, 2017; see Chap. 5). Astronomical alignments would also be included. The pyramids are almost perfectly orientated towards the north (Belmonte, 2001; Nell & Ruggles, 2014; Spence, 2000), and their access corridors were built with such a slope that this would facilitate the king's ascent to the northern skies (see Fig. 6.19), domain of the imperishable stars, the '*Ikhemu Sekiu*' (as also suggested in the Coffin Texts, Spell 115; Faulkner, 1994).

In addition, the pyramids could be associated with concomitant celestial phenomena possibly manifesting of the king's power after death. In particular, the zodiacal light, Venus (or both) for the case of the white crown and the northern lights or the dawn light for the red crown (Fig. 6.19). The pyramids could well be considered as petrified light. This unique astronomical symbolism would be put in writing two hundred years later in the *Pyramid Texts* of the fifth and sixth Dynasties. Sneferu presumably died in his thirty-first year of reign, leaving unfinished part of the structures related to the Red Pyramid, in particular his funerary temple that was possibly quickly finished by his eldest son, and successor, Khnum Khufu (Cheops) who, in order not to be less, would also try to develop his own particular cosmic landscape.

6.3.2 Akhet Khufu: The Sun at the Horizon

On June 21, 2006, the Egyptian–Spanish Mission, including the first author, travelled to the Giza Plateau in order to observe and document what we guessed had to be a spectacular astronomical hierophany (see Figure 6.22). The idea was to admire the sunset at the summer solstice behind the Sphinx, from a position where the statue was framed by the two largest pyramids that dominate the plateau. The Sphinx was known as Hor-em-akhet, 'Horus at the Horizon', during the New Kingdom. Egyptologists Richard Wilkinson and Mark Lehner had independently suggested that this name would have been inspired by the similarity between the position of the gigantic sculpture in the middle of the two pyramids and the compound hieroglyphic term for Hor-em-akhet (Fig. 6.22; Wilkinson, 1994; Lehner, 1985).

The phenomenon that was observed in Giza fully confirmed this idea, adding to the image proposed by Wilkinson and Lehner that of the setting Sun's disc exactly behind the head of the Sphinx, enveloping it with a brilliant halo of twilight light. The intuitive idea, reinforced from that day on, is that this phenomenology could certainly have been deliberate, and that the triad formed by the pyramid of Khefren, the Sphinx, and the pyramid of Cheops would have been at least designed with this goal in mind (Shalthout et al., 2007).

The idea of a single project for the two largest pyramids at Giza was independently developed in the 2000s by the first author and by Giulio Magli (2008). Conversations held over the years have signalled that there is apparently no straightforward contemporary evidence that the pyramid of Khaefre was built by this king. On the other hand, Colin Reader (2002) had even argued that work on the Khaefre causeway sector and other areas near the Sphinx would have started earlier, during at least the reign of Khufu. However, there is clear evidence that the Egyptians of subsequent dynasties had no doubt that the second Giza pyramid had to be assigned to Khaefre. This was beautifully illustrated in a frieze at the tomb of Qar, priest of the necropolis of Giza during the reign of Pepi I, where the names of the three

Fig. 6.22 Architecture and astral symbolism in Giza recreated in the Old Kingdom, through a hypothetical project of two pyramids (plus the Sphinx) imagined by Cheops, i.e. *ꜣḫt ḫfw*, the 'Horizon of Khufu', which was certainly partly appropriated and completed by his son Khaefre. This will be symbolically reinterpreted as the God Hor-em-akhet, 'Horus at the horizon', during the New Kingdom. This image could be a faithful reflection of the culmination of the king's solarization process during the reign of Khufu (*c.* 2550 BC). (This image was taken by the first author during sunset at the summer solstice at 21 June 2006)

pyramids of the plateau are clearly written, including Ur-Khaefre (Belmonte, 2012: fig. 6.3).

The hypothesis in this case would be the following. Since his father Sneferu had built a huge dual project at Dahshur, Khufu (*c.* 2550 BC) might have decided that he could not be less than his predecessor, planning another dual model but on an even larger scale. This unitary project would be made up of two huge pyramids (larger than any of his father's) and a huge statue of himself as an image of the solar deity (indeed the Sphinx), with which temples, roads, and a huge cemetery for members of the royal family would be associated. This is the first time that a large-scale orthogonal planning would be used in monumental construction. The new pyramid complex would have received the name , Akhet Khufu ('the Horizon of Cheops';(Fig. 6.22). In fact, the possibility that both Sneferu and Khufu imagined their funerary monuments on such a gigantic scale gives a dynastic meaning to the idea and offers additional, albeit circumstantial, support to the hypothesis of a common dual project in both cases.

A substantial difference would be that Sneferu would have had the time almost to end his project in Dahshur, while his son Khufu certainly left his unfinished after a long but slightly shorter reign of 27 years. This is what would have allowed his younger son, and second successor, Khaefre, to be in charge of carrying out the gigantic plan by simple 'usurpation' of almost half of his father's original project. He assumed it as his own under the name of Ur Khaefre, 'Khefren is a great one (from the sky)'. Consequently, the name Akhet Khufu would ultimately apply only to the Great Pyramid of Giza as later sources demonstrate.

Such an enormous dual construction project would have been part of a large-scale process of solarization of royalty during 4th Dynasty. This was initiated by

Khufu's grandfather, Huni, the first king of Egypt to write his name inside a cartouche, continued, by his father Sneferu, and reaching a climax during the reign of Cheops, who may even have identified himself with the sun god Re (Hawass, 2009). As a likely consequence of this process, his elder son and successor Djedefre would be the first king in Egyptian history to use the epithet 'Son of Re', a common term in the titling of the pharaohs hereafter. Therefore, Hor-em-akhet as the name for the sphinx would have been but a later New Kingdom reinterpretation of an idea that would already have been around for generations (Fig. 6.22).

In addition, the pyramid complex was orientated precisely to the north, possibly to the stars of Meskhetyu, the 'imperishable' *par excellence* (Belmonte, 2001). In this line of argument, Akhet Khufu complex could have been aligned in the interval between 2559 and 2541 BC approximately.

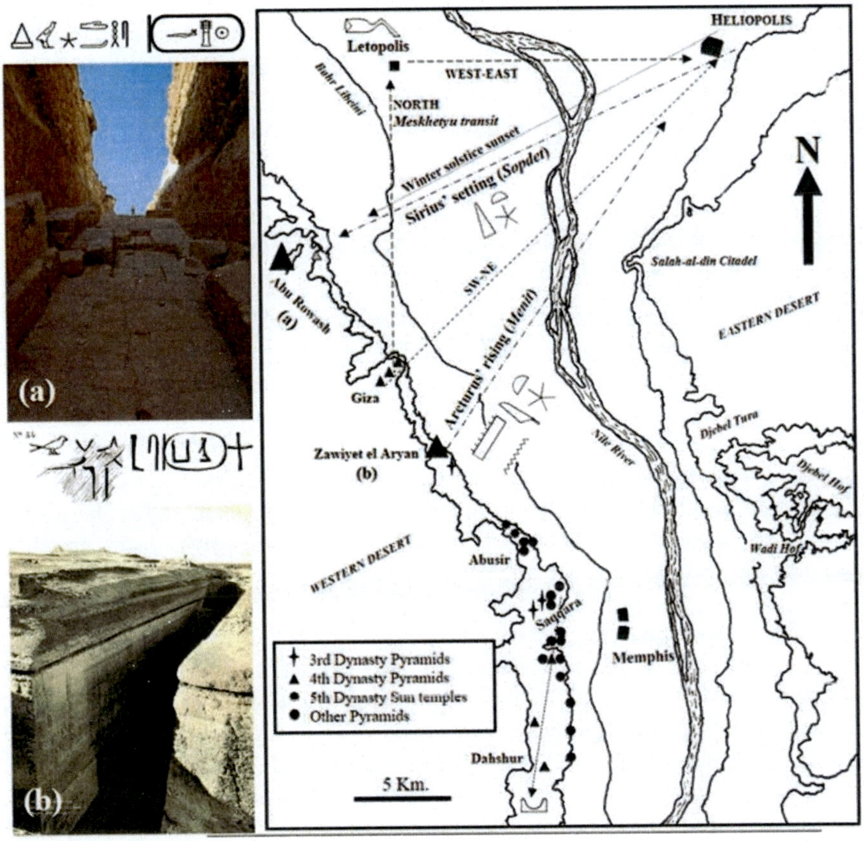

Fig. 6.23 Land- and skyscapes in the pyramid fields of the Old Kingdom in the Cairo area. Some suggestive astronomical and topographic alignments of 4th Dynasty pyramids and their geographic environment are shown. The graphic emphasizes the importance of Heliopolis, the city of the sun god. The images show the great trenches dug for the pyramids of Djedefre at Abu Rowash (**a**) and Setka at Zawiyet el Aryan (**b**), along with their 'stellar' names (certain in the first case, possible in the second). See text for further details. Authors' chart (graffiti 34 by A. Barsanti, 1906)

In Giza, however, not only astronomy, but also local topography and orography played very important roles (see Figure 6.23). As early as the 1960s, Georges Goyon had proposed that the plan of the three great pyramids of Giza suggested a relationship to the cities of Letopolis and Heliopolis (Goyon, 1970). The line that joins the southeast corners of the three main pyramids, with an azimuth slightly less than 45°, thus pointed towards the sacred city of the solar god, while the north direction led directly to Letopolis. Furthermore, the distance between Giza and Letopolis, and between the latter city and Heliopolis, was almost 30,000 royal cubits (or *c.* 100 stadia), a very round number.

This finding suggests as a possible corollary that the respective sites of the Giza necropolis with respect to the towns of Letopolis and Heliopolis might have been deliberately selected. A curious fact is that Letopolis was the capital of the Egyptian nome located just north of Giza, and whose emblem was a bull's foreleg, the same sign used as a determinative for the name of the constellation Meskhetyu (see above and Fig. 6.23). Was there a connection between celestial and geographical reality in ancient Egypt, similar to that found in the term 'arctic' (also *septentrional* in Latin languages) related to the Chariot asterism in many European languages?[1] If this were the case, in Giza and its geographical environment we would be really in the presence of a cosmic landscape on a mammoth scale (Belmonte, 2012: 217–22).

6.3.3 *Khufu's Successors and the Stellar Pyramids*

, the pyramid ('Djedefre is a Sehedu star'; see Fig. 6.23) is, as we saw, the name of the funerary monument of King Djedefre, elder son and successor of Khufu. This name suggests that this pharaoh, when he died, would have become a 'Sehedu' star, a term that seldom appears in the *Pyramid Texts* whose meaning is not yet certain. As we have discussed, it has been translated as 'Firmament' and it has even been suggested that it would be one of the possible Egyptian ways of naming the planets, although this is far from being proven (see Chap. 4). An alternative translation would suggest that is not Djedefre but his own pyramid, the one that would be a stone representation of the 'star Sehedu'. However, there is evidence that the first reading is perhaps the most probable.

Djedefre's apparently unfinished pyramid complex is located at Abu Rowash, on top of a rocky outcrop that, in ancient times, when the permanent haze of the Cairo megalopolis did not yet exist, would have been clearly visible from Heliopolis. In this sense, it is perhaps significant that sunset on the winter solstice, as observed from the northern limit of the sacred precinct of the Temple of Re in this city, would have been visible behind the southern corner of the pyramid, located more than 20 km away. This again suggests that the site of Abu Rowash could have been deliberately chosen (Fig. 6.23). We are thus confronting with a new solstitial relationship

[1] Arctic comes from the Greek 'Arctos', bear, while *septentrional* comes from *septem triones*, or the seven oxen the Romans saw in this important asterism.

in the context of the 4th Dynasty pyramids. This will add to those found in Dahshur and Giza (see Figs. 6.19 and 6.22; Belmonte & Magli, 2015).

However, a most interesting alternative could well have been the observation, from the sacred precinct of Heliopolis, of the setting of the star Sopdet, (our Sirius), the brightest in the Egyptian skies and herald of the Flooding (see Chaps. 4 and 5), in the surroundings of the pyramid of Djedefre. Could this fact be in relation to the stellar name of the pyramid? Without additional texts or other epigraphic evidence, both hypotheses are equally suggestive, but it is impossible to favour either of them.

However, the stellar hypothesis could be reinforced by an additional circumstance. It is accepted as a fact that another 4th Dynasty pyramid was planned at Zawyet el Aryan, south of Giza. Of this building, only a formidable substructure survives, similar to that of the Abu Rowash complex (see Fig. 6.23). This construction has been attributed to various alleged members of the royal family. The stonemasons' marks found at the site repeatedly mention the cartouche of a king '?Ka' (Barsanti, 1906). In particular, one of them (#34) mentions what could be the name of the pyramid complex (Fig. 6.23) since the inscription includes the text ('?Ka is his great star'), where the question mark could allow the reading of various names: Neferka , Setka , or , which could be read as Nebka or even Bika (perhaps Manetho's Biche[res]; Waddell, 1940). The last two, notably Setka, could correspond to known names of Djedefre's attested sons, so a certain dynastic continuity is verified (Dodson & Hilton, 2004).

Given the hypothetical 'stellar' name of the pyramid of Zawiyet el Aryan, a new astronomical connection between this building and Heliopolis could be envisaged. Either by chance or by deliberate choice, the star Arcturus, the fourth brightest in the Egyptian skies (indeed the closest bright star to the Pole and perhaps an 'imperishable' star), rose over Heliopolis around 2450 BC (Fig. 6.23). As discussed in Chap. 4, Arcturus may be identified as the most important star of the Mooring Post, *mnjt* (Lull & Belmonte, 2009; Belmonte, 2012: Table 3.1), a constellation of the Ramesside star clocks. The Mooring Post is also mentioned in the *Pyramid Texts* in the context of the astral journey of the deceased king to the imperishable stars, in which this astral entity, either a star or constellation, calls the king to take his place among the '*ikhemu sekiu*'. Therefore, there is a possible relationship that it is difficult to ignore.

On these premises, we would have two new pyramids—whose names are stellar ones (certain in one case, possible in another)—presumably started by father and son (not necessarily successors of each other) that would be located on both sides of the main necropolis of the 4th Dynasty at Giza. They would also show significant stellar alignments (Sirius and Arcturus, respectively) with the city of Heliopolis. Consequently, the importance of the skyscape in relation to the topography of the area occupied by the 4th Dynasty pyramid complexes (including that of Menkaure in Giza) seems increasingly evident.

The final act was going to be due to Shepseskaf, the last king of the dynasty. This king decided to build his funerary monument in the form of a gigantic mastaba instead of the traditional pyramid. The reason for this has always been an enigma.

Fig. 6.24 The group of monuments at Saqqara South, notably the pyramid of Pepi II and the mastaba of Shepseskaf with the drawback of the pyramids of Sneferu at Dahshur as seen from Djedkare Isesi's pyramid. Shepseskaf's monument seems to be somehow connected to his forefathers' behemoths. See the text for further discussion. (Photograph by the authors)

However, if this decision is analysed within the context of the landscape considered here, the reality becomes palpable. The two great pyramids of Sneferu in Dahshur are clearly visible in the distance, forming a very distinctive element of the southern horizon, as seen from the central nucleus of Saqqara (see Fig. 6.24; also Fig. 6.23), where this necropolis was accessed and where the monuments of the first dynasties are located. Shepseskaf's flat burial monument was built in such a way as to complete the space between both pyramids until the hieroglyphic sign *ḏw* was clearly visible (Figs. 6.23 and 6.24; Belmonte & Magli, 2015). If this was deliberate, this reality would complete our analysis of the relationships between astronomy, architecture, and landscape for this spectacular creative period in the history of Egypt, since it would establish a clear spatial relationship between the last king of the 4th Dynasty, Shepseskaf, to his great-great-grandfather Sneferu, founder of this dynastic line.

6.4 The Siblings of the Sun God: Temples, Pyramids, and Texts

First Shepseskaf, Khufu's great-grandson, and subsequently his successor Userkaf (*c.* 2470 BC) left the family cemeteries to the north of Memphis to build their pyramidal complex close to the capital, at Saqqara. As we have seen, Shepseskaf chose a monument in the form of gigantic mastaba that seemed to connect with the pyramids of the founder of his dynasty, Sneferu, in Dahshur and Userkaf manufactured

Fig. 6.25 Mortuary temple and pyramid of King Sahure at Abusir, the first pyramid to be built on site. The 5th Dynasty pyramids are mostly piles of rubble because of their construction techniques but the temples are beautifully preserved, high-reliefs included. (Image by the authors)

a small pyramid in the shadow of that of his glorious ancestor Netjerkhet. However, this last King also built at Abu Ghurob, in the area of Abusir, the first in a series of solar temples, two of which (his and Niuserre's) have been archaeologically uncovered. His successors, five 'solar' Kings of the 5th Dynasty (particularly Sahure, Neferirkare, Neferefre, Niuserre, and perhaps also Shepseskare) chose Abusir as their family necropolis (see Figure 6.25).

According to Stuart Jeffreys (1998), the location of the dynastic temple complexes devoted to the solar god Re would have been deliberately selected with the objective in mind of remaining in view of Heliopolis, the sacred city of the sun.[2]

In his opinion, this would also be the reason why the temples themselves were built a little to north of the pyramid complexes of the kings. This visibility would have been lost not only from Saqqara but also from the pyramid complex of Abusir, further north (see Fig. 6.23).

However, in the authors' opinion there is a complete series of astronomical and topographical connections between the pyramid fields of the 4th and 5th Dynasties (and indeed of the 6th Dynasty), the solar temples of Abu Ghurob and the sacred cities of the capital area as shown in Fig. 6.23 (see also, Belmonte et al., 2009b; Belmonte, 2012: 228–232).

Another highlight of Magli's work is the idea of several elaborate dynastic landscapes, including most suggestive topographic and astronomical alignments, notably what he has termed 'symbolic invisibility' (see Figure 6.26). One advantage of

[2] Most recent excavations in Abu Ghurob by an Italian team led by Massimiliano Nuzzolo have discovered the remains of an older sanctuary below the solar temple of Niuserre, perhaps either Sahure's or Neferirkare's lost solar temple.

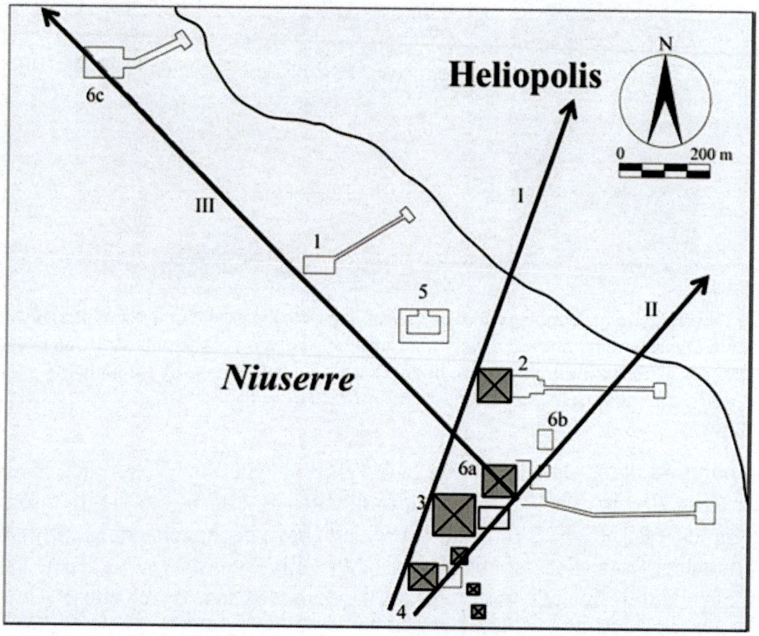

Fig. 6.26 The 'dynastic' diagonal of the 5th Dynasty (I) in the earlier pyramids of Abusir, pointing to Heliopolis. Niuserre's pyramid (6a) and solar temple (6c) were constructed later following related but alternative patterns (II and III). (Diagram adapted from Magli, 2013)

this hypothesis is that it includes a prediction (Magli, 2013: 119–143). The author forecasts where a few 'lost' pyramids of the Old Kingdom—notably Userkare's—could be located. If he were right on that, it would be a tremendous success and a confirmation of his ideas.

The pyramid complexes of the 5th Dynasty had an average orientation of half a degree (½°) to the south of east, if compared to those of their predecessors of the 4th Dynasty. This could mean that the usual method of orientation, which had been used for several generations since the reign of Sneferu, began to be imprecise. This might be suggested by the potential displacement in azimuth, presumably due to the precession of the equinoxes, when using the simultaneous transit of a couple of stars, if this was ever used as we discussed in the Introduction. Hence, a new alternative method needed to be developed. Perhaps, because of the process of solarization of the monarchy and the wandering nature of the calendar year with no leap years, there was also a change of paradigm in the cult, reflected in a change of methodology in the manner of aligning sacred monuments, notably the pyramids, but this is difficult to ascertain.

However, this change might also shed light on the mysterious sudden appearance of the *Pyramid Texts* in the burial chambers of all the royal pyramids, after the reign of Wenis, a custom that lasted until the end of the Old Kingdom. Perhaps, owing to precession and the wandering nature of the calendar year, at the end of the 5th Dynasty there was a loss of confidence in the classical procedures of orientation that had so well permitted the King to perform his post-mortem journey to the celestial realms of the 'imperishable' stars and the solar barque. For this reason, the guide to the afterlife was now written within the pyramid chambers for the exclusive use of the deceased kings (and queens).

6.5 The Solstice, the New Year, and the Nile: The Paradigm of Ipet Sut

The temple of the god Amun-Re at Karnak, or Ipet Sut in the ancient Egyptian sources, shows an impressive axis of symmetry that can certainly be interpreted in a context in which astronomy combines with religion, history, and landscape to produce one of the most sacred traditional spots on earth. The combination of the local course of the Nile, a solstitial orientation, the wandering aspect of the civil calendar and the nature of the deity worshipped in the temple might be considered as a paradigm of witnesses for the correct interpretation of the complex.

This magnificent religious complex located in ancient Thebes (modern Luxor) could have formed part of a relevant chapter in the history of archaeoastronomy. At the end of the nineteenth century, Sir Norman Lockyer argued that the main structure of the complex, the temple of Amun, would have been orientated towards sunset at the summer solstice, as the alignment of the main axis suggested (Lockyer, 1901). However, when he asked for this hypothesis to be checked on site, he learnt that the hills of Western Thebes precluded such an alignment (see Figure 6.27).

Fig. 6.27 The main axis of the Temple of Karnak, directed towards the west (left). The hills of Thebes, located on the horizon, prevent observation of sunset at summer solstice along the axis of the temple, as could have been anticipated by studying the master plan of the monument. This fact leads to the suggestion that it is the opposite direction (towards sunrise of the winter solstice) that could be relevant on site as demonstrated by sunrise (right) at winter solstice in December 2006 along the main axis of the Temple of Karnak, as observed from the old pier of the sanctuary. The phenomenon would have been much more precise 4000 years ago when the temple was first erected during the Middle Kingdom. (Images by the authors. Adapted from Belmonte, 2012)

Hence, the light of the setting sun never actually reached the interior chambers of the temple, unless the building had been constructed 56 centuries before, i.e. around 3600 BC (change of declination due to the variability of the ecliptic). In Lockyer's time, this date did sound problematic but still reasonable for the working chronology of the epoch.

However, when the old chronology was abandoned and a new one proposed at the turn of the nineteenth century—with a reduction of a complete Sothic cycle, i.e. a shortening by 1460 years—his hypothesis was severely questioned. Apparently, even though Lockyer was still alive and in his seventies, he never made the necessary effort to accommodate his proposals to the new chronology. Consequently, the potential solstitial alignment of Ipet Sut was forgotten for three quarters of a century.

Six decades later, a new argument was established (Barguet, 1952), namely that the inscriptions of the walls of the complex supported the idea that, although the main temple entrance was opened to the west and towards the river, the temple was somehow connected to the east, especially to sunrise. These ideas were later explored by Gerald Hawkins, who firstly reported on the winter solstice alignment of the 19th Dynasty temple of Re-Horakhty (included within Amun's precinct), but called particular attention to the so-called 'high-room' of the sun, a sanctuary associated with the Hall of Festivals built in the complex by Thutmose III (Hawkins, 1975; Krupp, 1988).

Two decades ago, it was demonstrated via archaeological excavations that the complex of the Amun temple at Karnak was surrounded during the Middle Kingdom by a village laid out in a Hyppodamian network whose main axis was that of the *dromos* connecting the temples of Mut and Amun (Carlotti, 2005). This was also the orientation of the axis of Mut temple. However, the east–west axis of this network diverged by more than 7° from the main axis of the Amun-Re temple (109¼° vs. 116¾°, respectively). This fact suggests that the orientation of this temple was chosen deliberately and not at all restricted by local urban necessities. This result is related to a couple of additional important facts.

It is important to notice that sunrise at the winter solstice may have had important mythological and/or calendrical implications. To be precise, because of the displacement of the civil calendar over the seasons, there were two occasions after the creation of the calendar when New Year's Eve or I Akhet 1 fell at the moment of the winter solstice owing to the wandering nature of the civil calendar (see Chap. 5). The first occurrence was in a four-year period centred on 2004 BC. This was a very interesting moment in Egyptian history. Mentuhotep II from Thebes had just re-unified the country, and new buildings, on a large monumental scale, were constructed for the first time in the very south of the country, including Karnak, where the most ancient register—a polygonal column possibly of a door jamb—is dated from the reign of his father Antef III (*c.* 2050 BC). A few years later, the first well documented temple of Amun, indeed a new aspect of the solar god, was re-erected by Senuseret I in Karnak, on a grand monumental scale on a river island, and orientated to the rising of the sun precisely at the winter solstice. This fact can hardly be ascribed to chance.

With the passage of time, the course of the river evolved as the Amun complex expanded westward until at some point, perhaps at the beginning of the New

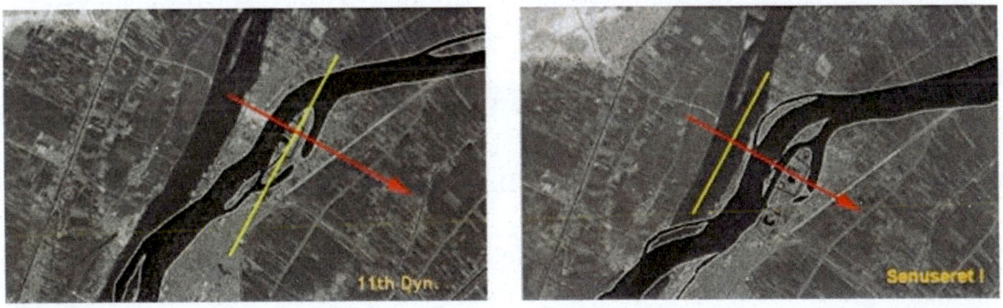

Fig. 6.28 Tentative orography around Karnak in the early Middle Kingdom. The temple of Amun, orientated towards sunrise at the winter solstice (arrow), was founded on a non-flooded turtleback on the banks of the Nile perpendicular to this alignment. As the temple expanded to the west, the riverbanks shifted, until at an indeterminate epoch they adapted to a channel (yellow dashed line) perpendicular to the solstice line, where it remains today. Whether this was deliberate or the result of chance is difficult to determine. (Authors' diagram on a series of images courtesy of Luc Gabolde)

Kingdom, it became practically parallel to that of that original island and, therefore, perpendicular to the solstice line (Fig. 6.28). It is difficult to know if this pattern, still recognizable today, was the result of pure chance, owing to the variations in the course of the Nile forced by the existing buildings after each rising of the waters, or if it was something deliberately sought after by the Egyptians. We know they preferred to have their sanctuaries orientated perpendicular to the river (Belmonte, 2012: 232–38; Belmonte, 2016, see Figs. 6.3 and 6.4).

In the first half of the fifteenth century BC, something extraordinary happened in Egypt. A woman, the royal wife Hatshepsut, proclaimed herself dual king of Egypt. To do so, she had to proclaim that her father had been none other than the god Amun-Re himself, who had elected her for royalty, usurping the rights of the ruling king Thutmose III. At this time, the great temple of Ipet Sut had been standing for at least half a millennium since the time of the early Middle Kingdom, when it had been precisely orientated towards sunrise at the winter solstice (Gabolde, 1998; see also, Gabolde, 1999). However, the Middle Kingdom temple, and later enlargements by the first kings of the 18th Dynasty, had faced west, towards the hill of Thebes. Instead, 'King' Hatshepsut built a new temple to 'Amun-Re-who-hears-the-prayers' exactly on the same axis but opening to the east and thus being the first structure at Karnak actually orientated towards sunrise at the winter solstice. Since this temple was in a court open to the public, we can only imagine the political revenues that such a divine hierophany would have accrued for Hatshepsut's coffers.

On the death of Hatshepsut, the actual legitimate sovereign, her nephew Thutmose III started his reign alone. Although it is not yet clear when the *dannatio memoriae* of Hatshepsut was performed, it is obvious that many monuments of the female 'king' were either usurped by the new king or somehow lost prominence. This was the case of the temple of 'Amun-Re-who-hears-the-prayers'. Thutmose erected a new structure in front of it, thus preventing the illumination by the Sun's rays of the statue of the Queen in close contact with the god Amun. The main focus of this new structure was a single huge obelisk, the highest ever to be erected in Egypt—but only by his grandson Thutmose IV—and which today adorns the Roman square of Saint John of Letran. This granite monolith was located exactly on the main axis of Ipet Sut. Afterwards, during the reign of Ramesses II, the obelisk was surrounded by the structures of the new temple of Re-Horakhty, and Hatshepsut's temple became sandwiched between two larger structures. This is the same situation in which we can see it today making it difficult to imagine how it would have been when it was the first temple in the Ipet Sut complex facing the winter rising of 'her father Amun'.

Following a different but convergent line of reasoning, Ipet Sut is located at the only site in the Nile Valley, above the first cataract, where the river flows in such a way that the average perpendicular direction to the water course is the solstitial line connecting winter solstice sunrise and summer solstice sunset. In fact, in a recent work, Luc Gabolde (2013: 3–12) suggests that Karnak was founded on a turtleback elongated island, free from Nile floods, that was practically perpendicular to the solstitial phenomenon (see Figure 6.28). When the temple was enlarged, the course of the Nile adapted to this situation so that it started to flow in a course that was virtually perpendicular to the solstitial line, nearly the same as we can see it today. The authors endorse the idea, previously outlined by Edwin Krupp (1984), that this natural accident could have been discovered by the Egyptians and may have helped to establish the sanctity of Thebes, and above all, the area of Karnak. We would then be witnessing an extraordinary case of a combination of topography and astronomy, and indeed another paradigm of cosmic landscape.

6.6 The Paradigm of Deir el-Bahari: Challenging Winter Solstice Alignments in Ancient Thebes

The temples of Deir el-Bahari, in particular those of Mentuhotep II and Hatshepsut, were studied in earlier campaigns carried out in the last decade within the framework of the Archaeoastronomy Mission of ancient Egypt. Although they seem to belong to the family of temples orientated to sunrise at the winter solstice, this orientation was not as precise as would be expected for temples of such importance.

Therefore, after resuming work in Egypt, it was decided to propose alternative hypotheses that could explain the deviations of these two temples to the south and north of the solstitial line, respectively. It is difficult to imagine that the deviations were mere design errors. In December 2017, further fieldwork was carried out in Thebes, in particular at Deir el-Bahari, and observations and measures were made on site during the winter solstice itself, the phenomenon being documented. The verifications carried out seem to confirm some new approaches. On the one hand, the burial temple of Mentuhotep II could be diverted *c.* 2° to the south so that there would be a lighting effect during the winter solstice in the chapel that closes the complex at the base of the cliff. This is an astronomical event that could be related to Wepet Renpet (New Year) at that period (11th Dynasty). On the other hand, the Million Years Temple of Hatshepsut could be orientated towards sunrise on another unique date of the civil calendar during the reign of that sovereign. All in all, astronomical orientations at the site could relate to key dates in the civil calendar at the time when the temples were built (Belmonte et al., 2020).

Landscape archaeology (including skyscape) has not traditionally enjoyed great prominence in Egyptological studies. As we have discussed, only in this century have specific works been developed in which the area of ancient Thebes has played a determining role beyond that of the temple of Karnak (Ullmann, 2007; Belmonte et al., 2009a, b; Magli, 2013: 194).

In the first decade of this century, the studies carried out within the framework of the Egyptian–Spanish Mission of Archaeoastronomy had already confirmed the existence of orientation patterns in the temples of Upper Egypt, both topographic (associated with the course of the Nile) and astronomical (Shaltout & Belmonte, 2005; Belmonte, 2012: 126–134). Among them, the pattern associated with the sunrise at the winter solstice stood out (Fig. 6.11). As we have discussed, the temple of Amun in Karnak may be considered the paradigm of this pattern of orientations.

Solstice orientations continued to be frequent throughout the history of Egypt, even when the winter solstice no longer coincided with Wepet Renpet, and became the second preferred form of orientation after the cardinal one, and the favourite one in Upper Egypt (Fig. 6.12). Later temples of Thebes such as the Millions of Year Mansion of Amenhotep III, preceded by the Colossi of Memnon, where astronomical symbolism played a determining role (Bryan, 1997), maintained this orientation.

However, the devil is sometimes in the details. Figure 6.11 shows that the peak of frequency distribution associated with the winter solstice has a width at mid-height of about 4°, which in part could be explained by intrinsic errors in the determination of the orientation or simply by misalignment. The latter may be due to the apparent 36′ diameter of the sun disc, the impossibility of determining the exact moment of sunrise (first ray, half sun on the horizon, visibility of the complete disc, etc.), or other purely physical effects, such as atmospheric extinction or refraction. However, even considering all these uncertainty margins, an amplitude of ±2° is relatively high. This means that there could be temples that although statistically counting as 'solstitial' may in reality not be such.

This has drawn our attention since the first campaign of the Egyptian–Spanish Mission in Thebes in 2004, and in particular during the second in 2006, when it was possible to observe sunrise on the winter solstice in various temples and sanctuaries of ancient Thebes. The situation of the temples at the foot of the cliffs of Deir el-Bahari, especially the best-preserved ones: the funerary temple of King Mentuhotep

Fig. 6.29 The two best-preserved great temples of the Deir el-Baharí circus, the funerary temple of Mentuhotep II (left) and the Million Year Mansion of Hatshepsut. The orientation of both temples (azimuth and declination, δ), which are not parallel, is close to that of sunrise at the winter solstice (text in white) but different enough to raise the possibility of alternative hypotheses. (Authors' diagram)

II of the 11th Dynasty, unifier of Egypt in the early Middle Kingdom; and the Million Year Mansion of Queen-Pharaoh Hatshepsut, known as Djeser-Djeseru, 'the most sacred of the holy places' (Szafranski, 2005).

However, both magnificent buildings, located side by side (Figure 6.29), were not parallel, but had an orientation of *c.* 118¼° and 115½°, respectively. These azimuths would correspond to a respective declination of −25½° for the sanctuary of Mentuhotep II and of −23° for that of Hatshepsut (Fig. 6.29), still within the margin of the solstice peak (−24° ± 2°, Fig. 6.29), but with differences marked enough to give food for thought.

Actually, what was most surprising was that they were not parallel, with a difference of almost 3° in their orientation, despite the fact that both 'looked' in a casual way towards the geographical location of the temple of Amun in Karnak.

There is no apparent orographic factor or feature forcing this divergence, and it was not logical to think that Egyptian planners, able to orientate pyramids with a precision of 3′ (see e.g. Belmonte, 2001, and references therein), would have made such great mistakes when selecting a privileged orientation. It was time critically to evaluate the results of the statistical analysis and focus on the details. To do this, our connection in Egypt with Prof. Magdi Fekri was once more re-engaged and, in collaboration with the European Union Project Stars4All, a new campaign was planned in Thebes. This again coincided with the winter solstice, in order to carry out new observations and document the phenomenon with more sophisticated equipment (Belmonte et al., 2020). Unfortunately, on this occasion we would not be joined by our greatly missed colleague, Professor Mosalam Shaltout, who had passed away 2 years earlier.

As will be analysed in later sections, the new data confirm the previous results on the orientation of the temples, but the images obtained, now of better quality, gave rise to new ideas that can explain the orientation of these sanctuaries. We delve into these new ideas in the next two sections, and venture to suggest alternative hypotheses to explain the detected anomalies.

6.6.1 The Funerary Temple of Mentuhotep II: An Enigma Wrapped in a Mystery

The temple of Nebhepetre Mentuhotep II at Deir el-Bahari is the oldest relatively well-preserved stone construction in ancient Thebes. Located in a spectacular setting, it was almost certainly the king's funerary temple and, at the same time, the site of his tomb and those of his closest relatives. It has a peculiar terraced structure—subsequently improved and enlarged in the Temple of Hatshepsut 500 years later—where architectural elements from the *Saff* Theban tombs of the 11th Dynasty and, to a lesser extent, from the royal funerary complexes of the Old Kingdom are combined. The complex had a clearly delimited axis of symmetry that begins at its lower ramp and ends at the chapel of the funerary cult, located in a space carved into the rock at the base of the cliffs.

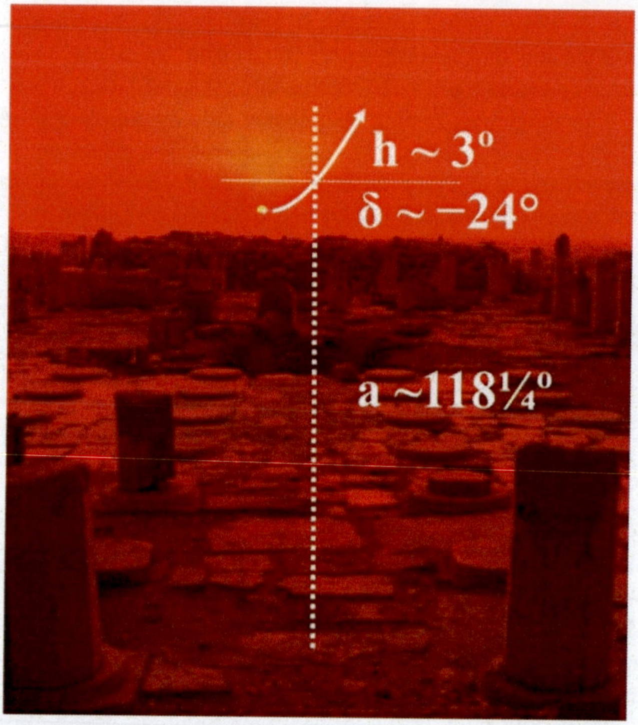

Fig. 6.30 Sunrise at the winter solstice of December 2017 at the temple of Mentuhotep II. The solar disc appears to be clearly misaligned with the axis of the sanctuary. For the alignment to be functional for Mentuhotep's epoch, the sun would have needed be located at a height of 3° above the horizon. (Authors' diagram)

Figure 6.30 shows sunrise on the day of the winter solstice right from the site of the chapel. The image shows that, even if we consider that in the time of Mentuhotep the sun would have risen more than ½° to the south (plus or minus a full solar diameter), the solar alignment does not work properly. In fact, the axis of the temple would be orientated to an object that had about −25½° declination (Fig. 6.29), beyond the solar range. At the time when this fact was first verified, in December 2006, a precise solstice alignment was ruled out without proposing alternatives, since the objective was a global statistically significant study of the orientation patterns of Egyptian temples (Fig. 6.11).

However, the new images obtained in December 2017, and above all, a more detailed inspection of the place, suggested a possible alternative explanation of the 'bad' orientation of the sanctuary. New measurements made in situ verified that, for a declination of −24°, corresponding to the sun at the winter solstice—and for the purpose, on New Year's Eve at the time of Mentuhotep II—the sun must have been at a height of about 3° above the horizon for the alignment to be functional (Fig. 6.30). This is where the most controversial element of the sanctuary enters the discussion.

The first surveys and excavations at the site discovered the presence of a large central quadrangular structure raised on a podium, located right on the axis of the building, which in Egyptological literature has received the German name of *Kernbau* ('nuclear structure'), as it seems to constitute the visible nucleus of the sanctuary (Arnold, 1974). The volume of stone preserved is not very large in reality (the place could have been used as a quarry for centuries), but this has not proved to be an obstacle for several reputed Egyptologists to have defended more or less realistic and suggestive hypotheses of what the purpose of this central structure might have been.

The most relevant and successful have been the three described below. The most traditional, and reproduced in almost all Egyptian architecture handbooks, is that of a pyramid with a slope similar to those of the Old Kingdom. This would have risen tens of metres above the complex (Naville, 1947: 28–29). Later, a great specialist in Egyptian architecture would propose a quadrangular mastaba about ten meters high (Arnold, 1974: 6). Finally, considering that this is a funerary temple, Stadelmann (1991: 232) has proposed that a burial mound would have been raised on the podium, perhaps made of rammed earth, even crowned with trees, which would have represented the primordial hill of Egyptian cosmogony, and that would simulate the Osiris tomb at Abydos.

Each scholar has defended one or another of these ideas (even a pointed pyramid of greater height has been postulated, similar in appearance to those that would later crown the tombs of minor kings and nobles in the Theban necropolis) without any of them having reached a majority acceptance. It is worth noting how Barry Kemp

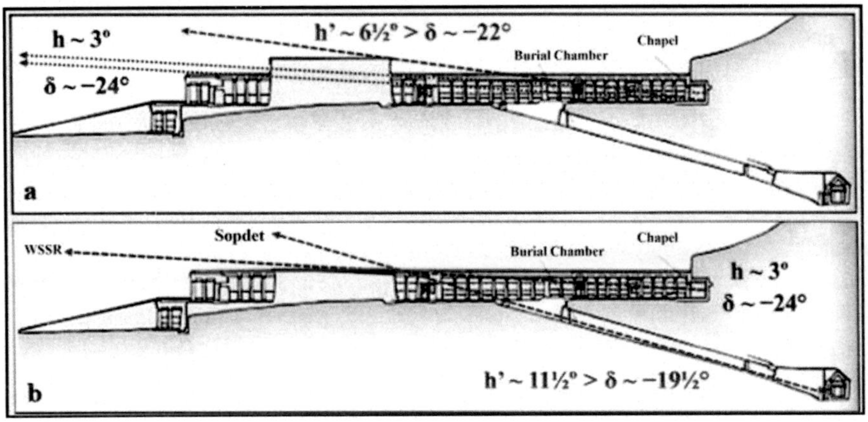

Fig. 6.31 Diagrams of possible astronomical alignments at the sanctuary of Mentuhotep II at Deir el Baharí. (**a**) On Arnold's hypothesis for the *Kernbau,* the solstice alignment would not be functional since sunlight would illuminate the funeral chapel on dates other than those expected (δ of −22°). (**b**) If, on the contrary, we suppose that this structure was much lower along the axis of the sanctuary, the image of the king in the chapel would have received sunlight at dawn on the winter solstice, which would also correspond to that of New Year's Eve or Wepet Renpet. Additionally, the access corridor to the king's tomb would have been orientated towards the acronical rising of the star Sopdet for the same epoch. See text for further details. (Authors' diagram)

(2006: 154–157) justifies the different interpretations of the *Kernbau* as sad examples of scholarly manipulation.

Herein lies the crux of the matter of the new hypothesis defended by Belmonte et al. (2020) lies. A pyramid, a large mastaba or a mound placed on the axis of the temple would have blocked the view of heaven from the funerary chapel of the complex. Figure 6.31 (Panel a) shows the least problematic case, that of a not-too-high mastaba. In this particular model, the structure would block the view of the sky up to an angular height of about 6½° (h' in Fig. 6.31a), thus preventing a view of sunrise on the winter solstice from the chapel on the axis of the sanctuary. The pyramid and mound cases are even worse, so if the building had been designed so that the alignment was functional, the central structure, the *Kernbau*, would have needed to be much lower, at least on the axis of symmetry of the temple, not being higher than 5 m at its western end.

This orientation and its corresponding alignment should not be interpreted as a functional aspect of the building, that is, the temple of Mentuhotep II would have not acted as an astronomical 'observatory': rather would have been of a purely symbolic character, in the same way that the winter solstice has always represented a moment of renewal or rebirth. In this case, this facet would be reinforced by the fact that the solstice coincided in time with Wepet Renpet, the Feast of the New Year.

However, it is remarkable how after decades in which the same patterns have constantly been repeated, a new hypothesis has recently emerged for the structure that would crown the central podium of the temple. It has been proposed by Luc Gabolde and, in reality, it is not one hypothesis but two with similar characteristics, inspired by astronomical phenomena occurring in the nearby Karnak temple, although half a millennium later (Gabolde, 2015). This has to do with obelisks and structures straddling pyramids, and obelisks that had been used as solar symbols since the 5th Dynasty.

The first is based on the 'unique obelisk' manufactured by Thutmose III and erected by his grandson Thutmose IV on the axis of Karnak temple, and behind which the sun would have risen on the winter solstice on the axis of the complex. This obelisk would have broken the astronomical alignment in Karnak by preventing the view of sunrise on the horizon, as seen in Fig. 6.27. We shall therefore not consider it in this analysis since any structure of this type located on the axis of the temple of Mentuhotep II would have blocked the view of the first sun of the New Year. The second hypothesis is based on the texts inscribed by Hatshepsut in Karnak, scripts where the solar disc is mentioned emerging between two obelisks 'on the horizon of the sky'.

the sun-disc rises between them as he rises on the horizon of the sky.

Figure 6.32 is a diagram of how Gabolde's hypothesis would work, also inspired, as has been said, in the solar temples of the 5th Dynasty, with their huge obelisks,

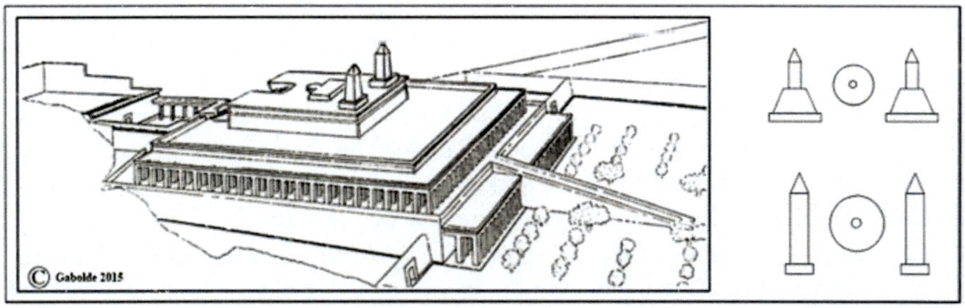

Fig. 6.32 Diagram showing the most recent reconstruction proposal for Mentuhotep II *Kernbau* inspired by Hatshepsut's text for her two obelisks erected at Karnak. If the access ramp to the upper terrace had been less inclined (Fig. 6.31b), this hypothesis would be perfectly compatible with the solar alignment at the New Year Festival (winter solstice for that time) in the reign of Mentuhotep II. (Authors' diagram on a graphic courtesy of Luc Gabolde)

and in the solar altars of the 18th Dynasty, whose best preserved prototype is precisely that of the neighboring temple of Hatshepsut (Szafranski, 2005). In this case, the solar disc would not appear 'on the horizon of the sky' between two monolithic obelisks, as reported by Hatshepsut, but between two representations of the Benben, similar to—although smaller than—the one found in Niuserre's temple at Abu Ghurob (Fig. 6.32). The *Kernbau* would now be a more complex structure in which the podium would house an access ramp or staircase (as in other solar altars), a high-place in the shape of a *hetep* sign and two pyramid trunks crowned by small obelisks, one on each side. The whole complex would embrace the rising sun.

In this case, the hypothesis proposed is perfectly compatible with the framework proposed here, which simply assumes that the podium and the access ramp were lower, allowing the observation of the solar disc precisely 3° above the horizon. This is shown in Panel (b) in Fig. 6.31 where the access ramp to the 'solar altar' located on the axis of the temple would allow the illumination of the chapel—and of any statue within—in just a few minutes after sunrise, when the sun would be right on the axis of the sanctuary at a height of 3°. For this to be acceptable, we have to assume that at least the central axis of the hypostyle hall that culminates in the chapel should not be completely roofed as proposed in different reconstructions of the complex. Under these premises, the alignment would therefore be functional. This is not to say that Gabolde's hypothesis is the correct one, but it would fit, both subtly and beautifully, with the astronomical hypothesis defended by Belmonte et al. (2020), and is accordingly subscribed to here.

Wepet Renpet, as the New Year Festival celebrated on I *Akhet* 1, or the first day of the Inundation season, which in the reign of Mentuhotep II occurred in accordance with the winter solstice, was the most important date in the civil calendar of ancient Egypt. However, this was not, as we have shown, the only important yearly festival. Another unique event was the heliacal rise of Sirius, or Peret Sopdet, whose trace is easily discerned in the Middle Kingdom (Belmonte, 2003a, b, c; see also Belmonte, 2012: 27–48 and 320–324; Hornung et al., 2006: Chaps. 9 and 10), perhaps even in the Old Kingdom (Habicht et al., 2015) as discussed in Chap. 5. This event acted as a herald of the true flooding (not the calendar season of the Flood) and occurred around July 19 in the Julian calendar, although it actually moved between the summer solstice and the second decade of July of the proleptic Gregorian calendar (the one linked to the climatic seasons) during the era of the Pharaohs. Therefore, the star Sopdet, our Sirius, the brightest in the Egyptian skies, was an astronomical reference of the first magnitude and her (she was considered a goddess) visibility must have been important.

Therefore, it is worth noting (as shown in Fig. 6.31b) that, if the *Kernbau* was much lower along the axis of the temple, as proposed here, a second astronomical alignment could have been functional, although once more certainly in a symbolic sense. The descending corridor to the burial chamber, which is located on the axis of the temple, shows an inclination of about 11½°, which for an azimuth of 118¼° implies a declination of −19½°, within the error margins. This declination corresponds to that of Sirius in the decades around 2000 BC, early in the Middle Kingdom, so a putative Sothic alignment could also be present in the funerary complex of Mentuhotep II at Deir el-Bahari.

However, we emphasize that this alignment would not be with the star's heliacal rising since it is only observable on that particular date for a few brief moments just at dawn, at a height of *c.* 2° before disappearing in the Sun's glare. For Sirius to be at 11½° at sunrise, you have to wait a fortnight later, at the end of July in the Julian calendar (end of the Peret III in the civil calendar at the time of Mentuhotep). Another interesting possibility is that what was relevant was not the heliacal rising of Sirius, but rather its acronical rising, which at that time occurred, interestingly, around the epoch of the winter solstice. We may remember that the acronical rising of a star occurs on the day when a star is seen to rise for the first time in the east just after sunset. It is an evening phenomenon, unlike the heliacal rising, which happens at dawn. Unfortunately, there is only the slightest evidence that Sirius was observed at other key moments in its annual cycle (see Chap. 5), so this possibility should remain just as a hypothesis, for the moment.

All in all, it is certainly curious that the mortuary temple of Mentuhotep II in Deir el-Bahari shows possible astronomical patterns related to renewal and rebirth through two first-order celestial objects in ancient Egypt, such as the sun and Sirius. This could perhaps be related to the name of the funerary complex: *ȝkh-swt* Nebhetepre (Fischer, 1996: 76; Belmonte et al., 2020). In the case of the former, in addition, the phenomenon happens at a key moment of the annual cycle, such as New Year's Eve. This perhaps has helped us to see with different eyes, and to interpret in a different way, the other great temple that occupies a most important site at the foot of the cliffs of Deir el-Bahari.

Fig. 6.33 Sunrise on the winter solstice in December 2017 in the Hathor Chapel of the Temple of Hatshepsut in Deir el-Bahari. This chapel has the same orientation as the rest of the sanctuary (115½°). The axis of the chapel (white arrow) is not aligned with the sunrise. Was it a construction mistake or a deliberate choice? A new interpretation suggests a deliberate alignment to crucial civil calendar dates during Hatshepsut's reign, notably to I Peret 1 or Nehebkau *c.* 1460 ± 16 BC. (Authors' diagram on an image adapted from Belmonte et al., 2020)

6.6.2 *The Million Year Mansion of Hatshepsut: A Source of Legitimacy*

Figure 6.33 shows sunrise on the winter solstice in December 2017. The image was obtained at the Hathor Chapel, in the southern sector of Hatshepsut's temple, but it could be valid for the main axis of the sanctuary, which has exactly the same orientation of 115½°. The solar disc (with a declination of *c.* −23½°) is clearly observed rising approximately ½° further south than the axis of the chapel (marked by a white arrow). In Hatshepsut's epoch, the discrepancy would have been even greater. Therefore, in spite of what has been published (Belmonte, 2012: fig. 13), Hatshepsut's master building is not solstitially aligned, unless we admit that the Queen allowed her architect, the great Senenmut, to making a bad mistake when orientating her temple; that is to say, the alignment is inaccurate.

This can also be seen in the images obtained inside the sancta sanctorum of the temple over several years, after considering the solstice orientation (Shaltout & Belmonte, 2005), by a team led by the tour specialist and excellent connoisseur of ancient Egypt David Furlong.[3] Therefore, either we admit a serious design error or we must think of the possibility that Djeser-Djeseru, the original name of the temple (Szafranski, 2005), was orientated with another objective in mind.

One could argue a topographical orientation towards the new buildings of Hatshepsut in Karnak, or even consider a small deliberate deviation that would allow accurate calculation of the proper date of the solstice by observing sunrise on the axis of the temple a few days before or after this astronomical phenomenon. However, as we have learned in the neighbouring temple of Mentuhotep II, the details are the key to allowing a new, alternative, serious and reasoned hypothesis. To do this, we should go back to what we have postulated for this temple and what we know about Egyptian civilization, particularly during the epoch of Hatshepsut's reign (18th Dynasty).

A clue may be gained from the texts of the nearby tomb of Senneferi (TT99; Strudwick, 2016), a high-rank civil servant and contemporary of Thutmose III—after the death of Hatshepsut—and his successors. Figure 6.34 shows a detail of the beautifully polychromed hieroglyphic texts where suggestive evidence for this work is collected. In this text, Senneferi's wife, children, siblings, and workers are mentioned as they come to make offerings to the tomb, apparently only on four key dates of the Egyptian calendar. These are the festivals of Wepet Renpet (or New Year's Eve), Nehebkau, deteriorated but still legible and on which we are going to focus, Tepy Renpet (or the 'First' of the Year, a still controversial date as we have discussed in Chap. 5) and Peret Sopdet, of whom we have already spoken at length.

The most interesting date in this case is the Festival of Nehebkau, dedicated to a deity of the same name that could be translated as 'the one who assigns the kas (or living spirits)' (see Figure 6.35). This was a very important holiday related to regeneration (its mention together with Wepet Renpet and Peret Sopdet proves this). In

[3] See: Last view, March 2021.

Fig. 6.34 Hieroglyph text from the tomb of Sennneferi, dated slightly after Hatshepsut's reign. This shows the most important dates or festivals in which offerings were brought to the tomb: Wepet Renpet, Nehebkau, Tepy Renpet, and Peret Sopdet. (Image by Juan A. Belmonte, courtesy of the West Bank office of the Supreme Council of Antiquities)

Fig. 6.35 There is a great deal of evidence suggesting a relationship between Hatshepsut and the Festival of Nehebkau (inscription in the upper left corner), celebrated on I Peret 1. Two such pieces of evidence are Hatshepsut's Name of Horus, beautifully written on the ceiling of Senenmut's tomb TT353, and the representation of the sovereign as Osiris on the façade of her temple. The festival of this deity was celebrated precisely in the month of Kaherka (Khoiak or IV Akhet), which culminated exactly in Nehebkau. (Diagram and images by the authors)

particular, Nehebkau was a feast serving to confirm royal power (something Hatshepsut was in great need of). It was celebrated on the first day of the Peret season, I Peret 1, and closed the cycle of festivities dedicated to Osiris that were celebrated throughout the month of Kaherka (Khoiak in Coptic) or IV Akhet (see Chap. 5). Additional proofs of this are Hatshepsut's titles, whose Name of Horus was Weserkaut ('that of powerful kas'), and the gallery of Osirian pillars of the queen, that imposingly decorate the façade of her Million Year Mansion (Fig. 6.35), along with other minor elements. Perhaps they tell us about an intentional relationship between the symbolism related to the female sovereign and the Nehebkau festival that guaranteed her much sought after legitimacy. The Theogamy that decorates the walls of the sanctuary is obviously another clear component of that particular message.

For this reason, it becomes remarkable when a new element, which could be considered controversial, comes into play: the orientation of the temple itself. In Thutmose III Year 1, usually set at 1479 BC (see Chap. 7), the winter solstice occurred on *c.* I Peret 6. This means that an orientation, with an azimuth of 115½° and a declination of −23°, such as that of Hatshepsut's temple (Fig. 6.29) would correspond to that of sunrise on I Peret 1, that is, on the day of the Nehebkau Festival. This would happen around a set of year centred on *c.* 1460 BC, with a margin of error of about 16 years due to different sources of uncertainty.

Thus, the axis of the Hatshepsut temple and its associated elements, including the Hathor Chapel, were aligned with sunrise on the Nehebkau Feast at some precise moment between *c.* 1473 and 1458 BC. This is a chronological framework that fits reasonably well within the dates when Hatshepsut rose to power and was certainly planning her Djeser-Djeseru, regardless of the chronology under consideration (Belmonte, 2012: Table 8.1, see also Chap. 7).

Therefore, the Million Year Mansion of the queen-pharaoh would follow a plan completely similar to that of her neighbour and source of inspiration, Mentuhotep II, but in a very different political and especially temporal framework. In the middle of the 18th Dynasty, Wepet Renpet occurred at an unimportant moment in the seasonal cycle. This was so until the full reign of Ramesses II, around 1253 BC, when it would coincide again with another key astronomical date such as, in this case, the summer solstice. On the other hand, Peret Sopdet occurred in III Shemu, a moment with minor significance in the civil calendar year. Interestingly, the arrival of the flooding was announced by Sirius' heliacal rising in the middle of the (civil) Drought season.

The anomaly of the orientation of Hatshepsut's temple could thus be explained by the coincidence of three factors: i) a close coincidence of a topographic orientation to Karnak; ii) the presence of an almost parallel temple of an illustrious ancestor orientated towards the winter solstice; and iii) Hatshepsut's alignment fitted a key date in the civil calendar that also gave legitimacy to her power seizure, that is, I Peret 1, the Feast of Nehebkau. Therefore, the hypothesis is that temple deviation is not an error! Senenmut deliberately planned the orientation to guarantee, and above all justify, the right to the throne of his beloved sovereign. Whether, in addition, given his knowledge and interest in astronomy, he used the near solstitial orientation of the building to predict the arrival of the winter solstice itself is another matter entirely.

6.6.3 A Solstice Delusion?

From the previous paragraphs, it follows that the apparent anomaly in the solstice orientation of the main temples of Deir el-Bahari is not really such an anomaly and instead may have a simple explanation.

In the first case, that of the funerary temple of Mentuhotep II, the 'solstice' orientation would be maintained by only taking into account a small factor in the reconstruction of the *Kernbau,* which should be slightly lower, on the axis of the temple itself, than considered so far. In this case, this solstice orientation would also have a markedly symbolic character, since it would also correspond to that of sunrise on the day of the New Year Festival, on I Akhet 1. This could follow a pattern that might have its origin in the temple of Amun in Karnak, whose oldest remains seem to date precisely from the beginning of the Middle Kingdom (Gabolde, 2018: 175).

In the second case, the Djeser-Djeseru of Queen-Pharaoh Hatshepsut, this relationship would be repeated on key dates in the civil calendar, although the Nehebkau Festival on I Peret 1, is in this case the relevant date. This was related to the confirmation of royal power and therefore to the, certainly questionable, legitimacy of the sovereign to the throne of Egypt. In this case, the near-solstice orientation could be either accidental or deliberate, since the alignment could be reliably used in a predictive role, although this point is difficult to prove.

However, a similar case is observed in the temple of Ramesses II at Abu Simbel, where there is a double alignment to I Peret 1 (actually also to I Shemu 1 for that time) and sunrise at the winter solstice (Belmonte, 2012: 146–153; see also Maravelia & Shaltout, 2003). The main sanctuary of this Nubian temple faces Nehebkau while the small lateral chapel dedicated to the solar cult, which has a different axis, faces sunrise at the winter solstice. We may perhaps speculate on a similar phenomenology in the solar sanctuary located to the north of the sancta sanctorum of Hatshepsut's temple if a window, or opening, had existed in the eastern wall for this purpose (Belmonte, 2012: fig. 13). The temple of Hatshepsut would not therefore be the only example with this phenomenology. The importance of key dates in the civil calendar of ancient Egypt is thus precisely shown.

We do not want to finish this discussion on the orientation of Deir el-Bahari temples without devoting, even if only briefly, a discussion to the third temple that stood at the foot of the cliffs, the Djeser-Akhet or memorial temple of Thutmose III, built later and straddling the structures of his predecessors, Mentuhotep II and Hatshepsut. This sanctuary has an azimuth of 118½° (divergent from the 127° of his Milliion Year Mansion) and points to a declination of −25¾° (Shaltout & Belmonte, 2005; Belmonte & Shaltout, 2009: Appendix II). It is, therefore, practically parallel to that of Mentuhotep II and diverges even more, perhaps deliberately, from that of his co-regent Hatshepsut. This temple is poorly preserved and little has survived of its superstructure, so it is impossible to guess if this orientation was deliberate and tried in some way to 'catch the solstice', or if, on the contrary, it simply imitates that of its glorious ancestor, Egypt's unifier Mentuhotep II. With no other options for now, this second possibility seems the most reasonable.

6.7 Tanis, a New Thebes in the North

One of the most interesting places in the Delta is the ancient city of Djanet, called Tanis by the Greeks, capital of Egypt during the 21st and 22nd Dynasties. The area was a turtleback in the low lands of the eastern Delta near the Tanititikos branch of the Nile and was thus normally free of flooding. There are no construction remains earlier than the 21st Dynasty, although a lot of movable monuments (sphinxes, statues, obelisks, steles, etc.) dating from the Middle and New Kingdoms were found in the city.

However, it is not known whether these monuments were the remains of earlier constructions of these dynasties on the site or whether they were brought to the city

Fig. 6.36 Principal axis of the Amun temple in the complex of the god at Tanis. The building was aligned at the beginning of the 21st Dynasty, either during the reign of Smendes or, more probably, that of Psusennes I. According to our analysis, this great sanctuary was erected as a lesser scale copy of the huge Amun complex at Karnak, in Thebes. The shrine offers a similar patter combining local landscape (Tanitic branch of the Nile) and skyscape (sunrise at Wepet Renpet). (Authors' image)

from the nearby area of Tell el Dabha and Qantir, where the cities of Avaris and Pi-Ramesses were located. Nothing better to express this shocking situation than reading the words of one of the former directors of excavations on site, Jean Yoyotte (1996: 31):

> The sudden transfer of the monarchy to Tanis and the creation of a Thebes of the North over a vast hill of sand, not occupied before by living people, near the great marshes, constitutes one of the greatest problems of Egyptian history.

Actually, it is believed that Tanis was converted into the new political and commercial capital of Egypt, perhaps from being a minor local village, at the beginning of the 21st Dynasty, thus replacing Pi-Ramesses, 22 km to the southeast, which had probably been abandoned owing to the Pelousiakos branch of the Nile silting up.

Almost from nothing, first the kings of the 21st and 22nd Dynasties, and later those of the 30th and Ptolemaic dynasties, built a splendid city that was a copy, on a smaller scale, of the old religious capital Thebes, 800 km upstream. This included an *Amen-em-opet* temple to the south of the main group in clear parallelism to the temple of Luxor (Aufrère & Golvin, 1997: 309–19). This *Waset Mehyt* (the 'Northern Thebes', was an alternative name of the city) became the showcase of the new kings of Egypt to the merchants of the Aegean and the Levant. The important sacred character of the city was emphasized by the erection of no less than five pairs of gigantic obelisks (more than in Karnak) along the main axis of the temple of

Amun (see Figure 6.36). Actually, Tanis, together with Thebes, had the privilege of being named as the city of the sun-god Re, *Iunu* (Heliopolis). These cities, Tanis and Thebes, were to be known as the (alternative) Northern and the Southern *Iunu*, respectively (Quirke, 2003: 79–80).

Our data (Belmonte & Shaltout, 2009: Appendix 2) supports this special sacred character. The main axis of the temple of Amun is open to the west, to a declination of −16° ± ¾°. This datum will be qualified later on. Initially we were tempted to include it as another, not too well orientated, member of the Sopdet family, since the declination of Sirius at the beginning of the 21st Dynasty (*c.* 1070 BC), when the temple could perhaps be aligned for the first time, was of −17½°.

However, when we noticed that the huge complex, integrating no less than five adjacent temples, was orientated according to the Nile, we thought of another striking possibility, apart from the very prosaic but reasonable one that the Nile, and it alone, was the sole reason for the alignment.

This consists of the idea that, if the opposite direction is considered (i.e. a declination of 16° ± ¾°, since the horizon is practically flat), then the main axis of the complex would have been orientated to sunrise at Wepet Renpet *c.* 1065 BC, with a margin of plus or minus 15 years (Shalthout et al., 2007). This would be in rough agreement with the reign of Smendes (*c.* 1069–1044 BC, see Table 7.1), founder of the 21st Dynasty and perhaps the first king to align the sacred complex. However, the earliest foundation deposit of the Amun temple bears the name of his successor Psusennes I. Actually, Smendes was not archaeologically attested at Tanis, although he is mentioned as living at this city in the famous story of Wenamun.

Interestingly, this hypothesis laid on the table a parallelism with what happened in Thebes. So now, in Waset Mehyt, there was a new temple of Amun with one of its axes perpendicular to the Nile and the other orientated to Wepet Renpet at the presumed moment of its construction. Indeed, one inference of this idea could be that the site of Tanis on the corresponding Nile branch was deliberately chosen to emphasize the sacred character of the city, even though there could be several other obvious strategic reasons. This was our proposal after fieldwork on site in June 2006.

However, Luc Gabolde and his collaborators has most recently produced a new in-depth analysis of the site (Gabolde et al., 2021). In this interesting work, they made an average of different GPS measurements of various remaining walls within the enclosure of the Amun temple and reached a more precise value of a slightly higher azimuth (*c.* 72½° vs. 71¾ ± ½°). This would suggest a slightly lower value for the declination, which, under the same premises, would fit sunrise at Wepet Renpet in the quadrennium centred on 1039 BC. This implies that the temple could be aligned at the beginning of the reign of Psusennes I (*c.* 1040–992 BC), in perfect agreement with archaeological data, if both research teams (Gabolde's and ours) wish to advocate a first alignment of the temple with New Year's Eve sunrise.

Fig. 6.37 Cosmic aspects of the Temple of the goddess Hathor at Dendara (**a**), Orientated perpendicularly to the Nile and, at the same time, to the rising (or 'akh') of Meskhetyu on the date of construction as suggested by local inscriptions (**b**). In the astronomical diagrams of the temple, Hathor was assimilated with Isis-Sopdet (**c**). The main axis of the adjacent Temple of Isis, included in the complex, and the monumental East Gate of the *temenos* of the Temple of Hathor – where a processional way ended (**d**)—faced the heliacal rising of Sirius (the brightest star of the sky, Egyptian Sopdet) in this area of the horizon during most of Egyptian history. Hence, the temple of Dendara might have been located on a site where the flow of the river aligned with the direction in which the heliacal rising of Sirius, Herald of the Flooding, would be observed. In the Graeco-Roman period, the perpendicular to this line also marked the rising of Meskhetyu. The temples erected on the site had their orientations established accordingly. (Diagram by the authors)

6.8 The Star of Isis, the Nile and Meskhetyu: Skyscaping at Dendara

We have so far dealt with sites which in some moment of Egyptian history became capitals of the country. We now analyse a singular site that did not attain such a distinction, but that has played a most important role in the history of cultural astronomy research in Egypt. We refer to Dendara. Throughout the 1990s, a team led by the French Egyptologist Sylvie Cauville made a detailed study of the Temple of the goddess Hathor at Dendara. Studying the texts, Cauville proposed, on the basis of certain epigraphic evidence, that the axis of the main temple, dedicated to Hathor (see Figure 6.37), was aligned on 16 July 54 BC, during the reign of Ptolemy

XII, father of Cleopatra VII. However, for the Temple of Isis, located behind the main temple, the situation was different (Cauville et al., 1992: 31–48; Cauville, 1997a, b).

This temple has no less than three major axes: the oldest one, associated with earlier foundations in the reign of Nectanebo (30th Dynasty) and subsequent constructions of the reigns of Ptolemy VI and Ptolemy X; a processional axis directed towards a monumental gate on the eastern wall of the temenos of the complex (see Fig. 6.37); and a festival hall dedicated to the birth of Isis, built in the time of Augustus. The former two diverge from one another by 4°, while the third was at an angle of 90°, so that the axis of the hall was parallel to the Temple of Hathor.

According to Cauville and her colleagues, the change of axis could have been assigned as a change in orientation towards the rising of the star Sirius—most prominently associated with Hathor in the celestial diagrams of her temple (Fig. 6.37)—caused by precession. The oldest axis (with an orientation 111°11′), from the building of Nectanebo, should have maintained the original orientation of a previous building, from which only some fragments have been preserved, and which would have been erected in the same location during the reign of Ramesses II (*c.* 1270 BC). The new one, with an azimuth of about 108°, would point to the rising of Sirius in 54 BC, when the axis of the new complex, including Hathor's temple was established in its entirety.

This would mean that this new alignment was determined in accordance with the orientation of the main axis of the Temple of Hathor, as one might expect, but in the perpendicular direction. From the symbolic, sociological, and even practical point of view, this solution seems reasonable, given that Isis, the owner of the smaller temple, is intermingled with Sopdet and therefore has Sirius as one of her major celestial manifestations. However, the inscriptions on the walls in the Temple of Hathor are crystal clear in this respect (Belmonte et al., 2009a). What they say is that the astronomical object observed in establishing its axis, and allegedly the plan of the entire complex, was the constellation of Meskhet(yu); in particular, towards the 'akh' of this constellation, a term with various significances (Fig. 6.37).

In 54 BC, the rising of Alkaid (η UMa)—the star located in the hoof of Meskhetyu's foreleg and one of its most conspicuous stars—was visible at an azimuth of 18° and at an angular height of about 2°, at the moment when all the constellation was fully visible after part of it had been hidden below the horizon. This was a celestial configuration that could be interpreted as the 'akh' (*ꜣḫ*, presumably 'appearance') of the constellation (Belmonte et al., 2009a). Consequently, it is almost certain that, according to the Egyptians and their texts, the main Temple of Hathor at Dendara was orientated towards the rising of a singular star of Meskhetyu and not of Sopdet. It therefore seems obvious that at Dendara Temple Complex a combination of stellar alignments was produced in such a form that Sopdet and Meskhetyu would be the focal point of reference for different buildings, such as the temple of Isis or the main temple of Hathor, respectively, at different epochs.

Up to this point, the opinion of the majority of specialists must be added: the idea that the Temple of Dendara is simply perpendicular to the Nile—which has a very peculiar course in that region of the Valley. The authors are particularly fascinated with these series of apparent contradictions because it should inevitably lead to another of those wonderful examples where astronomy and the local geography combined to create another possible example of a cosmic landscape (see Fig. 6.37).

The hypothesis is that the temple of Dendara could be chosen as an important place of worship, for both Hathor and Isis (considered as two different manifestations of a same divine power at least from the New Kingdom), because the Nile runs almost parallel to the direction in which the heliacal rising of Sopdet occurred during a substantial period of Egyptian history. We could even dare to speculate on the suggestive metaphor that the waters of the flood would be heralded by Sirius at the time of its earliest annually celestial manifestation: Peret Sopdet, precisely in this direction.

Later on, in the Hellenistic period, the obvious displacement of the position of Sirius' rising due to precession might introduce certain doubts as to the usefulness of the star as a precise reference for the orientation of the sanctuary. Hence, another procedure, already well known and used since ancient times (in Abydos, for example), would be selected and the new temple aligned with the stars of Meskhetyu. We would therefore be faced with a double stellar alignment that would additionally comply with the precept of its being in agreement with the Nile; once more, astronomy and landscape are seen to in close relationship.

6.9 Shrines of Ram-Headed Divinities and Canopus: Land- and Skyscapes of Herakleopolis Magna and Beyond

The Spanish Archaeological Mission at Herakleopolis Magna has for several decades been excavating in one of the most important ancient sites of Middle Egypt, and making extraordinary archaeological discoveries in what was once the capital of Egypt during the 9th and 10th Dynasties and again of a chiefdom for a short period during the Lybian epoch (Pérez Die, 2009). After our previous experience elsewhere in Egypt, it was hence decided that a new complete survey of the site should be made with an astronomical and landscape perspective in mind (Belmonte et al., 2016, 2017).

6.9.1 When Multidisciplinary Efforts Are Worthwhile: Skyscaping at Herakleopolis Magna

This section presents the result of such a survey where the relationship between land- and skyscape at the main monuments of the city is placed in the spotlight. One important outcome was the possible orientation to Canopus—the second brightest star in the Egyptian skies—of the main temple of the city, the one devoted to the patron divinity of Herakleopolis, the ram-headed god Heryshef. Epigraphy was integrated into the analysis and possible mythological relationships were explored, including the connections with other ram-headed deities of the Egyptian pantheon, such as Amun-Re and Banebdjedet. Interestingly, the temples of these divinities at Thebes and Mendes also show alignments that could be related to Canopus, thus offering a new challenge in the relationship between skyscaping and religion in the civilization of Pharaonic Egypt.

This collaboration with Carmen Pérez-Die, Head of the Spanish Mission, and the epigraphist Lucia Díaz Iglesias, the subject of whose PhD was Herishef (Díaz-Iglesias Llanos, 2012), represented a truly interdisciplinary effort.

The ancient Egyptian city of Henen-nesu, now Ehnasya el Medina in Beni Suef province, was the capital of the 20th Upper Egyptian nome called Naret Khentet (i.e. the Upper Naret tree). The town is located very close to the right bank of the Bahr Yusuf, an arm of the Nile that waters the Fayum Oasis. This province was one of the main centres in the history of Pharaonic and Coptic Egypt. Its origins go back to the first dynasties (as written sources demonstrate) but, for the moment, the archaeological evidence dates back only to the end of the Old Kingdom.

The Greeks called it Herakleopolis Magna, identifying the main ram-headed god of the city, Heryshef, with the Hellenic Herakles. In these paragraphs, the main results of an archaeoastronomical study (see Belmonte et al., 2015) carried out at Ehnasya in ancient religious and funerary structures are presented. This was carried out under the auspices of the Spanish Archaeological Mission (Pérez Die, 2012). Belmonte et al. (2016) also tried to contextualize these results within the framework of knowledge of other ram-headed divinities, such as Amun or Banebdjedet, and their temples in other sites of Egypt.

The most significant results focus on the orientation and location of the main temples of the city (see Figure 6.38), notably the temple of Heryshef.

This sacred structure, in addition to its undeniable relationship with the in-wall necropolises of the First and Third Intermediate Periods (hereafter FIP and TIP, respectively; Pérez Die, 2010), would have been astronomically orientated (azimuth of *c.* 201°) towards the setting of Canopus. This was the second brightest star in the Egyptian skies, at the time of the first monumental building of the temple during the Middle Kingdom (*c.* 1850 BC). This orientation could perhaps relate to the stellar epithets of the tutelary deity of the province, the ram-god Heryshef.

Such stellar epithets (Díaz-Iglesias Llanos, 2014: 472–473) placed Heryshef as the leader of the cosmic sphere or of some of its elements (see Fig. 6.39). Along with the more usual 'King of the Sky (*nsw pt*) and 'Lord of the Sky' (*nb pt*), written sources at Herakleopolis, and even at Thebes, mention the exceptional description of Heryshef as 'Pillar of the stars' (*ỉnw sb3w*) and 'Ruler of the stars' (*ḥḳ sb3w*). These were originally assigned (at least the first one) to the deceased monarch in his ascent to the sky and then, within the divine pantheon, only to the Herakleopolitan ram-headed divinity.

Other texts included the statement 'he [who] made his place in the sky as the Lone Star' (see Figure 6.39), linking an Osirianized Heryshef (he is often found in texts and reliefs in close connection with an Osiris of Naref, 'the One who is at the front of Naret') with a singular star. This Lone Star has occasionally been identified with Venus as Morning Star or Capella (Krauss, 1997), Canopus seems also a reasonable candidate, or perhaps the adjective 'lone' may apply to several distinct

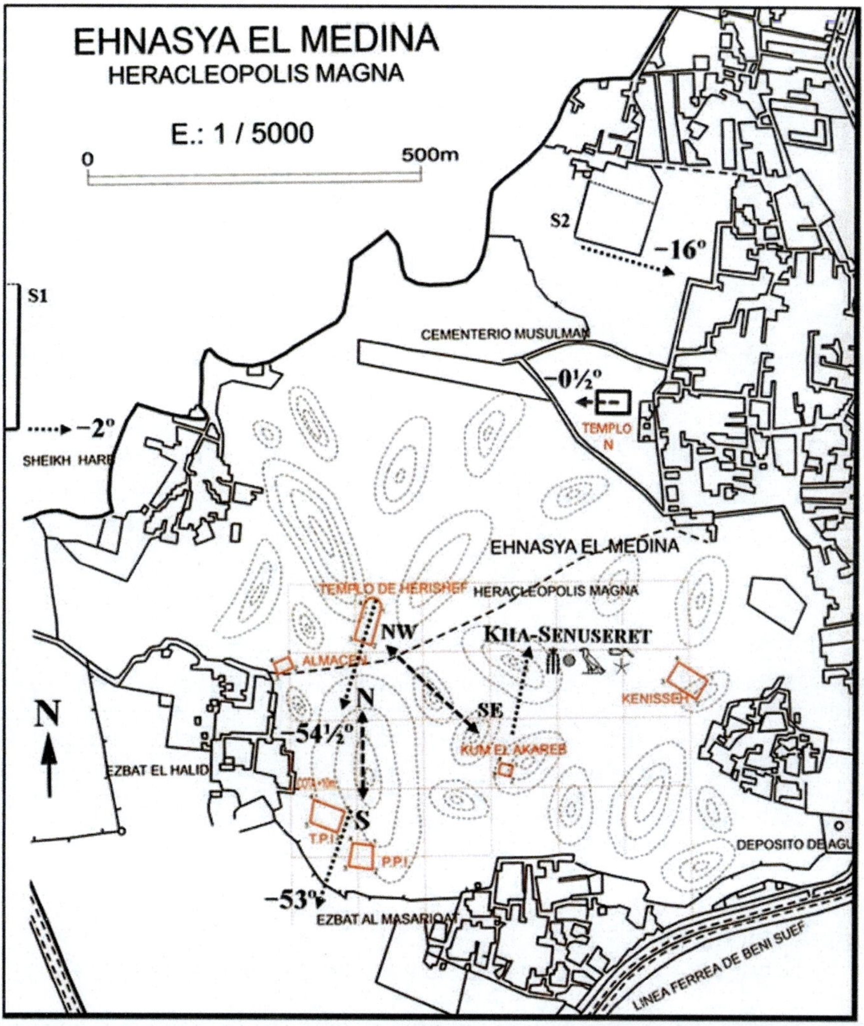

Fig. 6.38 Plan of Herakelopolis Magna within the context of the modern village of Ehnasya el Medina, including the main monuments of the city. The most significant topographical and astronomical (represented by the value of the declination) alignments are plotted. The temple of Heryshef could have originally been orientated towards the star Canopus as similarly to the chapel of the Third Intermediate Period necropolis (T.P.I. in the plot) twelve centuries later. The sanctuary at Kum el Akareb may have shown a double alignment to the (Imperishable) stars of the northern skies (notably those of Meskhetyu) which were orbiting above the pyramid complex of the Middle Kingdom Pharaoh Senuseret II in Lahun. The recently excavated—by an Egyptian team—'North Temple' is precisely equinoctially orientated. S1 and S2 stand for the two hypothetical saltpans that could be the modern counterparts of the classical Natron and Ma'at Lakes (described by many ancient sources) related to the city of Herakleopolis Magna. The expansion of the Muslim cemetery abuts on the northern sector of the site at the northern limit of the Spanish concession. See the text for further discussion. Figure elaborated upon a site plan of the excavation area by courtesy of the Archive of the Spanish Archaeological Mission at Herakleopolis Magna. (Adapted from Belmonte et al., 2016)

Fig. 6.39 Temple, iconography and 'celestial' epithets of Heryshef. Adapted aerial image of the temple of Heryshef showing where the data were collected and golden statue of the deity with his name in hieroglyphs, meaning 'The One who is upon his Lake'. The five celestial epithets from top to bottom are: 'King of the Sky', 'Lord of the Sky', 'Sovereign of the Stars', 'Pillar of the Stars', and 'He who is established in the Sky as the Lone Star' (Belmonte et al., 2016). (Diagram based on images by courtesy of José Javier Martínez, Boston MFA, and the Spanish Archaeological Mission at Herakleopolis Magna)

stars. In any case, all these epithets show a trend towards the stellar and solar transfiguration of Heryshef.

In the celestial sphere, it is important to take into account both the stars and the big luminaries, the sun and the moon, and their relationship to the calendar and festivals as possible frames of reference. The documented presence of stellar titles among the epithets of Heryshef justifies this claim (see Belmonte et al., 2015). In addition, the relevance of the stars is well characterized in the case of the bright Sirius (a reference star for another family of temple orientations and herald of the inundation). It was personalized in the goddess Sopdet, one of the manifestations of Isis (and of Hathor, Heryshef's partner). A certain Queen Shepensopdet has been reported at Heracleopolis in the excavations of the TIP Necropolis (Pérez Die, 2010: 443). In this context, we may include other important constellations such as Meskhetyu, widely analysed in previous studies as dominating the northern skies where the Egyptians recognized a most important part of the seat of immortality (Belmonte et al., 2009a).

Heryshef was the patron god of Herakleopolis Magna. Already witnessed by epigraphic sources from the early dynasties, his spheres of action are numbered

Fig. 6.40 The breath-taking geography of the former northern shore of Lake Fayum dwarfs the 12th Dynasty aniconic temple of Qsar As-Sagha, perhaps orientated to the rising of the star Canopus in that epoch. (Image by the authors)

four: water and purification, royal connections, reverential fear, and external manifestation or capacity for action (in Egyptian called 'ba', *bꜣ*, the same word used to denote rams). These spheres of action related him to defining attributes of ram-headed divinities. His stellar and royal facets are highlighted (Díaz-Iglesias Llanos, 2014: 467–475), as well as his relationship to Osiris (as his 'ba' or external manifestation), and perhaps to Amun.

These defining factors of Heryshef seem to be linked to the astronomical orientation of his sanctuary in Ehnasya. However, this case is not unique, since other parallels point in the same direction. For example, the unfinished and aniconic temple of Qsar As-Sagha, located north of the Fayum Lake some 60 km to the northwest of Herakleopolis as the crow flies was certainly built in the Middle Kingdom similarly to the first confirmed monumental complex of Heryshef (see Figure 6.40). It has an orientation that would be compatible with the rising of Canopus at the appropriate period. More examples will follow. However, we should go back to Herakleopolis since this is not all that the city has to offer.

The site included two splendid ancient necropolises of the FIP and TIP (periods when Herakleopolis rose to pre-eminence) excavated by the Spanish Mission (see Fig. 6.38). The former was located in a north–south (cardinal) relationship to the main temple, a fact that might be significant (the 'imperishable' stars would orbit above the temple). The latter was built on the axis of the temple and with nearly the same pattern of alignments. Interestingly, the chapel on the burial precinct (from the outside looking in) has an orientation that would fit the setting of Canopus *c.*

800 BC, adjusting the diachronic variation of declination of the star due to the precession of the equinoxes.

In the vicinity of the necropolis of the FIP, in a place called Kum el Akareb (see Fig. 6.38), there are remains of another building that must almost certainly have been an imposing temple. Nowadays, only part of the portico is well preserved consisting of several Cyclopean rectangular blocks, some columns and an ashlar with the name of the King Neferusobek, a woman who was the last sovereign of the glorious 12th Dynasty. Three colossi of Ramesses II, possibly reutilized, were found in three corners of the building. Two of them, exhibited today in the garden of the Cairo Museum, have been recently tentatively identified as portraits of the Middle Kingdom King Amenemhat IV, direct predecessor of Neferusobek and perhaps her brother. The third one, found in haphazard excavations during the 2011 'revolution', still unpublished and now in the Service of Antiquities Magazine at Ehnasya has a cartouche of Senuseret (III).

The probable topographical orientation of this large sacred structure to *Kha-Senuseret*, the pyramid complex of Senuseret II at Lahun, more than ten kilometres to the north, is very appealing. This could reinforce the idea of an original construction of this building in the Middle Kingdom. An astronomical orientation could also be suggested since the 'imperishable' constellation of Meskhetyu would have had its easternmost position in the sky above Senuseret II complex *c.* 2000 BC, precisely (Belmonte et al., 2017).

An additional monument, recently excavated by a team of Egyptian archaeologists of the Department of Antiquities led by Ahmed Galal, is a spectacular sanctuary located close to (and nearly surrounded by) the expanding Muslim cemetery of Ehnasya (Fig. 6.38). The construction reuses an enormous number of blocks of the Middle and New Kingdoms presumably from other temples, possibly including that of Heryshef and perhaps of the still unidentified temple of Hathor, his consort and partner. However, several additional blocks show the typical *opus quadratum* of Roman buildings. Hence, this third temple could have been built in the Roman or Byzantine periods.

The temple shows a very marked equinoctial orientation, which is exceptional in Egypt after the pyramid age (when it was the dominant one) but is frequent in Hellenistic buildings and Christian churches (González-García & Belmonte, 2014, 2015). This poses an interesting question mark on the kind and purpose of this building. Unfortunately, the outcomes and archaeological discoveries of this excavation have not been published yet, and remain completely unknown and cannot be used to extract further conclusions.

The archaeological site of Herakleopolis Magna therefore includes a series of temples and sanctuaries of different epochs, in a peculiar environment and likely to be interpreted within a general context of landscape archaeology, where the astronomical analysis of the skyscape plays a most relevant role.

6.9.2 *Beyond Herakleopolis: Ram-Headed Deities and Their Temples*

During the six campaigns of the Egyptian–Spanish Mission of Archaeoastronomy of ancient Egypt a total of some 350 temples were measured. This permitted a pioneering statistical study of Egyptian temple alignment patterns, which uncovered a total of seven families of astronomical orientations (Belmonte et al., 2009b). One of these families (numbered V) was assigned to Canopus. Nearly twenty temples across Egyptian geography could be related to it. Interestingly, several temples associated with ram-headed divinities—who were also somehow related to Osiris—were among them.

In this sense, some inscriptions recording stellar epithets of Heryshef have been found at the Amun-Re religious complex at Karnak (Thebes) in two reliefs dating to the reign of Ramesses III (Díaz-Iglesias Llanos, 2012). Amun-Re was normally represented in human form, although his relationship with the ram is well known. Suffice it to remember that his barque and the sphinx avenues were always decorated with ram-headed images and statues. The relationship between the nocturnal aspect of Re (who is depicted as a ram-headed creature) and Osiris is also a well-established fact as the famous relief of Nefertari's tomb (and in other royal tombs in the Valley of the Kings) demonstrates.

Following the same line of argument, numerous structures at Karnak had their axes presumably orientated to the setting of Canopus. These include the VIIth and VIIIth pylons creating a secondary axis of the complex (Gabolde, 1999) that was perpendicular to the main axis of the temple which was solstitially orientated (Hawkins, 1973; see Figure 6.41). The temple of Ramesses III and the TIP chapel of Osiris Heqadjet (although with their gates open to the north) could have had their axes orientated to the setting of Canopus as well (Belmonte et al., 2009b: Appendix II). This series of facts would reinforce this hypothesis.

It is worth noting that the temple of the Oracle at the Oasis of Siwa (see Figure 6.42), where Amun was conceived and represented as a ram-headed god (Fakhry, 2004), could also be orientated to Canopus (Shalthout et al., 2007). The Oasis of Siwa is one of the most important areas of Egypt outside the Nile Valley and, from our personal point of view, is by far the most fascinating of all the oases of the Western Desert. After the foundation of Cyrene by Dorian colonizers in 631 BC, the kings of Egypt became interested in the oasis, which was administered as a sort of vice-kingdom until Roman times.

The Oasis of Siwa is very rich in archaeological sites of Egyptian typology, with the remains of at least twelve temples of which the temple of the Oracle of Amun at Aghurmi is by far the best preserved and the most interesting. This temple has a splendid location in a low rocky outcrop above the palm-tree sea of the oasis with open views in all directions. It was built during the reign of Amasis (570–526 BC) perhaps over an earlier building, but the foundations of the temple have never been excavated in depth. The southeast horizon as seen from the sancta sanctorum of the temple is dominated by the hills of Djebel Takrur, where the quarry that was the

Fig. 6.41 The VIIth pylon of the temple of Amun-Re at Karnak, the first to be erected in the processional way to the temple of Mut during the reign of Thutmose III and Hatshepsut. This pylon, and the VIIIth one, defined a secondary axis of the temple that was not exactly perpendicular to the main axis, which was orientated to sunrise at the winter solstice. This axis, however, was aligned, either by chance or probably by design, to the setting of Canopus during the reign of Thutmose III. See the text for further discussion. (Adapted from Belmonte et al., 2016)

source of the stones to build the temple was located. This orography would have been dominated in Antiquity (sixth century BC) by the appearance of Canopus in the winter skies. The temples of Khamisa and Zeitun, both dated to late Ptolemaic or Roman times, would be other examples of a hypothetical orientation to the second brightest star in the Siwan skies (Shalthout et al., 2007).

Fig. 6.42 The splendid location of the temple of the Oracle of Amun on the acropolis of Aghurmi, in the middle of the palm-tree forest of the Oasis of Siwa. The temple itself is at the very centre of the fortified hill. (Photograph by the authors)

Heryshef had a possible Lower Egyptian counterpart in the ram-headed god Banebdjedet of the ancient city of Mendes (Djedet in Egyptian), capital of Egypt during the 29th Dynasty. Banebdjedet ('the Ram Lord of Djedet', also known as Mendes, as the city itself, in the Greek form of the name of the Egyptian king Nesobanebdjedet, i.e. Smendes) was considered as the 'ba' (note the homophony between *ba* for ram and *ba* for power of manifestation) of four deities. These were known to be four generations of father and son (Geb, Shu, Re, and Osiris), the latter being the most important of the quartet (Redford, 2010). A substantial temple of the god existed at Mendes whose most representative structures were four gigantic, monolithic naoi dedicated to each of these divinities of which only one survives today (see Figure 6.43).

Although the temple of Banebdjedet built by the kings of the 29th Dynasty was open to the north, there is archaeological evidence that this was not the case in earlier periods and that the temple, or parts of it, could have faced south in previous periods of Egyptian history. Hence, with an azimuth of 22° (202° if measured from the south, Shalthout et al., 2007; see Fig. 6.43), the temple would be nearly parallel to that of Heryshef at Herakleopolis Magna. Taking into account the different latitude, and possibly an alternative epoch (*c.* 600 BC or later), the temple axis could also have been orientated to the setting of Canopus. This similarity could indeed reinforce the association of Heryshef's temple at Herakleopolis, and even of the ram-headed god himself, with this bright star.

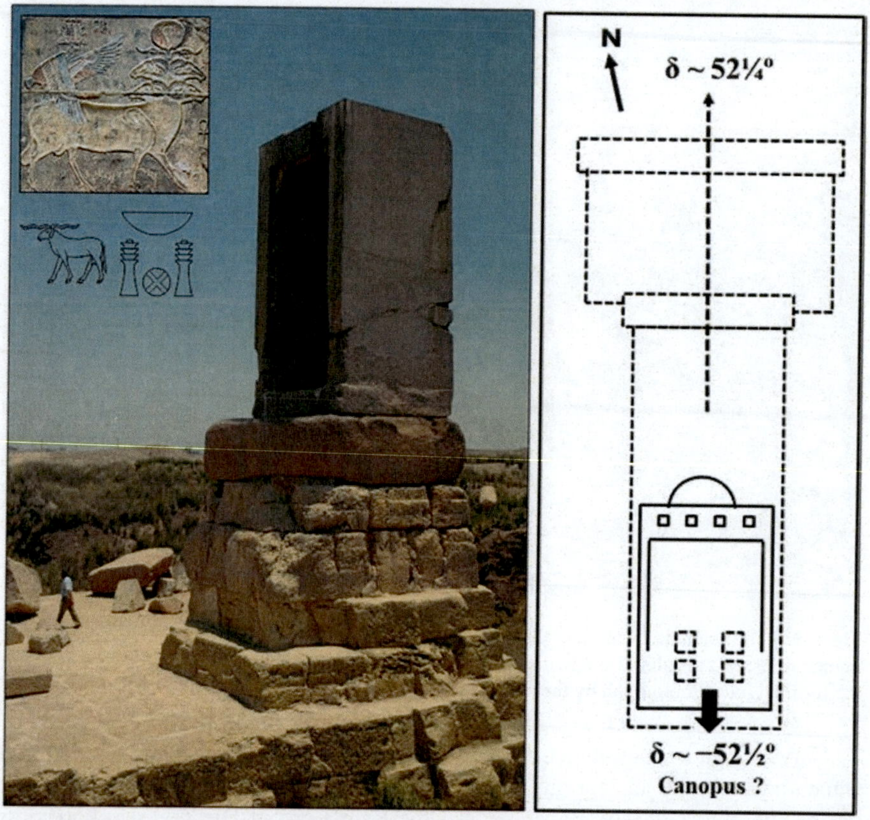

Fig. 6.43 The only surviving exemplar of the four huge naoi (note the human scale) once erected at the sancta sanctorum of the temple of the four ram-headed god Banebdjedet (image and glyph name within the diagram) at Mendes (ancient Egyptian Djedet), capital of Egypt during the 29th Dynasty. The temple plan shows that it is today orientated to the north. However, archaeological excavations on the site show that it might have originally been orientated to the south and hence to the setting of the star Canopus. See the text for further explanations. (Diagram by the authors)

However, it is worth mentioning that neither of the two best known temples of another ram-headed deity, the 'creator' god Khnum, at Elephantine and Esna, which are orientated closer to southeast, had any alignment compatible with the Canopus family (Belmonte et al., 2009b: Appendix II). A detailed comparison between the mythology associated with this divinity and his ram-headed counterparts could be interesting in establish similarities and discrepancies that may shed some light on this particular issue.

In contrast, a very interesting case of the Canopus Family would be that of the Isis temple complex at Philae (see Figure 6.44). The main axis of the Temple of Isis on the site was diverted to a declination of −53½° and hence could have been orientated to the setting of Canopus. We should not forget that Isis was the consort of Osiris. In this role, it might have been significant to orientate her temple to a star

Fig. 6.44 Western (**a**) and eastern (**b**) views of the splendid complex of Isis at Philae. The Temple of Isis could be orientated to Canopus, while the small Temple of Hathor (close-up in Panel b) could be a member of the Sopdet family of orientations. (Images by the authors)

which could have been associated with ram-deities, who themselves were so closely related to Osiris.

It is not self-evident whether the ancient Egyptians recognized Canopus or not. They probably did so since being the second brightest star in their skies, its presence would not have passed unnoticed, giving rise to mythological associations also reflected in architectural constructions. However, in Chap. 4 we were unable to identify its name, astronomical correlations, or possible religious connections beyond a reasonable doubt (see also Lull & Belmonte, 2009). There are, however, a couple of interesting exceptions. On the one hand, there is the very late reference by M. Capella, who called it *Ptolemaeus* in honour of King Ptolemy Lagos (Allen, 1963). It should be remembered that kingship is one of the main attributes of Heryshef.

On the other hand, a possible relation between Canopus and Osiris (and hence to the mythology of Isis) can be inferred from Plutarch's *Mysteries of Isis and Osiris*, lxvi, although it is not clear if the text refers to the star itself or to the pilot of the vessel Argo. However, Argo Navis, clearly the Greek constellation[4] (where Canopus is located at its helm), is related to the boat of Osiris within the same paragraph. The traditional Coptic name of the star may also relate it to sailing. The correlation could also be inferred from a controversial passage mentioning Osiris in the *Book of Day and Night*, where he is 'behind' Sah (Piankoff, 1942) and hence a celestial body that would be located behind this ancient Egyptian constellation which is often identified as Orion or, better, as parts of it (Chap. 4; Lull & Belmonte, 2009). The star Canopus, then, may be related in some way to the Osirian Triad and hence to the deities who were somehow connected to this charismatic god as a celestial manifestation of them.

This might be one explanation for the stellar attributes of the god Heryshef (see Fig. 6.39), his relationship with Osiris, and the orientation of his temple at Herakleopolis. The same might apply for other ram-headed divinities, such as Amun-Re and Banebdjedet who were somehow linked to Osiris and Re as well.

The land- and skyscape of Herakleopolis have proven to be deeply interconnected. The orientation of the main axis of the Temple of Heryshef allows confirmation of its membership to the Canopus family of orientations together with its visual connection with the TIP necropolis. The orientation of the chapel at this site seems to suggest a diachronic variability in the alignments that could perhaps be related to the precession of the equinoxes. Presumably, this will need further explorations by the members of the Spanish Mission.

Finally, the equinoctial orientation of the North 'Temple', possibly of the Roman period, suggests a completely different context (a sanctuary of the Egyptian divinities or perhaps a Christian church) that will also need to be explained once the results of the excavations by the local Egyptian team of archaeologists have been published, contrasted, and evaluated. The expectations for future research in Herakleopolis Magna are hence very promising.

[4] Argo Navis was broken up by N. L de Lacaille in the eighteenth century into three separate constellations (the present Carina, Puppis, and Vela). Canopus is now in Carina (α Car).

6.10 A Southern Reflection: The Kingdom of Kush

During the final years of the Egyptian–Spanish Mission, the project researchers decided that the Nubian temples and pyramids built by the kings of Kush would be an excellent laboratory to extend the analysis carried out in Egypt. This was a logical consequence of the study of the orientation of the temples of ancient Egypt in search of possible astronomical or topographic orientations, or—even better—those that could be contextualized in a global framework of land- and skyscapes (Belmonte et al., 2010).

In the case of Kush, we present the results of a preliminary study, based on bibliographic documentation, high quality plans (e.g., Edwards, 2004; Hinkel, 1997) and high-resolution Google Earth and satellite images. This study could be extended with still-pending future fieldwork. The study included data for fifty temples and more than a hundred pyramids erected by the kings of Kush in Upper Nubia, the Island of Meroë and the Butana region, closer to modern Khartoum, in what is now modern Sudan, over a period of more than a thousand years. This analysis became crucial, because it allows a contrast with the findings carried out in Egypt, within a culture highly influenced by the Pharaonic one (Welsby & Anderson, 2004). However, it also had an intrinsic interest:

> The excavations revealed that several times in its history the complex had been razed to the ground to be re-erected and expanded on the same spot, yet each time with a slight change of orientation ranging between 4° and 5° 20' In other words, the Great Enclosure had already existed in the Napatan period. The orientation of the temples must have been determined by certain stars, whose position in the sky changed over time, and this orientation was so quintaessential that the temples of the earlier complexes had to be re-erected several times. It was not dilapidation that motivated the repeated construction work, but a religious necessity to follow the stars in the orientation of the temples (Wening, 2001).[5]

This intriguing paragraph, written by one of the archaeologists who excavated the sacred complex of Musawwarat es-Sufra, in the region of Butana, one of the nuclear areas of the Kingdom of Kush during the Meroitic period, brings a number of very interesting ideas to the discussion. These include the presence of hypothetical stellar orientations of Kushite monuments and the reconstruction of certain buildings according to astronomical patterns that are closely related to religion. These ideas certainly need to be tested and verified.

There have been a number of cultural astronomy-themed essays on the ancient Kingdom of Kush, but until the publication of the work by Belmonte et al. (2010), there have been none explicitly devoted to archaeoastronomy. Thanks to these previous works, such as the writings of the Belgian Egyptologist Leo Depuydt (1998a, b, c) on the subject, we have some additional information. We know, for example, of the existence of some graffiti in Meroë suggesting the presence of possible astronomical instruments, such as a quadrant or a gnomon (see Chap. 3). Reports of alleged astronomical observations, such as those of the zenith passage of the sun in

[5] S. Wening, 'Musawwarat es-Sufra, interpreting the Great Enclosure', *Sudan & Nubia* 5, 2001, 72.

the enclosure of the old city of Meroë, having even postulated the presence of an observatory to control the Nile floods, as suggested by Logan and Williams (2000).

The encyclopedic work of Laszlo Török (1997), which presents numerous studies of archaeological landscapes of great interest, even informs us about the existence of a possible astronomer among the high-ranking officials of the Kingdom of Kush. This was a person named Wayekiye, son of a certain Hornakhtyotef (a prophet of Isis), who was a priest of Sopdet and of the 'five living stars', as well as supreme magician of the king of Kush. This is described in an inscription on the walls of Philae temple walls dated to a period (*c.* AD 227) when the Dodecasquenos border region was under Meroitic rule (Török, 1997: 472). The 'five living stars' are thought to be a euphemism for the five planets visible to the naked eye, while the mention of Sopdet, the brightest star in the sky and one of the celestial aspects of the goddess Isis, requires no further comment.

Thus, it may be reasonably stated that Sirius and the five naked-eye planets were diligently observed by a cast of priests with direct connections to royalty. Given our experience in Egypt, we may even hypothesize that the movements of the sun, perhaps the moon, and other important stars and asterisms were also followed and studied in detail. The presence in the region of celestial diagrams, such as the astronomical ceiling of pyramid number 26 of Djebel Barkal, dated in the reign of Amanislo (*c.* 270–260 BC), studied by a team from the Egyptian Museum from Barcelona (Berenguer & Diaz de Cerio, 2001), further substantiate these hypotheses. Therefore, the idea was to test whether this knowledge could be traced in the archaeological record and, if so, how to relate it to what has been learned further north, in the land of the Egyptian Pharaohs.

The double kings of Kush considered themselves as the legitimate heirs and *alter ego* of the kings of Egypt and, for a century or so (that is, during the 25th Dynasty)

Fig. 6.45 The plateau-shaped rock mass of Djebel Barkal (right), the so-called Pure Mountain of the ancient Egyptians and their successors, the inhabitants of Kush. The capital city of Napata was raised at its foot. It was a sacred place of the god Amun, whose temple (left), with several axes that denote a complicated construction history, was erected next to the base of the mountain. (Adapted from Belmonte, 2012)

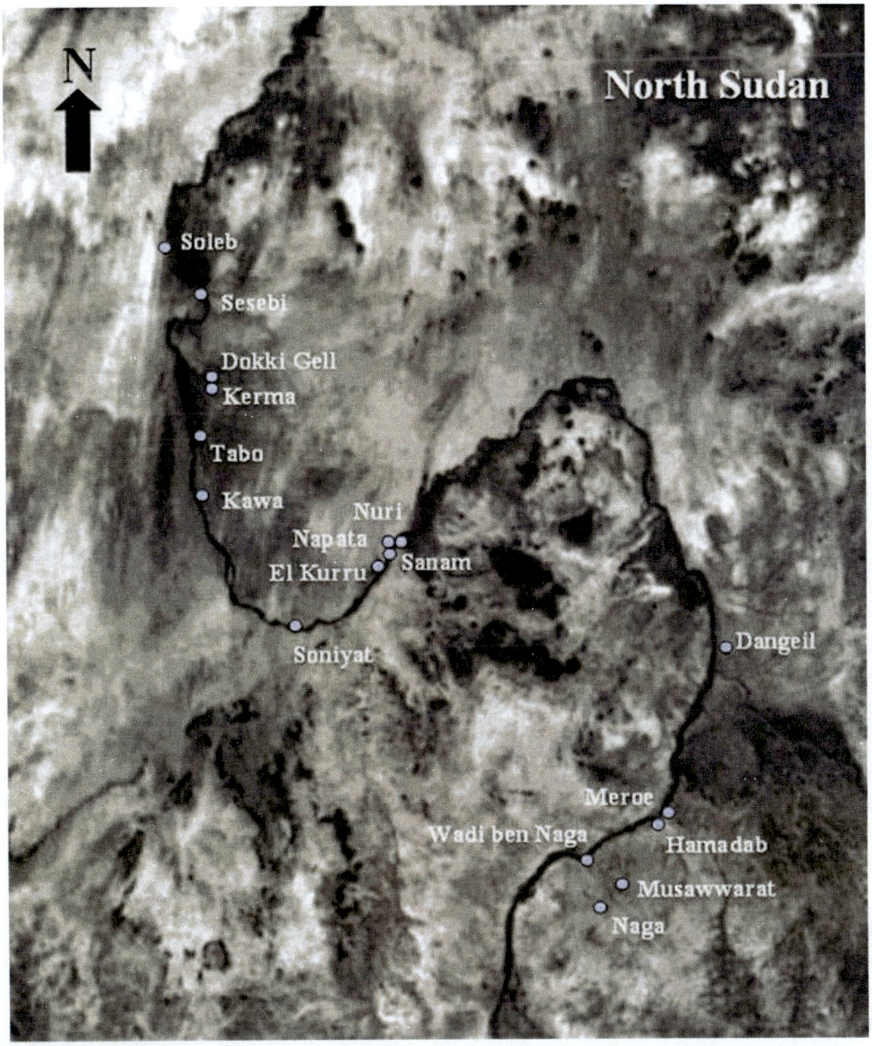

Fig. 6.46 Map of northern Sudan, showing the location of the sites discussed in the text. (Graphic adapted from an image, courtesy of Google Earth)

they really dominated the entire Nile Valley from Khartoum to the Mediterranean (Morkot, 2000; Welsby, 1996) Undoubtedly, this was an impressive tract of land for the time. Later on, after the Assyrian invasion, which reached Thebes, they retreated to their sheltered borders beyond the first or second cataract (depending on the period). However, they continued to rule a huge expanse of land, first from the capital city of Napata, at the foot of the 'Pure Mountain', current Djebel Barkal (see Figure 6.45), and after the reign of Aspelta (*c.* 580 BC), from the city of Meroë, in the Butana region, further inside the African continent (see Figure 6.46).

Török (1997: 201–6) provides us with a very detailed analysis of the series of monarchs of Napata and Meroë, their chronology, burial place, and full titles, among many other aspects. This analysis is generally accepted today by most specialists (e.g. Edwards, 2004) and allows placing this research in a more or less reliable spatio-temporal framework. However, there may still be large discrepancies in chronology, especially for the Meroitic period.

After Aspelta's transfer of the capital to Meroë, the area of Napata still retained a strong religious imprint and the royal cemetery would not be transferred to the Meroë sector until the reign of Arqamani(qo) at the beginning of the third century BC. This transfer is taken as the point of reference by which the history of Kush is normally divided between a previous Napatian and a later Meroitic period.

Figure 6.46 presents the map of the region showing the places where the data were obtained through the archaeoastronomical analysis of high-resolution images from Google Earth and updated highly accurate plans. On this occasion, given the prevailing insecure conditions in Sudan at the time of carrying out the analysis, Belmonte et al. (2010) opted to use this type of data. This decision prevented the team from carrying out an archaeological mission of the type required by our investigations, given the needed high mobility and a shortage of time and resources, even though standard missions dedicated to excavating in a single location, do operate on a regular basis in Sudan.

There are numerous studies on the architecture of Kush that we may use as references to supplement the data and complete the archaeological and historical context of the sites. Some of them have been basic in their analysis, such as the PhD thesis of the Canadian researcher Caroline Rocheleau on temples dedicated to the god Amun, where the importance of the 'dais' room is emphasized. This consists of an elongated room with a construction similar to an altar at one end that is accessed by a series of stairs. This was present in at least seven temples of special relief from the Napatian and Meroitic periods (Rocheleau, 2008: 25–54 and 77–78).

The analysis included more than fifty temples (data is presented in Belmonte et al., 2010: Table 2), covering the entire Sudanese geography and the complete chronology of the period. Some recently discovered temples, for which there are no detailed and reliable maps with which to compare, such as Hugeir Gubli, Awlib and Al Hassa (Rocheleau, 2008: 31–50), were discarded from the analysis. As can be seen on the map, most of the Napatian period sites are clustered in the Napata and Kerma areas, close to the Nile between the river's third and fourth cataracts, while most of the Meroitic sites are further south, between the fifth and sixth cataracts, and further inland in the Butana region. However, it is worth mentioning that most of the Napatian sites were still in use during the Meroitic phase, while several sites in this second phase show evidence of earlier settlement during the Napatian phase, and even during the imperial 25th Dynasty.

Several sites of the Egyptian New Kingdom, such as Sesebi, Dokki Gel, Kawa, and Napata underwent reconstructions, modifications, and additions, sometimes of great importance, during later periods. A good example of this is the Temple of Amun at Napata (Fig. 6.45), originally built (and therefore first orientated) during the 18th Dynasty, but to which was added a 'dais' hall during Peye's reign

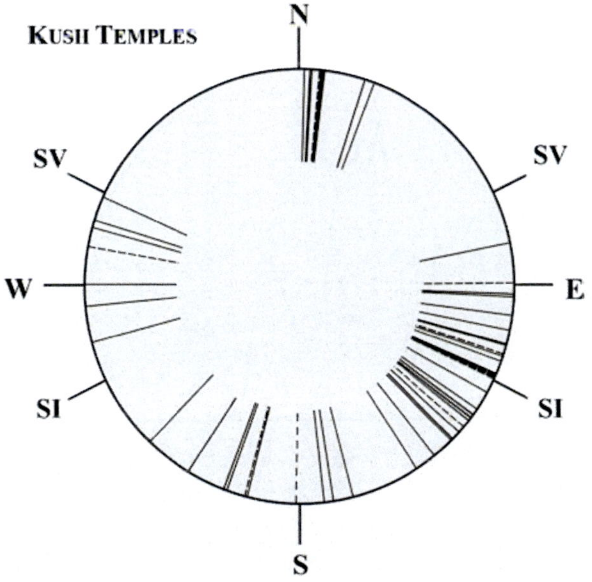

Fig. 6.47 Orientation diagram of the temples of Sudan, obtained from high-resolution images from Google Earth (solid lines) or from low-resolution images contrasted with bibliographic sources (dash lines). (Adapted from Belmonte et al., 2010)

(750–712 BC). This maintained the same orientation. Two large courtyards, with slightly different orientations, were also added at different times during the 25th Dynasty. The resulting curvature of the axis of the temple, from an original building with a typically Egyptian orientation towards the southeast (135°), could be due to successive attempts to accommodate the orientation of the temple to the local course of the Nile, which runs at an azimuth of *c.* 211°, although a fully perpendicular alignment was never achieved.

Figure 6.47 presents the orientation diagram of the temples. It is quite illustrative since, although there are temples orientated in almost all directions (with the notable exception of the northeast and northwest), there is a large concentration of buildings with azimuths between the east and the southeast, to which those with meridian orientations or those centred on the western solar arc would be added. These orientations are almost certainly the traces of an astronomical pattern that must be further explored.

6.10.1 South Versus North: Kush as a Mirror of Egypt

In previous sections, when we analysed the orientations of Egyptian temples, we emphasized the existence of a series of families of astronomical alignments. These reflected certain tendencies in the orientation of sacred spaces. These series were:

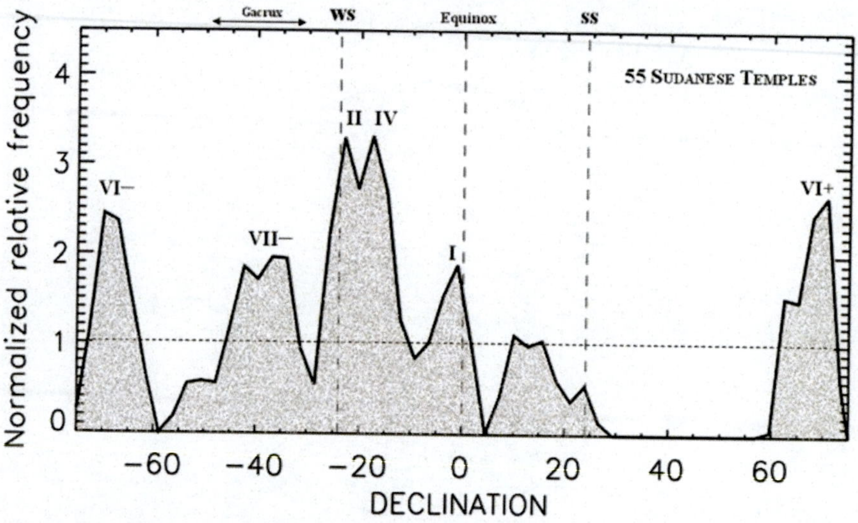

Fig. 6.48 Declination histogram of Sudanese temples. Six peaks are clearly representative, and two statistically significant. Interestingly, equivalences can easily be found for all of them with the set of orientation families that had previously been established for Egyptian temples. The most important ones, with confidence level higher than 99%, are related to the declination of the sun at the winter solstice (II) and with Sirius (IV), the brightest star in Sudanese skies

(I) eastern or 'equinoctial' (but certainly a byproduct of due-northern orientations), (II) winter solstice, (III) seasonal, (IV) Sothic, related to Sirius, (V) Canopus, (VI) meridian (somehow related to I), and (VII) inter-cardinal. These families were reflected in Fig. 6.11 and were often combined with the local course of the Nile, in a suggestive cross-correlation between land- and skyscape, or, in other words, astronomy and topography.

As we have argued on several occasions, the declination histogram has the advantage of being independent of latitude and therefore offers a first-hand view of overall possible astronomical patterns. Figure 6.48 shows the declination histogram of the 55 alignments that have been identified in Sudanese temples, precisely.

An analysis of the graph is most suggestive as there are several significant peaks. Interestingly, each of these peaks could be related to families already established in the case of Egypt. A fact that stands out is that the most important of them is a double peak with two maxima, at −23° and − 17½°, almost certainly related to the sun of the winter solstice and to Sopdet, that is, to Egyptian families II and IV, respectively. One possibility that was taken into account, recalling the supposed 'observatory' of Meroë, was that the second peak could perhaps be related to the zenith passage of the sun in the region, given the values of the latitudes of Napata (18°32′) and Meroë (16°56′). However, a detailed analysis of the data ruled out this possibility.

Solstice and Sothic orientations are, however, fully justified. On the one hand, solar orientations will be discussed extensively in the next section, when we deal with Amun temples, which are omnipresent in the region. On the other hand, the orientation to Sirius could be related to an Egyptian tradition imported into the region. However, this could be interpreted as directly related to a local interest in this star, as suggested by the Wayekiye titles discussed above.

Sirius heliacal rising and setting took place around May 18 and June 27, respectively, in the proleptic Gregorian calendar, for the latitude of Napata and the epoch of the 25th Dynasty. Both events could have been used as heralds of the arrival of the waters, taking into account that, in the Kush region, the rising waters of the Nile arrived between 2 weeks and a month (depending on the latitude) earlier than in Egypt. This is a very suggestive parallel to what was happening downstream, in the land of the Pharaohs.

Continuing with the histogram analysis, we now have to consider the accumulation peaks (VI + and VI-) that certainly represent the interest in meridian orientations, so common in Egypt, considering the influence of latitude, which is much more southerly in this area. The next one, a peak at −1°, is undoubtedly related to the cardinal pattern and will be discussed later. Finally, a wide peak with a maximum between −34° and −42°, which includes the different axes of the Temple of Amun in Napata, could be related to the inter-cardinal family.

The most fascinating aspect of this graph is that, without exception, all the relevant peaks have a reflection in the northern tradition, suggesting an Egyptian influence, which is perceived in many other aspects of the civilization of the Kingdom of Kush, especially with regard to religion. This should certainly be extended to the orientation practices of sacred buildings. Actually, this is not surprising since, as has been shown, religion and the orientation of sacred structures were already closely related in ancient Egypt.

Figure 6.11 also shows the aspect of the declination histogram for Egyptian and Sudanese temples when all alignments are taken into account. The diagrams with and without data for Sudan are practically identical except for the declination range of Family VI, which is logical considering that the complete set includes temples with much more southerly latitudes. This result confirms what was tentatively predicted. New data only qualify the proposed theory, which is positively falsified by the new data series from Sudan.

It is now the time to analyse the paragraph that leads into this section of the chapter, in order to discuss the great enclosure at Musawwarat es-Sufra. The excavations revealed that, on several occasions throughout the site's history, the complex had been razed in order later to be rebuilt and expanded in the same place. A set of eight construction stages have been identified between 500 BC and AD 350. However, on each reconstruction, the new building had a variation in the orientation that oscillates between 4° and 5°. According to Wening (2001), this must have to do with the variation of the position of a certain star, used to orientate the structures because of precession.

Unfortunately, we have no data obtained directly in situ that allow us to verify the different plans of the various structures built at the Musawwarat site over time. However, the set of orientations that has been determined through the detailed study of the plans and aerial images for the temples of the complex does not allow confirmation of Wening's hypothesis, even though, from our point of view, achieving this would have been fascinating. The main buildings of the complex show orientations that are either solstitial or with respect to the rising of Sirius; in either case, neither the variability of the obliquity of the ecliptic nor the precession justifies such large azimuthal variations.

It could be argued that the solstitial orientation of the main shrine (MS100, possibly dedicated to Amun, a solarized deity) is circumstantial and due to a later focus. Perhaps, earlier designs do indeed hide other orientations to stars or constellations. However, considering the statistical support for the solstice family and the reflection that we have found in the north, we think that this idea cannot be maintained. Maybe the severe changes in orientation were due to the search for a perfect alignment to a first order tradition. The architects did not stop tearing down and rebuilding on the same site until a perfect solstice alignment was achieved after several attempts.

A final comment to the temple of the lion god Apedemak in Naga; this had an azimuth of 119½°, pointing to an approximate declination of −28½° which is, within a certain error margin, the southernmost possible position of the moon, the lunastice. According to Török (1997: 506), Apedemak was a lunar god whose cult was first established in the area of Musawwarat in the third century BC by King Arnekhamani. This was later propagated by his successors, and it would be interesting to be able to establish a connection between the ascription of the temple and its orientation as we saw in the case of the Thoth temple at Seikh Abada (Belmonte, 2012: 246–7). However, this relationship must be considered with caution, because other temples of Apedemak, precisely those of Musawwarat and Meroë do not follow the same pattern.

6.10.2 Amun Temples, an Intriguing Case

Caroline Rocheleau (2008) has written an extremely interesting and novel study on the typology of the temples dedicated to Amun in ancient Nubia, particularly those of the Kushite period. Unfortunately, in this seminal work there is not a single line dedicated to a structural aspect as important as the orientations of the buildings, which, as has been shown, can offer important clues related to culture, festivals, general construction design, or even, as we shall see, to temple chronology.

The first structure that we should discuss is the temple of Amun at Napata, at the foot of the Djebel Barkal, also called the Ipet Sut after its northern namesake. Originally erected by Tutankhamun, the orientation was typically Egyptian. Another temple of Amun, dating from before the establishment of the Kingdom of Kush in the region, is that of Amun of Pnubs, a place that has recently been identified with the site of Dokki Gel, in Kerma, thanks to archaeological excavations carried out on the site by a team from the Sorbonne (Bonnet & Valbelle, 2005). The orientation of the oldest temple, originally erected by Thutmose IV, later rebuilt by Akhenaten as an Atonian temple, and completed in the Napatian period, points north, perhaps to the southernmost star of Meskhetyu (i.e. Alkaid).

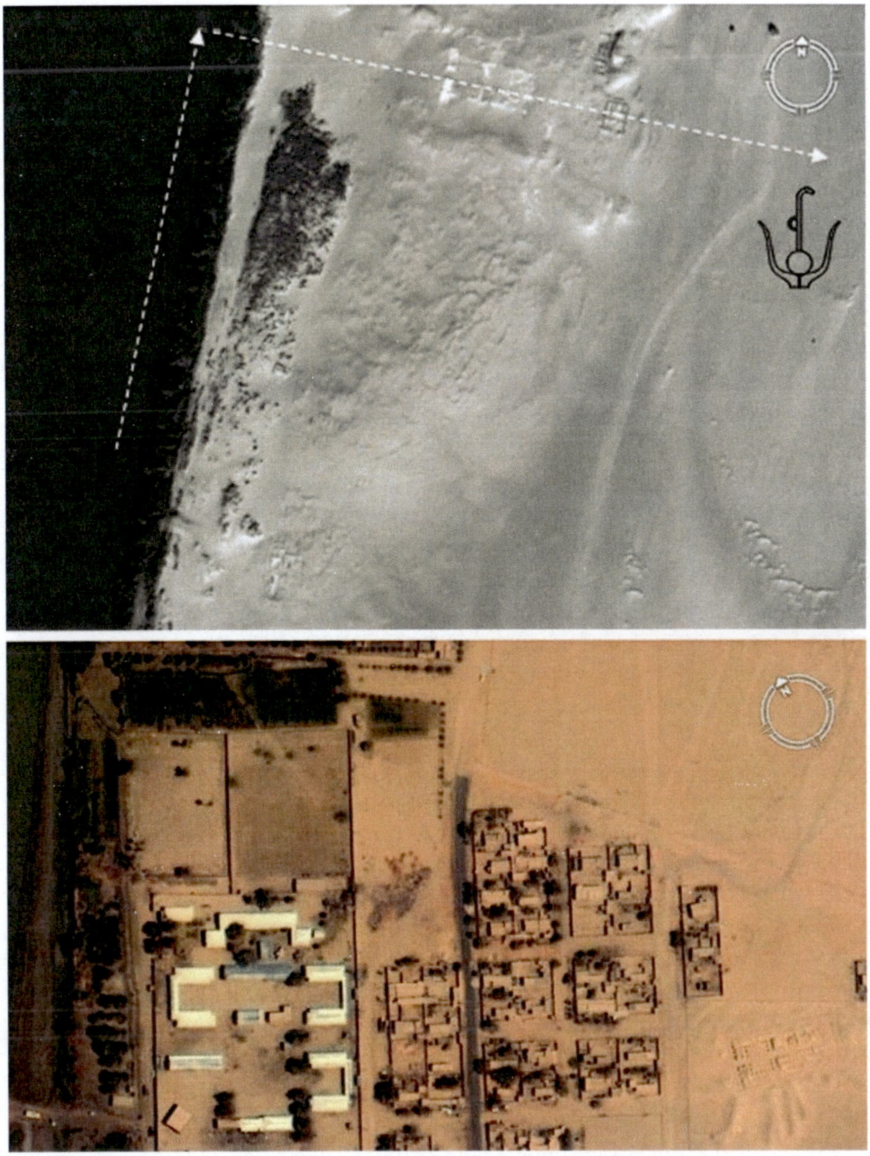

Fig. 6.49 Location and orientation of the temples of Amun of Gematen (above) and of Amun 'Bull of Nubia' (below), in Kawa and Sanam, respectively. The first faces the course of the Nile at the site. Kawa Temple is a reflection of Karnak Temple. Both shrines were allegedly erected by Taharqa (690–664 BC), double king of Egypt and Kush. See text for details. (Images courtesy of Google Earth, adapted from Belmonte, 2012)

Later temples erected on the site maintained almost the same alignment, so that when a new temple was built in parallel during the Meroitic period, it maintained a very similar orientation that can still be reasonably explained, as Alioth ceased to be circumpolar in the region around 500 BC. It is clear that this orientation towards the north was not the appropriate one for a solar deity, but we cannot be sure that this astral character was at the origin of the god Amun of Pnubs, not even for the Amun of Napata. In this later case, we do know that he was a chthonic deity residing inside the Pure Mountain; that is, the Djebel Barkal.

The situation changed radically with the arrival of the 25th Dynasty, particularly during the reign of Taharqa (690–664 BC). Both shrines were known during the Napatian period as distinguished members of a group of four temples, located between the third and fourth cataracts, which were visited during the pilgrimage that took place after the royal coronation, so that the ceremony was remembered at each one. The other two were the Temple of Amun at Gematen (undoubtedly the site of Kawa) and the Temple of Amun 'Bull of Nubia' at *Krtn* (identified with Sanam, opposite Napata). Both are illustrated in Figure 6.49.

These two temples, Kawa and Sanam, were conceived with a complex but remarkably similar architectural structure (including the location of the 'dais' hall) suggesting a common planning. Interestingly, both temples had almost the same orientation towards the western horizon, but while the Temple of Amun at Kawa, called Temple T by archaeologists, was perpendicular to the course of the Nile at the site, the Temple of Amun 'Bull of Nubia' was not (Fig. 6.49).

Could there be an explanation for this quirk? In fact, Temple T in Kawa is perpendicular to an earlier structure, Temple A, which was in ruins and abandoned on the site when Taharqa decided to build the impressive Temple of Amun from scratch on this barren land on the banks of the Nile. There is practically no cultivable land on the spot and it had been abandoned for several centuries. Nevertheless, there could be a very powerful reason for taking this decision.

We established early on that the Temple of Amun at Karnak was first aligned towards sunrise at the winter solstice when this astronomical phenomenon coincided with the Egyptian New Year; that is, with Wepet Renpet. Furthermore, by means of such an alignment, the temple also became perpendicular to the course of the Nile as time passed. The texts analysed in Kawa have shown that the feast of the god Amun of Gematen was celebrated in situ in I *Ahket* 1 in commemoration of the temple's foundation on New Year's Eve in 680 BC. At that time, the declination of the sun was about $-17°$, practically identical to that of the temple, if the opposite direction to that of the access door is considered (as in Karnak).

Consequently, the Temple of Amun at Kawa was orientated in precisely the same way as the Temple of Amun-Ra at Karnak thirteen centuries earlier. This would also imply that Taharqa, who was also a great builder in Thebes, would have deliberately selected Kawa owing to the dual alignment, both astronomical and topographic, that occurred at the site. Therefore, Kawa imitates Karnak (it would be a sort of echo or reflection in the south) while Sanam, further upstream possibly imitates Kawa. The orientation of the temple of Amun at Sanam, also built by Taharqa, is so similar to that of Kawa that this circumstance can hardly be attributed to chance.

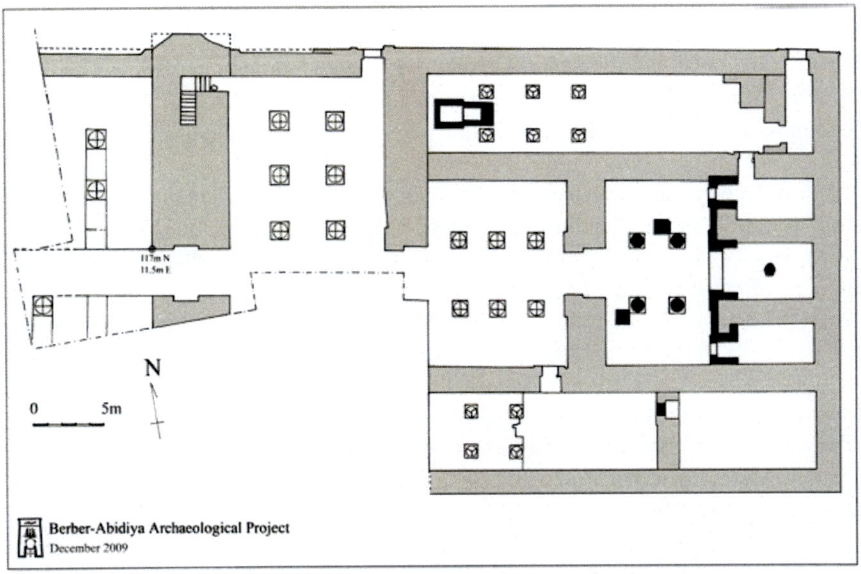

Fig. 6.50 Plan of the Temple of Amun in Dongeil, erected in the reign of the royal couple formed by Queen Amanitore and King Natakamani. The temple was orientated to New Year's Eve sunrise. Note the 'dais' room in the northeast corner of the building. (Image courtesy of the Berber-Abidiya Archaeological Project)

It is now the time to discuss the temple of Amun at Tabo (Jacquet-Gordon et al., 1969). This building has a close to equinoctial orientation and therefore conforms to the solar pattern. It is located on a seasonal island, so that it faces a channel of the Nile to the east when the waters are high. It is therefore a perfect orientation for a temple to the solar deity. This could support Rocheleau's (2008:76–7) hypothesis, based on certain structural aspects of the temple, including the possible existence of a 'dais' room in a similar location and an orientation that resembles that of the Kawa and Sanam temples. Tabo could have been the Pnubs temple of the royal pilgrimage and not that of Dokki Gel. The controversy is still open.

As we have seen, all *dais* rooms of the temples of the Napatian and Meroitic periods are orientated in such a way that any image on the altar, whether of a deity or a sovereign, gazes towards the east. This is also true for the Meroitic temples at Dangeil and Naga, although the 'dais' room is on the north side of these temples (see Figure 6.50).

In recent years, there has been a debate about the real purpose of dais rooms in Kushitic temples. Hypotheses range from the place for the coronation ceremony, to a sanctuary dedicated to the solar cult similar to those found in Egyptian temples (e.g. Abu Simbel). All in all, the most striking similarity is with the so-called 'high-room of the sun' at Karnak from where the winter solstice sunrise was observed precisely in the opposite direction to the general orientation of the temple (Krupp, 1988; see also, Belmonte et al., 2009b: 248–52).

It is probable that both hypotheses could be correct because they do not need to be mutually exclusive, but rather are complementary. Hence the king, whether sitting or standing in the right place in the 'dais' room, could have directed his gaze towards dawn at different times of the annual cycle, such as the 'equinox' (see, Belmonte, 2021), Wepet Renpet, or the winter solstice, depending on the temple selected for the occasion.

The currently visible remains of the Temple of Amun at Meroë, the second and ultimate capital of the Kingdom of Kush, are dated to the Meroitic period. The oldest archaeological record on site dates from the reign of Amanishakheto (*c.* AD 1), although there may have been another building on the site during the Napatian period. The temple faces sunrise on the winter solstice and the structure is also perpendicular to the Nile, a fact again reminiscent of Karnak. This shrine was not included in the list of temples of the coronation pilgrimage during this period. However, this fact could be understood as a proof that Meroë was already the capital of the kingdom at that time and therefore was nothing else than the starting point for the coronation journey. Of course, this could have been the case after the reign of Aspelta, the king who possibly moved the capital from Napata to Meroë, although the cartouche of Senkamanisken (643–623 BC), one of the immediate successors of the kings of the 25th Dynasty, has also been identified in the place.

Török (1997: 317 & 517) has suggested that the current plan of the temple has its origin in an extraordinary flood of the Nile that occurred in the year 6 of Taharqa (685 BC). The pylons of the original temple would face a branch of the Nile that existed at that time, which also implied an orientation towards the rising sun. This idea is based on old proposals that suggested that Meroë was originally an island in the middle of the river. However, recent excavations show that this idea should be abandoned, so that the similarity with Karnak increases. Some buildings in the area, where the alleged branch was located, have been dated to *c.* 800 BC. The rubble and pebble material found there seems more like evidence of a flood than of a dry branch of the river (Rocheleaau, 2004: 80–1). The temple of Amun therefore probably opened onto the Butana steppe from its inception and the building orientation is therefore extraordinary.

Aligned towards sunrise at the winter solstice, and in the opposite direction, perpendicular to the local course of the Nile, the temple orientation so closely resembles the orientation of the main temple of Amun at Karnak, that a corollary may well be that the site of Meroë could have been deliberately chosen. It would simultaneously fit the astronomical and topographical needs, land- and skyscape accomplishments, required for such an important structure. The parallelism with Karnak would be complete if the temple axis was firstly aligned around 500 BC, when Wepet Renpet occurred once more on dates coinciding with the winter solstice. However, this last possibility is unproven.

When a similar circumstance was analysed at Temple T at Kawa, we assumed that the Egyptian civil calendar was in common use in the Kingdom of Kush, at least during Taharqa's reign. This may certainly be the case for the reign of the sovereigns of the 25th Dynasty (including Taharqa), who were also dual kings of Egypt. For the new hypothesis to be true, we must accept that the civil calendar was thereafter adopted as the state calendar by the kings of Kush, a hypothesis that is far from proven.

However, there are interesting parallelisms. The sacred precinct of Amun at Meroë was completed with a processional avenue and a series of temples and chapels perpendicular to it, presumably dated to the reign of the royal couple of Queen Amanitore and King Natakamani (*c.* 20 BC), sovereigns who are well known in the history of Kush for being great builders.

One of the temples built by the royal couple is the Temple of Amun excavated by an Anglo-Sudanese mission coordinated by the British Egyptologist Julie Anderson in the village of Dangeil (Mohamed Ahmed & Anderson, 2005; see Fig. 6.50), which has a typical 'dais' room on the north side, a typical element in temples built in this epoch. The sanctuary doors open towards the valley, located to the west; however, the temple is not perpendicular to the Nile.

With a declination of *c.* 9½°, the temple faces sunset on dates close to August 28 or 29 in the proleptic Gregorian calendar. Interestingly, this has the consequence that the temple could have been orientated to sunset at Wepet Renpet around 20 BC, with an error margin of twenty years, provided we accept that the Egyptian civil calendar was in use during the Meroitic period. The decoration found at the site has shown that the temple was almost certainly built by King Natakamani and Queen Amanitore. This would favour a high chronology for these two sovereigns, since, according to the most accepted chronologies, this royal couple could have reigned, for a quarter of a century or so, sometime between the years 20 BC and AD 20.

However, there is another even more suggestive alternative. According to a majority of experts, around the year 25 BC, under the rule of Emperor Augustus and the influence of the Julian calendar, Egypt adopted the Alexandrian calendar as a new system of fixed time calculation, including leap years. Hence, New Year's Eve would always fall, in first approximation, at the same time of the seasonal cycle. The new era adopted Julian August 29 as the fixed date for I Akhet 1 in the years to come, although, owing to the mutual displacement between the Julian and Gregorian calendars, this date has gradually moved so that it now falls on September 11, a date of unpleasant memory nowadays. Copts still celebrate their New Year's Eve on this date.

What interests us here, however, is the fact that, according to this reasoning, the Temple of Amun at Dangeil could have been orientated to Wepet Renpet for any accepted chronology of the royal couple consisting of Natakamani and Amanitore, provided we accept that the Kingdom of Kush also implemented the reform of the Egyptian calendar imposed by Augustus. We are therefore faced with two very suggestive alternatives: either accept a high chronology for the period, or we consider the possibility that the Kushites adopted, in time and form, a calendar reform implemented by their worst enemies.

There is hardly any archaeological or documentary evidence supporting one alternative over the other. However, resorting to Ockham's razor, we would tend to favour the first because it is difficult to assume the second one. Just take into account that England lasted two centuries before adopting the Papal reform of the Gregorian calendar, despite its obvious virtues, and Tsarist Russia never did accept it.

Natakamani and Amanitore undoubtedly built another Temple of Amun of the same typology in the important holy city of Naga, to the south and in the interior of the Butana savannah. This place is far from the Nile and the river could not have commanded any of the local orientations. Nevertheless, once again, this temple is orientated to a singular moment of the solar seasonal cycle, the equinox on this occasion. During the Ptolemaic period, the Zodiac was accepted in Egypt, with Aries, the constellation of the Ram, at the head of the zodiacal signs, signaling the vernal equinox (Lull & Belmonte, 2009). The Temple of Amun in Naga has the largest collection of cryocephalic sphinxes in a Kushite temple. We may speculate that the Mediterranean tradition had reached Kush in the time of Natakamani, so that the temple was orientated towards the celestial ram, perhaps considered as an avatar of the god Amun (this is an idea that has haunted the first author for years, see Belmonte, 1999: 166–72), and therefore towards the equinox.

The dais rooms of both temples of Amun at Dangeil and Naga face east, in the opposite direction from the doors of the building, still following the same pattern established for the royal pilgrimage temples of the Napatian period (this is also the case for the Temple of Amun at Meroë). In this sense, there is a possibility that the three temples in the south sector of the country, separated from the northern core by the inhospitable Bayuda Desert, could be the distinguished members of a second set of temples related to the royal coronation pilgrimage, but for the Meroitic period (Rocheleau, 2008: 78).

This hypothesis is attractive, but we would like to emphasize that, once all the temples of Amun that have a 'dais' room have been taken into account (with the exception of Napata, forced by previous constructions), they all face a relevant time-mark of the seasonal cycle or of the civil calendar. These are the equinox and winter solstice sunrise, or Wepet Renpet, respectively, on occasions both at the same time. Consequently, the hypothetical solar aspect of this room would be reinforced. However, once again, we want to stress that the two possibilities under discussion (place of coronation or solar sanctuary) do not need to be mutually exclusive.

We wish to emphasize that the properties described so far could be used to try to ascribe a certain temple to the god Amun or to deny such an assignment. The pattern found for Musawwarat temple MS100, discussed earlier, seems to confirm that its ascription to the cult of Amun seems reasonable.

In contrast, the temple recently discovered by Bodgan Zurawski (1998) and his team at Saniyat, the ancient Tergedum of Pliny, has been classified, based on its plan, as a temple of Amun. However, the orientation of the sanctuary, almost perpendicular to the abnormal course of the Nile at the site, is so divergent from the rest of the Amun temples that there are serious doubts concerning this identification. Further work on the archaeological excavation of the sanctuary, where new epigraphic material could be recovered, would therefore be highly desirable.

All in all, the temples of Amun in Sudan have offered us a firm and suggestive clue about the relationship between astronomical practices, associated with the orientation of the temples, and various cultural aspects of the Kushite civilization, during both the Napatian and Meroitic periods. This confirms that Egyptian civilization had deep roots in the land of Kush. A final experiment will show us how true this statement is.

6.10.3 *Unveiling Nuri: Clues Towards Solving an Enigma*

Nowhere in Nubia is the relationship between archaeology and landscape more evident than in the area of ancient Napata, where the Nile, which flows in a southwesterly direction, and the Pure Mountain—on its right bank (west at the site)—integrate a very suggestive and evocative image of another peculiar landscape hundreds of kilometres to the north, in ancient Thebes.

In his manual of the Kushite civilization, Török (1997: 305) establishes a parallel between Napata and Thebes. In the Napata area, we have, on the one hand, the temples of Amun at the foot of the Djebel Barkal and the temple of Sanam. On the other hand, we encounter the necropolis of Nuri, on the left bank, where the royal pyramids were built (see Figure 6.51). In Thebes, we have the Karnak complex, the Temple of Amun at Medinet Habu, and the Valley of the Kings, respectively. Two paralleled sacred landscapes.

Along the same lines of argument, Thimothy Kendall (1997: 161–71 and note 78), who has excavated in the Djebel Barkal area for more than a decade, has presented an interesting hypothesis. He argues: 'From the top of Djebel Barkal, the pyramid of Taharqa (the first to be built and the largest of Nuri necropolis) marked the sunrise on the summer solstice and the heliacal rising of Sirius'. He continues: 'From Nuri, the sun descends on Djebel Barkal at the time of the winter solstice'. This would complement Török's ideas with suggestive astronomical relationships

Fig. 6.51 Satellite image of the Napata region in northern Sudan, showing the location of the Nuri necropolis and the holy mountain of Djebel Barkal on opposite banks of the Nile. The diagram also shows the image of Djebel Barkal from the base of Taharqa's pyramid (top left) and a detailed satellite image of the Nuri pyramid field (bottom right). The dashed line represents the orientation of the Pyramid of Taharqa (T). (Authors' diagram, on a series of images, courtesy of Google Earth)

such as the sun at the solstices or Sirius. Taking into consideration the facts we have established in the previous paragraphs of this chapter, this possibility obviously needs verification.

According to the data of Belmonte et al. (2010), the Pure Mountain covers an azimuth interval of 250° to 252½°, with an average angular height around 1°, as seen from Nuri (from Taharqa's pyramid to be precise). This represents an accurate declination range of −18¾° to −16¼°, respectively (Fig. 6.51). Kendall's hypotheses are therefore inaccurate, since the winter solstice sun sets a few degrees south of Pure Mountain as seen from Nuri. Consequently, the summer solstice sun never rises above Nuri as seen from Djebel Barkal, and neither is the heliacal rising of Sirius visible in that direction, but is several tens of degrees further to the south (although it is at approximately the same time of year, June 27 in the proleptic Gregorian during the reign of Taharqa).

Still, the data did show another couple of surprising possibilities. On the one hand, Sirius had a declination of −18°.3 to −17°.9 between the reigns of Tuthmose III and Tutankhamun, respectively, the first pharaohs who built at the foot of the Pure Mountain. This star had a declination of the order of −16°.8 during the reign of Taharqa, and of −16°.3 at the end of the Napatian period (*c.* 300 BC), reaching a minimum of −15°.8 in the middle of the first millennium. Therefore, Sirius setting over the Djebel Barkal plateau was always visible, from the site of Taharqa's pyramid, between the earlier settlement at Napata, in the New Kingdom, and the end of

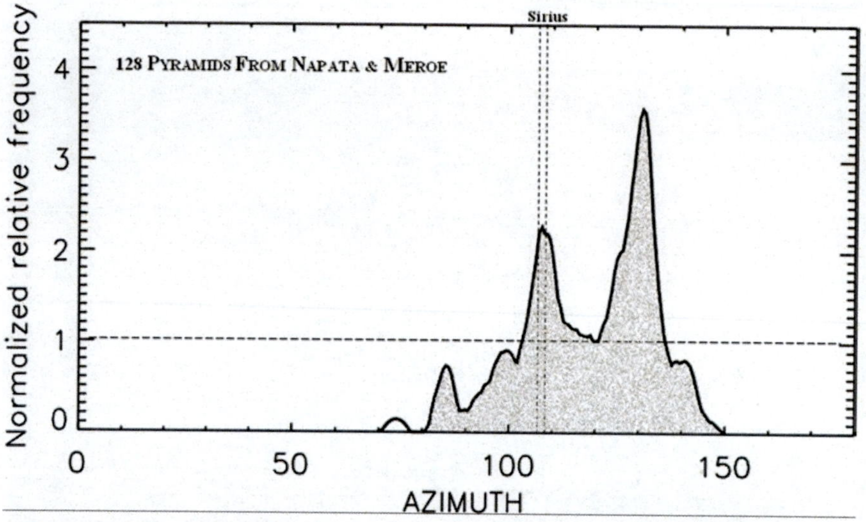

Fig. 6.52 Average orientation histogram of some 128 royal pyramids from the El Kurru, Djebel Barkal, Nuri and Meroë necropolises, with data extracted from the literature and high resolution images from Google Earth. Pending more precise in situ measurements, this diagram may be useful for analysing the orientation pattern of these monuments. Note the presence of two dominant patterns. The dashed vertical lines indicate the extreme azimuths of Sirius for the region and the corresponding period

the Napatian period, including of course the reign of Taharqa who would inaugurate the royal cemetery at Nuri.

In reality, as we have previously discussed, Sopdet heliacal setting would be visible over Djebel Barkal around May 18 in that historical period, which could suggest a hypothetical relationship with the arrival of the Nile flood. This fact alone could have singled out Nuri as a very important landmark in the Napata region and could have justified Taharqa's selection of the site as his, and his family's, burial site.

This connection would confirm the relationship between the orientation of the ancient pyramids of the 25th Dynasty kings and their immediate ancestors, downstream at El Kurru, and the rising of the star, as shown in Figure 6.52. Moreover, taking into account the course of the Nile at the site, El Kurru could also have been deliberately chosen as the site for the royal necropolis at the dawn of the Kingdom of Kush because it presented a double astronomical and topographic alignment (Belmonte et al., 2010, and note 57). Later, Taharqa, a fervent admirer of Thebes and certainly her landscape, would try to optimize that relationship by moving the royal cemetery to Nuri.

However, the location of Nuri additionally offers another astronomical connection; one which is even more suggestive in reinforcing the idea of its sacred character. Owing to the wandering nature of the Egyptian civil calendar (which was certainly in use during the 25th Dynasty), sunset at I Akhet 1, as observed from Nuri, began to happen on Djebel Barkal around 685 BC, early on the reign of Taharqa. This cosmic reality ceased around 648 BC during the reign of Atlanarsa, his son and second successor to the throne. Consequently, for most of Taharqa's reign, sunset at Wepet Renpet occurred over the flat surface of Pure Mountain as seen from the site selected for his pyramid.

Considering the effort invested by Taharqa in the construction of the Temple of Amun at Kawa, with the main axis aligned to sunrise on New Year's Eve, we can certainly raise the hypothesis that Taharqa's selection of Nuri as his resting place was not by chance. He could have been induced by this suggestive solar phenomenology, dramatically connecting his tomb in a symbolic way with the holiest place in his homeland. The casuistry associated with Sirius may well have reinforced that decision.

Interestingly, Taharqa selected for the orientation of his pyramid a pattern that undoubtedly broke with the tradition of his ancestors. However, this was nevertheless in accordance with a common tradition in the Theban area, the pattern related to inter-cardinal Family VII. The diagonal of the pyramid was thus orientated practically on the meridian line.

Figure 6.52 shows the tentative orientation histogram for 128 royal pyramids in the cemeteries of the Kushite Kingdom, located in El Kurru, Nuri, Djebel Barkal and Meroë. Data was obtained from HR Google Earth images and bibliographic sources (e.g. Lockyer, 1901). The figure clearly shows how the Kushite pyramids are concentrated into two preferred azimuth patterns, one centred on about 107½° and another on *c.* 131°. The former can tentatively be assigned to the rising of Sirius, and the latter with an orientation that probably mimics that of the pyramid of Taharqa. Sirius would not only be the main target in El Kurru, but also in the

Fig. 6.53 The northern necropolis of Meroë, which includes some of the best-preserved monuments from the Kingdom of Kush dating of the Meroitic period. Each pyramid was preceded by a funerary chapel from which there was access to the sepulchral chamber, located below the pyramid. On average, this group of pyramids is orientated, whether by chance or deliberate design, to the rising of Sopdet. (Image courtesy of Mosalam Shaltout)

northern cemetery of Meroë (see Figure 6.53) which has a distribution maximum centered at an azimuth of 107½°, precisely. In contrast, the maximum of the distribution at Nuri is about 130°, which would support the hypothesis of an inter-cardinal direction.

Taharqa's immediate successor, his nephew Tanwetamani, returned temporarily to El Kurru, selecting the cemetery of his ancestors as his resting place, but the next king, Atlanarsa and nineteen of his immediate successors returned to Nuri to be buried. Nastaseñ (*c.* 325 BC) was the last king to build a pyramid in this cemetery. One of the enigmas of Nubian archaeology is why the Nuri pyramid field was precisely abandoned after the funerals of Nastaseñ.

They first moved to Djebel Barkal, for just a couple of decades in the reigns of Aktisanes and Aryamani, and then, definitively, to Meroë around 295 BC, during the reign of Arkamani(qo), the king who would inaugurate the royal necropolis of Meroë (pyramid BrS6, built in the southern cemetery; Török, 1997: 442). He has been dated fairly accurately because he was a contemporary of Ptolemy II (282–246 BC).

According to Török (1997: 395), the abandonment of Nuri would have been motivated by the depletion of suitable sites for royal burials (and the construction of pyramids) in the elevated areas of the necropolis. However, taking into account the hypotheses analysed so far, which connect land- and skyscape as the main reason for the selection of Nuri as the royal cemetery during the reign of Taharqa, we can confront such a prosaic interpretation with an absolutely fascinating alternative.

Around 352 BC, because of the wandering nature of the civil calendar–with which the reader must be well acquainted by now—sunset at Wepet Renpet began to take place once more over the plateau summit of Djebel Barkal. This was after its almost three-centuries-long round-trip southern excursion. For the next 40 years, the sun on New Year's Eve continued to set over the Pure Mountain as seen from the site of Taharqa's pyramid, until it suddenly stopped doing so in *c.* 315 BC. This was the precise epoch when the royal cemetery was temporarily moved to the pyramid field near Djebel Barkal. Finally, around 298 BC, sunset on New Year's Eve ceased to occur above the Pure Mountain, wherever the place of the Nuri necropolis was chosen to visualize the phenomenon. Our suggestion, following Belmonte et al. (2010), is that this place lost its sacred character precisely because of this fact, perhaps combined with the almost simultaneous end of the casuistry related to Sirius at the same time.

During the reign of Taharqa, the sacred geography integrated by Nuri and Napata became a first order symbolic attractor (Fig. 6.51). Three and a half centuries later, the connection between heaven and earth came to an end and Nuri (and perhaps Djebel Barkal as well) lost a substantial part of their sacred aura. Consequently, the northern cemeteries were abandoned and the royal necropolis moved far south to Meroë. This movement culminated a dynamic that had begun almost three centuries earlier, with the change of capital during the reign of Aspelta. The Meroitic period of the Kushite civilization had definitely begun.

6.10.4 Kush: Egypt's Southern Mirror

As a summary of the experiment that was carried out with the monuments of Sudan, we may conclude that, as in the case of Egypt, only three astronomical orientation customs (cardinal, solar, and stellar) were present in ancient Kush, throughout the country and during the long course of her history.

The cardinal is perhaps the least representative, but the solar, formed by families II and I, has numerous paradigmatic examples such as the temples dedicated to the god Amun during the Napatian and Meroitic periods, in particular, those that include a 'dais' room in their structure. The stellar custom would be well represented by family IV, equally important for Kushite temples and pyramids. Once again, we have no doubts about the pertinence and relevance of Sirius alignments, although new epigraphic material, further confirming the importance of Sopdet in the history of Kush, would certainly be of great benefit.

On the other hand, it has also been established that the local topography and orography certainly command the orientation, and even the location, of certain sacred structures. Several temples were orientated almost perpendicular (or parallel) to the course of the Nile, as in Egypt, and in some cases the selection of the site could help towards having a double astronomical and topographic alignment. In this sense, the temples of Amun at Kawa and Meroë are highly illustrative and extremely suggestive.

In other cases, the selection of a specific place for causes related to the observation of solar and stellar phenomena in singular elements of the landscape was unambiguously recognized. The case of the Nuri necropolis is a paradigm in this resspect.

Consequently, we consider the experiment to have been prolific in its results, suggestive in ideas and hypotheses, and very positive as an element of contrast with what has been previously recognized for the ancient Egyptian civilization.

In this sense, we confirm that, Napatian sacred buildings, and their Meroitic heirs, followed identical, albeit more simplified, orientation patterns than those previously established in ancient Egypt, and that these can be related, without a doubt, to the tradition and religion of the Kush Realm, which was highly influenced by the Pharaonic culture. The mirror in the south concept is thus fully confirmed.

The study of the temples dedicated to Amun has been especially fruitful. We have shown that their standard plans were connected to solar orientations, and related to important milestones of the annual cycle, such as the winter solstice, the spring equinox, or the civil calendar New Year's Eve, I Akhet 1 or Wepet Renpet. This suggests a continuation of Egyptian traditions not only during the reign of the double kings of the 25th Dynasty, Taharqa in particular, but also during the Meroitic period, especially during the reign of the royal couple Natakamani and Amanitore. Moreover, this solar aspect was sometimes combined with the local topography to create a skyscape integrated into the landscape, as in the fascinating cases of Kawa and Meroë.

However, the skyscape paradigm has been found in the Napata region, where the location of the Nuri necropolis, with its dozens of pyramids, could have been deliberately selected by Taharqa to produce a link of clear celestial symbolism between his place of burial and the hill where the god Amun resided, the Pure Mountain of Djebel Barkal. With this idea in mind, the hypothesis of the abandonment of Nuri, when this celestial phenomenology ceased to occur at the end of the fourth century BC becomes a real and surprising possibility.

Most of the ideas devoted to Kush that have been reflected in this section are the result of the analysis of high-resolution images from Google Earth. However, being faithful to our own practices, it would be ideal to contrast them with data obtained from fieldwork (especially needed regarding the angular height).

Furthermore, a detailed visit to the sites may allow us a glimpse into aspects that surely escape our notice, given that we have not actually been to the sites. Our experience in Egypt tells us that this must be so. As Andrea Rodríguez-Antón often declares: 'Nothing compares to fieldwork'. However, we do think that the data are complete enough to enable us be reasonably certain that most of what has been said here is correct. Time will tell.

6.11 Epilogue

Pharaonic civilization temples in the Valley of the Nile were built in accordance with the local course of the river (either nearly perpendicular or parallel to it) and, at the same time, obeyed certain astronomical patterns.

This means that the ancient Egyptians should have dealt with particular situations to meet both requirements. This problem was solved by the selection of orientation patterns that would simultaneously comply with the constraints imposed by the Nile at different sites and, at the same time, obtain conspicuous astronomical orientations that were in accordance with the religious and symbolic character of the temple. The examples we have seen in this chapter are, in that sense, exceptional.

More cases, such as Akhetaten—the newly created city of Akhenaton at Tell el Amarna—could also have been explored and would have shown relevant examples of a similar phenomenology (Belmonte, 2012: 246–247; see Belmonte, 2022, and Chap. 7 for an update). As we have clearly shown, even cultures with strong Egyptian influences, such as the Kingdom of Kush (Belmonte et al., 2010), in ancient Sudan, show strong evidence of the same character, as we demonstrated in our last case study. Serabit el Khadim will probably need a new exploration before the role played on site by the precession of the equinoxes in relation to Sopdet could be affirmed (Belmonte et al., 2008). Many other sites have probably not yet unveiled all their secrets.

All in all, the results presented in this chapter illustrate, with little room for doubt, something that was not easy to imagine when our archaeoastronomical fieldwork in Egypt began two decades ago. We have shown how the sages of this exceptional culture scrutinized their environment, including the sky and their country's geographical framework, in their continuous fight against chaos and their quest for cosmic order. This is a reality, which has been reflected in the close relationship encountered between cosmogony, calendar, symbolism and land- and skyscapes in order to create what we may easily call cosmic landscapes.

Chapter 7
Astronomy and Chronology

This chapter has perhaps been the hardest one to write of the entire book. Since Belmonte (2012: 297) discussed this topic a decade ago, a dozen papers and a series of reviews and books on the topic have been published or are forthcoming. The resulting studies for any period of Egyptian history are seldom contradictory and scholars tend to favour that evidence with which they are more familiar or which better supports their preferred chronologies. Neither preconceived ideas are entirely absent. Examples of circularity occur more frequently than one would wish. Furthermore, a scholar, when trying to tackle the topic from an open and objective point of view, has to read such unfortunate phrases as: 'It would seem to be amply good enough [evidence] to do away once and for all with the revisionist low chronologies'. The label 'revisionist' has such unfortunate overtones that it would better to try to avoid such offensive terminology

The evidence is never as clear as chronologists would desire and scholars should be a little humbler when expounding their hypotheses, because most of the evidence is just that; simple hypotheses variously grounded in established facts.

7.1 Evidence, What Evidence?

In this, chapter we deal with the evidence related to astronomy. However, it is certain that we shall have to approach other dating techniques such as important textual evidence (for example, coeval references to kings that were contemporaneous), or the always controversial but most interesting models related to C14.

We can catalogue the astronomical evidence into three different classes depending on the celestial body, or bodies, in question, be it a star, the moon, or the sun. The first kind of evidence is the one that has been there since the beginning of Egyptology. This is Sothic dating, which relates to the yearly position of the bright star Sirius—identified with the goddess Sopdet, avatar of Isis and Seshat (see Figure 7.1)—notably her heliacal rising. The second has to do with the moon (*jꜥḥ*)

J. A. Belmonte, J. Lull, *Astronomy of Ancient Egypt*, Historical & Cultural Astronomy, https://doi.org/10.1007/978-3-031-11829-6_7

Fig. 7.1 Sopdet as Isis-Sopdet in the astronomical ceiling of Senenmut at TT353 in Deir el Bahari, western Thebes (left), and as Sopdet-[Isis]-Seshat in a cult chapel of King Amasis at Aïn al-Muftela in the Oasis of Bahariya (right). This multifaceted goddess was seen in the celestial vault as Sirius, the brightest star in the Egyptian skies. (Images by the authors)

and the synodic month, and the relationship established in ancient Egypt between the civil calendar and a series of lunar feasts or temple services embedded in a lunar time-reckoning system, as discussed in Chap. 5.

Finally, the third kind of evidence involves the most important of Egypt celestial objects, the sun, $r^{ʿ}(w)$, the celestial manifestation of the most important deities of the country. In this case, we deal with two kinds of events: eclipses or sunrises (rarely sunsets) at special moments of the civil year. The former rely on mere educated guesses since no direct textual evidence has been preserved in Egyptian sources. The latter has to do with the wandering nature of the civil calendar and relevant movable civil dates, such as Wepet Renpet, in the seasonal tapestry. This can be inferred from devoted archaeoastronomical research as shown in Chap. 6. Eventually, the changing coordinates of a star or asterism due to precession of the equinoxes can also be used to infer a date of when a sacred building might have been aligned for the first time.

It is not the purpose of this chapter to undertake a complete review of the role played by astronomical calculations in the determination of the chronology of ancient Egypt. Before entering into a discussion of a most relevant series of case studies, however, it is worth analysing the nature of astronomical evidence.

7.1.1 Stellar Inspirations

Sothic dates are not the fantastic evidence Egyptologists tend to believe they are. The different events identified in ancient sources could be classified as:

(a) The ideal one would be to have an undoubted observation of Sirius' heliacal rising on a definite civil date (including month and day) in an undisputed year of a well-identified sovereign. No such date exists for Pharaonic Egypt! The only one is Censorinus' account (n° 14 in Chap. 5) for the Roman Era, whose chronology is undisputed. The same can be argued for the inferred date of the Canopus Decree during the reign of Ptolemy III.

(b) The observation of the celestial phenomenon but in a case where some of the key information is lost. The Jebel Tjautti Sothic date is an example of this.

(c) Alternatively, we have cases where an observation is clearly involved, together with the civil date and the reign, but not the year. However, this occurs in a general religious context. The Ramesseum ceiling inscription (see Fig. 5.13) is an example of this.

(d) The prediction of the heliacal rising (or the related festival Peret Sopdet) in a clear civil-calendar and kingly context. This seldom happens, the great difficulty here lying in ascertaining that the event was observed at the relevant date and, above all, where it was predicted that the event would occur. Sothic date (2) is the best example.

(e) Mentions of the Peret Sopdet festival in a feast calendar context with a clear civil date related to it, but not the reign year. The Elephantine inscription (see Fig. 5.21) is an example. These serve as a chronological framework but not as a precise Sothic date.

(f) The same as above but with no civil date given for Peret Sopdet. The chronological use of these 'dates' is very limited unless strong assumptions are made. This is the case of the date in the stele of Buto (n° 7) discussed in Chap. 5.

(g) An implicit Sothic date from a difficult textual context. The Ebers calendar Sothic-inferred date of Amenhotep I's ninth year at III Shemu 9 is the best example. Prediction within a calendrical framework or observation, and the place related to the phenomenon are among the controversial issues related to this kind of evidence. The most recent, and perhaps most ancient, Sothic date, identified in Kaplony's collection (n° 1) is another example.

Under these premises, it is not rare for scholars and dedicated readers often to lose themselves in the endless discussions related to one kind of evidence or other.

However, this does not mean than Sothic dates should not be consider as precious jewels in establishing milestones in the intricate pathway of Egyptian chronology. Without them, we would be absolutely lost, and the wandering nature of the civil calendar would have remained an unsolved and inextricable mystery.

7.1.2 *Lunar Riddles*

Lunar dates have formed part of the chronological discussion since Brugsch (1885) first identified a series of them in his early collation of astronomical texts, while Borchardt (1935) qualified their possible chronological interest. The main aspect of lunar dates resides in the fact that 25 Egyptian civil years of 365 days are almost exactly equal to 309 average synodic months of 29.5306 days, with a small difference of *c.* 1 h. In principle, this means that after 25 years, lunar events, including lunar days, almost repeat themselves in an apparent endless cycle. Similarly, 11 civil years contain nearly 136 average synodic months (the error can amount to one day on this occasion). Lunar dates can therefore also repeat in shorter cycles, although less precisely.

This implies that, within a certain chronological framework, lunar dates (or LDs as defined in Chap. 5) can help distinguish between different alternatives. But life is not always easy, and the quality of lunar dates is as varied as human thought can devise and lunar phases permit (see Figure 7.2). There is a long list of possibilities stressing the quality of lunar dates. These can be catalogued as follows:

(i) The mention of a secure lunar day (e.g. a *psḏntyw*) within the context of a complete civil calendar date, i.e. including year of reign, month, and day precisely. These are by far the best lunar dates for chronology if the king's reign can be fixed within an adequate chronological framework.

Fig. 7.2 A representation of the 14 days of the waxing moon as symbolized by certain important deities until completion of the Wadjet Eye, at full-moon, presented by Thoth. The list in the lowest register below the stairway of divinities stands for the 30 days of the idealized lunar month. (Image by the astrophotographer Juan Carlos Casado, by courtesy of the EU project STARS4ALL)

(ii) A secure lunar day within a civil calendar context, which includes month and day but without the year of the king. Both, (i) and (ii) can be catalogued as explicit. However, (ii) is more problematic when the king's reign was very long or when the chronological uncertainty is greater than *c.* 25 years. This is due, as we have argued, to the repetition of lunar dates with cycles of some 11 or 14 years, and more closely, in cycles of 25 years.

(iii) A date in which a lunar event is cited, but for which we do not know if the accompanying complete civil date precisely refers to it or to some earlier or later moment in the lunar cycle. This therefore depends on the Egyptologist's preferences. On some occasions, scholars tend to confuse these with type (i) lunar dates but their quality is certainly inferior.

(iv) The mention of a lunar feast, which was probably celebrated on a certain lunar day (or days) within the context of a complete civil calendar date. This can be catalogued as implicit whenever the LD-to-feast relationship is clearly established. The mentions to the [beautiful] Feast of the Valley, or *ḥb jnt*, are clear examples of these because they probably refer to either LD1 or, even better, LD2. Indisputable mentions of the movable *w3gy* feast (see Chap. 5) can also be included here.

(v) A civil date probably associated with a certain festival occurring at a plausible lunar date because of the archaeological context. This may work quite well when there are series of them. Certain implicit *ḥb jnt* dates, or alternative festivals, such as a hypothetical feast of Ptah in LD4 in the Saqqara region are included in this range. These can be catalogued as extrapolated lunar dates.

(vi) An explicit lunar date within the general context of the reign of a certain sovereign, but with no mention of an explicit civil date. This can be useful if a series of these days are known because a statistical approach is then possible.

(vii) Similar to (vi), when a series of temple services (or *wrš*) can be determined within a clear civil calendar context in an almost secure historical framework. A great number of Illahun lunar dates (see below) are of this type. Egyptologists tend to include them in quality levels (i) or (ii) because they implicitly accept the *wrš* operational aspects, as discussed in Chap. 5.

(viii) A presumably lunar date within a general context of a guessed Pharaoh reign including a complete civil date. Educated guesswork play a role here (see below, Sect. 7.8).

Registers show examples of all these types. However, chronologists rarely mention the quality of the lunar date they are dealing with, so readers can hardly make up their minds as to the precision or reliability of the chronological estimates they are deriving. Once again, however, this does not mean that lunar dates have not played a most important role in the chronological debate. Actually, as we shall see, they are of substantial interest and are even crucial in distinguishing between clashing alternatives.

Disappointingly, the astronomer Bradley Schaefer (2000a) has discarded the utility of lunar dates in the chronological debate. He, his colleagues, and a team of observers have established that at least 15% of any lunar visibility (or invisibility) observation would be wrong owing simply to human (not atmospheric) failure. For example, as a conclusion of their experiments they argue: 'Current large uncertainties in predicting lunar visibility and in ancient Egyptian (astronomical) procedures do not allow for any possible astronomical solution of Egyptian absolute chronology with lunar dates' (Schaefer & Doggett, 1994).

However, Krauss, 2015b, and references therein) has shown, indeed correctly from our point of view, that this contention is overly pessimistic because modern untrained observers cannot and should not be compared with the dedicated astronomers of ancient Egypt (see Chap. 2).

Krauss's main argument is that the date of a certain lunar phase (for instance the last crescent visibility of the waning moon) depends not only on the synodic month, but also on the anomalistic and draconitic months, which have different lengths and thus do not exactly repeat in cycles of 25 civil years. Consequently, the interplay of these and other factors makes the behaviour of the moon over every 25-year period quite different from one period to those immediately before and after, resulting in a quite complicated pattern if Egyptian lunar dates are shifted in 25-year intervals (apart from the cumulative 1 h shift over 25 years). Cycles of 11 civil years behave similarly but offer less precision.

A basic trait of the pattern shows that only about 70% of a set of lunar dates repeat in case of a shift of 25 Egyptian years. However, because 150 years comprise an approximately common period of the synodic, anomalistic, and draconitic months, more dates repeat at this interval, but with a cumulative relative error of a quarter of a day, which still leads to detectable differences in the patterns. It then follows from these facts that a large enough statistically significant set of lunar dates ought to yield only one correct solution, i.e. a solution in which at least 85% of the recorded dates are astronomically correct.

7.1.3 The Wandering Wepet Renpet and Its Relatives

The wandering nature of the civil calendar offers a new kind of potential astronomical dating which had little or no impact on Egyptology until the beginning of the twenty-first century. The works by the members of the Egyptian–Spanish Mission (Shaltout & Belmonte, 2005; Belmonte & Shaltout, 2006) and the preliminary approach by Luc Gabolde to the temple of Karnak open up a new horizon for exploration (Gabolde, 1998).

Civil festivals such a Wepet Renpet (New Year's Eve) and Nehebkau, at I Peret 1, among many other relevant days within the civil calendar, wandered through the seasons of the tropical year in a cycle close to *c.* 1507 years (see Chap. 5). After such a period of time, the civil dates would return to the same position within the orbit of the earth around the sun (see Figure 7.3).

This implies the existence of certain moments, or milestones in time, when the time-marks of a tropical year would coincide with relevant moments of the civil calendar. Indeed, the most important such event would be Wepet Renpet. Events

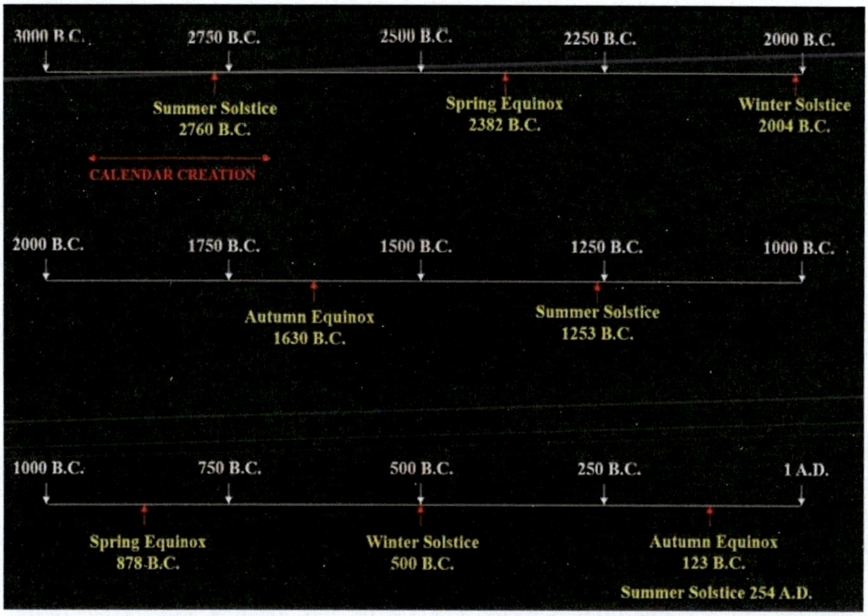

Fig. 7.3 Diagram showing the evolution of Wepet Renpet, or civil New Year's Eve, over time with respect to the corresponding seasons of the sun in the framework of the tropical year. Since the creation of the calendar in the early dynastic period, there were some important milestones in this evolution. See the text for further explanation. (Diagram by the authors)

like these happened during the 5th Dynasty, when for the first time in Egyptian history solar temples were systematically erected with a due-east orientation (and towards the spring equinox for a flat horizon) and furthermore orientated towards sunrise at New Year's Eve.

Even more suggestive is the fact that the *c.* 2004 BC sunrise at Wepet Renpet and the winter solstice coincided. This was at the beginning of the Middle Kingdom, when Thebes and its building solstitial alignments, including the great temple of Karnak, became important, as shown in Chap. 6. The same circumstance occurred in *c.* 500 BC under Persian rule.

A similar phenomenology eventually happened with the summer solstice of *c.* 1253 BC, during the reign of Ramesses II, when the civil and climatic seasons again coincided for the first time after the creation of the civil calendar. It is worth noting that this event was not linked to the simultaneity between Wepet Renpet and Peret Sopdet, which happened a few decades earlier because of the difference between the Sothic Year of *c.* 365¼ days and the tropical year of *c.* 365.2425 days in Antiquity.

This kind of astronomical dating is never explicit because there is not a single mention that a certain sacred building, temple, tomb, or shrine was orientated to sunrise at either a solar time-mark, such as a solstice or an equinox, or to Wepet Renpet or any other selective date of the civil calendar. Hence, the dates must be inferred from the data of the detected alignments. In Chap. 6 we discussed a few of these special events that will be appropriately recalled here. The credibility of any such date will strongly depend on how well justified and qualified the corresponding hypothesis is.

7.1.4 Aligning Stars

Finally, it is worth mentioning that, on certain occasions, astronomical alignments to stars and asterisms could be used to date a certain building whenever the relationship between star and structure could be clearly justified. This is of course related to precession.

However, the situation is not as clear as most specialists, including the authors, would desire. Actually, quite erroneous outcomes may come from badly justified assumptions. In the Introduction of this volume, we briefly discussed this topic in relation to the orientation of the Old Kingdom pyramids. As we show there, the topic is far from clear and there is not a single hypothesis that beats all the others, although we may have our own preferences.

Actually, the situation is seldom the other way round. If we have a temple with a certain presume stellar alignment and we know for sure the approximate date of the building, then we can justifiably claim that this alignment is real. The case of the temples of Hathor and Isis in Dendara and the risings of Alkaid and Sirius, respectively, as discussed in Chap. 6, would be an example of this. Indeed, if we had not had any idea of the temple builders, we would never have claimed an age for the temple based simply on these stellar alignments, even though texts may support such an assumption. Reality is more complex. We can use these data to infer cultural aspects but they should not be used in a chronological discussion.

Of course, there can always be a white blackbird when we have a temple or shrine of a certain deity (or related to him or her), whose celestial manifestation is well known and we do have an orientation to this celestial body (a star or asterism). In that case, we would be on firmer ground. However, we have not identified such a case so far. For example, we know that northern corridors of pyramids facilitate the ascent of the king's soul to the realm of the *Ikhemu Sekiu*, the Imperishable stars. However, this is a broad term that might include asterisms, such as the Plough (the Egyptian constellation of Meskhetyu) and individual stars very close to the pole, such as Thuban, or even bright stars such as Arcturus, the Egyptian Mooring Post (see Chap. 4).

We have identified four main typologies of astronomical dates which might help in establishing a coherent chronology. However, these would not have been necessary, provided the ancient Egyptians had developed a continuous system of time-reckoning from a delimited zero point, something they never, or even tried to, achieve.

7.2 Yearly Time-Reckoning in Ancient Egypt

To our misfortune, and the 'delight' of chronologists, the ancient Egyptians never used a linear dating system, with a certain origin or zero point, such as the one used today and that the West has exported to the rest of the world. For the Egyptians, the beginning of each new reign, the enthronement of a new Horus in the Double Land, was the beginning of an era, and the years began to be counted from that moment so that the year in which a king rose to the throne was year 1, and so on.

In particular, there is a significant lunar date of the reign of Ramesses II which can be illustrative of this phenomenology. The transcription in hieroglyphic writing is as follows (although the original text was written in the hieratic script):

,

Year 52, second month of Peret, day 27 in the House of Ramesses Meriamun, (is) psedjentiu [conjunction or LD1].

Here we can appreciate how the dates are recorded by the year of reign of the pharaoh, the season, the month and the day, as well as the information that this day was the date of the neomenia or psedjentiu.

Unfortunately, on some occasions, such as the following:

,

the name of the reigning pharaoh is not mentioned. It is perhaps possible to make an educated guess at which reign the text may refer to, but sometimes this is difficult as in the Sothic date of Jebel Tjautti. Besides, there are other occasions when the reign year appears but not the date itself, a fact which hinders the exact dating of the event.

To further complicate the issue, this dating system was not universal throughout Egyptian history, although it was practically the standard system from the New Kingdom onwards (Belmonte, 2012: 298). The same system, although the years of reign were counted from the date of the New Year and not the date of coronation, was in vogue during the Middle Kingdom. This means that, during that period, a full year was shared by a certain king and his deceased predecessor. Hence, the last year of a certain king was shared with year 1 of the successor, who counted year 2 from the following Wepet Renpet and not in the anniversary of his (seldom her) enthronement. This is a fact that must be taken into account in chronological analysis.

During the first dynasties, there is no proof that the years were counted from the coronation of the sovereign, although the partial first year was often called *sm3 t3wy*, Year of the unification of the country [by the Dual King]. However, there is certain evidence, such as that collected in the Stone of Palermo (see Figure 7.4), that each successive year was identified by a singular event. This important historical document also preserves the evidence that a certain count of various material elements—precious metals among them—was carried out in certain periods. This tally (*ṯnwt*) was produced in alternative years at least during the Second Dynasty (see Fig. 7.4).

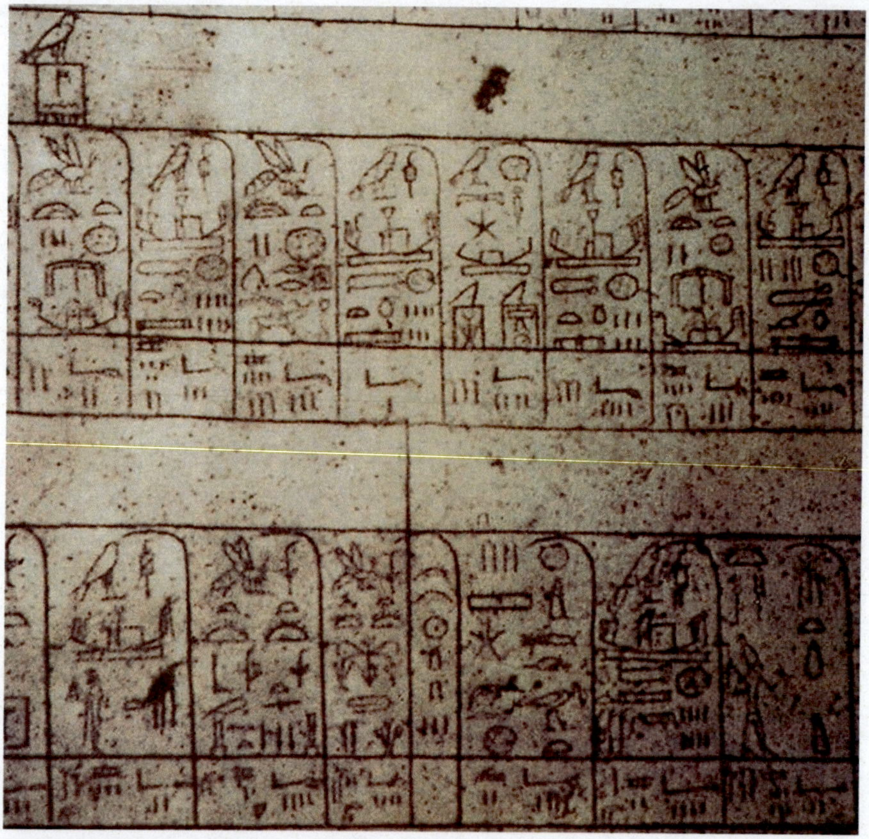

Fig. 7.4 The image shows a close-up of the Palermo Stone, in the sector occupied by the reign of Horus Ninetjer (upper row) and the beginning of the 3rd Dynasty (lower row). The upper row illustrates the consecutive system of counting in alternate years in columns 1, 3, 5, and 7, corresponding perhaps to years 10 (or 11 if the year of enthronement is separately counted) to 16 (17) of the reign of Ninetjer. The first register of the lower row shows a year identified by the 'creation' of a sculpture of King Khasekhemuy. Three registers later, a year is shared between this king and possibly his son Netjerkhet, the famous Djoser of later sources. (Image by the authors, courtesy of the Archaeological Museum of Palermo)

Perhaps with the arrival of the Old Kingdom, and certainly from the reign of Sneferu, the system was concretized in the so-called 'cattle count', where years of the count were often, but not always, followed by a year 'after'. Hence, we find years in the form:

year of the 14th occasion of the count (*rnpt sp ṯnwt*) of the cattle

and also in the form:

year after (*rnpt m-ẖt sp*) the sixteenth occasion.

The former formula was often abbreviated to 'year of the *n*th occasion'.

For decades it was thought that each 'occasion' was followed by a year after, so that the minimum years of reign for a given sovereign were calculated simply by doubling the greater of those two known counts. However, John Nolan's PhD work showed that this form of counting was not applied on a regular basis during the Old Kingdom, which suggests that the year after was counted only when an intercalary lunar month was needed (Nolan, 2003; he indeed is a supporter of Parker's Sirius-heralded lunar calendar, see Chap. 5). Thus, we find references, especially from Sneferu's own reign, that there were several instances of successive counts without intercalary years between them so that for each year 'after' there would be around three, eventually two, years of the 'occasion'. Therefore, the chronology of the Old Kingdom may well be overestimated.

As we shall discuss, Gautschy et al. (2017) have followed this hypothesis in a most interesting chronological analysis of the Old Kingdom. However, the number of years 'after' from King Djedkare Isesi to the end of the Old Kingdom during the reign of Pepy II almost equals the number of years of the 'occasion', strongly suggesting that a regular count was taken into account during the late 5th and the 6th Dynasties.

Krauss (2022a) has used this fact to produce an alternative chronological model for the Old Kingdom (see below). In short, all this is a huge headache with which Egyptologists dedicated to study the chronology have to deal before fixing the possible length of reign of a given ruler.

Once this is done, the most effective system of chronological dating, in the absence of an independent system, is what is called 'dead reckoning'. This is a term deriving from navigation, where it represents the determination of the distance travelled by a ship from the record of the courses sailed, without the direct aid of celestial navigation. These records can be estimated from velocity. The parallelism with Egyptology is astonishing.

The principle is simple, once the order of succession of the sovereigns has been established by independent and reliable methods, the years of individual reigns attested for them are added so that we can go back to the past. This system gives us a minimum interval for dating. This is the case because it is certain that there will not be records of all the reign years of a certain sovereign. Even if, although this is less likely, the information is so fragmentary that we do not know anything about a certain king attested by other sources. Egyptologist Kenneth Kitchen was the champion of dead reckoning, dismissing alternative 'scientific' methods as far-fetched (see Kitchen, 1993–2014).

However, with dead reckoning chronologists can only reach a minimum estimate that will have to be calibrated by other methods. The most reliable sources are royal lists, the genealogy of certain officials, synchronisms with the rulers of neighbouring countries, such as those of Mesopotamia or Anatolia, Kings of Hatti, Mittani, Assur, or Karduniash (that is Babylon), and others, archaeological stratigraphy, dating by carbon 14 (C14), and finally the most accurate astronomical dates (but whose reliability has often been questioned). The sources for these were clearly stressed earlier on this chapter.

The first three items may be classified as historiographical methods and have been pushed to unsuspected limits. The genealogy of officials can be useful as control information to confirm certain hypotheses about the duration of a given reign or the succession of a couple of contemporary kings concerning whom there are doubts, but it does not allow us to go much further. Much more important are the royal lists that have often been used to fix the reigns of certain problematic periods and which for certain epochs, such as the Second Intermediate Period (SIP), are fundamental. Among these we may highlight the lists of Abydos, Saqqara, and Karnak, and the Royal Papyrus of Turin.

It should be noted that the lists are highly politicized and often ignore the presence of sovereigns attested by other sources. For example, the table of Abydos ignores the presence of Shepsheskare Isi among the sovereigns of the 5th Dynasty (see Figure 7.5), the complete SIP, and all the sovereigns of the Amarna period, passing directly from Nebma'atre Amenhotep III to Djeserkheperure Horemheb. It also avoids the presence of female rulers, even if they were legitimized by contemporaneous sources, as in the case of Sobekkare Neferusobek, the last sovereign of the glorious 12th Dynasty.

The Royal Papyrus of Turin (Gardiner, 1959; see Figure 7.6) has been crucial in confirming the existence of certain groups of kings, such as Manetho's elusive 14th

Fig. 7.5 Extract from the table of Abydos showing part of the Old Kingdom (upper row) and the transition between the Middle and New Kingdoms (lower row). The listing of the kings of the 5th Dynasty ignores the reign of Shepsheskare [Isesi]. The table jumps directly from Ma'kherure [Amenemhat IV] to Nebphetyre [Ahmes] ignoring the whole SIP, including the female King Sobekkara. These omissions illustrate the limitations of this kind of source. (Image by the authors)

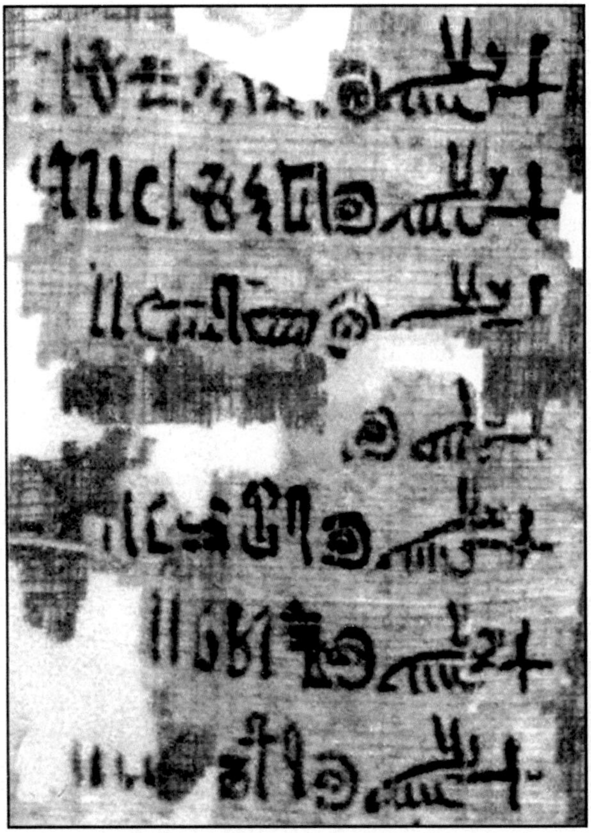

Fig. 7.6 Snapshot of the royal canon from the Royal Papyrus of Turin, presumably including the kings of Egypt up to the 19th Dynasty. This section of the royal papyrus, one of the best preserved, is of little use from the chronological point of view since it refers to reigns of little historical relevance of the SIP. The papyrus is unfortunately damaged in the parts covering most of the crucial periods of Egyptian history. (Image by the authors, by courtesy of Turin Egyptian Museum)

Dynasty, the existence of the Hyksos kings, or confirmation of the succession of certain monarchs (e.g. Ryholt, 1997). Among the royal lists, we may include some of an 'archaeological' nature such as the seals found in the necropolis of Abydos in Umm el Qab where the kings of the First Dynasty are collected in their respective order of succession (Kahl, 2006). Nor can we forget the fragments preserved in classical sources of Manetho's *Aegyptiaca*, the Egyptian priest who, despite his many inconsistencies, can still bring useful information and is, in addition, the basis of our system of counting the history of Egypt by dynasties (for these controversial classical sources, see Waddell, 1940, a comprehensive volume that has recently been re-edited).

Synchronisms with other rulers can be extremely useful if treated with due caution (e.g. Huber, 2011). In fact, there are researchers who give maximum weight to this type of information while disqualifying alternative solutions, such as C14 or even astronomy. For example, Kitchen (1993–2014) stated that 'science cannot solve in detail the intricate problems of the successions of Egypt's kings, and the chronological crossings with the neighboring regions of the Near East, only texts can do so'.

The Assyrian royal lists are quite reliable until the twelfth century BC and fix the reign of Assureshishi I, father of the great conqueror Tiglathpileser I, in 1132–1115 BC, with a margin of error of only two years (Krauss, 2015). For dates earlier than the twelfth and up to fifteenth centuries, things are no so clear. There are doubts about the reigns of Ashurdan I and his predecessor Ninurta-apil-ekur, which give us a margin of error of about 10 years. However, thanks to the synchronisms between Assyria and Babylon, the reign of Burnaburiash II in Karduniash (the name of the kingdom of Babylon in the Cassite era) can be fixed at *c.* 1349–1323 BC with a margin of ±5 years (Klinger, 2006). This king was a contemporary of both Amenhotep III and Amenhotep IV and this synchronism therefore allows an adequate upward correction of the dating system by dead reckoning, which at that time can amount to at least 20 years.

Two classical archaeometric methods (archaeological stratigraphy and C14 dating) offer only very general dating with relatively wide margins of error, and there are periods for which the corresponding information used to be contradictory. For example, there was a long-standing controversy between the data provided by the excavations at Tell el-Daba (site of ancient Avaris, capital of the Hyksos) the dates of the 18th Dynasty, and the dating of the explosion of the volcano at the island of Thera. Traditionally, this had been C14 dated at *c.* 1630 BC using the IntCal curve based on decadal and semidecadal blocks of tree measurements of bristlecone pines and European oak. This is the calibration curve often used for C14 dates in the Middle East, including Egypt. After the discovery of variabilities as large as 96 C14 years from one year to another, the paradigm ought to be upgraded. A new calibration curve was created for the 17th and 16th Dynasties based on annual tree ring measurements of California bristlecone pines and Irish oak, which has rendered a far better resolution for C14 dating (Pearson et al., 2018).

This new calibration curve has been applied to olive tree remains of the volcanic caldera and to the seeds from Akrotiri and offers a date for the eruption event between 1560 and 1540 BC, although even a date as low as 1510 BC would be acceptable. This offers a result much more compatible with the archaeological and historical evidence (Bietak, 2020). Hence, calibrated C14 dating cannot be considered as an exact science and needs to be subjected to improvements in the data and/or the analysis techniques. Before Present (BP) dates have a certain error margin, like any lab experiment, but are based on physics. However, the process of calibration needs a series of intermediate steps, including certain premises that have a greater degree of uncertainty and thus should be considered with a great deal of caution. This will be most important when we discuss the C14 dating of Egyptian remains. As Manfred Bietak has argued, this is especially meaningful considering recent attempts to raise the Bronze Age chronology by C14 dating alone (see below).

We finally arrive at astronomical dating in our search for a chronological framework. As we showed earlier, this method includes the so-called lunar dates, associated with the atavistic Egyptian custom of paying attention to the phases of the moon for the celebration of festivals, and the Sothic dates, related to the heliacal rising of Sirius, the Feast of Peret Sopdet. We shall see how the combination of both can be decisive in periods, such as the Middle Kingdom, when the margin of error of dating by dead reckoning is an insurmountable obstacle. The so-called archaeo-astronomical dating that we have been expounding throughout this work (notably in Chap. 6) will also be briefly discussed in relation to either stellar orientations, particularly those of Meskhetyu, or those related to the wandering nature of the Egyptian civil calendar and thus the New Year's Eve displacement on the tapestry of the seasons. These can be fundamental for periods when dead reckoning is of little use—except for the internal organization of the period itself—as is the case of the Old Kingdom.

But, before going into the study of a few study cases of these suggestive examples of how astronomy is useful to Egyptology, we give a few broad brushstrokes of those periods of Egyptian history in which the chronology is more controversial and for this we make use of Table 7.1 as a common thread. Table 7.1 contains suggestive, although occasionally outdated, information. Unlike in most Egyptology manuals where a certain chronology favoured by the author is chosen, here we present the most accepted and discussed chronologies of the last quarter century.

Table 7.1 includes the chronology of Jürgen von Beckerah, favoured by German scholars, and that of Ian Shaw, of Oxford University, preferred by many Anglo-Saxon researchers and the one deliberately selected in Bayesian approaches to C14 data by a lab project of the same university (see below). Finally, the table presents the chronology collected in their *Handbook of Egyptian Chronology* by the tandem formed by Eric Hornung, Rolf Krauss, and David Warburton (hereafter HKW); while revolutionary in many aspects, it is extremely controversial in others.

There are several periods of Egyptian history when the three chronologies differ widely. The beginning of 25th Dynasty King Taharqa's reign in 690 BC is the fixed point of all of them, and divergences are minimal until the beginning of the reign of the first sovereign of the Libyan dynasties Shoshenq I, between 943 and 948 BC. This discrepancy does not include the inner part of the period, where discrepancies can be substantial. Finally, the differences can be up to a decade for the end of the New Kingdom and the seizure of power of the 21th Dynasty, with Smendes in Tanis and Herihor in Thebes. This epoch, the beginning of the Third Intermediate Period (TIP) is a fascinating moment in Egyptian history, which is the specialty of one of the authors (Lull, 2002).

Another date that has tended to be considered a fixed milestone in the New Kingdom is year 1 of Ramesses II, established at 1279 BC by a set of astronomical data and calculations (see Fig. 7.7), while the most popular dates some years ago, such as 1290 BC (see Table 7.1) and 1304 BC had been brought into question. The former is questioned for not adjusting to every astronomical estimate of the 19th and 20th Dynasties, and the latter because it would mean that a total of 28 years would be absent from historical records, which seems impossible, and is in serious conflict with dead reckoning. However, in the last decade, some leading scholars, such as Peter Huber, David Aston, and Rita Gautschy have combined all their ample

Table 7.1 Comparative Chronology of Ancient Egypt, from the Proto-Dynastic Period to the Roman conquest, from three sources. B: **Jürgen von Beckerath**, *Chronologie des Pharaonischen Ägypten* (Mainz, 1997). S: **Ian Shaw** (editor), *The Oxford history of ancient Egypt* (Oxford, 2000). HKW: **Eric Hornung, Rolf Krauss and David A. Warburton** (editors, hereafter HKW), *Ancient Egyptian chronology* (Leiden, 2006). There is an additional one for the Ptolemaic Period, VV: V. Vanoyeke, *Les Ptolémées, derniers pharaons d'Egypte* (Paris, 1998). There has not been a completely new chronology produced in the last decade, but only individual approaches for various periods. Hence: (1) the Old Kingdom is still uncertain; (2) the dichotomy for the Middle Kingdom is not solved; and (3) all three chronologies presented here for the New Kingdom are outdated, and certainly wrong for most of the 18th Dynasty. Dates for SIP and TIP are not discussed in this volume. The only reliable dates are after Taharqa's reign in the 25th Dynasty

	B (1997)	S (2000)	HKW (2006)
1st Dynasty		**3000–2890**	
Narmer			2900–?$^{+25}$
Aha	2982–2950^{+50}	—	?–2870^{+25}
Iti	2950–2949^{+50}		
Djer	2949–2902^{+50}	—	2870–2823^{+25}
Djet	2902–2889^{+50}	—	2822–2815^{+25}
Merneith		—	
Den	2889–2842^{+50}	—	2814–2772^{+25}
Adjib	2842–2836^{+50}	—	2771–2764^{+25}
Semerkhet	2836–2828^{+50}	—	2763–2756^{+25}
Qa'	2828–2803^{+50}	—	2755–2732^{+25}
2nd Dynasty		**2890–2686**	
Hotepsekhemuy	2803–2775^{+50}	—	2730–?$^{+25}$
Nebre/Raneb	2775–2760^{+50}	—	?–2700^{+25} (= X?)
Ninetjer	2760–2717^{+50}	—	2700–2660^{+25}
Wenedj	2717–2710^{+50}	—	(= X?)
Senedj	2710–2699^{+50}	—	? –2610^{+25} ↓↓
Neferkare (I)	2699–2694^{+50}		
Neferkasokar	2694–2686^{+50}		
Hudjefa (*lacuna*)	2686–2684^{+50}		
Sekhemib	Eq. to Senedj		2650–?$^{+25}$ ↓
Peribsen	—	—	2660–2650^{+25}↑
Khasekhemuy	2684–2657^{+50}	—	2610–2593^{+25}
3rd Dynasty			
Netjerkhet [Djoser]	2640–2620^{+50}	2667–2648	2592–2566^{+25}
Sekhemkhet	2620–2613^{+50}	2648–2640	2565–2559^{+25}
Khaba	Id. to Huni	2640–2637	2559–?$^{+25}$
Sanekht Nebka	2657–2640^{+50} ↑↑↑	2686–2667 ↑↑↑	?–? $^{+25}$
Huni	2613–2589^{+50}	2637–2613	? –2544^{+25}
4th Dynasty			
Sneferu	2589–2554^{+50}	2613–2589	2543–2510^{+25}
Khufu	2554–2531^{+50}	2589–2566	2509–2483^{+25}
Djedefre	2531–2522^{+50}	2566–2558	2482–2475^{+25}

(continued)

Table 7.1 (continued)

	B (1997)	S (2000)	HKW (2006)
Bicheres (Bikka/Setka)	2496–2489^{+50}↓		2474–2473^{+25}
Khaefre	2522–2496^{+50}↑	2558–2532	2472–2448^{+25}
Menkaure	2489–2461^{+50}	2532–2503	2447–2442^{+25}
Shepseskaf	2461–2456^{+50}	2503–2498	2442–2436^{+25}
Dejedefptah	2456–2554^{+50}		
Khentkaus			---
5th Dynasty			
Userkaf	2454–2446^{+50}	2494–2487	2435–2429^{+25}
Sahure	2446–2433^{+50}	2487–2475	2428–2416^{+25}
Neferirkare Kakai	2433–2413^{+50}	2475–2455	2415–2405^{+25}
Shepseskare Isi	2413–2406^{+50}	2455–2448	2403^{+25} ↓
Neferefre	2406–2395^{+50}	2448–2445	2404^{+25} ↑
Neuserre Ini	2395–2364^{+50}	2445–2421	2402–2374^{+25}
Menkauhor	2364–2355^{+50}	2421–2414	2373–2366^{+25}
Djedkare Isesi	2355–2317^{+50}	2414–2375	2365–2322^{+25}
Unas/Wenis	2317–2297^{+50}	2375–2345	2321–2306^{+25}
6th Dynasty			
Teti	2297–2287^{+50}	2345–2323	2305–2279^{+25}
Userkare	2287–2285^{+50}	2323–2321	?–? $^{+25}$
Pepy I	2285–2235^{+50}	2321–2287	2276–2228^{+25}
Merenre	2235–2229^{+50}	2287–2278	2227–2217^{+25}
Pepy II	2229–2169^{+50}	2278–2184	2216–2153^{+25}
Nemtiemsaf II	2169–2168^{+50}		2152^{+25}
Nitocris/Nitiqret	2168–2166^{+50}	2184–2181	
7th Dynasty			
8th Dynasty	**2166–2120^{+50}**	**2181–2160**	**2150–2118**
Netjerkare			
Menkare			
Neferkare (II)			
Neferkare Neby			
Djedkare Shemay			
Neferkare Khendu			
Merenhor			
Nikare			
Neferkare (V)			
Neferkauhor (I)			
Neferkare (VI)			
Neferkamin Anu			
Qakare Ibi			
Neferkaure			2126–2123^{+25}
Neferkauhor (II)			2122–2120^{+25}
Neferirkare			2119–2118^{+25}

(continued)

Table 7.1 (continued)

	B (1997)	S (2000)	HKW (2006)
9th & 10th Dyns.	**2120^{+50}–2020^{+5}**	**2160–2025**	**2118–1980^{+25}**
Khety I Meribre		—	
Neferkare (VII)			
Khety Uahkare		—	
Senenen ?			
Khety Neferkare			
Khety Meri…			
Meyhathor			
Khety III Nebkaure		—	
Merikare		—	—
Iti			
11th Dynasty			
Mentuhotep I	2119–?		2080–$?^{+16}$
Intef I	? –2103	2125–2112	? –2067^{+16}
Intef II	2103–2054	2112–2063	2066–2017^{+16}
Intef III	2054–2046	2063–2055	2016–2009^{+16}
Mentuhotep II	2046–1995	2055–2004	2009–1959^{+16}
Mentuhotep III	1995–1983	2004–1992	1958–1947^{+16}
Mentuhotep IV	1983–1976	1992–1985	1947–1940^{+16}
12th Dynasty			
Amenemhat I	1976–1947	1985–1956	1939–1910^{+16}
Senuseret I	1956–1911/10	1956–1911	1920–1875^{+6}
Amenemhat II	1914–1879/76	1911–1877	1878–1843^{+3}
Senuseret II	1882–1872	1877–1870	1845–1837
Senuseret III	1872–1853/52	1870–1831	1837–1819
Amenemhat III	1853–1806/5	1831–1786	1818–1773
Amenemhat IV	1807/6–1798/97	1786–1777	1772–1764
Neferusobek	1797–1793	1777–1773	1763–1760
13th Dynasty	**1793–1645^{+3}**	**~1773–1650**	**~1759–1630**
Wegaf		—	~1759–1757
Senbuf			
Nerikare			
Amenemhat V			
Amenemhat VI			
Nebnuny			
Iufny			
Hornedjeriotef			
Swadjkare			
Nedjemibre			
Amenikemau			
Sobekhotep I		—	
Reniseneb			

(continued)

Table 7.1 (continued)

	B (1997)	S (2000)	HKW (2006)
Hori		—	
Amenemhat VII		—	~1753–1748
Sobekhotep II			~1737–1733
Khendjer		—	~1732–1728
Imy-ir-mesha			
Intef V			
Meribre Set			
Sobekhotep III		—	~1725–1722
Neferhotep I		—	~1721–1710
Sihathor		—	
Sobekhotep IV		—	~1709–1701
Sobekhotep V		—	~1700–1695
Sobekhotep VI			
Uahibre Ibiau			~1695–1685
Merneferre Ay		—	~1684–1661
Merhetepre Ini			~1660–1651
Mentuemsaf*			
Djedumose (I–II)*			
Sobekhotep VII			
Neferhotep II(I)			
Mentuhotep V			
Senebmiu			
14th Dynasty	**?–1645^{+3}**	**~1773–1650**	**?**
15th Dynasty		**~1650–1550**	**~?–1530**
Salitis/Sekerher	1645^{+3}	—	
Bnon	—		
Apachnas	–1587^{+3}		
Semqen			
Aper-Anat			
Sheshi			
Khyan		*c.* 1600	
Apophis	1587–1546^{+3}	*c.* 1555	~1575–1540
Khamudi	1546–1536^{+3}		
16th Dynasty	Lesser Hyksos	**~1650–1580**	
Djehuty		Theban kings	
Sobekhotep VIII*			
Neferhotep III			
Mentuhotepy			
Nebiriau I*			
Nebiriau II*			
Smenre			
Bebiankh*			

(continued)

Table 7.1 (continued)

	B (1997)	S (2000)	HKW (2006)
Shedwast			
Djedumose I*			
Djedumose II*			
Mentuemsaf*			
Mentuhotep VI			
Senuseret IV			
17th Dynasty	**1645–1550**	**~1580–1550**	**?–1540**
Intef V			—
Sobekhotep VIII*			—
Nebiriau (I & II)*			—
Bebiankh*			—
Rahotep		—	—
Sobekemsaf I		—	—
Intef VI		—	
Intef VII		—	
Intef VIII		—	
Sobekemsaf II		—	—
Tao I Senakhtenre		—	—
Tao II Sekenenre		*c.* 1560	—
Kamose		1555–1550	?–1540
18th Dynasty			
Ahmose	1550–1525	1550–1525	1539–1515
Amenhotep I	1525–1504	1525–1504	1514–1495
Thutmose I	1504–1492	1505–1492	1493–1483
Thutmose II	1492–1479	1492–1479	1482–1480
Thutmose III	1479–1425	1479–1425	1479–1425
Hatshepsut	1479/73–1458	1473–1458	1479–1458
Amenhotep II	1428–1397	1427–1400	1425–1400
Thutmose IV	1397–1388	1400–1390	1400–1390
	See Table 7.5		
Amenhotep III	1388–1351/50	1390–1352	1390–1353
AIV/Akhenaten	1351–1334	1352–1336	1353–1336
Semenkhkare	1337–1333	1338–1336	1336–1334
Neferneferuaten			1334–?
Tutankhamun	1333–1323	1336–1327	?–1324
Ay (II)	1323–1319	1327–1323	1323–1320
Horemheb***	1319–1292	1323–1295	1319–1292
19th Dynasty			
Ramesses I	1292–1290	1295–1294	1292–1291
Sethy/Seti I	1290–1279	1294–1279	1290–1279
Ramesses II	1279–1213	1279–1213	1279–1213
Merenptah	1213–1203	1213–1203	1213–1203

(continued)

Table 7.1 (continued)

	B (1997)	S (2000)	HKW (2006)
Sethy II	1200–1194 ↓	1200–1194 ↓	1202–1198
Amenmose	1203–1200 ↑	1203–1200 ↑	1202–1200
Siptah	1194–	1194–1188	1197–1193
Tousert /Tousre	–1186	1188–1186	1193–1191
20th Dynasty			
Sethnakht	1185–1182^{+1}	1186–1184	1190–1188
Ramesses III	1182–1151^{+1}	1184–1153	1187–1157
Ramesses IV	1151–1144^{+1}	1153–1147	1156–1150
Ramesses V	1144–1140^{+1}	1147–1143	1149–1146
Ramesses VI	1140–1132^{+2}	1143–1136	1145–1139
Ramesses VII	1132–1123^{+3}	1136–1129	1138–1131
Ramesses VIII	1123–1121^{+4}	1129–1126	1130
Ramesses IX	1121–1103^{+4}	1126–1108	~1129–1111
Ramesses X	1103–1099^{+4}	1108–1099	~1110–1107
Ramesses XI	1099^{+4}–1069^{+1}	1099–1069	~1106–1077
21st Dynasty			
Herihor			—
Smendes	1069–1043^{+1}	1069–1043	~1076–1052
Pinedjem			—
Amenemnisut	1043–1039^{+1}	1043–1039	~1005–1002 ↓
Psusennes I	1043–993^{+1}	1039–991	~1051–1006 ↑
Amenemope	995–984^{+1}	993–984	~1002–993
Osorkon the Elder	984–978^{+1}	984–978	992–987
Siamun	978–959^{+1}	978–959	986–968
Psusennes II	959–945^{+1}	959–945	~967–944
22nd Dynasty			
Shoshenq I	945–924^{+1}	945–924	943–923
Osorkon I	924–890	924–889	922–888
Takelot I	~890–877	889–874	~887–874
Shoshenq II	~877–875	*c.* 890	~873
Osorkon II	~875–837	874–850	~872–842
Takelot II*		850–825	
Shoshenq III	~837–798–785?	825–773	841–803
Shoshenq IIIa	~798–785		?–790
Pamay	~785–774	773–767	789–784
Shoshenq V	~774–736	767–730	783–746
Petubast II*			
Osorkon IV*/V		730–715	
23rd Dyn. UE-A			
Harsiese	~870–850		
Takelot II*	~841–816		845–821
Petubast I*	~830–800^{+5}		

(continued)

Table 7.1 (continued)

	B (1997)	S (2000)	HKW (2006)
Iuput I	~816–800	—	820–809
Shoshenq IV*	~800^{+5}–790		
Osorkon III	~790–762	—	~780±
Takelot III	~767–755	—	—
Rudamun*	~755–735		
Ini*	~735–730		
Pefdjauauibast*			
23rd Dyn. UE-B			
Petubast I*		—	834–812
Shoshenq IV*		—	—
Rudamun*		—	—
Ini*			—
Iuput II	~756–725 ?	—	
23rd Dyn. ME			
Pefdjauauibast*		—	—
23rd Dynasty LE			
Petubast II*	~756–730^{+2}		—
Osorkon IV*	~730–722		—
"Psammus"	~722–712		
24th Dynasty			
Tefnakht	740–717^{+2}		736–729
Bekerenef	717–712^{+2}	720–715	728–723
25th Dynasty			
Kashta	*c.* 746		
Piye	746–713^{+2}	747–716	753–723
Shabaka	713–698^{+2}	716–702	722–707
Shabitku	698^{+2}–690	702–690	706–690
Taharqa	690–664	690–664	690–664
Tanwentamani	664–655	664–656	664–655
26th Dynasty			
Nekao I		672–664	
Psametik I	664–610	664–610	664–610
Nekao II	610–595	610–595	610–595
Psametik II	595–589	595–589	595–589
Apries	589–570	589–570	589–570
Amasis	570–526	570–526	570–526
Psametik III	526–525	526–525	526–525
27th Dynasty			
Cambyses	525–522	525–522	525–522
Darius I	521–485^{+1}	522–486	521–486
Xerxes	485–464^{+1}	486–466	486–466
Artaxerxes I	464^{+1}–424	465–424	465–424

(continued)

Table 7.1 (continued)

	B (1997)	S (2000)	HKW (2006)
Xerxes II	424–423		
Darius II	423–404[+1]	424–405	424–404
Artaxerxes II	404–401(–358)[+1]	405–359	
28th Dynasty			
Amyrtaios	404–399	404–399	404–399
29th Dynasty			
Nepherites I	399–393	399–393	399–393
Psammuthis	393		393
Haker/Achoris	393–380	393–380	393–380
Nepherites II	380	380	380
30th Dynasty			
Nectanebo I	380–362	380–362	380–362
Teos	364–360	362–360	365–360
Nectanebo II	360–342	360–343	360–343
2nd Persian			
Artaxerxes III	342–338	343–338	343–338
Arses	338–336	338–336	338–336
Khababash (local)	337–335[+1]		
Darius III	336–332–(330)	336–332	335–332
Macedonians			
Alexander III		332–323	332–323
Philippos III		323–317	
Alexander IV		317–310	
Ptolemaic Dyn.			**VV**
Ptolemy I Soter I		305–285	305–283
PII Philadelphus		285–246	283–246
PIII Evergetes I		246–221	246–221
PIV Philopator		221–205	221–195
PV Epiphanes		205–180	193–181
PVI Philometor		180–145	173–145
PVII Evergetes II		170–116 (VIII)	169–116
PVIII Eupator		145 (VII)	
PIX Soter II		116–107 88–80	116–104 88–80 (VIII)
PX Alexander I		107–88	104–88 (IX)
Cleopatra			
PXI Alexander II		80	80 (X)
Berenike III			
PXII Auletes		80–51	80–51 (XI)
Berenike IV			
Ptolemy XIII		51–47	51–49 (XII)
Cleopatra VII		51–30	51–30

(continued)

Table 7.1 (continued)

	B (1997)	S (2000)	HKW (2006)
Ptolemy XIV		47–44	49–44 (XIII)
PXV Caesar		44–30	47–30 (XIV)
Upper Egypt			
Herwennefer			
Ankhwennefer			
Harsiese			
Roman Period			
Augustus		30– AD 14	

Notes:
—: Mentioned in the source but not assigned dates
*: Same king according to this source
↑ (↑): In upper position(s) with respect to the listed one
↓(↓): In lower position(s) with respect to the listed one
***: It is practically certain now that Horemheb would not have reigned for more than 15 years. Therefore, **all the chronologies of the 18th Dynasty must be reduced by between 11 and 14 years at least from the reign of Amenhotep III**. For a new chronology of the period, see **Table** 7.5

Fig. 7.7 From left to right, three kings of ancient Egypt whose astronomical dates are fundamental for discussions on Egyptian chronology. In all cases, top quality lunar data are available. These are: temple accounts at Illahun for Amenemhat III in the Middle Kingdom, a report of the battle of Meggido for Thutmose III, and a 'new moon' report at Pi-Ramesses for Ramesses II, during the New Kingdom. Sothic data for Amenemhat's father Senuseret III, for Thutmose III, and for Ramesses II are also relevant. Unfortunately, such precise evidence does not exist for the Old Kingdom. (Images by the authors, courtesy of the Museums of Munich, Luxor, and Turin, respectively)

knowledge in support of earlier chronologies. Huber (2011) begs for an unrealistically high chronology, settling an excessive trust in C14 Bayesian approaches. Aston (2013) goes back to 1290 in his most interesting discussion of wine sherds and C14, where he defends the maxim *in vino veritas*. Finally, Gautschy (2014) has performed a dedicated analysis with different alternatives and has chosen that model which situates Ramesses II year 1 in 1304 BC.

However, as we show later on (see Sect. 7.8), there are a series of impressive events supporting a low chronology for the Amarna period and the early 19th Dynasty. Hence, we agree with Krauss (2015a, 2016) and choose to accept 1279 BC as year 1 of Ramesses II, even if the odds are in favour of 1290 BC.

Doubts about the end of the 18th Dynasty suddenly appeared at the Xth International Congress of Egyptologists, held on the island of Rhodes in May 2008. There, Jacobus van Dijk reported—to the astonishment of the public and the authors—the discovery of a significant number of wine jar labels, or sherds, mentioning Horemheb's year 14 in the re-excavation of the interior of his tomb in the Valley of the Kings (van Dijk, 2009). No higher dates were reported. These jars were probably deposited at the time of his burial. Hence, this pharaoh's reign would have been 14 or at most 15 years long, much less than that hitherto assigned by most specialists. This obviously means that the dates of the 18th Dynasty must be significantly altered downwards. Hence, the chronology presented for the period in Table 7.1 would be inconsistent and certainly unrealistic.

A posteriori, several important chronological implications of this result will be analysed, showing that new findings can substantially alter the variables of a gear that seems firm.

The Second Intermediate Period (SIP) has its own discrepancies and lacunae owing to the absence of decisive information (Jebel Tjautti Sothic date is an example). However, it is during the Middle Kingdom, when relative information is again abundant, that the greatest discrepancies occur, with a high and a low chronology, depending on how the documentary information of the reigns of Senuseret III and Amenemhat III (see Figure 7.7) is interpreted. This issue is so important that we will devote an entire case study to it (Sect. 7.4).

On the other hand, the chronology of the Old Kingdom can be divided into a set of varied options of a high, medium, low, or very low chronologies This can locate the beginning of Khufu's reign as high as 2636 BC (Gautschy et al., 2017), or as low as 2470 BC (Dodson & Hilton, 2004: 288; see also Table 7.1), respectively. We shall soon come back to this suggestive controversy.

Finally, considering the enormous period of time elapsed, and the shortage of original sources, the discrepancies for the reign of Narmer and the beginning of Pharaonic history are not that substantial. Thus, the beginning of the proto-Dynastic period, the Thinite Dynasties, can be fixed around the year 3000 BC with a margin of approximately a century.

It is beyond the scope of this chapter to make a detailed analysis of every single period of Egyptian history. Several points have been tirelessly discussed and we do not want to bore the reader with endless discussion on such and such a Sothic, lunar, or C14 date, among others. This would probably need to be the subject of a dedicated specialized book. Hence, in this volume we restrict ourselves to a series of most interesting case studies that can give some hints to periods of special relevance in Egyptian history.

For example, despite the personal interest of the authors, we do not discuss the importance of the Feast of Tepy Shemu and the devilish chronology of the TIP

(Gautschy, 2015; Krauss, 2015). This is a fascinating period of Egyptian history, which, together with the SIP, can provoke more than one headache to the dedicated scholar. We shall instead concentrate in the elusive Old Kingdom lunar dates, the much-debated Middle Kingdom Sothic and lunar dates, and the varied nature of solar, lunar, and Sothic dates of the New Kingdom. We shall indeed discuss how the resulting chronologies fit with C14 analyses.

7.3 Astronomical Dates of the Old Kingdom and the Cattle Census Count

The Old Kingdom is a sort of nightmare with regard to its chronology. Manetho is basically useless, as in most periods of Egyptian history before the 21th Dynasty (Waddell, 1940). Apart from the 1st Dynasty, where the nesu-bity or nebty names of the kings were reported, and the preservation of otherwise unattested kings for other dynasties, one would better ignore his information for chronological purposes. The monumental royal lists provide long series of names and can be use with caution.

The Turin canon is only partially preserved, but it still offers a few hints, including information about the reign lengths of several sovereigns. The problem is that the data compiler was possibly puzzled by the cattle count census—as indeed we are (see below)—and we do not know how this person obtained the registers, whether from unknown direct sources, counting occasions as year counts, or doubling the maximum number of occasions to get a round number. Hence, we do not know the exact reign years of such important kings as Netjerkhet, Sneferu, Khufu, or even Pepy II, the pharaoh with the longest attested reign in Egyptian history according to Manetho (see Table 7.1 to check this dispersion). We still do not know if the king that saw the decline of the Old Kingdom was on the throne for a period longer than the also long-lived Ramesses II, i.e. at least 66 years. Radiocarbon dating also has problems as we shall show later on, and Dee's (2013d) dates are possibly too high.

Hence, the best evidence we have to get any clues are the astronomical dates. Although controversial, these may offer some hints of at least a general chronological framework for the period. There a few of varied nature and quality:

(a) Sothic date n° 1 (see Chap. 5). This is controversial in nature as we have seen. Even if the best option for a date, IV Akhet 1 (Habicht et al. 2015), were the correct one, the context of the piece and its finding in a private collection, renders it as quite useless for chronological purposes. To begin with, no king's name is preserved—or may be guessed at—and still less a reign year, although it was tentatively assigned to a 5th Dynasty king, perhaps Niuserre Ini.
(b) Two lunar dates implicit in the temple registers of the 5th Dynasty King Neferirkare Kakai temple at his pyramid complex in Abusir (Posener-Kriéger, 1976). Names of the kings and year dates are not explicit but can be derived from the context.

(c) A hypothetical mention of the lunar Wagy Feast in a papyrus fragment from the archive of King Neferefre, Kakai's son and heir, at his pyramid complex, also in Abusir (Posener Kriéger et al., 2006). The 'related' date was read by the excavators as III [?] 28, where the name of the season is not preserved but could only be either Akhet or Peret (see Chap. 5).
(d) The free interpretation of a *phyle* (i.e. a priest gang) roster date in Neferefre's archive as a first lunar date. A date of the 18th occasion, III Shemu 30, has been connected to a report 'when one phyle finishes a turn of duty and the other enters into its monthly service' (Krauss, 2022a). This has been assigned to the reign of Djedkara, the only verified king of the period to having reigned for so long.
(e) Three extrapolated lunar dates from those mentioned in royal decrees at the village of Coptos issued by Pepy II (two of them) and a barely attested king of the 8th Dynasty, Neferkauhor. Krauss (2022a) has conjectured that these dates were full-moon days, or LD15, precisely.
(f) The hypothesis of John Nolan (2015), where he suggests that a year after the occasions was only intercalated in the list following a lunar cycle, when an additional intercalary lunar moth was needed in the Sothic-based lunar calendar. Please remember that the authors do not endorse the existence of this phantasmagoric calendar. This year intercalation process might work following Parker's rule in his *Calendars*, or as Nolan now suggests, following the Metonic cycle of 19 years. To our knowledge, there is not single hint that ancient Egyptians ever knew this cycle prior to the Hellenistic period. Nolan's hypothesis would include a year after the occasion every three—rarely two—calendar years.
(g) The use of solar and stellar alignments as time-markers for dating selected events in the pyramid age (see the Introduction and Chap. 6). Belmonte (2012: 169–205; and references therein) made a full summary of all possibilities. We shall analyse some of these proposals later.

Rita Gautschy and her collaborators have produced a very sophisticated model for the period between Khufu and the end of the Old Kingdom (Gautschy et al., 2017). This was done with clues (a), (b) and (c), and by considering as presumably correct Nolan's hypothesis (f). It is worth noticing that only 105 dates of years of the occasion and years after have been preserved for a period of, at least, 350 years, less than 30 per cent. Still, these could be considered as an appropriately statistically significant sample to play with.

The main hypothesis of this fascinating research is that after a year 'sema-tawy', i.e. the fraction of the year when a king ascended to the throne, the following year could be counted as *rnpt m-ḫt sp tpy*, i.e. year after the first occasion, differentiating this from a year *rnpt m-ḫt sp 1*. This means than the adjective Tepy and the numeral 1 would have different meanings for this particular case, which is problematic. This strong hypothesis is needed for the model to work, but the existence during the reign of Djedkare of a *rnpt tpy*, perhaps following his year sema-tawy, impairs the strength of the hypothesis.

In any case, if this and Nolan's hypotheses are assumed, the Gautschy et al. model would offer two different possibilities for a chronology of the late Old Kingdom. These alternatives rely on the two different possibilities for the season of the Wagy Feast, i.e. clue (c): III Peret 28 or III Akhet 28. These render two distinct time frames: a Low and a High Chronology, respectively (Gautschy et al., 2017: Table 3). The model correctly reproduces 84% and 82% of years after of the cattle count, for the Low and High chronology, respectively. Truly a great achievement.

For this, other assumptions had to be considered such as the possibility of different reign lengths for the two models. A well-known king such as Pepy I is given 44 or 49 years, respectively. Problematic kings such as Userkare received either two, for the High Chronology, or even zero (a few months or even weeks) for the Low one. Another assumption is the presence of unattested sovereigns in the list, such as Manetho's elusive Thamphitis, who is given 4 and 2 years, for the Low and High chronology, respectively. All this may sound like some sort of selection effect, but as Gautschy has often argued, it is just a model, a working hypothesis, not the absolute truth.

An option apparently favoured by Gautschy et al. would be the High Chronology, closer to Jurgen von Beckerath's upper limit (see Table 7.1), perhaps because it is also closer to the upper boundary defined by the C14 Old Kingdom Bayesian Model, which offers the interval 2629–2558 BC for Khufu year 1 (Dee, 2013d). We shall discuss the pros and cons of C14 dating for this period later on. The High Chronology model locates Khufu's year one in 2636 BC, Djedkare's in 2468 BC, and Pepy II's in 2334 BC. This would appear to render too long a duration for the FIP of *c.* 170 years. However, they argue that, if a strictly biennial count is admitted for the 6th Dynasty (see below), this period could be lowered.

The Low Chronology model does not agree with any of the early chronologies (see Table 7.1) and would need a very short duration of 30–60 years—depending on Middle Kingdom dates—for the FIP, including the early years of Mentuhotep II before the conquest of Herakleopolis. However, this would offer a marginal chance for Spence's (2000) hypothesis for the orientation of the pyramids.

Rolf Krauss has recently questioned the model of Gautchy's team. Leaving aside the sema-tawy years—and the hypothesis that this could be followed by a *rnpt m-ḫt sp tpy*—ratios between existing years of the occasions and years after ascent to 17/9 for the 4th Dynasty (without Sneferu's reign), 30/15 for the 5th Dynasty, perhaps suggesting a year after every two years of the occasion, and 16/14 for the 6th Dynasty. This last ratio strongly suggests a biennial count during this period and perhaps even earlier. With data for the 5th and 6th Dynasties at hand, Nolan's (2003, 2015) intercalation model should hence be quarantined.

Importantly, Krauss (2021) has particularly questioned the value of the lunar date of the Wagy Feast. After a collation of the original image and not from Posener Kriéger's rendering of the event, Krauss realized that the date III [?] 28 is located in a different line from Wagy, which is instead preceded by the Feast of Thoth, strongly suggesting a civil feast framework (see Fig. 5.9). Hence, clue (c) should not be used for chronological purposes.

Bearing this in mind, Krauss (2022a) has developed another model for late Old Kingdom chronology based on evidence as reported in clues (b), (d), and (e), i.e. in implicit, inferred and extrapolated lunar dates, assuming a strict biennial count for the period. According to his, this method cannot be applied for earlier times because there are no astronomical [lunar] dates allowing a control and serving as anchors.

Kraus's (2022a) model offers also two alternatives, low and high, separated by a lunar cycle of 25 years. This would allocate Djedkare Isesi year 1 to either 2365 BC or 2390 BC, and the first year of Teti, and the beginning of the 6th Dynasty in either 2305 BC or 2330 BC. These dates are identical to those proposed by HWK years ago (see Table 7.1). Additionally, this model offers either 2119 BC or 2144 for the end of the 8th Dynasty. Consequently, a period of only 31 to 61 years would be allocated for the hegemony of the Herakleopolitan Dynasty (9th/10th).

In conversations with María del Carmen Pérez Die, Director of the Spanish Archaeological Mission at Herakleopolis Magna (Pérez Die, 2009, 2010: 16–17, 2012)—the authors have been members of this mission—we have tried to envisage for how long this hegemony could be sustained. The Mission earlier defended Ian Shaw's chronology for the period (see Table 7.1), but, as suggested by Bettina Bader, the team ceramic specialist, this is no longer sustainable. The excavation of the FIP cemetery, and recent analysis (by way of publication) of epigraphy and locally manufactured pottery, would suggest that this necropolis was in use not only in the Herakleopolitan Period, but also as late as the 12th Dynasty, and perhaps during the reign of Mentuhotep II.

This may suggest that the Herakleopolitan hegemony did not last very long, but left unexplained who was responsible for the destruction of the cemetery (originally thought to be the work of Mentuhotep II's army). The excavators' team is currently re-analysing the data in an attempt to solve the puzzle. The absence on site of a royal necropolis of the period does not help and this question is still open. In any case, we shall not advocate an extremely extended duration of this period, i.e. longer than a century.

Krauss also argued that there are no anchored dates for the early Old Kingdom. However, archaeoastronomy can offer some clues based on potentially datable alignments, either of a solar or stellar character. The former have to do with the wandering nature of the civil calendar, and hence of New Year's Eve or Wepet Renpet, and the latter with celestial objects in the realm of the Imperishable Stars (see Chap. 4).

As has already been discussed, the first of Thoth or I Akhet 1, the first day of the civil year, wandered through the cycle of the seasons, completing a whole round in just over 15 centuries, and a total of about two turns during the period of political validity of the civil calendar (see Fig. 7.3). Throughout these three millennia, sunrise and sunset in Wepet Renpet could well explain the orientation of several buildings of exceptional character, notably during the Old and the New Kingdoms (see Chap. 6, and Sect. 7.5 below).

The situation is a thorny one when we deal with the Old Kingdom. Here, Belmonte (2012: 176–205) has discussed four possible dates, two of them solar (for the New Year) and two stellar (for Meskhetyu), for the pyramid complexes of Teti, Khufu (two of them), and Netjerkhet, respectively.

The extremely anomalous orientation of Teti's pyramid could be explained if it had been orientated to sunset in Wepet Renpet around 2304 BC (with a margin of plus 28 or minus 18 years), in almost perfect agreement with one of Krauss's dates for the enthronement of Teti, as derived from lunar dates. However, this enters into conflict with C14 dates so we cannot be absolutely sure of the certainty of this hypothesis.

The results obtained on the Giza Plateau for Khufu's funerary complex are the most suggestive. Two independent archaeoastronomical dates have been reported: sunrise at Wepet Renpet along the monumental causeway (interval between 2558 and 2508 BC) and the lower meridian transit of Meskhetyu (interval between 2559 and 2541 BC)—for this we have to admit that Belmonte's model for pyramid orientation is valid (see the Introduction). Considering the results of the C14 Bayesian model and its limitations (Dee, 2013d), Belmonte has conjectured that Khufu's reign would have begun around 2550±10 BC. We believe that this could be a chronological milestone, indeed one of astronomical origin, for the early Old Kingdom that should be taken into account.

Finally, Belmonte (2012: 176–181) proposed a couple of dates for Netjerkhet's reign, following the outcomes of the Egyptian–Spanish Mission (Shaltout et al., 2007). These are grounded on the supposed orientation of the sight channels of the northern façade of the serdab at Djoser's pyramid complex at Saqqara towards the imperishable stars (see Figure 7.8). A wide time interval, between 2620 and 2485 BC, is derived with the former bordering the outcomes of C14 dating (see Sect. 7.6). We may reasonably conclude that the highest date (that is, around 2620 BC) is

Fig. 7.8 The 'serdab' of Netjerkhet, in the north-eastern corner of the Step Pyramid of Saqqara. This small chapel contained a statue of the king (now a replica) located in front of the two visible holes in its northern wall. The inclination of the two holes carved in the limestone is far from identical, since one is much stepper than the other (11½° versus almost 20°). When the serdab orientation is considered, the resulting declinations pointed to the lower culmination of the stars Dubhe and Kochab, respectively. This offers a possible dating interval: 2620–2485 BC. Dubhe and Kochab are in the precise position of the two celestial adzes (*nṯrwy*), used in the Opening of the Mouth ceremony, where the iron stellar (*sb3wy*) blades were located. The most important of these adzes was Meskhetyu (the asterism of the Plough). (Image by the authors)

certainly the most probable. Hence, 2620 BC should be included within the *c.* 27-year reign of Horus Netjerkhet, Dual King Djoser, thus confirming the most accepted chronologies. This would be compatible with 2550±10 BC as year 1 of Khufu.

Therefore, archaeoastronomical dating offers an interesting picture for the explanation of certain structural features and the orientation of a few emblematic buildings during the Old Kingdom. The chronological framework confirms that, and not the other way around. However, it also offers a reliable framework for dating the Giza complex, by providing redundant dating that could be effectively inserted as a prior into alternative dating methods, such as C14. Astronomy should still play a role in the dismaying, but fully alive, discussion about the chronology of the Old Kingdom.

7.4 Sothic and Lunar Dates at Sekhem Senuseret Ma'kheru: A Chronologists' Dream, and a Dilemma

7.4.1 Where Was Sopdet Observed (If It Ever Was)?

The 'Going Forth of Sirius', or in the Pharaonic sources:  (Peret Sopdet), was possibly a festival associated with the day of the heliacal rising of this singular star, the brightest in the Egyptian skies, associated with Isis from very early times. On that day, the star, after having been invisible for a theoretical period of about seven decades (variable according to latitude and epoch), being in conjunction with the sun, was again visible at dawn on the eastern horizon just before sunrise.

As we saw in the calendar chapter, the near coincidence of this astronomical phenomenon with the arrival of the life-giving waters of the Flood made Peret Sopdet the herald of the Inundation, enabling it to enjoy a recognizable importance throughout Egyptian history. As the heliacal rising of Sirius was related to the tropical year (and the proper movement of the star), Peret Sopdet was one day behind in the calendar every 4 years, providing a possibility (also theoretical, the reality is much more complicated) of dating the event accurately in the always hesitant framework of Egyptian chronology. As we anticipated in Chap. 5, these mentions of Peret Sopdet in a civil calendar context are known as Sothic dates.

A summary of the most important dates in their chronological framework was presented in Chap. 5 and the quality of these dates has been analysed at the beginning of this chapter (see Sect. 7.1). As has already been suggested, the case is especially dramatic for what has been considered the most important and suggestive of Sothic dates since the discovery of the Illahun archives more than a century ago (Luft, 1992a). These archives come from the temple of the pyramid complex of

Senuseret II at Illahun,[1] in the north bank of the Bahr Yusef in its course to Al-Fayoum Oasis, and date to the reigns of his son Senuseret III and his grandson Amenemhat III (see Fig. 7.7). A sketch of this date was presented in Fig. 5.19.

The problem is that the information collected in that archive is controversial in nature. On the one hand, in a letter dated III Peret 25 of year 7 of Senuseret III's reign, the chief of the temple overseers, Nebkaure, informed the chief of the reading priests Pepyhotep that Peret Sopdet would occur in that year on the date IV Peret 16, so he was therefore presumably making a prediction. However, in another section of the same document of the archive, someone reported that the offerings for the Feast of Peret Sopdet were transferred to the temple on IV Peret 17, a day after the heliacal rising of Sirius had supposedly occurred. This certainly created a controversy.

Normally, the offerings for a given festival arrived at the temple where it was to be held one or even two days before the date selected for the festival. Therefore, if the offerings came on IV Peret 17, Peret Sopdet should have been scheduled for IV Peret 18 or 19, or even later. Consequently, Luft (1992b) and Krauss (2003, confirmed in 2015) have argued that a correction should be applied to the original date of the document, where, by a scribal error, 16 would have been written where 18 should really appear (or 19, both entries are possible in hieratic).

After imposing this correction on the original hieratic text (recall Fig. 5.19), Luft obtained the date of 1866 BC for the corresponding heliacal rising, observed from the site, and, therefore, a date for the seventh year of Senuseret III. This would place the 19-year reign of this sovereign between 1872 and 1854 BC. The German Egyptologist and chronologist Jürgen von Beckerah (1997) has put the full weight of his authority behind this date by stating that July 17, 1866 BC would be the earliest absolute date in the entire history of Egypt (see Table 7.1).

However, as we will discuss later on, this date is in open contradiction with the one estimated by Krauss (2003), who places year 7 of Senuseret III in 1830 BC (see Table 7.1). He proposes that the astronomical phenomenon, i.e. the heliacal rising, was not actually observed but only predicted within the framework of a stable cycle of four calendar years calculated for the southern border of the country; that is, for Elephantine. According to his proposal, the heliacal rising would have taken place on IV Peret 18 (not 16 as the text mentions) in Elephantine on July 19, 1830 BC. Rolf Krauss is a supporter of this predictive character of astronomical phenomena within a regular basis time-reckoning system; unfortunately, there is no explicit textual evidence that this character ever existed.

To solve the dilemma, Belmonte (2003a) proposed an alternative hypothesis whereby prediction and actual observation could coincide in time and form through the possibility that the observation was predicted for Egypt's southern border during year 7 of Senuseret III, when the frontier of Egypt was not at Elephantine but more than two hundred kilometres to the south, at Buhen. In fact, the heliacal rising of Sirius would have taken place (under adequate atmospheric conditions and an appropriate *arcus visionis*) at the fortress of Buhen on IV Peret 16 in 1830 BC (see Figure 7.9).

[1] The pyramid complex received the name Kha-Senuseret. However, the valley temple city located to the east was called *sḫm snwsrt mꜣ'-ḫrw* ('Senuseret, voice justified [defunt], is powerful)', seldom referred as **Kahun** in the past.

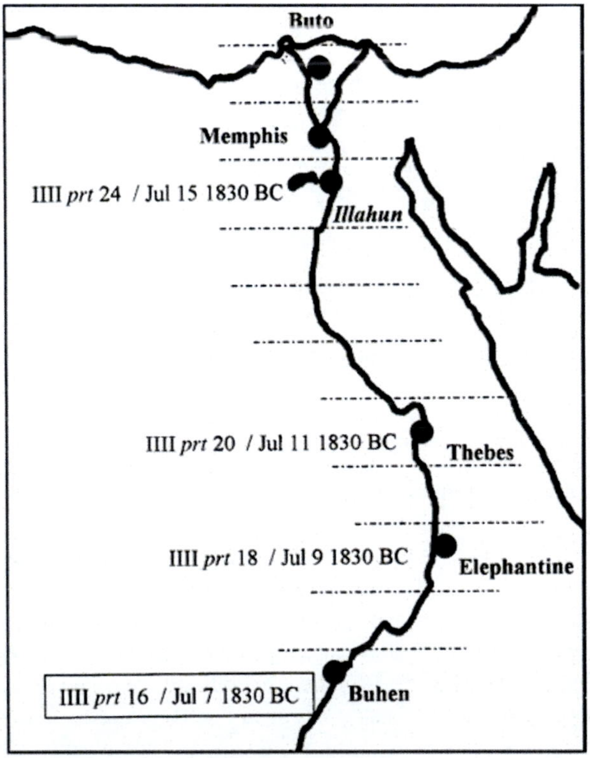

Fig. 7.9 Diagram of the Nile Valley in Egypt, showing the dates of the heliacal rising of Sirius (Egyptian Peret Sopdet) in July for different geographical locations in the country, in IV Peret of the seventh regnal year of Senuseret III (1830 BC under Krauss's model). At that stage, the southern limit of the country was fixed at Buhen, where Sirius' heliacal rising occurred 10 days earlier than in Buto, the sacred city on the Delta, and precisely at the date predicted by Senuseret II's cult priests. (Diagram by the authors, adapted from Belmonte 2003)

The *arcus visionis* is the angular distance between the rising star—or moon for lunar dates—and the sun below the horizon and is highly dependent in atmospheric conditions and also on latitude, since for lower latitudes it can be smaller because celestial objects rise at a greater angle to the horizon. An interesting discussion of the topic has been approached by Peter Huber, who has discussed different possibilities for a medium latitude at Egypt (Huber, 2011 and references therein). Table 7.2 presents a summary of the results.

Therefore, taking into account the local character of this astronomical event, Belmonte (2003) speculated that the date of the heliacal rising was predicted for the day when the phenomenon occurred at the southern end of the country for average atmospheric conditions (Buhen at that time). Instead, the feast would have been celebrated locally several days later and not necessarily on the exact date of the astronomical event. This could explain why the *ḥbjt nt prt spdt* (i.e. the offerings for Peret Sopdet) were moved to the temple of Illahun on IV Peret 17, thus avoiding the

Table 7.2 Path of change of the *arcus visionis* per day for an average latitude of *c.* 30°N, and estimated probabilities that a first sighting of Sirius heliacal rising occurs on a particular day

Day	*arcus visionis*	Probability
$n - 3$	6.6°	1%
$n - 2$	7.4°	6%
$n - 1$	8.2°	24%
n	9.0°	38%
$n + 1$	9.7°	24%
$n + 2$	10.5°	6%
$n + 3$	11.3°	1%

The average value is *c.* 9°, but a slightly wider range of ±1 day is an acceptable alternative. It is worth mentioning that these are probabilities. This does not mean that an event was not visible as much as 2 or 3 days, before predictions (day *n*), under extraordinary atmospheric conditions, or unobservable until 2 or 3 days after, under rough conditions such as clouds of Saharian dust, a common phenomenon in Egypt. However, the event should be expected within an appropriate date in *c.* 85% of the occasions. In any case, Sothic dates will not give a chronological accuracy better than ±10 years. Adapted from Huber (2011)

need to blame the scribe, a common practice in Egyptology when the data do not conform to the researcher's wishes (remember the Ebers).

This hypothesis has not received much attention in the Egyptological literature. Instead, new hypotheses have been produced that consider alternative explanations for Illahun lunar dates (Gautschy, 2011a). These hypotheses are based on a 30-year reign for Senuseret III (often, and reasonably, assigned just 19 years, see Table 7.1, but who was handed down 30+ years in the Turin royal papyrus) and give two alternatives for Senuseret III year 1 in 1872 BC and 1883 BC, respectively. The former is in agreement with an observation in Memphis or Heliopolis, in 1866 BC, one day after prediction (i.e. when the offerings arrived at the temple) under excellent atmospheric conditions. This coincides with von Beckerath's favourite date. The latter would fit the date of observation (IV Peret 16) in the year 1877 BC at Memphis under average atmospheric conditions (*arcus visionis* of 10°), thus raising the chronology of the period by 11 years. Both dates are compromised by Gautschy's alternative model for Illahun lunar dates.

Hence, the chronology of the Middle Kingdom is highly dependent on astronomical data but does not lead to a unique solution. This on-course 'fight' would not be complete without an analysis of the lunar dates and the different models used to approach them.

7.4.2 The Elusive and 'Damn' (But Fascinating) Lunar Dates of the Illahun Archive

As we have discussed, lunar dates have been considered problematic by the American astronomer Bradley Schaeffer, a specialist in the study of the visibility of astronomical events in the past (e.g. Schaefer, 2000b). In his analysis of the observations carried out with an ad hoc team of sky observers Schaefer conjectured that at least 15% of any observation of the lunar crescent, either its visibility or its invisibility, would be erroneous though simple human failure, rather than through atmospheric variability.

The Egyptologist Kenneth Kitchen has also described the moon as a futile chronological tool (Kitchen, 2007) because since the moon's phases are repeated in the civil calendar every 25 years, one must therefore know the approximate date of a document with an accuracy of at least half a century to be able even to begin to work. This preconception is deeply ingrained among Egyptologists. However, as we have earlier discussed, these prejudices are based on the most simplistic assumptions, calculations, and estimates. Lunar dynamics is far more complicated.

Synodic months, and the variables dependent on them such as lunar phases, are therefore not repeated in exact cycles of 25 calendar years. We should hence recall that, as a basic feature of the model, only about 70% of a lunar date set is repeated after a cycle of 25 calendar years (see, for example, Figure 7.10). Hence, a logical conclusion would be that a significant and coherent set of lunar dates should produce a single appropriate solution, that is, one where, considering the human factor, at least 85% of the recorded dates give us a solution that is astronomically correct.

The first person to recognize the presence of lunar dates in the letters of the Illahun archive was the great nineteenth century German Egyptologist Ludwig Borchardt, who properly identified an apparent cycle of synodic months in the service of the temple, although the discovery in the same archive of another document

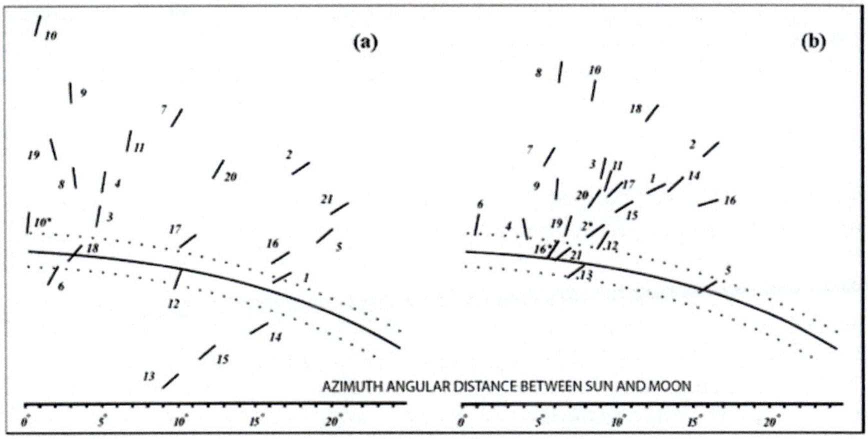

Fig. 7.10 Representation of the adjustment of the lunar dates in the Illahun archive to two models of visibility above the horizon (continuous line) adjusted for two possible alternatives for the year 1 of Amenemhat III: (**a**) if year 1 was 1844/43 BC, according to the high chronology, (**b**) if year 1 was 1819/18 BC, according to the low chronological model (see the 12th Dynasty in Table 7.1). Dotted lines represent intervals of reasonable uncertainties. The modelling strongly suggests that the low chronology, which would place year 1 of his father Senuseret III in 1837 BC, is the one that best fits—within the error margins—to the assumed astronomical observations, which are implicit in the archive. (Image adapted by the authors from two original diagrams, courtesy of Rolf Krauss 2009)

of this type in which the lunar festivals were framed within the framework of the civil calendar made him dismiss the possible chronological use of lunar dates as a whole (Borchardt, 1899). For several decades after that, the letters were inaccessible to specialists, including Richard Parker who could not use them in his Calendars. The correct reading of the documents could only be carried out by Rolf Krauss in the 1990s. The material with astronomical utility would later be published by Luft (1992), and indeed by Krauss (1985, 2003) himself.

Luft has collected 40 lunar dates from the Illahun papyri. He checked the respective dates, sometimes using photos instead of the originals. Sometimes he thought he could see a sign or the trace of a sign which was actually not there. Krauss rechecked the dates after Luft, using the papyri themselves, suggesting new alternative readings. Of the set of 40, Luft accepted 20 lunar dates as reliable and usable for chronological analysis (the ones with easy reading and dating); Krauss argued that there was one other reliable date, which brought the set to 21 dates.

This choice of 21 dates caused him more than one problem with absolutely unjustified criticisms such as 'questionable premises, faulty calculations, and unreliable self-correction of registered lunar data' from scholars who had never seen the original manuscripts. As we shall discuss later on, astronomer Rita Gautschy followed a different approach, in which all 40 lunar dates of the archive were considered, including those without dating. This was possibly dictated by a different philosophy in which she 'never throws available data away' (Gautschy, personal communication), looking at all of them, and using them in her analysis. This of course will produce a different chronological model (Gautschy, 2011), whose benefits were discussed by Huber (2011). We shall return to this intriguing conflict later on.

We should now analyse the basic chronological principles of the period, following Krauss's estimates. Between the year 1 of Ahmose and the year 23 of Thutmose III, in which the Battle of Megiddo occurred (see below), about 70 years are documented. Following a similar process for other earlier periods, such as the Hyksos era, or the 13th Dynasty, it can be estimated that between the year 1 Senuseret III and year 23 Thutmose III, a minimum of at least 197 calendar years is counted in the contemporary record.

As we have seen above, the parameters of the Illahun Sothic date are problematic. Consequently, Krauss (2003) has only worked on 21 reliable lunar dates from the Illahun archive, a vast majority from the reign of Amenemhat III (see Fig. 7.7). They span an interval of 42 years between year 9 of Senuseret III and year 32 of Amenemhat III. In his analysis, each 25-year period, from 2300 BC to 1300 BC (and some extra periods before and after) has been tested. As clearly shown in Figure 7.11, there is only one possibility that satisfies the condition that at least 85% of the recorded dates could be astronomically correct (there might be another, slightly above 80%, some 150 years later but this is too late, even with the shortest possible chronology). A date 25 years earlier would also be feasible if gross observational errors are admitted (see Fig. 7.10).

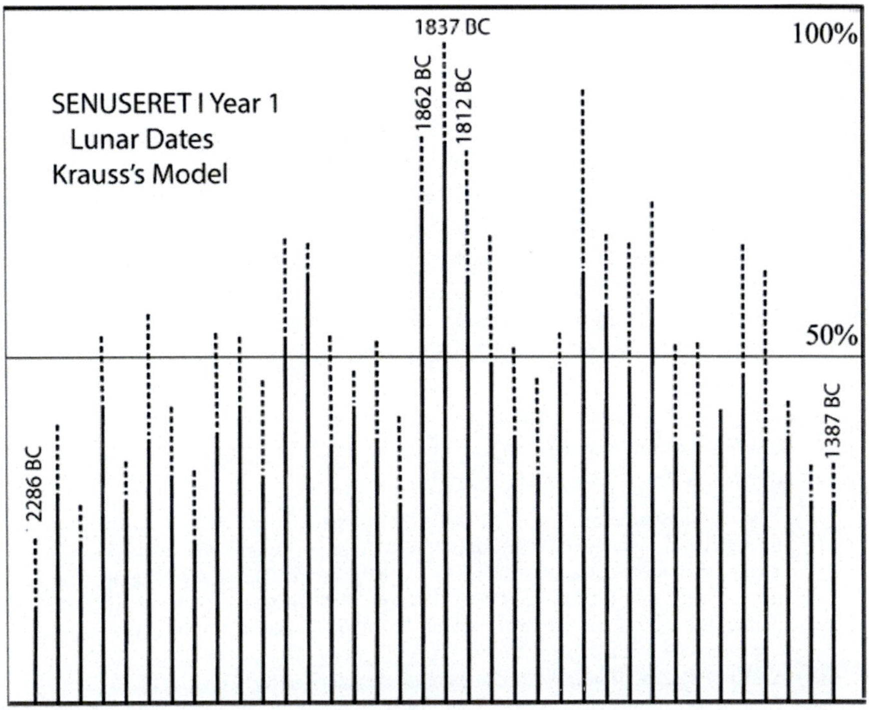

Fig. 7.11 Histogram of 21 lunar dates from the Illahun archive selected by Krauss (2003), from year 9 of Senuseret III to year 32 of Amenemhat III matching the actual behaviour of the moon as a function of time, in intervals of 25 calendar years between 2286 and 1387 BC. The only possibility that stands above an 85% threshold (the theoretically possible maximum related to human capability) would establish year 1 of Senuseret III in 1837/36 BC, regardless of the Sothic chronology. Solid lines indicate the percentage of those cases in which old crescent visibility would have been observable beyond reasonable doubt, and dashed lines those cases in which the calculated height of the crescent would have been within the zone of uncertain visibility (see Fig. 7.10). (Image adapted by the authors from an original, courtesy of Rolf Krauss 2009)

The seemingly inescapable result of the analysis is that the first year of Senuseret III straddles 1837 and 1836 BC, starting in November. This date would confirm his hypothesis for the Sothic date in the same archive.

However, life is not always as easy as chronologists would desire. Gautschy (2011) proposed a completely different model for the lunar dates of the Illahun archive. She decided to count on the basis of the 40 lunar dates: 26 of these correspond to the entries contained in explicitly dated documents (21 coincide with those used by Krauss). For the rest, the date is not given and should be derived from certain assumptions. Krauss has most recently argued that of the five additional lunar dates in dated documents only one would be possibly reliable. Gautschy's data set includes 14 and 26 lunar dates for the respective reigns of Senuseret III and Amenemhat III, which can be compared to the ratio 3/18 in Krauss's model.

However, Gautschy has honestly used all the data at hand and arrived to a completely different picture. She argues that if there are good reasons finally to exclude one or other lunar date, it should be clearly stated why. There are a few differences also in the approach to the data. A main difference between both analyses is the lunar day numbers for some of the feasts, namely the full moon feast and the Wagy feast; Gautschy strictly adhered to the Illahun material and made distance calculations based on well-known feast days. The latter leads to LD 16 for the full moon (the mean of values between 15 and 17) and LD 18 for the lunar Wagy feast. Krauss used LD 15 and 17 instead with the argument that full moon can only be on day 15, not 16, which is certain for Ptolemaic times (see Chap. 5). This is apparently not so evident for the Middle Kingdom where we lack a full list of moon days. Wagy is two days later than full moon, then LD 17, seldom 18, for Krauss.

Another difference is that Gautschy accepts, because the data analysis so suggests, an independent reign of 30 years for Senuseret III. Textual evidence does not offer a date higher that year 19, but he was assigned 30+ years in the Turin Canon. After her analysis, Gautschy produced two alternatives:

(A) Year one of Senuseret III in 1872 BC. A fit of 10/14 and of 14/26 is obtained for lunar dates of Senuseret III and Amenemhat III, respectively. This implies a set of 16 out of 40 lunar dates, or 40%, of wrong assignations. This is much higher than the one in Krauss's model (*c.* 15%). However, this will permit saving the Illahun Sothic date as observed from the temple in 1966 BC and hence the traditionally accepted high chronology.

(B) Year 1 Senuseret III in 1847 BC. In this case the match is 12/14 and 16/26, for lunar dates of the reigns of Senuseret III and Amenemhat III, respectively. There will be 12 out of 40 wrong assignations, i.e. an error of 30%. This is still quite high. This model does not adequately fit with Gautchy's analysis of Sothic dates. Interestingly, this alternative corresponds to year 1 Amenemhat III in 1817 BC, exactly the same as in Krauss's model, the difference being the 30-year reign assigned to Senuseret III.

Gautschy's analysis, when both Sothic and lunar dates in the Illahun archives are considered offers only one sound alternative, Year 1 in 1872, with the observation of the heliacal rising of Sirius in Memphis or Heliopolis in 1866 BC one day after prediction (i.e. when the offerings arrived at the temple). Consequently, Huber (2011) believed that the evidence presented in Gautschy's model provides a persuasive argument in favour of 1873/72 BC for the accession of Senuseret III, and a length of 30 years for his reign.

As we have seen, different scholars have come up with very different results for the lunar dates and the Illahun archive. It should be obvious that only one of these possibilities can be the right one (or perhaps all are wrong). What is certain, and we have been able to verify on several occasions, is the following. As Krauss affirms, every Egyptologist who deals with the chronology of ancient Egypt, using astronomical dates for it, is frequently stubborn to the limit, demonstrates prejudices towards the work of opponents, and has infinite difficulty in recognizing, much less admitting, the possibility of errors in his own calculations or assumptions. We believe that the situation is not so dramatic in this case. We personally know both Rolf Krauss and Rita Gautschy, both are honest and clever scholars. We prefer to think that their differences come from two different world-views, arising from different academic backgrounds. Krauss gives much credit to epigraphy. Gautschy tends to use all the data at hand.

The authors allied precisely to bridge this dichotomy. A last comment is perhaps relevant here. Michael Breger, the pope in the field where both Rita Gautschy and the first author started research, used to say, 'Bad data is worse than no data'. We are sure Krauss and Gautschy will certainly differ in what is bad data in the Illahun archive.

Hence, unless new irrefutable proof definitively tips the balance in one direction or the other, we believe that, despite evidence presented in however a convincingly honest manner, we shall continue to have discussions for the foreseeable future.

7.5 New Kingdom Astronomical Dates: Riddles for the Chronology of the 'Great' Pharaohs

The chronology of the New Kingdom in the first decade of the twenty-first century seemed quite clear. There was an almost general consensus in which year 1 of Thutmose III was precisely fixed in 1479 BC and year 1 of Ramesses II in 1279 BC, 200 years later. The dates of the former were obtained through two lunar dates (see below) and a Sothic date (n° 8, see Fig. 5.21).

For the time of the Ramesses II, the solution was apparently quite simple. As Krauss argued,

> The time of Ramesses II is fixed without any Sothic dates. There are synchronisms between Egypt, Assyria, Babylonia and Hatti. It is a fact that Ramesses II was a contemporary of certain Babylonian and Hittite rulers, who were in turn contemporary with certain Assyrian rulers. The uncertainty of Assyrian chronology in the 13th and 14th centuries BC amounts to 10 years only and, in consequence, year 1 of Rameses II lies at 1290 BC plus or minus 30 years. Within this interval, the lunar date from year 52 of Ramesses II allows only the following possible solutions: year 1 should be either 1304 BC, or 1290 BC or 1279 BC. If one takes into consideration the Festival-of-the-Valley dates from year 7 of Tewosre and year 7 of Ramesses III, which are lunar dates, it follows that only 1279 BC is compatible with these and other lunar dates. (In Belmonte, 2003, 2012: 298)

Unless in the Middle Kingdom, where lunar dates were extrapolated from either lunar feast or lunar temple services, the so-called civil-lunar double dates, or to be more concise, simple lunar dates of the New Kingdom, such as the one discussed at the beginning of Sect. 7.2 in relation to year 52 of the reign of Ramesses II, offer little scope for discussion. These may be considered explicit lunar date of top quality. Nevertheless, many Egyptologists are of the opinion that such dates are not entirely reliable given their apparent repetitive nature over 25, 14, or 11 year cycles.

In the last decade, a few papers have appeared which have questioned the 'revisionist' low chronology. Several arguments are based on C14 data (suggesting a need to raise the dates) that we have already challenged (the case of Thera eruption) and shall discuss in depth and qualify later on. Worth mentioning among these essays, are those by Huber (2011), already discussed for the Middle Kingdom,

Aston (2013), Quiles et al. (2013), Gautschy (2014), and a new updated insight by Krauss (2015a, see also 2016), which have enriched the discussion but offered proposals in flat contradiction between one and the other. However, for Krauss, lunar dates are extremely reliable information in helping to untangle the skein of chronology.

The evidence we have for dating the reigns of the great Pharaohs of the New Kingdom is varied and is the key for the chronology of the period. This evidence has to take into account that the process of re-excavation and improvement of KV57, Horemheb's tomb in the Valley of the Kings, has provided a large number of labels of wine jars dated to the 14th year of his reign, possibly deposited in the tomb on the occasion of the king's own burial (van Dijk, 2009). This has brought about a change of paradigm and, although not universally accepted, it is indeed a relevant clue to take into account. Under these considerations, the evidence at our disposal for chronological purposes may be classified as follows:

(i) From simple dead reckoning based on archaeological evidence, an interval of 186+X years should be counted between years 1 of Thutmose III and Ramesses II (Krauss, 2015), an interval *c.* 12 years lower than the earlier one considered from a longer reign of Horemheb, which ought to be abandoned.
(ii) Two explicit secure lunar dates of the reigns of Thutmose III and Ramesses II. These are LD1 in years 23 and 52, respectively.
(iii) A secure but not explicit lunar date in year 24 of Thutmose III, during the foundation ceremony of the Akhmenu in Karnak.
(iv) The explicit, but incomplete, Sothic date of Elephantine and the implicit Sothic date of the Ramesseum (n° 10, see Fig. 5.13) which documents the relationship of Peret Sopdet to New Year's Eve for the first time in Egyptian history.
(v) A series of implicit lunar dates derived from dated events almost certainly related to the Beautiful Feast of the Valley (*ḥb jnt*), which should correspond to either LD1 or LD2 (i.e. *psḏntyw* or *ꜣbd*) in the civil month of II Shemu. Most of them are graffiti in the Djeser-Akhet temple of Thutmose III below the cliffs of Deir el Bahari (Krauss 2015, Excursus 1).
(vi) Some events of similar status where the relationship to *ḥb jnt* is not so clear.
(vii) Civil dates of the Feast of Tepy Shemu during the TIP. A further-elaborated discussion of these dates is well beyond the scope of this volume and they will not be analysed as a case study in this work. Readers interested in greater detail concerning this fascinating but troublesome period may consult Gautschy's (2015) dedicated monograph and Krauss's extensive analysis on the topic (Krauss, 2015a).
(viii) Series of civil dates presumably related to a certain festival celebrated on a lunar day. The series of possible LD4 for a feast of Ptah in Saqqara that we discuss in later sections is an example of this (Krauss, 2016b). These may be catalogued as extrapolated lunar dates.
(ix) Hypothetical lunar dates related to the moment of ascension of a king to the throne. This is a sort of cherry-picking process with even two contradictory

choices: (a) full-moon (Edgerton, 1937), still discussed in certain academic cycles; and (b) first evening crescent (*3bd*), deliberately interpreted as LD1 by Derstine (2015 and 2016). This scholar is the proposer and lone defender of this speculative exercise. With such a controversial background, we shall not discuss this possibility further.

(x) The identification of solar eclipses as celestial phenomena related to certain remarkable events, as reported in the sources (see Sects. 7.7 and 7.8).

(xi) The archaeoastronomical dates of the *hwt-itn* in Akhetaten (see Sect. 7.8 for a discussion), and the Amun temple in Tanis (see Chap. 6). The latter would fit sunrise at Wepet Renpet in the quadrennium centred on 1039 BC. This implies that the temple could have been aligned at the beginning of the reign of Psusennes I (*c.* 1040–992 BC), in agreement with traditional chronological data (see Table 7.1). This datum alone precludes any attempt to substantially increase or reduce the dates for the end of the New Kingdom, because it adheres to standard chronologies as reported in Table 7.1.

This is the astronomical evidence at hand. Only numbers (i), (ii), and somehow (iv) are unquestionable. Numbers (iii), (v), and especially (vi) and (viii) are open to discussion. We shall not discuss either (vii) or (ix). Eclipse dates (x) are mostly educated guesses, whilst alignment dates (xi) are quite reliable.

C14 chronology is indeed a relevant source of information. Hence, we shall dedicate a whole section (7.6) to analysing its benefits and challenges. However, we now concentrate on dates related to astronomy.

We deal first with a singular case that Krauss has studied in detail, that of the Battle of Meggido. This act of arms confronted the Egyptian army commanded by Pharaoh Thutmose III in person and a coalition of Retenu peoples, who did not recognize Egyptian authority. The king was in his twenty-third year of reign, shortly after taking the reins of absolute power after the death of his aunt and stepmother Hatshepsut. He was then eager to achieve a crushing victory. Sources inform us that the battle took place in, 'Year 23, First month of Shemu, day 21, psedjentiu precisely' (see Figure 7.12).

In the past, various specialists have argued, on the basis of historical events and following Parker's *Calendars*, that the battle could not have taken place on the 21st, but on the 20th, since the army arrived at the gates of Megiddo after a skillful ploy on I Shemu 19. Therefore, both possibilities, that LD1 was either I Shemu 20 or 21 should be taken into account. This option is untenable today.

Using dead reckoning, a date around 1450 BC can be reached for year 23 of Thutmose III. The Sothic date of Elephantine also suggests that the interval 1433½±4½ BC could be between years 33 and 54 of this king (Quiles et al., 2013). Besides, using synchronisms with other states in the Middle East, a date for the battle in the interval between 1458 and 1443 BC can be inferred. If the date of I Shemu 20 is considered, we see that it corresponds to an LD1 in the years 1468 or 1443 BC, which does not fit well with the information of the sources. If, on the other hand, we admit that the Egyptian source is correct, we get two possible dates, 1457 and 1446 BC as the most reasonable ones, corresponding to years 1479 and 1468 BC for Thutmose year 1.

Fig. 7.12 Image by the hill of Megiddo and the remains of the Bronze Age city whose capture by the army of Thutmose III in his 23rd year provides the earliest chronological milestone of the Egyptian New Kingdom in most accepted chronologies. The date was recorded in a register of the battle where I Shemu 21, year 23 was written (upper hieroglyph register). This also precisely informs us that this date was an LD1 (psedjentiu). (Diagram and image by the authors)

Traditionally, the first had been admitted and the beginning of the reign of Thutmose III in 1479 BC was therefore 'astronomically' fixed. However, as we have seen the process of re-excavation and improvement of Horemheb's tomb suggests that this king's reign should be shortened to 14 or, at most, 15 years (van Dijk, 2009).

Therefore, if this reasoning is correct, the entire chronology of the 18th Dynasty has to be reduced by around a dozen years with spectacular consequences. In the particular case at hand, the date of the Battle of Megiddo would have to be moved according to Krauss (2009) from the year 1457 to 1446 BC, lowering in turn the date of the beginning of the reign of Thutmose III to the year 1468 BC. However, astronomical dating is not always so easy.

There is a second implicit lunar date from the reign of Thutmose III that is most relevant. This was a foundation ritual in the Karnak temple, possibly related to the erection of the Akhmenu, the festival hall of the king in the eastern quarters of the complex. The text informs us that the king ordered preparations to be made for the ritual on II Peret 30, while 'waiting for the day of pesedjentyu', i.e. *ḥr s3wt hrw n psḏntyw* in the original Egyptian language (Krauss, 2015).

This has been variously interpreted as one of three possible options. The most favoured one (a) is that the day of the order is the same as the day of the lunar event, and hence II Peret 30 should be an LD1. A second possibility (b) is that the day of the ceremony in LD1 was the day after the order, a plausible possibility. Then, LD1 would be III Peret 1. These two alternatives (a) and (b) are compatible with 1479 BC as Thutmose III year 1. The third possibility (c) would be that the day or the order was II Peret 30. The day of preparation for the ceremony would be III Peret 1, an LD29, the day of 'the standing guardian' waiting to detect the last dawn crescent visibility or invisibility, and LD1 would be the next day III Peret 2, when the ceremony would be performed. This would be compatible with 1468 BC as Thutmose III year 1, still compatible with Ramesses II year 1 in 1279 BC, 189 years apart. However, Krauss has recently stressed to the authors that this third option is 'out of the question'.

This ambivalence when setting safe 'astronomical' dates depending on the moon is what makes some specialists distrust them. However, we must bear in mind that the possibilities are not endless. It is perhaps worth make a summary of the most important analyses performed in the decade before this book was planned. We start with the in-depth study performed by Peter Huber (2011), who has analysed all possible options, including some which would certainly be outdated. Table 7.3 summarizes his analysis.

Of the several options he presented, 1 and 3 should be discarded because all the evidence points to a short reign for Horemheb. Huber's favourite hypothesis is Option 4, representing a very high chronology, with assumed 'errors' in the lunar dates. This is a little distressing. Hence, Option 2 would indeed be a much more reliable possibility, saving 1479 BC as Thutmose III year 1, but then Ramesses II year 1 has to be brought forward to 1290 BC. As we shall discuss in Sect. 7.8, both options for an eclipse, close to the spring equinox, should probably be discarded when Hittite religion is taken into account (González-García & Belmonte, 2011).

Huber's is just one of this series of analyses. Table 7.4 presents a comprehensive summary (centred on Tuthomse III's and Ramesses II's enthronement dates) of the chronologies defended in the last decade based on more or less solid hypotheses.

Table 7.3 Outcomes for the reigns of the 'great' Pharaohs of Huber's (2011) analysis of lunar dates of Thutmose III with alternatives (a) and (b) as discussed before, and considering two possible lengths for Horemheb's reign

Input	Option 1	Option 2	Option 3	Option 4
Horemheb's reign	Long (*c.* 27Y)	Short (*c.* 14/5Y)	Long (*c.* 27Y)	Short (*c.* 14/5Y)
Thutmose III Lunar dates (LD1)	Y23 I Shemu 21 Y24 II Peret 30	Y23 I Shemu 21 Y24 II Peret 30	Y23 I Shemu 20 Y23 III Peret 1	Y23 I Shemu 20 Y23 III Peret 1
Thutmose III Y1	1479 BC	1479 BC	1504 BC	1504 BC
Ramesses II Y1	1279 BC	1290 BC	1304 BC	1315 BC
Eclipse in Murshili II Y9/10	April 13 1308 BC *c.* 50% at sunrise	April 13 1308 BC *c.* 50% at sunrise	March 13 1335 BC >75% at *c.* noon	March 13 1335 BC >75% at *c.* noon

The last row shows his proposal for the solar omen interpreted as an eclipse in the reign of Murshili II (dates in the Julian proleptic calendar). See the text for further details. Y stands for Year

Table 7.4 Different 2010s models for the chronology of the New Kingdom, matching the first year of both Thutmose III and Ramesses II

Researcher	Option	Thutmose III Year 1	Ramesses II Year 1
Schneider (2010)	Sethy I 10 year reign Sethy I 13 year reign	1476 BC 1479 BC	1290 BC 1290 BC
Huber (2011)	Option 2 Option 4	1479 BC 1504 BC	1290 BC 1315 BC
Aston (2013)	High Ultra high	1493 BC 1504 BC	1290 BC 1290 BC
Gautschy (2014)	P3 P1 P_{AG}*	1493 BC 1468 BC 1479 BC	1304 BC 1279 BC 1279 BC
Krauss (2015a)	'Best LD1s fit' Qualified in (2016)	1479 BC 1479 BC	1290 BC 1279 BC
This volume Sethy I 12 years	11 years to AII or TIV* Karnak LD1 in III Peret 2	1479 BC 1468 BC	1279 BC 1279 BC

The option with an asterisk (*) supposes that 11 years must be searched for in the reigns of Amenhotep II and/or Thutmose IV. The last row presents the working hypotheses that we shall defend in this volume. See the text for further discussions

These use dead reckoning (and hence archaeology), the lunar dates of the period and eventually Elephantine Sothic date along with other evidence, and a bit of subjectivity depending on the author preferences. All of them accept a short reign for Horemheb.

A first look at the table clearly shows that the odds are in favour of a higher chronology for Ramesses II year 1. Aston's (2013) model relies on his analysis not only of astronomical dates, on which he does not put too much emphasis, but on wine jars dockets and C14 dates. For C14 he assumed that dates must in general be raised because this is what the Bayesian models show or 'demonstrate', despite individual analyses suggesting the contrary (see Sect. 7.6). Most important are his lengthy arguments based on wine labels. He argues that the same weight must be given to the sherds found in KV57 as to those date labels in amphora sherds found in a pottery deposit that apparently (evidently for him) derived from the interior of KV17, the tomb of Sethy I. Some 500 labels were identified by a team led by Nicholas Sartori, all inscribed with Sethy I's year 8, thus suggesting (following Horemheb's precedent) a reign of 8—at most 9—years for this king.

We shall return to this argument late on. Notwithstanding, the most important of Aston's argument is that the reigns of both Amenhotep II and Thutmose IV should be increased by at least 34 (or even 36) and 19 (or even as much as 34) years, respectively, hence his high and ultra-high chronologies for Thutmose III (see Table 7.4). This last hypothesis has not received much acceptance, even if grounded on apparently firm evidence (Aston, 2013).

Gautschy (2014) made a dedicated analysis of all kinds of evidence, including the solar eclipse as the solar omen in Murshili II's annals. After checking different possibilities, she favoured the eclipse seen at sunrise in Hattusha in April 13 1308 BC (Gautschy, 2017). Her best model is P3 (see Table 7.4), which implies a rising in the dates of both Thutmose III and Ramesses II, although P1 is also possible, with the issue of Karnak lunar date. An alternative solution (P_{AG}) she offered, to keep 1479 BC as Thutmose III year 1, was to add 11 years to either Amenhotep II or Thutmose IV, as suggested by Aston (2013). In any case, she personally argued during a visit to the first author's institution, weeks before the Covid 19 pandemics, that the New Kingdom chronology is far from solved and that new evidence would be greatly welcomed.

Finally, Krauss (2015a) initially thought that Ramesses II year 1 in 1290 BC would probably be the best fit to the data, with 1279 BC as an alternative. However, he recently changed his mind, using newly extrapolated lunar data for the reign of Ay (Krauss, 2016b) and other evidence that will later be more deeply analysed. He now thinks that there are impressive arguments supporting 1279 BC as Ramesses II year 1, even if the odds were in favour of 1290 BC. When analysing the Amarna period, the end of the 18th Dynasty and the beginning of the 19th Dynasty in Sect. 7.8, we shall show that his bet for 1279 BC would certainly be in full agreement with at least one of our working hypotheses (see Table 7.4). However, prior to this, we make two much-needed excursus to deal with C14 dating and the elusive solar eclipses.

7.6 C14 Versus Astronomy: A Scholar's Dilemma

Carbon 14 dating is one of the most widespread and reliable systems of dating ancient remains. Its archaeological importance is indisputable. Interestingly, however, in the case of ancient Egypt this importance has always been relative, when it has not been absolutely absent from the chronological debate. In the Valley of the Nile, C14 specialists have encountered numerous problems. There was an angry debate on the 'ancient wood' issue for the dating of the remains of C14 of the Old Kingdom (Bonani et al., 2001), a controversy that continues so far. Sturt Manning's earlier work is illustrative: using only sophisticated techniques of statistical sampling, and through simulations, he postulated a date for the reign of Khufu around 2580±50 BC (Manning, 2006).

The same could be said for other epochs of Egyptian history, such as the Amarna period, which was dated in two very different time intervals: 1368–1360 BC and 1336–1320 BC, depending on the type of material considered: wood or animal bone samples, respectively. This was not of much help in solving the problem of the chronology of the period, and the situation has not improved, as we shall discuss later on.

Another unique case was that of the destructive eruption of the Thera volcano (see Figure 7.13). This was dated by C14 in an interval between 1720 and 1640 BC, within a 2σ margin, the one commonly accepted by specialists, while the archaeological-historical dating of the same event would place it apparently between 1530 and 1480 BC. The dichotomy was therefore evident, as there was a margin of more than a century between both estimates.

Fig. 7.13 Panoramic view of the islands forming the gigantic caldera of the Thera archipelago from the cliffs of Santorini at its northern rim. The Island of Thera exploded in a huge volcanic event, which nearly destroyed it. This had been C14 dated within an interval between 1720 and 1640 BC, while the archaeological-historical dating of the same event placed it between 1530 and 1480 BC. However, a new C14 calibration curve has offered a date for the eruption event between 1560 and 1540 BC, or even as low as 1510 BC, in complete agreement with the archaeological evidence (Bietak, 2020). This has obvious consequences for C14 dates in Pharaohs' Egypt. (Image by the authors)

However, as we discussed earlier on, such a dichotomy never really existed, as a new calibration curve has offered a date for the eruption event between 1560 and 1540 BC, or even as low as 1510 BC, in complete agreement with the archaeological evidence (Bietak, 2020). We wonder how the development of new, more sophisticated, and upgraded calibration curves for other earlier and later periods might alter the results we are going to discuss in the following paragraphs and which has been part of the debate for the last decade.

Aware of a series of problems, but not yet of the possibility that their calibration curve could be wrong, a team of specialists in various fields led by Christopher Bronk Ramsey of the University of Oxford decided to take a totally different approach to the problem. They would use samples from various museums and collections but only of short-lived plant materials that had been cut shortly before their use, such as flower necklaces deposited inside a tomb at the time of the burial of their owner.

This would supposedly eliminate problems, such as that of the old wood, and other polluting effects that were difficult to assess. The work was embodied in an article in the prestigious journal *Science* (Bronk Ramsey et al., 2010) as a triumph of the collaboration between natural scientists and Egyptologists, and was fully introduced into the chronological debate in the following years. The sample included in total 188 useful dates—about twenty samples were discarded as outliers for giving illogical values—of which 128 came from the New Kingdom, 43 from the Middle Kingdom, and only 17 from the Old Kingdom. Of these, 48, 7, and 7, respectively, could not be assigned to a precise reign. This means that only ten dates cover reigns of the Old Kingdom (seven of them of Netjerkhet's) so their reliability should be relativized.

On the chronological framework imposed by the C14 dates—obtained using the IntCal (09 version) calibration curve—and therefore upon the time scale frame of the model, 'historical' data were superimposed. These included series of reigns and their duration, although not the dates themselves. For this, the Oxford chronology model proposed by Ian Shaw (see Table 7.1) was selected. Hence, it is worth noting that this model includes a few undocumented reigns, such as that of Shepseskare, and far from obvious reign lengths for several kings, such as a clear overestimate of 39 years for Senuseret III. Moreover, it did not include the lowering of Horemheb's reign from 27 to 15/4 years (already reported and widely known in 2010). A statistical method known as Bayesian modelling of the data was subsequently applied.to the whole set of C14 date framework, kings' succession, and reigns duration.

As deftly and succinctly described by Anita Quiles,

> Bayesian modelling is based on subjective hypotheses that represent the state of the established knowledge on a studied event or object when the dating is performed. Such an approach highlights the combination of two systems of time: the first is relative time (represented by the archaeological evidence) and incorporated as *a priori* to a second time, which is absolute time represented by [C14] measurement and called the *likelihood*. The age deduced by combining archaeological and dating information is the *a posteriori* law, and this distribution has to be accepted or rejected depending on particular criteria that we have to define in advance. (Quiles et al., 2013: 424)

The key point here is subjectivity. The priors have to be decided in advance and are subjected to the researcher's wishes or guesses. The order of reigns and their duration is clearly a prior. The sample is also a prior, and last decade of research has demonstrated that short-lived remains of living beings are the best objectives. The likelihood is also not as secure as scholars would desire. C14 dating is a precise science when BP (i.e. before the present or AD 1950) are obtained with a certain margin of physical error. However, the calibration curve used is also a prior and improved version, or even completely different curves (remember Thera), may offer quite different results. Hence, C14 dating should be observed, not with skepticism, but with a certain degree of caution. Miracles should not be expected as in astronomical dating.

Bronk Ramsey et al. (2010) divided the sample into three groups according to the three main historical periods of Egyptian history, obtaining very suggestive results. These were compared with three chronological models, those of Shaw, Hornung-Krauss-Warburton (see Table 7.1) and Kate Spence (2000), although the last only for the Age of the Pyramids. This last model, based on alignments and astronomy, which postulated an ultra-short chronology for the Old Kingdom with a reduction of about 80 years in the dates of Khufu's reign, seemed to be outright discarded.

However, the situation is not so clear in the other two cases, although the authors conclude that Ian Shaw's model, with a high chronology, seems to fit their data better. In this regard, we would like to stress that there has obviously been a certain kind of circularity of argument. Using the data provided by Shaw for the order of, and especially the length of reigns, it is logical that his approximation is closer to that provided by the model—otherwise it would be illogical. However, Shaw's sample includes manifest biases, such as, for example, an excessively short duration for Netjerkhet's, Sneferu's, or Khufu's reigns (19, 24, and 24, compared to about 26, more than 30, and 27 years, respectively, according to the most confronted current historical information; see Table 7.1 and Chap. 6).

Likewise, the reduction to 14 or 15 years for the duration of Horemheb's reign, made public in 2008, two years before the publication of the article, had not been taken into consideration. This would include an obsolescence clause for the dates proposed by the authors for the 18th Dynasty. In subsequent analyses, the research team decided to take into account other options, producing a series of most interesting works that were finally compiled and edited in a single volume (Shortland & Bronk Ramsey, 2013). Other priors were considered, including one that eliminated the obvious outliers presented in the C14 data. It is hence worth studying these updated analyses, where the contributions of Michael W. Dee have been fundamental.

A first challenging approach has to do with the analysis of a historically dated (with a margin of ±2 years) sample of short-lived plants collected in Egypt between AD 1700 and 1900, a crucial period prior to the damming of the Nile. A total of 77 dates were obtained on 66 samples. Interestingly, there was an unexpected average difference of 19±5 between the known ages and the C14 calibrated dates, in the sense that the remains seemed younger than they were. This little but significant 'offset' was ascribed to a local growing season effect (Dee, 2013a). This offset would subsequently be applied to the C14 calibrated dates in every period of ancient Egypt history from the New Kingdom to the Archaic period.

However, there are two facts to be taken into account. The period analysed was that of the Little Ice Age between the fourteenth and the nineteenth centuries, with three temperature minima in 1650, 1770, and 1850, precisely. This caused alterations in harvests and plant growth in Europe and the Mediterranean basin that may have nothing to do with a local growing season effect. Worst still, there is not a single shred of evidence supporting the idea that this offset may be happily applied to the past, to periods thousands of years earlier than this local effect. Its application to the Egypt of the Pharaohs, offering higher dates that further support certain preconceptions is a kind of circular argument. We will soon discuss a clear example of this reality related to the Amarna period.

Having said that, we should indeed try to analyse some results of the model. If we take into account the 2σ precision, which, in theory, provides a precise chronological margin of 95%, we see that the ascension to the throne of Ramesses II, which we have considered as a chronological milestone, would be between 1301 and 1272 BC, according to the NKM5 model avoiding outliers (Dee, 2013d). This would encompass both dates provided by the astronomical evidence (1290 and 1279

BC), which is good. This stresses the capacity of the model when the sample is big enough, as in the case of the New Kingdom, and illustrates what can be achieved before the reliability of the calibration curve itself could be called into question.

We shall nevertheless hypothesize later that 1279 BC is the best candidate. This suggests that, within the interval considered, the lowest dates seem to be the most likely. Another example of this happens when the accession date of Tutankhamun is analysed. Evidence from C14 data only from his reign offers the interval 1359–1319 BC for Nebkheperure year 1. However, when these data are included in a broader Bayesian model with samples of several other kings' reigns of the 18th Dynasty, the interval is 1365–1328 BC, with a decalage of a decade to older dates (Quiles et al., 2013).

The situation is still more challenging when the Amarna period is taken into consideration. The analysis of Amarna C14 samples (with a 5-year resolution curve) shows that this period ought to be dated between 1345–1315 BC with *c.* 68% probability, with a mean 1333±17 BC and a median of 1331 BC (Manning et al., 2013). This chronological framework agrees with Belmonte (2013) and Krauss (2016a) and the updated and upgraded model that is defended in this volume, but not with the traditional chronologies (see Table 7.1). Only when the 19±5 year omnipresent—and annoying—offset is taken into account, do the old chronologies fit within the 68% probability range of 1352–1314 BC.

If we analyse the Middle Kingdom, where the model should not show large errors, the situation is (apparently) clearer. In this case, the kings' succession is quite well determined, with minor differences in the reign lengths, such as with Senuseret III (19, 30, or even 39 years). Also, a significant sample of C14 dates is available, but mostly form the reigns of Mentuhotep II, Senuseret III and Amenemhat III (Shortland & Bronk Ramsey, 2013: Table 2). Dee (2013c) showed that C14 cannot discriminate between the two dates of the reign of Senuseret III that have been discussed in previous sections of this chapter.

The 2σ MKM4 Bayesian model, which eliminates outliers but still includes the 19-year offset and a 39-year reign, situates the enthronement of this pharaoh between 1890 and 1837 BC (Dee, 2013c). Interestingly, model MKM2 which uses data from HKW (see Table 7.1), and a 19-year reign, gives an interval 1894–1828 BC for Senuseret year 1. Hence, models cannot distinguish which of the two dates, 1872 BC of von Beckerah, defended by Gautschy, or 1937 BC defended by Krauss (see above, and Fig. 7.11), is the most accurate. Dee (2013c) tends to favour the higher chronology, although he states that if consensus could be reached, and new data were available, the MKM models could perhaps establish the definite position of the period in absolute time.

As far as the Old Kingdom is concerned, the dates obtained from the Bayesian models are again much closer to the models proposed by Shaw and von Beckherah (high chronologies) than to the low chronologies (notably HKW, see Table 7.1). It is worth noting that model OKM4, where outliers are cancelled, does not converge owing to data shortage (Dee, 2013d). Hence, Bayesian statistics is not of much help in this particular case, so caution is advised.

Only Netjerkhet's sample analysis converges, fixing Djoser's year 1 and the beginning of the pyramid era in the interval 2690–2625 BC, a quite reasonable time framework. The OKM1 model, using Shaw's data, gives the intervals 2649–2582 BC and 2629–2558 BC, for year 1 of Sneferu and Khufu, respectively; but OKM2, where HKW's data (see Table 7.1) are considered, offers a slightly lower range for Khufu, 2618–2549 BC, because a logical longer reign for Sneferu is considered. Dates yielded for Djedkare Isesi—for whom only one C14 date is reported—(2486–2400 BC for year 1) are still far from those reported from astronomical dating (year 1 in 2365 or 2390 BC, see Sect. 7.3). The discrepancy would not be as large if the ubiquitous 19-year offset were ignored.

It is certainly possible that the valid dates are the closest to the lower values of the interval, taking into account what was mentioned regarding the reign of Ramesses II, and the scantiness of the sample, which is not statistically significant. However, Michael Dee is optimistic and considers that 'the inclusion of more data may enable C14 dating to fully resolve this important period of Egyptian history' (Dee, 2013d: 216). We hope he is correct.

Finally, dates for the proto-dynastic period are scarce but still offer a hint. The dates produced by the Oxford project are 3019–2893 BC, for textile from Tarkhan, and 3023–2891 BC, from seeds from Abydos necropolis at Umm el Qab (Rowland, 2013). Once more, these intervals are close to those offered by high chronologies (see Köhler, 2013 and Table 7.1). Doubts, however, still persist.

In any case, we wish to highlight the substantial efforts carried out by Bronk Ramsey and his team. In the future, new investigations of the Oxford and other teams could lead to suggestive new results, with due precautions and perhaps with the proper improvements in the models and calibration curves. These outcomes could be very useful in solving, once and for all, the dichotomy between radiocarbon dating and the chronology of ancient Egypt, in particular that based on astronomical data.

7.7 Hiding the Aten: Was the Sun Eclipsed in Ancient Egypt?

We have often been asked about the observation of non-cyclical phenomena in ancient Egypt. Among these, supernovae, comets, and solar and lunar eclipses are certainly the most striking ones (see Chap. 4). The last two in particular are in fact the most frequent it being (for a non-urban culture of the pre-industrial era) virtually impossible not to notice them when close to totality under clear skies. However, there is not a single explicit or obvious reference to any of these phenomena in the entire history of Pharaonic Egypt, with Perhaps one exception (Parker, 1957b). This is the case of an eclipse implicitly mentioned in a papyrus just before the death of Psametik I (see Chap. 4). The solar event happened on September 30 610 BC, posterior to the fixed date of Taharqa in 664 BC (see Table 7.1). Hence, this eclipse is not needed into fixing the chronology of the period (Shortland, 2013).

Fig. 7.14 Detailed photograph of a total solar eclipse showing in a perceptible way the filamentary structure of the solar corona, extending several radii beyond the hidden disc of the sun. This structure has occasionally been related to the wings of the god Horus-Behedety, and is ubiquitous in pylon gates and temple ceilings. (Image courtesy of the IAC Multi Media Service)

This absence of sources is especially dramatic in the case of solar eclipses, especially total ones. This apparent absence of information in our knowledge of Egyptian astronomy has puzzled many historians and scientists. Did Egypt suffer from a shortage of total solar eclipses in her history? The answer is a resounding no! The solar corona, clearly noticeable during a total eclipse (see Figure 7.14), was visible from somewhere in the Nile Valley up from Aswan on average once every 75 years or so, in the period between the year 3000 BC and the Roman era.

It is impossible, and at the same time difficult to imagine, that this spectacular series of solar eclipses went unnoticed by a culture that worshipped the sun as the supreme divinity in its pantheon and orientated its buildings accordingly (see Chap. 6). Consequently, it is probable that there was a certain degree of censorship in the way this very unique information was treated (see Chap. 4).

However, according to some researchers, the appearance of the sun during the phase of totality was preserved in a symbolic form. For example, Jane Sellers (1992: 74) has proposed that part of the myth of the struggle between Seth and Horus could be reinterpreted in light of a phenomenology related to a solar eclipse, although her ideas are very controversial and have not been widely accepted. However, there is another certainly suggestive hypothesis, regarding the solar corona and its distinctive aspect during eclipses, an image that varies with the 11-year cycle of solar activity (see Fig. 7.14).

During the minimal phase of this cycle, the brightness of the corona is less intense, so it extends on both sides of the solar disc in the form of extensive flares of light that are clearly visible to the naked eye during the phase of totality. According to Brewer (1991), it is not difficult to see the similarity between this image and that offered by the wings of the celestial god Behedety, spreading about the solar disc, one of the most frequent and undoubtedly most symbolically charged representations of the solar deity. Although quite speculative and impossible to prove, this hypothesis is quite suggestive.

Another idea proposed by Sellers (1992: 277) is that total solar eclipses may have had something to do with certain significant dynastic changes in ancient Egypt, especially those of a dramatic nature. This idea has been furthest developed in the Internet by a series of researchers such as the Leo Duval in France, the American Willian McMurray and the Egyptian Ahmed Ibrahim.[2] The latter has even speculated on the controversial idea of the existence of a hypothetical link between the observation of certain solar eclipses and the erection of the largest obelisks, at certain critical moments in Egyptian history. None of these ideas has received wider audience.

One of the problems with solar eclipses is that it is extremely difficult to determine the precise date, the nature of the eclipse, and especially the exact path of totality for events prior to 900 BC. This is due to various uncertainties concerning the dynamics of the orbits of earth and the moon that impose a degree of uncertainty in the value of the acceleration of the rotational speed of our planet (ΔT, see Figure 7.15). This is a well-known problem, widely publicized by one of the most serious eclipse hunters, NASA researcher Fred Espenak, who frequently updates and optimizes his own hypotheses in his extraordinarily managed webpage.[3] To this problem, we must add the discrepancies in Egyptian chronology that we have lengthily discussed throughout this chapter.

Most of the proposals refer to eclipses that occurred during the New Kingdom. Thus, for example, the total eclipse that presumably covered the city of Akhetaten in shadow in May 1338 BC (see Sect. 7.8) has been variously interpreted.

Duval, for example, mentioned it as either a terrible harbinger of the imminent death of Akhenaten or as the inspiration of this king for the foundation of the city in that precise location of Middle Egypt. McMurray has suggested that the eclipse would have been the ill omen that convinced Tutankhamun to leave the city in his year 4. These three interesting hypotheses are at the same time mutually exclusive and certainly unprovable, and will remain so until the chronology of the period is clarified once and for all. Neither is there any historiographical confirmation that the eclipse might have been observed (a fact concerning which we should have no doubts) or indeed recorded by the ancient sources. We shall soon return to this point.

Interestingly, we could try to verify, or refute, these kinds of hypotheses by making a comparison with other key moments, or even complete periods, of Egyptian history. Figure 7.15 shows, as an example, the solar eclipse that occurred at sunset

[2] See, for example, L. Duval, 'Timing Akhenaten', http://www.archaeometry.org/nefertiti.htm; W. McMurray, 'Dating the Amarna period in Egypt: did a solar eclipse inspire Akhenaten?' and 'Towards an absolute chronology of ancient Egypt' http://www.egyptomania.com/EEF/EEFLibrary.html; and A. Ibrahim, 'An eclipse-based Hittite chronology?' http://www.eclipse-chasers.com/egypt4.htm

[3] https://eclipse.gsfc.nasa.gov/

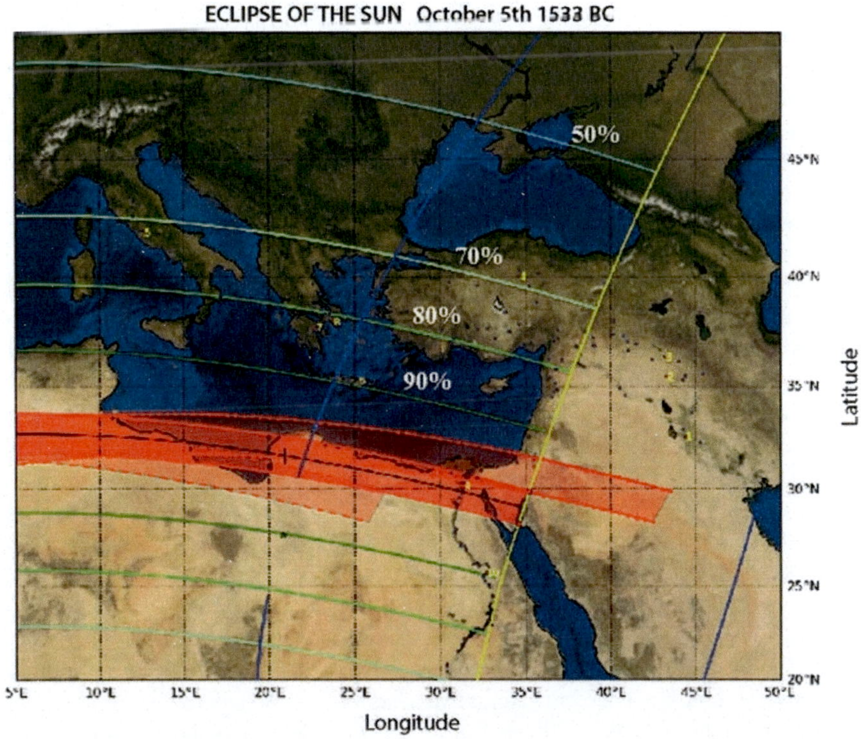

Fig. 7.15 Possible trajectories of a total solar eclipse that presumably occurred on October 5, 1533 BC in the Julian proleptic calendar, according to the mean value of the acceleration of the speed of rotation of the earth (ΔT, continuous line), and its maximum (red dotted line) and minimum (red stroke line) values accepted today. Totality may or may not have been visible in the Delta depending on the values of ΔT considered. Image by the authors, adapted from a graphic courtesy of Rita Gautschy. [http://gautschy.ch/~rita/archast/archast.html]

on October 5 1533 BC. This was in theory observable from the Nile Delta. This date is according to the various chronologies of the period (see Table 7.1) and almost coincides with the beginning of the war of independence from Hyksos rule in Avaris. The Theban sovereigns Kamose and Ahmose waged the war. Perhaps, the solar eclipse could have been interpreted as a bad omen for the future of the northern enemy. A circumstance that might have encouraged the southern conquerors to start the campaign.

However, this is one of those eclipses in which the uncertainties in ΔT prevent a clear conclusion from being arrived at. This raises further doubts as to whether the eclipse was observable as total, or simply as partial, from the Delta because the strip of totality would not have reached Egypt if the maximum possible values of this variable were considered. Therefore, such doubts far outweigh the certainties and thus render idle any such speculative exercises.

In contrast, we have a pair of annular solar eclipses, with the maxima over Memphis, in the years 1646 BC and 1635 BC. These were observable as partial throughout Egypt. This, we may speculate, could relate to the end of the Middle Kingdom and the seizure of power in Avaris by the Hyksos rulers, a historical event fixed at around 1640 BC by most chronologists (see Table 7.1). However, once again, the uncertainties concerning this enigmatic period of Egyptian history make this hypothesis a mere speculative exercise.

We end the experiment by mentioning that there were three total solar eclipses in a row, within a short span of time, in the years 1935, 1902. and 1884 BC. These were possibly observable as total from various places in Egypt and hence presumably did not go unnoticed. At this time, Egypt was ruled by the great sovereigns of the 12th Dynasty. Thus, the situation was especially stable so it seems that these eclipses did not cause any dramatic effect that can be identified in the sources (even if we might have expected any). If, however, they resulted in other kinds of political, social, or religious decisions, such as the founding of a temple or the beginning of a regency, again our lack of accuracy in the records prevents us from being more precise.

The problem is that Egyptian history is so rich that we could certainly find satisfying events that coincided with almost every total solar eclipse that occurred at the time. Even more if partial or lunar eclipses are considered. However, such sources are scarce, if not simply non-existent, so that we cannot go further in our conclusions. Therefore, astronomy does not always have the definitive answer to a certain historical or cultural problem. Actually, a serious approach to the problem of eclipses in Egypt through a dedicated statistical analysis of all solar eclipses (and—why not?—lunar eclipses as well) observable throughout the 3000 years of Egyptian history is still pending. This could certainly be a very stimulating experiment.

7.8 Lunar Dates, Solar Alignments, Wine Dockets, C14 and Eclipses: Dating the Amarna Period

A key issue for a comprehensive understanding of the dynamic years, which closed the Amarna period up to the reign of Ramesses II, is the time frame. This section incorporates the most recent approach to this amazing era performed by the authors as gift contributions for the *Festschrifts* of Professors Jiro Kondo (Belmonte, 2022) and, eventually, John Taylor (Lull & Navarro-López 2022), which were under discussion at the moment of writing this chapter.

Table 7.5 reports an updated chronology of the one defended in Belmonte (2022), where the Lull and Navarro-López (2022) outcomes have been added. This is mostly based on the standard low chronology as first developed by Krauss and Warburton (2009), including a reduction of Horemheb's reign to *c.* 14 years, with subsequent improvements. The most important modification is that related to a series of lunar dates associated with a Feast of Ptah in LD4. This strongly suggests that Ay's first

Table 7.5 Chronological framework for the end of the 18th early 19th Dynasties defended in this volume

King	Years	Dates BC	Great royal wife
Nebma'atre Amenhotep III	38	(1379–1342)	*Tiyi*
Neferkheperure Akhenaten	17	(1341–1325)	*Neferneferuaten Nefertiti* *
Ankhkheperure Semenkhkare**	<1	(?–?)	*Meritaton*
Ankh(et)kheperure Neferneferuaten*	1+2	(1325–1323)	*Meritaten ?*
Nebkheperure Tutankhaten/Tutankhamun	9	(1322–1314)	*Ankhesenamun*
Kheperkheperure Ay	5	(1313–1309)	*Tey*
Djeserkheperure Horemheb	15	(1308–1293)	*Mutnedjemet*
Nebphetire Ramesses I	2	(1293–1292)	
Menma'atre Sethy I	12	(1291–1280)	
Usirma'atre Ramesses II	67	(1279–1213)	

It reflects the latest findings (e.g. 1313 BC as Year 1 of Ay; Krauss, 2016b), for the last sovereigns of the 18th Dynasty and the first of the 19th, based on the most recent genetic, forensic, historical, epigraphic, and archaeological evidence, correlation with Hittite sovereigns and various astronomical dates: several moon dates, a solar alignment and an eclipse. The reigns of the Hittite kings Shuppiluliuma *c.* (1348–1322), Arnuwanda II (1322–1321), Murshili II (1321–1295), and Muwatalli II *c.* (1295–1272) run in parallel (e.g. Bryce, 2001:19). See text for further details
(*) The same person. A minor part of her reign could coincide with her husband's
(**) Very short coregency between years 13 and 15 of Akhenaten, or, even better, as late as year 17

year was 1313 BC (Krauss, 2016b). As we have previously discussed, C14 dates are not in conflict with this chronological framework (Manning et al., 2013).

This chronology also supports the total solar eclipse of June 24, 1312 BC as the ill solar omen reported in Murshili II's chronicles when he was campaigning in Azzi lands in his year 9/10 (see Figure 7.16), as we had argued (Belmonte 2012: 381). Rolf Krauss has recently supported this hypothesis, although under certain conditions (Krauss, 2016b). Many other eclipses have been proposed for the omen, if indeed it was an eclipse at all. An updated discussion on this topic can be followed in Gautschy (2017, and references therein). However, Gautschy's favourite option is instead the partial eclipse at sunrise on April 13, 1308 BC, as observed from Yerkapi, in Hattusha, because this better agrees with her new proposal for New Kingdom chronology (Gautschy, 2014).

Although this hypothesis would still be within the appropriate chronological frame, historically, it is very difficult to imagine that a Hittite king would be campaigning abroad in early April when he had to be at Hattusha celebrating the AN.TAH.SUM and Purulli festivals. These festivals were celebrated for a period of 40 days when the land 'prospers and thrives' and were possibly timed with the new crescent moon of the first month after the spring equinox. This event happened on *c.* April 4 in the Julian proleptic calendar in this epoch (González-García & Belmonte, 2011).

A different but very suggestive issue is the foundation ceremonies and the alignment of Akhetaten's first structures in Akhenaten's year 5. Both Gabolde (2009) and Belmonte (Belmonte 2012: 339; see also Belmonte & Shaltout, 2009: Appendix II)

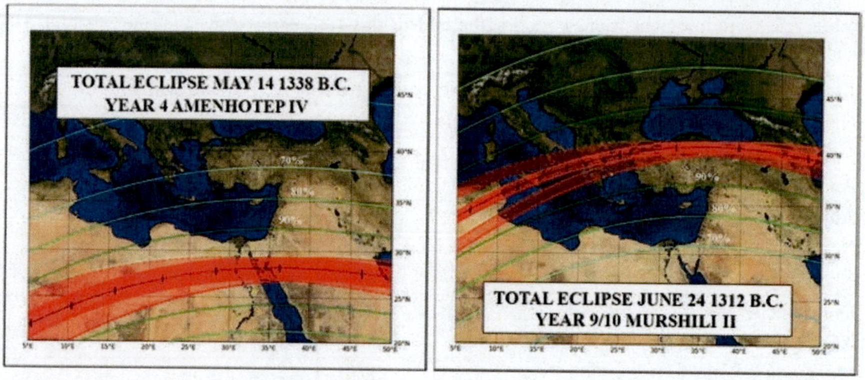

Fig. 7.16 Graphical simulation of the totality bands of two interesting solar eclipses that occurred during the end of the Eighteenth Dynasty. The first was on May 14 1338 BC when the disc was 100% hidden in the Akhetaten region shortly before the city was founded. The second, which occurred in June 24 1312 BC, was visible in northern Anatolia, and might have been the bad omen mentioned in Hittite sources in the year 9 or 10 of Murshili II. Note that both statements are independent of the value considered for ΔT (the earth's acceleration rate). (Image adapted from Belmonte (2022) on graphics courtesy of Rita Gautschy). [http://gautschy.ch/~rita/archast/archast.html]

had defended the hypothesis that the first alignment ceremony, presumably a 'stretching of the cord', was celebrated in IV Peret 13 in year 5 at the Small Aten Temple (the *ḥwt-jtn*) in Amarna. This would offer a suggestive time milestone for year 5 in 1335±4 BC in perfect agreement with the proposed chronology which would locate Neferkheperure's year 5 in 1337 BC (see Table 7.5).

However, a little later on, Edward Castle resumed the question, arguing that the foundation stelae texts could reveal that the ceremony happened at the Large Aten Temple (the *pr-jtn*) instead of in the smaller one. He argued that the place where this happens, *pꜣy* [*pr*] *jtn*, incompletely preserved in the stelae texts, but which could be reconstructed, already referred to the *pr-jtn* at this so early stage. He also provided circumstantial archaeological evidence that the act was made on an altar in the middle of the desert that was preserved and included within the limits of the Large Aten Temple when this structure was built (Castle, 2015). It is worth noting that neither this altar nor the Large Temple faced any conspicuous landmark on Amarna's eastern horizon. The Small Temple does! Apart from the doubts that might be generated concerning the idea of equating *pꜣy* [*pr*] *jtn* to *pr-itn*, where the first could refer at this early stage to the whole area of Akhetaten and not to a very specific place, there are other important considerations.

The centre of the city of Akhetaten follows a nearly Hypodamic plan, consisting of a square grid where the most important urban buildings, such as the two temples, the main promenade, and the royal palace, were later enclosed. In the authors' experience, the alignment of this kind of grid often follows either a relevant landmark or a conspicuous astronomical phenomenon. In this, Egypt is a paradigm (see Belmonte, 2016; and references therein). In Akhetaten we may have both, as

Fig. 7.17 The entrance of the royal tomb wadi forms a characteristic notch in the eastern horizon of Akhetaten where the sun rises in perfect—and indeed suggestive considering the name of the site—alignment with the enclosure north wall of the Mansion of the Aten (*ḥwt-jtn*), the so-called Small Temple. The horizon is so close that the southern wall would not fit the alignment. (Image by the authors, adapted from Belmonte 2012)

illustrated in Figure 7.17. The rising sun and a conspicuous notch in the eastern horizon, where the valley of the royal necropolis opens, form an inspiring sight resembling the *ꜣḫt* sign defining the proper name of the city *ꜣḫt-jtn*, i.e. Akhetaten. If either of the authors had been there, they would certainly have chosen this splendid view.

However, the cliffs are near enough so that a large rectangular structure built in the centre of the city would not have had both walls pointing in the appropriate direction. This certainly happens in the Small Temple, where the north wall of the enclosure faces the 'akhet', but not the southern one, as anyone on site can appreciate. Not even the axis of the temple does so.

Hence, we stick to the hypothesis that the 'stretching on the cord' ceremony in the foundation act of the city was performed in IV Peret 13 year 5, with the king located in the northwest corner of the future *ḥwt-ỉtn* enclosure facing towards the east. At this moment the four corners of the sanctuary would have been established (Belmonte et al., 2009a), and with them the Hypodamic city grid. Chronology, aesthetics, and even logic support this reasonable possibility.

Belmonte (2022) finally included a working hypothesis as a gift for Prof. Kondo. This was a brief discussion of a so far unresolved obscure situation, concerning the Amarna period. First, why did Amenhotep IV decide to change her name to Akhenaten (the last attestation of 'Amenhotep, divine ruler of Thebes' is in year 5,

II Peret 19, at Ghurob; Lull, 2019b)? Second, and most significantly, why did he decide to create a whole new city from scratch in a barren, isolated place of Middle Egypt?

The boundary stelae marking the northern and southern limit of the town mention that the king had heard that something terrible had happened, something never heard of—or seen—before during his earlier years or during the reigns of his father Nebma'atre or his grandfather Menkheper[u]re (perhaps his ancestor Menkheperre Thutmose III, the text is damaged here; Ridley, 2019: 32).

Several Egyptologists have wondered what could have been the reason for such a drastic decision and have even suggested that Amenhotep IV's action was the result of some sort of theophany (Hoffmeier, 2014: 154–7). In earlier works, Belmonte (2012: 408) has speculated that this 'theophany' could have been a total solar eclipse occurring in May 14, 1338 BC (see Fig. 7.16). The eclipse was almost total (> 90%) in Thebes and Memphis, but was certainly total in Amarna, Akhmin being almost on the borderline of totality, depending slightly on the value of ΔT. This impressive phenomenon occurred in III Shemu 23 in Amenhotep IV's year 4, according to the chronology defended in Table 7.5.

Total solar eclipses are normally repeated at a certain location on earth in intervals of a few centuries, while an appreciable degree of totality of a nearby total eclipse may on average take a century or so to be repeated.

This time frame would match what the king stressed in his proclamation. He could have observed the phenomenon in Thebes, but he was presumably advised that, further north, the disc of the sun was completely obscured. Considering the relevance that the sun disc cult had already reached in his early years, such a situation might have been dramatic for him. Hence, a few months later, he changed his name to Akhenaten and looked for a perfect place for a new capital in the area signalled by the Aten. Nine months after the eclipse, the place had been selected and a city founded 'at the place of the first event'.

These different astronomical possibilities (an alignment, two potential solar eclipses, and a series of LD4; besides the lunar date in year 52 of Ramesses II) are reflected in Table 7.5 (an upgrade of Belmonte, 2022). This new chronology is also reliant on discarding a co-regency between Amenhotep III and Amenhotep IV, as demonstrated by Lull (2019b). In this context, (Ta)-Hemet-Nesu Neferneferuaten Nefertiti, either as a regent queen or as King Ankh(et)kheperure Neferneferuaten, a possible throne ascension name for her, would in all likelihood be the Dakhamunzu of Hittite texts. Semenkhkare (KV55 male and father of Tutankhamun) was a shadowy co-regent of Akhenaten. Interestingly, Amenhotep III year 1 would be 1379 BC. We shall come back to this fascinating point later on this section.

Accepting these chronological milestones implies a gap of 3 years that Krauss (2016) tentatively, and with much caution, assigned to Sethy I, being initially followed by Belmonte (2022) on that point. This could be a problem. However, Lull and Navarro-López (2022) have most recently proposed an alternative solution to fill this gap.

Thanks especially to the work of Brand (2000), the co-regency hypothesis between Sethy I and Ramesses II has been totally discarded. However, there is still no unanimous agreement on the length and chronology of Sethy I's reign (e.g. Dodson, 2019), although in recent years, as we have seen, the latest advances in archaeoastronomy and astronomy, combined with improvements in the relative chronology of the New Kingdom, have served to improve our knowledge of the period.

The festival calendar of Ramesses III in his Mansion of a Million Years at Medinet Habu indicates that the Beautiful Feast of the Valley (*nfr ḥb jnt*) began with LD1 (*pšdntyw*) of II Shemu (El-Sabban, 2000: 67). This was the moment when the statue of the god Amun was taken to the west bank of the Nile, in Thebes, where in LD2 (*ꜣbd*) the offerings were made to the god in the king's funerary temple. Thanks to this information, the texts referring to the Beautiful Feast of the Valley can provide, even though indirectly as we have shown, astronomical data of great importance for the absolute chronology.

The coronation date of Sethy I is not known with certainty. However, according to Helck (1990), the enthronement took place on III Shemu 24, a date that in ostracon Gardiner 11 (from year 6 of Ramesses VI) is remembered as *ẖnw n Sty*, a term that has been linked to the celebration of the king's coronation. Also, the coronation of Ramesses II took place on III Shemu 27, as Helck (1959) and Krauss (1977) originally concluded. This implies that the last year of Sethy I's reign was short-lived, no more than three days.

The length of Sethy I's reign has also been the subject of discussion. The highest date ever found, a year 11, comes from a stela found by Reisner in the temple of Amun at Gebel Barkal, in Nubia (Reisner & Reisner, 1933). However, Van Dijk (2011) questioned the reading, estimating instead a year 3. We have already noticed that the highest dates written on amphorae sherds of Sethy I found in a deposit at the Valley of the Kings refer to year 8. Aston (2016) reinforced the idea that they could refer to the last vintage before the death of the king, making year 9 the last of Sethy I. However, these series of amphorae of year 8 could be part of the progressive provisioning for the tomb, as happened in Tutankhamun's burial, where the number of wine jars with labels from year 5 doubled the number of year 9, presumably Nebkheperure's last year. The dilemma is thus still open to debate.

It is at this precise moment when we must introduce a document (ostracon DM 21) that, despite the uncertainties that surround it—also discussed by Krauss (2016)—can help to establish the chronological match we need to better define the reign of Sethy I. The hieratic inscription offered by oDM 21 (Černý, 1935: pl. 5A 21),[4] reads, in two lines: II *ꜣbd šmw* 25 (II Shemu 25), and *ḥnḳt jnḥt* 51 (Beer: 51 *inhet*-jars).

The huge amount of beer mentioned (about 637 litres!), being delivered in II Shemu, suggested to Kitchen[5] an association with offerings linked to the Beautiful Feast of the Valley and, stilistically, with a group of ostraca from year 3 of Sethy I.

[4] Černý, *Catalogue des ostraca*, https://www.ifao.egnet.net/bases/archives/ostraca/?inv=DM+21&os=0

[5] *RITANC* (1993–2014), volume 1; 257.

Even considering these series of premises, Lull and Navarro-López (2022) have suggested that oDM 21 may have an obvious chronological utility. They have calculated when II Shemu 25 coincided with a LD1 between 1295 BC and 1280 BC; that is, within the most probable years of the reign of Sethy I, considering 1279 BC as Ramesses II year 1. Lull and Navarro found that on May 1, 1288 BC, in the Julian proleptic calendar, there was the lone coincidence of II Shemu 25 with an LD1. That means that if oDM 21 is related to *ḥb jnt* of year 3 of Sethy I, the reign of Sethy I should have a length of at least 12 full years, beginning in 1291 BC and ending almost in the middle of 1279 BC, when Ramesses II ascended to the throne. This is a fascinating possibility (see Table 7.5).

Interestingly, this will also explain the inscription found in the hemi-speos shrine built by Sethy I in El Kanayis, in the Eastern Desert, 50 km east of Edfu, where year 9 is attested as the moment of the first visit of the king to this mining area (see Figure 7.18). The temple, decorated with reliefs, a well, and other minor facilities for improving miners' lives, could hardly have been built in just a few months. Besides, the king made a second visit to the site to consecrate the sanctuary. Indeed, a period of nearly 3 years seems much more reasonable for the completion of this infrastructure in the middle of nowhere.

Fig. 7.18 The tiny hemi-speos temple built by Sethy I at El Kanayis, on the desert road connecting Edfu to the Red Sea. The decision to build the temple—and auxiliary structures such as a well—was taken in III Shemu 20 of his year 9, just 4 days before regnal year 10. Once everything was finished, the king made a second visit to the site to consecrate the shrine. This certainly suggests a reign of at least 10 years, perhaps as much as 12. This great builder was responsible for the recovery of Egyptian international prestige after the Amarna episode. (Photograph by the authors)

In this case, how might this chronology fit with the immediate reigns (accepting the correspondence Ay year 1 to year 1313 BC (see Table 7.5)? Lull and Navarro-López (2022) suggest a solution.

On the one hand, if Ay's reign began in III or IV Peret 1313 BC (Julian year 1313 BC began in II Peret 18) and ended in II Akhet 1308 BC, Ay would have been on the throne 5 full years and 5 or 6 months of his sixth year. On the other hand, If Horemheb's reign began in II Akhet 1308 BC and ended in the middle of 1293 BC, Horemheb would have reigned 14 full years and about 9 or 10 months of his year 15. That is, up to III–IV Shemu, so near the beginning of the grape harvest.

Nowadays in Egypt, the grape harvest season begins in May and ends in September, but low temperatures during the growing season can slow down the plant growth and high temperatures can accelerate ripening. The wine jars could be docketed at the end of the harvest season, especially during the first months of Akhet, suggesting almost 15 years for Horemheb. In that case, the reign of Sethy's father Ramesses I would have lasted 2 full years or almost 2 full years. Interestingly, this is the chronological framework proposed in Table 7.5 as a working hypothesis. This would fix the beginning of the reign of Amenhotep III in 1379 BC.

These creates a sort of rule of thumb for the chronology of the New Kingdom. The reigns of the three most important sovereigns of the period (not necessarily the most famous) Menkheperre Thutmose III, Nebma'atre Amenhotep III, and Weserma'atre Ramesses II, would have started in 1479 BC, 1379 BC and 1279 BC, respectively: a truly easy mnemonic.

However, as we have argued, the chronologist's life is not always an easy one. If Thutmose III ascended to the throne in 1479 BC and died in 1425 BC, and Amenhotep III was enthroned in 1379 BC, *c.* 46 years later, we have a problem. Shaw's and HKW's chronologies assigned 26 and 9 years to Amenhotep II and Thutmose IV (see Table 7.1), making a total of 35. This means that at least 11 years are lost if we accept the chronological framework of Table 7.5. Two alternatives are possible to solve the gap (see Table 7.4).

Aston (2013) gave a long list of arguments, proposing that Amenhotep II should be assigned at least 30 years (even as much as 34) and Thutmose IV at least 19 years (or perhaps as much as 34). These include possible commemorative *ḥb-sd* festivals after year 30 (probable for Amenhotep II, less likely for Thutmose IV), the number of elite tombs, wine dockets, the number of great royal spouses, even pottery, among other minor reasons. Hence Aston proposed very high chronologies (see Table 7.4). The ratio of preserved scarabs from these two reigns also suggested raising the numbers of Thutmose IV. This irritated (*sic*) Rolf Krauss for years and he felt tempted to accept a longer reign for this king (Krauss, 2015, and references therein). However, he no longer seems to be in favour of such a possibility.

Indeed, a reign of 18 years for Thutmose IV, as suggested by a wine label from the workers' village of Deir el Medina (Aston, 2013), and a small increase of regnal years for Amenhotep II would suffix to fill the 11-year gap.

A second possibility has to do with the Karnak lunar date (LD1) related to an order of King Thutmose III in II Peret 30 (see Sect. 7.5). We have discussed 1479 BC as year 1 of Thutmose III as the most acceptable solution for this date and that of the Battle of Meggido (see Table 7.4). However, we have argued that 1468 BC as Thutmose year 1, with LD1 in III Peret 2 could have a chance. The first invisibility of the waning moon crescent occurs on *c.* 82% of occasions in the day of astronomical conjunction. However, in *c.* 10% of cases this can happen one day before and in *c.* 8% one day after (Huber, 2011). Hence, other observing variables are possible.

We cannot objectively evaluate whether the best alternative for a solution of the 11-year gap is raising the dates of, mostly, Thutmose IV and keep 1479 as year 1 of Thutmose III, or accepting 1468 BC as year 1 of Thutmose III. The former has all the weight of Aston's somewhat controversial arguments. The latter means accepting a bad prediction from the side of royal advisors or a sequence of events different from what is often assumed. This was considered a sound possibility in the past.

However, we would love to keep the 1479-1379-1279 BC mnemonic rule for dating the 'great' pharaohs of imperial Egypt. Readers may decide for themselves.

7.9 Conclusion

Astronomy has proven to be a serious and crucial auxiliary science to ancient chronology. Without Sothic, lunar and even solar dates, Egyptian history would not have an adequate time frame beyond that provided by dead reckoning (Kitchen, 2013) and contemporaneity with foreign rulers, although even the chronology of these sovereigns relies on astronomical calculations of reported observed phenomena, such as lunar eclipses (e.g. Huber, 2011; Krauss, 2015b). In earlier sections of this chapter we have analysed how astronomy helped fix the chronology of the most important periods (the Old, Middle and New kingdoms) and stated that this is the case even for the TIP (Gautschy, 2015; Krauss, 2015a). Radiocarbon (C14) dating has also proven to be a premier league player in this extremely sophisticated game (Shortland & Bronk Ramsey, 2013), but it still suffers from certain deficiencies that may be set right in the future (see Bietak, 2020)

Before concluding this chapter on chronology and astronomy we recall, albeit briefly, a number of dates, obtained by considering the orientation of sacred structures that we have been expounding throughout this work, especially in Chap. 6. We could well catalogue them as archaeoastronomical, either solar or stellar. Perhaps even lunar ones in the future. Who knows!

We have no doubt about the validity of the orientation of the temple of Hathor in Dendara towards Meskhetyu in 54 BC, particularly to the raising of Alkaid. The chronological and astronomical framework obtained is more than reasonable. If we can confirm their use beyond doubt, as is the case, the stars can offer an accurate chronological framework.

Another, much more substantial example is that of the great temple of Amun at Karnak. Orientated, as we discussed in Chap. 6, to sunrise at the winter solstice and perpendicular to the Nile, our proposal was that it would also be aligned at the time when the winter solstice coincided with Wepet Renpet, towards the year 2004 BC, with a margin of a quarter of a century approximately.

This certainly suggests that the first platform identified in the excavations, which seems to maintain the same orientation of the great temple later erected on the site by Senuseret I, could have been built (and orientated) during the reign of the great king Mentuhotep II. Perhaps even under Amenemhat I, although this would only be valid within the framework of the high chronology. Again, the solution seems reasonable. This would compare perfectly with the similar orientation of the funerary complex of Mentuhotep II at Deir el-Bahari, which followed a similar pattern (Chap. 6). For this, we had to assume a certain architectural model (Gabolde, 2015) for this impressive temple.

Further examples, but related to the important festival of Nehebkau, at the beginning of the Growing Season (I Peret 1), can be found at Hatshepsut's Djeser-Djeseru at Deir el-Bahari (Chap. 6) and the great temple of Ramesses II at Abu Simbel (Belmonte et al., 2009b). Here, the king, speaking in the third person, dictated the following inscription:

He made these great and perfect monuments for all eternity. He knew how to orientate them towards the Horizon of the Sky.

Words are unnecessary!

The land- and skyscapes of ancient Egypt provided the ideal framework for the development of Egyptian history, for establishing a comprehensive chronology, and, most importantly, for the understanding of the ancient Egyptian cosmic order.

Appendix: Glossary of Ancient Egyptian Astronomical Terms and Related Words

Fig. A.1 Limestone blocks decorated on both sides with five pointed stars with which, according to R. Stadelmann, the manoeuvre chamber (located above the granite burial chamber) and the lateral galleries of the stepped pyramid of Netjerkhet were walled up. The first example of this short of astronomical decoration in Egyptian history. (Image by the authors)

Abd(u) [*3bd*]: The lunar month, although from very early on, it was written in a similar way as the second day of the lunation (LD2) in which the lunar crescent was normally seen for the first time, so many Egyptologists identify it simply with the crescent. It is a typical Egyptian term of multiple meanings. However, it is highly possible that the vocalization was different for both terms, as in certain Coptic dialects, in which month is pronounced and written 'abot' (ⲁⲃⲟⲧ), and Monday, the second day (of the week), 'abit' (ⲁⲃⲓⲧ).

J. A. Belmonte, J. Lull, *Astronomy of Ancient Egypt*, Historical & Cultural Astronomy, https://doi.org/10.1007/978-3-031-11829-6

Akh [*ꜣḫ*]: Spirit, bright, beneficial. This term can have all these meanings and more. In particular, it applied to beneficial spirits who became recognizable stars in the sky. One of the stellar decans was called Akhuy, the two spirits. It is a spiritual concept that, like *ka* and *ba*, was related to man. It can represent transfiguration and, as a solar principle, divine cosmic energy. In the *Pyramid Texts* (PT 305 §474), it is said that 'the "akh" belongs to heaven, the corpse to earth'.

Akhet [*ꜣḫt*, , the name is included in hieroglyphic to clearly differentiate it from the following term, although vocalization could be different, as in *ꜣbd*] **1:** The Season of the Flood, the first of the Egyptian civil calendar. At the time of the founding of this calendar, it extended more or less to the months when the benefactor waters covered Egypt, between the end of June and October. Later it wandering ahead on the tapestry of the climatic seasons owing to the non-existence of the leap year in the Egyptian system of time reckoning. The seasons of the calendar year and the tropic year fitted again after a period of just over 1500 years.

Akhet [*ꜣḫt*, , in the Old Kingdom, or later on] **2:** The Horizon. Perhaps homophone of the previous entry with which it should never be confused. The 'akhet' was one of the parts that constituted the Egyptian cosmos. It was the place where the sun rose every morning and set every evening. It appears in the name of cities, such as Akhetaten ('The Horizon of the Disk') or divinities such as Hor-em-akhet ('Horus at the Horizon') or Re-Horakhty ('Re, the Horus of the two horizons'). It appears with its ancient spelling in the name of the Great Pyramid, Akhet Khufu, that could be translated as Khufu's Horizon, or as Khufu belongs to the horizon.

Ankh [*ʿnḫ*]: Cross-shaped symbol, which, as a hieroglyph, means life. It is very common to see it in the hands of a divinity.

Anuqet [*ʿnḳt*]: (Anukis in the papyri of the Greek period): Goddess of the First Cataract, companion of the god Khnum. Together with him and the goddess Satet, she was supposed to control the arrival of the Flood. She had her temple on the island of Sehel, a few kilometres south of Elephantine. She is represented in the Zodiac of Dendara but it has not been possible to establish unambiguously whether she represents a constellation of the Egyptian skies on her own or its presence is merely symbolic.

Aphelion: Furthest point from the Sun's centre in orbit of a body that revolves around the sun.

Apokatastasis: [fr. Greek, ἀποκατάστασις]: The moment at which the first day of the Egyptian civil calendar, or Wepet Renpet, and the heliacal rising of the star Sirius, or Peret Sopdet, again concurs.

Apotropaic: say of what serves to nullify or limit the influences of evil beings.

Arcus visionis (**β**)**:** Angle of the sector of a major circle connecting a celestial object above the horizon and the sun below it. It represents the minimum angle for the object to be visible against the solar glare. It is most relevant for calculating the date of the heliacal rising of Sirius or the visibility of a first waxing lunar crescent at dusk or a last waning lunar crescent at dawn.

Atmospheric extinction: Dimming of the light from celestial objects due to the greater thickness of the atmosphere at the level of the horizon. The lower the height of a star, the dimmer the filter of the atmosphere makes it look, even rendering it virtually impossible to observe. This effect can be dramatic under Saharan dusty conditions.

Azimuth: The clockwise arc between due north and the **vertical** of the star,

Ba [*bꜣ*]: plural, bau. The personal and animated manifestation of the deceased, with which he can travel through the sky or in his own grave. In a way, it personifies the consciousness of the individual, usually depicted as a bird with a human head. It is sometimes translated as soul although no complete parallelism exists.

Benben [*bnbn*]: The primordial hill, the first piece of land that emerged from the waters of the primaeval ocean. By extension, it was the name of a stone representing the primordial hill. The most famous was the one at Heliopolis.

Cenotaph: False, symbolic tomb for the funerary cult.

Colour index (B-V): The result of the subtraction between the blue magnitude of a star and its visual magnitude.

Colossus: Statue of greater than natural proportion, usually dedicated to a pharaoh and placed in most cases on the outside of the doors or pylons of temples.

Cosmogony: Mythological theory about the creation of the Universe, elaborated at different religious centres such as Heliopolis, Memphis, Hermopolis, Esna, and others.

Cosmovision or world-view: A given culture's way of understanding, seeing and interpreting the world and the universe.

Culmination: The passage of a celestial body across the local meridian. Circumpolar stars have two culminations, an upper one above the celestial pole and a lower below it.

Declination [δ]: In the equatorial coordinate system, the meridian arc in degrees from the celestial equator to the star.

Demiurge: The creator of the world, who comes to life in the depths of the primordial ocean.

Demotic: Egyptian writing of highly cursive nature, written mainly on papyrus. The demotic was born as a development of the hieratic script around the 7th century BC and was used at least until AD 452, when it was finally substituted by the Coptic alphabet.

Duat [*dw3t*]: In the *Pyramid Texts*, a region of the afterlife reached by the dead on their journey to heaven. It seems to be closely related to the stars, the sky, and the celestial vault.

Ecliptic: The apparent path of the sun on the celestial sphere. It is the projection of the plane of the orbit of the earth around the sun. Its inclination to the equator, ε, is today *c.* 23½°, but was closer to 24° in the era of the pyramids. Today, it decreases at a rate of 0.46 seconds of arc per year.

Elongation: Angular distance from the sun. This can be west, when the planet is observed in the morning, or east in the afternoon. The smaller the elongation, the harder the planet becomes to detect because of its greater proximity to the sun.

Ennead: Group of nine divinities. The best known comes from Heliopolis, where the sun god Re was followed by Shu and Tefnut, Geb and Nut and their children Osiris, Isis, Seth, Nephthys, and Haroeris.

Equinox: One of the two points at which the sun crosses the celestial equator and its declination is 0°. There are two equinoxes, one in the spring and the other in autumn. The name derives from the day on which light and darkness are approximately equal. By extension, it is applied to the day when this phenomenon occurs. It is not clear whether, prior to the Hellenistic period, the ancient Egyptians had a similar concept or not.

Heliacal setting: The moment at which a star is seen for the last day before entering into conjunction with the sun.

Heliacal rising: The first day on which a star can be observed at dawn after its conjunction with the sun.

5 Heryu Renpet [*5(djw) ḥrjw rnpt*]: The 'five' upon the year. This is the name given by the Egyptians to the five supernumerary days that were added to the year of 12 months of 30 days each to complete the Egyptian year of 365 days. The Greeks gave them the name 'epagomenoi', the name by which they are usually known. They are still used today in the liturgical calendar of the Coptic church in which they are called the 'little month'.

Hieratic: Egyptian cursive script born as a development of hieroglyphic writing, usually employed on papyri and ostraca. This script was used for official documents, accountability, or in religious contexts.

[Astronomical] Horizon: The great circle of the celestial sphere that is perpendicular to the vertical of the place. It divides the celestial sphere into two hemispheres, the upper (visible) one, and the lower (invisible) one. In standard conditions, the astronomical horizon is often below the apparent visual horizon, due to orographic elements such as mountains, trees, and buildings.

Hypostyle hall: Room of columns. It was the most outstanding and amazing part of the temples that, at times, had a row of such halls. The most famous are those of the temple of Amon-Re in Karnak and the temple of Hathor in Dendara.

Ikhemu sekiu. [*jḫmw skj(w)*, 'those who do not know destruction']: A term applied to a set of stars that Egyptologists have called 'imperishable'. They are often mentioned in the *Pyramid Texts* as one of the deceased king's favourite celestial destinies. They are usually identified with the circumpolar stars, because they never set or rise, a fact that could be interpreted as never having been born or destined never to die; in other words, they were eternal. Some authors suggest as an alternative that the imperishable ones would be those stars that were visible every night of the year even if they rise or set during the night. Meskhetyu was the imperishable constellation *par excellence*, while Arcturus would be an imperishable star according to the latter possibility.

Ka [*kꜣ*]: plural, *kꜥu*. The vital force of a human being; the spiritual, invisible, and intangible element which imbues men and women with life, and distinguishes them from corpses. The ka was born with the person, and accompanied him or her as a double all through life. However, when that person died, the ka continued in existence. The ka would then need, even if only symbolically, food and drink to continue living.

Logogram: A sign that represents an entire word. Egyptian hieroglyphs were a combination of logograms and phonograms (either mono-, bi- or trilateral), resulting in a logo-syllabic writing system.

Lunation or synodic month: The interval of time that has elapsed between two analogous positions of the earth, moon, and sun. Hence, it is the period of time between two identical phases. It has an average value of 29.5306 days. Often called a lunar month, although in ancient Egypt this was ambiguous.

Lunar apogee: The point in the lunar orbit that is farthest from earth, **perigee** being the closest point.

Lunar cycle: A 19-year period in which the phases of the moon are repeated on the same calendar dates; also known as the **Metonic Cycle**. Less precise alternatives were the 8-year cycle, or **Octaeteris**.

Ma'at or Maat [*mꜥꜣt*]: The personification of law, governing everything that exists. The concepts of law, order, and truth merge in her. As a concept, ma'at is equally related to justice, truth, and cosmic order.

Magnitude: Unit of measurement of the brightness of the stars observed from earth. It is measured on a logarithmic scale, reflecting the response of the human eye. The dimmer a star is, the higher magnitude is. There are very bright stars, such as Sirius and Canopus, that even have negative values.

Mastaba: word of Arabic origin meaning bench for sitting. It is used to designate the exempt tombs of the Archaic Period and the beginning of the Old Kingdom. Its shape is that of an elongated cube with a smooth ceiling and slightly sloping walls.

Maximum Digress(ion) [aka maximum elongation]: In astronomy, the maximum separation of a star from the local meridian in its nightly journey. Often called elongation, there are two, one eastern and the other western. It applies particularly to circumpolar or almost circumpolar stars.

Meridian line: The horizontal line joining the points of intersection of the local meridian with the horizon. It is commonly known as the N–S line.

Meridian: A great circle that intersects the rotational axis of the world. The **local meridian** passes through the celestial pole and the zenith, intersecting the horizon in the north and south cardinal points.

Merkhet [*mrḫt* o *mrḫyt*]: The name of an astronomical instrument frequently mentioned in inscriptions, especially in the Ptolemaic Period, associated with the computation of time and the orientation of temples. Traditionally, the merkhet has usually been related to certain L-shaped devices arranged with the long arm lying horizontally that served to measure the hours of the day, since this is the determinative that is used to write its name (a plumb line is often added in the hieroglyph). However, its precise form and method of use still remain to be clarified.

Meskhetyu [*msḫtjw*]: One of the most important constellations of the firmament of ancient Egypt. It was the 'imperishable star' (see *Ikhemu Sekiu*) *par excellence*. It is most certainly equivalent to the asterism of the Plough (or Big Dipper) in the constellation of the Big Bear (Ursa Major). The Egyptians recognized in it either a bull's foreleg (sometimes a complete bull), or one of the celestial adzes used in the ceremony of the opening of the mouth. Two of its stars (now called Phecda and Megrez) marked the direction of the pole (and the pole star of the time, Thuban or α Draconis), in the era of the great pyramids.

Naos: Sanctuary in which statues representing the deities were preserved inside the sancta sanctorum of a temple. They were usually made of wood and were occasionally placed inside another monolithic stone naos. The term naos is sometimes used to designate the temple itself.

Nehaha: Instrument that symbolizes the royal power of Pharaoh. It is a scourge.

Nome [Greek *nomos*, plural *nomoi*]: Territorial district, comparable to a province. In Upper Egypt there were 22 and in Lower Egypt 20.

Nut [*nwt*]: The goddess of the sky *par excellence*, daughter of Shu (the ethereal space) and Tefnut (the wet space) and sister-wife of the earth god Geb. She is the mother of Osiris, Isis, Nephthys, Seth, and Haroeris (Horus the Elder), the five divinities of the epagomenal days. She is often represented with the arched body simulating the celestial vault, through which the stars and the sun travel. According to some authors, the goddess should be more precisely identified with the Milky Way, the true backbone of the sky. This is not universally accepted.

Obelisk: Monolithic squared column, usually of pink granite, and crowned by a pyramidon which was covered with electron, an alloy of gold and silver. Obolisks were basic elements of the solar cult.

Ogdoad: Group of eight divinities, the best known of which is that of the city of Hermopolis.

Omphalos: (plural *omphaloi*), rounded sacred stone in the shape of half a primordial egg. This was related to creation.

Opposition: A planet is said to be in opposition when it is on opposite side of the sky to the sun; that is, when the sun, earth, and the planet are configured in an approximately straight line. The instant of opposition is defined as occurring when the apparent geocentric celestial longitude of the body differs by 180° from the apparent geocentric longitude of the sun.

Osireion: Name designating an underground megalithic construction behind the temple of Sethy I in Abydos, symbolically related to the regeneration of Osiris.

pedj-sesh(er) [*pḏ-sšr*]: Literally, the stretching, or tensioning, of the cord. It is the name of a ceremony performed during the founding rites of the sacred buildings of Pharaonic Egypt. During this rite, the goddess Seshat and the king himself stretched a cord or rope between two poles to establish the plan of the structure. According to the sources, the ceremony began to be performed in the Archaic Period, perhaps earlier, and was in use until the Roman rule of the country.

Peret [*prt*, 'Going Forth']: Second season of the civil calendar that refers to the going forth (*prt*) of the earth after the inundation period. It has sometimes been associated with the rebirth of vegetation after the life-giving flooding of the waters. Both possibilities are reasonable. It originally corresponded to the months between November and February. However, in the Graeco–Roman period it was translated as winter, when being approximately adjusted to the coldest months during that time (December to March).

Peret Sopdet [*prt spdt*, 'Going Forth of Sopdet']: The name of a feast, and of an associated astronomical event (the heliacal rising of Sirius, i.e. Sopdet) that marked an important milestone in the Egyptian calendar. This festival acted as a herald of the rising waters of the flooding and the arrival of the actual inundation. Throughout the history of the pharaohs, the event wandered between the end of June (summer solstice) and mid-July in the proleptic Gregorian calendar, but stayed almost fixed at July 19 in the Julian proleptic calendar. The delay of the phenomenon, of one day every four years, on the dates of the civil calendar has been used, not without great controversy, to fix some periods of Egyptian chronology. These are the so-called Sothic dates. The theoretical displacement of Peret Sopdet through the 365-day civil calendar year over a period of approximately 1460 Julian years (1461 calendar years) is known as the 'Sothic cycle'.

Peribolos: Wall that delimits the sacred space of a temple or temenos.

Perihelion: The point closest to the centre of the sun of a body in the orbit around it. Throughout recorded history, in the case of the earth, this has happened in January.

Phonogram: Written sign that recalls a sound.

Precession of the equinoxes: Retrograde movement of the intersection of the celestial equator and the ecliptic (measured from the first point of Aries), caused by the movement of the earth's polar axis, like that of a spinning top. As a consequence, the equatorial coordinates of the stars (including the pole star) change with time.

Pronaos: Room that precedes the naos of a temple. Sometimes it coincides with (one of) the hypostyle halls.

Pesedjentyu. [*psḏntjw*]: 'Neomenia', new moon, conjunction, or lunar day one (LD1). This name was applied to the first day of the Egyptian lunation, the next to the day when the crescent waning moon was visible at dawn for the last time. On that day, the new moon was supposedly begotten. It used to be an important festival day and is recorded on many occasions in the sources, associated with certain dates of the civil calendar, which has allowed the so-called 'lunar dates' to be used to fix, not without controversy, certain moments of the Egyptian chronology.

Pylon: Monumental entrance to a temple, formed by a pair of massive towers with an opening in the centre, usually worked in the form of a portal. The pylon with the disc of the sun in the middle resembled the hieroglyph 'akhet', i.e. horizon.

Pyramid Texts: Collection of utterances written in the hieroglyphic script inside the burial chamber of Wenis and several kings and queens of the 6th Dynasty. They represent a sort of astral eschatology, helping the deceased to reach his or her celestial destiny either among the imperishable stars or the unwearing stars at the solar bark. It is the oldest registered religious corpus in the history of the world. Several Egyptian constellations were mentioned in them for the first time.

Quadrature: A special position of a planet in its orbit. This occurs when the planet forms an angle of 90° in the sun–planet–earth system.

Ramesseum: Name given to the Million Year Mansion (aka funerary temple) of Ramesses II in western Thebes. It includes one of the finest celestial diagrams of ancient Egypt.

Ramesside: Relating to the time of the Ramesses, Egyptian 19th and 20th dynasties, applied in particular to the Ramesside star clocks of the tombs of Ramesses IV, VII, and IX.

Reret [*rrt*]: The female Hippopotamus, one of the most suggestive constellations of the Egyptian skies. Of enormous proportions, it probably stretched from the bright star Vega (presumably, a hypostasis of Isis), and the head of Draco to Boötes and Corona Borealis. She is represented in the vast majority of celestial diagrams of ancient Egypt, including the Zodiac of Dendara, and is depicted next to one or more mooring posts (*mnjt*). Frequently associated with the goddess Ipet, protector of childbirth and pregnant women, she was one of the celestial manifestations of the goddess Isis. Several stars in the constellation were used in the Ramesside star clocks to calculate the hours of the night.

Retrograde motion: The retrograde movement or retrogression of a planet is a transient phenomenon that, with respect to the superior planets, occurs when the earth overtakes the planet in its orbit. The loop that originates in retrograde motion is due to the different orbital inclination of these planets with respect to the plane of the ecliptic.

Right ascension [RA or α]: The arc measured anticlockwise along the equator, from the vernal point or First Point of Aries to the meridian of a star. It is one of the coordinates of the equatorial system.

Sah [*sꜣḥ*]: An important constellation of the Egyptian firmament, mentioned in the *Pyramid Texts*, in which he pairs with Sopdet and sponsors the deceased pharaoh on his journey to heaven as Horus-Sopdu or as the Morrning Star. Later on, he was part of the decan stars. Sah was associated from an early date with the god Osiris. Usually identified with the constellation of Orion, it is very likely that he occupied only the southernmost sector of this asterism, extending from Orion's Belt southwards to the constellation of Lepus.

Sancta sanctorum: The most sacred part of the temple, where the statue of the deity is located.

Satet [*sṯt* or *sṯjt*]: Goddess of the First Cataract, forming a triad with Anuqet and the god Khnum. Her temple on the island of Elephantine is one of the oldest recorded in Egypt, as excavations have shown that, by 3600 BC, there was already a sanctuary on the site. Generator of the rising flood waters next to Khnum and Anuqet, she was therefore identified with Sopdet (Sothis) as a herald of the inundation. She appears in the Zodiac of Dendara located behind Sopdet. However, it is impossible to specify whether she represents a different star or an individualized constellation or, on the contrary, she should simply be identified as an alternative manifestation of Sothis.

Seba [*sbꜣ*]: Star. This is the standard term for a celestial body in Egypt. The sign for it (Gardiner's N14) was a five pointed star formed by an internal dot and five rays. The universal five-pointed star symbol presumably originated in Egypt in pre-Dynastic times, when we have a first attestation in an ivory knife handle of unknown provenance (now at MET), and a series of them in the names of royal domains of the kings of the first dynasties, starting by Andjib whose domain was called *ḥr-sbꜣ-ẖt*. The sign was standardized during the Old Kingdom (Fig. A.1). The origin of this sign is uncertain. A possibility is the Egyptian detailed observation of nature, notably the fossils of Miocene/Pliocene clypeasterid echinoderms in the Egyptian deserts. A most singular case would be the *clypeaster aegyptiacus* present in the Giza plateau, which strongly resembles the sign for *dwꜣt*, a *sbꜣ* surrounded by a circle (N15). This subject would be an objective of future studies.

Seqed [*sḳd*] or seked: An Egyptian unit of measurement that was used to indicate the inclination of a wall, including that of the faces of a pyramid. The ancient Egyptians did not use angular but linear units for this magnitude. The seked was fundamentally the number of horizontal palms (and fingers) that the wall had to be tilted for each royal cubit in height. Since a cubit had 7 palms, a seked 5/7, or simply 5, implied a slope equivalent to an angle of almost 55°.

Semedet [*smdt*]: Day of the full moon. It received this name from the fifteenth day of the moon (LD15) when the moon was full regardless of whether the corresponding lunation duration was 29 or 30 days. It is worth noting that the moon could looked full as early as LD14 or as late as LD16. It was an important date for the celebration of festivals.

Serdab: From the Arabic 'basement'. In order to protect the statues of the deceased, a necessary precaution for their continued existence in the new life, the Egyptians placed them in closed chambers called serdabs by Egyptologists. In ancient Egyptian, it was called *pr twt*, 'house of the image'. On many occasions, the serdab was related to a place of offerings through wall openings, as in Djoser's pyramid complex.

Serekh [*srẖ*]: Of the five titles or royal names by which a pharaoh was called, the name of Horus is found in a rectangular space on which a falcon, an image of Horus, rests. This space is the serekh, which symbolizes the façade of a palace. The kings of the proto-Dynastic Period and the early 3rd Dynasty were mostly known by their serekhs.

Serqet [*srḳt*]: Selkis in the papyri of the Greek period. One of the four protective goddesses of the Egyptian pantheon, she had her equivalent in the firmament of ancient Egypt. However, given the ambiguities in her representation in the celestial diagrams and that it was not used in any of the star clocks, her precise identification in the firmament is very difficult. This volume defends her equivalence to the constellation of Virgo or even with part of Ursa Minor, but both possibilities are far from being verified.

Seshat [*sš3t*]: Goddess of writing and time reckoning of ancient Egyptians. She was a very ancient divinity, already mentioned in the Palermo Stone in close relationship to the ceremony of the stretching of the cord. Her hieroglyphic sign, which is depicted above the goddess's head, has been the subject of debate and speculation. In this volume, a hypothesis is raised: this sign could be a very stylized representation of a transit instrument, similar to the Roman groma, presumably used by the Egyptians for the orientation of their sacred buildings.

Shemu [*šmw*]: Third season of the Egyptian civil calendar. It could be translated according its origins as 'drought', when it covered the months of low waters between March and June. Afterwards, in the papyri of the Graeco–Roman period, the term was translated as 'summer', because in that epoch the calendar season coincided with the hottest months of the year (May to August). Its usual translation in ancient Egypt manuals as 'harvest' should, however, be abandoned.

Shendjut [*šnḏwt*]: Piece of men's dress made of linen cloth, which was girded at the waist like a skirt.

Sidereal period: The time, in average days, taken by a planet to describe its orbit around the sun (or the moon around the earth). That is, the time in which the sun–planet or earth–moon vector radius describes a 360° angle. It should not be confused with the synodic period.

Sidereal year: The time that the sun needs to travel 360° in the ecliptic, going back to the same position relative to the background stars. Today, it is 365.2563 days long.

Solar conjunction: A star or planet is said to be in conjunction with the sun when it is aligned with it as seen from earth. In conjunction, stars and planets are invisible owing to the glare of sunlight. A solar eclipse is a best example of a conjunction.

Solstice: Point where the apparent motion of the sun in declination reverses. This is either a point on the horizon or a point in the ecliptic. There are two solstices, one in winter and the other in summer. In the ecliptic, the line of the solstices is perpendicular to that of the equinoxes. The solstices correspond to the points of the ecliptic whose angular distance from the celestial equator is maximum and equals ε, the obliquity of the ecliptic.

Sopdet [*spdt*]: Sothis in the papyri of the Graeco–Roman period, hence the adjective Sothic. The brightest star in the Egyptian skies, equivalent to our Sirius (α Canis Majoris). Sopdet was apparently part of a constellation or asterism of the same name associated with a goddess of the greatest importance. Already mentioned in the *Pyramid Texts*, she was related to the goddess Isis from an early age, of whom this star would be her most important celestial manifestation. Sirius' heliacal rising (see Peret Sopdet) was certainly recognized by the time of the Middle Kingdom, perhaps earlier, as a herald of the arrival of the true water flooding (not of the civil season Akhet, aka Inundation).

Sothic cycle: A 1461 calendar year long period of time. This period takes into account two factors: the 365 days of the civil calendar year, and the observation of the heliacal rising of Sirius. According to Censorinus, a Sothic cycle began when Sirius' heliacal rising coincided with the first day of the Egyptian calendar year (I Akhet 1). Because of the difference between the 365-day civil year and the *c.* 365½ days Sothic year, this coincidence theoretically occurred 1460 Julian years later, which is the duration of that period.

Syncretism: Fusion, convergence or union of ideological elements in the sphere of religious thought.

Synodic period: Time for a planet to orbit the sun (or the moon around the earth), but relative to the sun itself as appreciated by a terrestrial observer. Hence, the period of a synodic revolution is an apparent quantity related to our geocentric perception of celestial movements.

Temenos: Sacred space that surrounds a temple, delimited by the peribolos.

Tepy Renpet: [*tpj rnpt*]: The first (literally the one in front) of the year, it was the name of a festival. It should refer to a feast that took place at the beginning or on dates close to the beginning of a year. However, Tepy Renpet's precise time framework is unknown. According to some authors, it could be the first day of the first lunation that followed the beginning of the year. According to others, it would be the name of the first day of the hypothetical lunar calendar, while some propose that it was the term used for Sirius' heliacal rising during the Old Kingdom (this latter option should probably be abandoned today). In this book, we support the hypothesis that it would be the first day of the first lunation that followed the rising of the beneficial waters, and thus the arrival of the flooding. This could be the date of the beginning of the year in an original lunar calendar related to the behaviour of the Nile.

Tropical year: Also called solar year, the time elapsed between two successive passages of the sun through the vernal equinox or First Point of Aries. It is shorter than the sidereal year because of the backwards movement of the vernal equinox due to precession. It has a length of 365.2422 days today, but was 365.2425 days long in Antiquity. The 365½ days Julian calendar and the 365.2425 days Gregorian calendar try to accommodate its numbers.

Uraeus: [plural, *uraei*]: Cobra snake placed on the king's forehead to protect him by throwing fire at his enemies. The cobra is associated with the goddess Wadjet or with the eye of the sun.

Uroboros: A snake biting its tail.

Vernal Point or First Point of Aries: The point at which the sun ascending along the ecliptic intersects with the celestial equator. This marks the moment of the equinox. The First Point of Aries is the one that corresponds to the passage from a negative to a positive declination, and the First Point of Libra from positive to negative declination. Both are known as equinoxes.

Vertical: The planes, or great circles, that pass through the zenith-nadir line. The First Vertical in particular is the one passing through east and west, defining on the ground, the perpendicular E–W line. Angular height is always measured on a vertical, while azimuth is measured on the horizon.

Was: [*wꜣs*]: The characteristic sceptre of the gods and goddesses of Egypt that has the shape of a stylized canid head in its upper part. It also meant 'power'.

Wepet Renpet: [*wpt rnpt*, 'the one that opens the year']: Name of New Year's Eve of the Egyptian civil calendar (I Akhet 1, or first day of Thoth), from the invention of this calendar until the sunset of the Pharaonic period. It should never be confused with Peret Sopdet. By extension, it is also the original name of the twelfth month of the calendar, IV Shemu, later known as Misore (aka *mswt rꜤw*, the 'birth of Re').

Weresh: [*wrš*]: Temple service cycle. This was a period of time equivalent to a synodic month, in which a phyle, or priest gang, served in a temple. They typically began at LD2 *ꜣbd* and ended at LD1 *psḏntjw*. Hence, they were later assimilated as *pꜣ wrš* to the lunar months of the Babylonian lunar calendar of the Assyrian and Persian invaders.

Zenith: The point directly overhead, in the vertical of a place: The zenith is directly above the observer and the **nadir** is in the opposite direction

Zodíac: Zone limited by two planes parallel to the ecliptic, of respective celestial latitudes ±9°. This is the zone containing the orbits of the moon and the most important planets. As an extension, this name was given to the 12 constellations (or the 12 zodiacal signs) identified in that region of the sky. Hence, those Egyptian celestial diagrams including these constellations are also known as Zodiacs.

Bibliography

Abbreviations

ÄA: Ägyptologische Abhandlungen
AnBib: Analecta Biblica
AoF: Archiv für Orientforchung
ASAE: Annales du Service des antiquités de l'Egypte
ÄuL: Ägypten und Levante.
BAR IS: British Archaeological Reports. International Series
BdÉ: Bibliothèque d'étude
BIFAO: Bulletin de l'Institut français d'archéologie orientale
BSÉG: Bulletin de la Société d'Égyptologie, Genève
BSFÉ: Bulletin de la société française d'égyptologie
CdÉ: *Chronique d'Égypte*
CENiM: Cahiers Égypte Nilotique et Méditérranéenne
CNIP: Carsten Niebuhr Institute of Near Eastern Studies Publications
CNRS: Centre national de la recherche scientifique
CRIPEL: Cahiers de recherches de l'institut de papyrologie et d'egyptologie de Lille
EAT: Neugebauer, O. and Parker, R. A., 1960–1969. *Egyptian Astronomical Texts*. Vols. I–III. Providence: Brown University.
EEF: Egypt Exploration Fund
ENiM: Égypte nilotique et méditerranéenne
HÄB: Hildesheimer Ägyptologische Beiträge
HdO: Handbuch der Orientalistik / Handbook of Oriental Studies
IFAO: Institut français d'archéologie orientale
JAOS: Journal of the American Oriental Society
JARCE: Journal of the American Research Center in Egypt
JCS: Journal of Cuneiform Studies
JEA: Journal of Egyptian Archaeology
JEGH: Journal of Egyptian History
JNES: Journal of Near Eastern Studies
KRI: Kitchen, K.A., 1969–1990. *Ramesside Inscriptions, Historical and Biographical*. Vols. I–VIII. Oxford: Blackwell.
MAIBL NS: Mémoires de l'Académie des Inscriptions et Belles-Lettres, Nouvelle Série, Paris.
MÄS: Münchner Ägyptologische Studien
MDAIK: Mitteilungen des Deutschen Archäologischen Institus Abteilung Kairo
MEEF: Memoir of the Egypt Exploration Fund

J. A. Belmonte, J. Lull, *Astronomy of Ancient Egypt*, Historical & Cultural Astronomy, https://doi.org/10.1007/978-3-031-11829-6

MMAF: Mémoires de la Mission Archéologique Française au Caire
NINO: The Netherlands Institute for the Near East
OBO: Orbis Biblicus et Orientalis
OIP: Oriental Institute Publications
OLA: Orientalia Lovaniensa Analecta
RdÉ: Revue d'Égyptologie
RecTrav: Recuil de Travaux Relatifs a la philologie et a l'archéologie égyptiennes et assyriennes
SA: Studia Aegyptiaca
SAK: Studien zur altägyptischen Kultur
SAOC: Studies in Ancient Oriental Civilization
Urk, III: Schäfer, H. (1905). *Urkunden der älteren Äthiopenkönige*. Leipzig: J. C. Hinrichs.
Urk, IV: Helck, W., 1955. *Urkunden der 18. Dynastie. Historische Inschriften Thutmosis' III. und Amenophis II*. Berlin: Akademie Verlag.
Urk, VI: Schott, S. (1929). *Urkunden Mythologischen Inhalts. Bücher und Spruche gegen den Gott Seth*. Leipzig: J.C. Hinrichs.
Wb: Erman, A. and Grapow, H. (Eds.), 1926–1961. *Wörterbuch der Aegyptischen Sprache*, Vols. I–VII. Berlin: Akademie-Verlag.
ZÄS: Zeitschrift für Ägyptische Sprache und Altertumskunde

List of Works Consulted

Abry, J. H. (Ed.). (1993). *Les tablettes astrologiques de Grand (Vosges) et l'astrologie en Gaule Romaine*. De Boccard.

Aja Sánchez, J. R. (2015). *Aguas mágicas. El Nilo en la memoria y la religiosidad del mundo antiguo*. Universidad Nacional de Educación a Distancia.

Allen, E. B. (1947). A coptic solar eclipse record. *JAOS, 67*, 267–269.

Allen, T. G. (1960). *The Egyptian Book of the Dead documents in the Oriental Institute Museum at the University of Chicago* (OIP, 82). University of Chicago Press.

Allen, R. H. (1963). *Star names, their lore and meaning*. Dover Inc. Pub.

Allen, T. G. (1974). *The Book of the Dead or Going Forth by Day. Ideas of the ancient Egyptians concerning the hereafter as expressed in their own terms*. University of Chicago Press.

Allen, J. P. (1988). *Genesis in ancient Egypt. The philosophy of ancient Egyptian creation accounts*. Yale University Press.

Allen, J. P. (1989). The cosmology of the pyramid texts. In W. K. Simpson et al. (Eds.), *Religion and philosophy in ancient Egypt* (pp. 1–28). Yale University Press.

Allen, J. P. (1993). Reading a pyramid. In C. Berger, G. Clerc, & N. Grimal (Eds.), *Hommages à Jean Leclant, Vol. 1: Études pharaoniques* (BdÉ, 106) (pp. 5–28). IFAO.

Allen, J. P. (2000). *Middle Egyptian. An introduction to the language and culture of hieroglyphs*. Cambridge University Press.

Allen, J. P. (2005). The ancient Egyptian Pyramid Texts. In P. Der Manuelian (Ed.), *Writings from the ancient world*. Society of Biblical Literature.

Allen, J. P. (2013). *A new concordance of the pyramid texts* (Vol. 1–6). Brown University.

Alliot, M. (1949). *Le culte d'Horus à Edfou au temps des Ptolémées*. IFAO.

Almansa-Villatoro, V. (2019). The Cultural Indexicality of the N41 Sign for *bjꜣ*: The metal of the sky and the sky of metal. *JEA, 105*, 73–81.

Altenmüller, H. (1965). *Die Apotropaïa und die Götter Mittlelägyptens*. Thesis dissertation.

Altmann-Wendling, V. (2019). Shapeshifter. Knowledge of the Moon in Graeco-Roman Egypt. In J. Althoff, D. Berrens, & T. Pommerening (Eds.), *Finding, inheriting or borrowing? The construction and transfer of knowledge in antiquity and the middle ages* (pp. 213–252). Transcript Verlag.

Altmann-Wendling, V., & Stockhusen, M. (2020). Die Statue eines ägyptischen Priesters und Astronomen (JE 38545). In R. Färber & R. Gautschy (Eds.), *Zeit in den Kulturen des Altertums. Antike Chronologie im Spiegel der Quellen* (pp. 103–115). Böhlau Verlag.

Andrews, C. A. R. (1992). Unpublished demotic texts in the British Museum. In J. H. Johnson (Ed.), *Life in a multi-cultural society: Egypt from Cambyses to Constantine and beyond* (pp. 1–14). University of Chicago Press.

Anonymous. (1991). *The Greek Alexander Romance* (R. Stoneman, Trans.). Penguin.

Anonymous of 379. (1993). *The treatise on the bright fixed stars* (R. Schmidt, Trans.). The Golden Hind Press.

Antelme, R. S., & Rossini, S. (2007). *Nout. Le cosmos des pharaons*. Ed. Du Rocher.

Anthes, R. (1983). *Ägyptische Theologie im dritten Jahrtausend v. Chr* (SA, 9). Université Loránd Eötvös.

Arnold, D. (1974). *Der Tempel des Können Mentuhotep von Deir el Bahari I*. Archäologische Veröffentlichungen 08.

Aristotle. (1922). *De Caelo* (J. L. Stocks, Trans.). Oxford University Press.

Aristotle. (1952). *Aristotle Meteorologica* (H. D. P. Lee, Trans.). Harvard University Press/William Heinemann.

Arlt, C. (2008). The Egyptian Notary Offices in the Ptolemaic Fayyum. In S. Lippert & M. Schentuheit (Eds.), *Graeco-Roman Fayum. Texts and archaeology* (pp. 15–25). Harrasowitz.

Assmann, J. (1970). *Der König als Sonnenpriester: ein kosmographischer Begleittext zur kultischen Sonnenhymnik in thebanischen Tempeln und Gräbern*. J. J. Augustin.

Assmann, J. (1972). Die Inschrift auf dem äußeren Sarkophagdeckel des Merenptah. *MDAIK, 28*, 47–73.

Assmann, J. (1975). *Zeit und Ewigkeit im alten Ägypten. Ein Beitrag zur Geschichte der Ewigkeit*. C. Winter.

Assmann, J. (1977). *Das Grab der Mutirdis*. Philipp von Zabern.

Assmann, J. (1983). Das Doppelgesicht der Zeit im altägyptischen Denken. In A. Mohler & A. Peisl (Eds.), *Die Zeit*. (Schriften der Carl Friedrich von Siemens-Stiftung Nr. 6) (pp. 189–223). Oldenbourg Wissenschaftsverlag

Assmann, J. (1999). *Ägyptische Hymnen und Gebete*. Universitätsverlag / Göttingen: Vandenhoeck Ruprecht.

Assmann, J. (2003). La notion d'éternité dans l'Egypte ancienne. In V. Pirenne-Delforge & Ö. Tunca (Eds.), *Représentations du temps dans les religions* (pp. 111–122). Université de Liège.

Assmann, J. (2011). *Steinzeit und Sternzeit. Altägyptische Zeitkonzepte*. Wilhelm Fink.

Assmann, J. (2019). Ouroboros. The Ancient Egyptian Myth of the Journey of the Sun, Aegyptiaca. *Journal of the History of Reception of Ancient Egypt, 4*, 19–32.

Aston, D. (2013). Radiocarbon, wine jars and New Kingdom chronology. *Ägypten und Levante, 22*, 289–315.

Aston, D. (2016). *In Vino Veritas*. A Docketed History of the New Kingdom between Year 1 of Tuthmosis III and Year 1 of Ramesses II. In J. van Dijk (Ed.), *Another mouthful of dust. Egyptological Studies in Honour of Geoffrey Thorndike Martin* (OLA, 246) (pp. 1–42). Peeters.

Aubourg, É. (1995). La date de conception du zodiaque du temple d'Hathor à Dendera. *BIFAO, 95*, 1–10.

Aubourg, É., & Cauville, S. (1998). En ce matin du 28 Décembre 47 .... In W. Clarysse, A. Schoors, & H. Willems (Eds.), *Egyptian religion the last thousand years: Studies dedicated to the memory of Jan Quaegebeur* (Vol. I–II, pp. 767–772). Peeters.

Aufrère, S. H., & Golvin, J. C. (1997). *L'Égypte restituée, Volume 3. Sites, temples et pyramides de Moyenne at basse Égypte*. Gallimard.

Aufrère, S. H., & Marganne, M. H. (2019). Encounters between Greek and Egyptian science. In K. Vandorpe (Ed.), *A companion to Greco-Roman and late antique Egypt* (pp. 501–518). Wiley.

Aufrère, S. H., Golvin, J. C., & Goyon, J. C. (1991). *L'Égypte restituée, Volume 1, Sites et temples de l'Haute Égypte*. Gallimard.

Aveni, A. F. (1989). *Empires of time*. Harper and Row. Revised Edition 2002, University Press of Colorado.

Aveni, A. F. (1990). *Skywatchers of ancient Mexico*. University of Texas Press.

Baines, J., Van Heel, K. D., & Fischer-Elfert, H.-W. (1998). Abnormal hieratic in Oxford: Two new papyri*. *JEA, 84*, 235–237.

Bakir, A. E.-M. (1966). *The Cairo calendar no. 86637*. General Organisation for Govt. Print. Offices.

Baratta, N. C., & Magli, G. (2021). The role of astronomy and FengSui in the planning of Ming Beijing. *Nexus Network Journal, 23*, 767–787.

Bard, K. A. (1998). *Encyclopaedia of the archaeology of ancient Egypt*. Rotledge.

Barguet, P. (1950). Les stèles du Nil au Gebel Silsileh. *BIFAO, 50*, 49–63.

Barguet, P. (1952). Le rituel archaïque de fondation des temples de Medinet-Habou et de Louxor. *RdÉ, 9*, 1–22.

Barguet, P. (1977). Le cycle lunaire d'après deux textes d'Edfou. *RdÉ, 29*, 14–20.

Barsanti, M. A. (1906). Fouilles de Zaouiét el-Aryan. *ASAE, 7*, 257–286.

Barta, W. (1973). *Untersuchungen zum Götterkreis der Neunheit* (MÄS 28). Deutscher Kunstverlag.

Barta, W. (1980). Thronbesteigung und Krönungsfeier als unterschiedliche Zeugnisse königlicher Herrschaftsübernahme. *SAK, 8*, 33–53.

Barta, W. (1981a). *Die Bedeutung der Pyramidentexte für den verstorbenen König* (MÄS, 39). Deutscher Kunstverlag.

Barta, W. (1981b). Der Dekankalender des Nutbildes und das Sothisdatum aus dem 7. Regierungsjahr Sesostris' III. *SAK, 9*, 85–103.

Bauval, R. (1989a). A master plan for the three pyramids of Giza based on the three stars of the belt of Orion. *Discussions in Egyptology, 13*, 7–18.

Bauval, R. G. (1989b). Investigation on the origins of the Benben Stone: Was it an iron stone? *Discussions in Egyptology, 14*, 5–17.

Bauval, R. (2006). *The Egypt code*. Century.

Bauval, R., & Brophy, T. (2013). *Imhotep the African. Architect of the cosmos*. Disinformation.

Bauval, R., & Gilbert, A. (1994). *The Orion mystery*. Crown.

Beatty, M. H. (1998). *The image of celestial phenomena in the book of coming forth by day: An astronomical and philological analysis*. PhD thesis, Temple University.

Bedier, S. (1994). Ein Stiftungsdekret Thutmosis III. *Bulletin of the Center of Papyrological Studies, 10*, 1–23.

Beinlich, H. (1991). *Das Buch vom Fayum. Zum religiösen Eigenverständnis einer ägyptischen Landschaft*. Otto Harrassowitz.

Belmonte, J. A. (1999). *Las leyes del cielo*. Temas de Hoy.

Belmonte, J. A. (2000). Astronomía y arquitectura: el papel de los astros en la cultura y el arte del antiguo Egipto. In M. A. Molinero & D. Sola (Eds.), *Arte y Sociedad del Antiguo Egipto* (pp. 109–136). Encuentro.

Belmonte, J. A. (2001). On the orientation of the Old Kingdom pyramids. *Archaeoastronomy, 26*, S1–S20.

Belmonte, J. A. (2002). The Decans and the Ancient Egyptian Skylore: An astronomer's approach. *Memorie della Societa Astronomica Italiana, 73*, spec. 1, 43–57.

Belmonte, J. A. (2003a). Some open questions on the Egyptian calendar: An astronomer's view. *Papers on Ancient Egypt (TdE), 2*, 7–56.

Belmonte, J. A. (2003b). A celestial map of the ancient Egyptian firmament. In A.-A. Maravelia (Ed.), *European archaeoastronomy and the orientation of monuments in the Mediterranean basin* (BAR IS, 1154) (pp. 31–36). Archaeopress.

Belmonte, J. A. (2003c). The Ramesside star clocks and the ancient Egyptian constellations. In M. Blomberg, P. E. Blomberg, & G. Henriksson (Eds.), *Calendars, symbols, and orientations: legacies of astronomy in culture* (Uppsala Astronomical Observatory report, 59) (pp. 57–65). Ocarina Books.

Belmonte, J. A. (2006). Astronomy on the horizon and dating, a tool for ancient Egyptian chronology? In E. Hornung, R. Krauss, & D. A. Warburton (Eds.), *Ancient Egyptian chronology* (HdO, 83) (pp. 380–385). Brill.

Belmonte, J. A. (2009). The Egyptian calendar: Keeping Ma'at on Earth. In J. A. Belmonte & M. Shaltout (Eds.), *In search of cosmic order, selected essays on Egyptian archaeoastronomy* (pp. 75–132). Supreme Council of Antiquities.

Belmonte, J. A. (2010). Ancient Egypt. In C. Ruggles & M. Cotte (Eds.), *Heritage sites of astronomy and archaeoastronomy in the context of the UNESCO World Heritage Convention* (pp. 117–136). Icomos.

Belmonte, J. A. (2012). *Pirámides, templos y estrellas: astronomía y arqueología en el Egipto antiguo*. Crítica.

Belmonte, J. A. (2013). DNA, wine and eclipses: The Dakhamunzu 'affaire'. *Anthropological Notebook, 19*, 419–441.

Belmonte, J. A. (2016). Cosmic landscapes of ancient Egypt: A diachronic perpective. *Culture and Cosmos, 20*, 3–30.

Belmonte, J. A. (2021). What equinox? In E. Boutsikas, S. C. McCluskey, & J. Steele (Eds.), *Advancing cultural astronomy: Studies in honour of Clive Ruggles* (pp. 11–31). Springer.

Belmonte, J. A. (2022). Nefertiti strikes back! A comprehensive multidisciplinary approach for the end of the Amarna Period. In N. Kawai and B.G. Davies (Eds.), The star who appears in thebes: studies in Honour of Jiro Kondo (pp. 54–86). Abercromby Press.

Belmonte, J. A., & Magli, G. (2015). Astronomy, architecture and symbolism: The global project of Sneferu at Dahshur. *Journal for the History of Astronomy, 46*, 173–205.

Belmonte, J. A., & Magli, G. (2015). *Kha-Sneferu*, ¿un proyecto singular en Dahshur? astronomía, arquitectura y simbolismo en la IV Dinastía, in *V Congreso Ibéricco de Egiptología*, Burgos Bernal L., Pérez Largacha, A. y Vivas sanz, I. (eds.) Ediciones de la UCLM, 157 (pp.1583–1598).

Belmonte, J. A., & Shaltout, M. (2006). On the orientation of ancient Egyptian temples: (2) new experiments at the oases of the Western Desert. *Journal for the History of Astronomy, 37*, 173–192.

Belmonte, J. A., & Shaltout, M. (2007). The astronomical ceiling of Senenmut: A dream of mystery and imagination. In M. Zedda & J. A. Belmonte (Eds.), *Light and shadows in cultural astronomy* (pp. 145–154). Associazione Archeofila Sarda.

Belmonte, J. A., & Shaltout, M. (2009). *In search of cosmic order, selected essays on Egyptian archaeoastronomy*. Supreme Council of Antiquities.

Belmonte, J. A., & Urrutia-Aparicio, M. (2022). The solstice delusion! Challenging winter solstice alignments in ancient Egypt. In A. Stoev & P. Maglova (Eds.), *28th SEAC Conference on Cultural astronomy and ancient skywatching, Stara Zagora, September 2021*, in the press.

Belmonte, J. A., & Zedda, M. P. (2007). Light and shadows on the pyramids. In M. Zedda & J. A. Belmonte (Eds.), *Light and shadows in cultural astronomy* (pp. 188–196). Associazione Archeofila Sarda.

Belmonte, J. A., Shaltout, M., & Fekri, M. (2005). Astronomy and landscape in ancient Egypt: Challenging the enigma of the minor step pyramids. *Papers on ANCIENT EGYPt (TdE), 4*, 7–18.

Belmonte, J. A., Shaltout, M., & Fekri, M. (2008). On the orientation of ancient Egyptian temples: (4) epilogue at Serabit el Khadem and overview. *Journal for the History of Astronomy, 39*, 181–211.

Belmonte, J. A., Molinero Polo, M. A., & Miranda, N. (2009a). Unveilling Seshat: New Insights into the stretching of the Cord Ceremony. In J. A. Belmonte & M. Shaltout (Eds.), *Search of cosmic order. Selected essays on Egyptian archaeoastronomy* (pp. 197–212). Supreme Council of Antiquities.

Belmonte, J. A., Shaltout, M., & Fekri, M. (2009b). Astronomy, landscape and symbolism: A study on the orientations of ancient Egyptian temples. In J. A. Belmonte & M. Shaltout (Eds.), *In search of cosmic order, selected essays on Egyptian archaeoastronomy* (pp. 213–284). Supreme Council of Antiquities.

Belmonte, J. A., Fekri, M., Abdel-Hadi, Y., Shaltout, M., & González-García, A. C. (2010). On the orientation of ancient Egyptian temples: (5) testing the theory in Middle Egypt and Sudan. *Journal for the History of Astronomy, 41*, 65–93.

Belmonte, J. A., Pérez Die, M. C., & Díaz-Iglesias, L. (2015). Astronomía y paisaje en Heracleopolis Magna: un estudio de los templos. In N. Castellano, M. Mascort, C. Piedrafita, & J. Vivó (Eds.), *Ex Aegypto lux et sapientia, Homenatge al professor Josep Padró Parcerisa* (Nova Studia Aegyptiaca, 9) (pp. 107–123). Universitat de Barcelona.

Belmonte, J. A., Pérez Die, M. C., & Díaz-Iglesias, L. (2016). Shrines of Ram-Headed Divinities and Canopus: Skyscaping at Herakleópolis Magna. *Mediterranean Archaeology and Archaeometry, 16*(4), 125–132.

Belmonte, J. A., Fekri, M., & Serra, M. (2020). ¿Atrapando el Solsticio? Un Análisis crítico de la orientación de los templos de Deir el Bahari. *Papers on ANCIENT EGYPt, TdE, 10*, 11–26.

Bennett, C. (2011). *Alexandria and the Moon: An investigation into the lunar calendar of Ptolemaic Egypt* (Studia Hellenistica, 52). Peeters.

Benson, M., & Gourlay, J. (1899). *The Temple of Mut in Asher*. John Murray.

Berenguer, F. (1997). Report of the first season at Gebel Barkal Necropolis, 1995 Sector 2000. *Kush, 17*, 108–136.

Berenguer, F., & Diaz de Cerio, M. (2001). *En busca de los faraones negros*. Fundació Arqueològica CLOS.

Berlev, O. (1999). Precisely two Sothic periods between year 18 of King Sénw/Tosorthos and Year 2 of Pharaoh Antoninus Pius. In O. I. Paylove (Ed.), *Ancient Egypt: Language, culture, consciousness* (pp. 87–120). Priscel's.

Bickel, S. (1994). *La cosmogonie égyptienne: avant le Nouvel Empire* (*OBO*, 134). Éditions Universitaires/Vandenhoeck and Ruprecht.

Bickel, S. (2003). Temps liminaires, temps meilleurs? Qualifications de l'origene et de la fin du temps en Égypte ancienne. In V. Pirenne-Delforge & Ö. Tunca (Eds.), *Représentations du temps dans les religions* (pp. 43–53). Université de Liège.

Bickel, S., & Gautschy, R. (2014). Eine ramessidische Sonnenuhr im Tal der Könige. *ZÄS, 141*, 3–14.

Biegel, R. A. (1921). *Zur Astrognosie der alten Ägypter*. Universität Zürich.

Bietak, M. (2020). Why radiocarbon dates from Egyptian funerary contexts are approximately accurate while those from stratified settlements are precisely wrong. In J. Kamrin, M. Bárta, S. Ikram, M. Lehner, & M. Megahed (Eds.), *Guardian of ancient Egypt, studies in honor of Zahi Hawass* (pp. 235–245). Charles University.

Billing, N. (2002). *Nut. The goddess of life in text and iconography*. Uppsala University Press.

Birk, R. (2014). Titel-Bilder. Zur amtsspezifischen Ikonographie thebanischer Priester der Ptolemäerzeit. In G. Neunert, A. Verbovsek, & K. Gabler (Eds.), *Bild: Ästhetik – Medium – Kommunikation. Beiträge des dritten Münchner Arbeitskreises Junge Aegyptologie* (MAJA 3) 7. bis 9.12.2012 (pp. 79–101). Harrassowitz Verlag.

Blackman, A. M. (1972). *Middle-Egyptian Stories: 1. The story of Sinuhe – 2. The shipwrecked sailor. Bibliotheca Aegyptiaca, 2*. Edition de la fondation égyptologique Reine Elisabeth.

Bland, P. A., & Artemieva, N. A. (2006). The rate of small impacts on Earth. *Meteoritics and Planetary Science, 41*, 607–631.

Bleeker, C. J. (1973). *Hathor and Thoth. Two key figures of the ancient Egyptian religion*. Brill.

Bohleke, B. (1996). In terms of fate: a survey of the indigenous Egyptian contribution to ancient astrology in light of Papyrus CtYBR inv.1132 (B). *SAK, 23*, 11–46.

Böker, R. (2007). Über Namen und Identifizierung der ägyptischen Dekane. *Centaurus, 27*, 189–217.

Boll, F. (1903). *Sphaera. Neueu griechische Texte und Untersuchungen zur Geschichte der Sternbilder*. Teubner.

Bonani, G., Herbert, H., Hawass, Z., Lehner, M., Nakhla, S., Nolan, J. S., Wenke, R., & Wölfli, W. (2001). Radiocarbon dates of Old and Middle Kingdom monuments in Egypt. *Radiocarbon, 43*, 1297–1320.

Bonano, M. (2017). *kkw* y *kkw-zmAw*. Un estudio del concepto de oscuridad en la Duat de los Libros del Más Allá del Reino Nuevo Egipcio. *Sociedades Precapitalistas, 6*, 1–16.
Bonnet, H. (1952). *Reallexikon der aegyptischen Religionsgeschichte*. Walter de Gruyter and Co.
Bonnet, C., & Valbelle, D. (2005). *The Nubian pharaohs: Black kings of the Nile*. American University in Cairo Press.
Boraik, M. (2007). Stela of Bakenkhonsu, High Priest of Amun-Re. *Memnonia, 18*, 119–126.
Borchardt, L. (1899). Ein altägyptisches astronomisches Instrument. *ZÄS, 37*, 10–17.
Borchardt, L. (1910). Altägyptische Sonnenuhren. *ZÄS, 48*, 9–17.
Borchardt, L. (1911). Eine Reisesonnenuhr aus Ägypten. *ZÄS, 49*, 66–68.
Borchardt, L. (1917). *Die Annalen und die zeitliche Festlegung des Alten Reiches der ägyptischen Geschichte*. Behrend & Co.
Borchardt, L. (1920). *Die altägyptische Zeitmessung*. B. Walter de Gruyter and Co.
Borchardt, L. (1925). *Statuen und Statuetten von Königen und Privatleuten im Museum von Kairo Nr. 1-1294, II*. Reichsdruckerei.
Borchardt, L. (1935). *Die Mittel zur zeitlichen Festlegung von Punkten der ägyptischen Geschichte und ihre Anwendung*. Selbstverlag.
Borrego Gallardo, F.L. (2014) Lord of Maat: Innovation and Change of the Egyptian Divine Kingship Under the Reign of Snefru. *Arys, 12*, 87–127.
Borghouts, F. (1989). A New Middle Kingdom Netherworld Guide. In S. Schoske (Ed.), *Akten des vierten internationalen Ägyptologen Kongresses. München 1985* (SAK Beiheft, 3) (pp. 131–139). Helmut Buske Verlag.
Bosch Puche, F. (2009). *Alexandre el gran a Egipte: documentació, protocol onomàstic i legitimació*. PhD thesis, Universitat de Barcelona.
Bosticco, S. (1965). *Museo Archeologico di Firenze. La stele egiziane del Nuovo Regno*. Istituto poligrafico dello Stato.
Bosticco, S. (1972). *Museo Archeologico di Firenze. Le stele egiziane di epoca tarda*. Istituto poligrafico dello Stato.
Botti, G. (1967). *L'archivio demotico da Deir el-Medineh. Catalogo del Museo Egizio di Torino Seria Prima (Monumenti e Testi)*. Felice Le Monnier.
Bouché-Leclercq, A. (1899). *L'astrologie grecque*. Ernest Leroux.
Boudreaux, P. (1921). *Catalogus Codicum Astrologorum Graecorum VIII. Codicum Parisinorum IV*. Mauritii Lamertin.
Brady, B. (2015). Star phases: The naked-eye astronomy of the old kingdom pyramid texts. In F. Silva & N. Campion (Eds.), *Skyscapes. The role and importance of the sky in archaeology*. Oxbow Books.
Brand, P. (2000). *The monuments of Sety I. Epigraphic, historical and art historical analysis. Probleme der Ägyptologie, 16*. Brill.
Brandt, J. C., & Chapman, R. D. (2004). *Introduction to comets*. Cambridge University Press. [1st edn 1981].
Breasted, H. (1906). *Ancient records of Egypt, IV*. The University of Chicago Press.
Brennan, C. (2017). *Hellenistic astrology. The study of fate and fortune*. Amor Fati Publications.
Brewer, B. (1991) *Eclipse*. Clinton.
Briggs, R. E. (1952). Astronomy in the pyramid texts. In S. A. B. Mercer (Ed.), *The pyramid texts in translation and commentary, vol. IV Excursues* (pp. 38–49). Longmans, Green and Co.
Broekman, G. (2012). Once again the Piankhy-blocks from the Temple of Mut at Karnak. *CdÉ, 87*, 233–258.
Bronk Ramsey, Ch., Dee, M. W., Rowland, J. M., Higham, T. F. G., Harris, S. A., Borck, F., Quiles, A., Wild, E. M., Marcus, E. S., & Shortland, A. J. (2010). Radiocarbon-based chronology for Dynastic Egypt. *Science, 238*, 1554–1557, plus supplementary materials.
Brotto, A. (2016). *The death of ancient Egyptian Gods. Terminological and semantic analysis of literary evidence*. Università degli Studi di Pisa, Tesi di Laurea.
Brovarski, E. (1977). The doors of heaven. *Orientalia, 46*, 107–115.
Brown, D., Fermor, J., & Walker, C. (1999–2000). The water clock in Mesopotamia. *Archiv für Orientforschung, 46(47)*, 130–148.

Brugsch, H. K. (1862). *Recueil de monuments égyptiens dessinés sur lieux et publiés sous les auspices de son Altesse le vice-roi d'Egypte Mohammed-Said-Pacha*. J. C. Hinrichs.

Brugsch, H. K. (1883, re-edition 1968). *Thesaurus Inscriptionum Aegyptiacarum altaegyptischen Inschriften. I./II. Abteilung. Astronomische und Astrologische Inschriften altaegyptische Denkmaeler*. J.C. Hinrichssche Buchhandlung [1]. Akademischer Drück [2].

Brugsch, H. K. (1891). *Die Aegyptologie. Abriss der Entzifferungen und Forschungen auf dem Gebiete der aegyptischen Schrift, Sprache und Alterthumskunde*. Wilhelm Friedrich.

Bruins, E. M. (1965). The Egyptian shadow clock. *Janus, 52*, 123–137.

Brunner, H. (1988). Zeichendeutung aus Sternen und Winden in Ägypten. In H. Brunner (Ed.), *Das hörende Herz: Kleine Schriften zur Religions- und Geistesgeschichte Ägyptens* (*OBO*, 80) (pp. 224–229). Universitätverlag.

Brunner-Traut, E., & Brunner, H. (1981). *Die ägyptische Sammlung der Universität Tübingen*. Philipp von Zabern.

Bryan, B. M. (1997). The statue program for the mortuary temple of Amenhotep III. In S. Quirke (Ed.), *The temples in ancient Egypt, new discoveries and recent research* (pp. 57–81). British Museum Press.

Bryce, T. (2001). *The Kingdom of the Hittites*. Oxford University Press (Spanish edition, Madrid, Cátedra, 2001).

Buchwald, J. Z., & Greco Josefowicz, D. (2010). *The Zodiac of Paris*. Princeton University Press.

Budge, E. A. W. (1899). *The Book of the Dead. Papyri of Hunefer, Anhai, Kerasher and Netchemet with Supplementary Text from the Papyrus of Nu*. British Museum.

Budge, E. A. W. (1913). *The Papyrus of Ani* (Vol. 3). G. P. Putnam's Sons/Philip Lee Warner.

Bull, C. H. (2018). *The tradition of Hermes Trismegistus: The Egyptian priestly figure as a teacher of Hellenized Wisdom*. Brill.

Burke, J. G. (1991). *Cosmic debris: Meteorites in history*. University of California Press.

Calder, N. (1985). *¡Que viene el cometa!* (Biblioteca científica Salvat, 10). Salvat. [Spanish translation of *The comet is coming*, 1980].

Calmettes, M.-A. (2003). *La création du monde dans l'Égypte ancienne*. Paris: Éditions Louvre.

Calmettes, M.-A. (2008). Une représentation cosmologique du temple de Kom Ombo. *Acta Orientalia Belgica, 21*, 1–11.

Calmettes, M.-A. (2017). La représentation cosmologique du sarcophage d'Ourechnefer. *ENIM, 10*, 25–50.

Caminos, R. A. (1958). *The Chronicle of Prince Osorkon* (Analecta Orientalia, 37). Pontificium Institutum Biblicum.

Caminos, R. A. (1974). *The new-kingdom temples of Buhen*. Egypt Exploration Fund.

Caminos, R. A. (1975). *Ei*. In W. Helck & E. Otto (Eds.), *Lexikon der Ägyptologie* (Vol. I, cols. 1185–1188). Otto Harrassowitz.

Caminos, R. A., et al. (1954). *Reliefs and Inscriptions at Karnak, III: The Bubastite Portal* (*OIP*, 74). University of Chicago Press.

Carlotti, J.-F. (2005). Considérations architecturales sur l'orientation, la composition et les proportions des structures du temple d'Amon-Rê á Karnak. In P. Jánosi (Ed.), *Structure and significance: Thoughts on ancient Egyptian architecture* (pp. 169–191). Verlag der Österreichischen Akademie der Wissenschften.

Castle, E. W. (2015). The foundation ceremony at Akhetaten. *SAK, 44*, 43–82.

Cauville, S. (1997a). *Dendara, Le Zodiaque d'Osiris*. Peeters.

Cauville, S. (1997b). *Dendara. Les chapelles osiriennes*. IFAO.

Cauville, S., Aubourg, E., Deleuze, P., & Lecler, A. (1992). Le temple d'Isis à Dendera. *BSFÉ, 123*, 31–48.

Černý, J. (1935). *Catalogue des ostraca hiératiques non littéraires de Deir el-Médineh* (Nos 1 à 113). IFAO.

Černý, J. (1943). The origin of the name of the month Tybi. *ASAE, 43*, 173–181.

Černý, J. (1948). Note on *aAwy pt* 'Shrine'. *JEA, 34*, 120.

Černý, J. (1963). A possible Egyptian word for 'astronomer'. *JEA, 49*, 173.

Černý, J. (1973). *The valley of the kings. Fragments d'un manuscrit inachevé* (*BdÉ*, 61). IFAO.

Cervelló-Autuori, J. (2011). The sun-religion in the Thinite Age: Evidence and political significance. In R. F. Friedman & P. N. Fiske (Eds.), *Egypt and its origins* 3 (OLA, 205) (pp. 1125–1150). Peeters.

Champollion, J. F. (1842). Mémoire sur les signes employés par les anciens égyptiens à la notation des divisions du temps, dans leurs trois systems d'écriture. In *Mémoires de l'Institut royal de France, Académie des inscriptions et belles-lettres* 215 (pp. 73–134).

Champollion, J.-F. (1845). *Monuments de l'Egypte et de la Nubie, 3*. Firmin Didot Frères.

Champollion, J.-F. (1868). *Lettres écrites d'Égypte et de Nubie en 1828 et 1829*. Didier et C.

Chassinat, E. (1931). *Le temple d'Edfou VI* (MMAF, 23). IFAO.

Chassinat, É. (1935). *Le temple de Dendara*, III. Cairo: IFAO.

Chatley, H. (1940). Egyptian astronomy. *JEA, 26*, 120–126.

Chévrier, H. (1964). Techniques de construction dans l'ancienne Égypte. I. Murs en briques crues. *RdÉ, 16*, 11–17.

Cintron, D. A. (2003). A new angle on Sneferu's pyramids. In *54th meeting of the American Research Centre of Egypt*, unpublished.

Clagett, M. (1992). *Ancient Egyptian science I: Knowledge and order*. American Philosophical Society.

Clagett, M. (1995). *Ancient Egyptian Science II: Calendars, clocks and astronomy*. American Philosophical Society.

Clark, R. T. R., 1949–1950. The origin of the phoenix *University of Birmingham Historical Journal*, 2, 1–29; 105–140.

Claudiani, C. (1824). *Opera Omnia, N. L. Artaud*. Lemaire.

Cleomedes. (2004). *Cleomedes' lectures on astronomy. A translation of the heavens with an introduction and commentary by Alan C. Bowen and Robert B. Todd*. University of California Press.

Clère, J. J. (1949). *Un texte astronomique de Tanis. Kêmi, 10*, 3–27.

Clère, J. J. (1958). Fragments d'une Nouvelle Représentation Égyptienne du Monde. *MDAIK, 16*, 30–46.

Comellas, J. L. (2016). *El cielo de Colón. Técnicas navales y astronómicas en el viaje del Descubrimiento*. Athenaica.

Comelli, D., D'Orazio, M., Folco, L., El-Halwagy, M., Frizzi, T., Alberti, R., Capogrosso, V., Elnaggar, A., Hassan, H., Nevin, A., Porcelli, F., Rashed, M. G., & Valentini, G. (2016). The meteoritic origin of Tutankhamun's Iron Dagger Blade. *Meteoritics and Planetary Science, 51*(7), 1301–1309.

Congdom, L. O. (2000). A rare solar display depicted in the Tomb of Meryre at El Amarna. *Amarna Letters, 4*, 45–59.

Conman, J. (2002). *The Round Zodiac ceiling of the Temple of Hathor at Denderah*. https://saturniancosmology.org/files/denderah/dendera.round.html.txt. Last viewed 11 Jan 2021.

Conman, J. (2003). It's about time: Ancient Egyptian cosmology. *SAK, 31*, 33–71.

Conman, J. (2013). *Ancient Egyptian Sky Lore. Rethinking the conventional wisdom*. Amazon Italia.

Cooper, G. M. (2018). Astrology. The science of signs in the heavens. In P. T. Keyser & J. Scarborough (Eds.), *The Oxford handbook of science and medicine in the classical world* (pp. 381–407). Oxford University Press.

Cotterell, B., Dickson, F. P., & Kamminga, J. (1986). Ancient Egyptian water-clocks: A reappraisal. *Journal of Archaeological Science, 13*(1), 31–50.

Couprie, D. (2011). *Heaven and Earth in ancient Greek cosmology. From Thales to Heraclides Ponticus*. Springer.

Couprie, D. L. (2013). The Qumran Roundel and the mrḫyt: A comparative approach. *Dead Sea Discoveries, 20*, 264–306.

Cramer, F. H. (1954). *Astrology in Roman law and politics*. American Philosophical Society.

Crum, W. E. (2000). *A coptic dictionary*. Oxford University Press.

D'Orazio, M., Folco, L., Zeoli, A., & Cordier, C. (2011). Gebel Kamil: The iron meteorite that formed the Kamil crater (Egypt). *Meteoritics and Planetary Science, 46*, 1179–1196.

Daoud, K. (1993). An early Ramesside stela of a chief of hour watchers in the Memphite temple of Ptah. *JEA, 79*, 261–265.

Daressy, G. (1896). Une inondation à Thèbes sous le règne d'Osorkon (III). *RecTrav, 18*, 181–186.

Daressy, G. (1900). Une ancienne liste des décans égyptiens. *ASAE, 1*, 79–90.

Daressy, G. (1902). Grand vase en pierre avec graduations. *ASAE, 3*, 236–239.

Daressy, G. (1916). L'estatue d'un astronome. *ASAE, 16*, 1–4.

Daressy, G. (1919). Antiquités trouvées à Fostat. *ASAE, 18*, 275–278.

Darnell, J. C., & Darnell, D. (2002). Gebel Tjauti rock inscription 11. In *Theban desert road survey in the Egyptian Western Desert, Vol. 1 Gebel Tjauti and Wadi El-Hol rock inscriptions 1–45* (pp. 49–52). Chicago University Press.

Dash, G. (2015). On the field testing of methods the Egyptians may have used to find cardinal directions. *Journal for the History of Astronomy, 46*, 351–364.

Dash, G. (2017). Occam's Egyptian Razor: The Equinox and the Alignment of the Pyramids. *The Journal of Ancient Egyptian Architecture, 2*, 1–8.

Daumas, F. (1987). *Le temple de Dendara IX*. IFAO.

Davies, N. D. G. (1908). *The Rock Tombs of El Amarna, part VI: Tombs of Parennefer, Tutu, and Aÿ* (Archaeological Survey of Egypt, 18). Egypt Exploration Fund.

Davies, N. D. G. (1917). *The Tomb of Nakht at Thebes*. The Metropolitan Museum of Art.

Davies, N. D. G. (1953). *The Temple of Hibis in El Khargeh Oasis. Part III, The Decoration*. Metropolitan Museum of Art.

Davis, V. L. (1985). Identifying Ancient Egyptian constellations. *Journal of the History of Astronomy, 16*, 102–104.

Dee, M. W. (2013a). Investigating the accuracy of radiocarbon dating in Egypt: Checks with samples of known age. In A. J. Shortland & C. Bronk Ramsey (Eds.), *Radiocarbon and the chronologies of ancient Egypt* (pp. 53–64). Oxbow Books.

Dee, M. W. (2013b). A radiocarbon based chronology for the Old Kingdom. In A. J. Shortland & C. Bronk Ramsey (Eds.), *Radiocarbon and the chronologies of ancient Egypt* (pp. 209–217). Oxbow Books.

Dee, M. W. (2013c). A radiocarbon based chronology for the Old Kingdom. In A. J. Shortland & C. Bronk Ramsey (Eds.), *Radiocarbon and the chronologies of ancient Egypt* (pp. 174–181). Oxbow Books.

Dee, M. W. (2013d). A radiocarbon based chronology for the Old Kingdom. In A. J. Shortland & C. Bronk Ramsey (Eds.), *Radiocarbon and the chronologies of ancient Egypt* (pp. 65–75). Oxbow Books.

Depauw, M. (1997). *A companion to Demotic studies*. Fondation Égyptologique Reine Élisabeth.

Depauw, M. A. (2001). A Demotic Stela of an Astronomer. *Enchoria, 27*, 1–2.

Depuydt, L. (1995a). Regnal years and civil calendars in Achaemenid Egypt. *JEA, 81*, 151–173.

Depuydt, L. (1995b). On the consistency of the Wandering Year as a backbone of Egyptian Chronology. *JARCE, 32*, 43–58.

Depuydt, L. (1996). The function of the Ebers Calendar concordance. *Orientalia, 65*, 61–88.

Depuydt, L. (1997). *Civil and lunar calendar in ancient Egypt* (OLA, 77). Peeters.

Depuydt, L. (1998a). Gnomons at Meroë and early trigonometry. *JEA, 84*, 171–180.

Depuydt, L. (1998b). The demotic mathematical astronomical papyrus Calsberg 9 reinterpreted. In W. Clarysse, A. Schoors, & H. Willems (Eds.), *Egyptian religion. Studies Ian Quaegebeur* (pp. 1277–1297). Peeters.

Depuydt, L. (1998c). The Hieroglyphic Representation of the Moon's Absence (*Psḏntyw*). In L. H. Lesko (Ed.), *Ancient Egyptian and Mediterranean studies in memory of William A. Ward* (pp. 71–85). Brown University.

Depuydt, L. (1999). The two problems of the month names. *RdÉ, 50*, 107–133.

Depuydt, L. (2000). Sothic chronology and the old kingdom. *JARCE, 37*, 167–186.

Depuydt, L. (2007). Calendars and years in ancient Egypt: The soundness of Egyptian and west Asian chronology in 1500-500 B.C. and the consistency of the Egyptian 365-day wandering year. In J. M. Steele (Ed.), *Calendars and years: Astronomy and time in the ancient Near East* (pp. 35–82). Oxbow Books.

Depuydt, L. (2009). From twice helix to double helix: A comprehensive model for Egyptian calendar history. *JEGH, 2*, 115–147.
Depuydt, L. (2010). Ancient Egyptian star tables: A reinterpretation of their fundamental structure. In A. Imhausen & T. Pommerening (Eds.), *Writings of early scholars in the ancient near East, Egypt, Rome, and Greece* (pp. 235–270). De Gruyter.
Depuydt, L. (2016). The Regnal Years of the "Mathematical Astronomical" Demotic Papyrus Carlsberg 9 Reinterpreted. In R. Landgráfová & J. Mynářová (Eds.), *Rich and great. Studies in honour of Anthony J. Spalinger on the occasion of his 70th Feast of Thoth* (pp. 39–60). Charles University.
Depuydt, L. (2017). The calendars and the year-counts of ancient Egypt. *CdÉ, 92*, 271–294.
Derchain, P. (1956). A propos de Claudien. Eloge de Stilichon, II, 424–436. *ZÄS, 81*, 4–6.
Derchain, P. (1989). Harkhébis, le Psylle-Astrologue. *Chronique d'Égypte, 64*, 74–89.
Derchain, P. (2006). La clepsydre de la derniere nuit. *Göttinger Miszellen, 210*, 19–21.
Derchain-Urtel, M.-T. (1974). Die Schlange des "Schiffbrüchigen". *SAK, 1*, 83–104.
Derstine, P. (2016). The start of the Egyptian lunar month in light of early Eighteenth Dynasty Sothic and lunar events. *Göttinger Miszellen, 249*, 39–57.
Derstine, P. (2017). Early Eighteenth Dynasty chronology and Thutmoside succession. *Göttinger Miszellen, 252*, 41–59.
Desroches-Noblecourt, C. (1976). *Ramses le Grand.* Galeries Nationales du Grand Palais.
Detoma, E. (2014). On two star tables on the lids of two cofins in the Egyptian Museum of Turin. *Atti Sc Mor, 148*, 117–169.
Dhennin, S. (2012). Une Stèle de Mefkat (Montgeron 2007.4). *RdE, 63*, 67–82.
Díaz-Iglesias Llanos, L. E. (2012). *La proyección mitológica de Heracleópolis Magna: continuidades, transformaciones y contextualización de los mitemas heracleopolitanos en fuentes funerarias y cultuales.* PhD thesis, Universidad de La Laguna.
Díaz-Iglesias Llanos, L. (2014). *El ciclo mítico de Heracleópolis Magna: continuidad y reelaboración a partir de las fuentes funerarias y cultuales* (Aula Aegyptiaca Studia, 7). Universitat Autònoma de Barcelona.
Dieleman, J. (2003a). Claiming the stars. Egyptian priests facing the sky. *Aegyptiaca Helvetica, 17*, 277–289.
Dieleman, J. (2003b). Stars and the Egyptian priesthood in the Graeco-Roman period. In S. Noegel, J. Walker, & B. Wheeler (Eds.), *Prayer, magic, and the stars in the ancient and late antique world* (pp. 137–153). The Pennsylvania State University.
Diodorus Siculus. (1989). *Diodorus Siculus, I. Books I–II.34* (Loeb Classical Library) (C. H. Oldfather, Trans.). Harvard University Press [1933].
Diodorus Siculus, Library of History (Texts/Diodorus_Siculus/1D*.html). Last seen 26 May 2020.
Diógenes Laercio. (2007). *Vidas y opiniones de los filósofos ilustres. Traducción, introducción y notas de Carlos García Gual.* Alianza Editorial.
Dodson, A. (2019). *Sethy I King of Egypt. His life and afterlife.* American University in Cairo Press.
Dodson, A., & Hilton, D. (2004). *The complete royal families of ancient Egypt.* Thames and Hudson.
Dorman, P. F. (1991). *The Tombs of Senenmut: The architecture and decoration of tombs 71 and 353.* The Metropolitan Museum of Art Egyptian Expedition.
Dorner, J. (1981). *Die Absteckung Und Astronomische Orientierung Aegyptischer Pyramiden.* Universitaet Innsbruck.
Dorner, J. (1985). Tagesuhr, Nachtuhr. Ein Beitrag zur altägyptischen Zeitmessung. *Jahreshefte des Österreichischen Archäologischen Institutes in Wien, 56*, 27–37.
Dreyer, G. (1998). *Umm el-Qaab 1: Das prädynastische Königsgrab U-j und seine frühen Schriftzeugnisse. DAI Archäologische Veröffentlichungen*, 86.
Edgar, C. C. (1906). *Sculptors studies and unfinished works: Catalogue général des antiquités égyptiennes du Musée du Caire (33301-33506).* IFAO.
Edgerton, W. F., (1937). On the chronology of the early eighteenth dynasty (Amenhotep I to Thutmose III), *The American Journal of Semitic Languages and Literatures, 53*, 188–197.
Edwards, I. E. S. (1960). *Oracular Amuletic decrees of the late new kingdom. Hieratic Papyri in the British Museum. Fourth Series,* 2 vols. Trustees of the British Museum.

Edwards, I. E. S. (1993). *The pyramids of Egypt*. Penguin [3rd ed.].

Edwards, D. N. (2004). *The Nubian past: An archaeology of the Sudan*. Routledge.

Eigner, D. (1984). *Die Monumentalen Grabbauten der Spätzeit in der Thebanischen Nekropole*. Österreichischen Akademie.

El-Hawary, A. (2010). *Wortschöpfung: die Memphitische Theologie und die Siegesstele des Pije – zwei Zeugen kultureller Repräsentation in der 25. Dynastie* (OBO, 243). Universitätsverlag / Göttingen: Vandenhoeck Ruprecht.

El-Sabban, S. (2000). *Temple festival calendars of ancient Egypt*. Liverpool University Press.

El-Sayed, R. (1980). Au sujet de la statue CG. 680 du musée du Caire de l'époque ptolémaïque et provenant de Thèbes-Ouest. *BIFAO, 80*, 233–248.

El-Sebaie, S. M. (2000). *The destiny of the world: A study on the end of the universe in the light of ancient Egyptian texts*. University of Toronto. PhD Thesis.

Epigraphic Survey. (1930). *Medinet Habu, Volume I: Earlier Historical Records of Ramses III. OIP, 8*. University of Chicago Press.

Epigraphic Survey. (1934). *The Calendar, the "Slaughter-House", and Minor Records of Ramses III, OIP, 23 – Medinet Habu vol. III*. University of Chicago Press.

Epigraphic Survey (1954). *Reliefs and Inscriptions at Karnak, III: The Bubastite Portal. OIP, 74. University of Chicago Press.*

Epigraphic Survey. (1963). *Medinet Habu vol. 6. The Temple Proper, II. OIP, 84*. University of Chicago.

Epigraphic Survey. (1980). *The Tomb of Kheruef: Theban Tomb 192. OIP, 102*. Chicago: University of Chicago Press.

Erichsen, W. (1954). *Demotisches Glossar*. Ejnar Munksgaard.

Erichsen, W. (1956). *Eine neue demotische Erzählung, Akademie der Wissenschaften und der Literatur in Mainz*. Franz Steiner Verlag.

Erichsen, W., & Schott, S. (1954). *Fragmente memphitischer Theologie in demotischer Schrift (Pap. Demot. Berlin 13603)*. Akademie der Wissensschaften und der Literatur in Mainz.

Erman, A. (1911). Ein Denkmal memphitischer Theologie. *Sitzungsberichte der Preußischen Akademie der Wissenschaften zu Berlin, 43*, 916–950.

Escolano-Poveda, M. (2018–2019). Astronomica Montserratensia I. A Demotic monthly almanac with synodic phenomena (P. Monts.Roca inv. 314). *Enchoria, 36*, 1–36.

Escolano-Poveda, M. (2020). *The Egyptian priests of the Graeco-Roman Period. An analysis on the basis of the Egyptian and Graeco-Roman literary and paraliterary sources*. Harrassowitz Verlag.

Escolano-Poveda, M. (2022). Astrologica athribitana: Four demotic-hieratic horoscopes from Athribis (O. Athribis 17-36-5/1741 and ANAsh. Mus.D.O.633 reedited). *Journal for the History of Astronomy, 53*(1), 49–87.

Étienne, M. (2009). *Les portes du ciel. Visions du monde dans l'Egypte ancienne*. Somogy éditions d'art.

Etz, D. V. (1997). A new look at the constellation figures in the celestial diagram. *JARCE, 34*, 143–161.

Evans, J. (2004). The astrologer's apparatus: A picture of professional practice in Greco-Roman Egypt. *Journal for the History of Astronomy, 35*, 1–44.

Evans, J., & Marée, M. (2008). A Miniature Ivory Sundial with Equinox Indicator from Ptolemaic Tanis, Egypt. *Journal for the History of Astronomy, 39*, 1–17.

Fakhry, A. (1959). *The monuments of Sneferu at Dahshur*. General Organization for Government Printing Offices.

Fakhry, A. (2004). *Siwa Oasis* (5th ed.). American University of Cairo Press.

Färber, R., & Gautschy, R. (Eds.). (2020). *Zeit in den Kulturen des Altertums. Antike Chronologie im Spiegel der Quellen*. Böhlau Verlag.

Faulkner, R. O. (1933). *The Papyrus Bremner-Rhind, BA III*. Fondation Égyptologique Reine Élisabeth.

Faulkner, R. O. (1937a). The Bremner-Rhind Papyrus-III. D. The book of overthrowing Apep. *JEA, 23*, 166–185.

Faulkner, R. O. (1937b). The Bremner-Rhind Papyrus-IV. D. The Book of Overthrowing Apep. *JEA, 24*, 41–53.
Faulkner, R. O. (1962). *Concise dictionary of middle Egyptian*. Oxford Univeristy Press.
Faulkner, R. O. (1966). The King and the Star-religion in the pyramid texts. *JNES, 25*(3), 153–161.
Faulkner, R. O. (1969). *The ancient Egyptian pyramid texts*. Oxford University Press.
Faulkner, R. O. (1988). *A Concise Dictionary of Middle Egyptian*. Oxford: Griffith Institute.
Faulkner, R. O. (1994). *The ancient Egyptian coffin texts*, vol. I–III. Aris and Phillips.
Fermat, A. (2010). *Le rituel de la Maison de Vie. Papyrus Salt 825*. MdV Éditeur.
Fermor, J. (1993). Perceived night length ratios in ancient Egypt. *Vistas in Astronomy, 36*, 363–373.
Fincke, J. C., 2003/2004. The Babylonian Texts of Nineveh: Report on the British Museum's "Ashurbanipal Library" Project. *Archiv für Orientforschung, 50*, 111–149.
Finnestad, R. B. (1985). *Image of the world and symbol of the creator. On the cosmological and iconological values of the Temple of Edfu*. Otto Harrassowitz.
Firmicus Maternus. (1975). *Ancient astrology theory and practice. Matheseos Libri VIII* (J. R. Bram, Trans.). Noyes Press.
Fischer, H. G. (1978). Five inscriptions of the old kingdom. *ZÄS, 105*, 42–59.
Fischer, H. G. (1996). *On the interpretation of names of the pyramids. Egyptian studies III – Varia Nova* (pp. 73–77). Metropolitan Museum of Art.
Fissolo, J.-L. (2001). Les astronomes égyptiens. *Égypte. Afrique and Orient, 21*, 15–24.
Folco, L., Di Martino, M., El Barkooky, A., D'Orazio, M., Lethy, A., Urbini, S., Nicolosi, I., Hafez, M., Cordier, C., van Ginneken, M., Zeoli, A., Radwan, A. M., El Khrepy, S., El Gabry, M., Gomaa, M., Barakat, A. A., Serra, R., & El Sharkawi, M. (2010). The Kamil Crater in Egypt. *Science, 329*, 804.
Forenbaher, S., & Jones, A. (2011). The Nakovana Zodiac: Fragments of an Astrologer's Board from an Illyrian-Hellenistic Cave Sanctuary. *Journal for the History of Astronomy, 42*, 425–438.
Frankfort, H. (1933). *The Cenotaph of Seti I at Abydos, II*. Egypt Exploration Fund.
Friedman, R. (1996). The ceremonial centre at Hierakonpolis locality HK29A. In J. Spencer (Ed.), *Aspects of early Egypt* (pp. 16–35). University College.
Fukaya, M. (2019). *The festivals of Opet, the valley, and the new year. Their socio-religious functions*. Archaeopress.
Gabolde, L. (1998). *Le 'Grand Château' d'Amon de Sésostris Ier à Karnak* (MAIBL NS, 17). De Boccard.
Gabolde, L. (1999). Canope et les orientations nord-sud de Karnak étables par Thoutmosis III. *RdÉ, 50*, 278–282.
Gabolde, L. (2009). L'horizon d'Aton, exactement? *Cahiers de l'Égypte Nilotique et Mediterranéenne, 2*, 145–157.
Gabolde, L. (2013). L'implantation du temple: contingencies religieuses et contraintes géomorphologigues. *Les Cahiers de Karnak, 14*, 3–12.
Gabolde, L. (2015). The "Kernbau" of the temple of Mentuhotep II at Deir el Bahari: A monumental sun altar? In R. Jasnow, K. M. Cooney, & K. E. Davis (Eds.), *Joyful in Thebes: Egyptological studies in honor of Betsy Bryan* (pp. 145–154). Lockwood Press.
Gabolde, L. (2018). *Karnak, Amon-Rê. La genèse d'un temple, la naissance d'un dieu* (BdÉ, 167). IFAO.
Gabolde, L., & Gabolde, M. (2015). Les textes de la paroi sud de la salle des Annales de Thoutmosis III à Karnak: Reconstitution architecturale et restitution épigraphique. *Kyphi. Cercle lyonnais d'égyptologie Victor Loret*, 45–110.
Gabolde, L., Laisney, D., Leclère, F., & Payraudeau, F. (2021). L'orientation de grand temple d'Amon à Tanis. Données topographiques et archéologiques, hypothèses astronomiques et conséquences historiques. In P. Collombert, L. Coulon, I. Guermeur, & C. Thiers (Eds.), *Questionner le Sphinx. Mélanges offerts à Christiane Zivie-Coche* (BdÉ, 178) (pp. 309–349). IFAO.
Gadré, K. (2008). *Conception d'un modèle de visibilité d'étoile à l'oeil nu. Application à l'identification des décans égyptiens*. PhD thesis, Université Paul Sabatier, Toulouse III.

Gadré, K., & Roques, S. (2008a). L'année civile égyptienne et les horloges stellaires. *Revista de la Sociedad Urugaya de Egiptologia RSUE, 25.*

Gadré, K., & Roques, S. (2008b). Catalogue d'étoiles peuplant le ciel méridional de l'Égypte ancienne. *Cahiers Caribéens d'Egyptologie, 11*, 1–7.

Gadré, K., & Roques, S. (2009). Astronomical dating proposals of the ancient Egyptian stellar clocks. *Revista de la Sociedad Urugaya de Egiptologia RSUE, 26.*

Galán, J. M. (2002). *El imperio egipcio. Inscripciones, ca. 1550-1300 a.C.* Trotta.

Gallo, P. (1997). *Ostraca demotici e ieratici dall'archivio bilingue di Narmouthis* (Vol. II). Edizioni ETS.

Gardiner, A. H. (1906). Mesore as first month of the Egyptian year. *ZÄS, 43*, 136–144.

Gardiner, A. H. (1909). *The Admonitions of an Egyptian Sage from a Hieratic Papyrus in Leiden (Pap. Leiden 344 recto).* Hinrichs'sche Buchhandlung.

Gardiner, A. H. (1935). *Hieratic Papyri in the British Museum. Third Series. Chester Beatty Gift.* British Museum.

Gardiner, A. H. (1944). Horus the Behdetite. *JEA, 30*, 23–60.

Gardiner, A. H. (1947). *Ancient Egyptian Onomastica, vol. I.* Oxford University Press.

Gardiner, A. H. (1955). The problem of the month-names. *RdE, 10*, 9–31.

Gardiner, A. H. (1957). *Egyptian grammar being an introduction to the study of Hieroglyphs.* Griffith Institute.

Gardiner, A. H. (1959). *The Royal Canon of Turin.* Griffith Institute.

Garstang, J. (1916). Fifth interim report on the excavations at Meroë. *Liverpool Annals of Archaeology and Anthropology, 7*, 1–24.

Gautschy, R. (2011a). Monddaten aus dem Archiv von Illahun: Chronologie des Mittleren Reiches. *ZÄS, 138*, 1–19.

Gautschy, R. (2011b). Lunar and Sothic data from the archive of el-Lahun revisited: Chronology of the Middle Kingdom. In M. Horn, J. Kramer, D. Soliman, N. Staring, C. van den Hoven, & L. Weiss (Eds.), *Current research in Egyptology 2010: Proceedings of the eleventh annual symposium which took place at Leiden University* (pp. 53–61). Oxbow Books.

Gautschy, R. (2014). A reassessment of the absolute chronology of the Egyptian New Kingdom and its 'brotherly' countries. *ÄuL, 24*, 143–160.

Gautschy, R. (2015). The Tepi Shemu Feast: A tool for testing chronologies of Dynasty 21 to 25? *JEGH, 8*, 81–114.

Gautschy, R. (2017). Remarks concerning the alleged solar eclipse of Murshili II. *AoF, 44*(1), 23–29.

Gautschy, R. (2020a). Die Dekansternuhr des Idy. In R. Färber & R. Gautschy (Eds.), *Zeit in den Kulturen des Altertums. Antike Chronologie im Spiegel der Quellen* (pp. 43–53). Köln, Böhlau Verlag.

Gautschy, R. (2020b). Eine Sonnenuhr aus dem Tal der Könige. In R. Färber & R. Gautschy (Eds.), *Zeit in den Kulturen des Altertums. Antike Chronologie im Spiegel der Quellen* (pp. 95–102). Böhlau Verlag.

Gautschy, R. (2020c). The Karnak Clepsydra: Votive Gift or Utilitarian Object? In K. Gabler, R. Gautschy, L. Bohnenkämper, H. Jenni, C. Reymond, R. Zillhardt, A. Loprieno-Gnirs, & H.-H. Münch (Eds.), *Text-Bild-Objekte im archäologischen Kontext Festschrift für Susanne Bickel. Lingua Aegyptia Studia Monographica* (pp. 171–183). Widmaier Verlag.

Gautschy, R. (2020d). Eine Wasseruhr aus der Zeit des Pharaos Amenophis III. In R. Färber & R. Gautschy (Eds.), *Zeit in den Kulturen des Altertums. Antike Chronologie im Spiegel der Quellen* (pp. 87–94). Böhlau Verlag.

Gautschy, R. (2020e). Die astronomische Decke des Senenmut Grabes. In R. Färber & R. Gautschy (Eds.), *Zeit in den Kulturen des Altertums. Antike Chronologie im Spiegel der Quellen* (pp. 65–75). Böhlau Verlag.

Gautschy, R., Habicht, M. E., Galassi, F. M., Rutica, D., Hannig, R., & Rühli, F. J. (2017). A new astronomically based chronological model for the Egyptian Old Kingdom. *JEGH, 10*, 69–108.

Gayet, A. (1886). *Musée Du Louvre Stèles de la XII Dynastie* (Bibliothèque de l'Ecole des Hautes Etudes, 68). Vieweg.

Gayet, A. (1894). *Le temple de Louxor* (MMAF XV fasc. 1). Ernest Leroux ed.
George, B. (1974). Die Bahn der Sonne am Tage und in der Nacht: altägyptische Sonnenuhren und Königsgräber. In L. Kákosy (Ed.), *Recueil d'études dédiées à Vilmos Wessetzky* (SA, 1, pp. 101–116).
Gilhou, N. (2004). L'Héritage culturel de l'Égypte. L'Égypte et la science grecque: Planetes et constellations. In *Cultural heritage of Egypt and Christian Orient*, vol. 2 (pp. 25–36). Institute of Oriental Studies.
Goebs, K. (2008). *Crowns in Egyptian funerary literature: Royalty, rebirth, and destruction.* Griffith Institute.
González-García, A. C., & Belmonte, J. A. (2011). Thinking Hattusha: Astronomy and landscape in the Hittite lands. *Journal for the History of Astronomy, 42*, 461–494.
González-García, A. C., & Belmonte, J. A. (2014). Sacred architecture orientation across the Mediterranean: A comparative statistical analysis. *Mediterranean Archaeology and Archaeometry, 14*, 95–113.
González-García, A. C., & Belmonte, J. A. (2015). The orientation of pre-Romanesque churches in the Iberian Peninsula. *Nexus Network Journal, 17*, 353–377.
Goodenow, J., Orr, R., & Ross, D. (2015). *Mathematical models of water clocks.* https://www.simiode.org/resources/1002
Goyon, G. (1970). Nouvelle observations relatives à l'orientation de la pyramide de Khéops. *RdÉ, 22*, 85–98.
Goyon, G. (1974). Kerkasôre et l'ancien observatoire d'Eudoxe. *BIFAO, 74*, 135–147.
Gozzoli, R. B. (2009). Kawa V and Taharqo's Byjwt: Some aspects of Nubian Royal Ideology. *JEA, 95*, 235–248.
Graefe, E. (1982). Morgenstern. In W. Helck & W. Westendorf (Eds.), *Lexikon der Ägyptologie IV (col. 206)*. Otto Harrassowitz.
Grandet, P. (2005). *Le Papyrus Harris I*, vols. 1–2 (BdÉ, 109/1-2). IFAO.
Greenbaum, D. G. (2016). *The Daimon in Hellenistic astrology: Origins and influence*. Brill.
Greenbaum, D. G., & Ross, M. T. (2010). The role of Egypt in the development of the horoscope. In L. Bareš, F. Coppens, & K. Smoláriková (Eds.), *Egypt in transition. Social and religious development of Egypt in the First Millennium BCE* (pp. 146–181). Charles University in Prague.
Greenbaum, D. G., & Ross, M. T. (2015). Various Renderings of Πίναξ in Greek and Demotic at Medīnet Māḍi. In N. Campion & D. G. Greenbaum (Eds.), *Astrology in time and place: Cross-cultural questions in the history of astrology* (pp. 109–129). Cambridge Scholars Publishing.
Grenfell, B. P., & Hunt, A. S. (1903). *The Oxyrhynchus Papyri, part III*. EEF.
Griffith, F. L. (1909). *Catalogue of the Demotic papyri in the John Rylands Library Manchester, vol. III: Key-list, translations, commentaries and indices*. Manchester University Press.
Griffith, F. L. (1937). *Catalogue of the Demotic Graffiti of the Dodecaschoenus,* vols. I–II. Oxford University Press.
Griffith, F. L., Petrie, W., & Flinders, M. (1889). *Two Hieroglyphic Papyri from Tanis* (EEF). Trübner and Co.
Grimal, N.-C. (1981a). *La stèle triomphale de Pi('ankh)y au Musée du Caire*. IFAO.
Grimal, N.-C. (1981b). *Quatre Stèles Napaténnes au Musée du Caire. JE 48863-48866*. IFAO.
Grimm, A. (1994). Die altägyptischen Festkalender in den Tempeln der griechisch-römischen Epoche. In *Ägypten und Altes Testament, 15*. Harrassowitz.
Grusson, H. (1895). *Im Reiche des Lichtes. Sonnen, Zodiakallichte, Kometen*. George Westermann.
Guermeur, I. (2003). Glanures (§ 1–2). *BIFAO, 103*, 281–296.
Guermeur, I. (2005). *Les cultes d'Amon hors de Thèbes. Recherches e Géographie religieuse*. Brepols.
Guilhou, N. (2007). La naissance du soleil et ses voyages d'Esna à Saïs: une lecture astronomique de la Cosmogonie d'Esna. In E. Kormysheva (Ed.), *Cultural heritage of Egypt and Christian Orient 4* (pp. 13–24). Moscow.

Gundel, W. (1936a). *Dekane und Dekansternbilder. Ein Beitrag zur Geschichte der Sternbilder der Kulturvölker*. J. J. Augustin.

Gundel, W. (1936b). *Neue astrologische Texte des Hermes Trismegistos: Funde und Forschungen auf dem Gebiet der antiken Astronomie und Astrologie*. Verlag der Bayerischen Akademie der Wissenschaften.

Gundel, W. (1969). *Dekane und Dekansternbilder*. Wissenschaftliche Buchgesellschaft.

Gundel, W., & Gundel, H. G. (1966). *Astrologumena. Die astrologische Literatur in der Antike und ihre Geschichte*. Franz Steiner Verlag.

Gunn, B. (1921). A note on the verb *wrš*. *RecTrav, 39*, 108–109.

Gunn, B. (1926). The coffins of Heny. *ASAE, 26*, 166–171.

Güterbock, H. G. (1956). The Deeds of Suppiluliuma as told by his son Mursili II. *JCS, 10*, 41–130.

Gysembergh, V. (2018). Hipparchus and the "Ancients:" Nechepsos-Petosiris? *Journal of Ancient Civilizations, 33*, 165–179.

Haack, S. C. (1984). The astronomical orientation of the Egyptian Pyramids. *Archaeoastronomy, 7*, S119–S125.

Habachi, L., & Habachi, B. (1952). The Naos with the decades (Louvre D 37) and the discovery of another fragment. *JNES, 11*, 251–263.

Habicht, M. A., Gautschy, R., Siegmann, R., Rutica, D., & Hannig, R. (2015). A New Sothis rise on a small cylindrical jar from the old kingdom (Collection of Prof. Peter A. Kaplony). *Göttinger Miszellen, 247*, 41–50.

Hall, E. S. (1986). *The Pharaoh smites his enemies* (MÄS, 44). Deutscher Kunstverlag.

Handoussa, T. (1979). A propos de l'offrande šbt. *SAK, 7*, 65–74.

Hannig, R. (1997). *Die Sprache der Pharaonen. Großes Handwörterbuch Ägyptisch–Deutsch (2800 bis 950 v. Chr.)*. Philipp von Zabern.

Hannig, R. (2003). *Ägyptisches Wörterbuch I. Altes Reich und Erste Zwischenzeit*. Philipp von Zabern.

Hannig, R., & Vomberg, P. (1998). *Wortsschatz der Pharaonen in Sachgruppen. Hannig-Lexica 2*. Philipp von Zabern.

Hardy, P. A. (2002–2003). The Cairo calendar as a stellar Almanac. *Archaeoastronomy, 17*, 48–63.

Harris, J. R. (1961). *Lexicographical studies in ancient Egyptian minerals*. Akademie Verlag.

Hawass, Z. (2009). *Mountains of the pharaohs*. Routledge.

Hawkins, G. S. (1973). *Beyond Stonehenge*. Routledge.

Hawkins, G. S. (1975). Astroarchaeology: The unwritten evidence. In A. Aveni (Ed.), *Archaeoastronomy in pre-Columbian America* (pp. 131–162). Austin University Press.

Heilen, S. (2011). Some metrical fragments from Nechepsos and Petosiris. In I. Boehm & W. Hübner (Coords.), *La poésie astrologique dans l'Antiquité* (pp. 23–93). De Boccard.

Helck, W. (1954). *Untersuchungen zu den Beamtentiteln des ägyptischen alten Reiches* (Ägyptologische Forschungen, 18). J. J. Augustin.

Helck, W. (1959). Bemerkungen zu den Thronbesteigungsdaten im Neuen Reich. *AnBib, 12*, 113–129.

Helck, W. (1990). Drei Ramessidische Daten. *SAK, 17*, 205–214.

Helck, W. (1992). *Die Prophezeiung des Nfr.tj*. Harrassowitz Verlag.

Henne, W. (1991). Bemerkungen zur altägyptischen Streiflicht-Uhr. *ZÄS, 118*, 21–31.

Henriksson, G. (2020). Aristotle, King David, King Zhou and Pharaoh Thutmosis III have seen comet Encke. *Mediterranean Archaeology and Archaeometry, 20*, 29–44.

Hephaistio of Thebes. (1994). *Apotelesmatics, Book I* (R. Schmidt, Trans.). The Golden Hind Press.

Herodotus. (1920). *Herodotus, the Histories* (A. D. Godley, Trans.). Harvard University Press.

Hinkel, W. (1997). L'architecture Méroïtique. In D. Wildung (Ed.), *Soudan, royaumes sur le Nil* (pp. 391–417). Institut du monde arabe.

Hoffmann, F. (1995). Astronomische und astrologische Kleinigkeiten I: Pap. Wien D6005. *Enchoria, 22*, 22–26.

Hoffmann, F. (2000). *Ägypten. Kultur und Lebenswelt in griechisch-römischer Zeit. Eine Darstellung nach den demotischen Quellen*. Akademie Verlag.

Hoffmann, F. (2010). Astronomische und astrologische Kleinigkeiten VI: Neumonddaten aus dem Jahre 184/185 n. Chr. In C. Leitz, D. von Recklinghausen, & H. Knuf (Eds.), *Honi soit qui mal y pense: Studien zum pharaonischen, griechisch-römischen und spätantiken Ägypten zu Ehren von Heinz-Josef Thissen* (OLA, 194) (pp. 233–236). Peeters.

Hoffmann, F. (2016). Beispiele für Übersetzung und Kommentierung ägyptischer astronomischer Texte: Sonnenuhr, Sonnenaufgang und Dekansterne. In A. Imhausen & T. Pommerening (Eds.), *Translating writings of early scholars in the ancient near East, Egypt, Greece and Rome* (Beiträge zur Altertumskunde, 344) (pp. 335–378). De Gruyter.

Hoffmann, F. (2017). Astronomische und astrologische Kleinigkeiten VII: Die Inschrift zu Tages und Nachtlängen aus Tanis. In R. Jasnow & G. Widmer (Eds.), *Illuminating Osiris. Egyptological studies in honor of Mark Smith* (pp. 135–153). Lockwood Press.

Hoffmann, F., & Jones, A. (2009). Astronomische und astrologische Kleinigkeiten V: die Mondephemeride des P. Carlsberg 638. *Enchoria, 30*, 10–20.

Hoffmann, F., & Quack, J. (2014). Pastophoros. In A. M. Dodson, J. J. Johnston, & W. Monkhouse (Eds.), *A good scribe and an exceedingly wise man. Studies in Honour of W. J. Tait* (Golden House Publications Egyptology, 21) (pp. 127–155). Golden House.

Hoffmeier, J. K. (2014). *Akhenaten and the origins of monotheism*. Oxford University Press.

Hölbl, G. (1984). Die ägyptische Wasseruhr Ptolemaios'II. In G. Langmann, M. Firneis, and G. Hölbl (Eds.), Die ägyptische Wasserauslaufuhr aus Ephesos. *Jahreshefte des Österreichischen Archäologischen Institutes, 55*, 21–68.

Holden, J. H. (2006). *A history of horoscopic astrology. From the Babylonian period to the modern age*. AFA.

Hollenback, G. M. (2004). Gnomon graffito at Meroe. *JEA, 90*, 223–224.

Homer. (1924). *The Iliad* (A. T. Murray, Trans.). Harvard University Press/William Heinemann.

Horapolo. (1840). *The Hieroglyphics of Horapollo Nilous* (A. T. Cory, Ed.). William Pickering.

Horapolo. (2011). *Hieroglyphica, traducción de María José García Soler*. Akal.

Hornung, E. (1956). Chaotische Bereiche in der geordneten Welt. *ZÄS, 81*, 28–32.

Hornung, E. (1963). *Das Amduat, die Schrift des verborgenen Raumes, vol. 2*. Otto Harrassowitz.

Hornung, E. (1965). Die sonnenfinsternis nach dem tode Psammetichs I. *ZÄS, 92*, 38–39.

Hornung, E. (1975). *Dat, in Lexikon der Ägyptologie, I. (col. 994-995)*. Otto Harrassowitz.

Hornung, E. (1982). *Der ägyptische Mythos von der Himmelskuh: Ein Ätiologie des Unvollkommenen*. Universitätsverlag/Göttingen: Vandenhoeck Rup-recht.

Hornung, E. (1983). *Conceptions of god in ancient Egypt: The one and the many*. Routledge and Kegan Paul Ltd. (Spanish Ed. 1999. *El uno y los múltiples: concepciones egipcias de la divinidad*. Trotta).

Hornung, E. (1990). *Zwei ramessidische Königsgräber: Ramses IV. und Ramses VII*. Philip von Zabern.

Hornung, E. (1992). *Texte zum Amduat, vol. 2. Langfassung 4. bis 8. Stunde*. Editions de belles-lettres.

Hornung, E. (1997). *Die Unterweltsbücher der Ägypter*. Artemis Verlag.

Hornung, E. (1999). *The ancient Egyptian books of the afterlife*. Cornell University Press.

Hornung, E. (2001). Time in the Egyptian Netherworld. *Cadmos, 11*, 7–14.

Hornung, E., Krauss, R., & Warburton, D. A. (Eds.). (2006). *Ancient Egyptian chronology* (HdO, 83). Brill.

Huber, P. J. (2001). The solar omen of Mursili II. *JAOS, 121*, 640–644.

Huber, P. J. (2011). The astronomical basis of Egyptian chronology of the Second Millennium BC. *JEGH, 4*, 172–227.

Hübner, W. (1995). *Grade und Gradbezirke der Tierkreiszeichen, II. Kommentar*. De Gruyter.

Hughes, G. R. (1951). A demotic astrological text. *JNES, 10*(4), 256–264.

Hughes, G. R. (1986). An astrologer's handbook in demotic Egyptian. In L. H. Lesko (Ed.), *Egyptological studies in honor of Richard A. Parker, presented on the occasion of his 78th birthday* (pp. 53–69). Brown University Press.

Hughes, G. R., & Jasnow, R. (1997). *Oriental Institute Hawara Papyri. Demotic and Greek Texts from an Egyptian family archive in the Fayum (Fourth to Third Century B.C.)* (OIP, 113). University of Chicago Press.

Hunger, H. (1992). *Astrological reports to Assyrian Kings*. Helsinki University Press.

Hunger, H., & Pingree, D. (1989). *MUL.APIN. An Astronomical Compendium in Cuneiform. AfO*, 24 suplement. Ferdinand Berger and Söhne.

Hunger, H., & Pingree, D. (1999). *Astral sciences in Mesopotamia*. Brill.

Iamblichus. (1818). *Life of Pythagoras* (T. Taylor, Trans.). J. M. Watkins.

Ifrah, G. (2005). *The universal history of numbers I*. Pengüin Books.

Ingham, M. F. (1969). The length of the Sothic cycle. *JEA, 55*(1969), 36–40.

Isler, M. (1989). An ancient method of finding and extending direction. *JARCE, 26*, 191–206.

Isler, M. (1991). The Gnomon in Egyptian antiquity. *JARCE, 28*, 155–185.

Isler, M. (2001). *Sticks, stones and shadows: Building the Egyptian pyramids*. University of Oklahoma Press.

Jacquet-Gordon, H., Bonnet, C., & Jacquet, J. (1969). Pnubs and the temple of Tabo on Argo island. *JEA, 60*, 103–111.

Jambon, E. (2012). Les signes de la nature dans l'Égypte Pharaonique. In S. Georgoudi, R. Koch Poettre, & F. Schmidt (Eds.), *La Raison des Signes. Présages, rites, destin dans les sociétés de la Méditerranée ancienne* (pp. 131–156). Brill.

Jambon, E. (2013). Calendriers et prodige: remarques sur la divination égyptienne d'après Hérodote II, 82. In *Hérodote et l'Égypte. Regards croisés sur le Livre II de l'Enquête d'Hérodote. Actes de la journée d'étude organisée à la Maison del'Orient et de la Méditerranée* – Lyon, le 10 mai 2010 (pp. 145–166). Maison de l'Orient et de la Méditerranée Jean Pouilloux.

James, T. G. H. (1962). *The Hekanakhte papers and other early Middle Kingdom documents* (MET Pub. No 12). Metropolitan Museum of Art.

Jansen-Winkeln, K. (1998). Beiträge zu den Privatinschriften der Spätzeit. *ZÄS, 125*, 1–13.

Janssen, J. J. (1961). *Two ancient Egyptian ship's logs*. E. J. Brill.

Jasnow, R. (2010). On Sothis and the So-Called Clepsydra (?) in the Book of Thoth (B02, 3/16 – 4/11). In Ch. Leitz, D. von Recklinghausen, and H. Knuf (Eds.), *Honi soit qui mal y pense. Studien zum pharaonischen, griechisch-römischen und spätantiken Ägypten zu Ehren von Heinz-Josef Thissen*. OLA, 194 (pp. 237–243). Peeters.

Jeffreys, D. (1998). The topography of Heliopolis and Memphis: Some cognitive aspects. In R. Stadelmann (Ed.), *Stationen. Beitrage zur Kulturgeschichte Ägyptens* (pp. 63–71). Philipp von Zabern.

Jéquier, G. (1935). *La Pyramide d'Aba. Fouilles à Saqqarah*. IFAO.

Johnson, J. H. (Ed.). (2001). *The demotic dictionary of the oriental institute of the University of Chicago*. The Oriental Institute.

Johnson, J. D. M., Roberts, C. H., & Hunt, A. S. (Eds.). (1915). *Catalogue of the Greek papyri in the John Rylands Library Manchester, vol. II. Documents of the Ptolemaic and Roman Periods (Nos. 62-456)*. The University Press.

Johnson, D., Tyldesley, J., Lowe, T., Withers, P. J., & Grady, M. M. (2013). Analysis of a prehistoric Egyptian iron bead with implications for the use and perception of meteorite iron in ancient Egypt. *Meteoritics and Planetary Science, 48*, 997–1006.

Jones, D. (1988). *A glossary of ancient Egyptian nautical titles and terms*. Kegan Paul.

Jones, A. (1994). The place of astronomy in Roman Egypt. *Apeiron. A Journal for Ancient Philosophy and Science, 27*, 25–52.

Jones, A. (1997). Studies in the astronomy of the Roman period. I. The standard lunar scheme. *Centaurus, 39*, 1–36.

Jones, A. (1998). Three astronomical tables from Tebtunis. *Zeitschrift für Papyrologie und Epigraphik, 121*, 211–218.

Jones, A. (1999a). *Astronomical Papyri from Oxyrhynchus: (P. Oxy. 4133–4300a)*. American Philosophical Society.

Jones, A. (1999b). Classification of astronomical tables on Papyrus. In N. M. Swerdlow (Ed.), *Ancient astronomy and celestial divination* (pp. 299–340). MIT Press.

Jones, A. (2000a). Calendrica II: Date equations from the reign of Augustus. *Zeitschrift für Papyrologie und Epigraphik, 129*, 159–166.

Jones, D. (2000b). *An index of ancient Egyptian titles, epithets and phrases of the old kingdom* (v. I,. BAR IS, 866 (I)). Archaeopress.
Jones, A. (2007). Astrologers and their astronomy. In A. Bowman, R. A. Cles, N. Gonis, & P. J. Parsons (Eds.), *Oxyrhynchus: A city and its texts* (pp. 307–314). Egypt Exploration Society.
Jones, A. (2009). Mathematics, science, and medicine in the Papyri. In R. S. Bagnall (Ed.), *The Oxford handbook of Papyrology* (pp. 338–357). Oxford University Press.
Jones, A., & Perale, M. (2012). Greek Astronomical Tables in the Papyrus Carlsberg Collection (with associated fragments from other collections). *Archiv für Papyrusforschung, 58*, 308–343.
Junge, F. (1973). Zur Fehldatierung des sog. Denkmals memphitischer Theologie oder Der Beitrag der ägyptischen Theologie zur Geistesgeschichte der Spätzeit. *MDAIK, 29*, 195–204.
Kahl, J. (1993). Textkritische Bemerkungen zu den Diagonalsternuhren des Mittleren Reiches. *SAK, 20*, 95–107.
Kahl, J. (2006). Inscriptional evidence for the relative chronology of Dyns. 0–2. In E. Hornung, R. Krauss, & D. A. Warburton (Eds.), *Ancient Egyptian chronology* (HdO, 83) (pp. 94–115). Brill.
Kahl, J. (2007). *Ra is my lord: searching for the rise of the sun god at the dawn of Egyptian history*. Harrassowitz.
Kahl, J., Kloth, N., & Zimmermann, U. (1995). *Die Inschriften der 3 Dynastie* (ÄA, 56). Harrassowitz.
Kákosy, L. (1963). Schöpfung und Weltuntergang in der ägyptischen Religion. *Acta antiqua Academiae Scientiarum Hungaricae, 11*, 17–30.
Kákosy, L. (1982). Decans in Late Egyptian Religion. *Oikumene, 3*, 163–191.
Kamal, A. (1906). Rapport sur quelques localités de la Basse-Égypte. *ASAE, 7*, 232–240.
Kaper, O. (1995). The astronomical ceiling of Deir el-Haggar in the Dakhleh Oasis. *JEA, 81*, 175–195.
Kees, H. (1942). Die Feuerinsel in den Sargtexten und im Totenbuch. *ZÄS, 78*, 41–53.
Kees, H. (1949). Ein Sonnenheiligtum im Amonstempel von Karnak. *Orientalia, 18*, 427–442.
Kemp, B. J. (2006). *Ancient Egypt. Anatomy of a civilization*. Routledge.
Kendall, T. (1997). Les souverains de la montagne sacrée, Napata et la dynastie des Koushites. In D. Wildung (Ed.), *Soudan, royaumes sur le Nil* (pp. 161–171). Institut Monde Arabe.
Kitchen, K. A. (1993–2014). *Ramesside inscriptions translated and annotated: Translations* (7 volumes). Blackwell/Wiley-Blackwell.
Kitchen, K. (2013). Establishing chronology in Pharaonic Egypt and the ancient Near east: interlocking textual sources relating to c. 1600-664 BC. In A. J. Shortland & C. Bronk Ramsey (Eds.), *Radiocarbon and the chronologies of ancient Egypt* (pp. 1–19). Oxbow Books.
Klinger, J. (2006). Chronological links between the Cuneiform world of the ancient Near East and ancient Egypt. In E. Hornung, R. Krauss, & D. A. Warburton (Eds.), *Ancient Egyptian chronology* (HdO, 83) (pp. 305–325). Brill.
Klotz, D. (2012). Articulata Forma Dei. A Cosmic Epithet from Esna and Medinet Habu. *ENIM, 5*, 31–37.
Klug, A. (2002). *Königliche Stelen in der Zeit von Ahmose bis Amenophis III* (Monumenta Aegyptiaca, 8). Brepols.
Knigge, C. (2006). *Das Lob der Schöpfung: Die Entwicklung ägyptischer Sonnen- und Schöpfungshymnen nach dem Neuen Reich* (OBO, 219). Universitätsverlag.
Koch, R. (1990). *Die Erzählung des Sinuhe*. Fondation égyptologique reine Élisabeth.
Koch-Westenholz, U. (1995). *Mesopotamian astrology. An introduction to Babylonian and Assyrian Celestial Divination*. Museum Tusculanum Press.
Köhler, E. C. (2013). Early dynastic Egyptian chronologies. In A. J. Shortland & C. Bronk Ramsey (Eds.), *Radiocarbon and the chronologies of ancient Egypt* (pp. 224–234). Oxbow Books.
Kolbe, G. (2008). A test of the 'Simultaneous transit method'. *Journal for the History of Astronomy, 39*, 515–517.
Kozloff, A. P., & Berman, L. M. (Eds.). (1993). *Aménophis III le Pharaon-Soleil*. Réunion des musées nationaux.
Krauss, R. (1977). Untersuchungen zu König Amenmesse. *SAK, 5*, 131–174.

Krauss, R., 1985. *Sothis und Monddaten*. HÄB, 20. : Gerstenberg Verlag.

Krauss, R. (1992a). Das Kalendarium des Papyrus Ebers und seine chronologische Verwertbarkeit. *ÄuL, 3*, 75–96.

Krauss, R. (1992b). Lässt sich die astronomische Decke im Senenmut-Grab für die absolute Thutmoside-Chronologie auswerten? In Das *Kalendarium des Papyrus Ebers und seine chronologische Verwertbarkeit (Exkurs 3). ÁuL, 3* (pp. 75–96).

Krauss, R. (1995). Nochmals die ägyptische Nacht vom 14./15. November. *Göttinger Miszellen, 146*, 61–70.

Krauss, R. (1996). The length of Sneferu's reign and how long it took to build the 'red pyramid'. *JEA, 82*, 43–50.

Krauss, R. (1997). *Astronomische Konzepte und Jenseitsvorstellungen in den Pyramidentexten* (ÄA, 59). Harrassowitz Verlag.

Krauss, R. (1998a). Altägyptische Sirius- und Monddaten aus dem 19 und 18 Jahrhundert vor Christi Geburt (Berliner Illahun-Archiv). *ÄuL, 8*, 113–123.

Krauss, R. (1998b). Wenn und aber: das Wag-Fest und die Chronologie des Alten Reiches. *Göttinger Miszellen, 162*, 53–63.

Krauss, R. (1999). Nähere Mitteilungen über Seth/Merkur und Horusauge/Venus im grossen Tagewählkalender. *SAK, 27*, 233–254.

Krauss, R. (2002). *Egyptian calendars and astronomy* (Cambridge history of science). Cambridge University Press.

Krauss, R. (2003). Arguments in favor of a low chronology for the Middle and New Kingdom in Egypt. In M. Bietak (Ed.), *The synchronisation of civilisations in the eastern Mediterranean in the second millennium B.C. II* (pp. 175–197). Austrian Academy of Sciences.

Krauss, R. (2006a). Dates relating to seasonal phenomena and miscellaneous astronomical dates. In E. Hornung, R. Krauss, & D. A. Warburton (Eds.), *Ancient Egyptian chronology* (pp. 369–379). Brill.

Krauss, R. (2006b). La orientación de los corredores en las pirámides egipcias según John Herschel. In J. Lull (Ed.), *Trabajos de Arqueoastronomía. Ejemplos de África, América, Europa y Oceanía* (pp. 185–192). Agrupación Astronómica de La Safor.

Krauss, R. (2006c). Egyptian Sirius/Sothic Dates, and the question of the Sothis-based Lunar calendar. In E. Hornung, R. Krauss, & D. A. Warburton (Eds.), *Ancient Egyptian chronology* (pp. 439–457). Brill.

Krauss, R. (2007). Die Bubastiden-Finsternis im Licht von 150 Jahren Forschungsgeschichte. *MDAIK, 63*, 211–223.

Krauss, R. (2009). Astronomical chronology. In J. A. Belmonte & M. Shaltout (Eds.), *In search of cosmic order, selected essays on Egyptian archaeoastronomy* (pp. 133–154). Supreme Council of Antiquities.

Krauss, R. (2011). Stellar and solar components in ancient Egyptian mythology and royal ideology. In M. Rappenglück, B. Rappenglück, N. Campion, & F. Silva (Eds.), *Astronomy and power* (BAR IS, 2794) (pp. 137–142). Archaeopress.

Krauss, R. (2015). Zur Interpretation der Wasseruhren aus Karnak und im Museo Barracco. In K. Finneiser & J. Helmbold-Doyé (Eds.), *Der andere Blick. Forscherlust und Wissensdrang. Museumsgabe zum 80. Geburtstag von Karl-Heinz Priese* (pp. 132–165). E-B Verlag.

Krauss, R. (2015a). Egyptian chronology: Ramesses II through Shoshenq III, with analysis of the lunar dates of Thutmoses III. *ÄuL, 25*, 335–382.

Krauss, R. (2015b). Astronomy and chronology – Babylonia, Assyria, and Egypt. In C. L. N. Ruggles (Ed.), *Handbook of archaeoastronomy and ethnoastronomy* (pp. 31–41). Springer.

Krauss, R. (2016a). Aussagen über die Ungenauigkeit der ägyptischen Auslauf-Wasseruhren. *Göttinger Miszellen, 250*, 119–132.

Krauss, R. (2016b). The Azzi eclipse and Hittite-Egyptian chronology. In J. van Dijk (Ed.), *Another mouthful of dust. Egyptological studies in honour of Geoffrey Thorndike Martin* (OLA, 246) (pp. 335–379). Peeters.

Krauss, R. (2018). Über die L-förmigen Schattenuhren und die Schlacht von Megiddo. *SAK, 47*, 149–175.

Krauss, R. (2019). Die Kanäle in der Cheops-Pyramide: Luftschächte, Modellkorridore oder Leitwege zu den Sternen? *SAK, 48*, 151–181.
Krauss, R. (2021). The assumed movable *wAgy* feast in the Raneferef archive. *Göttinger Miszellen, 263*, 91–96.
Krauss, R. (2022a). Late Old Kingdom chronology – Another model. *ÄuL*, forthcoming.
Krauss, R. (2022b). The layout of the Ebers calendar. *Orientalia*, forthcoming.
Krauss, R. (forthcoming). The *dwAt* as part of the visible sky. *JNES*.
Krauss, R., & Reijs, V. (2021). Do ancient Egyptian almanacs show evidence of celestial recurrence? In A. C. González-García, R. Frank, L. D. Sims, M. Rappenglueck, G. Zotti, J. A. Belmonte, & I. Sprajc (Eds.), *Beyond paradigms in cultural astronomy* (BAR international series, 3033) (pp. 3–10). Archeopress.
Krauss, R., & Warburton, D. A. (2009). The basis for the Egyptian dates. In D. A. Warburton (Ed.), *Time's up! Dating the Minoan eruption of Santorini* (pp. 15–138). Monographs of the Danish Institute.
Kronk, G. W. (1999). *Cometography. A catalog of comets, I. Ancient – 1799.* Cambridge University Press.
Krupp, E. C. (1984). Egyptian astronomy: A tale of temples, tradition, and tombs. In E. C. Krupp (Ed.), *Archaeoastronomy and the roots of science* (AAAS Selected Symposium, 71) (pp. 289–320). Routledge.
Krupp, E. C. (1988). Light in the temples. In C. L. Ruggles (Ed.), *Records in stone: Papers in memory of Alexander Thom* (pp. 473–499). Cambridge University Press.
Kurth, D. (2004a). *Edfu Temple: A guide by an ancient Egyptian priest.* American University in Cairo Press.
Kurth, D. (2004b). *Edfou VII.* Otto Harrassowitz.
Lange, H. O., & Neugebauer, O. (1940). *Papyrus Carlsberg No. I. Ein hieratisch-demotischer kosmologischer text.* E. Munksgaard.
Lange, H. O., & Schäfer, H. (1902). *Catalogue général des antiquités égyptiennes du Musée du Caire, N. 20001-20399. Grab- und Denksteine des Mittleren Reichs, I.* Reichdrückerai.
Lange, H. O., & Schäfer, H. (1908). *Catalogue général des antiquités égyptiennes du Musée du Caire, N. 20400–20780. Grab- und Denksteine des Mittleren Reichs, II.* Reichdrückerai.
Larcher, C. (2016). Quand Thot devient babouin. Essai de datation et d'interprétation de l'apparition de la forme simiesque du dieu. *ZÄS, 143*(1), 60–76.
Larson, J. A. (1992). The Tutankhamen astronomical instrument. *Amarna Letters, 2*, 77–86.
Le Page Renouf, P. (1874). Calendar of astronomical observations found in royal tombs of the XXth dynasty. *Transactions of the Society of Biblical Archaeology, 3*, 400–421.
Leemans, C. (1843). *Monumens égyptiennes du Musée d'Antiquités des Pays-Bas à Leide.* Brill.
Lefebvre, G. (1923). L'oeuf divin d'Hermopolis. *ASAE, 23*, 65–67.
Lefebvre, G. (1924). *Le tombeau de Petosiris.* IFAO.
Lehner, M. (1985). Giza, a contextual approach to the pyramids. *AfO, 32*, 139–159.
Lehner, M. (1997). *The complete pyramids.* Thames and Hudson.
Leitz, C. (1991). *Studien zur Ägyptischen Astronomie* (ÄA, 49). Harrassowitz.
Leitz, C. (1992). Bemerkungen zur astronomischen Chronologia. *ÄuL, 3*, 97–102.
Leitz, C. (1994a). Der Mondkalender und der Beginn des ägyptischen Kalendartages. *BSÉG, 18*, 49–60.
Leitz, C. (1994b). Tagewählerei. In *Das Buch ḥ3t nḥḥ pḥ.wy ḏt und verwandte Texte.* Harrassowitz.
Leitz, C. (1995). *Altägyptische Sternuhren* (OLA, 62). Peeters.
Leitz, C. (1999). *Magical and Medical Papyri in the British Museum, Hieratic papyri in the British Museum, VII.* British Museum Press.
Leitz, C. (2006). Die Sternbilder auf den rechteckigen und runden Tierkreis von Dendara. *SAK, 34*, 285–318.
Leitz, C. (2008). Zu einigen astronomischen Aspekten im sogenannten Nutbuch oder Grundriß des Laufes der Sterne. *Enchoria, 31*, 1–21.
Lepsius, K. R. (1849–1859). *Dankmäler aus Aegypten und Aethiopien.* Nicolaische Buchhandlung.

Lesko, L. H. (1991). Ancient Egyptian cosmogonies and cosmologies. In B. E. Shafer (Ed.), *Religion in ancient Egypt: Gods, myths, and personal practice* (pp. 88–122). Cornell University Press.
Lesko, L. H. (2002). *A dictionary of late Egyptian*. Fall River Publication.
Lichtheim, M. (1975). *Ancient Egyptian literature, vol. I: The old and middle kingdoms*. University of California Press.
Lichtheim, M. (1976). *Ancient Egyptian literature, vol. 2: The new kingdom*. University of California Press.
Lightboy, D. I. (2020). Moving heaven and earth for Khufu: Were the Trial Passages at Giza components of a rudimentary stellar observatory? *The Journal of Ancient Egyptian Architecture, 4*, 29–53.
Locher, K. (1981). A conjecture concerning the early Egyptian constellation of the sheep. *Journal for the History of Astronomy, 12*, 73–75.
Locher, K. (1985). Probable identification of the ancient Egyptian circumpolar constellations. *Journal for the History of Astronomy, 16*, 151–152.
Locher, K. (1990). The ancient Egyptian Constellation Group 'The Lion Between the Two Crocodiles' and the bird. *Journal for the History of Astronomy, 21*, 49–51.
Locher, K. (1993). New arguments for the celestial location of the Decanal Belt and for the origins of the S3h-Hieroglyph. In G. M. Zaccone & T. R. Netro (Eds.), *VI Congresso Internazionale di Egittologia vol. II* (pp. 279–284). Società Italiana.
Locher, K. (1998). Middle Kingdom astronomical coffin lids, extension of the corpus from 12 to 17 Specimens Since Neugebauer and Parker. In *Transactions of the 7th international congress of Egyptologists* (OLA, 82) (pp. 697–702). Peeters.
Lockyer, J. N. (1901). *The dawn of astronomy*. Garland Publishing, Inc., 1993 new edition.
Lodomez, G. (2007). Les fragments de clepsydre de la dynastie des Argéades (332–304 av. J. C.). *CdÉ, 82*, 57–76.
Logan, T. J., & Williams, B. (2000). On the Meroe observatory. *Beitrage zur Sudanforschung, 7*, 59–84.
Loprieno, A. (2003). Temps des dieux et temps des hommes en ancien Égypte. In V. Pirenne-Delforge & Ö. Tunca (Eds.), *Représentations du temps dans les religions* (pp. 123–142). Université de Liège.
Loprieno-Behlmer, H. (1986). Stern. In W. Helck & W. Westendorf (Eds.), *Lexikon der Ägyptologie, VI (cols. 11–14)*. Otto Harrassowitz.
Luft, U. (1992a). *Die chronologische Fixierung des ägyptischen Mittleren Reiches nach dem Tempelarchiv von Illahun*. Verlag der Österreichischen Akademie der Wissenschaften.
Luft, U. (1992b). Remarks of a philologist on Egyptian chronology. *ÄuL, 3*, 109–114.
Luft, U. (1994). The date of the *wAgy* feast: Considerations on the chronology of the Old Kingdom. In A. J. Spalinger (Ed.), *Revolutions in time: Studies on ancient Egyptian calendrics* (pp. 39–44). Van Sieclen Books.
Lull, J. (2002). *Las tumbas reales egipcias del Tercer Período Intermedio (dinastías XXI–XXV): tradición y cambios* (BAR international series, 1045). Archaeopress.
Lull, J. (2003). Posible identificación de un instrumento de observación astronómica. *Huygens, 42*, 7–13.
Lull, J. (2004a). *La astronomía en el antiguo Egipto*. Publicacions de la Universitat de València.
Lull, J. (2004b). Identification of an Ancient Egyptian Star: *Tms n xntt* 'the red (star) of Khentet'. *Göttinger Miszellen, 199*, 73–77.
Lull, J. (2006–2009). A possible ancient Egyptian astronomical instrument for positioning the Central Meridian. *Discussions in Egyptology, 64*, 47–56.
Lull, J. (2006a). El eclipse solar del papiro demótico Berlín 13588. *Huygens, 59*, 16–18.
Lull, J. (2006b). Los cúmulos estelares en las listas decanales y relojes ramésidas del antiguo Egipto. In J. Lull (Ed.), *Trabajos de Arqueoastronomía. Ejemplos de África, América, Europa y Oceanía* (pp. 193–208). Agrupación Astronómica de La Safor.
Lull, J. (2006c). La constelación de Mesjetiu en el antiguo Egipto. *Astronomía, 84*, 24–31.
Lull, J. (2008a). El astrónomo en el antiguo Egipto. *Astronomía, 103*, 72–79.

Lull, J. (2008b). El planisferio egipcio del templo de Dendera. *Astronomía, 113*, 78–83.
Lull, J. (2008c). Meschetiu in der Mythologie und der Orientierung der ägyptischen Tempel. *Acta Praehistorica et Archaeologica, 40*, 85–92.
Lull, J. (2018a). The ancient Egyptian constellation of *Wjꜣ* (Boat) and its link to Sagittarius in the Ptolemaic and Roman era. *Aula Orientalis, 36*(2), 57–77.
Lull, J. (2018b). Wia. Una barca en el cielo egipcio. *Astronomía, 223*, 32–37.
Lull, J. (2019a). El León divino, el león y el Leo zodiacal en el antiguo Egipto. *Astronomía, 235*, 32–39.
Lull, J. (2019b). A chronological perspective on the transition from Amenhotep III to Amenhotep IV/Akhenaten'. *Aula Orientalis, 37*, 61–89.
Lull, J. (2022). Cometas, meteoros y meteoritos en el antiguo Egipto. *Astronomía, 271*, 38–44. Forthcoming.
Lull, J. (2023). El reloj estelar de Kom Abu Yasin. Un precedente egipcio al Reloj de las Guardas. Forthcoming.
Lull, J., & Belmonte, J. A. (2006). A firmament above Thebes: Uncovering the constellations of ancient Egyptians. *Journal for the History of Astronomy, 37*, 373–392.
Lull, J., & Belmonte, J. A. (2009). The constellations of ancient Egypt. In J. A. Belmonte & M. Shaltout (Eds.), *In search of cosmic order: Selected essays on Egyptian archaeoastronomy* (pp. 155–195). Supreme Council of Antiquities Press.
Lull, J., & Belmonte, J. A. (2015). Egyptian constellations. In C. L. N. Ruggles (Ed.), *Handbook of archaeoastronomy and ethnoastronomy* (pp. 1477–1487). Springer.
Lull, J., & Navarro-López, D. (2022). The reign of Sety I from the absolute chronology, forthcoming.
Macalister, R. A. S. (1912). *The excavation of Gezer 1902–1905 and 1907–1909, vol. 2*. John Murray.
Mace, A. C. (1922). A group of Scarabs found at Lisht. *JEA, 8*, 13–15.
Magdolen, D. (2001). An astronomical inscription on the Berlin Merkhet. *Asian and African Studies, 10*(1), 80–87.
Magli, G. (2008). Akhet Khufu: Archaeoastronomical hints at a common project of the two main pyramids of Giza. *Nexus Network Journal, 11*, 35–50.
Magli, G. (2013). *Architecture, astronomy and sacred landscape in ancient Egypt*. Cambridge University Press.
Manley, B. (2002). Some images of the king and queen together in stele of Ahmose I. *Papers on Ancient Egypt (TdE), 1*, 35–44.
Manning, S. W. (2006). Radiocarbon dating and Egyptian chronology. In E. Hornung, R. Krauss, & D. A. Warburton (Eds.), *Ancient Egyptian chronology* (HdO, 83) (pp. 327–355). Brill.
Manning, S. W., Kromer, B., Dee, M. W., Friedrich, M., Higham, T. F. G., & Bronk Ramsey, C. (2013). Radiocarbon calibration in the mid to later 14th Century BC and radiocarbon dating Tell al-Amarna, in Egypt. In A. J. Shortland & C. Bronk Ramsey (Eds.), *Radiocarbon and the chronologies of ancient Egypt* (pp. 121–145). Oxbow Books.
Manassa, C. (2007). *The Late Egyptian Underworld: Sarcophagi and Related Texts from the Nectanebid Period. ÄAT, 72*. Wiesbaden: Harrassowitz Verlag.
Maravelia, A. (2019). The Conception of the Cosmic Egg (*swHt*) in the ancient Egyptian and in the Orphic Cosmovision. *Shodoznavstvo, 83*, 25–52.
Maravelia, A. A., & Shaltout, M. (2003). Illumination of the sacrarium in the great temple of Abu Simbel, its astronomical explanation and some hints on the possible stellar orientation of the small temple. In A. A. Maravelia (Ed.), *Ad astra per aspera et per ludum: European archeoastronomy and the orientation of monuments in the Mediterranean basin* (BAR IS, 1154) (pp. 7–30). Archaeopress.
Mariette, A. (1889). *Les mastabas de l'ancien Empire*. F. Vieweg.
Martin, G. T. (1971). *Egyptian administrative and private-name seals: Principally of the middle kingdom and second intermediate period*. Griffith Institute.
Martin, G. T. (1979). *The Tomb of Hetepka and Other Reliefs and Inscriptions from the Sacred Animal Necropolis North Saqqâra, 1964–1973*. Egypt Exploration Society.

Mathieu, B. (1986). Les hommes de larmes: à propos d'un jeu de mots mythique dans les textes de l'ancienne Égypte. In: *Hommages à François Daumas* (pp. 499–505). OrMonsp 3.

Mathieu, B. (1987). Le voyage de Platon en Egypte. *ASAE, 71*, 153–167.

Mathieu, B. (2008). Les Enfants d'Horus, théologie et astronomie. *ENIM, 1*, 7–14.

Matthey, P. (2012). *Pharaon, magicien et filou: Nectanébo II entre l'histoire et la légende*. PhD thesis, University of Genève.

McMurray, W. (2003). Dating the Amarna period in Egypt: did a solar eclipse inspire Akhenaten? http://egyptologyforum.org/EMP/DAPE.pdf. Last consulted 5 Mar 2021.

Meeks, D. (1982). Meteor. In W. Helck & W. Westendorf (Eds.), *Lexikon der Ägyptologie, IV (cols. 117–118)*. Otto Harrassowitz.

Meeks, D., & Favard-Meeks, C. (1994). *La vida cotidiana de los dioses egipcios*. Ediciones Temas de Hoy.

Menchetti, A. (2009). Un aperçu des textes astrologiques de Médinet Madi. In: G. Widmer, D. Devauchelle (Ed.): *Actes du IXe congrès international des études démotiques, Paris, 31 août - 3 septembre 2005. Bibliothèque d'Étude, 147* (pp. 223–241).

Mengoli, P. (2006). *Astronomia Egizia*. Budrio.

Mengoli, P. (2019). *L'astronomia degli antichi egiziani*. A. Pontecorboli editore.

Mercer, S. A. B. (1952). *The pyramid texts in translation and commentary. Vol. IV*. Longman, Green and Co.

Meyer, E. (1904). *Ägyptische Chronologie*. Verlag der Königl. Akademie der Wissenschaften.

Mogensen, M. (1918). *Inscriptions Hiéroglyphiques du Musée National de Copenhague*. A. F. Höst et fils.

Mohamed Ahmed, S., & Anderson, J. (2005). Le temple d'Amon à Dangeil (Soudan). *BSFÉ, 162*, 10–17.

Molinero Polo, M. Á. (2000). Templo y cosmos. In M. Á. Molinero Polo & D. Sola Antequera (Eds.), *Arte y sociedad del Egipto antiguo* (pp. 69–94). Encuentro.

Molinero Polo, M. Á. (2003). *Realeza y concepción del universo en los textos de las pirámides*. PhD thesis, UCM.

Molinero Polo, M. Á. (2014). A bright sky over Karakhamun; the astronomical ceiling of the main burial chamber in TT2233. In E. Pischikova (Ed.), *Tombs of the south Asasif necropolis Thebes, Karakhamun (TT223), and Karabasken (TT391) in the 25th dynasty* (pp. 201–238). American University in Cairo Press.

Möller, G. (1918). Ägyptische Abteilung. Eine neue demotische Erzählung. *Amtliche Berichte aus den Königlichen Kunstsammlungen, 39*, 12–14.

Möller, G. (1923). G. Hieratische Lesestücke für akademischen Gebrauch 1. *Bibliolife* 2009.

Montet, P. (1951). *Les constructions et le tombeau de Psousennès à Tanis. Fouilles de Tanis, La nécropole royale de Tanis,* 2. Paris.

Morenz, L. D. (2004). Apophis: On the origin, name, and nature of an ancient Egyptian anti-God. *JNES, 63*, 201–205.

Moret, A. (1917). Chartes d'immunité dans l'Ancien Empire égyptien. *Journal Asiatique, 11*, 359–447.

Morkot, R. G. (2000). *The black pharaohs: Egypt's Nubian rulers*. Rubicon Press.

Morrow, S. B. (2015). *The dawning of the mind. Unlocking the pyramid texts*. Ferraus, Straus and Giroux.

Moursi, M. I. (1972). *Die Hohenpriester des Sonnengottes von der Frühzeit Ägyptens bis zum Ende des Neuen Reiches* (MÄS, 26). Deutcher Kunstverlag.

Moyer, I. S. (2011). *Egypt and the Limits of Hellenism*. Cambridge University Press.

Murray, M. A. (1904). *The Osireion at Abydos*. B. Quaritch.

Myśliwiec, K. (1979). *Studien zum Gott Atum, II*. Gerstenberg Verlag.

Naether, F., & Ross, M. (2008). Interlude: A series containing a Hemerology with Lengths of Daylight. *Egitto e Vicino Oriente, 31*, 51–90.

Naville, E. (1894). *Ahnas el Medineh (Heracleopolis Magna): with chapters on the Mendes, the Nome of Thoth, and Leontopolis* (MEEF, 11). Egypt Exploration Fund.

Naville, E. (1903). La Pierre de Palermo. *RecTrav, 25*, 64–75.
Naville, E. (1947). *The XIth Dynasty temple at Deir el-Bahari I* (MEEF, 28). EEF.
Nell, E., & Ruggles, C. L. (2014). The orientation of the Giza pyramids and associated structures. *Journal for the History of Astronomy, 45*, 308–369.
Nelson, H. H. (1913). *The Battle of Megiddo*. University of Chicago Libraries.
Nemes, G. W. (2020). The mythological importance of the constellation Msḫtjw in mortuary representations until the end of the New Kingdom. *ENIM, 13*, 1–61.
Nenna, M.-D. (2003). De Douch (oasis de Kharga) à Grand (Vosges). Un disque en verre peint à représentations astrologiques. *BIFAO, 103*, 355–376.
Neugebauer, O. (1938). Die Bedeutungslosigkeit der Sothisperiode für die älteste ägyptische Chronologie. *Acta Orientalia, 17*, 169–195.
Neugebauer, O. (1942). The origin of the Egyptian calendar. *JNES, 1*(4), 396–403.
Neugebauer, O. (1943). Demotic horoscopes. *JAOS, 63*, 115–127.
Neugebauer, O. (1947). Studies in ancient astronomy. VIII. The water clock in Babylonian astronomy. *Isis, 37*, 37–43.
Neugebauer, O. (1980). On the orientation of the pyramids. *Centaurus, 24*, 1–3.
Neugebauer, O. (1983). Die Bedeutungslosigkeit der "Sothisperiode" für die älteste ägyptische Chronologie (1938). In O. Neugebauer (Ed.), *Astronomy and history. Selected essays* (pp. 169–195). Springer.
Neugebauer, O., & Parker, R. A. (1960–1969). *Egyptian astronomical texts*, vol. I–III. Brown University.
Neugebauer, O., & Van Hoesen, H. B. (1987). *Greek Horoskopes*: American Philosophical Society [1959].
Neugebauer, O., Parker, R. A., & Zauzich, K.-T. (1981). A demotic lunar eclipse text of the first century, B.C. *Proceedings of the American Philosophical Society, 125*, 312–327.
Niwiński, A. (1981). Noch einmal über zwei Ewigkeitsbegriffe. Ein Vorschlag der graphischen Lösung in anlehung an die Ikonographie der 21. Dynastie. *Göttinger Miszellen, 48*, 41–53.
Niwinski, A. (1989). *Studies on the illustrated Theban Funerary Papyri of the 11th and 10th Centuries B.C. Freiburg* / Göttingen: Universitätsverlag / Vandenhoeck Ruprecht.
Nolan, J. F. (2003). The original lunar calendar and cattle counts in old kingdom Egypt. *Aegyptiaca Helvetica, 17*, 75–98.
Nolan, J. F. (2015). Kings and priests: Phyle rotations and Old Kingdom civil dates. In P. der Manuelian & T. Schneider (Eds.), *Towards a new history for the Egyptian old kingdom: Perspectives on the Pyramid Age* (pp. 337–365). Brill.
Nuzzolo, M. (2020). La pierre de Palermo at les fragments associés. Nouvelles découvertes sur les plus anciennes annales royales égyptiennes. *BSFÉ, 202*, 55–82.
Nuzzolo, M., & Krejčí, J. (2017). Heliopolis and the solar cult in the third millennium BC. *ÄuL, 27*, 357–380.
Ogden, J. (2000). Metals. In I. Shaw & P. Nicholson (Eds.), *Ancient Egyptian materials and technology* (pp. 148–176). Cambridge University Press.
Onasch, H.-U. (1994). *Die Assyrischen Eroberungen Ägyptens I*. Harrassowitz.
Ossendrijver, M., & Winkler, A. (2018). Chaldeans on the Nile: Two Egyptian astronomical procedure texts with Babylonian Systems A1 and A2 for Mercury. In C. J. Crisostomo et al. (Eds.), *The scaffolding of our thoughts. Essays on Assyriology and the history of science in honor of Francesca Rochberg* (pp. 382–419). Brill.
Otto, W. (1905). *Priester und Tempel im hellenistischen Ägypten: ein Beitrag zur Kulturgeschichte des Hellenismus*, vols. 1–2. Teubner.
Park, R. (2008). The First Decan. In K. Griffin (Ed.), *Current research in Egyptology 2007* (pp. 103–111). Oxbow Books.
Park, R., & Eccles, B. (2012). Dating the Dendara zodiac. Egypt's famous Greco-Roman 'Zodiac'. *Aram, 24*, 175–192.
Parker, R. A. (1950). *The calendars of ancient Egypt* (Studies in Ancient Oriental Civilization, 26). Chicago Universsity Press.

Parker, R. A. (1957a). The problem of the month-names: a reply. *RdE, 11*, 85–107.

Parker, R. A. (1957b). The length of reign of Amasis and the beginning of the 26th Dynasty. *MDAIK, 15*, 208–212.

Parker, R. A. (1959). *A Vienna Demotic Papyrus on Eclipse- and Lunar-Omina. Edited with translation and commentary*. Brown University Press.

Parker, R. A. (1974). Ancient Egyptian astronomy. *Philosophical Transactions of the Royal Society of London Series A, Mathematical and Physical Sciences, 276*, 51–65.

Parker, R. A. (1981). Some reflections on the lunar dates of Thutmose III and Ramesses II. In W. K. Simpson & W. M. Davis (Eds.), *Studies in ancient Egypt, the Aegean, and the Sudan. Essays in Honor of Dows Dunham on the occasion of his 90th Birthday, June 1, 1980* (pp. 146–148). Museum of Fine Arts.

Parker, R. A., & Lesko, L. H. (1988). The Khonsu Cosmogony. In J. Baines, T. G. H. James, A. Leahy, & A. F. Shore (Eds.), *Pyramid studies and other essays presented to I. E. S. Edwards* (pp. 168–175). EES.

Parpola, S. (2007). Letters from Assyrian Scholars to the Kings Esarhaddon and Assurbanipal, *part II: Commentary and Appendices*. Eisenbrauns.

Pearson, Ch. Brewer, P., Brown, D., Heaton, T., Hodgins, G.W.L., Jull, A.J., Lange, T. and Salzer, W.S. (2018) Annual Radiocarbon record indicates 16th Century BCE date for the Thera eruption. *Sience Advances, 4*, eaar8241

Pepin, J.-F. (1989). Quelques aspects de Nouou dans les textes des pyramides et les textes des sarcophages. In S. Schoske (Ed.), *Akten des vierten internationalen Ägyptologen Kongresses, München 1985* (SAK Beiheft 3) (pp. 339–345). Helmut Buske Verlag.

Pepirakis, S. (2017). Decanal Iconography and Natural Materials in the Sacred Book of Hermes to Asclepius. *Greek, Roman, and Byzantine Studies, 57*, 136–161.

Pérez Die, M. C. (2009). Excavaciones de la Misión Arqueológica Española en Heracleopolis Magna (Ehnasya el Medina). In M. C. pérez Die & W. Al-Sadiik (Eds.), *120 years of Spanish archaeology in Egypt* (pp. 54–66). Ministerio de Cultura.

Pérez Die, M. C. (Ed.). (2010). *Heracleópolis Magna (Ehnasya el Medina, Egipto). La necrópolis "real" del Tercer Periodo Intermedio y su reutilización*. Ministerio de Cultura.

Pérez Die, M. C. (2012). El proyecto de Investigación "Heracleópolis Magna" (Ehnasya el Medina). Trabajos 2008–2009. In L. M. de Araujo & J. Das Candeias Sales (Eds.), *Novos Trabalhos de Egiptologia Ibérica, IV Congreso Ibérico de Egiptología* (pp. 923–935). Universidade de Lisboa.

Petrie, W. M. F. (1883). *The pyramids and temples of Gizeh*. Kegan Paul.

Petrie, W. M. F. (1900). *Dendereh. 1898*. The Egypt Exploration Fund.

Petrie, W. M. F. (1924). *Ancient Egypt*. Macmillan.

Petrie, W. M. F. (1940). *Wisdom of the Egyptians*. British School of Archaeology. in Egypt.

Petrie, W. M. F., Wainwright, G. A., & Mackay, E. (1912). *The Labyrinth, Gerzeh and Mazghuneh*. British School of Archaeology in Egypt.

Piankoff, A. (1933). Le naos D 29 du Musée du Louvre. *RdE, 1*, 161–179.

Piankoff, A. (1942). *Le livre du jour et de la nuit*. IFAO.

Piankoff, A. (1946). *Le livre des Querets*. IFAO.

Piankoff, A. (1949). Une representation rare sur l'une des chapelles de Toutankhamon. *JEA, 35*, 113–116.

Piankoff, A. (1951). *Les Chapelles de Tout-ankh-Amon, vol. 2*. IFAO.

Piankoff, A. (1954). *The Tomb of Ramesses VI*. Pantheon Books.

Piankoff, A., & Rambova, N. (1957). *Mythological Papyri, vols. I-2*. Bollingen Foundation.

Pilcher, E. J. (1923). Portable Sundial From Gezer. *Palestine Exploration Quarterly, 55*, 85–89.

Pingree, D. (1997). *From Astral Omens to Astrology: from Babylon to Bikaner*. Istituto italiano per l'Africa et l'Oriente.

Pires, G. B. (2019). Before time, after time: Existential time markers in Ancient Egypt – Beginning, end and restart. A preliminary approach (with a special focus on the Heliopolitan conception). *RES Antiquitatis, 1*, 143–157.

Pliny the Elder. (1855). *Natural history* (Vol. I) (J. Bostock & H. T. Riley, Trans.). Henry G. Bohn.
Plutarch. (1970). *De Iside et Osiride* (J. G. Griffiths, Trans.). University of Wales Press.
Plutarch. *Isis and Osiris*. https://penelope.uchicago.edu/Thayer/E/Roman/Texts/Plutarch/Moralia/Isis_and_Osiris*/A.html. Last viewed 26 May 2020.
Pogo, A. (1930). The Astronomical Ceiling-decoration in the Tomb of Senmut (XVIII Dynasty). *Isis, 14*, 301–325.
Pogo, A. (1932). Calendars on coffin lids from Asyût (second half of the third millenium). *Isis, 17*, 4–24.
Pogo, A. (1936a). Der kalender auf dem Sargdeckel des Idy in Tübingen. *Studien der Bibliothek Warburg, 19*, 22–26.
Pogo, A. (1936b). Egyptian water clocks. *Isis, 25*, 403–425.
Polák, B. (1952). Astronomická orientace egyptsých chrámú a pyramid. *Říše hvězd, 9–10*, 209–223.
Popielska-Grzybowska, J. (2017). Nu, continuity and everlastingness in the Pyramid Texts. In M. Tomorad & J. Popielska-Grzybowska (Eds.), *Egypt 2015: Perspectives of research proceedings of the seventh European conference of Egyptologists 2nd–7th June 2015, Zagreb, Croatia* (pp. 17–29). Archaeopress.
Popko, L. (2006). *Untersuchungen zur Geschichtsschreibung der Ahmosiden- und Thutmosidenzeit*. Ergon.
Porceddu, S., Jetsu, L., Markkanen, T., & Toivari-Viitala, J. (2008). Evidence of periodicity in ancient Egyptian calendars of lucky and unlucky days. *Cambridge Archaeological Journal, 18*, 327–339.
Porceddu, S., Jetsu, L., Markkanen, T., Lyytinen, J., Kajatkari, P., Lehtinen, J., & Toivari-Viitala, J. (2018). Algol as Horus in the Cairo Calendar: The possible meaning and the motives for the observation. *Open Astronomy, 27*, 232–264.
Porphyry. (1823). *Select works of Porphyry* (Translated from the Greek by T. Taylor). T. Rodd.
Posener-Kriéger, P. (1976). *Les archives du temple funéraire de Neferirkare-Kakai (Les papyrus d'Abousir)*, 2 vols. BdÉ, 65. IFAO.
Posener-Kriéger, P., Verner, M. and Vymazalová, H. (2006). *Abusir X: The Pyramid Complex of Raneferef : the Papyrus Archive*. Czech Institute of Egyptology, Charles University in Prague.
Priskin, G. (2015a). The Dendera zodiacs as narratives of the myth of Osiris, Isis, and the child Horus. *ENIM, 8*, 133–185.
Priskin, G. (2015b). The encounter between the sun and the moon on hypocephali. *Birmingham Egyptology Journal, 3*, 24–41.
Priskin, G. (2016). The depiction of the entire lunar cycle in Graeco-Roman temples. *JEA, 102*, 111–144.
Priskin, G. (2017). The Depictions of Astral Rebirth in the 26th Dynasty Tomb of Benaty. *Chronica: Annual of the Institute of history University of Szeged, 17*, 4–18.
Priskin, G. (2018). *The ancient Egyptian Book of the Moon: Coffin texts spells 154–160*. PhD thesis.
Priskin, G. (2019a). *The ancient Egyptian Book of the Moon: Coffin texts spells 154–160*. Archaeopress.
Priskin, G. (2019b). The constellations of the Egyptian astronomical diagrams. *ENIM, 12*, 137–180.
Quack, J. F. (1995). Dekane und Gliedervergottung. Altägyptische Traditionen im Apokryphem Johannis. *Jahrbuch für Antike und Christentum, 38*, 97–122.
Quack, J. F. (1996). Review of F. R. Herbin, Le livre de parcourir l'éternité, OLA 58 (Leuven, 1994). *Orientalistische Literaturzeitung, 91*, 152–158.
Quack, J. F. (1999). Frühe ägyptische Vorläufer der Paranatellonta? *Sudhoffs Archiv, 83*, 212–223.
Quack, J. F. (2000a). Das Buch vom Tempel und verwandte Texte. Ein Vorbericht. *Archiv für Religionsgeschichte, 2*, 1–20.
Quack, J. F. (2000b). Kollation und Korrekturvorschläge zum Papyrus Carlsberg 1. In P. J. Frandsen & K. Ryholt (Eds.), *The Carlsberg Papyri 3: A miscellany of demotic texts and studies* (pp. 165–171). Museum Tusculanum Press.
Quack, J. F. (2002a). Die Dienstanweisung des Oberlehrens aus dem Buch vom Tempel. In H. Beinlich (Ed.), *5. Ägyptologische Tempeltagung, Würzburg, 23.-26. September 1999. Akten*

der ägyptologischen Tempeltagung (Ägypten und Altes Testament 33/3) (pp. 159–171). Otto Harrassowitz.

Quack, J. F. (2002b). Die Spur des Magiers Petese. *Chronique d'Égypte, 77*, 76–92.

Quack, J. F. (2003). Le manuel du temple une nouvelle source sur la vie des prêtres égyptiens. *Égypte, Afrique and Orient, 29*, 11–18.

Quack, J. F. (2004). Organiser le culte idéal. Le Manuel du temple. *BSFÉ, 160*, 9–25.

Quack, J. F. (2008). Demotische magische und divinatorische Texte. In B. Janowski & G. Wilhelm (Eds.), *Omina, Orakel, Rituale und Beschwörungen* (pp. 331–385). Gütersloher.

Quack, J. F. (2010). The Naos of the decades and its place in Egyptian astrology. In D. Robinson & A. Wilson (Eds.), *Alexandria and the North-Western Delta. Joint conference proceedings of Alexandria: City and Harbour, Oxford 2004, and The Trade and Topography of Egypt's North-Western Delta, 8th Century BC to 8th Century AD, Berlín, 2006* (pp. 175–181). Oxford Centre for Maritime Archaeology.

Quack, J. F. (2016). On the concomitancy of the seemingly incommensurable, or why Egyptian Astral tradition needs to be analyzed within its cultural context. In J. M. Steele (Ed.), *The Circulation of the Astronomical Knowledge in the Ancient World* (pp. 230–244). Brill.

Quack, J. F. (2017). "Assur Will Suffer:" Predicting disaster in ancient Egypt. In G. Schenk (Ed.), *Historical disaster experiences. Towards a comparative and transcultural history of disasters across Asia and Europe* (pp. 189–206). Springer.

Quack, J. F. (2018a). *Denkmal memphitischer Theologie, in Das wissenschaftliche Bibellexikon im Internet*. http://www.bibelwissenschaft.de/stichwort/16329/

Quack, J. F. (2018b). Egypt as an astronomical-astrological centre between Mesopotamia, Greece and India. In D. Brown (Ed.), *The interactions of ancient Astral science* (pp. 69–123). Hempen.

Quack, J. F. (2018c). Astronomy in ancient Egypt. In P. T. Keyser & J. Scarborough (Eds.), *The Oxford handbook of science and medicine in the classical world* (pp. 61–70). Oxford University Press.

Quack, J. F. (2019). The planets in ancient Egypt. In *Oxford research encyclopedia of planetary science* (pp. 1–33). Oxford University Press.

Quaegebeur, J. (1975). *Le dieu égyptien Shaï dans la religion et l'onomastique* (OLA, 2). Leuven University Press.

Quiles, A., Aubourg, E., Berthier, B., Delque-Loliç, E., Pierrat-Bonnefois, G., Dee, M. W., Andreu-Lanoë, G., Bronk Ramsey, C., & Moreau, C. (2013). Bayesian modelling of an absolute chronology forr Egypt's 18th Dynasty by astrophysical and radiocarbon methods. *JAS, 40*, 423–432.

Quintus Curtius. (1971). *Quintus Curtius. Books I-IV.* (J. C. Rolfe, Trans.). Harvard University Press/William Heinemann.

Quirke, S. (2001). *The cult of Ra: Sun-worship in ancient Egypt.* Thames and Hudson.

Quirke, S. (2003). *Ra, el Dios del Sol.* Oberón.

Quirke, S. (2004). *Titles and bureaux of Egypt 1850–1700 BC.* Golden House Publications.

Quirke, S. (2013). *Going out in Daylight – prt m hrw. The Ancient Egyptian Book of the Dead, translations, sources, meanings.* Golden House Publications.

Ransom, C. L. (1914). A Late Egyptian Sarcophagus. *The Metropolitan Museum of Art Bulletin, 9*(4), 112–120.

Ray, J., & Gilmore, G. (2006). A fixed point in Coptic Chronology: The Solar Eclipse of 10 March, 601. *Zeitschrift für Papyrologie und Epigraphik, 158*, 189–193.

Razanajao, V. (2006). *D'Imet à Tell Farâoun: recherches sur la géographie, les cultes et l'histoire d'une localité de Basse-Égypte orientale.* PhD thesis, Université Paul Valéry Montpellier III.

Reader, C. (2002). Giza before the Fourth Dynasty. *Journal of the Ancient Chronology Forum, 9*, 1–5.

Redford, D. B. (2010). *City of the Ram-Man: The story of ancient Mendes.* Princeton University Press.

Reemes, D. M. (2015). *The Egyptian Ouroboros: An iconological and theological study.* PhD thesis, University of California.

Régen, I. (2015). Le faucon, rtḥ-qꜣb.t et le lever du soleil. Trois extraits inédits du Livre de Nout dans l'Assassif (TT 34, TT 33, TT 279). In C. Thiers (Ed.), *Documents de Théologies Thébaines Tardives (D3T 3)* (CENiM, 13) (pp. 217–246) Montpellier.
Rehren, T., Belgya, T., Jambon, A., Káli, G., Kasztovszky, Z., Kis, Z., Kovács, I., Maróti, B., Martinón-Torres, M., Miniaci, G., Pigott, V. C., Radivojević, M., Rosta, L., Szentmiklósi, L., & Szőkefalvi-Nagy, Z. (2013). 5,000 years old Egyptian iron beads made from hammered meteoritic iron. *Journal of Archaeological Science, 40*, 4785–4792.
Reisner, G., & Reisner, M. (1933). Inscribed monuments from Gebel Barkal. Part 3. The Stela of Sety I. *ZÄS, 69*, 73–78.
Renberg, G. H. (2016). *Where dreams may come: Incubation sanctuaries in the Greco-Roman world*. Brill.
Reymond, E. A. E. (1962). The Primeval Djeba. *JEA, 48*, 81–88.
Reymond, E. A. E. (1969). *The mythical origin of the Egyptian temple*. Manchester University Press.
Richter, B. A. (2012). *The theology of Hathor of Dendera: Aural and visual scribal techniques in the Per-Wer sanctuary*. PhD thesis, University of California.
Ridley, R. (2019). *Akhenaten, a historian's view*. American University in Cairo Press.
Riess, E. (1891). Nechepsonis et Petosiridis Fragmenta Magica. *Philologus, Supplementband, 6*, 325–394.
Ritner, R. K. (1993). *The mechanics of ancient Egyptian magical practice*. The University of Chicago.
Ritner, R. K. (2003). The legend of Isis and the name of Re (l. 22) (P. Turin 1993). In W. W. Hallo & K. L. Younger (Eds.), *The context of scripture* (Vol. 1, pp. 33–34). Brill.
Ritner, R. K. (2009). *The Libyan Anarchy: Inscriptions from Egypt's third intermediate period*. Brill.
Ritner, R. K. (2016). Oriental Institute Museum Notes 16: Two Egyptian Clepsydrae (OIM E16875 and A7125). *JNES, 75*, 361–389.
Roberts, A., Donalson, J., & Coxe, A. C. (2007). *The Ante-Nicene Fathers, II. Fathers of the Second Century: Hermas, Tatian, Theophilus, Athenagoras, and Clement of Alexandria*. Cosimo Classics.
Robinson, J. (Ed.). (1999). *The Oxford companion to wine*. Oxford University Press.
Rochberg, F. (1989). Babylonian horoscopes and their sources. *Orientalia NS, 58*, 102–123.
Rochberg, F. (1998). *Babylonian horoscopes*. American Philosophical Society.
Rochberg, F. (2009). The hellenistic transmission of Babylonian astral sciences. *Mélanges de l'Université Saint-Joseph, 61*, 13–32.
Rochberg, F. (2010). *In the path of the moon: Babylonian celestial divination and its legacy*. Brill.
Rocheleau, C. M. (2008). *Amon temples in Nubia: a typological study of New Kingdom, Napatan and Meroitic temples* (BAR IS, 1850). Archeopress.
Rochemonteix, M. D. (1892). *Le temple d'Edfou I*. MMAF.
Rodríguez-Antón, A., González-García, A. C., & Belmonte, J. A. (2018). Astronomy in Roman urbanism: A statistical analysis of the orientation of Roman towns in the Iberian Peninsula. *Journal for the History of Astronomy, 49*, 363–387.
Roeder, G. (1913). *Aegyptische Inschriften aus den Staatlichen Museen zu Berlin, I*. J. C. Hinrich.
Roeder, G. (1917). Die Himmelsbeobachtung der alten Ägypter. *Sirius, 50*, 1–11.
Rogers, J. (1998). Origins of the ancient constellations. *Journal of the British Astronomical Association, 108*, 9–28. (part 1) and 79–89 (part 2).
Rondot, V., & Török, L. (2010). La maison du Dieu: le temple. In M. Baud (Ed.), *Méroé. Un empire sur le Nil* (pp. 227–233). Musée du Louvre.
Ross, M. (2006). An introduction to the horoscopic Ostraca of Medinet Madi. *Egitto e Vicino Oriente, 29*, 147–163.
Ross, M. (2007). A survey of demotic astrological texts. *Culture and Cosmos, 11*, 1–25.
Ross, M. T. (2019). Philology and science at Dendera. In M. Brose, P. Dils, F. Naether, L. Popko, & D. Raue (Eds.), *En détail – Philologie und Archäologie im Diskurs. Festschrift für Hans-W. Fischer-Elfert* (Zeitschrift für ägyptische Sprache und Altertumskunde Beiheft 7) (pp. 1011–1019). De Gruyter.

Rossi, C. (2004). *Architecture and mathematics in ancient Egypt*. Cambridge University Press.

Rossi, C., & Magli, H. (2019). Wind, sand and water: The orientation of teh late Roman forts in the Kharga Oasis (Egyptian Western Desert). In G. Magli, A. C. González-García, J. Belmonte Avilés, & E. Antonello (Eds.), *Archaeoastronomy in the Roman world* (pp. 153–168). Springer.

Rossi, F., & Pleyte, W. (1869–1876). *Papyrus de Turin*. Brill.

Roth, A. M. (1993). Fingers, stars, and the opening of the mouth: The nature and function of the *NTrwj*-blades. *JEA, 79*, 57–79.

Rotsch, H. (2005). The Primeval Ocean Nun and the terminology of water in ancient Egypt. In A. Amenta, M. M. Luiselli, & M. Novella Sordi (Eds.), *L'acqua nell'antico Egitto: vita, rigenerazione, incantesimo, medicamento. Proceeding of the first international conference for young Egyptologists* (pp. 229–239). L'Erma di Bretschneider.

Rotsch, H. (2019). *Der Gott Nun und die mythologische Topographie der Unterwelt. Der Nun am Westhorizont*. PhD thesis.

Roulin, G. (1996). *Le Livre de la Nuit. Une composition égyptienne de l'au-delà* (OBO 147/1s). Ed. Universitaire.

Roullet, A. (1972). *Egyptian and Egyptianizing Monuments of Imperial Rome. Études préliminaires aux religions orientales dans l'Empire romain* (Vol. 20). Brill.

Rowland, J. M. (2013). Problems and possibilities for achieving absolute dates from prehistoric and early historic contexts. In A. J. Shortland & C. Bronk Ramsey (Eds.), *Radiocarbon and the chronologies of ancient Egypt* (pp. 235–249). Oxbow Books.

Ryholt, K. S. B. (1997). *The political situation in Egypt during the Second Intermediate Period, c. 1800-1550 BC* (CNIP, 20). Museum Tusculanum Press.

Ryholt, K. (2006). *The Petese Stories II (P Petese II)*. Museum Tusculanum Press.

Ryholt, K. (2011). New Light on the Legendary King Nechepsoso of Egypt. *JEA, 97*, 61–72.

Ryholt, K. (2013). *Narrative literature from the Tebtunis Temple Library*. Museum Tusculanum Press.

Ryholt, K. (2019). Libraries from late period and Graeco-Roman Egypt, c. 800 BCE–250 CE. In K. Ryholt & G. Barjamovic (Eds.), *Libraries before Alexandria. ancient near Eastern traditions* (pp. 390–472). Oxford University Press.

Sachs, A. (1952). Babylonian horoscopes. *Journal of Cuneiform Studies, 6*, 49–75.

Salmas, A.-C. (2014). La mesure du temps de la journée (II). Modules et fonctionnement des horloges à ombre tardives et des cadrans solaires. *BIFAO, 114*, 419–446.

Salmas, A.-C. (2013). La mesure du temps de la journée (1). Modules et fonctionnement des premières horloges à ombre. *BIFAO, 113*, 353–380.

Sambin, C. (1986). Les horloges hydrauliques dans l'Égypte ancienne. État des questions. In P. Louis (Ed.), *L'Homme et l'Eau en Méditerranée et au Proche-Orient. III. L'eau dans les techniques* (Séminaire de recherche 1981–1982) (pp. 75–83). Maison de l'Orient et de la Méditerranée Jean Pouilloux.

Sambin, C. (1987). Les objets *Sbt* des musées. *BIFAO, 87*, 275–292.

Sambin, C. (1988). *L'offrande de la soi-disant "Clepsydre". Le symbole šbt/wnšb/wtt* (Studia Aegyptiaca, XI). Université Eötvos Lorand.

Sartori, N. (forthcoming). *Jarres inscrites de la vallée des rois.*

Sauneron, S. (1959). Le prêtre astronome du temple d'Esna. *Kêmi, 15*, 36–41.

Sauneron, S. (1960). *The priests of ancient Egypt*. Evergreen Books.

Sauneron, S. (1962). *Les fêtes religieuses d'Esna aux derniers siècles du paganisme*. IFAO.

Sauneron, S. (1963). *Le temple d'Esna II*. IFAO.

Sauneron, S. (1968). *Le temple d'Esna III*. IFAO.

Sauneron, S. (1969). *Le temple d'Esna IV*. IFAO. [2nd ed. 2012].

Sauneron, S. (1983). *La porte ptolémaïque de l'enceinte de Mout à Karnak* (MIFAO, 107). IFAO.

Sauneron, S., & Yoyotte J. (1959). La naissance du monde selon l'Égypte ancienne. In *La naissance du monde, Sources orientales, 1* (pp. 17–91). Editions du Seuil.

Schaefer, B. (2000a). Astronomy and the limits of vision. *Vistas in Astronomy, 36*, 311–361.

Schaefer, B. (2000b). The heliacal rise of Sirius and ancient Egyptian chronology. *Journal for the History of Astronomy, 31*, 149–155.

Schaefer, B., & Doggett, L. (1994). Lunar crescent visibility. *Icarus, 107*, 338–403.
Schlögl, H. A. (1980). *Der Gott Tatenen: Nach Texten und Bildern des Neuen Reiches* (OBO, 29). Universitätsverlag/Göttingen: Vandenhoeck Ruprecht.
Schmidt, R. (1995). *The astrological record of the early sages in Greek*. Golden Hind Press.
Schneider, T. (2010). Contributions to the chronology of the New Kingdom and Third Intermediate Period. *ÄuL, 20*, 307–403.
Schomberg, A. (2017). 'To amaze the world'. A contribution to the shape and meaning of the water clock in antiquity. In K. Wellbrock (Ed.), *Cura Aquarum in Greece, Proceedings of the 16th international conference on the history of water management and hydraulic engineering in the Mediterranean Region Athens* (Schriften der Deutschen Wasserhistorischen Gesellschaft (DWhG) e. V. Vol. 27, Part 1) (pp. 301–340). DWhG.
Schott, S. (1938). Das blutrünstige Keltergerät. *ZÄS, 74*, 88–93.
Schott, S. (1950). *Altägyptische Festdaten*. Akademie der Wissenschaften und der Literatur in Mainz.
Schott, S. (1959). Altägyptische Vorstellungen vom Weltende. *Analecta Biblica, 12*, 319–330.
Scott, G. D. (1986). *Ancient Egypt art at Yale*. Yale University Press.
Scott, G. D. (1992). *Temple, Tomb and Dwelling: Egyptian Antiquities from the Harer Family Trust Collection*. University of California Press.
Sellers, J.B. (1992). *The Death of the Gods in Ancient Egypt* . Penguin.
Seneca, L. A. (1972). *Seneca. Naturales quaestiones* (T. H. Corcoran, Trans.). Harvard University Press.
Seneca. *Naturales Quaestiones*.http://naturalesquaestiones.blogspot.com/2009/08/book-vii-tr-john-clarke.html. Last viewed 21 Feb 2019.
Servajean, F. (2007). *Djet et Neheh. Une histoire du temps égyptien*. Presses universitaires de la Méditerranée.
Sethe, K. (1908). *Die Altaegyptischen Pyramidentexte, I*. J.C. Hinrichsche Buchhandlung.
Sethe, K. (1918). Zu den mit wr »der Große« beginnenden alten Titeln. *ZÄS, 55*, 65–67.
Sethe, K. (1919–1920). *Die Zeitrechnung der alten Aegypter im Verhältnis zu der andern Völker. Eine entwicklungsgeschichtliche Studie*.
Sethe, K. (1928). *Dramatische Texte zu altägyptischen Mysterienspielen*. Hinrichs.
Sethe, K. (1929). *Amun und die acht Urgötter von Hermopolis. Eine Untersuchung über Ursprung und Wesen des ägyptischen Götterkönigs*. Akademie der Wissenschaften.
Sethe, K. (1962). *Übersetzung und Kommentar zu den altägyptischen Pyramidentexten. Band V (PT 507-582)*. J. J. Augustin.
Shalthout, M., Belmonte, J. A., & Fekri, M. (2007). On the orientation of ancient Egyptian temples: (3) key points at Lower Egypt and Siwa Oasis. *Journal for the History of Astronomy, 38*, 141–160. (PI) and 413–442 (PII).
Shaltout, M., & Belmonte, J. A. (2005). On the orientation of ancient Egyptian temples: (1) Upper Egypt and Lower Nubia. *Journal for the History of Astronomy, 36*, 273–298.
Sharpe, S. (1864). *The Alabaster Sarcophagus of Oimenepthah I, King of Egypt*. Longman.
Shmakov, T. T. (2012). *Critical analysis of J. P. Allen's "The Ancient Egyptian Pyramid Texts"*. A. K. Eyma.
Shortland, A. J. (2013). An introduction to Egyptian historical chronology. In A. J. Shortland & C. Bronk Ramsey (Eds.), *Radiocarbon and the chronologies of ancient Egypt* (pp. 19–28). Oxbow Books.
Shortland, A. J., & Bronk Ramsey, C. (Eds.). (2013). *Radiocarbon and the chronologies of ancient Egypt*. Oxbow Books.
Sidebotham, S. E., Hense, M., & Nouwens, H. M. (2008). *The red land: The illustrated archaeology of Egypt's eastern desert*. American University in Cairo Press.
Simpson, W. K. (1974). *The terrace of the great god at Abydos the offering chapels of dynasties 12 and 13*. Peabody Museum of Natural History of Yale University.
Simpson, W. K. (1984). Schiffbrüchiger. In W. Helck & W. Westendorf (Eds.), *Lexikon der Ägyptologie, V (cols. 619–622)*. Otto Harrassowitz.
Simpson, W. K. (2005). *The literature of ancient Egypt. An anthology of stories, instructions, stelae, autobiographies, and poetry*. American University in Cairo Press.

Sloley, R. W. (1931). Primitive methods of measuring time with special reference to Egypt. *JEA, 17*, 166–178.

Smith, M. (1991). Did Psammetichus I die abroad? *OLP, 22*, 101–109.

Smith, M. (2002). *On the Primaeval Ocean, The Carlsberg Papyri 5*. Museum Tusculanum Press.

Smith, D. G. (2007). *Total solar eclipses in Ancient Egypt – A new interpretation of some New Kingdom texts*. http://www.egiptomania.com/EEF/Blindness.pdf. Last viewed 13 Oct 2021.

Smyth, C. P. (1858). *Teneriffe, an Astronomer's experiment: Or, specialities of a residence above the clouds*. Cambridge University Press, re-edited 2010.

Smyth, C. P. (1864). *Our Inheritance in the Great Pyramid*. Cambridge University Press, re-edited 2012.

Sourouzian, H. (1989). *Les Monuments du roi Merenptah*. Verlag Philipp von Zabern.

Sousa, R. (Ed.). (2018). *The Tomb of the Priests of Amun. Burial Assemblages in the Egyptian Museum of Florence* (Gate of the Priests Series Volume 1). Brill.

Spalinger, A. J. (1990). A mark of renewal. *SAK, 17*, 289–294.

Spalinger, A. J. (1992). *Three studies on Egyptian Feasts and their Chronological Implications*. Halgo Press, Johns Hopkins.

Spalinger, A. J. (1993a). Parallelism of thought: Hommages à Jean Leclant. *BdÉ, 106*(4), 363–377.

Spalinger, A. J. (1993b). A chronological analysis of the feast of *txy*. *SAK, 20*, 289–303.

Spalinger, A. J. (Ed.). (1994a). *Revolutions in time: Studies on ancient Egyptian calendrics*. Van Sieclen Books.

Spalinger, A. J. (1994b). Thoth and the calendars. In A. J. Spalinger (Ed.), *Revolutions in time: studies on ancient Egyptian calendrics* (pp. 45–60). Van Sieclen Books.

Spalinger, A. J. (1995a). Month representations. *CdÉ, 70*, 110–122.

Spalinger, A. J. (1995b). Some remarks on the Epagomenal days in ancient Egypt. *JNES, 54*, 33–47.

Spalinger, A. J. (1995c). Notes on the ancient Egyptian Calendars. *Orientalia, 64*, 17–32.

Spalinger, A. J. (1995d). The lunar system in festival calendars: From the New Kingdom onwards. *BSÉG, 19*, 25–40.

Spalinger, A. J. (1996a). Some times. *RdÉ, 47*, 67–77.

Spalinger, A. J. (1996b). *The private feast lists of ancient Egypt*. Harrassowitz.

Spalinger, A. J. (1996c). The festival structure of Thutmose III's Buto Stele. *JARCE, 33*, 69–76.

Spalinger, A. J. (2002a). Ancient Egyptian calendars: How many were there? *JARCE, 34*, 241–250.

Spalinger, A. J. (2002b). Egyptian festival dating and the moon. In J. M. Steele & A. Imhausen (Eds.), *Under one sky* (pp. 379–404). Ugarit Verlag.

Spalinger, A. J. (2010). The beginning of the civil calendar. In M. Barta, F. Coppens, & J. Malek (Eds.), *Abusir and Saqqara, edited by in the year 2010* (pp. 723–735). Charles University.

Spalinger, A. J. (2012). Nut and the Egyptologists. *SAK, 41*, 353–377.

Spalinger, A. J. (2018). *Feasts and fights, essays on time in ancient Egypt* (Yale Egyptological Studies, 10). Yale Egyptological Institute.

Spalinger, A. J. (2020a). Perceiving the days of the month. In J. Kamrin, M. Bárta, S. Ikram, M. Lehner, & M. Megahed (Eds.), *Guardian of ancient Egypt, studies in honor of Zahi Hawass* (pp. 1528–1538). Charles University.

Spalinger, A. (2020b). Notes on months names. In S.-W. Hsu, V. P.-M. Laisney, & J. Moje (Eds.), *Ein Kundinger, der in die Gottesworte eingedrungen ist. Festchrift für den Ägyptologen Karl Jansen-Winkeln zum 65. Geburstag* (Ägyten und Altes Testament, 99) (pp. 273–282). Zaphon.

Spence, K. (2000). Ancient Egyptian chronology and the astronomical orientation of the pyramids. *Nature, 408*, 320–324.

Spiegelberg, W. (1904). Demotische Miscellen. *RecTrav, 26*, 154–165.

Spiegelberg, W. (1911). Die ägyptische Namen und Zeichen der Tierkreisbilder in demotischer Schrift. *ZÄS, 48*, 146–151.

Spiegelberg, W. (1912). Demotische Inschrift auf einem Sargbrett. *ZÄS, 50*, 44–45.

Spiegelberg, W. (1922). *Eine neue Spur des Astrologen Petosiris. Sitzungsberichte der Heidelberger Akademie der Wissenschaften*. Carl Winters Universitätsbuchhandlung.

Šprajc, I. (2018). Astronomy, architecture, and landscape in Prehispanic Mesoamerica. *Journal of Archaeological Research, 26*, 197–251.

Stadelmann, R. (1991). *Die äegyptischen Pyramiden, von Ziegelbau zum Weltwunder*. Phillip von Zabern.
Stadelmann, R. (2011). The Heb-Sed Temple of Snefru at Dahshur. In M. Barta, F. Voppens, & J. Krejci (Eds.), *Abusir and Saqqara in the Year 2010* (pp. 738–739). Institute of Archaeology.
Stadler, M. A. (2018). Eine demotische Version des Töpferorakels?: Der Papyrus Berlin P 23888+Wien D 9906b, c recto. In: P. Davoli and N. Pellé (Eds), *Polymatheia: Studi classici offerti a Mario Capasso*. Lecce: Pensa MultiMedia Editore (pp. 395–411).
Steele, J. M. (2000). *Observations and predictions of eclipse times by early astronomers*. Kluver.
Steele, J. M. (2018). The development of the Babylonian zodiac: Some preliminary observations. *Mediterranean Archaeology and Archaeometry, 18*(4), 97–105.
Steinmann, F. (2006). Eine Stundenuhr aus Tuna el-Gebel. *Imago Aegypti, 1*, 125–127.
Stockhusen, M. (2020). Die Grabinschrift des »Uhrmachers« Amenemhet. In R. Färber & R. Gautschy (Eds.), *Zeit in den Kulturen des Altertums. Antike Chronologie im Spiegel der Quellen* (pp. 55–63). Böhlau Verlag.
Strabo. *The geography*. http://penelope.uchicago.edu/Thayer/E/Roman/Texts/Strabo/17A3*.html. Last viewed 21 Feb 2019.
Strudwick, N. (1985). *The administration of Egypt in the old kingdom, the highest titles and their holders*. KPI.
Strudwick, N. (2016). *The Tomb of Pharaoh's chancellor Senneferi at Thebes (TT99), Part I, The new kingdom*. Oxbow Books.
Symons, S. (1999). *Ancient Egyptian astronomy. Timekeeping and cosmography in the new kingdom*. PhD thesis, University of Leicester.
Symons, S. (2002). The 'Transit Star Clock' from the Book of Nut. In J. M. Steele & A. Imhausen (Eds.), *Under one sky: Astronomy and mathematics in the ancient near East* (pp. 429–446). Ugarit-Verlag.
Symons, S. (2007). A star's year: The annual cycle in the ancient Egyptian sky. In J. M. Steele (Ed.), *Calendars and years: Astronomy and time in the ancient world* (pp. 1–33). Oxbow Books.
Symons, S. (2015). Egyptian 'Star Clocks'. In C. L. N. Ruggles (Ed.), *Handbook of archaeoastronomy and ethnoastronomy* (pp. 1495–1500). Springer.
Symons, S. (2020). Sun and stars: Astronomical timekeeping in ancient Egypt. In K. J. Miller & S. Symons (Eds.), *Down to the hour: Short time in the ancient Mediterranean and near East* (pp. 14–51). Brill.
Symons, S., & Khurana, H. (2016). A catalogue of Ancient Egyptian sundials. *Journal for the History of Astronomy, 47*, 375–385.
Symons, S., & Tasker, E. (2015). Stars of the dead. *Scientific American*, October 2015, 32–37.
Symons, S., Cockcroft, R., Bettencourt, J., & Koykka, C. (2013). *Ancient Egyptian astronomy*. [Online database]. Available at: http://aea.physics.mcmaster.ca/. Last viewed 20 Aug 2020.
Szafranski, Z. E. (2005). Deir el Bahari: The temple of Hatshepsut season 2004/2005. *Polish Archaeology in the Mediterranean, 17*, 241–254.
Taller, P., & Lehner, M. (2021). *Les papyrus de la mer Rouge. L'inspecteur Merer, un témoin oculaire de la construction de la pyramide de Chéops*. Errance.
te Velde, H. (1977). The theme of the separation of heaven and earth in Egyptian mythology. *Studia Aegyptiaca, 3*, 161–170.
Thausing, G. (1939). Der ägyptische Schicksalsbegriff. *MDAIK, 8*, 46–70.
Thiers, C. (2009). Le ciel septentrional ghr.t et le ciel méridional gb.t. *ENIM, 2*, 53–58.
Thomas, S. E. (2014). The 'Pastophorion' revisited: Owners and users of 'priests houses' in Ptolemaic Pathyris and elsewhere in Egypt. *JEA, 100*, 111–132.
Thuault, S. (2020). L'herminette et la cuisse, histoire d'un taureau parmi les étoiles. *BIFAO, 120*, 411–449.
Török, L. (1984). Meroitic architecture: Contributions to problems of chronology and style. *Meroitica, 7*, 351–366.
Török, L. (1997). *The kingdom of Kush. Handbook of the Napatan-Meroitic civilization*. Brill.

Török, L. (2009). *Between two worlds: The frontier region between ancient Nubia and Egypt, 3700 BC–500 AD* (Probleme der Ägyptologie, Vol. 29). Brill.

Trauckener, C. (1993). Les résidents des rives du Lac Sacré. Le cas d'Ankhefenkhonsou. *CRIPEL, 15*, 83–93.

Trevisan, C. (2011). La rappresentazione delle costellazioni nello zodiaco circolare di Dendera. In *Ikhnos, Annale di Analisi grafica e Storia della Rappresentazione* (pp. 9–35). Lombardi Editori.

Troy, L. (1986). *Patterns of kingship in ancient Egyptian myth and history*. University of Uppsala.

Tupikova, I., & Soffel, M. (2012). Modelling sundials: Ancient and modern errors. In K. Geus & M. Geller (Eds.), *Productive errors: Scientific concepts in antiquity* (pp. 93–114). Max-Planck-Institut für Wissenschaftsgeschichte.

Turner, E. G., & Neugebauer, O. (1949). Gymnasium debts and new moons. *Bulletin of the John Rylands Library, 32*(1), 80–96.

Ulanowski, K. (2016). A comparison of the role of Barû and Mantis in ancient warfare. In K. Ulanowski (Ed.), *The religious aspects of war in the ancient near East, Greece, and Rome* (pp. 65–98). Brill.

Urbini, S., Nicolosi, I., Zeoli, A., El Khrepy, S., Lethy, A., Hafez, M., El Gabry, M., El Barkooky, A., Barakat, A., Gomaa, M., Radwan, A. M., El Sharkawi, M., D'Orazio, M., & Folco, L. (2012). Geological and geophysical investigation of Kamil crater, Egypt. *Meteoritics & Planetary Science, 47* (11), 1842–1868.

Ullmann, M. (2007). Thebes: Origins of a ritual landscape. In P. F. Dorman & B. M. Bryan (Eds.), *Sacred space and sacred function in ancient Thebes* (SAOC, 61) (pp. 3–26). The Oriental Institute.

van der Baerden, B. L. (1972). Ägyptische Planetenrechnung. *Centaurus, 16*, 65–91.

van der Plas, D. (1986). *L'Hymne a la crue du Nil, II*. NINO.

van der Spek, R. J. (2008). Berossus as a Babylonian chronicler and Greek Historian. In R. J. van der Spek (Ed.), *Studies in ancient Near Eastern World View and Society presented to Marten Stol on the occasion of his 65th birthday, 10 November 2005, and his retirement from the Vrije Universiteit Amsterdam* (pp. 277–318). CDL Press.

van Dijk, J. (2009). New evidence of the length of the reign of Horemheb. *JARCE, 44*, 193–200.

van Dijk, J. (2011). The date of the Gebel Barkal Stela of Sety I. In D. Aston, B. Bader, C. Gallorini, P. Nicholson, & S. Buckingham (Eds.), *Under the Potter's Tree. Studies on ancient Egypt presented to Janine Bourriau on the cccasion of her 70th birthday* (OLA, 204) (pp. 325–332). Peeters.

Vandier, J. (1961). *Le papyrus Jumilhac*. CNRS.

Vandorpe, K., & Clarysse, W. (2019). Cults, creeds, and clergy in a multicultural context. In K. Vandorpe (Ed.), *A companion to Greco-Roman and late antique Egypt* (pp. 407–427). Wiley Blackwell.

Vercoutter, J., Malinine, M., & Posener, G. (1968). *Catalogue des stèles de Sérapéum de Memphis*. Editions des Musées Nationaux.

Vettius Valens. *Anthologies* (M. Riley, Trans.). https://www.csus.edu/indiv/r/rileymt/Vettius%20Valens%20entire.pdf. Last viewed 3 Jan 2020.

Vitruvius Pollio, M. (1874). *The architecture* (J. Gwilt, Trans.). George Phipps.

Vittmann, G. (1998). *Der demotische Papyrus Rylands 9* (Vol. I). Harrassowitz.

Vodolazhskaya, L. (2014). Reconstruction of vertical and L-shaped ancient Egyptian sundials and methods for measuring time. *arXiv*: 1408.0987v2 [physics.hist-ph]. Available at https://arxiv.org/abs/1408.0987. Last viewed 30 Aug 2020.

Volten, A. (1958). Das Harpunierergestirn. *MDAIK, 16*, 346–366.

von Beckerah, J. (1997). *Chronologie des Pharaonischen Ägypten: Die Zeitbestimmung der Ägyptischen Gerchichte von der Vorzeit bis 332 v. Chr* (MÄS, 46). Philipp von Zabern.

von Beckerath, J. (1993). Bemerkungen zum ägyptischen Kalender III: Zum Kalendarium des Papyrus Ebers. *ZÄS, 120*, 131–136.

von Bomhard, A.-S. (1999). *The Egyptian calendar, a work for eternity*. Periplus Publishing Ltd.

von Bomhard, A.-S. (2008). *The Naos of the Decades. From the observation of the sky to mythology and astrology*. Oxford Centre for Maritime Archaeology.
von Bomhard, A.-S. (2011). The Naos of the Decades, puzzle archéologique et thématique. *ENIM, 4*, 107–136.
von Bomhard, A.-S. (2012). Ciels d'Égypte. Le "ciel du sud" et le "ciel du nord". *ENIM, 5*, 73–102.
von Lieven, A. (1999). Divination in Ägypten. *Altorientalische Forschungen, 26*, 77–126.
von Lieven, A. (2000). *Der Himmel über Esna: eine Fallstudie zur religiösen Astronomie in Ägypten* (ÄA, 64). Harrassowitz.
von Lieven, A. (2001). Scheiben am Himmel – Zur Bedeutung von ı'tn und ı'tn.t. *Studien zur Altägyptischen Kultur, 29*, 277–282.
von Lieven, A. (2007). *Grundriss des Laufes der Sterne: Das Sogenannte Nutbuch*. Museum Tusculanum Press.
von Lieven, A. (2010). Translating the fundamentals of the course of the stars. In A. Imhausen & T. Pommerening (Eds.), *Writings of early scholars in the ancient near East, Egypt, Rome and Greece* (pp. 139–150). Walter de Gruyter.
von Lieven, A. (2014). Father of the fathers, mother of the mothers. God as father (and mother) in ancient Egypt, themes in Biblical narrative 18. In F. Albrecht & R. Feldmeier (Eds.), *The divine father: Religious and philosophical concepts of divine parenthood in antiquity* (pp. 17–36).
von Lieven, A. (2016a). Tagewählerei im Alten Ägypten. Mythos. *Rivista di Storia delle Religioni, 10*, 31–41.
von Lieven, A. (2016b). The movement of time. News from the "Clockmaker" Amenemhet. In R. Landgráfová & J. Mynářová (Eds.), *Rich and great. Studies in honour of Anthony J. Spalinger on the occasion of his 70th Feast of Thoth* (pp. 207–231). Charles University in Prague Faculty of Arts.
von Lieven, A. (2021). Astromie und Zeit im Alten Ägypten. In H. Meller, A. Reichenberger, & R. Risch (Eds.), *Time is power. Who makes time?* (pp. 199–215). Landesammt für Denkmalpflege und Archälogie Sachsen-Anhalt.
von Spaeth, O. (2000). Dating the oldest star map. *Centaurus, 42*, 159–179.
Waddell, W. G. (1940). *Manetho* (Loeb Classical Library, 350). HUP.
Wainwright, G. A. (1932a). Iron in Egypt. *JEA, 18*, 3–15.
Wainwright, G. A. (1932b). A pair of constellations. In *Studies presented to F.Ll. Griffith* (pp. 375–379). Egypt exploration society, H. Milford, OUP.
Wainwright, G. A. (1935). Amun's Meteorite and Omphaloi. *ZÄS, 71*, 41–44.
Waitkus, W. (2003). Zur möglichen Identifizierung einer weiteren Konstellation des nördlichen altägyptischen Sternhimmels. In N. Kloth, K. Martin, & E. Pardey (Eds.), *Es werde niedergelegt als Schriftstück. Festschrift für Hartwig Altenmüller* (SAK, Beihefte 9) (pp. 453–470). Buske.
Waitkus, W. (2010). Anmerkungen zum Gründungsritual und seinen astronomischen Bezügen in Edfu und Dendera. In D. Kurth & W. Waitkus (Eds.), *Edfu: Materialien und Studien. Die Inschriften des Tempels von Edfu, Begleitheft 6* (pp. 173–183). PeWe-Verlag.
Ward, W. A. (1982). *Index of Egyptian administrative and religious titles of the Middle Kingdom*. American University of Beirut.
Wells, R. A. (1985). Sothis and the Satet temple on Elephantine: A direct connection. *SAK, 12*, 255–302.
Wells, R. A. (1992). The mythology of nut and the birth of Ra. *SAK, 19*, 305–321.
Wells, R. A. (1994). Re and the calendars. In A. J. Spalinger (Ed.), *Revolutions in time: Studies on ancient Egyptian calendrics* (pp. 1–37). Van Sieclen Books.
Welsby, D. A. (1996). *The kingdom of Kush: The Napatan and Meroitic empires*. British Museum Press.
Welsby, D. A., & Anderson, J. R. (2004). *Sudan, ancient treasures*. British Museum Press.
Wening, S. (2001). Musawwarat es-Sufra, interpreting the Great Enclosure. *Sudan and Nubia, 5*, 71–86.
Wessetzky, G. (1954). Les problemes des "amulettes de nouvel an" égyptiennes. *Bulletin du Musée Hongrois des Beaux-Arts, 5*, 3–10.

Wilcken, U. (1903). Die berliner Papyrusgrabungen in Herakleopolis Magna im Winter 1898/9. *Archiv für Papyrusforschung, 3*, 294–336.
Wild, H. (1969). Quatre statuettes du Moyen Empire dans une collection privée de Suisse. *BIFAO, 69*, 89–130.
Wilkinson, R. H. (1991). New kingdom astronomical paintings and methods of finding and extending direction. *JARCE, 28*, 149–156.
Wilkinson, R. H. (1994). *Symbol and magic in ancient Egypt*. Thames and Hudson.
Wilkinson, T. H. A. (1999). *Early dynastic Egypt*. Rotledge.
Wilkinson, R. H. (2000a). *The complete temples of ancient Egypt*. Thames and Hudson.
Wilkinson, T. A. H. (2000b). *Royal annals of ancient Egypt: The palermo stone and its associated fragments*. Kegan Paul International.
Wilkinson, R. H. (2003). *The complete gods and goddesses of ancient Egypt*. Thames and Hudson.
Willems, H. (1988). *Chests of life: A study of the typology and conceptual development of Middle Kingdom standard class coffins*. Ex Oriente Lux.
Wilson, J. V. K. (1972). *The Nimrud Wine Lists. A Study of men and administration at the Assyrian capital in the Eighth Century, B.C.* British School of Archaeology in Iraq.
Wilson, P. (1997). *A Ptolemaic Lexikon: A lexicographical study of the texts in the Temple of Edfu* (OLA, 78). Peeters.
Winkler, A. (2016). Some astrologers and their handbooks in demotic Egyptian. In J. M. Steele (Ed.), *The circulation of the astronomical knowledge in the ancient world* (pp. 245–286). Brill.
Winkler, A. (2018). A starry summer night in AD142: A Theban Horoscope (Griffith MSS 3.59) in context. In K. D. van Heel, F. A. J. Hoogendijk, & C. J. Martin (Eds.), *Hieratic, demotic and Greek studies and text editions of making many books there is no end: Festschrift in honour of Sven P. Vleeming* (P. L. Bat. 34) (pp. 297–308). Brill.
Wolf, W. (1929). Der Berliner Ptah-Hymnus (P 3048, II-XII). *ZÄS, 64*, 17–44.
Yellin, J. W. (1984). An astronomical text from Beg. South 503. *Meroitica, 7*, 577–582.
Yoyotte, J. (1996). L'Égypte du Delta: les capitals du nord. *Dossiers d'Archéologie, 213*, 6–11.
Zaccagnini, C. (1983). Patterns of mobility among ancient near Eastern craftsmen. *JNES, 42*, 245–264.
Zayed, A. H. (1968). Painted wooden stelae in the Cairo museum. *RdÉ, 20*, 149–170.
Zenihiro, K. (2009). *The complete funerary cones*. Ancient Egyptian Museum.
Zitman, M. (2010). *The Necropolis of Assiut: A case study of local Egyptian funerary culture from the old kingdom to the end of the middle kingdom*. Peeters.
Zivie-Coche, C. (2008). Religion de l'Égypte ancienne. *Annuaire de l'École pratique des hautes études (EPHE), Section des sciences religieuses, 115*, 73–83.
Zivie-Coche, C. (2016). L'ogdoade d'Hermopolis à thèbes et ailleurs ou l'invention d'un mythe. *Egitto e Vicino Oriente, 39*, 57–90.
Zurawski, B. (1998). Pliny's 'Tergedum' discovered. *Sudan and Nubia, 2*, 74–81.

Index

J. A. Belmonte, J. Lull, *Astronomy of Ancient Egypt*, Historical & Cultural Astronomy, https://doi.org/10.1007/978-3-031-11829-6

N

O

P

T

U